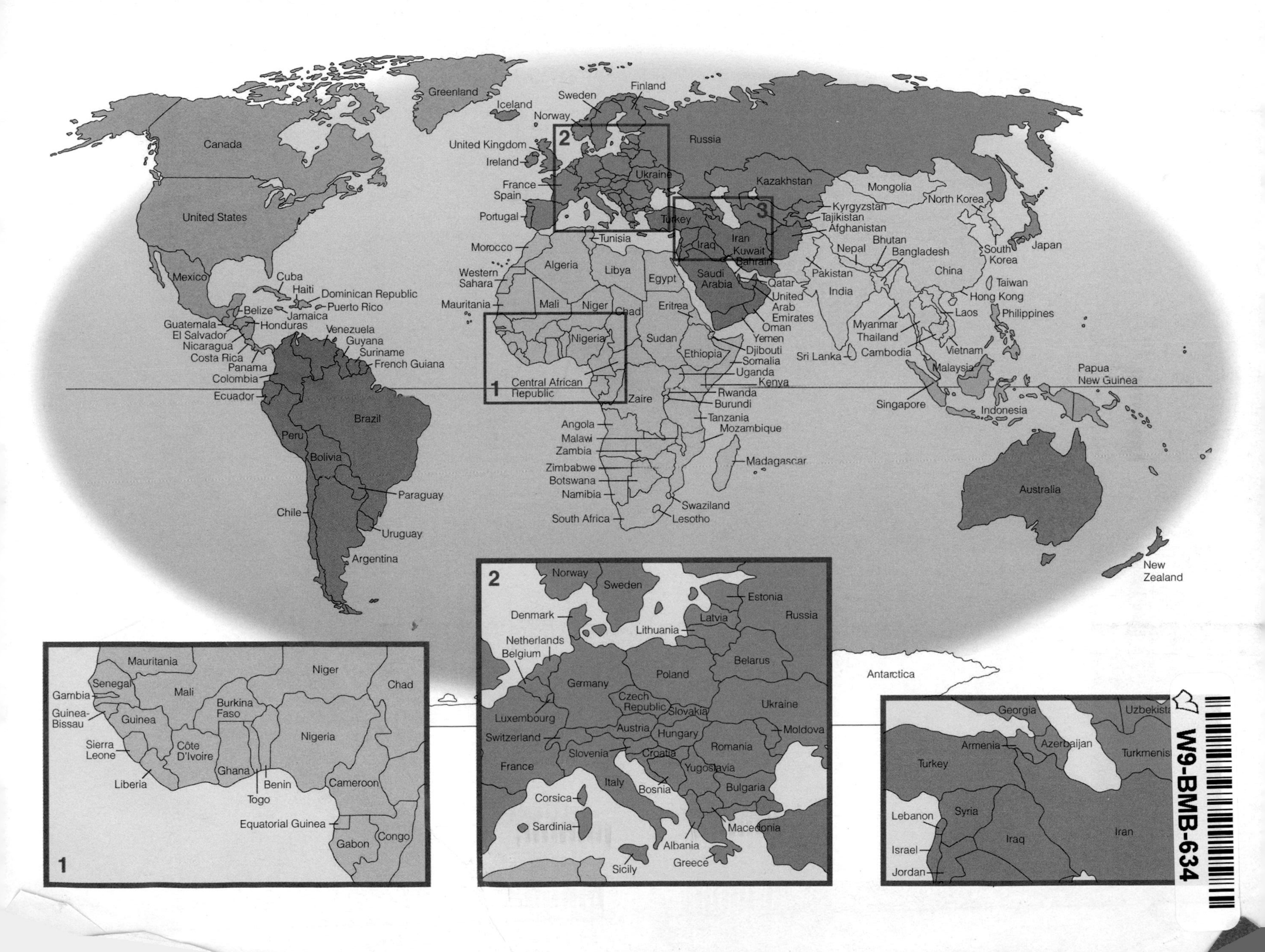

Greenland
Iceland
Sweden
Finland
Norway
Canada
United Kingdom
Ireland
Russia
France
Spain
Ukraine
Kazakhstan
Mongolia
North Korea
Portugal
United States
Turkey
Kyrgyzstan
Tajikistan
Afghanistan
Tunisia
Iran
Iraq
Kuwait
Bahrain
Nepal
Bhutan
Bangladesh
South Korea
Japan
Morocco
Algeria
Libya
Egypt
Saudi Arabia
Pakistan
China
Mexico
Cuba
Haiti
Dominican Republic
Puerto Rico
Western Sahara
Qatar
United Arab Emirates
India
Taiwan
Hong Kong
Belize
Jamaica
Mauritania
Mali
Niger
Chad
Eritrea
Laos
Philippines
Guatemala
El Salvador
Nicaragua
Costa Rica
Panama
Colombia
Honduras
Venezuela
Guyana
Suriname
French Guiana
Nigeria
Sudan
Ethiopia
Oman
Yemen
Djibouti
Somalia
Uganda
Kenya
Myanmar
Thailand
Cambodia
Vietnam
Sri Lanka
Malaysia
Singapore
Papua New Guinea
Indonesia
Central African Republic
Ecuador
Zaire
Rwanda
Burundi
Tanzania
Mozambique
Brazil
Peru
Bolivia
Angola
Malawi
Zambia
Zimbabwe
Botswana
Namibia
Madagascar
Paraguay
Swaziland
South Africa
Lesotho
Chile
Uruguay
Argentina
Australia
New Zealand
Antarctica
1
Mauritania
Senegal
Gambia
Guinea-Bissau
Guinea
Sierra Leone
Liberia
Mali
Burkina Faso
Côte D'Ivoire
Ghana
Togo
Benin
Niger
Nigeria
Chad
Cameroon
Equatorial Guinea
Gabon
Congo
2
Norway
Sweden
Estonia
Russia
Denmark
Latvia
Lithuania
Netherlands
Belgium
Germany
Poland
Belarus
Czech Republic
Slovakia
Ukraine
Luxembourg
Switzerland
Austria
Hungary
Moldova
Romania
Slovenia
Croatia
Yugoslavia
France
Italy
Bosnia
Bulgaria
Corsica
Sardinia
Macedonia
Albania
Greece
Sicily
Georgia
Uzbekistan
Armenia
Azerbaijan
Turkmenistan
Turkey
Syria
Lebanon
Iraq
Iran
Israel
Jordan

Understanding Social Problems

Fourth Edition

Linda A. Mooney

David Knox

Caroline Schacht

East Carolina University

Australia • Canada • Mexico • Singapore • Spain • United Kingdom • United States

Sociology Editor: Robert Jucha
Development Editor: Julie Sakaue
Assistant Editor: Stephanie Monzon
Editorial Assistant: Melissa Walter
Technology Project Manager: Dee Dee Zobian
Marketing Manager: Matthew Wright
Marketing Assistant: Tara Pierson
Advertising Project Manager: Linda Yip
Project Manager, Editorial Production: Cheri Palmer
Print/Media Buyer: Karen Hunt
Permissions Editor: Joohee Lee

Production: Mary Douglas, Rogue Valley Publications
Text Designer: Jeanne Calabrese
Photo Researcher: Roberta Broyer
Copy Editor: Patterson Lamb
Illustrator: G&S Typesetters
Cover Designer: Hiroko/Cuttriss & Hambleton
Cover Image: Corbis and Getty Images
Cover Printer: Phoenix Color Corp
Compositor: G&S Typesetters
Printer: R.R. Donnelley/Willard

Printed in the United States of America
2 3 4 5 6 7 08 07 06 05

Library of Congress Control Number: 2004102002

Student Edition: ISBN 0-534-62514-2

Instructor's Edition: ISBN 0-534-62515-0

0-534-62537-1

Thomson Wadsworth
10 Davis Drive
Belmont, CA 94002-3098
USA

Asia
Thomson Learning
5 Shenton Way #01-01
UIC Building
Singapore 068808

Australia/New Zealand
Thomson Learning
102 Dodds Street
Southbank, Victoria 3006
Australia

Canada
Nelson
1120 Birchmount Road
Toronto, Ontario M1K 5G4
Canada

Europe/Middle East/Africa
Thomson Learning
High Holborn House
50/51 Bedford Row
London WC1R 4LR
United Kingdom

Latin America
Thomson Learning
Seneca, 53
Colonia Polanco
11560 Mexico D.F.
Mexico

Spain/Portugal
Paraninfo
Calle Magallanes, 25
28015 Madrid, Spain

Brief Contents

Contents

Thinking About Social Problems 1

Illness and the Health Care Crisis 27

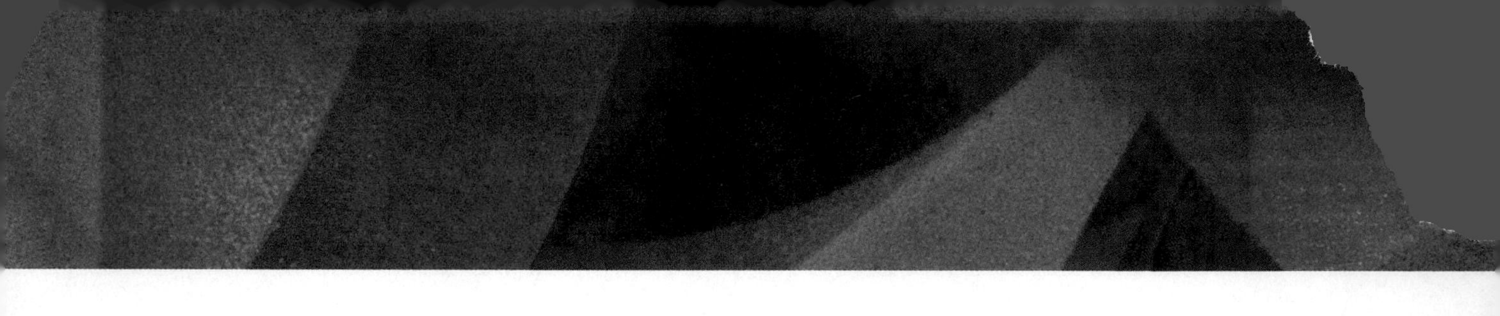

Problems of Youth and Aging 265

Poverty 295

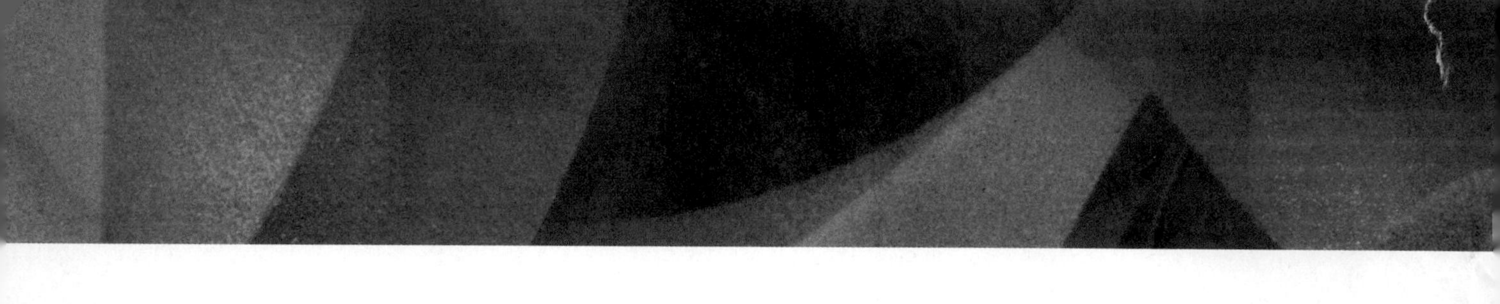

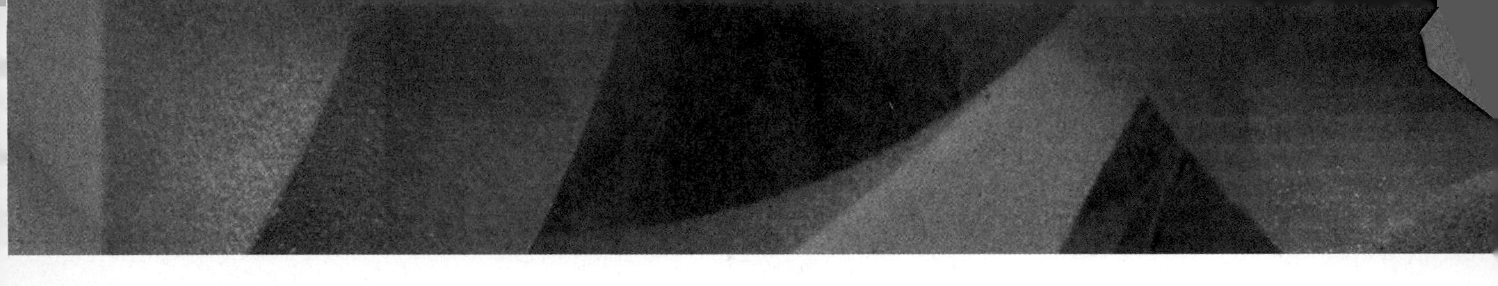

Conflict, War, and Terrorism 488

Preface

Understanding Social Problems is intended for use in a college-level sociology course. We recognize that many students enrolled in undergraduate sociology classes are not sociology majors. Thus, we have designed our text with the aim of inspiring students—*no matter what their academic major or future life path may be*—to care about the social problems affecting people throughout the world. In addition to providing a sound theoretical and research basis for sociology majors, *Understanding Social Problems* also speaks to students who are headed for careers in business, psychology, health care, social work, criminal justice, and the nonprofit sector, as well as those pursuing degrees in education, fine arts, the humanities, and those who are "undecided." Social problems, after all, affect each and every one of us, directly or indirectly. And everyone, whether a leader in business or politics, a stay-at-home parent, or a student, can become more mindful of how his or her actions (or inactions) perpetuate or alleviate social problems. It is our hope that *Understanding Social Problems* not only informs, but also inspires planting seeds of social awareness that will grow no matter what academic, occupational, and life path students choose.

New to This Edition

As societies change at an ever-increasing pace, issues and events once defined as "social problems" are solved or redefined and replaced by new, emerging social problems. The Fourth Edition of *Understanding Social Problems* reflects this evolution not only with updates to important data—such as the global poverty statistics, population growth rates, and the percentage of U.S. residents without health insurance—but also with the incorporation of important new research findings and changing social policies, laws, and programs. Updates and revisions are included throughout the book's chapters, as detailed below.

Chapter 1 (Thinking About Social Problems) now includes a discussion of social constructionism and how it helps our understanding of the origin and evolution of social problems. The Self and Society feature has been updated to reflect the most recent "freshman norms." A debate over the legal protection of sociologists' sources is included at the end of the chapter.

In **Chapter 2** (Illness and the Health Care Crisis), a new opening section explores ways in which globalization affects health. We have also updated information on two major problems: the global HIV/AIDS crisis and the problem of inadequate health insurance coverage. The section on mental illness has been expanded, highlighting men's experiences with depression, attitudes toward seeking psychological help, and educational strategies to combat the stigma of mental illness. The concept of medicalization has also been expanded, and we introduce a new related concept: biomedicalization.

Chapters 3 (Alcohol and Other Drugs) now includes a discussion of oxycontin and methadone use along with an expanded discussion of "date rape" and club drugs. Results from the most recently available National Household Survey on Drug Abuse are presented throughout the chapter. A discussion of prevention techniques has been added along with a discussion of California's Proposition 36. The priorities of our national drug control strategy are also new to this chapter.

Chapter 4 (Crime and Violence) presents the most recent crime statistics available from the National Crime Victimization Survey and the Uniform Crime Reports. The section on white-collar crime has been expanded to include recent (e.g., WorldCom) accusations of securities fraud, tax invasion, and insider trading, and a new section on corrections has been added. Further, the chapter ends with a debate over whether or not sex offenders should be required to register with local or state agencies.

In **Chapter 5** (Family Problems), information on changing structures and patterns in U.S. families and households has been reorganized in a more concise fashion. We incorporate feminist perspectives on the family, address issues concerning DNA paternity testing, and discuss the Community Marriage Policy—a faith-based strategy to strengthen marriage. Information on child abuse and intimate partner violence has been updated, including the cycle of abuse that characterizes many chronically abusive relationships.

Chapter 6 (Race and Ethnic Relations) provides increased coverage of prejudice and discrimination against Muslims and people of Middle Eastern descent—a topic that has gained widespread attention since the September 11, 2001 terrorist attacks. We have also expanded and updated the discussion of affirmative action, including the 2003 Supreme Court ruling on affirmative action.

Chapter 7 (Gender Inequality) now includes a discussion of feminism in the theory section and an expanded discussion of media, language, and cultural sexism. A review of the relationship between gender and health has also been expanded and now more completely addresses the issue of women and HIV/AIDS. The chapter now ends with a "take-a-stand" debate over gender differences and the nature-versus-nurture argument.

In **Chapter 8** (Issues in Sexual Orientation), we present recent U.S. census data on same-sex cohabiting couples in the United States and provide updated information on the legal status of sexual orientation issues, including the Massachusetts court ruling on same-sex marriages and the 2003 Supreme Court ruling in *Lawrence v. Texas* that struck down sodomy laws. The findings from a national survey on the campus climate for GLBT are discussed, and new material on campus "safe zones" is included in the section on reducing antigay prejudice and discrimination.

Chapter 9 (Problems of Youth and Aging) has been significantly revised. Adding to its strength, we have included several new sections in the chapter. Discussions of children's health, orphans and street children, child labor, and child trafficking and prostitution are now included in this chapter. A new section on "fighting age discrimination" has also been added, including the 2003 case of *EEOC and Arnett et al. v. CalPERS* (California Public Employees' Retirement System). This updated chapter also includes new information on physician-assisted suicide.

Chapter 10 (Poverty) includes updated information on the extent of poverty and economic inequality worldwide, patterns of poverty in the United States, public assistance programs and myths and realities of welfare in the United States, and "corporate welfare." Important new research findings on the use of genetically engineered crops, which its advocates promote as a strategy for feeding the world's poor populations, have been added to the "Focus on Technology" feature in this chapter.

New information on modern slavery has been added to **Chapter 11** (Work and Unemployment), along with expanded coverage of the globalization of trade and free trade agreements and transnational corporations. In a new "Human Side" feature, Barbara Ehrenreich, author of the best-selling *Nickel and Dimed,* talks about her experiences working in low-wage jobs. This chapter also includes new survey data on corporate and government repression of labor union efforts from the International Confederation of Free Trade Unions as well as updated statistics on job health and safety issues.

In **Chapter 12** (Problems in Education), we have expanded the sections on school segregation and have added a discussion of a new trend—"school resegregation." The section on school violence has also been expanded, and we have added a section on President Bush's "No Child Left Behind" initiative and its four organizational principles. Finally, we have added a discussion of "learning while black," the educational equivalent of racial profiling.

Chapter 13 (Population and Urbanization), which was titled "Cities in Crisis" in the third edition, has been merged with discussions of problems associated with population growth. New material in this chapter includes recent United Nations population projections and a discussion of the effects of urbanization on the demographic transition. The "Human Side" and "Social Problems Research Up Close" features in this chapter are also new, and both deal with the effects of family planning on the lives of women and men in several countries.

In the previous edition of this text, environmental problems shared a chapter with population problems. New to this edition, population problems are discussed elsewhere (Chapter 13, Population and Urbanization), and all of **Chapter 14** (Environmental Problems) is now devoted to environmental problems. New topics added to this chapter include worldwide energy use (including a new "Self and Society" feature, "Energy IQ Test"), the environmental effects of mining (including a new "Human Side" about mountain-top mining in Appalachia), the vulnerability of children to environmental chemical exposure, and the effects of "bioinvasion."

Chapter 15 (Science and Technology) contains an expanded section on partial birth abortions and the legal challenges to them as well as a discussion of feminists' opposition to the new ban. A discussion of several laws that advocate and protect "fetal rights" is also new to this chapter, including a discussion of the *Unborn Victims of Violence Act.* In addition to the topics listed, Chapter 15 now includes a discussion of the problem of pornography on the Internet.

Chapter 16 (Conflict, War, and Terrorism) has been renamed to better reflect its new content. A significant portion of the chapter is now devoted to the types, patterns, and roots of terrorism and to responses to terrorism (for instance, USA Patriot Act). Further, a feminist perspective of war is presented as well as an analysis of September 11, 2001, media coverage. Other topics new to this chapter include the war with Iraq, the conflict in Afghanistan, and a discussion of Samuel P. Huntington's *The Clash of Civilizations and the Remaking of World Order* (1996).

Features and Pedagogical Aids

We have integrated a number of features and pedagogical aids into the text to help students learn to think about social problems with a sociological perspective. It is our mission not only to help students apply sociological concepts to observed situations in their everyday life and to think critically about social problems and their implications, but to learn to assess how social problems relate to their lives on a personal level.

Study Aids in Each Chapter That Are New to This Edition

Taking a Stand. Because writing and critical thinking skills are so important, we've added the "Taking a Stand" exercise at the end of each chapter to help students hone these specific skills. New to the Fourth Edition, the "Taking a Stand" exercise begins by presenting a single question addressing a particular issue that's relevant to the chapter topic (for example, "Should Social Security be privatized?"). We follow the question with a short paragraph that introduces the major issues associated with the question. We then refer students to Wadsworth's many sociology online resources to help them research and write their response to the question posed. With this new exercise, students will not only sharpen their understanding of specific social issues, but they will also hone their analytical and writing skills in general.

End-of-Chapter Reviews in Question-and-Answer Format. To the Fourth Edition, we've added chapter summaries to help students study and review each chapter. To make the summaries particularly effective, we've formatted them using the easy-to-read, question-and-answer approach. This approach highlights and reinforces the most important concepts while helping students to understand rather than just simply memorize the material.

Exercises and Boxed Features

Self and Society. Each chapter includes a social survey designed to help students assess their own attitudes, beliefs, knowledge, or behavior regarding some aspect of the social problem under discussion. In Chapter 5 (Family Problems), for example, the "Abusive Behavior Inventory" invites students to assess the frequency of various abusive behaviors in their own relationships.

The Human Side. In addition to the "Self and Society" boxed features, each chapter includes a boxed feature entitled "The Human Side," which further personalizes the social problems under discussion by describing personal experiences of individuals who have been affected by them. The "Human Side" feature in Chapter 6 (Race and Ethnic Relations), for example, profiles several post-9/11 hate crime victims, revealing the senseless violence against Muslims and people of Middle Eastern descent that peaked after 9/11 and remains a problem today.

Social Problems Research Up Close. In every chapter, boxed features called "Social Problems Research Up Close" present examples of social science research. These features demonstrate for students the sociological enterprise from theory and data collection to findings and conclusions. Through the "Social Problems Research Up Close" feature, students are thus exposed to a variety of studies and research methods.

Focus on Technology. Because technology is a pervasive part of our society, we have added a "Focus on Technology" boxed feature to every chapter, in addition to dedicating an entire chapter to the topic (Chapter 15, Science and Technology). The "Focus on Technology" features present information on how technology may contribute to the specific social problems and their solutions that are under discussion in any given chapter. In Chapter 10, Poverty, for example, the "Focus on Technology" feature examines the controversy surrounding the use of agricultural biotechnology, weighing its potential benefits (such as increasing food supplies and nutrition for the poor) against the potential risks (such as adverse effects on human and environmental health).

In-Text Learning Aids

Vignettes. Each chapter begins with a vignette designed to engage students and draw them into the chapter by illustrating the current relevance of the topic under discussion. Chapter 2, Illness and the Health Care Crisis, for instance, begins with the case of 57-year-old Larry Causey. Mr. Causey staged a federal crime (post-office robbery) because he wanted to be arrested and put in prison where he would receive medical treatment for cancer—treatment he could not otherwise afford due to the escalating cost of health care.

Critical Thinking. Each chapter ends with a brief section called "Critical Thinking," which raises several questions related to the chapter topic. These questions invite students to apply their critical thinking skills to the information discussed in the chapter.

Key Terms. Important terms and concepts are highlighted in the text where they first appear. To re-emphasize the importance of these words, they are listed at the end of every chapter and included in the glossary at the end of the text.

Glossary. All of the key terms are defined in the end-of-text glossary.

Understanding [specific social problem] Sections. All too often students, faced with contradictory theories and study results, walk away from social problems courses without any real understanding of their causes and consequences. To address this problem, chapter sections entitled "Understanding . . . [specific social problem]" cap the body of each chapter just before the chapter summaries. Unlike the chapter summaries, these sections synthesize the material presented in the chapter, summing up the present state of knowledge and theory on the chapter topic.

Supplements

The Fourth Edition of *Understanding Social Problems* comes with a full complement of supplements designed with both faculty and students in mind.

Supplements for the Instructor

Instructor's Resource Manual with Test Bank (with Multimedia Manager CD-ROM). This supplement offers the instructor learning objectives, key terms, lecture outlines, student projects, classroom activities, and Internet and InfoTrac® College Edition exercises. Test items include 60–100 multiple-choice and true/false questions with answers and page references, as well as short answer and essay questions for each chapter. Concise user guides for InfoTrac College Edition and WebTutor™ are included as appendixes. An all-new Multimedia Manager Instructor Resource CD-ROM is now packaged with the IRM/TB. This new instructor resource includes book-specific PowerPoint lecture slides, graphics from the book, the IRM/TB Word documents, CNN video clips, and links to many of Wadsworth's impressive sociology resources. All of your media teaching resources in one place!

ExamView® Computerized Testing. Create, deliver, and customize tests and study guides (both print and online) in minutes with this easy-to-use assessment and tutorial system. ExamView offers both a quick test wizard and an online test wizard that guide you step by step through the process of creating tests. The test appears on screen exactly as it will print or display online. Using ExamView's complete word processing capabilities, you can enter an unlimited number of new questions or edit existing questions included with ExamView.

Videos. Adopters of *Understanding Social Problems* have a couple of different video options available with the text.

CNN Today: Social Problems/Issues and Solutions Video

Social Problems: Volume I: 0-534-54119-4
Social Problems: Volume II: 0-534-54124-0
Social Problems: Volume III: 0-534-61898-7
Social Problems: Volume IV: 0-534-61933-9

Issues and Solutions Volume I: 0-534-54122-4
Issues and Solutions: Volume II: 0-534-54123-2

Integrate the up-to-the-minute programming power of CNN and its affiliate networks right into your classroom! Updated yearly, *CNN Today* videos are course-specific videos that can help you launch a lecture, spark a discussion, or demonstrate an application—using the top-notch business, science, consumer, and political reporting of the CNN networks. Produced by Turner Learning, Inc., these 45-minute videos show your students how the principles they learn in the classroom apply to the stories they see on television. Special adoption conditions apply.

Wadsworth Sociology Video Library. This large selection of thought-provoking films is available to adopters based on adoption size.

Supplements for the Student

Study Guide. The study guide includes learning objectives, chapter outlines, key terms, Internet activities, completion exercises, practice tests consisting of 25 multiple-choice and true/false questions with answers and page references, and 5 short answer questions with page references to enhance and test student understanding of chapter concepts.

Wadsworth's Sociology Online Resources and Writing Companion for Mooney/Knox/Schacht's *Understanding Social Problems,* Fourth Edition. NEW! This valuable guide shows students how they can use Wadsworth's exclusive Online Resources—InfoTrac College Edition, the Opposing Viewpoints Resource Center (OVRC), and MicroCase Online—to assist them in their study of sociology and to build essential research and writing skills. Part One provides informative user guides that introduce each of these powerful research tools, while Part Two contains chapter-by-chapter directed exercises designed to develop research and critical thinking proficiency for each of the core topics in social problems. Part Three encourages students to "put it all together" and apply their newly acquired research skills to respond to the specific sociological questions posed in the "Taking a Stand" feature in the main text. Part Four then provides an overview of some of the research and writing tools available online, such as InfoWrite and the OVRC research guide, and shows students how they can effectively integrate their research findings into class assignments.

***Investigating Social Problems: An Introduction Using MicroCase®,* First Edition.** Written by David J. Ayers of Grove City College, this supplementary workbook and statistical package takes a skills-based approach to teach students how to perform social research into common social problems using real, sociological data such as that found in the GSS. Each chapter contains perforated worksheets with quizzes and exercises using MicroCase. By using the workbook, students develop skills using empirical data in order to examine real world problems.

***Social Problems: Readings with Four Questions,* First Edition.** Edited by Joel M. Charon of Minnesota State University–Moorhead, this reader is a unique and groundbreaking collection of 58 articles, organized in 13 thematic sections, that takes a structural-conflict approach yet lets the voices of those impacted by social problems be heard. The articles are a mix of classic and contemporary readings that cover a wide range of issues in the United States and the world. The introductory article, written by the author, focuses on four questions that students are urged to apply throughout the reader: What is the problem? What makes the problem a "social problem"? What causes the problem? What can be done? This "four-questions" approach gives students a consistent sociological framework within which to analyze social problems.

***Social Problems of the Modern World: A Reader,* First Edition.** Edited by Frances V. Moulder of Three Rivers Community College, this unique anthology includes readings from both American and global perspectives for each problem area, while also covering all key issues typically addressed in social problems courses. With all readings chosen for their interest and accessibility, this wonderful collection includes articles written by sociologists as well as fascinating accounts representing a diversity of voices from different nationalities and backgrounds. No other anthology so effectively helps readers understand social problems on both national and global scales and think critically and deeply about each issue.

***Experiencing Poverty: Voices from the Bottom,* First Edition.** Edited by D. Stanley Eitzen, Colorado State University, and Kelly Eitzen Smith, University of Arizona, this reader is unique in its comprehensive understanding of poverty. The readings incorporate the experiences of the impoverished in different situations, including people who are homeless, housed, working, nonworking, single, married, urban, rural, born in the United States, first-generation Americans, young, and old. Most significantly, the crucial dimensions of inequality—race and gender—are interwoven through the selected readings.

***Researching Sociology on the Internet,* Third Edition.** Prepared by D.R. Wilson of Houston Baptist University, this guide is designed to assist sociology students with doing research on the Internet. Part One contains general information necessary to get started and answers questions about security, the type of sociology material available on the Internet, the information that is reliable and the sites that are not, the best ways to find research, and the best links to take students where they want to go. Part Two looks at each main topic in the area of sociology and refers students to sites where they can obtain the most enlightening research and information.

***Applied Sociology: Terms, Topics, Tools, and Tasks,* First Edition.** This concise, user-friendly book by award-winning sociology professor Stephen F. Steele of Anne Arundel Community College and Jammie Price of the University of North Carolina–Wilmington addresses a common question among many introductory sociology students: "What can I do with sociology?" The book introduces students to sociology as an active and relevant way to understand human social interaction by offering a clear, direct linkage between sociology and its practical use. It focuses on the core concepts in sociology (terms and topics), presents contemporary and practical skills used by sociologists to investigate these concepts (tools), and contains concrete exercises for learning and applying these skills (tasks). The book also includes brief sections on using sociology to make a difference in the community and on developing a career in sociology. *Applied Sociology* is an ideal supplement to traditional sociology texts to add an applied component to the course.

***Six Steps to Effective Writing in Sociology,* First Edition.** Written by Judy H. Schmidt, Pennsylvania State University, Michael K. Hooper, California Department of Justice, and Diane Kholos Wysocki, University of Nebraska–Kearney, this compact resource is intended to embed strong writing skills in students and prepare them for their academic and professional pursuits. The authors approach writing as a series of skills to be applied at each stage of the writing process: generating ideas, developing and planning, drafting, revising, and editing. Sample writing topics, examples, formats, and sample papers reflect the discipline, providing a complement to classroom instruction and discussion.

Online Resources

Wadsworth's Virtual Society: The Wadsworth Sociology Resource Center (http://www.wadsworth.com/sociology). Here you will find a wealth of sociology resources such as Census 2000: A Student Guide for Sociology, Breaking News in Sociology, a Guide to Researching Sociology on the Internet, Sociology in Action, and much more. Contained on the home page is the companion Web site for *Understanding Social Problems,* Fourth Edition.

Mooney/Knox/Schacht's *Understanding Social Problems* Companion Web Site (http://sociology.wadsworth.com/mooney_knox_schacht/problems4e). Access to useful learning resources for each chapter of the book. And for instructors, the site includes password-protected Instructor's Manuals, PowerPoint lectures, and important sociology links. Click on the companion Web site to find useful learning resources for each chapter of the book. Some of these resources include:

- Tutorial Practice Quizzes that can be scored and e-mailed to the instructor
- Web Links
- Internet Exercises
- Video Exercises
- InfoTrac College Edition Exercises
- Flashcards of the text's glossary
- Crossword Puzzles
- Essay Questions
- Learning Objectives
- Virtual Explorations

And much more!

WebTutor™ Toolbox for WebCT or Blackboard. Preloaded with content and available free via pincode when packaged with this text, WebTutor Toolbox pairs all the content of this text's rich book companion Web site with all the sophisticated course management functionality of a WebCT or Blackboard product. You can assign materials (including online quizzes) and have the results flow automatically to your grade book. Toolbox is ready to use as soon as you log on—or you can customize its preloaded content by uploading images and other resources, adding Web links, or creating your own practice materials. Students have access only to student resources on the Web site. Instructors can enter a pincode for access to password-protected instructor resources.

InfoTrac College Edition. With each purchase of a new copy of the text comes a free four-month passcode to InfoTrac College Edition, the online library that gives students access to reliable resources anytime, anywhere. This fully searchable database offers 20 years' worth of full-text articles from almost 5000 diverse sources, such as academic journals, newsletters, and up-to-the-minute periodicals including *Time, Newsweek, Science, Forbes,* and *USA Today.* This incredible depth and breadth of material—available 24 hours a day from any computer with Internet access—makes conducting research so easy that your students will want to use it to enhance their work in every course! Through InfoTrac's InfoWrite, students now also have instant access to critical-thinking and paper-writing tools. Both adopters and their students receive unlimited access for four months.

Opposing Viewpoints Resource Center (OVRC). Newly available from Wadsworth, this online center presents varying perspectives on today's most compelling issues. OVRC draws on Greenhaven Press's acclaimed Social Issues series, as well as core reference content from other Gale and Macmillan Reference USA sources. The result is a dynamic online library of current events topics—the facts as well as the arguments of each topic's proponents and detractors. Special sections focus on critical thinking—walking students through the steps involved in critically evaluating point-counterpoint arguments—and researching and writing papers.

Interactions CD-ROM for Mooney/Knox/Schacht's Fourth Edition. This free CD-ROM comes automatically packaged inside the text and is a dynamic, colorful, and exciting all-new interactive study tool. It includes Multimedia Chapter Summaries that contain interest-generating multimedia presentations of chapter topics with audio, photos, and selected videos; chapter-by-chapter quizzes with scoring and feedback; a direct link to the InfoTrac College Edition; and links to other online sociology resources and the book's companion Web site!

Acknowledgments

This text reflects the work of many people. We would like to thank the following for their contributions to the development of this text: Eve Howard, Bob Jucha, Julie Sakaue, Dee Dee Zobian, Stephanie Monzon, Cheri Palmer, Mary Douglas, Joohee Lee, and Karen Hunt.

We would also like to acknowledge the support and assistance of Carol Jenkins (thanks CJ), Susan Gatewood, Allison Van Wyke, Lorraine Rollins, Marieke Van Willigen, Bob Edwards, Bill Robbins, Lee Maril, Sandy Alessio, Ellen Kittredge, and Alexander Aman. To each our heartfelt thanks.

Additionally, we are indebted to those who read the manuscript in its various drafts and provided valuable insights and suggestions, many of which have been incorporated into the final manuscript:

Wendy Beck
Eastern Washington University

Deanna Chang
Indiana University of Pennsylvania

Margaret Choka
Pellissippi State Technical Community College

Robert R. Cordell
West Virginia University at Parkersburg

Kim Davies
Augusta State University

Katheryn Dietrich
Blinn College

William Feigelman
Nassau Community College

Roberta Goldberg
Trinity College

Roger Guy
Texas Lutheran University

Sandra Krell-Andre
Southeastern Community College

Pui-Yan Lam
Eastern Washington University

Dale Lund
University of Utah

JoAnn Miller
Purdue University

Frank J. Page
University of Utah

Barbara Perry
Northern Arizona University

Donna Provenza
California State University at Sacramento

Carl Marie Rider
Longwood University

Jeffrey W. Riemer
Tennessee Technological University

Cherylon Robinson
University of Texas at San Antonio

Rita Sakitt
Suffolk County Community College

D. Paul Sullins
The Catholic University of America

Robert Turley
Crafton Hills College

Robert Weaver
Youngstown State University

Robert Weyer
County College of Morris

We are also grateful to the reviewers of the first, second, and third editions: David Allen, *University of New Orleans;* Patricia Atchison, *Colorado State University;* Walter Carroll, *Bridgewater State College;* Roland Chilton, *University of Massachusetts;* Verghese Chirayath, *John Carroll University;* Kimberly Clark, *DeKalb College–Central Campus;* Anna M. Cognetto, *Dutchess Community College;* Barbara Costello, *Mississippi State University;* William Cross, *Illinois College;* Doug Degher, *Northern Arizona University;* Katherine Dietrich, *Blinn College;* Jane Ely, *State University of New York Stony Brook;* Joan Ferrante, *Northern Kentucky University;* Robert Gliner, *San Jose State University;* Julia Hall, *Drexel University;* Millie Harmon, *Chemeketa Community College;* Sylvia Jones, *Jefferson Community College;* Nancy Kleniewski, *University of Massachusetts, Lowell;* Daniel Klenow, *North Dakota State University;* Mary Ann Lamanna, *University of Nebraska;* Phyllis Langton, *George Washington University;* Cooper Lansing, *Erie Community College;* Tunga Lergo, *Santa Fe Community College, Main Campus;* Lionel Maldonado, *California State University, San Marcos;* Judith Mayo, *Arizona State University;* Peter Meiksins, *Cleveland State University;* Madonna Harrington-Meyer, *University of Illinois;* Clifford Mottaz, *University of Wisconsin–River Falls;* Lynda D. Nyce, *Bluffton College;* James Peacock, *University of North Carolina;* Ed Ponczek, *William Rainey Harper College;* Cynthia Reynaud, *Louisiana State University;* Rita Sakitt, *Suffolk County Community College;* Mareleyn Schneider, *Yeshiva University;* Paula Snyder, *Columbus State Community College;* Lawrence Stern, *Collin County Community College;* John Stratton, *University of Iowa;* Joseph Trumino, *St. Vincent's College of St. John's University;* Alice Van Ommeren, *San Joaquin Delta College;* Joseph Vielbig, *Arizona Western University;* Harry L. Vogel, *Kansas State University;* Rose Weitz, *Arizona State University;* Bob Weyer, *County College of Morris;* Oscar Williams, *Diablo Valley College;* Mark Winton, *University of Central Florida;* Diane Zablotsky, *University of North Carolina.*

Finally, we are interested in ways to improve the text and invite your feedback and suggestions for new ideas and material to be included in subsequent editions. You can contact us at mooneyl@mail.ecu.edu, knoxd@mail.ecu.edu, or schachtc@mail.ecu.edu.

About the Authors

Linda A. Mooney, Ph.D., is an Associate Professor of Sociology at East Carolina University in Greenville, North Carolina. In addition to social problems, her specialties include law, criminology, and juvenile delinquency. She has published over thirty professional articles in such journals as *Social Forces, Sociological Inquiry, Sex Roles, Sociological Quarterly,* and *Teaching Sociology.* She has won numerous teaching awards, including the University of North Carolina Board of Governor's Distinguished Professor for Teaching Award.

David Knox, Ph.D., is Professor of Sociology at East Carolina University. He has taught Social Problems, Introduction to Sociology, and Sociology of Marriage Problems. He is the author or co-author of ten books and more than sixty professional articles. His research interests include deception in relationships and how blacks and whites differ in their expectations of romantic partners.

Caroline Schacht, M.A., is an Instructor of Sociology at East Carolina University. She has taught Introduction to Sociology, Deviant Behavior, Individuals in Society, and Courtship and Marriage. She has co-authored several textbooks in the areas of social problems, introductory sociology, courtship and marriage, and human sexuality. Her areas of interest include mediation and conflict resolution, alternative education, social inequality, and environmental problems.

"Unless someone like you cares a whole awful lot, nothing is going to get better. It's not." *Dr. Seuss, The Lorax*

Thinking About Social Problems

esearchers at Fordham University conducted a study called "The Social Health of the States" that evaluates the cumulative effect on Americans of 16 major social problems including crime, unemployment, drug abuse, suicide rates, homicide rates, and child abuse. According to analyses of these 16 social indicators, many states are "experiencing significant social deterioration" characterized by or at risk of "social recession" (Fordham University 2003). Moreover, a Gallup Poll reveals that "satisfaction with the way things are going in the country" is at its lowest point since 1996 (Newport 2003).

A global perspective on social problems is also troubling. In 1990 the United Nations Development Programme published its first annual Human Development Report, which measured the well-being of populations around the world according to a "human development index" (HDI). This index measures three basic dimensions of human development — longevity, knowledge (i.e., educational attainment), and a decent standard of living. The most recent report reveals that "the world is facing an acute development crisis with many poor nations suffering severe and continuous social-economic reversals." Such reversals indicate "an urgent call for action to address health and education as well as income levels in these countries." Of the 34 lowest ranking countries, 30 are in sub-Saharan Africa (Human Development Report 2003).

Problems related to poverty and malnutrition, inadequate education, acquired immunodeficiency syndrome (AIDS) and other sexually transmitted diseases (STDs), inadequate health care, crime, conflict, oppression of minorities, environmental destruction, and other social issues are both national and international concerns. Such problems present both a threat and a challenge to our national and global society.

The primary goal of this text is to facilitate increased awareness and understanding of problematic social conditions in U.S. society and throughout the world. Although the topics covered in this text vary widely, all chapters share common objectives: to explain how social problems are created and maintained; to indicate how they affect individuals, social groups, and societies as a whole; and to examine programs and policies for change. We begin by looking at the nature of social problems.

Only relatively recently have suicide bombers been considered a social problem to the American public. More specifically, since the horror of September 11, 2001, terrorism in the United States has taken on new meaning. Here airport security guards inspect vehicles approaching the terminals.

© AP/Wide World Photos

What Is a Social Problem?

There is no universal, constant, or absolute definition of what constitutes a social problem. Rather, social problems are defined by a combination of objective and subjective criteria that vary across societies, among individuals and groups within a society, and across historical time periods.

Objective and Subjective Elements of Social Problems

Although social problems take many forms, they all share two important elements: an objective social condition and a subjective interpretation of that social condition. The **objective element** of a social problem refers to the existence of a social condition. We become aware of social conditions through our own life experience, through the media, and through education. We see the homeless, hear gunfire in the streets, and see battered women in hospital emergency rooms. We read about employees losing their jobs as businesses downsize and factories close. In television news reports we see the anguished faces of parents whose children have been killed by violent youths.

The **subjective element** of a social problem refers to the belief that a particular social condition is harmful to society, or to a segment of society, and that it should and can be changed. We know that crime, drug addiction, poverty, racism, violence, and pollution exist. These social conditions are not considered social problems, however, unless at least a segment of society believes that these conditions diminish the quality of human life.

© AP/Wide World Photos

Whereas some individuals view homosexual behavior as a social problem, others view homophobia as a social problem. In 2003, the U.S. Supreme Court held that laws banning intimate sexual relations between same sex partners are unconstitutional. Here, participants carry a giant rainbow flag during the gay pride parade in Toronto, Canada, Sunday June 29, 2003. The parade is the final event of Pride Week, Canada's largest gay and lesbian celebration.

By combining these objective and subjective elements, we arrive at the following definition: A **social problem** is a social condition that a segment of society views as harmful to members of society and in need of remedy.

Variability in Definitions of Social Problems

Individuals and groups frequently disagree about what constitutes a social problem. For example, some Americans view the availability of abortion as a social problem, while others view restrictions on abortion as a social problem. Similarly, some Americans view homosexuality as a social problem, while others view prejudice and discrimination against homosexuals as a social problem. Such variations in what is considered a social problem are due to differences in values, beliefs, and life experiences.

Definitions of social problems vary not only within societies, but across societies and historical time periods as well. For example, prior to the nineteenth century, it was a husband's legal right and marital obligation to discipline and control his wife through the use of physical force. Today, the use of physical force is regarded as a social problem rather than a marital right.

Tea drinking is another example of how what is considered a social problem can change over time. In seventeenth- and eighteenth-century England, tea drinking was regarded as a "base Indian practice" that was "pernicious to health, obscuring industry, and impoverishing the nation" (Ukers 1935, cited in Troyer & Markle 1984). Today, the English are known for their tradition of drinking tea in the afternoon.

Because social problems can be highly complex, it is helpful to have a framework within which to view them. Sociology provides such a framework. Using a sociological perspective to examine social problems requires a knowledge of the basic concepts and tools of sociology. In the remainder of this chapter, we discuss some of these concepts and tools: social structure, culture, the "sociological imagination," major theoretical perspectives, and types of research methods.

Elements of Social Structure and Culture

Although society surrounds us and permeates our lives, it is difficult to "see" society. By thinking of society in terms of a picture or image, however, we can visualize society and, therefore, better understand it. Imagine that society is a coin with two sides: on one side is the structure of society, and on the other is the culture of society. Although each "side" is distinct, both are inseparable from the whole. By looking at the various elements of social structure and culture, we can better understand the root causes of social problems.

Elements of Social Structure

The *structure* of a society refers to the way society is organized. Society is organized into different parts: institutions, social groups, statuses, and roles.

Institutions. An **institution** is an established and enduring pattern of social relationships. The five traditional institutions are family, religion, politics, economics, and education, but some sociologists argue that other social institutions, such as science and technology, mass media, medicine, sports, and the military, also play important roles in modern society.

Many social problems are generated by inadequacies in various institutions. For example, unemployment may be influenced by the educational institution's failure to prepare individuals for the job market and by alterations in the structure of the economic institution.

Social Groups. Institutions are made up of social groups. A **social group** is defined as two or more people who have a common identity, interact, and form a social relationship. For example, the family in which you were reared is a social group that is part of the family institution. The religious association to which you may belong is a social group that is part of the religious institution.

Social groups may be categorized as primary or secondary. **Primary groups,** which tend to involve small numbers of individuals, are characterized by intimate and informal interaction. Families and friends are examples of primary groups. **Secondary groups,** which may involve small or large numbers of individuals, are task-oriented and characterized by impersonal and formal interaction. Examples of secondary groups include employers and their employees, and clerks and their customers.

Statuses. Just as institutions consist of social groups, social groups consist of statuses. A ***status*** is a position a person occupies within a social group. The statuses we occupy largely define our social identity. The statuses in a family may consist of mother, father, stepmother, stepfather, wife, husband, child, and so on. Statuses may be either ascribed or achieved. An **ascribed status** is one that society assigns to an individual on the basis of factors over which the individual has no control. For example, we have no control over the sex, race, ethnic background, and socioeconomic status into which we are born. Similarly, we are assigned the status of "child," "teenager," "adult," or "senior citizen" on the basis of our age—something we do not choose or control.

An **achieved status** is assigned on the basis of some characteristic or behavior over which the individual has some control. Whether you achieve the status of college graduate, spouse, parent, bank president, or prison inmate depends largely on your own efforts, behavior, and choices. One's ascribed statuses may affect the likelihood of achieving other statuses, however. For example, if you are born into a poor socioeconomic status, you may find it more difficult to achieve the status of "college graduate" because of the high cost of a college education.

Every individual has numerous statuses simultaneously. You may be a student, parent, tutor, volunteer fund-raiser, female, and Hispanic. A person's **master status** is the status that is considered the most significant in a person's social identity. Typically, a person's occupational status is regarded as his or her master status. If you are a full-time student, your master status is likely to be "student."

"When I fulfill my obligations as a brother, husband, or citizen, when I execute contracts, I perform duties that are defined externally to myself. . . . Even if I conform in my own sentiments and feel their reality subjectively, such reality is still objective, for I did not create them; I merely inherited them."

Emile Durkheim
Sociologist

Roles. Every status is associated with many **roles,** or the set of rights, obligations, and expectations associated with a status. Roles guide our behavior and allow us to predict the behavior of others. As a student, you are expected to attend class, listen and take notes, study for tests, and complete assignments. Because you know what the role of teacher involves, you can predict that your teacher will lecture, give exams, and assign grades based on your performance on tests.

A single status involves more than one role. For example, the status of prison inmate includes one role for interacting with prison guards and another role for interacting with other prison inmates. Similarly, the status of nurse involves different roles for interacting with physicians and with patients.

Elements of Culture

Whereas social structure refers to the organization of society, culture refers to the meanings and ways of life that characterize a society. The elements of culture include beliefs, values, norms, sanctions, and symbols.

Beliefs. **Beliefs** refer to definitions and explanations about what is assumed to be true. The beliefs of an individual or group influence whether that individual or group views a particular social condition as a social problem. Does secondhand smoke harm nonsmokers? Are nuclear power plants safe? Does violence in movies and on television lead to increased aggression in children? Our beliefs regarding these issues influence whether we view the issues as social problems. Beliefs not only influence how a social condition is interpreted, but they also influence the existence of the condition itself. For example, police officers' beliefs about their supervisors' priorities impacted officers' problem-solving behavior and the time devoted to it (Engel & Worden 2003). The *Self and Society* feature on the next page allows you to assess your own beliefs about various social issues and compare your beliefs with a national sample of first-year college students.

Values. **Values** are social agreements about what is considered good and bad, right and wrong, desirable and undesirable. Frequently, social conditions are viewed as social problems when the conditions are incompatible with or contradict closely held values. For example, poverty and homelessness violate the value of human welfare; crime contradicts the values of honesty, private property, and nonviolence; racism, sexism, and heterosexism violate the values of equality and fairness.

Values play an important role not only in the interpretation of a condition as a social problem, but also in the development of the social condition itself. Sylvia Ann Hewlett (1992) explains how the American values of freedom and individualism are at the root of many of our social problems:

> There are two sides to the coin of freedom. On the one hand, there is enormous potential for prosperity and personal fulfillment; on the other are all the hazards of untrammeled opportunity and unfettered choice. Free markets can produce grinding poverty as well as spectacular wealth; unregulated industry can create dangerous levels of pollution as well as rapid rates of growth; and an unfettered drive for personal fulfillment can have disastrous effects on families and children. Rampant individualism does not bring with it sweet freedom; rather, it explodes in our faces and limits life's potential. (pp. 350–351)

Absent or weak values may contribute to some social problems. For example, many industries do not value protection of the environment and thus contribute to environmental pollution.

Norms and Sanctions. **Norms** are socially defined rules of behavior. Norms serve as guidelines for our behavior and for our expectations of the behavior of others.

There are three types of norms: folkways, laws, and mores. **Folkways** refer to the customs and manners of society. In many segments of our society, it is customary to shake hands when being introduced to a new acquaintance, to say "excuse me" after sneezing, and to give presents to family and friends on their birthdays. Although no laws require us to do these things, we are expected to do them because they are part of the cultural traditions, or folkways, of the society in which we live.

Self and Society

Personal Beliefs About Various Social Problems

Indicate whether you agree or disagree with each of the following statements:

Statement	Agree	Disagree
1. Federal military spending should be increased.	______	______
2. Colleges should prohibit racist/sexist speech on campus.	______	______
3. There is too much concern in the courts for the rights of criminals.	______	______
4. Abortion should be legal.	______	______
5. The death penalty should be abolished.	______	______
6. The activities of married women are best confined to the home and family.	______	______
7. Marijuana should be legalized.	______	______
8. It is important to have laws prohibiting homosexual relationships.	______	______
9. People should not obey laws that violate their personal values.	______	______
10. The federal government should do more to control the sale of handguns.	______	______
11. Racial discrimination is no longer a major problem in America.	______	______
12. Realistically, an individual can do little to bring about changes in our society.	______	______
13. Wealthy people should pay a larger share of taxes than they do now.	______	______
14. Affirmative action in college admissions should be abolished.	______	______
15. Same-sex couples should have the right to legal marital status.	______	______

Percentage* of First-Year College Students Agreeing with Belief Statements

	Percentage Agreeing in 2002		
Statement Number	**Total**	**Women**	**Men**
1. Military spending	45	40	51
2. Prohibit speech on campus	60	65	54
3. Too much concern for criminals' rights	64	63	66
4. Abortion rights	54	53	54
5. Abolishment of death penalty	32	35	28
6. Women's activities confined to home	22	17	28
7. Legalization of marijuana	40	35	46
8. Laws prohibiting gay relationships	25	19	33
9. Personal values	35	32	40
10. Federal control of handgun sales	78	85	69
11. Racial discrimination not a problem	22	18	26
12. Individuals can't influence social change	28	24	32
13. Wealthy should pay higher taxes	50	50	50
14. Affirmative action abolished in college	49	44	55
15. Legal right of same-sex couples to marry	59	66	51

*Percentages are rounded.

Source: The American Freshman: *National Norms for Fall 2002.* Los Angeles: Higher Education Research Institute, UCLA. Copyright © 2002 by the Regents of the University of California. Used by permission.

Table 1.1
Types and Examples of Sanctions

	Positive	Negative
Informal	Being praised by one's neighbors for organizing a neighborhood recycling program.	Being criticized by one's neighbors for refusing to participate in the neighborhood recycling program.
Formal	Being granted a citizen's award for organizing a neighborhood recycling program.	Being fined by the city for failing to dispose of trash properly.

Laws are norms that are formalized and backed by political authority. It is normative for a Muslim woman to wear a veil. However, in the United States, failure to remove the veil for a driver's license photo is grounds for revoking the permit. Such is the case of a Florida woman who has brought suit against the state claiming that her religious rights are being violated because she is required to remove her veil for the driver's license photo (Canedy 2003).

Some norms, called **mores,** have a moral basis. Violations of mores may produce shock, horror, and moral indignation. Both littering and child sexual abuse are violations of law, but child sexual abuse is also a violation of our mores because we view such behavior as immoral.

All norms are associated with **sanctions,** or social consequences for conforming to or violating norms. When we conform to a social norm, we may be rewarded by a positive sanction. These may range from an approving smile to a public ceremony in our honor. When we violate a social norm, we may be punished by a negative sanction, which may range from a disapproving look to the death penalty or life in prison. Most sanctions are spontaneous expressions of approval or disapproval by groups or individuals—these are referred to as informal sanctions. Sanctions that are carried out according to some recognized or formal procedure are referred to as formal sanctions. Types of sanctions, then, include positive informal sanctions, positive formal sanctions, negative informal sanctions, and negative formal sanctions (see Table 1.1).

Symbols. A **symbol** is something that represents something else. Without symbols, we could not communicate with each other or live as social beings.

The symbols of a culture include language, gestures, and objects whose meaning is commonly understood by the members of a society. In our society, a red ribbon tied around a car antenna symbolizes Mothers Against Drunk Driving, a peace sign symbolizes the value of nonviolence, and a white hooded robe symbolizes the Ku Klux Klan. Sometimes people attach different meanings to the same symbol. The Confederate flag is a symbol of southern pride to some, a symbol of racial bigotry to others.

The elements of the social structure and culture just discussed play a central role in the creation, maintenance, and social response to various social problems. One of the goals of taking a course in social problems is to develop an awareness of how the elements of social structure and culture contribute to social problems. Sociologists refer to this awareness as the "sociological imagination."

The Sociological Imagination

The **sociological imagination,** a term developed by C. Wright Mills (1959), refers to the ability to see the connections between our personal lives and the social world in which we live. When we use our sociological imagination, we are able to distinguish between "private troubles" and "public issues" and to see connections between the events and conditions of our lives and the social and historical context in which we live.

For example, that one man is unemployed constitutes a private trouble. That millions of people are unemployed in the United States constitutes a public issue. Once we understand that personal troubles such as HIV infection, criminal victimization, and poverty are shared by other segments of society, we can look for the elements of social structure and culture that contribute to these public issues and private troubles. If the various elements of social structure and culture contribute to private troubles and public issues, then society's social structure and culture must be changed if these concerns are to be resolved.

Rather than viewing the private trouble of being unemployed as a result of an individual's faulty character or lack of job skills, we may understand unemployment as a public issue that results from the failure of the economic and political institutions of society to provide job opportunities to all citizens. Technological innovations emerging from the Industrial Revolution led to individual workers' being replaced by machines. During the economic recession of the 1980s, employers fired employees so the firm could stay in business. Thus, in both these cases, social forces rather than individual skills largely determined whether a person was employed.

Theoretical Perspectives

Theories in sociology provide us with different perspectives with which to view our social world. A perspective is simply a way of looking at the world. A theory is a set of interrelated propositions or principles designed to answer a question or explain a particular phenomenon; it provides us with a perspective. Sociological theories help us to explain and predict the social world in which we live.

Sociology includes three major theoretical perspectives: the structural-functionalist perspective, the conflict perspective, and the symbolic interactionist perspective. Each perspective offers a variety of explanations about the causes of and possible solutions for social problems.

Structural-Functionalist Perspective

The structural-functionalist perspective is based largely on the works of Herbert Spencer, Emile Durkheim, Talcott Parsons, and Robert Merton. According to **structural-functionalism,** society is a system of interconnected parts that work together in harmony to maintain a state of balance and social equilibrium for the whole. For example, each of the social institutions contributes important functions for society: family provides a context for reproducing, nurturing, and socializing children; education offers a way to transmit a society's skills, knowledge, and culture to its youth; politics provides a means of governing members of society; economics provides for the production, distribution, and consumption of goods and services; and religion provides moral guidance and an outlet for worship of a higher power.

The structural-functionalist perspective emphasizes the interconnectedness of society by focusing on how each part influences and is influenced by other parts. For example, the increase in single-parent and dual-earner families has contributed to the number of children who are failing in school because parents have become less available to supervise their children's homework. As a result of changes in technology, colleges are offering more technical programs, and many adults are returning to school to learn new skills that are required in the workplace. The increasing number of women in the workforce has contributed to the formulation of policies against sexual harassment and job discrimination.

Structural-functionalists use the terms "functional" and "dysfunctional" to describe the effects of social elements on society. Elements of society are functional if they contribute to social stability and dysfunctional if they disrupt social stability. Some aspects of society may be both functional and dysfunctional for society. For example, crime is dysfunctional in that it is associated with physical violence, loss of property, and fear. But according to Durkheim and other functionalists, crime is also functional for society because it leads to heightened awareness of shared moral bonds and increased social cohesion.

Sociologists have identified two types of functions: manifest and latent (Merton 1968). **Manifest functions** are consequences that are intended and commonly recognized. **Latent functions** are consequences that are unintended and often hidden. For example, the manifest function of education is to transmit knowledge and skills to society's youth. But public elementary schools also serve as baby-sitters for employed parents, and colleges offer a place for young adults to meet potential mates. The baby-sitting and mate selection functions are not the intended or commonly recognized functions of education—hence, they are latent functions.

Structural-Functionalist Theories of Social Problems

Two dominant theories of social problems grew out of the structural-functionalist perspective: social pathology and social disorganization.

Social Pathology. According to the social pathology model, social problems result from some "sickness" in society. Just as the human body becomes ill when our systems, organs, and cells do not function normally, society becomes "ill" when its parts (i.e., elements of the structure and culture) no longer perform properly. For example, problems such as crime, violence, poverty, and juvenile delinquency are often attributed to the breakdown of the family institution; the decline of the religious institution; and inadequacies in our economic, educational, and political institutions.

Social "illness" also results when members of a society are not adequately socialized to adopt its norms and values. Persons who do not value honesty, for example, are prone to dishonesties of all sorts. Early theorists attributed the failure in socialization to "sick" people who could not be socialized. Later theorists recognized that failure in the socialization process stemmed from "sick" social conditions, not "sick" people. To prevent or solve social problems, members of society must receive proper socialization and moral education, which may be accomplished in the family, schools, churches, workplace, and/or through the media.

Social Disorganization. According to the social disorganization view of social problems, rapid social change disrupts the norms in a society. When norms become weak or are in conflict with each other, society is in a state of **anomie** or normlessness. Hence, people may steal, physically abuse their spouse or children, abuse drugs, commit rape, or engage in other deviant behavior because the norms regarding these behaviors are weak or conflicting. According to this view, the solution to social problems lies in slowing the pace of social change and strengthening social norms. For example, although the use of alcohol by teenagers is considered a violation of a social norm in our society, this norm is weak. The media portray young people drinking alcohol, teenagers teach each other to drink alcohol and buy fake identification cards (IDs) to purchase alcohol, and parents model drinking behavior by having a few drinks after work or at a social event. Solutions to teenage drinking may involve strengthening norms against it through public education, restricting media depictions of youth and alcohol, imposing stronger sanctions against the use of fake IDs to purchase alcohol, and educating parents to model moderate and responsible drinking behavior.

Conflict Perspective

Whereas the structural-functionalist perspective views society as comprising different parts working together, the **conflict perspective** views society as comprising different groups and interests competing for power and resources. The conflict perspective explains various aspects of our social world by looking at which groups have power and benefit from a particular social arrangement.

The origins of the conflict perspective can be traced to the classic works of Karl Marx. Marx suggested that all societies go through stages of economic development. As societies evolve from agricultural to industrial, concern over meeting survival needs is replaced by concern over making a profit, the hallmark of a capitalist system. Industrialization leads to the development of two classes of people: the bourgeoisie, or the owners of the means of production (e.g., factories, farms, businesses), and the proletariat, or the workers who earn wages.

The division of society into two broad classes of people—the "haves" and the "have-nots"—is beneficial to the owners of the means of production. The workers, who may earn only subsistence wages, are denied access to the many resources available to the wealthy owners. According to Marx, the bourgeoisie use their power to control the institutions of society to their advantage. For example, Marx suggested that religion serves as an "opiate of the masses" in that it soothes the distress and suffering associated with the working-class lifestyle and focuses the workers' attention on spirituality, God, and the afterlife rather than on such worldly concerns as living conditions. In essence, religion diverts the workers so that they concentrate on being rewarded in heaven for living a moral life rather than on questioning their exploitation.

Conflict Theories of Social Problems

There are two general types of conflict theories of social problems: Marxist and non-Marxist. Marxist theories focus on social conflict that results from economic inequalities; non-Marxist theories focus on social conflict that results from competing values and interests among social groups.

"Underlying virtually all social problems are conditions caused in whole or in part by social injustice."

Pamela Ann Roby
Sociologist, University of California, Santa Cruz

Marxist Conflict Theories. According to contemporary Marxist theorists, social problems result from class inequality inherent in a capitalistic system. A system of "haves" and "have-nots" may be beneficial to the "haves" but often translates into poverty for the "have-nots." As we shall explore later in this text, many social problems, including physical and mental illness, low educational achievement, and crime, are linked to poverty.

In addition to creating an impoverished class of people, capitalism also encourages "corporate violence." Corporate violence may be defined as actual harm and/or risk of harm inflicted on consumers, workers, and the general public as a result of decisions by corporate executives or managers. Corporate violence may also result from corporate negligence; the quest for profits at any cost; and willful violations of health, safety, and environmental laws (Reiman 2003). Our profit-motivated economy encourages individuals who are otherwise good, kind, and law-abiding to knowingly participate in the manufacturing and marketing of defective brakes on American jets, fuel tanks on automobiles, and contraceptive devices (i.e., intrauterine devices [IUDs]). The profit motive has also caused individuals to sell defective medical devices, toxic pesticides, and contaminated foods to developing countries. As Eitzen and Baca Zinn note, the "goal of profit is so central to capitalistic enterprises that many corporate decisions are made without consideration for the consequences" (Eitzen & Baca Zinn 2000, 483).

Marxist conflict theories also focus on the problem of **alienation,** or powerlessness and meaninglessness in people's lives. In industrialized societies, workers often have little power or control over their jobs, a condition that fosters in them a sense of powerlessness in their lives. The specialized nature of work requires workers to perform limited and repetitive tasks; as a result, the workers may come to feel that their lives are meaningless.

Alienation is bred not only in the workplace but also in the classroom. Students have little power over their education and often find the curriculum is not meaningful to their lives. Like poverty, alienation is linked to other social problems, such as low educational achievement, violence, and suicide.

Marxist explanations of social problems imply that the solution lies in eliminating inequality among classes of people by creating a classless society. The nature of work must also change to avoid alienation. Finally, stronger controls must be applied to corporations to ensure that corporate decisions and practices are based on safety rather than profit considerations.

Non-Marxist Conflict Theories. Non-Marxist conflict theorists such as Ralf Dahrendorf are concerned with conflict that arises when groups have opposing values and interests. For example, anti-abortion activists value the life of unborn embryos and fetuses; pro-choice activists value the right of women to control their own body and reproductive decisions. These different value positions reflect different subjective interpretations of what constitutes a social problem. For anti-abortionists, the availability of abortion is the social problem; for pro-choice advocates, restrictions on abortion are the social problem. Sometimes the social problem is not the conflict itself but rather the way that conflict is expressed. Even most pro-life advocates agree that shooting doctors who perform abortions and blowing up abortion clinics constitute unnecessary violence and lack of respect for life. Value conflicts may occur between diverse categories of people, including non-whites versus whites, heterosexuals versus homosexuals, young versus old, Democrats versus Republicans, and environmentalists versus industrialists.

Solving the problems that are generated by competing values may involve ensuring that conflicting groups understand each other's views, resolving differences through negotiation or mediation, or agreeing to disagree. Ideally, solutions should be win-win, with both conflicting groups satisfied with the solution. However, outcomes of value conflicts are often influenced by power; the group with the most power may use its position to influence the outcome of value conflicts. For example, when Congress could not get all states to voluntarily increase the legal drinking age to 21, it threatened to withdraw federal highway funds from those that would not comply.

Symbolic Interactionist Perspective

Both the structural-functionalist and the conflict perspectives are concerned with how broad aspects of society, such as institutions and large social groups, influence the social world. This level of sociological analysis is called **macro sociology:** It looks at the "big picture" of society and suggests how social problems are affected at the institutional level.

> "Each to each a looking glass, Reflects the other that doth pass."
>
> **Charles Horton Cooley**
> **Sociologist**

Micro sociology, another level of sociological analysis, is concerned with the social psychological dynamics of individuals interacting in small groups. **Symbolic interactionism** reflects the micro sociological perspective and was largely influenced by the work of early sociologists and philosophers such as Max Weber, George Simmel, Charles Horton Cooley, G. H. Mead, W. I. Thomas, Erving Goffman, and Howard Becker. Symbolic interactionism emphasizes that human behavior is influenced by definitions and meanings that are created and maintained through symbolic interaction with others.

Sociologist W. I. Thomas (1931/1966) emphasized the importance of definitions and meanings in social behavior and its consequences. He suggested that humans respond to their definition of a situation rather than to the objective situation itself. Hence, Thomas noted that situations we define as real become real in their consequences.

Symbolic interactionism also suggests that our identity or sense of self is shaped by social interaction. We develop our self-concept by observing how others interact with us and label us. By observing how others view us, we see a reflection of ourselves that Cooley calls the "looking glass self."

Lastly, the symbolic interaction perspective has important implications for how social scientists conduct research. The German sociologist Max Weber (1864–1920) argued that to understand individual and group behavior, social scientists must see the world through the eyes of that individual or group. Weber called this approach *Verstehen,* which in German means "empathy." *Verstehen* implies that in conducting research, social scientists must try to understand others' view of reality and the subjective aspects of their experiences, including their symbols, values, attitudes, and beliefs.

Symbolic Interactionist Theories of Social Problems

A basic premise of symbolic interactionist theories of social problems is that a condition must be defined or recognized as a social problem for it to be a social problem. Based on this premise, Herbert Blumer (1971) suggested that social problems develop in stages. First, social problems pass through the stage of "societal recognition"—the process by which a social problem, for example, drunk driving, is "born." Second, "social legitimation" takes place when the social problem achieves recognition by the larger community, including the media, schools, and churches. As the visibility of traffic fatalities associated with alcohol increased, so did the legitimation of drunk driving as a social problem. The next stage in the development of a social problem involves "mobilization for action," which occurs when individuals and groups, such as Mothers Against Drunk Driving, become concerned about how to respond to the social condition. This mobilization leads to the "development and implementation of an official plan" for dealing with the problem, involving, for example, highway checkpoints, lower legal blood-alcohol levels, and tougher drunk driving regulations.

Blumer's stage development view of social problems is helpful in tracing the development of social problems. For example, although sexual harassment and date rape have occurred throughout this century, these issues did not begin to receive recognition as social problems until the 1970s. Social legitimation of these problems was achieved when high schools, colleges, churches, employers, and the media recognized their existence. Organized social groups mobilized to develop and implement plans to deal with these problems. For example, groups successfully lobbied for the enactment of laws against sexual harassment and the enforcement of sanctions against violators of these laws. Groups also mobilized to provide educational seminars on date rape for high school and college students and to offer support services to victims of date rape.

Some disagree with the symbolic interactionist view that social problems exist only if they are recognized. According to this view, individuals who were victims of date rape in the 1960s may be considered victims of a problem, even though date rape was not recognized at that time as a social problem.

Labeling theory, a major symbolic interactionist theory of social problems, suggests that a social condition or group is viewed as problematic if it is labeled as such. According to labeling theory, resolving social problems sometimes involves changing the meanings and definitions that are attributed to people and situations. For example, as long as teenagers define drinking alcohol as "cool" and "fun," they will continue to abuse alcohol. As long as our society defines providing sex education and contraceptives to teenagers as inappropriate or immoral, the teenage pregnancy rate in our country will continue to be higher than in other industrialized nations.

Social constructionism is another symbolic interactionist theory of social problems. Similar to labeling theorists and symbolic interactionism in general, social constructionists argue that reality is socially constructed by individuals who interpret the social world around them. Society, therefore, is a social creation rather than an objective given. As such, social constructionists often question the origin and evolution of social problems. For example, most Americans define "drug abuse" as a social problem in the United States but rarely include alcohol or cigarettes in their discussion. A social constructionist would point to the historical roots of alcohol and tobacco use as a means of understanding their legal status. Central to this idea of the social construction of social problems are the media, universities, research institutes, and government agencies that are often responsible for the public's initial "take" on the problem under discussion.

Table 1.2 summarizes and compares the major theoretical perspectives, their criticisms, and social policy recommendations as they relate to social problems. The study of social problems is based on research as well as theory, however. Indeed, research and theory are intricately related. As Wilson (1983) states,

> Most of us think of theorizing as quite divorced from the business of gathering facts. It seems to require an abstractness of thought remote from the practical activity of empirical research. But theory building is not a separate activity within sociology. Without theory, the empirical researcher would find it impossible to decide what to observe, how to observe it, or what to make of the observations. (p. 1)

Table 1.2
Comparison of Theoretical Perspectives

	Structural-Functionalism	Conflict Theory	Symbolic Interactionism
Representative Theorists	Emile Durkheim Talcott Parsons Robert Merton	Karl Marx Ralf Dahrendorf	George H. Mead Charles Cooley Erving Goffman
Society	Society is a set of interrelated parts; cultural consensus exists and leads to social order; natural state of society—balance and harmony.	Society is marked by power struggles over scarce resources; inequities result in conflict; social change is inevitable; natural state of society—imbalance.	Society is a network of interlocking roles; social order is constructed through interaction as individuals, through shared meaning, make sense out of their social world.
Individuals	Individuals are socialized by society's institutions; socialization is the process by which social control is exerted; people need society and its institutions.	People are inherently good but are corrupted by society and its economic structure; institutions are controlled by groups with power; "order" is part of the illusion.	Humans are interpretative and interactive; they are constantly changing as their "social beings" emerge and are molded by changing circumstances.
Cause of Social Problems?	Rapid social change: social disorganization that disrupts the harmony and balance; inadequate socialization and/or weak institutions.	Inequality; the dominance of groups of people over other groups of people; oppression and exploitation; competition between groups.	Different interpretations of roles; labeling of individuals, groups, or behaviors as deviant; definition of an objective condition as a social problem.
Social Policy/Solutions	Repair weak institutions; assure proper socialization; cultivate a strong collective sense of right and wrong.	Minimize competition; create an equitable system for the distribution of resources.	Reduce impact of labeling and associated stigmatization; alter definitions of what is defined as a social problem.
Criticisms	Called "sunshine sociology"; supports the maintenance of the status quo; needs to ask "functional for whom?" Does not deal with issues of power and conflict; incorrectly assumes a consensus.	Utopian model; Marxist states have failed; denies existence of cooperation and equitable exchange. Can't explain cohesion and harmony.	Concentrates on micro issues only; fails to link micro issues to macro-level concerns; too psychological in its approach; assumes label amplified problem.

Social Problems Research

Most students taking a course in social problems will not become researchers or conduct research on social problems. Nevertheless, we are all consumers of research that is reported in the media. Politicians, social activist groups, and organizations attempt to justify their decisions, actions, and positions by citing research

Social Problems Research Up Close

The Sociological Enterprise

Each chapter in the book contains a *Social Problems Research Up Close* box that describes a research report or journal article that examines some sociologically significant topic. Some examples of the more prestigious journals in sociology include the *American Sociological Review*, the *American Journal of Sociology*, and *Social Forces*. Journal articles are the primary means by which sociologists, as well as other scientists, exchange ideas and information. Most journal articles begin with *an introduction and review of the literature.* It is here that the author examines previous research on the topic, identifies specific research areas, and otherwise "sets the stage" for the reader. It is often in this section that research hypotheses, if applicable, are set forth. A researcher, for example, might hypothesize that the sexual behavior of adolescents has changed over the years as a consequence of increased fear of sexually transmitted diseases, and that such changes vary on the basis of sex.

The next major section of a journal article is entitled *sample and methods.* In this section the author describes the characteristics of the sample, if any, and the details of the type of research conducted. The type of data analysis used is also presented in this section (see Appendix A). Using the research question above, a sociologist might obtain data from the Youth Risk Behavior Surveillance Survey collected by the Centers for Disease Control. This self-administered questionnaire is distributed biennially to over 10,000 high school students across the United States.

The final section of a journal article includes the *findings and conclusions.* The findings of a study describe the results, that is, what the researcher found as a result of the investigation. Findings are then discussed within the context of the hypotheses and the conclusions that can be drawn. Often research results are presented in tabular form. Reading tables carefully is an important part of drawing accurate conclusions about the research hypotheses. In reading a table you should follow the steps below (see table on next page):

1. *Read the title of the table and make sure that you understand what the table contains.* The title of the table indicates the unit of analysis (high school students), the dependent variable (sexual risk behaviors), the independent variables (sex and year), and what the numbers represent (percentages).
2. *Read the information contained at the bottom of the table including the source and any other explanatory information.* For example, the information at the bottom of this table indicates that the data are from the Centers for Disease Control, that sexually active was defined as having intercourse in the last three months, and that data on condom use were only from those students who were defined as currently sexually active.
3. *Examine the row and column headings.* This table looks at the percentage of males and females, over four years, that reported ever having sexual intercourse, having four or more sex partners in a lifetime, being currently sexually active, and using condoms during the last sexual intercourse.
4. *Thoroughly examine the data contained within the table carefully looking for patterns between variables.* As indicated in the table, the first three columns indicate that "risky" sexual behaviors of both males and females, in general, have decreased between the years surveyed. There are several exceptions, however. For example, between 1997 and 1999 the percentage of males ever having sexual intercourse, having four or more sex partners, or currently being sexually active increased.
5. *Use the information you have gathered in step 4 to address the hypotheses.* Clearly sexual practices, as hypothesized, have changed over time. Not only

results. As consumers of research, we need to understand that our personal experiences and casual observations are less reliable than generalizations based on systematic research. One strength of scientific research is that it is subjected to critical examination by other researchers (see this chapter's *Social Problems Research Up Close* feature). The more you understand how research is done, the better able you will be to critically examine and question research rather than to passively consume research findings. The remainder of this section discusses the stages of conducting a research study and the various methods of research used by sociologists.

have "risky" sexual practices, in general, declined over the time period study, but there has also been a general increase in condom use during sexual intercourse for both males and females.

6. *Draw conclusions consistent with the information presented.* From the table can we conclude that sexual practices have changed over time? The answer is probably yes although the limitations of the survey, the sample, and the measurement techniques used always should be considered. Can we conclude, however, that the observed changes are a consequence of the fear of sexually transmitted diseases? Although the data may imply it, having no measure of fear of sexually transmitted diseases over the time period studied, we would be premature to come to such a conclusion. More information, from a variety of sources, is needed. The use of multiple methods and approaches to study a social phenomenon is called **triangulation.**

Percentage of High School Students Reporting Sexual Risk Behaviors, by Sex and Survey Year

Survey year	Ever had sexual intercourse	Four or more sex partners during lifetime	Currently sexually active[a]	Condom used during last intercourse[b]
Male				
1995	54.0	20.9	35.5	60.5
1997	48.8	17.6	33.4	62.5
1999	52.2	19.3	36.2	65.5
2001	48.5	17.2	33.4	65.1
Female				
1995	52.1	14.4	40.4	48.6
1997	47.7	14.1	36.5	50.8
1999	47.7	13.1	36.3	50.7
2001	42.9	11.4	33.4	51.3

[a] Sexual intercourse during the three months preceding the survey.

[b] Among currently sexually active students.

Sources: Youth Risk Behavior Survey, Centers for Disease Control. 1999, Tables 30 and 32; "Trends in Sexual Risk Behaviors among High School Students—United States, 1991–1997." 1998. *Morbidity and Mortality Weekly Report* 47, September 18. Youth Risk Behavior Surveillance System. Center for Disease Control. 2002. "Sexual Behaviors." http://apps.nccde.cdc.gov/yrbss

Stages of Conducting a Research Study

Sociologists progress through various stages in conducting research on a social problem. This section describes the first four stages: formulating a research question, reviewing the literature, defining variables, and formulating a hypothesis.

Formulating a Research Question. A research study usually begins with a research question. Where do research questions originate? How does a particular researcher come to ask a particular research question? In some cases, researchers have

"In science (as in everyday life) things must be believed in order to be seen as well as seen in order to be believed."

Walter L. Wallace
Social scientist

a personal interest in a specific topic because of their own life experience. For example, a researcher who has experienced spouse abuse may wish to do research on such questions as "What factors are associated with domestic violence?" and "How helpful are battered women's shelters in helping abused women break the cycle of abuse in their lives?" Other researchers may ask a particular research question because of their personal values—their concern for humanity and the desire to improve human life. Researchers who are concerned about the spread of human immunodeficiency virus (HIV) infection and AIDS may conduct research on such questions as "How does the use of alcohol influence condom use?" and "What educational strategies are effective for increasing safer sex behavior?" Researchers may also want to test a particular sociological theory, or some aspect of it, to establish its validity or conduct studies to evaluate the effect of a social policy or program. Research questions may also be formulated by the concerns of community groups and social activist organizations in collaboration with academic researchers. Government and industry also hire researchers to answer questions such as "How many children are victimized by episodes of violence at school?" and "What types of computer technologies can protect children against being exposed to pornography on the Internet?"

Reviewing the Literature. After a research question is formulated, the researcher reviews the published material on the topic to find out what is already known about it. Reviewing the literature also provides researchers with ideas about how to conduct their research and helps them formulate new research questions. A literature review serves as an evaluation tool, allowing a comparison of research findings and other sources of information, such as expert opinions, political claims, and journalistic reports.

Defining Variables. A **variable** is any measurable event, characteristic, or property that varies or is subject to change. Researchers must operationally define the variables they study. An **operational definition** specifies how a variable is to be measured. For example, an operational definition of the variable "religiosity" might be the number of times the respondent reports going to church or synagogue. Another operational definition of "religiosity" might be the respondent's answer to the question, "How important is religion in your life?" (1 = not important, 2 = somewhat important, 3 = very important.)

Operational definitions are particularly important for defining variables that cannot be directly observed. For example, researchers cannot directly observe concepts such as "mental illness," "sexual harassment," "child neglect," "job satisfaction," and "drug abuse." Nor can researchers directly observe perceptions, values, and attitudes.

"My latest survey shows that people don't believe in surveys."

Laurence Peter
Humorist

Formulating a Hypothesis. After defining the research variables, researchers may formulate a **hypothesis,** which is a prediction or educated guess about how one variable is related to another variable. The **dependent variable** is the variable that the researcher wants to explain; that is, it is the variable of interest. The **independent variable** is the variable that is expected to explain change in the dependent variable. In formulating a hypothesis, the researcher predicts how the independent variable affects the dependent variable. For example, Kmec (2003) investigated the impact of segregated work environments on minority wages, concluding that "minority concentration in different jobs, occupations, and establishments than whites is a considerable social problem because it perpetuates racial wage inequality"

(p. 55). In this example, the independent variable is workplace segregation and the dependent variable is wages.

In studying social problems, researchers often assess the effects of several independent variables on one or more dependent variables. Jekielek (1998) examined the impact of parental conflict and marital disruption (two independent variables) on the emotional well-being of children (the dependent variable). Her research found that both parental conflict and marital disruption (separation or divorce) negatively affect children's emotional well-being. However, children in high-conflict intact families exhibit lower levels of well-being than children who have experienced high levels of parental conflict but whose parents divorce or separate.

Methods of Data Collection

After identifying a research topic, reviewing the literature, defining the variables, and developing hypotheses, researchers decide which method of data collection to use. Alternatives include experiments, surveys, field research, and secondary data.

Experiments. Experiments involve manipulating the independent variable to determine how it affects the dependent variable. Experiments require one or more experimental groups that are exposed to the experimental treatment(s) and a control group that is not exposed. After the researcher randomly assigns participants to either an experimental or a control group, she or he measures the dependent variable. After the experimental groups are exposed to the treatment, the research measures the dependent variable again. If participants have been randomly assigned to the different groups, the researcher may conclude that any difference in the dependent variable among the groups is due to the effect of the independent variable.

An example of a "social problems" experiment on poverty would be to provide welfare payments to one group of unemployed single mothers (experimental group) and no such payments to another group of unemployed single mothers (control group). The independent variable would be welfare payments; the dependent variable would be employment. The researcher's hypothesis would be that mothers in the experimental group would be less likely to have a job after 12 months than mothers in the control group.

The major strength of the experimental method is that it provides evidence for causal relationships; that is, how one variable affects another. A primary weakness is that experiments are often conducted on small samples, usually in artificial laboratory settings; thus, the findings may not be generalized to other people in natural settings.

Surveys. Survey research involves eliciting information from respondents through questions. An important part of survey research is selecting a sample of those to be questioned. A **sample** is a portion of the population, selected to be representative so that the information from the sample can be generalized to a larger population. For example, instead of asking all abused spouses about their experience, you could ask a representative sample of them and assume that those you did not question would give similar responses. After selecting a representative sample, survey researchers either interview people, ask them to complete written questionnaires, or elicit responses to research questions through computers.

1. *Interviews.* In interview survey research, trained interviewers ask respondents a series of questions and make written notes about or tape-record the respondents'

answers. Interviews may be conducted over the telephone or face-to-face. A recent Gallup Poll (Newport 2003) involved telephone interviews with a randomly selected national sample of over 1,000 U.S. adults. One of the questions interviewers asked was for respondents to "name the most important problem facing the country." The top three responses were fear of war (see Chapter 16), the economy (see Chapter 10), and unemployment (see Chapter 11) (Newport 2003).

One advantage of interview research is that researchers are able to clarify questions for the respondent and follow up on answers to particular questions. Researchers often conduct face-to-face interviews with groups of individuals who might otherwise be inaccessible. For example, some AIDS-related research attempts to assess the degree to which individuals engage in behavior that places them at high risk for transmitting or contracting HIV. Street youth and intravenous drug users, both high-risk groups for HIV infection, may not have a telephone or address because of their transient lifestyle. These groups may be accessible, however, if the researcher locates their hangouts and conducts face-to-face interviews. Research on drug addicts may also require a face-to-face interview survey design (Jacobs 2003).

The most serious disadvantages of interview research are cost and the lack of privacy and anonymity. Respondents may feel embarrassed or threatened when asked questions that relate to personal issues such as drug use, domestic violence, and sexual behavior. As a result, some respondents may choose not to participate in interview research on sensitive topics. Those who do participate may conceal or alter information or give socially desirable answers to the interviewer's questions (e.g., "No, I do not use drugs.").

"When I was younger I could remember anything— whether it happened or not."

Mark Twain
American humorist and writer

2. *Questionnaires.* Instead of conducting personal or phone interviews, researchers may develop questionnaires that they either mail or give to a sample of respondents. Questionnaire research offers the advantages of being less expensive and time-consuming than face-to-face or telephone surveys. In addition, questionnaire research provides privacy and anonymity to the research participants. This reduces the likelihood that they will feel threatened or embarrassed when asked personal questions and increases the likelihood that they will provide answers that are not intentionally inaccurate or distorted.

The major disadvantage of mail questionnaires is that it is difficult to obtain an adequate response rate. Many people do not want to take the time or make the effort to complete and mail a questionnaire. Others may be unable to read and understand the questionnaire.

3. *"Talking" Computers.* A new method of conducting survey research is asking respondents to provide answers to a computer that "talks." Romer et al. (1997) found that respondents rated computer interviews about sexual issues more favorably than face-to-face interviews and that the former were more reliable. Such increased reliability may be particularly valuable when conducting research on drug use, deviant sexual behavior, and sexual orientation as respondents reported the privacy of computers as a major advantage.

Field Research. Field research involves observing and studying social behavior in settings in which it occurs naturally. Two types of field research are participant observation and nonparticipant observation.

In participant observation research, the researcher participates in the phenomenon being studied so as to obtain an insider's perspective of the people and/or behavior being observed. Palacios and Fenwick (2003), two criminologists, attended dozens of raves over a 15-month period to investigate the south Florida drug culture. In nonparticipant observation research, the researcher observes the phenomenon being studied without actively participating in the group or the activity. For example, Dordick (1997) studied homelessness by observing and talking with homeless individuals in a variety of settings, but she did not live as a homeless person as part of her research.

Sometimes sociologists conduct in-depth detailed analyses or case studies of an individual, group, or event. For example, Fleming (2003) conducted a case study of young auto thieves in British Columbia. He found that unlike professional thieves, the teenagers' behavior was primarily motivated by thrill-seeking—driving fast, the rush of a possible police pursuit, and the prospect of getting caught.

The main advantage of field research on social problems is that it provides detailed information about the values, rituals, norms, behaviors, symbols, beliefs, and emotions of those being studied. A potential problem with field research is that the researcher's observations may be biased (e.g., the researcher becomes too involved in the group to be objective). In addition, because field research is usually based on small samples, the findings may not be generalizable.

Secondary Data Research. Sometimes researchers analyze secondary data, which are data that have already been collected by other researchers or government agencies or that exist in forms such as historical documents, police reports, school records, and official records of marriages, births, and deaths. Caldas and Bankston (1999) used information from Louisiana's 1990 Graduation Exit Examination to assess the relationship between school achievement and television viewing habits of over 40,000 tenth graders. The researchers found that, in general, television viewing is inversely related to academic achievement for whites but has little or no effect on school achievement for African-Americans. A major advantage of using secondary data in studying social problems is that the data are readily accessible so researchers avoid the time and expense of collecting their own data. Secondary data are also often based on large representative samples. The disadvantage of secondary data is that the researcher is limited to the data already collected.

"Feminists in all disciplines have demonstrated that objectivity has about as much substance as the emperor's new clothes."

Connie Miller
Feminist scholar

Goals of the Text

This text approaches the study of social problems with several goals in mind.

1. *Provide an integrated theoretical background.* The book reflects an integrative theoretical approach to the study of social problems. More than one theoretical perspective can be used to explain a social problem because social problems usually have multiple causes. For example, youth crime is linked to (1) an increased number of youths living in inner-city neighborhoods with little or no parental supervision (social disorganization), (2) young people having no legitimate means of acquiring material wealth (anomie theory), (3) youths being angry and frustrated at the inequality and racism in our society (conflict theory), and (4) teachers regarding youths as "no good" and treating them accordingly (labeling theory).
2. *Encourage the development of a sociological imagination.* A major insight of the sociological perspective is that various structural and cultural elements of society have far-reaching effects on individual lives and social well-being. This insight, known as the sociological imagination, enables us to understand how social forces underlie personal misfortunes and failures as well as contribute to personal successes and achievements. Each chapter in this text emphasizes how structural and cultural factors contribute to social problems. This emphasis encourages you

to develop your sociological imagination by recognizing how structural and cultural factors influence private troubles and public issues.

3. *Provide global coverage of social problems.* The modern world is often referred to as a "global village." The Internet and fax machines connect individuals around the world, economies are interconnected, environmental destruction in one region of the world affects other regions of the world, and diseases cross national boundaries. Understanding social problems requires an awareness of how global trends and policies affect social problems. Many social problems call for collective action involving countries around the world; efforts to end poverty, protect the environment, control population growth, and reduce the spread of HIV are some of the social problems that have been addressed at the global level. Each chapter in this text includes coverage of global aspects of social problems. We hope that attention to the global aspects of social problems broadens students' awareness of pressing world issues.

4. *Provide an opportunity to assess personal beliefs and attitudes.* Each chapter in this text contains a section called *Self and Society,* which offers you an opportunity to assess your attitudes and beliefs regarding some aspect of the social problem discussed. Earlier in this chapter, the *Self and Society* feature allowed you to assess your beliefs about a number of social problems and compare your beliefs with a national sample of first-year college students.

5. *Emphasize the human side of social problems.* Each chapter contains a feature called *The Human Side,* which presents personal stories of how social problems have affected individual lives. By conveying the private pain and personal triumphs associated with social problems, we hope to elicit a level of understanding and compassion that may not be attained through the academic study of social problems alone. This chapter's *The Human Side* feature presents stories about how college students, disturbed by various social conditions, have participated in social activism.

6. *Social Problems Research Up Close.* In every chapter there are boxes called *Social Problems Research Up Close,* which present examples of social science research. These boxes demonstrate for students the sociological enterprise from theory and data collection to findings and conclusions. Examples of research topics covered include "National College Health Risk Behavior Survey" (Chapter 2), "The Social Construction of the Hacking Community" (Chapter 15), and "Family Adjustment to Military Deployment" (Chapter 16).

7. *Focus on Technology.* Boxes called *Focus on Technology* also appear in every chapter. These boxes present information on how technology may contribute to social problems and their solutions. For example, in Chapter 4, Crime and Violence, the Focus on Technology feature highlights the use of DNA testing in criminal investigations. In Chapter 7, Gender Inequality, issues relating to "women, men, and computers" are discussed.

8. *Challenge students to "take a stand."* New to the fourth edition is a feature called *Take a Stand,* which challenges students to take and defend a position on a current issue. In doing so, students are encouraged to use reason, scientific evidence, and logic rather than emotionality and anecdotal evidence in making a cohesive argument.

9. *Encourage students to take pro-social action.* Individuals who understand the factors that contribute to social problems may be better able to formulate interventions to remedy those problems. Recognizing the personal pain and public costs associated with social problems encourages some to initiate social intervention.

"Activism pays the rent on being alive and being here on the planet. . . . If I weren't active politically, I would feel as if I were sitting back eating at the banquet without washing the dishes or preparing the food. It wouldn't feel right."

Alice Walker
Novelist

Individuals can make a difference in society by the choices they make. Individuals may choose to vote for one candidate over another, demand the right to re-

© Paul Fusco/Mangum Photos

One way to affect social change is through demonstrations. A U.S. survey of first-year college students revealed that 47 percent reported having participated in organized demonstrations in the last year (Higher Education Research Institute 2002). Here students march against the war in Iraq.

productive choice or protest government policies that permit it, drive drunk or stop a friend from driving drunk, repeat a racist or sexist joke or chastise the person who tells it, and practice safe sex or risk the transmission of sexually transmitted diseases. Individuals can also "make a difference" by addressing social concerns in their occupational role, as well as through volunteer work.

"In a certain sense, every single human soul has more meaning and value than the whole of history."

Nicholas Berdyaev
Philosopher

Although individual choices make an important impact, collective social action often has a more pervasive effect. For example, while individual parents discourage their teenage children from driving under the influence of alcohol, Mothers Against Drunk Driving contributed to the enactment of national legislation that potentially will influence every U.S. citizen's decision about whether to use alcohol and drive.

Schwalbe (1998) reminds us that we don't have to join a group or organize a protest to make changes in the world.

> We can change a small part of the social world single-handedly. If we treat others with more respect and compassion, if we refuse to participate in re-creating inequalities even in little ways, if we raise questions about official representation of reality, if we refuse to work in destructive industries, then we are making change. (p. 206)

Understanding Social Problems

At the end of each chapter, we offer a section entitled *Understanding* in which we reemphasize the social origin of the problem being discussed, the consequences, and the alternative social solutions. It is our hope that the reader will end each chapter with a "sociological imagination" view of the problem and how, as a society, we might approach a solution.

Sociologists have been studying social problems since the Industrial Revolution at the turn of the twentieth century. Industrialization brought about massive social changes: the influence of religion declined and families became smaller and moved from traditional, rural communities to urban settings. These and other

The Human Side

College Student Activism

Some people believe that in order to promote social change one must be in a position of political power and/or have large financial resources. However, the most important prerequisite for becoming actively involved in improving levels of social well-being may be genuine concern and dedication to a social "cause." The following vignettes provide a sampler of college student activism—college students making a difference in the world:

- In May 1989, hundreds of Chinese college students protested in Tiananmen Square in Beijing, China, because Chinese government officials would not meet with them to hear their pleas for a democratic government. These students boycotted classes and started a hunger strike. On June 4, 1989, thousands of students and other protesters were massacred or arrested in Tiananmen Square.
- Students at several colleges are petitioning their university administrations to buy "fair trade coffee"—coffee that is certified by monitors to have come from farmers who were paid a fair price for their beans. Many of these students are members of Students for Fair Trade. As one student said, "This is easy activism." Students make their voices heard by buying coffee with a fair-trade certified label or not buying coffee at all (Batsell 2002).
- Students at the University of California-Berkeley recently came together to protest cutbacks in the campus's Ethnic Studies Department. The protest lasted over a month with over 100 arrests and six hunger strikes. As a consequence of the students' activism, the Administration agreed to reopen a multicultural student center, hire eight tenure-track ethnic studies faculty over the next five years, and invest $100,000 in an Ethnic Research Center (Alvarado 2000).
- While a student at George Washington University, Ross Misher started an organization called Students Against Handgun Violence. When Ross was 13, his father was shot and killed by a coworker who had purchased a handgun during his lunch hour and returned to shoot Ross's father before killing himself (Lewis 1991).
- While a zoology major at the University of Colorado, Jeff Galus began the Animal Rights Student Group. This organization focuses on informing the public about how animals are treated in research and what corporations use animals in testing their products.
- Students at over 150 campuses are members of the anti-sweatshop movement, many belonging to the Worker's Rights Consortium (WRC). WRC is a student-run watchdog organization that inspects factories worldwide, monitoring the monitors, as part of the anti-sweatshop movement. WRC requires that member schools agree to closely scrutinize manufacturers of collegiate apparel. WRC also mandates "the protection of workers' health and safety, compliance with local labor laws, protection of women's rights, and prohibition of child labor, forced labor, and forced overtime" (Boston College 2003).

Students who are interested in becoming involved in student activism, or who are already involved, might explore the web site for the Center for Campus Organizing (2003)—a national organization that supports social justice activism and investigative journalism on campuses nationwide. The organization was founded on the premise that students and faculty have played critical roles in larger social movements for social justice in our society, including the Civil Rights movement, the anti-Vietnam War movement, the Anti-Apartheid movement, the women's rights movement, and the environmental movement. In 2002, almost half of all college students participated in an organized demonstration (Sax, Lindholm, Astin, Korn, & Mahoney 2002).

Sources: Alvarado, Diana. 2000. "Student Activism Today." *Diversity Digest.* http://www.inform.umd.edu/DiversityWeb/Digest/sm99/activism.html; Batsell, Jake. 2002 "USA: Students Campaign for Coffee in Good Conscience." *The Seattle Times.* March 17; Boston College. 2003. "The Anti-Sweatshop Movement and Boston College." http://www.bc.edu/bc_org/cas/soc/justice; Center for Campus Organizing. 2003. http://www.cco.org/; Jeff Galus, http://www.colorado.edu/StudentGroups/animalrights/Galus@UCSU; Lewis, Barbara A. 1991. *The Kid's Guide to Social Action,* pp. 110–11. Minneapolis, MN: Free Spirit Publishing.

changes have been associated with increases in crime, pollution, divorce, and juvenile delinquency. As these social problems became more widespread, the need to understand their origins and possible solutions became more urgent. The field of sociology developed in response to this urgency. Social problems provided the initial impetus for the development of the field of sociology and continue to be a major focus of sociology.

There is no single agreed-upon definition of what constitutes a social problem. Most sociologists agree, however, that all social problems share two important elements: an objective social condition and a subjective interpretation of that condition. Each of the three major theoretical perspectives in sociology—structural-functionalist, conflict, and symbolic interactionist—has its own notion of the causes, consequences, and solutions of social problems.

Chapter Review

- **What is a social problem?**
  Social problems are defined by a combination of objective and subjective criteria. The objective element of a social problem refers to the existence of a social condition; the subjective element of a social problem refers to the belief that a particular social condition is harmful to society, or to a segment of society, and that it should and can be changed. By combining these objective and subjective elements, we arrive at the following definition: a social problem is a social condition that a segment of society views as harmful to members of society and in need of remedy.

- **What is meant by the structure of society?**
  The structure of a society refers to the way society is organized.

- **What are each of the components of the structure of society?**
  Institutions are an established and enduring pattern of social relationships and include family, religion, politics, economics, and education. Social groups are defined as two or more people who have a common identity, interact, and form a social relationship. A status is a position a person occupies within a social group and may be achieved or ascribed. Every status is associated with many roles, or the set of rights, obligations, and expectations associated with a status.

- **What is meant by the culture of society?**
  Whereas social structure refers to the organization of society, culture refers to the meanings and ways of life that characterize a society.

- **What are each of the components of the culture of society?**
  Beliefs refer to definitions and explanations about what is assumed to be true. Values are social agreements about what is considered good and bad, right and wrong, desirable and undesirable. Norms are socially defined rules of behavior. Norms serve as guidelines for our behavior and for our expectations of the behavior of others. Finally, a symbol is something that represents something else.

- **What is the sociological imagination and why is it important?**
  The sociological imagination, a term developed by C. Wright Mills (1959), refers to the ability to see the connections between our personal lives and the social world in which we live. It is important because when we use our sociological imagination, we are able to distinguish between "private troubles" and "public issues" and to see connections between the events and conditions of our lives and the social and historical context in which we live.

- **What are the differences between the three sociological perspectives?**
  According to structural-functionalism, society is a system of interconnected parts that work together in harmony to maintain a state of balance and social equilibrium for the whole. The conflict perspective views society as comprising different groups and interests competing for power and resources. Symbolic interactionism reflects the micro sociological perspective and emphasizes that human behavior is influenced by definitions and meanings that are created and maintained through symbolic interaction with others.

- **What are the first four stages of a research study?**
  The first four stages of a research study are: (1) formulating a research question, (2) reviewing the literature, (3) defining variables, and (4) formulating a hypothesis.

- **How do the various research methods differ from one another?**
  Experiments involve manipulating the independent variable in order to determine how it affects the dependent variable. Survey research involves eliciting information from respondents through questions. Field research involves observing and studying social behavior in settings in which it occurs naturally. Secondary data are data that have already been collected by other researchers or government agencies, or exist in forms such as historical documents, police reports, school records, and official records of marriages, births, and deaths.

Critical Thinking

1. People increasingly are using information technologies as a means of getting their daily news. As a matter of fact, some research indicates that news on the Internet is beginning to replace television news as the primary source of information among computer users (see Chapter 15). What role does the media play in our awareness of social problems, and will definitions of social problems change as sources of information change?
2. Each of you occupies several social statuses, each one carrying an expectation of role performance—that is, what you should and shouldn't do given your position. List five statuses you occupy, the expectations of their accompanying roles, and any role conflict that may result. What types of social problems are affected by role conflict?
3. Definitions of social problems change over time. Identify a social condition, now widely accepted, that might be viewed as a social problem in the future.

Key Terms

achieved status
alienation
anomie
ascribed status
beliefs
conflict perspective
dependent variable
experiment
field research
folkway
hypothesis
independent variable
institution
labeling theory
latent function
law
macro sociology
manifest function
master status
micro sociology
mores
norm
objective element
operational definition
primary group
role
sanction
sample
secondary group
social constructionism
social group
social problem
sociological imagination
status
structural-functionalism
subjective element
survey research
symbol
symbolic interactionism
triangulation
values
variable

Taking a Stand

Should sociologists be required by law to reveal their sources?

In a free society there must be freedom of information. That is why journalists' sources are protected by the Constitution and, more specifically, the First Amendment. If journalists are compelled to reveal their sources, their sources may be unwilling to share information, and this would jeopardize the public's right to know. A journalist cannot reveal information given in confidence without permission from the source or a court order. Sociologists, in some circumstances, have been given the same legal rights as journalists. However, not all states protect sociologists' confidential information.

Use Wadsworth's exclusive online resources—InfoTrac College Edition, MicroCase Online, and OVRC—to formulate a position on this topic.

The Wadsworth's Sociology Online Resources and Writing Companion will help you get started. This valuable guide will show you how to use Wadsworth's exclusive online resources when studying social problems. It will also help you to build essential research and writing skills. InfoTrac College Edition, MicroCase Online, OVRC, and an electronic copy of portions of this companion are available at http://sociology.wadsworth.com/mooney_knox_schacht/problems4e, the companion Web site for *Understanding Social Problems,* Fourth Edition.

Media Resources

The Companion Web Site for *Understanding Social Problems,* Fourth Edition

http://sociology.wadsworth.com/mooney_knox_schacht/problems4e

Supplement your review of this chapter by going to the companion Web site to take one of the Tutorial Quizzes, use the flash cards to master key terms, and check out the many other study aids you'll find there. You'll also find special features such as *Wadsworth's Sociology Online Resources and Writing Companion,* GSS Data, and Census 2000 information, data, and resources at your fingertips to help you complete that special project or do some research on your own.

Interactions CD-ROM

Go to the Interactions CD-ROM for *Understanding Social Problems,* Fourth Edition, to access additional interactive learning tools, such as in-depth review materials, corresponding practice quizzes, and other engaging resources and activities to help you study the concepts in this chapter.

"The health of each person affects the health of our families, our workplaces, our communities, our economy, and our society."
Linda Peeno, M.D.

Illness and the Health Care Crisis

The Global Context: Effects of Globalization on Health

Societal Measures of Health and Illness

Sociological Theories of Illness and Health Care

HIV/AIDS: A Global Health Concern

Mental Illness: The Invisible Epidemic

Social Factors Associated with Health and Illness

Problems in U.S. Health Care

Strategies for Action: Improving Health and Health Care

Understanding Illness and Health Care

Chapter Review

Larry Causey, age 57, called the FBI and told them he was going to rob the post office in West Monroe, LA. Then he went to the post office and handed a note to a teller demanding money. He left empty-handed and sat in his car until officers arrested him. Larry had no intention of committing robbery. Larry had cancer and could not afford cancer treatment, so he staged a robbery in order to get arrested and be put in jail, where he would receive cancer medical treatment ("Access to Free Health Care Means Crime Does Pay for Some Sick Inmates" 2001). Larry Causey's story is not uncommon. "Sheriffs nationwide say they're also arresting people willing to trade their freedom for a free visit to the doctor" (n.p.).

In this chapter, we review problems of illness and health care throughout the world and in the United States. Taking a sociological look at health issues, we examine how social forces affect and are affected by health and illness and why some social groups suffer more illness than others.

While significant gains in global health have been made over the last several decades, there have also been serious setbacks. Old diseases thought to be under control, including cholera, tuberculosis, and malaria, have increased in number of cases or geographical spread. And new diseases and health threats have emerged in recent decades, such as the HIV/AIDS epidemic, growing resistance to antibiotics, the threat of anthrax and other bioterrorist attacks, and the emergence of Severe Acute Respiratory Syndrome (SARS). One of the most significant social forces influencing health is the increasing globalization of the world.

The Global Context: Effects of Globalization on Health

Globalization, broadly defined as the growing economic, political, and social interconnectedness among societies throughout the world, has eroded the boundaries that separate societies, creating a "global village." Globalization has had both positive and negative effects on health. On the positive side, globalized communication technology enhances the capacity to monitor and report on outbreaks of disease, disseminate guidelines for controlling and treating disease, and share scientific knowledge and research findings (Lee 2003). Globalization also provides opportunities for establishing international health programs and agreements. In 2003, for example, the World Health Organization (made up of 192 member countries) voted unanimously to adopt the Framework Convention on Tobacco Control, which urges countries to eliminate tobacco advertising, establish stronger warning labels, raise cigarette prices, and adopt smoke-free workplace laws around the world (SmokeFree Educational Services 2003). On the negative side, features of globalization, including expansion of trade and transnational corporations, and the growth of travel and information technologies, have been linked to a number of health problems.

Increased Travel and Information Technology

As discussed later in this chapter, modern communication technology has useful applications in health and medicine. But such technology is also linked to a number of health problems. Following the terrorist attacks of September 11, 2001, the world was reminded that global communication systems and international travel enable terrorist groups to form well-organized networks that can move around the globe. The terrorist attacks of September 11 resulted not only in direct deaths and injuries but also raised awareness of the prospect of biological and chemical weapons attacks as a major public health threat. Increased business travel and tourism have encouraged the spread of disease, such as the potentially fatal West Nile virus, which first appeared in the United States in 1999, and the spread of Severe Acute Respiratory Syndrome (SARS), which was first diagnosed in Asia in 2002 and within months spread to 30 countries, infecting thousands of individuals and killing more than 800.

The Expansion of Trade and Transnational Corporations

Increased international trade, another feature of globalization, has expanded the range of goods available to consumers, but at a cost to global health. The increased transportation of goods by air, sea, and land contributes to the pollution caused by the burning of fossil fuels. The expansion of international trade of harmful products such as tobacco, alcohol, and processed or "fast" foods are associated with a rise in cancer, heart disease, stroke, and diabetes (World Health Organization 2002).

Expanding trade has also facilitated the growth of transnational corporations (see also Chapter 11). For example, 75 percent of the world cigarette market is controlled by just four companies: Philip Morris, British American Tobacco, Japan Tobacco/R. J. Reynolds, and the China National Tobacco Corporation. Each company has plants in at least 50 countries throughout the world (Collin 2003). Transnational corporations profit from the new global system of production that allows companies to set up shop in developing countries in order to take advantage of lower labor costs and lax environmental and labor regulations. Due to lax labor and human rights regulations, factory workers in transnational corporations—typically low status, uneducated women—are often exposed to harmful working conditions that increase the risk of illness, injury, and mental anguish (Hippert 2002). These workers often suffer exposure to toxic substances, lack safety equipment such as gloves and goggles, are denied bathroom breaks (which leads to bladder infections), and are assaulted at the workplace, as "physical brutality is frequently used as a mechanism of control on the production floor of the factory" (p. 863).

Due to weak environmental laws in developing countries, transnational corporations are responsible for high levels of pollution and environmental degradation, which negatively impacts the health of entire populations. Finally, the movement of factories out of the United States to other countries has resulted in significant losses of U.S. jobs in manufacturing and the textile and apparel industry, which is significant because of the relationship between physical and mental illness and unemployment (Bartley, Ferrie, & Montgomery 2001).

The economic benefits of expanding trade opportunities have not been shared equally. As the gap between rich and poor countries and individuals within them has grown, so have the disparities in the health of populations (Feachum 2000; Schaeffer 2003). A United Nations Population Fund (2002a) report suggests that "to make the most of globalization, part of the economic gains must be ploughed back into social programmes that directly help the poor" (p. 24).

Societal Measures of Health and Illness

Measures of health and illness reveal striking disparities among countries and regions. The following sections describe various measures that provide indicators of the health of populations, including measures of morbidity, mortality, and life expectancy.

Morbidity

Morbidity refers to illnesses, symptoms, and impairments they produce. Measures of morbidity are often expressed in terms of the incidence and prevalence of specific health problems. *Incidence* refers to the number of new cases of a specific health problem with a given population during a specified time period. *Prevalence* refers to the total number of cases of a specific health problem within a population that exist at a given time. For example, the *incidence* of HIV infection worldwide was 5 million in 2003; meaning that there were 5 million people newly infected with HIV in 2003. In the same year, the worldwide *prevalence* of HIV was 40 million, meaning that a total of 40 million people worldwide were living with HIV infection in 2003 (National Institutes of Health 2004).

Rates of morbidity in a population provide one measure of the health of that population. As we discuss later in this chapter, patterns of morbidity vary according to social factors such as social class, education, sex, and race. Morbidity patterns also vary according to the level of development of a society and the age structure of the population. In the less-developed countries, infectious and parasitic diseases such as HIV disease, tuberculosis, diarrheal diseases (caused by bacteria, viruses, or parasites), measles, and malaria are the major health threats (Weitz 2004). In the industrialized world, where infectious and parasitic diseases have been largely controlled by advances in sanitation, immunizations, and antibiotics, noninfectious diseases have emerged as major sources of morbidity. These include chronic and degenerative conditions such as heart disease, cancer, stroke, arthritis, diabetes, mental disorders, and respiratory diseases.

Patterns of Longevity

Another measure of the health of a population is the average number of years individuals born in a given year can expect to live, referred to as **life expectancy.** Globally, average life expectancy has increased from 46.5 years in 1950–1955 to 65.2 years in 2002 (World Health Organization 2003); however, wide disparities exist between societies. In 2003, Japan had the longest life expectancy (81 years) while 13 countries (primarily in Africa) had life expectancies of less than 50 years. As shown in Table 2.1, several countries have life expectancies that are greater than life expectancy in the United States.

Patterns of Mortality

Rates of **mortality,** or death—especially those of infants, children, and women—provide sensitive indicators of the health of a population. Worldwide, the leading cause of death is infectious and parasitic diseases (Weitz 2004). In the United States, the three leading causes of death for both women and men are heart disease, cancer, and stroke (see Figure 2.1). Later, we discuss how patterns of mortality are

Table 2.1
Countries with Low and High Life Expectancies,* 2004

Low Life Expectancies

Countries	Life Expectancy
Botswana	31
Angola; Malawi; Mozambique	37
Swaziland; Zimbabwe	38
Rwanda	39
Ethiopia	41
Cote d'Ivoire; Niger	42
Sierra Leone	43
Burkina Faso; South Africa	44
Kenya; Mali; Uganda	45
Afghanistan	47
Somalia	48

High Life Expectancies

Countries	Life Expectancy
Cuba; Denmark; United States	77
Belgium; Finland; United Kingdom	78
Austria; France; Germany; Greece; Israel; Netherlands; Norway; New Zealand; Spain	79
Australia; Canada; Iceland; Italy; Sweden	80
Japan	81

*Rounded to the nearest whole number.

Source: U.S. Census Bureau. 2004. International Data Base.

related to social factors, such as social class, sex, race/ethnicity, and education. Mortality patterns also vary by age, as shown in Table 2.2.

Infant and Childhood Mortality Rates. Infant mortality rates, the number of deaths of live-born infants under 1 year of age per 1,000 live births (in any given year), provide an important measure of the health of a population. In 2001, infant mortality rates ranged from 182 in Sierra Leone to only 3 in Ireland, Japan, Singapore, and Sweden. In 2001, the U.S. infant mortality rate was 7; 31 countries had infant mortality rates that were lower than that of the United States (UNICEF 2003).

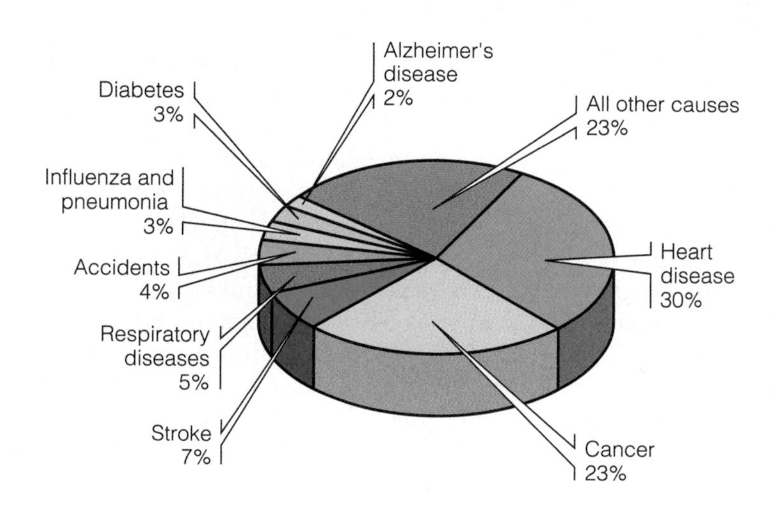

Figure 2.1
Leading Causes of Death in the United States: 2000.*

Source: Based on Anderson, Robert N. 2002. "Deaths: Leading Causes for 2000." *National Vital Statistics Reports* 50(16): all.

*Rounded to the nearest percent

Table 2.2
Top Three Causes of Death by Selected Age Groups: United States, 2000

Age	Leading Causes of Death		
	1	2	3
10–14	accidents	cancer	suicide
15–24	accidents	homicide	suicide
25–34	accidents	suicide	homicide
35–44	cancer	accidents	heart disease
45–54	cancer	heart disease	accidents
55–64	cancer	heart disease	respiratory disease
65+	heart disease	cancer	stroke

Source: Based on Anderson, Robert N. 2002. "Deaths: Leading Causes for 2000." *National Vital Statistics Reports* 50(16):all.

"The state of the world's children is the best measure of human well-being, and the health of children is the best measure of the health of our planet."

Carol Bellamy
Director, UNICEF

Under-5 mortality rates, another useful measure of child health, refer to the rate of deaths of children under age 5. Under-5 mortality rates range from 316 in Sierra Leone, to 4 in Denmark. In 2001, the U.S. under-5 mortality rate was 8; 32 countries had infant mortality rates that were lower than that of the United States (UNICEF 2003).

Most deaths of infants and children under age 5 occur in developing countries, where underweight is a contributing factor in 60 percent of childhood deaths (World Health Organization 2002). Although mortality among infants and children has been declining in most developing countries from the mid-1980s through the 1990s, this decline has recently slowed, stopped, or reversed itself in some countries of sub-Saharan Africa, largely as a result of the rate of HIV infection among infants and children (Rustein 2000).

Maternal Mortality Rates. Maternal mortality rates, a measure of deaths that result from complications associated with pregnancy, childbirth, and unsafe abortion, also provide a sensitive indicator of the health status of a population. Maternal mortality is the leading cause of death and disability for women ages 15 to 49 in developing countries (Ransom & Yinger 2002). The three most common causes of maternal death are hemorrhage (severe loss of blood), infection, and complications related to unsafe abortion.

For every woman who dies from pregnancy-related causes, about 30 suffer from serious health problems such as infections, anemia, infertility, and damage to the uterus and reproductive tract. One of the most devastating pregnancy-related injuries occurs when a hole develops between a woman's vagina and her bladder or rectum, or both, usually as a result of trauma during childbirth. Known as *obstetric fistulas*, these pregnancy-related injuries cause permanent incontinence (leakage of urine or feces or both) if not treated. The estimated 2 million women living with obstetric fistulas are shamed, ostracized, divorced, abandoned, and left without support (United Nations Population Fund 2002b).

Rates of maternal mortality show a greater disparity between rich and poor countries than any of the other societal health measures (United Nations Population Fund 2002b). Women's lifetime risk of dying from pregnancy or childbirth is highest in sub-Saharan Africa, where 1 in 13 women dies of pregnancy-related causes during her lifetime, compared to 1 in 4,085 women in industrialized countries (Ransom & Yinger 2002).

High maternal mortality rates in less-developed countries are related to poor quality and inaccessible health care, malnutrition and poor sanitation, and higher rates of pregnancy and childbearing at early ages. Pregnancy before the age of 18 is several times more risky than for women over 20 (United Nations Population Fund 2002a). In developing countries, only 53 percent of all births are attended to by professionals and nearly 30 percent of women who give birth in developing countries receive no care after the birth (United Nations Population Fund 2000). Women in many countries also lack access to family planning services and/or do not have the support of their male partners to use contraceptive methods such as condoms. Consequently, many women resort to abortion to limit their childbearing, even in countries where abortion is illegal.

"It is not uncommon for women in Africa, when about to give birth, to bid their older children farewell."

United Nations Population Fund

Illegal abortions in less-developed countries have an estimated mortality risk of 100 to 1,000 per 100,000 procedures (Miller & Rosenfield 1996). In contrast, the U.S. mortality risk for legal abortion is very low: 0.6 per 100,000. Unsafe abortion represents a serious threat to the health and lives of women. Each year, women undergo an estimated 50 million abortions, 20 million of which are unsafe, resulting in the deaths of 78,000 women (United Nations Population Fund 2000a).

The Epidemiological Transition. Life expectancy and rates and causes of mortality and morbidity vary dramatically between developing and developed nations. As societies develop and increase the standard of living for their members, life expectancy increases and birth rates decrease. At the same time, the causes of death and disability shift from infectious disease and maternal and infant mortality and morbidity to chronic, noninfectious causes. This shift is referred to as an **epidemiological transition** whereby low life expectancy and predominance of parasitic and infectious diseases shifts to high life expectancy and predominance of chronic and degenerative diseases. As societies make the epidemiological transition, birthrates decline and life expectancy increases, so diseases that need time to develop, such as cancer, heart disease, Alzheimer's disease, arthritis, and osteoporosis become more common, and childhood illnesses, typically caused by infectious and parasitic diseases, become less common, as do pregnancy-related deaths and health problems.

In the *World Health Report 2002,* the World Health Organization (2002) notes that changes in patterns of consumption, particularly of food, alcohol, and tobacco, around the world are creating a "risk transition." Changes in food production and processing and in agricultural and trade policies have led to increased consumption of alcohol, tobacco, salt, sugar, and fat and subsequent increases in noninfectious diseases related to these substances, such as cancer, respiratory disease, heart disease, and diabetes. For low- and middle-income countries that are still dealing with high rates of infectious diseases and malnutrition, this creates a "double burden."

Patterns of Burden of Disease

A new approach to measuring the health status of a population provides an indicator of the overall burden of disease on a population through a single unit of measurement that combines not only the number of deaths but also the impact of premature death and disability on a population (Murray & Lopez 1996). This comprehensive unit of measurement, called the *disability-adjusted life year* (DALY), reflects years of life lost to premature death and years lived with a disability. More simply, 1 DALY is equal to 1 lost year of healthy life.

Worldwide, tobacco is the leading cause of burden of disease (World Health Organization 2002). Hence, tobacco has been called "the world's most lethal weapon of mass destruction" (SmokeFree Educational Services 2003, 1). The top ten risk factors that contribute to the global burden of disease are underweight; unsafe sex; high blood pressure; tobacco; alcohol; unsafe water, sanitation, and hygiene; high cholesterol; indoor smoke from solid fuels; iron deficiency; and overweight (World Health Organization 2002).

Sociological Theories of Illness and Health Care

The sociological approach to the study of illness, health, and health care differs from medical, biological, and psychological approaches to these topics. Next, we discuss how three major sociological theories—structural-functionalism, conflict theory, and symbolic interactionism—contribute to our understanding of illness and health care.

Structural-Functionalist Perspective

The structural-functionalist perspective is concerned with how illness, health, and health care affect and are affected by other aspects of social life. For example, rather than look at individual reasons for suicide, a structural-functionalist approach looks for social patterns that may help explain suicide rates. Durkheim (1897/1951) conducted one of the first scientific sociological research studies that found that suicide rates were higher in countries characterized by lower levels of social integration and regulation—a finding that has been replicated in recent studies (Stockard & O'Brien 2002).

Many health behaviors and outcomes can be understood by examining the influence of social patterns and social changes. The women's movement and changes in societal gender roles have led to more women smoking and drinking, and experiencing the negative health effects of these behaviors. Increased modernization and industrialization throughout the world has resulted in environmental pollution—a major health concern (see Chapter 15).

Just as social change affects health, health concerns may lead to social change. The emergence of HIV and AIDS in the U.S. gay male population was a force that helped unite and mobilize gay rights activists. Concern over the effects of exposure to tobacco smoke—the greatest cause of disease and death in the United States and other developed countries—has led to legislation banning smoking in workplaces, restaurants, and bars in at least five states (California, Delaware, New York, Connecticut, and Maine) (SmokeFree Educational Services 2003).

According to the structural-functionalist perspective, health care is a social institution that functions to maintain the well-being of societal members and, consequently, of the social system as a whole. Illness is dysfunctional in that it interferes with people's performing needed social roles. To cope with nonfunctioning members and to control the negative effects of illness, society assigns a temporary and unique role to those who are ill—the sick role (Parsons 1951). This role assures that societal members receive needed care and compassion, yet at the same time, it carries with it an expectation that the person who is ill will seek competent medical advice, adhere to the prescribed regimen, and return as soon as possible to normal role obligations.

Structural-functionalists explain the high cost of medical care by arguing that society must entice people into the medical profession by offering high salaries. Without such an incentive, individuals would not be motivated to endure the rigors of medical training or the stress of being a physician.

Finally, the structural-functionalist perspective draws attention to latent functions, or unintended and often unrecognized consequences of social patterns or behavior. For example, a latent function of widespread use of some drugs is the emergence of drug resistance, which occurs when drugs kill the weaker disease-causing germs while allowing variants resistant to the drugs to flourish. For generations the drug chloroquine was added to table salt to prevent malaria. But overuse led to drug-resistant strains of malaria, and now chloroquine is useless in preventing malaria (McGinn 2003). The most alarming example of the development of a drug-resistant germ is tuberculosis, which kills more people yearly than any other infectious disease. Tuberculosis is caused by bacilli that attack and destroy lung tissue and is spread when infected individuals cough or sneeze. The World Health Organization estimates that one-third of the world's population is infected, although only about 10 percent of infected persons ever develop symptoms (Weitz 2004).

Conflict Perspective

The conflict perspective focuses on how wealth, status, power, and the profit motive influence illness and health care. Worldwide, populations living in poverty, with little power and status, experience more health problems and have less access to quality medical care (Feachum 2000).

The conflict perspective criticizes the pharmaceutical and health care industry for placing profits above people. For example, pharmaceutical companies' research and development budgets are spent not according to public health needs, but rather according to calculations about maximizing profits. Because the masses of people in developing countries lack the resources to pay high prices for medication, pharmaceutical companies do not see the development of drugs for diseases of poor countries as a profitable investment. This explains why about 90 percent of the $70 billion invested annually in health research and development by pharmaceutical companies and Western governments focuses on the health problems of the 10 percent of the global population living in developed industrialized countries (Thomas 2003). This allocation of health-related funding clearly benefits the wealthier citizens of the world and neglects the needs of poor populations. Consider, for example, malaria—a tropical disease that kills about 3 million people annually, killing more people than AIDS. Malaria remains one of the world's leading health threats, yet it is rarely covered in the news and is a relatively low public health priority. Between 1975 and 1999, only 4 of the 1,393 new drugs developed worldwide were antimalarials. McGinn (2003) explains, "the reality is that malaria is a disease of poor countries. If it were a constant threat in industrial countries, the story would be completely different" (p. 63).

Conflict theorists also point to the ways in which health care and research are influenced by male domination and bias. When the male erectile dysfunction drug Viagra made its debut in 1998, women across the United States were outraged by the fact that some insurance policies covered Viagra (or were considering covering it), although female contraceptives were not covered. The male-dominated medical research community has also been criticized for neglecting women's health issues and excluding women from major health research studies (Johnson & Fee 1997). Bias in medical research also results from sponsorship of research by industry. A recent study found that industry-sponsored research is 3.6 times more likely to produce results favorable to the sponsoring company (Bekelman, Li, & Gross 2003).

Corporations may influence health-related policies and laws through contributions to politicians and political candidates. In a study on the influence of tobacco industry campaign contributions on state legislators in six states, researchers found that legislators who received higher tobacco industry campaign contributions had more pro-tobacco policy positions (Monardi & Glantz 1998).

Conflict theorists argue that the high costs of medical care in the United States are a result of a capitalistic system in which health care is a commodity, rather than a right. The conflict perspective views power and concern for profits as the primary obstacles to U.S. health care reform. Insurance companies realize that health care reform translates into federal regulation of the insurance industry. In an effort to buy political influence to maintain profits, the insurance industry has contributed millions of dollars to congressional candidates.

Symbolic Interactionist Perspective

Symbolic interactionists focus on (1) how meanings, definitions, and labels influence health, illness, and health care and (2) how such meanings are learned through interaction with others and through media messages and portrayals. According to the symbolic interactionist perspective of illness, "there are no illnesses or diseases in nature. There are only conditions that society, or groups within it, have come to define as illness or disease" (Goldstein 1999, 31). Psychiatrist Thomas Szasz (1970) argued that what we call "mental illness" is no more than a label conferred on those individuals who are "different," that is, who don't conform to society's definitions of appropriate behavior.

Defining or labeling behaviors and conditions as medical problems is part of a trend known as **medicalization.** Initially, medicalization was viewed as occurring when a particular behavior or condition deemed immoral (e.g., alcoholism, masturbation, homosexuality) was transformed from a legal problem to a medical problem that required medical treatment. The concept of medicalization has expanded to include (1) any new phenomena defined as medical problems in need of medical intervention, such as post-traumatic stress disorder (PTSD), premenstrual syndrome (PMS), and attention deficit hyperactivity disorder (ADHD) and (2) "normal" biological events or conditions that have come to be defined as medical problems in need of medical intervention, including childbirth, menopause, and death.

Conflict theorists view medicalization as resulting from the medical profession's domination and pursuit of profits. A symbolic interaction perspective suggests that medicalization results from the efforts of sufferers to "translate their individual experiences of distress into shared experiences of illness" (Barker 2002, 295). In her study of women with fibromyalgia (a pain disorder that has no identifiable biological cause), Barker (2002) suggests that the medicalization of symptoms and distress through a diagnosis of fibromyalgia gives sufferers a framework for understanding and validating their experience of distress.

Recent theorists have observed a shift from medicalization to *biomedicalization.* Spurred by technoscientific innovation, **biomedicalization** refers to the view that medicine can not only control particular conditions, but can also *transform* bodies and lives. "Such transformations range from life after complete heart failure to walking in the absence of leg bones, to giving birth a decade or more after menopause, to the capacity to genetically design life itself" (Clarke, Mamo, Fishman, Shim, & Fosket 2003).

The concepts of medicalization and biomedicalization suggest that conceptions of health and illness are socially constructed. It follows then that definitions of health and illness vary over time and from society to society. In some countries, being fat is a sign of health and wellness; in others it is an indication of mental illness or a lack of self-control. Among some cultural groups, perceiving visions or voices of religious figures is considered normal religious experience, whereas such "hallucinations" would be indicative of mental illness in other cultures. In eighteenth- and nineteenth-century America, masturbation was considered an unhealthy act that caused a range of physical and mental health problems. Individuals caught masturbating were often locked up in asylums, treated with drugs (such as sedatives and poisons), or subjected to a range of interventions designed to prevent masturbation by stimulating the genitals in painful ways, preventing genital sensation, or deadening it. These physician-prescribed interventions included putting ice on the genitals; blistering and scalding the penis, vulva, inner thighs, or perineum; inserting electrodes into the rectum and urethra; cauterizing the clitoris by anointing it with pure carbolic acid; circumcising the penis; and surgically removing the clitoris, ovaries, and testicles (Allen 2000). Today, most health professionals agree that masturbation is a normal, healthy aspect of sexual expression.

Symbolic interactionists also focus on the stigmatizing effects of being labeled "ill." A **stigma** refers to any personal characteristic associated with social disgrace, rejection, or discrediting. (Originally, the word "stigma" referred to a mark burned into the skin of a criminal or slave.) Individuals with mental illnesses, drug addictions, physical deformities and impairments, and HIV and AIDS are particularly prone to being stigmatized. Stigmatization may lead to prejudice and discrimination and even violence against individuals with illnesses or impairments.

Finally, symbolic interactionism draws attention to the effects that meanings and labels have on behaviors that affect our health. For example, as tobacco sales have declined in developing countries, transnational tobacco companies have looked for markets in developing countries, using advertising strategies that depict smoking as "an inexpensive way to buy into glamorous lifestyles of the upper or successful social class" (Egwu 2002, 44). Meanings and labels also affect social policy. Dennis Altman (2003) suggests that we conceptualize HIV/AIDS as a global security issue, rather than a health issue, "because how we conceptualize the pandemic will impact on the extent of political commitment governments bring to dealing with it. . . . Redefining the disease to encompass security issues almost inevitably pushes it higher on government agendas" (pp. 40–41). We examine HIV/AIDS in the following section.

HIV/AIDS: A Global Health Concern

One of the most urgent worldwide public health concerns is the spread of the human immunodeficiency virus (HIV), which causes acquired immunodeficiency syndrome (AIDS). HIV/AIDS is the fourth leading cause of death in the world and the major cause of death in Africa (United Nations Population Fund 2002a; World Health Organization 2002). The epidemic is rapidly expanding into Eastern Europe and Central Asia.

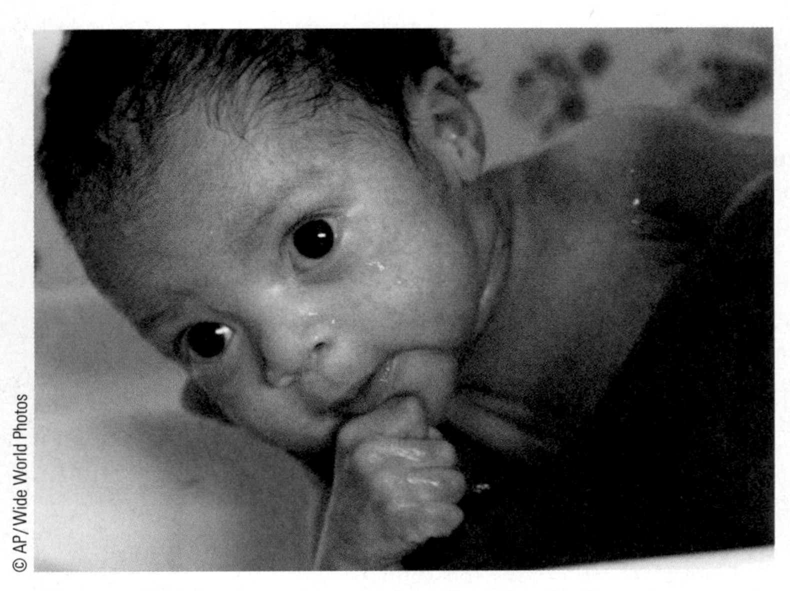

© AP/Wide World Photos

Although HIV/AIDS first emerged as an adult health problem, the disease is now a major killer of children in countries with high HIV prevalence rates.

HIV is transmitted through sexual intercourse; through sharing unclean intravenous needles; through perinatal transmission (from infected mother to fetus or newborn); through blood transfusions or blood products; and, rarely, through breast-milk. Worldwide, the predominant mode of HIV transmission is through heterosexual contact.

Approximately two-thirds percent of people living with HIV/AIDS live in sub-Saharan Africa, where 8 percent of all adults (ages 15 to 49) are HIV-infected (National Institutes of Health 2004). More than half of infected adults in sub-Saharan Africa are women, and among 15- to 24-year-olds, HIV is twice as prevalent among females as in males (6%–11% versus 3%–6%) (Stephenson 2003). African girls are much more likely to be HIV-infected than African boys of their same age because girls' sex partners are more likely to be older and thus more likely to be HIV-infected than boys of their own age. The low status of women makes it difficult for them to refuse sex or demand condom use.

The high rates of HIV in developing countries, particularly sub-Saharan Africa, are having alarming and devastating effects on societies. The HIV/AIDS epidemic creates an enormous burden on the limited health care resources of poor countries. Economic development is threatened by the HIV epidemic, which diverts national funds to health-related needs and reduces the size of a nation's workforce. In countries that are affected by HIV/AIDS as well as a food shortage, the growing rate of HIV infection in women is exacerbating the food shortage, as women and girls are responsible for up to 80 percent of food production in developing countries (Stephenson 2003). Finally, AIDS deaths have left over 13 million children without one or both parents—a third of all orphans (United Nations Population Fund 2002a).

HIV/AIDS in the United States

An estimated 850,000 to 950,000 U.S. residents are infected with HIV, one-quarter of whom are unaware of their infection (National Institutes of Health 2004). About 70 percent of new HIV infections in the United States each year are among men and 30 percent among women, and half of new infections are in people younger than 25 years of age. Of new infections among U.S. men, about 60 percent were infected through same-sex sexual contact, 25 percent through injection drug use, and 15 percent through heterosexual sex. Of new infections among women, 75 percent were infected through heterosexual sex, and 25 percent through injection drug use (National Institutes of Health 2004).

Despite the widespread concern about HIV, many Americans—especially adolescents and young adults—engage in high-risk behavior. Although a national survey of U.S. teens found that most teens (81%) say that AIDS is a serious problem for people their age (Kaiser Family Foundation 2000), one study of sexually active U.S. high school students found that only 58 percent reported that either they or their partner had used a condom during the last sexual intercourse (Kann et al. 2000).

Mental Illness: The Invisible Epidemic

What it means to be mentally healthy varies across and within cultures. **Mental health** has nevertheless been defined as the successful performance of mental function, resulting in productive activities, fulfilling relationships with other people, and the ability to adapt to change and to cope with adversity (U.S. Department of Health and Human Services 2001). **Mental illness** refers collectively to all mental disorders, which are health conditions that are characterized by alterations in thinking, mood, and/or behavior associated with distress and/or impaired functioning, and that meet specific criteria (such as level of intensity and duration) specified in the classification manual used to diagnose mental disorders: *The Diagnostic and Statistical Manual of Mental Disorders* (APA Online 2000) (see Table 2.3).

Mental illness is a "hidden epidemic" because the shame and embarrassment associated with mental problems discourages people from acknowledging

Table 2.3
Mental Disorders Classified by the American Psychiatric Association

Classification	Description
Anxiety Disorders	Disorders characterized by anxiety that is manifest in phobias, panic attacks, or obsessive–compulsive disorder
Dissociative Disorders	Problems involving a splitting or dissociation of normal consciousness such as amnesia and multiple personality
Disorders First Evident in Infancy, Childhood, or Adolescence	Including mental retardation, attention-deficit hyperactivity, and stuttering
Eating or Sleeping Disorders	Including anorexia, bulimia, and insomnia
Impulse Control Disorders	Including the inability to control undesirable impulses such as kleptomania, pyromania, and pathological gambling
Mood Disorders	Emotional disorders such as major depression and bipolar (manic-depressive) disorder
Organic Mental Disorders	Psychological or behavioral disorders associated with dysfunctions of the brain caused by aging, disease, or brain damage (such as Alzheimer's disease)
Personality Disorders	Maladaptive personality traits that are generally resistant to treatment, such as paranoid and antisocial personality types
Schizophrenia and Other Psychotic Disorders	Disorders with symptoms such as delusions or hallucinations
Somatoform Disorders	Psychological problems that present themselves as symptoms of physical disease, such as hypochondria
Substance-Related Disorders	Disorders resulting from abuse of alcohol and/or drugs such as barbiturates, cocaine, or amphetamines

The Human Side

Men Talk About Depression

As part of the National Institute of Mental Health campaign "Real Men, Real Depression," the following men went public with their experiences of depression. In the following interview excerpts, these men reveal their personal struggles with depression.

NIMH

Jimmy Brown, Firefighter

"CAN YOU DESCRIBE A TYPICAL DAY?"
Many days . . . I just didn't wanna get out of bed. Honestly, the only reason I got out of bed . . . was because the dog had to get walked and my wife had to go to work. So I walked the dog, take her to work, and come back. Some days I'd get back in bed, some days I'd just sit on the couch and wonder what I was going to do next.

"DO YOU BELIEVE IT WILL EVER END?"
No . . . you just don't know if it's gonna end. . . . You don't know if you're ever gonna be the person that you were before. There were days when I thought I'd never be myself again.

"WAS THERE ANYTHING YOU LIKED TO DO?"
I pretty much lost interest in just about everything. Every aspect of your life the interest level just goes. You're just kinda there.

"CAN YOU PULL OUT OF IT ON YOUR OWN?"
They think I'm a big, tough fireman. I'm supposed to be able to deal with anything, I'm supposed to be able to just pick up, carry on. . . . It's not that easy. . . . I don't know if I'd be a firefighter today if I didn't get help."

NIMH

Patrick McCathern, Retired First Sergeant, U.S. Air Force

"HOW IS DEPRESSION DIFFERENT FROM THE BLUES?"
Everybody gets the blues. I call depression the super blues. . . . When you have the super blues, you can't find your way back cause you've gotten so far in. It's like a hole that closes up behind you and you just get lost in your own mind. You literally get lost.

"WHY DIDN'T YOU TALK TO PEOPLE ABOUT YOUR DEPRESSION?"
Here I am in the Air Force and I'm one of the senior leaders in the enlisted ranks. And that would be a sign that, well, maybe I'm not a leader. And then my career's derailed or maybe I'll lose my security clearance. I can't let anybody know. I've got to gut it out, I've got to fake my way through it. . . . You don't want to be perceived as weak, you finally get to a point where . . . you don't care how you're perceived, because you are barely breathing, you're barely getting up. . . .

"WHAT DID YOU DO TO RELIEVE THE PAIN?"
I'd drink and I'd just get numb. . . . We're talking many, many beers to get to that state where you could shut your head off,

and talking about it. Because male gender expectations associate masculinity with emotional strength, men are particularly prone to deny or ignore mental problems. In an effort to raise awareness that depression in men is a major public health problem, the National Institutes of Mental Health (NIMH) launched a public health education campaign called "Real Men, Real Depression," which includes print, television, and radio public service announcements (see this chapter's *Human Side* feature).

Extent and Impact of Mental Illness

Although transnational estimates of the prevalence of mental disorders vary, one study found a 40 percent lifetime prevalence of any mental disorder in Netherlands and the United States; 12 percent lifetime prevalence in Turkey, and 20 percent in Mexico (WHO International Consortium in Psychiatric Epidemiology 2000). On an

but then you wake up the next day and it's still there. Because you have to deal with it, it doesn't just go away.

NIMH

Rene Ruballo, Retired Police Officer

"CAN YOU DESCRIBE HOW IT STARTED?"
It started with my loss of interest in basically everything that I like doing. . . . I just didn't really feel like doing anything any more. . . . Sometimes I didn't even want to get out of bed.

"WHAT DID YOU THINK WAS HAPPENING TO YOU?"
I am thinking there's got to be something wrong because . . . I feel like nothing matters. My children, my family, nothing matters.

"WHY DIDN'T YOU SEEK HELP RIGHT AWAY?"
I was hoping that it would just go away . . . but it didn't. It just got worse. . . . Every day was a struggle, just to do minor things.

"HOW DO THE CHILDREN FEEL NOW THAT YOU ARE BETTER?
Well, they feel they got Daddy back the way he used to be. I'm doing more things with them and . . . taking more interest in things and school.

NIMH

Rodolf Palma-Lulion, Recent College Graduate

"WHAT WERE THE FIRST SYMPTOMS OF DEPRESSION?"
I just felt terrible and I didn't know why. . . . I didn't want to face anyone. I didn't want to talk to anyone. I didn't really want to do anything for myself because I felt . . . like I was such an awful person that there was no real reason for me to do anything for myself.

"DESCRIBE HOW YOU FELT."
I just didn't feel any emotions. I just couldn't feel. My real feeling was just pure numbness. . . .

"HOW DID DEPRESSION AFFECT YOU AT SCHOOL?"
I didn't read a book. I barely went to class. I just couldn't wake up in time for class. If I had class at two, I'd sleep till three. . . .

"DID BEING LATINO MAKE A DIFFERENCE?"
Yeah, I totally think that being a Latino made it harder. . . . My little brother went through depression before me . . . and we never even really talked about it because . . . there's just things you don't talk about. . . . When I told my parents I had depression . . . my mom was like you're not depressed! . . . You're gonna get over it. You just got to be strong. You just gotta finish school. . . .

Source: National Institute of Mental Health. 2003. Real Men, Real Depression. Personal Stories. http://menanddepression.nimh.nih.gov/personal.asp

annual basis, about one in five (21%) of U.S. adults and children experience mental illness (U.S. Department of Health and Human Services 2001).

In 1998, major depression was the leading cause of disability in developed nations, including the United States (World Health Organization 1999). Untreated mental disorders can lead to poor educational achievement, lost productivity, unsuccessful relationships, significant distress, violence and abuse, incarceration, and poverty. On any given day, about 150,000 people with severe mental illness are homeless, living on the streets or in public shelters. In addition, 60 to 75 percent of incarcerated youth and 16 percent of adult inmates have a mental health disorder (National Council on Disability 2002).

Mental disorders also contribute to mortality, with suicide being the fourth leading cause of death worldwide among 15- to 44-year-olds (Mercy, Krug, Dahlberg, & Zwi 2003). In the United States, suicide is the third leading cause of death among people aged 10 to 24, and the second leading cause of death for ages

25 to 34 (Anderson 2002). Most suicides in the United States (60% to 90%) are committed by individuals with mental illness (Ezzell 2003). Suicides outnumber homicides two to one every year in the United States (Ezzell 2003). In 2000, about half of the estimated 1.7 million violent deaths that occurred in the world were the result of suicide, about one-third resulted from homicide, and one-fifth were from war injuries (Mercy et al. 2003).

Causes of Mental Disorders

Biological and social factors are linked to mental illness. Some mental illnesses are caused by genetic or neurological pathology. However, social and environmental influences, such as poverty, history of abuse, or other severe emotional trauma, also affect individuals' vulnerability to mental illness and mental health problems. For example, iodine deficiency, common in poor countries, is believed to be the single most common preventable cause of mental retardation and brain damage (World Health Organization 2002). War within and between countries also contributes to mental illness (see Chapter 16 for a discussion of combat-related post-traumatic stress disorder). Garfinkel and Goldbloom (2000) explain, "The radical shifts in society towards technology, changes in family and societal supports and networks and the commercialization of existence . . . may account for the current epidemic of depression and other psychiatric disorders" (p. 503). It may be safe to conclude that the causes of most mental disorders lie in some combination of genetic, biological, and environmental factors (U.S. Department of Health and Human Services 2001).

Social Factors Associated with Health and Illness

Public health education campaigns, articles in popular magazines, college-level health courses, and health professionals emphasize that to be healthy, we must adopt a healthy lifestyle. In response, many people have at least attempted to quit smoking, eat a healthier diet, and include exercise in their daily or weekly routine. However, health and illness are affected by more than personal lifestyle choices. In the following section, we examine how social factors such as social class, poverty, education, race, and gender affect health and illness.

Social Class and Poverty

In an address to the 2001 World Health Assembly, UN Secretary-General Kofi Annan remarked, "The biggest enemy of health in the developing world is poverty" (United Nations Population Fund 2002a). Approximately one-fifth of the world's population live on less than U.S. $1 per day and nearly one-half live on less than U.S. $2 per day (World Health Organization 2002). Poverty is associated with malnutrition, unsafe water and sanitation, indoor air pollution, hazardous working conditions, and lack of access to medical care (see also Chapter 10).

In the United States, low socioeconomic status is associated with higher incidence and prevalence of health problems, disease, and death (Malatu & Schooler 2002). The percentage of Americans reporting fair or poor health is nearly four times as high for persons living below the poverty line as for those with family income at least twice the poverty threshold (National Center for Health Statistics 2000).

Poverty is both a cause and an effect of poor health. For example, malaria deepens the poverty of people who are already barely getting by, as households spend as much as one-third of their income on malaria treatment and prevention and lose income when a wage-earning member is sick (McGinn 2003). In the United States, poverty is associated with higher rates of health-risk behaviors such as smoking, alcohol drinking, being overweight, and being physically inactive. The poor are also exposed to more environmental health hazards and have unequal access to and use of medical care (Lantz et al. 1998). In addition, the lower class tends to experience high levels of stress, while having few resources to cope with it (Cockerham 1998). Stress has been linked to a variety of physical and mental health problems, including high blood pressure, cancer, chronic fatigue, and substance abuse.

Low socioeconomic status is also associated with increased risk of a broad range of psychiatric conditions (Williams & Collins 1999). Rates of depression and substance abuse, for example, are higher in the lower socioeconomic classes (Kessler et al. 1994). One explanation suggests that lower-class individuals experience greater stress as a result of their deprived and difficult living conditions. Others argue that members of the lower class are simply more likely to have their behaviors identified and treated as mental illness.

Education

In general, lower levels of education are associated with higher rates of health problems and mortality, largely because of the link between education and income (Lantz et al. 1998). Education level and family income are closely tied, for example, to the likelihood of being uninsured (Nelson 2003). Individuals with low levels of education are more likely to engage in health-risk behaviors such as smoking and heavy drinking. Women with less education are less likely to seek prenatal care and are more likely to smoke during pregnancy, which helps explain why low birthweight and infant mortality are more common among children of less-educated mothers (Children's Defense Fund 2000). In some cases, lack of education means that individuals do not know about health risks or how to avoid them. A national survey in India found that only 18 percent of illiterate women had heard of AIDS, compared with 92 percent of women who had completed high school (Ninan 2003). Despite the association between education and better health, many college students engage in high-risk health behaviors (see this chapter's *Social Problems Research Up Close* feature).

Family and Household Factors

The household in which one lives has consequences for one's physical and mental health. A study of adults in their fifties found that married people who live only with their spouse or with spouse and children had the best physical and mental health, while single women living with children had the lowest measures of health (Hughes & Waite 2002). Other research findings concur that married adults have lower levels of depression and anxiety compared to adults who are single, divorced, cohabiting, or widowed (Mirowsky & Ross 2003). Better health among married individuals results from the economic advantages of marriage, and from

Social Problems Research Up Close

The National College Health-Risk Behavior Survey

In 1995, the first national college based survey was conducted to measure a broad range of health-risk behaviors among U.S. college students. A brief description of the methods and selected findings from the National College Health Risk Behavior Survey follows.

Methods and Sample

A nationally representative sample of two- and four-year colleges and universities were selected, resulting in 136 institutions participating in the survey. A random sample of full- and part-time undergraduate students aged 18 and older was selected from the 136 participating colleges and universities. A total of 7,442 students were selected and eligible for the study, of which 4,838 (65%) completed the questionnaire. Students aged 18 to 24 represented 64 percent of the sample. The survey questionnaire, developed by the Centers for Disease Control and Prevention, was sent by mail to students in the sample. The questionnaire, available in both English and Spanish, consisted of a booklet that could be scanned by a computer and contained 96 multiple-choice questions. Responses to the survey were voluntary and confidential.

Findings and Conclusions*

- Use of safety belts: Nationwide, 10 percent of college students rarely or never used safety belts when riding in a car driven by someone else. Of those students who had driven a car, 9 percent said they rarely or never used safety belts when driving a car.
- Riding with a driver who had been drinking alcohol and driving after drinking alcohol: During the 30 days preceding the survey, more than one-third (35%) of college students nationwide had ridden with a driver who had been drinking alcohol and 27 percent of students had driven a vehicle after drinking alcohol. Male students were more likely than female students to report these behaviors, and white students were more likely than black students to report these behaviors.
- Suicide: Nationwide, 10 percent of college students had seriously considered attempting suicide during the 12 months preceding the survey; 7 percent had made a specific plan to attempt suicide, and 2 percent had attempted suicide.
- Tobacco use: Nearly one-third (32%) of college students nationwide reported either current cigarette use or current smokeless tobacco use. Male students (37%) were significantly more likely than female students (29%) to report current tobacco use. White students (36%) were significantly more likely than black (15%) and Hispanic (26%) students to report this behavior.
- Sexual behavior and HIV testing: More than one-third (35%) of college students nationwide had had sexual intercourse with six or more sex partners during their lifetime. Students aged 25 and older (50%) were significantly more likely to report this behavior than students aged 18 to 24 years. Among currently sexually active college students nationwide, 30 percent reported that either they or their partner had used a condom during last intercourse. Nationwide, 39 percent of college students had ever had their blood tested for HIV infection.
- Dietary behaviors: About one-fourth (26%) of college students had eaten five or more servings of fruits and vegetables during the day preceding the survey.
- Weight: Although 21 percent of college students were classified as being overweight based on body mass index calculations, 42 percent of college students believed themselves to be overweight. Black students (34%) were significantly more likely than white (20%) and Hispanic (21%) students to be overweight. Female students (49%) were significantly more likely than male students (32%) to perceive themselves as overweight. Nationwide, 4 percent of female students and less than 1 percent of male students had either vomited or taken laxatives to lose weight or to keep from gaining weight. Female students (7%) were significantly more likely than male students (1%) to have taken diet pills to either lose weight or keep from gaining weight.
- Physical activity: More than one-third (38%) of college students had participated in activities that had made them sweat and breathe hard for at least 20 minutes on at least 3 of the 7 days preceding the survey. Male students (44%) were significantly more likely than female students (33%) to report vigorous physical activity.

The results of the National College Health-Risk Behavior Survey indicate that many U.S. college students engage in behaviors that place them at risk for serious health problems. The survey results provide important baseline data for college leaders and health officials to use in reducing health-risk behaviors among college students.

Source: Centers for Disease Control and Prevention.1997. "Youth Risk Behavior Surveillance: National College Health-Risk Behavior Survey—United States, 1995." www.cdc.gov/ nccdphp/dash/ MMWRFile/ss4606.htm

*Percentages are rounded.

the emotional support provided by most marriages—the sense of being cared about, loved, and valued (Mirowsky & Ross 2003).

For children, living in a two-parent household is associated with better health outcomes. A Swedish study found that children living with only one parent have a higher risk of death, mental illness, and injury than those in two-parent families, even when their socioeconomic disadvantage is taken into account (Hollander 2003). For both children and adults, psychosocial stress involved in living in a single-parent household appears to have a negative effect on health.

Gender

Gender affects the health of both women and men. Gender discrimination and violence against women produce adverse health effects in girls and women worldwide. Violence against women is a major public health concern: at least one in three women has been beaten, coerced into sex, or abused in some way—most often by someone she knows (United Nations Population Fund 2000). "Although neither health care workers nor the general public typically thinks of battering as a health problem, woman battering is a major cause of injury, disability, and death among American women, as among women worldwide" (Weitz 2004, 56).

In Africa, where the leading cause of death is HIV/AIDS, HIV-positive women outnumber men by 2 million, in part because African women do not have the social power to refuse sexual intercourse and/or to demand that their male partners use condoms (United Nations Population Fund 2002a). As noted earlier, women in developing countries suffer high rates of mortality and morbidity due to the high rates of complications associated with pregnancy and childbirth. The low status of women in many less-developed countries results in their being nutritionally deprived and having less access to medical care than do men (Murphy 2003).

In the United States before the twentieth century, the life expectancy of U.S. women was shorter than that of men because of the high rate of maternal mortality that resulted from complications of pregnancy and childbirth. In the United States today, life expectancy of women (79.5) is greater than that of men (74.1) (National Center for Health Statistics 2004). Lower life expectancy for U.S. men is due to a number of factors. Men tend to work in more dangerous jobs than women, such as agriculture and construction. In addition, "beliefs about masculinity and manhood that are deeply rooted in culture . . . play a role in shaping the behavioral patterns of men in ways that have consequences for their health" (Williams 2003, 726). Men are socialized to be strong, independent, competitive, and aggressive, and to avoid expressions of emotion or vulnerability that could be construed as weakness. These male gender expectations can lead men to take actions that harm themselves or to refrain from engaging in health-protective behaviors. For example, socialization to be aggressive and competitive leads to risky behaviors (such as dangerous sports, fast driving, and violence) that contribute to men's higher risk of injuries and accidents. Men are more likely than women to smoke cigarettes and abuse alcohol and drugs, but are less likely than women to visit a doctor and to adhere to medical regimens (Williams 2003).

Although there are no gender differences in the overall prevalence of mental disorders, women have higher rates of internalizing disorders (feelings are focused on the self) such as depression and anxiety, while men have higher rates of externalizing disorders (emotions are expressed in outward behavior) such as alcohol and drug abuse, and antisocial behavior (Williams 2003). Although women are

Table 2.4
Life Expectancy* in United States by Race and Sex

Black		White	
Female	Male	Female	Male
74.9	68.2	80.0	74.8

*For persons born in 2000

Source: National Center for Health Statistics. 2004. *Health, United States, 2003.* Hyattsville, MD: U.S. Government Printing Office.

Table 2.5
U.S. Infant Mortality Rate by Mothers' Race and Ethnicity: 2000

Black or African American	13.5
American Indian or Alaska Native	8.3
White non-Hispanic	5.7
Hispanic or Latino	5.6
Asian or Pacific Islander	4.9
All mothers	6.9

Source: National Center for Health Statistics. 2004. *Health, United States, 2003.* Hyattsville, MD: U.S. Government Printing Office.

more likely to attempt suicide, men are more likely to succeed at it because they use deadlier methods.

Racial and Ethnic Minority Status

In the United States, racial and ethnic minorities tend to suffer higher rates of mortality and morbidity. Black U.S. residents, especially black men, have a lower life expectancy than whites (see Table 2.4) and also have the highest rate of infant mortality of all racial and ethnic groups, largely because of higher rates of prematurity and low birthweight (see Table 2.5). Black Americans are more likely than white Americans to die from stroke, heart disease, cancer, HIV infection, unintentional injuries, diabetes, cirrhosis, and homicide.

Compared with white Americans, Native Americans have higher death rates from motor vehicle injuries, diabetes, and cirrhosis of the liver (caused by alcoholism) (Weitz 2004). Compared with non-Hispanic whites, Hispanics have more diabetes, high blood pressure, and lung cancer, and are at higher risk of dying from violence, alcoholism, and drug use (Weitz 2004). Because Asian Americans have the highest levels of income and education of any racial/ethnic U.S. minority group, they typically have high levels of health.

Socioeconomic differences between racial and ethnic groups are largely responsible for racial and ethnic differences in health status (Weitz 2004; Williams & Collins 1999). Racial and ethnic minorities are less likely to have insurance coverage for health care and are more likely than whites to live and work in environments where they are exposed to hazards such as toxic chemicals, dust, and fumes. In addition, discrimination contributes to poorer health among oppressed racial and ethnic populations by restricting access to the quantity and quality of public education, housing, and health care. In one study, after controlling for symptoms and insurance coverage, researchers found that doctors were more likely to offer white patients life-saving treatments (such as by-pass surgery and the most effective drugs for HIV) and more likely to offer minorities less desirable treatments (such as leg amputations for diabetes) (Nelson, Smedley, & Stith 2002).

"Of all the forms of inequality, injustice in health care is the most shocking and inhumane."

Martin Luther King, Jr.
Civil Rights Leader

Racial and ethnic minorities also bear a disproportionate burden of mental illness, in part because of the higher rates of poverty among minorities. Racism and discrimination also adversely affect physical and mental health, and they place minorities at higher risk for mental disorders, such as anxiety and depression (U.S. Department of Health and Human Services 2001). Minorities also have less access to mental health services, are less likely to receive needed mental health services, often receive lower quality mental health care, and are underrepresented in mental health research (U.S. Department of Health and Human Services 2001).

Problems in U.S. Health Care

The United States boasts of having the best physicians, hospitals, and advanced medical technology in the world, yet problems in U.S. health care remain a major concern on the national agenda. A World Health Organization report on the first-ever analysis of the world's health systems noted that although the United States spends a higher portion of its gross domestic product on health care than any other country, it ranks 37 out of 191 countries according to its performance (World Health Organization 2000). The report concluded that France provides the best overall health care among major countries followed by Italy, Spain, Oman, Austria, and Japan. After presenting a brief overview of U.S. health care, we address some of the major health care problems in the United States—inadequate health insurance coverage, the high cost of medical care and insurance, inadequate mental health care, and the managed care crisis.

U.S. Health Care: An Overview

In traditional health insurance plans, the insured choose their health care provider, who is reimbursed by the insurance company on a fee-for-service basis. The insured individual typically must pay an out-of-pocket "deductible" (perhaps $350.00 per year per person or per family) and then is often required to pay a percentage of medical expenses (e.g., 20%) until a maximum out-of-pocket expense amount is reached (after which insurance will cover 100 percent of medical costs).

Health maintenance organizations (HMOs) are prepaid group plans in which a person pays a monthly premium for comprehensive health care services. HMOs attempt to minimize hospitalization costs by emphasizing preventive health care.

Preferred provider organizations (PPOs) are health care organizations in which employers who purchase group health insurance agree to send their employees to certain health care providers or hospitals in return for cost discounts. In this arrangement, health care providers obtain more patients but charge lower fees to buyers of group insurance.

Managed care refers to any medical insurance plan that controls costs through monitoring and controlling the decisions of health care providers. In many plans, doctors must call a utilization review office to receive approval before they can hospitalize a patient, perform surgery, or order an expensive diagnostic test. Although the terms "HMO" and "managed care" are often used interchangeably, HMOs are only one form of managed care. About three-fourths of all Americans with private insurance belong to some form of managed care plan (Iglehart 1999). Recipients of Medicaid and Medicare may also belong to a managed care plan.

Medicare, Medicaid, and the state Children's Health Insurance Program (SCHIP) are the major publicly funded health programs. **Medicare** is funded by the federal government and reimburses the elderly and the disabled for their health care (see also Chapter 9). Medicare consists of two separate programs: a hospital insurance program and a supplementary medical insurance program. The hospital insurance program is free, but enrollees may pay a deductible and a copayment, and there is limited coverage of home health nursing and hospice care. Medicare's medical insurance program is not free; enrollees must pay a monthly premium as well as a copayment for services.

Medicare does not cover long-term nursing home care, dental care, eyeglasses, and other types of services, which is why many individuals who receive Medicare also purchase supplementary private insurance known as *medigap* policies. The Medicare Prescription Drug, Improvement, and Modernization Act of 2003, signed into law by President G. W. Bush, contains a prescription drug benefit for the disabled and seniors; it will become available in 2006 and will be offered by private insurers and health plans under contract with the government. Critics of the Medicare Prescription bill argue that the drug coverage is inadequate and complicated, denies beneficiaries the right to purchase supplemental coverage, and fails to lower the cost of prescription drugs. The Medicare Prescription Drug legislation provides billions of dollars in subsidies to HMOs and other managed care plans, paying them much more than it costs regular Medicare to provide the same services. These private plans can elect to cover a limited number of drugs and deny coverage for other drugs. Finally, critics point out that the legislation is projected to cause more than 2 million retirees to lose the drug coverage they previously had (Park, Nathanson, Greenstein, & Springer 2003).

Medicaid, which provides health care coverage for the poor, is jointly funded by the federal and state governments (see also Chapter 10). Eligibility rules and benefits vary from state to state and in many states Medicaid provides health care only for the very poor who are well below the federal poverty level. Even so, Medicaid covers one in seven Americans and is the second-largest item in state budgets, surpassed only by education (Toner & Stolberg 2002). In response to large budget deficits, two-thirds of states cut Medicaid benefits in 2003, and nearly half froze or reduced Medicaid payments to providers (PNHP Data Update 2003). Because of the low reimbursement from Medicaid, many health care providers do not accept Medicaid patients.

In 1997, the state Children's Health Insurance Program (SCHIP) was created to expand health coverage to uninsured children, many of whom come from families with incomes too high to qualify for Medicaid but too low to afford private health insurance (Health Care Financing Administration 2000). Under this initiative, states receive matching federal funds to provide medical insurance to uninsured children. In 2002, the Bush administration issued a regulation extending coverage under SCHIP to "unborn children," thus covering children from conception up to age 19. For the first time, the United States recognizes a zygote, embryo, or fetus as a "person" eligible for government aid (International Women's Health Coalition 2003).

Inadequate Health Insurance Coverage

Virtually all elderly Americans have Medicare coverage and most non-elderly Americans receive health insurance coverage through their employers. But in 2002, more than 43 million persons in the United States—15.2 percent of the U.S. population—lacked health insurance (Institute of Medicine 2004). Contrary to the belief that Medicaid covers all poor people, it does not. In 35 states, even a part-time minimum wage job with earnings equal to the poverty line could disqualify a single mother from Medicaid (Harrell & Carrasquillo 2003).

Individuals who lack health insurance are much more likely than individuals with insurance to delay or forgo needed medical care. A survey of U.S. adults found that of those who were uninsured, 20 percent needed but did not get care for a serious medical problem, 39 percent skipped a recommended test or treatment, 30 percent did not get a prescription filled, and 13 percent had problems getting needed mental health care (Kaiser Commission on Medicaid and the Uninsured 2000). Indeed, individuals lacking insurance during an entire year received about half as much care as the privately insured (Hadley & Holahan 2003). An estimated 18,000 deaths in the United States annually are attributable to lack of health insurance (Institute of Medicine 2004).

The High Cost of Health Care

The United States spends more on health care per person than any other country in the world. Most industrialized countries spend between 6 and 10 percent of gross domestic product (GDP) on health care, while in 2002 the United States spent 14.9 percent of GDP ($5,440 per person), and is projected to spend 17.7 percent of GDP by 2012 (Pear 2004; PNHP Data Update 2003). In 2001, health care spending rose 9.3 percent, the fastest rate in 11 years (Pear 2004).

Several factors have contributed to escalating medical costs. First, the U.S. population is aging. Because people over age 65 use medical services more than younger individuals, the growing segment of the elderly population means more money is spent on medical care (Conrad & Brown 1999).

Health care administrative expenses are higher in the United States than in any other nation. Insurance companies and for-profit health maintenance organizations (HMOs) spend between 20 to 30 percent of their budgets to cover the costs of stockholder dividends, lobbyists, huge executive salaries, marketing, and wasteful paperwork (Conyers 2003).

The high cost of drugs also contributes to health care costs. In 2001, the top 10 U.S. pharmaceutical companies increased profits by 33 percent, while the overall profits of Fortune 500 companies declined by 53 percent (Public Citizen 2002). In 2001, the prices of the 50 prescription drugs most commonly used by the elderly rose by 7.8 percent—nearly three times the rate of inflation (2.7%). Medicare beneficiaries' average out-of-pocket spending on medications increased from $644 in 2000 to $866 in 2002 (PNHP Data Update 2003). While the pharmaceutical industry claims that profits are needed to develop new drugs, most large drug companies

In 2003, 18,000 General Electric workers in 23 states went on strike to protest an increase in health insurance co-pays.

spend more than twice as much on marketing, advertising, and administration than on research and development of new drugs (PNHP Data Update 2003). Top executives at the nine largest pharmaceutical companies received an average annual income of $21 million in 2001 with an average of $48 million additional compensation in stock options (PNHP Data Update 2003).

High doctors' fees and hospital costs are also factors in the rising costs of health care. The use of expensive medical technology, unavailable just decades ago, also contributes to high medical bills. Increasingly, hospitals are purchasing more expensive equipment than is needed.

High costs of public and private insurance have also contributed to escalating health care expenditures. The average premium of a family health plan is about $500 per month, or $6,000 per year (Harrell & Carrasquillo 2003). Workers who are covered by insurance provided by their employer pay about one-quarter of the cost of a family plan. With the rising cost of medical insurance, companies are increasing the employees' share of the cost, decreasing the benefits, or not providing insurance at all. Facing a 12.7 percent increase in health premiums in 2002, followed by a 15 percent increase in 2003, it is no surprise that the number of small businesses providing employee coverage dropped from 67 percent in 2001 to 61 percent in 2002 (PNHP Data Update 2003).

The Managed Care Crisis

Increasingly, Americans are concerned about the reduced quality of health care resulting from the emphasis on cost containment in managed care. In a survey of physicians' views on the effects of managed care, the majority responded that managed care has negative effects on the quality of patient care because of limitations on diagnostic tests, length of hospital stay, and choice of specialists (Feldman, Novack,

& Gracely 1998). One former director of a health maintenance organization described the situation:

> I've seen from the inside how managed care works. I've been pressured to deny care, even when it was necessary. I have seen the bonus checks given to nurses and doctors for their denials. I have seen the medical policies that keep patients from getting care they need . . . and the inadequate appeal procedures. (Peeno 2000, 20)

Inadequate Mental Health Care

Since the 1960s, U.S. mental health policy has focused on reducing costly and often neglectful institutional care and, instead, providing more humane services in the community. This movement, known as **deinstitutionalization,** had good intentions but has largely failed to live up to its promises. Only one in five U.S. children with emotional disturbance received any mental health services, and fewer than half of adults with a serious mental illness received treatment or counseling for a mental health problem during the past year (National Council on Disability 2002; Substance Abuse and Mental Health Services Administration 2003).

Reasons for not seeking treatment include the stigma associated with mental illness (and thus with treatment for mental illness), fear and mistrust of treatment, cost of care (which is often not covered by health insurance), and lack of access to services. Most private health insurance plans do not provide the same benefits for mental health care as they do for other health care. Compared to coverage for other illnesses, insurance coverage for mental health care often involves higher copayments and deductibles, and allows fewer doctor visits or days in the hospital. Families of children with mental disorders often face the heart-wrenching choice of not receiving adequate mental health services for their children or relinquishing custody of the children in order to qualify for Medicaid (Lehmann 2003).

The mental health system is also plagued by inadequate funding. Medicaid provides more than half of funding for public mental health services, and every state has cut or plans to cut Medicaid funds (Mulligan 2003). Inadequate federal and state funding of public mental health centers results in rationing care to those most in need. Thus, people must "hit bottom" before they can receive services. Mental health services are often inaccessible, especially in rural areas. In most states, services are available from "9 to 5"; the system is "closed" during evenings and weekends when many people with mental illness experience the greatest need. Across the nation, people with severe mental illness end up in jails and prisons, homeless shelters, and hospital emergency rooms. Many children with untreated mental disorders drop out of school or end up in foster care or the juvenile justice system. Given the increasing growth of minority populations, another deficit in the mental health system is the inadequate numbers of mental health clinicians who speak the client's language and who have awareness of cultural norms and values of minority populations (U.S. Department of Health and Human Services 2001).

Strategies for Action: Improving Health and Health Care

There are two broad approaches to improving the health of populations: selective primary health care and comprehensive primary health care (Sanders & Chopra 2003). **Selective primary health care** uses technocratic solutions to target a specific health problem, such as using immunization and oral rehydration therapy to promote child survival. In contrast, **comprehensive primary health care** focuses on the broader social determinants of health, such as poverty and economic inequality, gender inequality, environment, and community development.

In recent decades, health agendas have focused on selective primary health care approaches, such as child immunization, which increased from 20 percent (of children under age 1) in 1980 to 80 percent by 1990 (Sanders & Chopra 2003). However, more recent declines in immunization coverage in some regions, along with rising infant mortality rates, provide evidence that selective primary health care produces "short term but unsustainable results" (p. 107). Targeting specific health problems may be necessary, but not sufficient, for achieving long-term health gains. Sanders and Chopra (2003) emphasize that "only where health interventions are embedded within a comprehensive health care approach, including attention to social equity, health systems and human capacity development, can real and sustainable improvements in health status be seen" (p. 108).

As you read the following sections on improving maternal and infant health, and on preventing and alleviating HIV/AIDS, see if you can identify which strategies represent selective primary health care approaches and which strategies are comprehensive. Also, be mindful that strategies to alleviate social problems discussed in subsequent chapters of this text are also important elements to a comprehensive primary health care approach (e.g., see Chapters 6, Race and Ethnic Relations; 7, Gender Inequality; 10, Poverty; 12, Problems in Education; 14, Environmental Problems).

Improving Maternal and Infant Health

As discussed earlier, pregnancy and childbirth are major causes of mortality and morbidity among women of reproductive age in the developing world. In 1987, the Safe Motherhood Initiative was launched to improve maternal health. This global initiative involves a partnership of governments, nongovernmental organizations, agencies, donors, and women's health advocates working to protect women's health and lives, especially during pregnancy and childbirth.

In many developing countries, women's lack of power and status means that they have little say over their reproductive health (Murphy 2003). Men make the decisions about whether or when their wives (or partners) will have sexual relations, use contraception, or use health services. Thus, improving the status and power of women is an important strategy in improving their health. Promoting women's education increases the status and power of women to control their reproductive lives, exposes women to information about health issues, and also delays marriage and childbearing.

Access to family planning services, skilled birth attendants, affordable methods of contraception, and safe abortion services are important determinants of the well-being of mothers and their children (Save the Children 2002). Family planning reduces maternal mortality simply by reducing the number of unintended pregnancies. Spacing births two to three years apart decreases infant mortality significantly (Murphy 2003). Implementing these strategies requires funding, but the cost is not prohibitive. The price of ensuring that women in low-income countries get health care during pregnancy, delivery, and after birth; family planning services; and newborn care is estimated at only $3 (U.S.) per person per year (Family Care International 1999). The question is, do the rich countries of the world have the political will to support efforts to protect the health and lives of women and infants in the developing world?

HIV/AIDS Prevention and Alleviation Strategies

As of this writing, there is no vaccine to prevent HIV infection. As researchers continue to work on developing such a vaccine, a number of other strategies are available to help prevent and treat HIV/AIDS.

Alleviating HIV/AIDS requires educating populations about how to protect against HIV and providing access to male and female condoms. The need for education about HIV/AIDS is evident in a statement by Libyan leader Moammar Gadhafi, who told African leaders that heterosexuals do not have to worry about getting HIV/AIDS, implying that only gays get AIDS ("Libya Leader: Only Gays Get AIDS" 2003). At least 30 percent of young people in a survey of 22 countries had never heard of AIDS, and in 17 countries surveyed, over half of adolescents could not name a single method of protecting themselves against HIV (United Nations Population fund 2002a).

Many HIV/AIDS public educational campaigns target young adults—the age group that is most at risk for acquiring HIV.

Providing access to male and female condoms is an important HIV-prevention strategy. Yet, because of widespread concern that sex education and access to condoms will encourage young people to become prematurely sexually active, many sex education programs in the United States have focused solely upon abstinence. Several studies have concluded, however, that sex education programs that combine messages about abstinence and safer sex practices (e.g., condom use) may delay the initiation of sexual behavior as well as increase preventive behaviors among young people who are already sexually active.

To reduce transmission of HIV among injection drug users, their sex partners, and their children, some countries and U.S. communities have established **needle exchange programs,** which provide new, sterile syringes in exchange for used, contaminated syringes. Many needle exchange programs also provide drug users with a referral to drug counseling and treatment. Research suggests that needle exchange programs result in lower rates of HIV transmission (U.S. Department of Health and Human Services 1998). In Canada, sterile injection equipment is available to drug users in pharmacies and through numerous needle exchange programs. In contrast, most U.S. states prohibit the sale or possession of sterile needles or syringes without a medical prescription, and only a small number of communities have legal needle exchange programs.

Another strategy to curb the spread of HIV involves encouraging individuals to get tested for HIV so they can modify their behavior (to avoid transmitting the virus to others) and so they can receive early medical intervention, which can slow or prevent the onset of AIDS. However, individuals who have been diagnosed with

HIV often continue to engage in risky behaviors such as unprotected anal, genital, or oral sex and needle sharing (Diamond & Buskin 2000).

Developing countries—those hardest hit by the HIV/AIDS—depend on aid from wealthier countries to fight HIV/AIDS. HIV/AIDS education programs, condoms, and drugs used to treat HIV cost money that many poor countries do not have. The U.S. Leadership Against HIV/AIDS, Tuberculosis and Malaria Act of 2003, signed into law by President George W. Bush, authorized the expenditure of $15 billion over five years to fight the global pandemic of HIV/AIDS. However, the law requires one-third of the funds for HIV prevention to be spent on abstinence-only-until-marriage programs, which critics argue are not as effective as programs that provide sexuality education and condom distribution.

Commenting on HIV/AIDS prevention, Altman (2003) notes, "the great irony is that we know how to prevent HIV transmission and it is neither technically difficult nor expensive. Most HIV transmission can be stopped by the widespread use of condoms and clean needles" (p. 41). Efforts to promote condom use require empowering women, so that they have a voice in negotiating condom use in their relationships. Implementing these strategies, however, conflicts with religious and cultural beliefs and they threaten the political power structure. Altman explains that "effective HIV prevention requires governments to acknowledge a whole set of behaviours—drug use, 'promiscuity,' homosexuality, commercial sex work—which they would often rather ignore, and a willingness to support, and indeed empower, groups practicing such behaviours" (p. 42).

U.S. Health Care Reform

"Healthcare is one of our most basic needs. Making it available is perhaps our greatest test of humanity."

Linda Peeno, M.D.

The United States is the only country in the industrialized world that does not have any mechanism for guaranteeing health care to its citizens. Other countries such as Canada, Great Britain, Sweden, Germany, and Italy have national health insurance systems, also referred to as **socialized medicine** and **universal health care.** Despite differences in how socialized medicine works in various countries, in all systems of socialized medicine the government (1) directly controls the financing and organization of health services, (2) directly pays providers, (3) owns most of the medical facilities (Canada is an exception), (4) guarantees equal access to health care, and (5) allows some private care for individuals who are willing to pay for their medical expenses (Cockerham 1998).

Since 1912, when Theodore Roosevelt first proposed a national health insurance plan, the idea of health care for all Americans has been advocated by the Truman, Nixon, Carter, and Clinton administrations. A 2002 Gallup poll found that nearly two-thirds (62%) of U.S. adults believe that the federal government is responsible for guaranteeing health care coverage for all Americans (Gallup Poll Social Series 2002). Currently, advocates are promoting a national health insurance bill that would create an improved and expanded Medicare for All program. This program would create a single-payer system in which a single tax-financed public insurance program replaces private insurance companies. Under this plan, every U.S. resident would be issued a national health insurance card; would receive all medically necessary services (including prescription drugs and long-term care); have no copayments or deductibles; and see the doctor of his or her choice (Conyers 2003). If this plan is adopted, it is estimated to save approximately $100 billion per year on administrative costs—enough to provide coverage for all the uninsured and substantially help the underinsured. The proposed single-payer program would use the already existing Medicare program by expanding it to cover all U.S. residents and would provide comprehensive coverage (including dentistry, eye care, mental health services, substance abuse treatment, prescription drugs, and long-term care) with no copays or deductibles.

The insurance industry, not surprisingly, opposes the adoption of such a system because the private health insurance industry would be virtually eliminated. The health insurance industry's opposition to a single-payer universal health plan is matched only by the persistent efforts of those who advocate such a plan. Bills for single-payer health programs have been introduced in a number of states, including California, Hawaii, Minnesota, New Mexico, Delaware, Massachusetts, Wisconsin, and Maine (PNHP Data Updates 2003).

Strategies to Improve Mental Health Care

In 1999 the first White House conference on Mental Health and the first Surgeon General's Report on Mental Health established mental health care as a national health priority. Two areas for improving mental health care in the United States are eliminating the stigma associated with mental illness and improving health insurance coverage for treating mental disorders. Efforts in these areas will, hopefully, promote the delivery of treatment to individuals suffering from mental illness. Although most mental disorders may be successfully treated with medications and/or psychotherapy or counseling, nearly half of all Americans who have a severe mental illness do not seek treatment (U.S. Department of Health and Human Services 1999).

"Americans assign high priority to preventing disease and promoting personal well-being and public health; so too must we assign priority to the task of promoting mental health and preventing mental disorders."

Mental Health: A Report of the Surgeon General

Eliminating the Stigma of Mental Illness. The first White House Conference on Mental Health called for a national campaign to eliminate the stigma associated with mental illness. Fearing the negative label of "mental illness" and the social rejection and stigmatization associated with mental illness, individuals are reluctant to seek psychological services (Komiya, Good, & Sherrod 2000). In addition, "stigma deters the public from wanting to pay for care and, thus, reduces consumers' access to resources and opportunities for treatment and social services" (U.S. Department of Health and Human Services 1999, viii). To assess your attitudes toward seeking professional psychological help, see this chapter's *Self and Society* feature.

Reducing stigma associated with mental illness may be achieved through encouraging individuals to seek treatment and making treatment accessible and affordable. The Surgeon General's Report on Mental Health explains,

> Effective treatment for mental disorders promises to be the most effective antidote to stigma. Effective interventions help people to understand that mental disorders are not character flaws but are legitimate illnesses that respond to specific treatments, just as other health conditions respond to medical interventions. (U.S. Department of Health and Human Services 1999, viii)

The National Alliance for the Mentally Ill (NAMI) has a StigmaBusters campaign, whereby the public submits instances of media content that stigmatize individuals with mental illness to StigmaBusters, which then investigates and conveys their concerns to media organizations, urging them to avoid stigmatizing portrayals of mental illness.

Another tool used to reduce the stigma of mental illness is a school curriculum called *Breaking the Silence,* available through NAMI in elementary, middle, and high school levels. Using true stories, activities, a board game, and posters, *Breaking the Silence* debunks myths about mental illness and sensitizes students to the pain that words like "psycho" and "schizo," as well as frightening or comic media images of mentally ill people, can cause (Harrison 2002).

Self and Society

Attitudes Toward Seeking Professional Psychological Help

Directions: For each of the following statements, indicate your level of agreement according to the following scale:

Agree = 3 Partly agree = 2 Partly disagree = 1 Disagree = 0

	Statement	Scoring
1.	If I believed I was having a mental breakdown, my first inclination would be to get professional attention.	S
2.	The idea of talking about problems with a psychologist strikes me as a poor way to get rid of emotional conflicts.	R
3.	If I were experiencing a serious emotional crisis at this point in my life, I would be confident that I could find relief in psychotherapy.	S
4.	There is something admirable in the attitude of a person who is willing to cope with his or her conflicts and fears *without* resorting to professional help.	S
5.	I would want to get psychological help if I were worried or upset for a long period of time.	S
6.	I might want to have psychological counseling in the future.	S
7.	A person with an emotional problem is not likely to solve it alone; he or she *is* likely to solve it with professional help.	S
8.	Considering the time and expense involved in psychotherapy, it would have doubtful value for a person like me.	R
9.	A person should work out his or her own problems; getting psychological counseling would be a last resort.	R
10.	Personal and emotional troubles, like many things, tend to work out by themselves.	R

Scoring: Straight items (S) are scored 3-2-1-0, and reversal items (R) 0-1-2-3, respectively, for the response alternatives *agree, partly agree, partly disagree,* and *disagree.* Sum the responses for each item to find your total score, which will range from 0 to 30. The higher your score, the more positive your attitudes toward seeking professional psychological help; the lower your score, the more negative your attitudes toward seeking professional psychological help.

In a study of 389 undergraduate students (primarily 18-year-old freshmen), the average score on this scale was 17.45 (Fischer & Farina 1995). Women were more likely to have higher scores, reflecting more positive attitudes. For women in the study, the average score was 19.08 compared to 15.46 for men.

Source: Fisher, Edward, and Amerigo Farina. 1995. "Attitudes Toward Seeking Professional Psychological Help: A Shortened Form and Considerations for Research." *Journal of College Student Development* 36(4):368–373.

Eliminating Inequalities in Health Care Coverage for Mental Disorders. Another priority on the agenda to improve the nation's mental health care system involves eliminating the inequalities in health care coverage for mental disorders. The Mental Health Parity Act of 1996 was an important step in ending health care discrimination against individuals with mental illnesses by requiring equality between mental health care insurance coverage and other health care coverage—a concept known as **parity.** However, this federal law applies only to annual or lifetime cost limits, but not to substance abuse, copayments, deductibles, or inpatient/outpatient treatment limits. This law was set to expire at the end of 2002, although the Senate extended it for one year following the death of Senator Paul Wellstone, who pushed for expanding the law to prevent insurance companies from charging higher deductibles and copayments for mental health care (Vastag 2003). Some states have enacted mental health parity laws, which vary in their scope and application: Many do not address substance abuse, are limited to the more serious mental illnesses, or apply only to government employees.

Computer Technology in Health Care

Computer technology has revolutionized many aspects of modern life, including health care. **Telemedicine** involves using information and communication technologies to deliver a wide range of health care services, including diagnosis, treatment, prevention, health support and information, and education of health care workers. Many hospitals and clinics store their medical records electronically, allowing doctors to access information about their patients very quickly and to send medical records, vital signs (such as heart rate and blood pressure), and radiological images (such as x-ray and ultrasound images) to another location for the purpose of interpretation or consultation.

Telemedicine offers increased access to health care in rural and remote areas. In addition, telemedicine increases the public's access to specialists and expensive equipment and enables medical specialists to spread their skills across continents.

Computer technology also enables health consumers to access health information and support services on the Internet, which helps empower individuals in managing their health concerns. Through e-mail, bulletin boards, and chat rooms, individuals with specific health problems can network with other similarly affected individuals. This social support assists in patient recovery, reduces the number of visits to physicians and clinics, and provides networks of social support to individuals with disabilities.

One of the most fascinating uses of computer technology in health care is biofeedback. This chapter's *Focus on Technology* feature provides an overview of the types, uses, and benefits of biofeedback in health care.

Understanding Illness and Health Care

Human health has probably improved more over the past half century than over the previous three millennia (Feachum 2000). Yet the gap in health between rich and poor remains very wide and the very poor suffer appallingly. Health problems are affected not only by economic resources, but also by other social factors such as aging of the population, family structure, gender, education, and race/ethnicity.

U.S. cultural values and beliefs view health and illness as determined by individual behavior and lifestyle choices, rather than by social, economic, and political forces. We agree that an individual's health is affected by the choices he or she makes—choices such as whether to smoke, exercise, engage in sexual activity, use condoms, wear a seatbelt, and so on. However, the choices individuals make are influenced by social, economic, and political forces that must be taken into account if the goal is to improve the health not only of individuals, but also of entire populations. By focusing on individual behaviors that affect health and illness, we often overlook not only social causes of health problems but social solutions as well. For example, at an individual level, the public has been advised to rinse and cook meat, poultry, and eggs thoroughly and to carefully wash hands, knives, cutting boards, and so on in order to avoid illness caused by Escherichia coli and Salmonella bacteria. But whether one becomes ill from contaminated meat, eggs, or poultry is also affected by governmental action in the 1980s, which reduced the number of government food inspectors and deregulated the meat-processing industry (Link & Phelan 2001). Just as governmental actions created the need for individuals to use caution in food preparation, governmental actions can also offer solutions by providing for more food inspectors and stricter regulations on food industries.

"The defense this nation seeks involves a great deal more than building airplanes, ships, guns and bombs. We cannot be a strong nation unless we are a healthy nation."

U.S. President Franklin Roosevelt 1940

Although certain changes in medical practices and policies may help to improve world health, "the health sector should be seen as an important, but not the

Focus on Technology

The Uses of Biofeedback in Health Care

Imagine sitting in a chair facing a computer monitor that displays an image of a rocket ship traveling through space. You are controlling the speed of the rocket ship, but instead of using a joystick, you are using the electrical activity of your brain, or brain waves, to control the rocket ship. This is not a science fiction scenario; this is actually a very powerful and effective health intervention known as biofeedback.

Biofeedback is a process in which information that is fed back to the brain enables a person to change his or her biological functioning. Biofeedback treatment teaches a person to influence biological responses such as heart rate, nervous system arousal, muscle contractions, and even brain wave functioning. Biofeedback is used at about 1,500 clinics and treatment centers worldwide. A typical session lasts about an hour and costs $60 to $150.

Types and Uses of Biofeedback

Several types of biofeedback are used to treat a wide variety of physical and mental health problems.

Electromyographic (EMG) Biofeedback, which measures electrical activity created by muscle contractions, is often used for relaxation training as well as stress and pain management. EMG biofeedback is also used to treat urinary and fecal incontinence, muscle incoordination (e.g., for persons with cerebral palsy or multiple sclerosis), and repetitive strain injuries such as carpal tunnel syndrome.

Thermal or Temperature Biofeedback uses a temperature sensor to detect changes in temperature of the fingertips or toes. Stress causes blood vessels in the fingers to constrict, reducing blood flow, leading to cooling. Thermal biofeedback, which trains people to quiet the nervous system arousal mechanisms that produce hand and/or foot cooling, is often used for stress and pain management, arthritis, anxiety, Raynaud's disease, and irritable bowel and other applications.

Galvanic Skin Response (GSR) Biofeedback utilizes a finger electrode to measure sweat gland activity. This measure is very useful for relaxation and stress management training, and is also used in the treatment of attention deficit/hyperactivity disorder.

Neurofeedback, also called *Neurobiofeedback* or *Electroencephalogram (EEG) Biofeedback,* trains people to enhance their brain wave functioning and has been found to be effective in treating a wide range of conditions, including attention deficit/hyperactivity disorder, depression, anxiety and panic disorder, sleep disorders, obsessive-compulsive disorder, closed-head injury, tension and migraine headaches, addictions, chronic pain, epilepsy, anger, and high blood pressure.

As neurofeedback is the fastest growing field in biofeedback, with new applications being developed every year, let's take a closer look at this treatment modality. Neurofeedback involves a series of sessions in which a client sits in a comfortable chair facing a specialized game computer. Small sensors are placed on the scalp to detect brain wave activity and transmit this information to the computer. The neurofeedback therapist also sits in front of a computer that displays the client's brain wave patterns in the form of an electroencephalogram (EEG). Based on a clinical assessment of the client's functioning, the therapist determines what kinds of brainwave patterns are optimal for the client. During neurofeedback sessions, clients learn to produce desirable brain waves by controlling a computerized game or task, similar to playing a videogame, but instead of a joystick, it is the client's brain waves that control the game.

Neurofeedback is like an exercise of the brain, helping it to become more flexible and effective. But unlike body exercise, which when training is stopped will lose its benefits over time, brain wave training generally does not. Once the brain is trained to function in its optimal state (which may take an average of 20 to 25 sessions), it generally remains in this more healthy state. Neurofeedback therapists liken the process to that of learning to ride a bicycle—once you learn to ride a bike, you can do so even if it has been years since you have ridden one. One neurofeedback therapist explains that "clients speak often of their disorders—panic attacks, chronic pain, and so on—as if they were stuck in a certain pattern of response. Consistently in clinical practice, EEG biofeedback helps 'unstick' people from these unhealthy response patterns" (Carlson-Catalano, 2003).

Individual and Social Benefits of Biofeedback

Biofeedback treatment has been shown to be effective in alleviating a variety of physical and mental health problems in about 80 percent of individuals who receive treatment. Biofeedback reduces the cost of health care as it eliminates or reduces the need for medications, surgery, caretaking costs, and other medical treatments. It also increases the client's sense of control and self-efficacy, as biofeedback is essentially a process of self-regulation. For example, biofeedback provides victims of domestic violence and abuse a mechanism for increasing self-esteem and sense of power and control over one's life (Carlson-Catalano 2003).

As biofeedback provides effective and long-term treatment for such ills as substance addictions, depression, rage disorders, and attention deficit/hyperactivity disorder, it also promises to remedy social problems associated with such conditions. Thus, widespread use of biofeedback could potentially reduce violence, divorce, crime, and incarceration, and could increase educational success and attainment and work productivity.

The Future of Biofeedback

In the 1960s and 1970s, when research demonstrated that biofeedback was effective in preventing seizures associated with epilepsy, biofeedback was considered a "fringe" therapy. Advances in biotechnology, a growing body of research and clinical evidence supporting its efficacy, and growing public interest in alternative therapies have contributed to growing recognition of biofeedback by the mainstream medical and professional community.

Yet biofeedback is not widely embraced by insurance companies, health care professionals, educators, policy makers, and judges. Many insurance companies do not cover biofeedback at all, while others cover it for some problems but not others. For example, an insurance company may cover the use of biofeedback to treat headaches, but not high blood pressure or attention deficit disorder (despite the voluminous research evidence showing that biofeedback is useful in treating both). And while judges often order individuals convicted of spousal abuse, substance abuse, road rage, or violent crimes to participate in treatment programs for anger management or substance abuse, they overlook biofeedback as a potential treatment, despite research that finds a significantly lower recidivism rate among felons who received biofeedback during their incarceration (von Hilsheimer & Quirk 2001).

From a conflict perspective, lack of support for biofeedback can be explained by the fact that biofeedback threatens the profits of pharmaceutical companies and mainstream health professionals (e.g., doctors, psychiatrists, psychologists). For example, attention deficit/hyperactivity disorder (AD/HD) generates enormous profits for pharmaceutical companies that make drugs such as Ritalin and Adderall (commonly prescribed for AD/HD), and for the doctors who prescribe them. These stimulant drugs, known as "speed" to drug abusers, can have serious side effects such as sleep problems, weight loss, jitters, stomach upset, and potential for abuse by the patient (or the patient's family). Long-term effects of stimulant use are unknown. While stimulant drugs treat only the symptoms of AD/HD, biofeedback goes to the root of the problem by correcting dysregulation of the brain, so drugs and doctors are no longer needed to manage the symptoms.

The authors of *Getting Rid of Ritalin: How Neurofeedback Can Successfully Treat Attention Deficit Disorder without Drugs* claim that "intelligent and widespread use of neurofeedback by clinicians and educators has the potential to heal some of the deepest and most devastating wounds of our children and our society" (p. xi). Biofeedback is one of the best examples of how technology can alleviate social problems. Unfortunately, it is also an example of how, in our society, profits take precedence over people.

Sources: Carlson-Catalano, Judy. 2003 (June 9). Director, Clinical Biofeedback Services. *Health Innovations.* Greenville, NC 27878. www.healthinnovationsinc.com. Personal communication; Hill, Robert W., and Eduardo Castro. 2002. *Getting Rid of Ritalin: How Neurofeedback Can Successfully Treat Attention Deficit Disorder without Drugs.* Charlottesville, VA: Hampton Roads Publishing; Kirchheimer, Sid. 2002 (December 20). "Biofeedback Enhances ADHD Treatments." WebMD Medical News. http://webmd.lycos.com/content/article/57/66030.htm; Monastra, Vincent J., Donna M. Monastra, and Susan George. 2002. "The Effects of Stimulant Therapy, EEG Biofeedback, and Parenting Style on the Primary Symptoms of Attention-Deficit/Hyperactivity Disorder." *Applied Psychophysiology and Biofeedback,* 27(4):231–249; Oubre, Alondra. 2002. "EEG Neurofeedback for Treating Psychiatric Disorders." *Psychiatric Times* 19(2). www.psychiatrictimes.com/p020268.html; Robbins, Jim. 2000. *A Symphony in the Brain: The Evolution of the New Brain Wave Biofeedback.* New York: Grove Press; Samilo, Kathleen, and Lela Carlson. 2003. "Integrating Biofeedback in Community Mental Health Settings: Experiences from a Clinical Demonstration Project." *Biofeedback* 31(1):30–34; von Hilsheimer, George, and Douglas A. Quirk. 2001. "Using Neurofeedback to Correct the Incorrigible." The Biofeedback Center. Maitland, FL 32751. www.drbiofeedback.com/sections/library/articles/incorrigible.html

sole, force in the movement toward global health" (Lerer, Lopez, Kjellstrom, & Yach 1998, 18). Improving the health of a society requires addressing diverse issues, including poverty and economic inequality, gender inequality, population growth, environmental issues, education, housing, energy, water and sanitation, agriculture, and workplace safety. Improving health worldwide calls for cooperation in the international community. At a 2003 meeting of the Commission on Human Rights in Geneva, 42 countries supported a resolution urging countries to commit to realizing the universal right to the highest attainable standards of physical and mental health. The resolution calls on all countries to increase efforts to eliminate discrimination in health care, prevent violence, promote sexual and reproductive health, and assist developing countries in achieving higher standards of health. The United States was the only country to vote against the resolution (International Women's Health Coalition 2003).

Improving the health of the world also means seeking nonmilitary solutions to international conflicts. In addition to the deaths, injuries, and illnesses that result from combat, war diverts economic resources from health programs, leads to hunger and disease caused by the destruction of infrastructure, causes psychological trauma, and contributes to environmental pollution (Sidel & Levy 2002). Thus, "the prevention of war . . . is surely one of the most critical steps mankind can make to protect public health" (White 2003, 228).

The World Health Organization (1946) defines **health** as "a state of complete physical, mental, and social well-being" (p. 3). Based on this definition, we conclude this chapter with the suggestion that the study of social problems *is,* essentially, the study of health problems, as each social problem is concerned with the physical, mental, and social well-being of humans and the social groups of which they are a part. As you read the remaining chapters in this text, consider how the problems in each chapter affect the health of individuals, families, populations, and nations.

Chapter Review

- **What three features of globalization have contributed to health problems?**
  Three features of globalization that have been linked to problems in health are increased transportation and information technology, the expansion of international trade, and the growth of transnational corporations.

- **What are three measures that serve as indicators of the health of populations? Which health measure reveals the greatest disparity between developed and developing countries?**
  Measures of health that serve as indicators of the health of populations include morbidity (often expressed as incidence and prevalence rates), life expectancy, and mortality rates (including infant and under-5 childhood mortality rates and maternal mortality rates). Maternal mortality rates reveal the greatest disparity between developed and developing countries.

- **Which theoretical perspective criticizes the pharmaceutical and health care industry for placing profits above people?**
  The conflict perspective criticizes the pharmaceutical and health care industry for placing profits above people. For example, pharmaceutical companies' research and development budgets are spent not according to public health needs, but rather according to calculations about maximizing profits. Because the masses of people in developing countries lack the resources to pay high prices for medication, pharmaceutical companies do not see the development of drugs for diseases of poor countries as a profitable investment.

- **Worldwide, what is the most common means by which HIV is transmitted?**
  Most cases of HIV worldwide are transmitted through heterosexual contact.

- **Why do the authors of your text refer to mental illness as a "hidden epidemic"?**
  Mental illness is a "hidden epidemic" because the shame and embarrassment associated with mental problems discourage people from acknowledging and talking about mental illness. Because male gender expectations associate masculinity with emotional strength, men are particularly prone to deny or ignore mental problems.
- **What, according to UN Secretary-General Kofi Annan, is the "biggest enemy of health in the developing world"?**
  In an address to the 2001 World Health Assembly, UN Secretary-General Kofi Annan remarked, "the biggest enemy of health in the developing world is poverty." Approximately one-fifth of the world's population live on less than U.S. $1 per day and nearly one-half live on less than U.S. $2 per day. Poverty is associated with malnutrition, unsafe water and sanitation, indoor air pollution, hazardous working conditions, and lack of access to medical care.
- **According to a World Health Organization report analysis of the world's health systems, which country provides the best overall health care?**
  The World Health Organization found that France provides the best overall health care among major countries followed by Italy, Spain, Oman, Austria, and Japan. The United States ranked 37 out of 191 countries, despite the fact that the United States spends a higher portion of its gross domestic product on health care than any other country.
- **What is the difference between "selective primary health care" and "comprehensive primary health care"?**
  Selective primary health care uses technocratic solutions to target specific health problems, such as using immunization to promote child survival or condoms to prevent the spread of HIV. In contrast, comprehensive primary health care focuses on the broader social determinants of health, such as poverty and economic inequality, gender inequality, environment, and community development.
- **How does the World Health Organization define "health"?**
  "Health," according to the World Health Organization is "a state of complete physical, mental, and social well-being." Based on this definition, the authors suggest that the study of social problems *is,* essentially, the study of health problems, as each social problem is concerned with the physical, mental, and social well-being of humans and the social groups of which they are a part.

Critical Thinking

1. For nearly a century, campaigns to bring about universal health care in the United States have failed. Do you envision universal health care in the United States in your lifetime? Why or why not?
2. Why do you think the American Psychiatric Association (2000) avoids the use of such expressions as "a schizophrenic" or "an alcoholic" and instead uses the expressions "an individual with schizophrenia" or "an individual with alcohol dependence"?
3. As noted in this chapter, suicides outnumber homicides two to one every year in the United States (Ezzell 2003). Why do you think media coverage of homicides is much more common than media coverage of suicides?

Key Terms

biomedicalization
comprehensive primary health care
deinstitutionalization
epidemiological transition
globalization
health
health maintenance organizations (HMOs)
life expectancy
managed care
maternal mortality rates
Medicaid
medicalization
Medicare
mental health
mental illness
morbidity
mortality
needle exchange programs
parity
preferred provider organizations (PPOs)
selective primary health care
socialized medicine
stigma
telemedicine
universal health care

Taking a Stand

Should prison inmates receive costly medical treatments at taxpayers' expense?

In 2002, a California inmate became the first person to receive an organ transplant while in state prison. A 31-year-old felon serving a 14-year sentence for robbery, was given a new heart, financed by taxpayers. The cost of the transplant operation was over $150,000, added to the security costs, aftercare, and post-transplant medication for a total cost of up to $1 million (Wiegand 2002). When the inmate received the heart, there were more than 4,000 people nationwide waiting for a heart to become available for a transplant. Should the heart have been given to someone else who was not in prison? In 1976, the U.S. Supreme Court held that "deliberate indifference" to a prison inmate's health problems constituted cruel and unusual punishment and thus violated the Eighth Amendment of the Constitution. What are the implications of a policy that gives inmates a Constitutional right to health care, whereas the rest of the population has no such guarantee?

Use Wadsworth's exclusive online resources—InfoTrac College Edition, MicroCase Online, and OVRC—to formulate a position on this topic.

The *Wadsworth's Sociology Online Resources and Writing Companion* will help you get started. This valuable guide will show you how to use Wadsworth's exclusive online resources when studying social problems. It will also help you to build essential research and writing skills. InfoTrac College Edition, MicroCase Online, OVRC, and an electronic copy of portions of this companion are available at http://sociology.wadsworth.com/mooney_knox_schacht/problems4e, the companion Web site for *Understanding Social Problems,* Fourth Edition.

Media Resources

The Companion Web Site for *Understanding Social Problems,* Fourth Edition

http://sociology.wadsworth.com/mooney_knox_schacht/problems4e

Supplement your review of this chapter by going to the companion Web site to take one of the Tutorial Quizzes, use the flash cards to master key terms, and check out the many other study aids you'll find there. You'll also find special features such as *Wadsworth's Sociology Online Resources and Writing Companion,* GSS Data, and Census 2000 information, data and resources at your fingertips to help you complete that special project or do some research on your own.

Interactions CD-ROM

Go to the Interactions CD-ROM for *Understanding Social Problems,* Fourth Edition, to access additional interactive learning tools, such as in-depth review materials, corresponding practice quizzes, and other engaging resources and activities to help you study the concepts in this chapter.

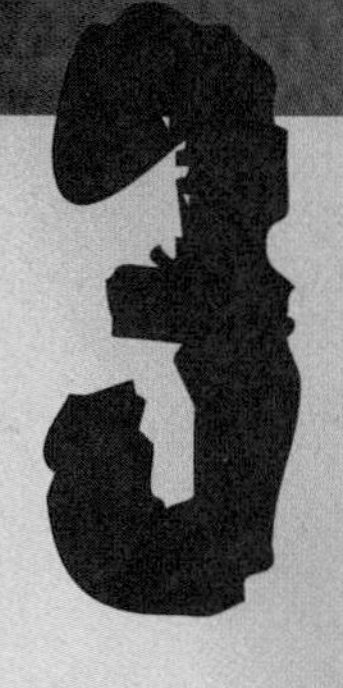

"Substance abuse, the nation's number one preventable health problem, places an enormous burden on American society, harming health, family life, the economy, and public safety, and threatening many other aspects of life." *Robert Wood Johnson Foundation, Institute for Health Policy, Brandeis University*

Alcohol and Other Drugs

Scott Krueger was athletic, intelligent, handsome, and what you'd call an all-around "nice guy." A freshman at Massachusetts Institute of Technology, he was a three-letter athlete and one of the top 10 students in his high school graduating class of over 300. He was a "giver" not a "taker," tutoring other students in math after school while studying second-year calculus so he could pursue his own career in engineering. While at MIT he rushed a fraternity and celebrated his official acceptance into the brotherhood. The night he celebrated he was found in his room, unconscious, and after three days in an alcoholic coma he died. He was 18 years old (Moore 1997). In September of 2000, MIT agreed to pay Scott's parents, Bob and Darlene Krueger, $4.75 million in a settlement over the death of their son and to establish a scholarship in his name (AP 2000, 11). In 2002, an out-of-court settlement was reached with Phi Gamma Delta and the fraternity officers paying the Kruegers $1.75 million. The defendants also agreed to produce an educational video on the dangers of drinking (Crittenden 2002).

Drug-induced death is just one of many negative consequences that can result from alcohol and drug abuse. The abuse of alcohol and other drugs is a social problem when it interferes with the well-being of individuals and/or the societies in which they live—when it jeopardizes health, safety, work and academic success, family, and friends. But managing the drug problem is a difficult undertaking. In dealing with drugs, a society must balance individual rights and civil liberties against the personal and social harm that drugs promote—crack babies, suicide, drunk driving, industrial accidents, mental illness, unemployment, and teenage addiction. When to regulate, what to regulate, and who should regulate are complex social issues. Our discussion begins by looking at how drugs are used and regulated in other societies.

The Global Context: Drug Use and Abuse

Pharmacologically, a **drug** is any substance other than food that alters the structure or functioning of a living organism when it enters the bloodstream. Using this definition, everything from vitamins to aspirin constitutes a drug. Sociologically, the term "drug" refers to any chemical substance that (1) has a direct effect on the user's physical, psychological, and/or intellectual functioning, (2) has the potential to be abused, and (3) has adverse consequences for the individual and/or society. Societies vary in how they define and respond to drug use. Thus, drug use is influenced by the social context of the particular society in which it occurs.

Drug Use and Abuse Around the World

According to estimates, the prevalence of drug use around the world varies dramatically. For example, the lifetime prevalence of illicit drug use varies from 46 percent of the adult population in the United States, to 36 percent in England, 26 percent in Italy, 18 percent in Poland, and 9 percent in Sweden (ODCCP 2003). In Europe, lifetime prevalence of illegal drug use *excluding* cannabis ranges from 12 percent in England to 2 percent in Finland. Moreover, 23 percent of European 15- to 16-year-olds have used marijuana at least once in their lifetime compared to 35 percent of 15- to 16-year-olds in Canada, 43 percent in Australia, and 41 percent in the United States.

In England, illegal drug use continues to spread, particularly among those under 25 (BBC 2003). Whereas in a recent survey 10 percent of the 11- to 15-year-old

Self and Society

U.S. Social Policy

Indicate whether you agree or disagree with the following policy-oriented statements and then compare your answers to those from a national sample of U.S. adults 18 and over.

	Agree	Disagree
1. Restrict drinking on city streets.	______	______
2. Ban youth-oriented packaging.	______	______
3. Ban keg sales to individuals.	______	______
4. Ban home delivery of alcohol.	______	______
5. Restrict drinking in parks.	______	______
6. Ban beer/wine ads on TV.	______	______
7. Require server training.	______	______
8. Ban alcohol marketing with athletes.	______	______
9. Restrict drinking at sports stadiums.	______	______
10. Ban happy hours.	______	______
11. Restrict drinking on college campuses.	______	______
12. Require legal age for alcohol servers.	______	______
13. Check everyone's ID.	______	______
14. Require beer keg registration.	______	______
15. Punish adult providers.	______	______

Results of a National Survey, 2001

	Percent Support
1. Restrict drinking on city streets.	93
2. Ban youth-oriented packaging.	70
3. Ban keg sales to individuals.	31
4. Ban home delivery of alcohol.	64
5. Restrict drinking in parks.	91
6. Ban beer/wine ads on TV.	59
7. Require server training.	90
8. Ban alcohol marketing with athletes.	62
9. Restrict drinking at sports stadiums.	74
10. Ban happy hours.	38
11. Restrict drinking on college campuses.	89
12. Require legal age for alcohol servers.	78
13. Check everyone's ID.	80
14. Require beer keg registration.	62
15. Punish adult providers.	87

Source: Alcohol Epidemiology Program. 2002. "Public Support for Alcohol Policies: 1997–2001." Youth Access to Alcohol Survey. Robert Wood Johnson Foundation. University of Minnesota.

population reported smoking at least one cigarette a week, 18 percent reported regularly taking drugs, and 24 percent reported having an alcohol drink in the previous week. Further, the average age of a caller to the British government drug information hotline fell from 29 to 20 in the last year. Statistics also indicate that drug users in the UK are more likely to be male, unemployed, and living in or around London when compared with nonusers.

Some of the differences in international drug use may be attributed to variations in drug policies. The Netherlands, for example, has had an official government policy of treating the use of such drugs as marijuana, hashish, and heroin as a health issue rather than a crime issue since the mid-1970s. In the first decade of the policy, drug use did not appear to increase. However, increases in marijuana use were reported in the early 1990s with the advent of "cannabis cafes." These coffee shops sell small amounts of marijuana for personal use and, presumably, prevent casual marijuana users from coming into contact with drug dealers (MacCoun &

U.S. citizens visiting the Netherlands may be shocked or surprised to find people smoking marijuana and hashish openly in public. Pictured here is a tourist using a water pipe to smoke marijuana in a coffee shop.

Reuter 2001; Drug Policy Alliance 2003a). Some evidence suggests that marijuana use among Dutch youth is decreasing (Sheldon 2000).

Great Britain has also adopted a "medical model," particularly in regard to heroin and cocaine. As early as the 1960s, English doctors prescribed opiates and cocaine for their patients who were unlikely to quit using drugs on their own and for the treatment of withdrawal symptoms. By the 1970s, however, British laws had become more restrictive making it difficult for either physicians or users to obtain drugs legally. Today, British government policy (this chapter's *Self and Society* deals with different policy approaches) provides for limited distribution of drugs by licensed drug treatment specialists to addicts who might otherwise resort to crime to support their habits. Further, in 2002 a report from the Advisory Council on the Misuse of Illicit Drugs recommended that marijuana be downgraded to a Class C drug, the lowest ranking possible, noting that it is less dangerous to the individual or to society than tobacco or alcohol (Drug Policy Alliance 2003b).

In stark contrast to such health-based policies, other countries execute drug users and/or dealers, or subject them to corporal punishment. The latter may include whipping, stoning, beating, and torture. Such policies are found primarily in less-developed nations such as Malaysia, where religious and cultural prohibitions condemn the use of any type of drug, including alcohol and tobacco.

Drug Use and Abuse in the United States

According to government officials there is a drug crisis in the United States—a crisis so serious that it warrants a multibillion-dollar-a-year "war on drugs." Americans' concern with drugs, however, has varied over the years. Ironically, in the 1970s when drug use was at its highest, concern over drugs was relatively low. Recently, when a sample of Americans were asked whether they considered drugs to be absolutely wrong, mostly wrong, sometimes wrong, or not wrong at all, over half of the respondents selected absolutely wrong, and 32 percent said sometimes or mostly wrong. Less than 10 percent responded not wrong at all (Harris Poll 2003).

As Table 3.1 indicates, use of illicit drugs in one's lifetime is fairly common. Of persons 12 years old and older, 41.7 percent report using an illicit drug sometime in their life. Marijuana and hashish have the highest occurrence in a lifetime (36.9%)

Table 3.1

Percentages Reporting Lifetime, Past Year, and Past Month Use of Illicit and Licit Drugs Among Persons Aged 12 or Older: 2001

	TIME PERIOD		
Drug	**Lifetime**	**Past Year**	**Past Month**
Any Illicit Drug[1]	41.7	12.6	7.1
Marijuana and Hashish	36.9	9.3	5.4
Cocaine	12.3	1.9	0.7
Crack	2.8	0.5	0.2
Heroin	1.4	0.2	0.1
Hallucinogens	12.5	2.0	0.6
LSD	9.0	0.7	0.1
PCP	2.7	0.1	0.0
Ecstasy	3.6	1.4	0.3
Inhalants	8.1	0.9	0.2
Nonmedical Use of Any Psychotherapeutic[2]	16.0	4.9	2.1
Pain Relievers	9.8	3.7	1.6
Tranquilizers	6.2	1.6	0.6
Stimulants	7.1	1.1	0.5
Methamphetamine	4.3	0.6	0.3
Sedatives	3.3	0.4	0.1
Cigarettes	67.2	29.1	24.9
Alcohol	81.7	63.7	48.3

[1] Any Illicit Drug includes marijuana/hashish, cocaine (including crack), heroin, hallucinogens, inhalants, or any prescription-type psychotherapeutic used nonmedically.

[2] Nonmedical use of any prescription-type pain reliever, tranquilizer, stimulant, or sedative; does not include over-the-counter drugs.

Source: HHS (U.S. Department of Health and Human Services). 2002. "2001 National Household Survey on Drug Abuse." Substance Abuse and Mental Health Service Administration. Washington, DC: U.S. Government Printing Office.

with heroin being the lowest (1.4%). Despite these relatively high numbers, particularly for marijuana and hashish, use of alcohol and tobacco are much more widespread than use of illicit drugs. Almost half of Americans 12 and over report being *current* alcohol drinkers and an estimated 66.5 million Americans over the age of 12 reported *current* use of a tobacco product (HHS 2002). In the United States, cultural definitions of drug use are contradictory—condemning it on the one hand (e.g., heroin), yet encouraging and tolerating it on the other (e.g., alcohol). At various times in U.S. history, many drugs that are illegal today were legal and readily available. In the 1800s and the early 1900s, opium was routinely used in medicines as a pain reliever, and morphine was taken as a treatment for dysentery and fatigue. Amphetamine-based inhalers were legally available until 1949, and cocaine was an active ingredient in Coca-Cola until 1906 when it was replaced with another drug—caffeine (Witters, Venturelli, & Hanson 1992).

Sociological Theories of Drug Use and Abuse

Most theories of drug use and abuse concentrate on what are called psychoactive drugs. These drugs alter the functioning of the brain, affecting the moods, emotions, and perceptions of the user. Such drugs include alcohol, cocaine, heroin, and marijuana. **Drug abuse** occurs when acceptable social standards of drug use are violated resulting in adverse physiological, psychological, and/or social consequences. For example, when an individual's drug use leads to hospitalization, arrest, or divorce, such use is usually considered abusive. Drug abuse, however, does not always entail drug addiction. **Drug addiction**, or **chemical dependency**, refers to a condition in which drug use is compulsive—users are unable to stop because of their dependency. The dependency may be psychological, in that the individual needs the drug to achieve a feeling of well-being, and/or physical, in that withdrawal symptoms occur when the individual stops taking the drug. For example, withdrawal from marijuana includes depression, anger, decreased appetite, and restlessness (Zickler 2003). Of those Americans who define themselves as drug dependent, 46 percent said their problem was alcohol, 22 percent said alcohol and drugs were equal problems, and 19 percent said their chief problem was drugs (Willing 2002).

Various theories provide explanations for why some people use and abuse drugs. Drug use is not simply a matter of individual choice. Theories of drug use explain how structural and cultural forces, as well as biological factors, influence drug use and society's responses to it.

Structural-Functionalist Perspective

Functionalists argue that drug abuse is a response to the weakening of norms in society. As society becomes more complex and rapid social change occurs, norms and values become unclear and ambiguous, resulting in **anomie**—a state of normlessness. Anomie may exist at the societal level, resulting in social strains and inconsistencies that lead to drug use. For example, research indicates that increased alcohol consumption in the 1830s and the 1960s was a response to rapid social change and the resulting stress (Rorabaugh 1979). Anomie produces inconsistencies in cultural norms regarding drug use. For example, while public health officials and health care professionals warn of the dangers of alcohol and tobacco use, advertisers glorify the use of alcohol and tobacco and the U.S. government subsidizes alcohol and tobacco industries. Further, cultural traditions, such as giving away cigars to celebrate the birth of a child and toasting a bride and groom with champagne, persist.

Anomie may also exist at the individual level as when a person suffers feelings of estrangement, isolation, and turmoil over appropriate and inappropriate behavior. An adolescent whose parents are experiencing a divorce, who is separated from friends and family as a consequence of moving, or who lacks parental supervision and discipline may be more vulnerable to drug use because of such conditions. Thus, from a structural-functionalist perspective, drug use is a response to the absence of a perceived bond between the individual and society, and to the weakening of a consensus regarding what is considered acceptable. Consistent with this perspective, a national poll of Americans 18 years and older found that peer pressure and lack of parental supervision were the two most common responses given as to why teenagers take drugs (Pew 2002).

Conflict Perspective

Conflict perspectives emphasize the importance of power differentials in influencing drug use behavior and societal values concerning drug use. From a conflict perspective, drug use occurs as a response to the inequality perpetuated by a capitalist system. Societal members, alienated from work, friends, and family, as well as from society and its institutions, turn to drugs as a means of escaping the oppression and frustration caused by the inequality they experience. Further, conflict theorists emphasize that the most powerful members of society influence the definitions of which drugs are illegal and the penalties associated with illegal drug production, sales, and use.

"There are but three ways for the populace to escape its wretched lot. The first two are by route of the wine-shop or the church; the third is by that of the social revolution."

Mikhail A. Bakunin
Anarchist and revolutionary

For example, alcohol is legal because it is often consumed by those who have the power and influence to define its acceptability—white males (HHS 2002). This group also disproportionately profits from the sale and distribution of alcohol and can afford powerful lobbying groups in Washington to guard the alcohol industry's interests. Since tobacco and caffeine are also commonly used by this group, societal definitions of these substances are also relatively accepting.

Conversely, crack cocaine and heroin are disproportionately used by minority group members, specifically, blacks and Hispanics (HHS 2002). Consequently, the stigma and criminal consequences associated with the use of these drugs are severe. The use of opium by Chinese immigrants in the 1800s provides a historic example. The Chinese, who had been brought to the United States to work on the railroads, regularly smoked opium as part of their cultural tradition. As unemployment among white workers increased, however, so did resentment of Chinese laborers. Attacking the use of opium became a convenient means of attacking the Chinese, and in 1877 Nevada became the first of many states to prohibit opium use. As Morgan (1978, 59) observes:

> The first opium laws in California were not the result of a moral crusade against the drug itself. Instead, it represented a coercive action directed against a vice that was merely an appendage of the real menace—the Chinese—and not the Chinese per se, but the laboring "Chinamen" who threatened the economic security of the white working class.

The criminalization of other drugs, including cocaine, heroin, and marijuana, follows similar patterns of social control of the powerless, political opponents, and/or minorities. In the 1940s, marijuana was used primarily by minority group members and carried with it severe criminal penalties. But after white middle-class college students began to use marijuana in the 1970s, the government reduced the penalties associated with its use. Though the nature and pharmacological properties of the drug had not changed, the population of users was now connected to power and influence. Thus, conflict theorists regard the regulation of certain drugs, as well as drug use itself, as a reflection of differences in the political, economic, and social power of various interest groups.

Symbolic Interactionist Perspective

Symbolic interactionism, emphasizing the importance of definitions and labeling, concentrates on the social meanings associated with drug use. If the initial drug use experience is defined as pleasurable, it is likely to recur, and over time, the individual may earn the label of "drug user." If this definition is internalized so that the individual assumes an identity of a drug user, the behavior will likely continue and may even escalate.

Drug use is also learned through symbolic interaction in small groups. In a recent study of binge drinking researchers found that students who believed their friends were binge drinking were more likely to binge drink themselves (Weitzman, Wechsler, & Nelson 2003). First-time users learn not only the motivations for drug use and its techniques, but also what to experience. Becker (1966) explains how marijuana users learn to ingest the drug. A novice being coached by a regular user reports the experience:

> I was smoking like I did an ordinary cigarette. He said, "No, don't do it like that." He said, "Suck it, you know, draw in and hold it in your lungs . . . for a period of time." I said, "Is there any limit of time to hold it?" He said, "No, just till you feel that you want to let it out, let it out." So I did that three or four times. (Becker 1966, 47)

Marijuana users not only learn how to ingest the smoke, but also learn to label the experience positively. When certain drugs, behaviors, and experiences are defined by peers as not only acceptable but pleasurable, drug use is likely to continue.

> Because they (first-time users) think they're going to keep going up, up, up till they lose their minds or begin doing weird things or something. You have to like reassure them, explain to them that they're not really flipping or anything, that they're gonna be all right. You have to just talk them out of being afraid. (Becker 1966, 55)

Interactionists also emphasize that symbols may be manipulated and used for political and economic agendas. The popular DARE (Drug Abuse Resistance Education) program, with its anti-drug emphasis fostered by local schools and police, carries a powerful symbolic value that politicians want the public to identify with. "Thus, ameliorative programs which are imbued with these potent symbolic qualities (like DARE's links to schools and police) are virtually assured wide-spread public acceptance (regardless of actual effectiveness) which in turn advances the interests of political leaders who benefit from being associated with highly visible, popular symbolic programs" (Wysong, Aniskiewicz, & Wright 1994, 461).

Biological and Psychological Theories

Drug use and addiction are likely the result of a complex interplay of social, psychological, and biological forces. Biological research has primarily concentrated on the role of genetics in predisposing an individual to drug use. Research indicates that severe, early onset alcoholism may be genetically predisposed, with some men having 10 times the risk for addiction as those without a genetic predisposition.

Interestingly, other problems such as depression, chronic anxiety, and attention-deficit disorder are also linked to the likelihood of addiction. Nonetheless, researchers warn, "Nobody is predestined to be an alcoholic" (Firshein 2003).

Biological theories of drug use hypothesize that some individuals are physiologically predisposed to experience more pleasure from drugs than others and, consequently, are more likely to be drug users. According to these theories, the central nervous system, which is composed primarily of the brain and spinal cord, processes drugs through neurotransmitters in a way that produces an unusually euphoric experience. Individuals not so physiologically inclined report less pleasant experiences and are less likely to continue use (Jarvik 1990; Alcohol Alert 2000).

Psychological explanations focus on the tendency of certain personality types to be more susceptible to drug use. Individuals who are particularly prone to anxiety may be more likely to use drugs as a way to relax, gain self-confidence, or ease tension. For example, research indicates that female adolescents who have been sexually abused or who have poor relationships with their parents are more likely to have severe drug problems (NIDA 2000). Psychological theories of drug abuse also emphasize that drug use may be maintained by positive and negative reinforcement.

Frequently Used Legal and Illegal Drugs

Over 16 million people in the United States are illicit drug users, which represents 7.1 percent of the 12-and-older population. Users of illegal drugs, although varying by type of drug used, are more likely to live in metropolitan areas, to be male, young, and minority group members (HHS 2002). Social definitions regarding which drugs are legal or illegal, however, have varied over time, circumstance, and societal forces. In the United States, two of the most dangerous and widely abused drugs, alcohol and tobacco, are legal.

Alcohol

Americans' attitudes toward alcohol have had a long and varied history. Although alcohol was a common beverage in early America, by 1920 the federal government had prohibited its manufacture, sale, and distribution through the passage of the Eighteenth Amendment to the Constitution. Many have argued that Prohibition, like the opium regulations of the late 1800s, was in fact a "moral crusade" (Gusfield 1963) against immigrant groups who were more likely to use alcohol. The amendment had little popular support and was repealed in 1933. Today, the United States is experiencing a resurgence of concern about alcohol. What has been called a "new temperance" has manifested itself in federally mandated 21-year-old drinking-age laws, warning labels on alcohol bottles, increased concern over fetal alcohol syndrome and underage drinking, and stricter enforcement of drinking and driving regulations. Such practices may have had an impact on drinking norms. For example, in 2002, the number of college freshmen who reported that they drank beer "frequently" or "occasionally" was at a record low, part of a downward trend of two decades (Sax et al. 2002).

Nonetheless, alcohol remains the most widely used and abused drug in America. Although most people who drink alcohol do so moderately and experience few negative effects, alcoholics are psychologically and physically addicted to alcohol and suffer various degrees of physical, economic, psychological, and personal harm.

The National Household Survey on Drug Abuse (now called the National Survey on Drug Use and Health) conducted by the Department of Health and Human Services reports that 109 million Americans 12 and older consumed alcohol at least once in the month preceding the survey, that is, were current users (HHS 2002). Of this number, 5.7 percent reported being heavy drinkers (defined as drinking 5 or more drinks per occasion on 5 or more days in the survey month) and 20.5 percent were binge drinkers (defined as drinking 5 or more drinks on at least one occasion during the survey month).

Even more troubling were the 10 million current users of alcohol who were 12 to 20 years old, a quarter of whom reported being heavy or binge drinkers (HHS 2002). Binge drinking in college has attracted media attention and thus the public's attention. The alcohol consumed by binge drinkers represents 70 percent of all alcohol consumed by college students (Schemo 2002). Further, the money spent annually by college students on alcohol, $5.5 billion, is more than they spend on milk, soft drinks, coffee, tea, and books combined (MADD 2003a). Many binge drinkers began in high school, with almost one-third having their first drink before age 13. Research indicates that the younger the age of onset, the higher the probability that an individual will develop a drinking disorder at some time in his or her life. For example, an individual's chance of becoming dependent on alcohol, as defined by the National Household Survey, is 40 percent if the person's drinking began before the age of 13. The chances of being alcohol dependent also increase if an individual's parents (1) are alcoholics, (2) drink, (3) have a positive attitude about drinking, or (4) use discipline sporadically (*Affects Child* 2003). Heavy teenage drinkers, as with their adult counterparts, are more likely to be white, non-Hispanic, and male (HHS 2002).

Additional results from the National Household Survey on Drug Abuse (HHS 2002) include the following:

- The highest levels of both heavy and binge drinking are among 18- and 25-year-olds peaking at age 21.
- Rates of alcohol use are higher among the employed than the unemployed; however, patterns of heavy or binge drinking are highest among the unemployed.
- College graduates are less likely to be binge drinkers than high school graduates but more likely to report alcohol use in the past month.
- Drinking is higher among 12- to 17-year-olds who live in rural counties than for those who live in large metropolitan counties; however, those 18 and above who live in large metropolitan areas report higher rates than their counterparts in rural areas.
- Binge drinking was least likely to be reported by Asians and most likely to be reported by Native Americans/Alaska Natives.

Many alcohol users report using other controlled substances, but the more frequently a student binges, the higher the probability of reporting other drug use. Some evidence suggests that certain combinations of drugs—for example, alcohol and cocaine—may heighten the negative effects of either drug separately, that is, there is a negative drug interaction.

Tobacco

Although nicotine is an addictive psychoactive drug and tobacco smoke has been classified by the Environmental Protection Agency as a Group A carcinogen, tobacco continues to be among the most widely used drugs in the United States. According to the U.S. Department of Health and Human Services survey, 66.5 million

Americans continue to smoke cigarettes—30 percent of the 12-and-older population. Thirteen percent of the 12- to 17-year-old population reporting smoking, a decline from 14.9 percent in 1999, and 13.4 in 2000 (HHS 2002). Interestingly, among 12- to 17-year-olds, three brands account for over 50 percent of the tobacco market—Marlboro, Newport, and Camel. Use of all tobacco products including smokeless tobacco (7.3 million users), cigars (12.1 million users), pipe tobacco (2.3 million users), and cigarettes is higher for high school compared with college graduates, males, and Native Americans/Alaska Natives (NHSDA Report 2003).

Although teenage cigarette use decreased between 2000 and 2001, nearly 25 percent of high school seniors smoke everyday (Update 2003). Cigarette advertising that targets youth such as the now defunct "Joe Camel" campaign are often blamed. There is also evidence that cigarette advertisers target minorities.

> Recent studies have shown a higher concentration of tobacco advertising in magazines aimed at African Americans, such as *Jet* and *Ebony,* than in similar magazines aimed at a broader audiences, such as *Time* and *People*. . . . From 1992 to 2000 smoking rates increased among African American 12th graders from 8.7 percent to 14.2 percent. (AHA 2003a)

Similarly, the tobacco industry is accused of developing advertising campaigns that target Hispanics and women. Advertising is but one venue criticized for the positive portrayal of tobacco use. In this chapter's *Social Problems Research Up Close* feature, images of tobacco and alcohol use in children's animated films are examined.

© John Kobal Foundation/Hulton/Archive

Up until the 1960s, the harmful effects of tobacco were not part of the American collective consciousness. Tobacco was glamorized in the movies and on television, and it was not unusual to use "stars" to advertise tobacco products. Countless numbers of television and movie personalities have died from tobacco-related illnesses including actor Humphrey Bogart.

Tobacco was first cultivated by Native Americans, who introduced it to the European settlers in the 1500s. The Europeans believed it had medicinal properties, and its use spread throughout Europe, assuring the economic success of the colonies in the New World. Tobacco was initially used primarily through chewing and snuffing, but in time smoking became more popular even though scientific evidence that linked tobacco smoking to lung cancer existed as early as 1859 (Feagin & Feagin 1994). However, it was not until 1989 that the U.S. Surgeon General concluded that tobacco products are addictive and that it is nicotine that causes the dependency. Today, the hazards of tobacco use are well documented and have resulted in federal laws that require warning labels on cigarette packages and prohibit cigarette advertising on radio and television. By the year 2030, tobacco-related diseases will be the number one cause of death worldwide, killing one of every six people. Eighty percent of the deaths will take place in poor nations where many smokers are unaware of the health hazards associated with their behavior (Mayell 1999).

"Everybody knew it's addictive. Everybody knew it causes cancer. We were all in it for the money."

Victor Crawford
Former tobacco lobbyist and smoker who developed lung cancer

Marijuana

Marijuana is the most commonly used and most heavily trafficked illicit drug in the world. When just the top of the plant is sold, it is called hashish and is much more potent than marijuana, which comes from the entire plant (Thio 2004). Globally, there are 147 million marijuana users representing 3.1 percent of the world's population and 4.3 percent of the global population 15 years old and older. Marijuana use, in general, has increased in Europe, Africa, Oceania, and the Americas. It has decreased in South and Southwest Asia (ODCCP 2003). It is higher among males in

Social Problems Research Up Close

Images of Alcohol and Tobacco Use in Children's Animated Films

The impact of media on drug and alcohol use is likely to be recursive—media images affect drug use while, alternatively, societal drug use helps define media presentations. Previous research has documented the rate of tobacco and alcohol use in print media, advertising, and Hollywood movies. In the present research, Goldstein, Sobel, and Newman (1999) use content analysis to investigate the prevalence of tobacco and alcohol use in children's animated films as one step in assessing the growing concern with media influence on children's smoking and drinking behavior.

Sample and Methods

The researchers examined all G-rated animated films released between 1937 (*Snow White and the Seven Dwarfs*) and 1997 (*Hercules, Anastasia, Pippi Longstocking,* and *Cats Don't Dance*). Criteria for sample inclusion included that the film be at least 60 minutes in length and before video distribution was released to theaters. The resulting sample included all of Disney's animated children's films produced during the target years with the exception of three that were unavailable on videocassette. The remaining films included all children's animated films produced by MGM/United Artists, Universal, 20th Century Fox, and Warner Brothers since 1982. Variables coded included the (1) presence of alcohol or tobacco use, (2) length of time of use on screen, (3) number of characters using alcohol or tobacco, (4) value of the character using tobacco or alcohol (i.e., good, neutral, or bad), (5) any implied messages about the drug use, and (6) the type of tobacco or alcohol being used.

Findings and Conclusions

Of the 50 films analyzed, at least one episode of alcohol and/or tobacco use was portrayed in 34 (68%) with tobacco use (N = 28) slightly exceeding portrayals of alcohol use (N = 25). Tobacco was used by 76 different characters with an onscreen time of 45 minutes—an average of 1.62 minutes per movie. Characters were most likely to use cigars followed by cigarettes, and pipes. Of the 76 characters using tobacco, 28 (37%) were classified as good. Surprisingly, the use of tobacco products by "good" characters has increased rather than decreased over time.

Sixty-two characters, averaging 2.5 per film, were shown using alcohol with a total duration of 27 minutes across all films. Characters were most likely to consume wine followed by beer, spirits, and champagne. The number of good characters using alcohol was similar to the number of characters classified as bad. In 19 of the 25 films in which alcohol use was portrayed, tobacco use was also pictured. Although several films portrayed the physical consequences of smoking (N = 10) (e.g., coughing) or drinking (N = 7) (e.g., passing out), no film verbally referred to the health hazards of either drug.

One particularly interesting finding of the research concerned the use of alcohol and tobacco as a visual prop in character development. For example, although cigar smokers were portrayed as tough and powerful (e.g., Sykes in *Oliver and Company*), pipe smokers were most often older, kindly, and wise (e.g., Geppetto in *Pinocchio*), and cigarette smokers independent, witty, and intelligent (e.g., the Genie in *Aladdin*). There was also a tendency for alcohol and tobacco use to be portrayed together. When one, the other, or both are associated with positively defined characters the impact may be detrimental to the lifestyle choices of viewers.

Although this study cannot assess the "impact question," advertising campaigns such as Joe Camel and the Budweiser frogs have been linked to detrimental results. Although in each of these cases the motivation for the use of such appealing characters is clear, the presentation of "good" characters using alcohol and tobacco products in children's animated films remains unexplained. Interpretation of the results is further complicated by the lack of change over time, that is, as our knowledge of the harmful effects of these products increased, their presence in children's films did not, as expected, decrease. In light of these results, the researchers call for an end to the portrayal of alcohol and tobacco use in all children's animated films and associated products (e.g., posters, books, games).

Source: Adam Goldstein, Rachel Sobel, and Glen Newman. 1999. "Tobacco and Alcohol Use in G-rated Children's Animated Films." *Journal of the American Medical Association* 281 : 1121–1136.

most countries, as high as 90 percent in traditional Asian cultures. The gender difference in the United States is much narrower, with 56 percent of marijuana users being male (ODCCP 2003).

Marijuana's active ingredient is THC (delta-9-tetrahydrocannabinol) which, in varying amounts, may act as a sedative or as a hallucinogen. Marijuana use dates back to 2737 B.C. in China and has a long tradition of use in India, the Middle East, and Europe. In North America, hemp, as it was then called, was used for making rope and as a treatment for various ailments. Nevertheless, in 1937 Congress passed the Marijuana Tax Act, which restricted its use; the law was passed as a result of a media campaign that portrayed marijuana users as "dope fiends" and, as conflict theorists note, was enacted at a time of growing sentiment against Mexican immigrants (Witters et al. 1992, 357–59).

There are more than 12 million current marijuana users representing 5.4 percent of the U.S. population. According to the 2002 *Monitoring the Future Study,* among twelfth graders, 47.8 percent have used marijuana/hashish at least once in their lifetime, 36.2 percent in the last year, 21.4 percent in the last month, and 6.0 percent daily. This is despite the fact that 53 percent of the seniors responded that smoking marijuana regularly is harmful. The study further shows that 10.4 percent of eighth graders and 20.8 percent of tenth graders reported use of an illicit drug in the past month with marijuana being the most commonly reported (MTF 2002; ONDCP 2003a).

Not surprisingly, perceived risk of marijuana use is associated with prevalence rates. As perceived risk of marijuana use goes up, the likelihood of use goes down. In recent years the perception that smoking marijuana is a risky behavior has decreased. For example, among 12- to 17-year-olds, the percentage of those who believe that smoking marijuana is a "great risk" decreased from 56.0 percent in 2000 to 53.5 percent in 2001 (HHS 2002).

Although the effects of alcohol and tobacco are, in large part, well known, the long-term psychological and physiological effects of marijuana are less understood. According to the Office of National Drug Control Policy (ONDCP 2003b), marijuana is associated with many negative health effects including impaired memory, anxiety, panic attacks, and increased heart rate. Another important concern is that marijuana may be a **gateway drug** that causes progression to other drugs such as

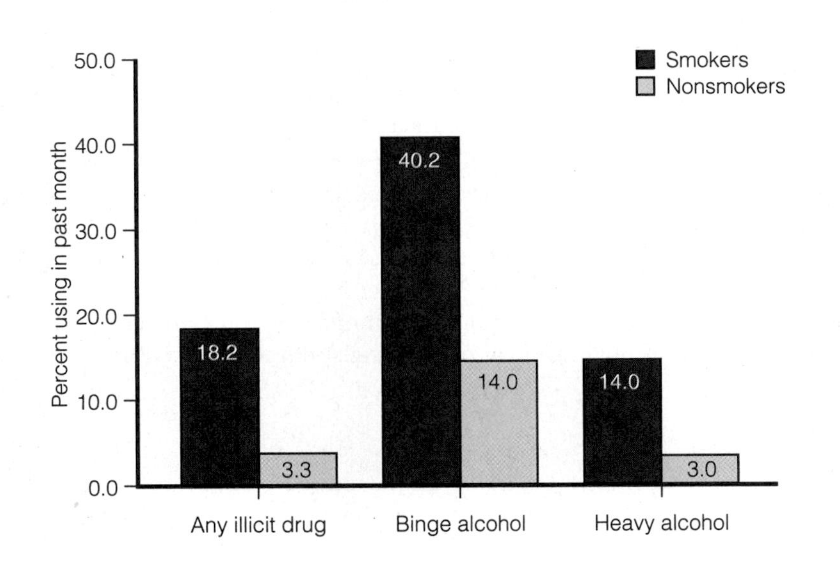

Figure 3.1
Past Month Any Illicit Drug, Binge Alcohol, and Heavy Alcohol Use Among Smokers and Nonsmokers Aged 12 or Older: 2001

Source: 2001 National Household Survey on Drug Abuse, 2002. "Chapter 4: Tobacco Use." U.S. Department of Health and Human Services. Washington, D.C.

cocaine and heroin. More likely, however, is that persons who experiment with one drug are more likely to experiment with another. Indeed, most drug users are polydrug users with the most common combination being alcohol, tobacco, and marijuana.

Cocaine

Cocaine is classified as a stimulant and, as such, produces feelings of excitation, alertness, and euphoria. Although such prescription stimulants as methamphetamine and dextroamphetamine are commonly abused, over the last 10 to 20 years societal concern over drug abuse has focused on cocaine. Such concerns have been fueled by its increased use, addictive qualities, physiological effects, and worldwide distribution. More than any other single substance, cocaine has led to the present "war on drugs."

"Crack is a drug peddler's dream: it is cheap, easily concealed and provides a short-duration high that invariably leaves the user craving more."

Tom Morganthau
Journalist

Cocaine, which is made from the coca plant, has been used for thousands of years, but anti-cocaine sentiment in the United States did not emerge until the early 1900s, when it was primarily a response to cocaine's heavy use among urban blacks, poor whites, and criminals (Witters et al. 1992, 260; Thio 2004). Cocaine was outlawed in 1914 by the Harrison Narcotics Act, but its use and effects continued to be misunderstood. For example, a 1982 *Scientific American* article suggested that cocaine was no more habit forming than potato chips (Van Dyck & Byck 1982). As demand and then supply increased, prices fell from $100 a dose to $10 a dose, and "from 1978 to 1987 the United States experienced the largest cocaine epidemic in history" (Witters et al. 1992, 256, 261).

Cocaine use in recent years has decreased in the United States although increasing in South America, Africa, and Europe (ODCCP 2003). According to the National Household Survey, 27.8 million people—12.3 percent of the U.S. population 12 and over—have tried cocaine once in their lifetime (ONDCP 2003c). Forty-two percent of high school seniors surveyed in the 2002 *Monitoring the Future Study* indicated that getting powdered cocaine is "fairly easy"or "very easy." Almost half of the high school seniors surveyed reported that use of powdered cocaine or crack cocaine once or twice is a "great risk" (MTF 2002).

Crack is a crystallized product made by boiling a mixture of baking soda, water, and cocaine. The result, also called rock, base, and gravel, is relatively inexpensive and was not popular until the mid-1980s. Crack is one of the most dangerous drugs to surface in recent years. Crack dealers often give drug users their first few "hits" free, knowing the drug's intense high and addictive qualities are likely to produce returning customers. An addiction to crack can take six to ten weeks; to pure cocaine, three to four years (Thio 2004).

According to the National Household Survey, 2.8 percent of Americans over the age of 12 (6.2 million people) have tried crack cocaine once in their lifetime. In 2002 there were 406,000 current users. This is a slight reduction from the 413,000 crack users in 1999 (ONDCP 2003d). Crack use may be decreasing as young people begin to associate crack use with "burnouts" and "junkies." This chapter's *The Human Side* feature graphically describes conditions in a crack house and associated criminal behaviors through the eyes of sociologist-ethnographer Terry Williams.

Other Drugs

Other drugs abused in the United States include "club drugs" (e.g., LSD, ecstasy), heroin, prescription drugs (e.g., tranquilizers, amphetamines), and inhalants (e.g., glue).

The Human Side

An Excerpt from *The Cocaine Kids*

The door opens a crack before I can knock, a tall African-American man brusquely thrusts his palm toward me and asks, "You got three dollars?" He motions excitedly, "If you ain't got three dollars you can't come in here." The entrance fee. I pay and walk in.

The establishment is desolate, uninviting, dank, and smoky. The carpet in the first room is shit-brown and heavily stained, pockmarked by so many smoke burns that it looks like an abstract design. In the dim light, all the people on the scene seem to be in repose, almost inanimate, for a moment.

As my eyes adjust to the smoke, several bodies emerge. I see jaws moving, hear voices barking hoarsely into walkie-talkies—something about money; their talk is jagged, nasal, and female. One woman takes out an aluminum foil packet, snorts some of its contents, passes it to her partner then disappears into another room. In a corner near the window, a shadowy figure moans. One woman sits with her skirt over her head, while a bobbing head writhes underneath her. In an adjacent alcove, I see another couple copulating. Somewhere in the corridor a man and woman argue loudly in Spanish. Staccato rap music sneaks over the grunts and hollers.

The smell is a nauseating mix of semen, crack, sweat, other human body odors, funk and filth. Two men dicker about who took the last "hit" (puff); two others are on their hands and knees looking for crack particles they claim they have lost in the carpet.

In the crack houses, the sharing rituals associated with snorting are being supplanted by more individualistic, detached arrangements where people come together for erotic stimulation, sexual activity, and cocaine smoking. They may be total strangers, seeking only brief and superficial physical contact, encounters designed to heighten sensations; the smoking act is a narcissistic fix—there is little thought for the other person. The emotional content is largely due to the momentary excitation of the setting and the cocaine. Much of the sexual behavior is performed to acquire more cocaine.

Nothing better exemplifies the new attitude than the act of Sancocho (a word meaning to cut up in little pieces and stew). To sancocho is to steal crack, drugs, or money from a friend or other person who is not alert, a regular practice in the crack houses. Another example is the "hit kiss" ritual: after inhaling deeply, basers literally "kiss"—put their lips together and exhale the smoke into each other's mouths. This not only saves all the valuable smoke, but also stimulates the other sexually. Other versions of the kiss extend to other orifices.

Source: Terry Williams. 1990. From *The Cocaine Kids: The Inside Story of a Teenage Drug Ring,* by Terry Williams.

Club Drugs. Club drugs is a general term used for illicit, often synthetic drugs commonly used at nightclubs or all-night dances called raves. Club drugs include ecstasy (MDMA), ketamine ("Special K"), LSD ("acid"), GHB ("liquid ecstasy"), and Rohypnol ("roofies"). Ecstasy, manufactured in and trafficked from Europe, is the most popular of the club drugs ranging in price from $20 to $30 a dose (Leshner 2003). Use of ecstasy is growing, with 3.6 percent of Americans, 8.1 million people, having tried ecstasy once in their lifetime (HHS 2002). Ecstasy is associated with feelings of euphoria and inner peace, yet critics argue that as the "new cocaine," it can produce both long- (e.g., permanent brain damage) and short-term (e.g., hyperthermia) negative effects (Cloud 2000; DEA 2000; ONDCP 2003e; Thio 2004).

Ketamine and LSD (lysergic acid diethylamide) both produce visual effects when ingested. Use of ketamine, an animal tranquilizer, can also cause loss of long-term memory, respiratory problems, and cognitive difficulties. In 2002, 2.6 percent of high school seniors reported past year ketamine use (MTF 2002). LSD is a synthetic hallucinogen, although many other hallucinogens are produced naturally (e.g., peyote). There are currently over 320,000 LSD users in the United States equaling .01 percent of the 12-and-over population. However, of those 12 to 17 years old, lifetime use rises to 3.1 percent and for 18- to 25-year-olds, 15.1 percent

(HHS 2002; ONDCP 2003f). Use of LSD is three to four times higher in North American and Australia than in Europe (ODCCP 2003).

GHB (gamma hydroxybutyrate) and Rohypnol (flunitrazepam) are often called **date-rape drugs** because of their use in rendering victims incapable of resisting sexual assaults. GHB, a central nervous system depressant, was banned by the Food and Drug Administration in 1990 although kits containing all the necessary ingredients to manufacture the drug continued to be available on the Internet (ONDCP 2003e). On February 18, 2000, President Clinton signed a bill that made GHB a controlled substance and, thus, illegal to manufacture, possess, or sell. Nonetheless, 1.5 percent of twelfth graders report past-year use of GHB (MTF 2002).

Rohypnol, presently illegal in the United States, is lawfully sold by prescription in over 50 countries for the short-term treatment of insomnia (ONDCP 2003g). It belongs to a class of drugs known as benzodiazepines, which also includes such common prescription drugs as Valium, Halcion, and Xanax. Rohypnol is tasteless and odorless. The effects of rohypnol begin within 15 to 20 minutes; 1 mg of the drug can incapacitate a victim for up to 12 hours (NIDA 2000; DEA 2000; ONDCP 2003g). "Roofies" are popular with high school and college students, costing only $5 a tablet.

"The only livin' thing that counts is the fix . . . Like I would steal off anybody— anybody, at all, my own mother gladly included."

Heroin Addict

Heroin. Heroin is an analgesic, that is, a painkiller. Most heroin comes from the poppy fields of Myanmar, Thailand, Afghanistan, Pakistan, Iran, Mexico, and Colombia (Thio 2004). Overall use of heroin is higher in Europe than in the United States (ODCCP 2003). The number of Americans over the age of 12 who report lifetime heroin use is 3.1 million. The number of Americans who are heroin dependent is, of course, much lower, ranging from an estimated 850,000 to 1 million addicts. The highest use of heroin is among 18- to 25-years-olds with 1.6 percent lifetime heroin use (ONDCP 2003h). Heroin is a highly addictive drug that is increasingly popular among school-aged youth. It can be injected, snorted, or smoked. If intravenous injection is used the effects are felt within 7 to 8 seconds; if snorted or smoked, the effects are felt within 10 to 15 minutes. While crack cocaine has become less fashionable among youthful offenders, heroin, an opium-based narcotic, has increased in acceptability to the point of being glamorized in recent motion pictures and song lyrics (Heroin Drug Conference 1997; NIDA 2000). In 2001, the rate of heroin use for 12- to 17-year-olds decreased slightly (ONDCP 2003h). In addition to experiencing the negative repercussions heroin shares with all other drugs, heroin users are subjected to the risks of HIV/AIDS if using intravenously (HHS 2002; NIDA 2000).

In recent years methadone, a synthetic drug often used in the treatment of heroin, has been responsible for an increasing number of deaths. In Florida, for example, there were 209 methadone-related deaths in 2000, 357 deaths in 2001, and 254 deaths in the first six months of 2002—the latest available data. Such numbers prompted a Florida official to announce that methadone "is the fastest rising killer drug" (Belluck 2003, p. 1). The increase in methadone abuse is somewhat puzzling given its sedating effect and delayed reaction, sometimes up to an hour or two. The increase in methadone use may be linked to the growing abuse of heroin and OxyContin, a prescription pain killer (Belluck 2003).

Psychotherapeutic Drugs. Use of psychotherapeutic drugs, that is, nonmedical use of any prescription pain reliever, stimulant, sedative, or tranquilizer tends to be high at either end of the age continuum. In 2001, nearly 3 million 12- to 17-year-olds and 7 million 18- to 25-year-olds reported using a prescription drug for nonmedical reasons at least once in their lifetime. Highest abuse was among 12- to 25-year olds. However, those over the age of 60, because they take so many prescription drugs, also have high rates of misuse. An estimated 17 percent of those over 60 years old misuse prescription drugs, many because of a misunderstanding or uncertainty about the proper dose (Snider 2003).

Methamphetamine, a stimulant, is one example of a popular psychotherapeutic drug. Although occasionally prescribed for legitimate medical reasons, "meth" is often made in clandestine laboratories in Mexico and the United States. Recent increases in the use of methamphetamine have alarmed international authorities with use in some areas of the world rivaling that of cocaine. It is notable that such increases are not just in industrialized nations but in many developing countries as well. For example, in Thailand methamphetamine has replaced heroin as the most commonly used drug. U.S. use is highest among 18- 25-year-olds with 1.2 percent lifetime use. Methamphetamine is linked to violent behavior and is often used in combination with other drugs (Final Report 2000; International Narcotics Control Strategy Report 2000; HHS 2002).

Recently, there has been an increase in the use of psychotherapeutic drugs, particularly pain killers (HHS 2002). One such drug is OxyContin. The pharmacological characteristics of OxyContin make it a suitable heroin substitute. As a prescription pain killer, OxyContin is often covered by health insurance plans, which contributes to its appeal. When insurance will no longer pay, it is not uncommon for people to switch to heroin, which is less expensive on the street than OxyContin. In 1999, the number of nonmedical lifetime OxyContin users was 221,000; in 2001, that number was 957,000 (HHS 2002).

Inhalants. Common inhalants include adhesives (e.g., rubber cement), food products (e.g., vegetable cooking spray), aerosols (e.g., hair spray, air fresheners), anesthetics (ether), and cleaning agents (e.g., spot remover). In total, over 1,000 household products are currently abused (ONDCP 2003k). According to the Youth Risk Behavior Surveillance Survey, 14.7 percent of high school students nationwide "have sniffed glue, breathed the contents of aerosol spray cans, or inhaled paints or spray to get high at least once in their lifetime" (YRBSS 2002). Young people often use inhalants believing they are harmless or that prolonged use is necessary for any harm to result. In fact, inhalants are very dangerous because of their toxicity and may result in what is called Sudden Sniff Death Syndrome.

Societal Consequences of Drug Use and Abuse

Drugs are a social problem, not only because of their adverse effects on individuals, but also as a result of the negative consequences their use has for society as a whole. Everyone is a victim of drug abuse. Drugs contribute to problems within the family and to crime rates, and the economic costs of drug abuse are enormous. Drug abuse also has serious consequences for health at both the individual and societal level.

Family Costs

The cost to families of drug use is incalculable. When one or both parents use or abuse drugs, needed family funds may be diverted to purchasing drugs rather than necessities. Children raised in such homes have a higher probability of neglect,

This poster from the Office of National Drug Control Policy National Youth Anti-Drug Media Campaign emphasizes the importance of a close relationship between parent and child in the fight against youthful drug use.

behavioral disorders, and absenteeism from school as well as lower self-concepts and increased risk of drug abuse (Easley & Epstein 1991; AP 1999; ONDCP 2000j; *Affects Child* 2003). It is estimated that 6 million children—9 percent of those under the age of 18—live in homes where drugs are being abused (SAMHSA 2003).

Drug abuse is also associated with family disintegration. For example, alcoholics are seven times more likely to separate or divorce than nonalcoholics, and as much as 40 percent of family court problems are alcohol related (Sullivan 2003, 336). Close to 50 percent of American adults have a close family member who is or was drug dependent (*Affects Child* 2003).

Abuse between intimates is also linked to drug use. Alcohol misuse, for example, is the single most common trait associated with wife abuse (Charon 2002). The more violent the interaction, the more likely it is that the husband has been drinking excessively. Further, serious violence in the first year of marriage is twice as high among heavy drinkers than among social drinkers (Johnson 2003). In a study of 320 men who were married or living with someone, twice as many reported hitting their partner only after they had been drinking compared with those who reported the same behavior while sober (Leonard & Blane 1992). Additionally, Straus and Sweet (1992) found that alcohol consumption and drug use were associated with higher levels of verbal abuse among spouses.

Crime Costs

"... every $1 invested in addiction treatment programs yields a return of between $4 and $7 in reduced drug-related crime ..."

Principles of Drug Addiction Treatment
National Institute of Drug Abuse

The drug behavior of persons arrested, those incarcerated, and those in drug treatment programs provides evidence of a link between drugs and crime. Drug users commit a disproportionate number of crimes. For example, it is estimated that at the time of arrest 63.3 percent of males and 63.9 percent of females test positive for cocaine, opiates (e.g., heroin), marijuana, methamphetamine, and/or PCP (ONDCP 2003l). Further, the majority of crimes that take place on college campuses—over 90 percent—are alcohol related (Thio 2004).

The relationship between crime and drug use, however, is a complex one. Sociologists disagree as to whether drugs actually "cause" crime or whether, instead, criminal activity leads to drug involvement. Further, because both crime and drug use are associated with low socioeconomic status, poverty may actually be the more powerful explanatory variable. After extensive study of the assumed drug-crime link, Gentry (1995, 491) concludes that "the assumption that drugs and crime are causally related weakens when more representative or affluent subjects are considered."

In addition to the hypothesized crime–drug use link, some criminal offenses are drug defined: possession, cultivation, production, and sale of controlled substances; public intoxication; drunk and disorderly conduct; and driving while intoxicated. Driving while intoxicated is one of the most common drug-related crimes.

> Nationwide, drunk drivers account for 14 percent of all probationers, 7 percent of local jail inmates, and 2 percent of state prisoners. Of drunk driving offenders, most (89 percent) were on probation; only 11 percent were sentenced to jail time. The average length of incarceration for those who did serve time was 11 months, and nearly half were sentenced to at least six months in jail. About two-thirds of those incarcerated for drunk driving were repeat offenders (State Legislatures 2000).

In 2002, 41 percent of all traffic crashes were alcohol related and 17,419 Americans were killed in drunk driving accidents—an average of one every 30 seconds (NHTSA 2003; MADD 2003b). A recent federal law requires states to adopt a .08 blood alcohol content limit by 2004 or lose highway funds (also, see Collective Action).

Economic Costs

The economic costs of drug use are high, over $143 billion, the majority of which comes from loss of "productivity due to drug related illnesses and deaths, as well as from incarcerations and work hours missed by crime victims" (ONDCP 2003b, 1). Billions of dollars in insurance are also lost. For example, alcohol-related car accidents account for 18 percent of the over $100 billion in U.S. auto insurance payments (MADD 2003c). There too is the revenue lost in the underground economy as Americans spend an estimated $36 billion on cocaine, $11 billion on marijuana, $10 billion on heroin, and $5.4 billion on methamphetamine (ONDCP 2003l). Concern that on-the-job drug use may impair performance and/or cause fatal accidents has led to drug testing in many industries. For many employees, such tests are routine both as a condition for employment and as a requirement for keeping their jobs. This chapter's *Focus on Technology* feature reviews some of the issues related to technologies, privacy rights, and drug testing.

Other economic costs of drug abuse include the cost of homelessness, the cost of implementing and maintaining educational and rehabilitation programs, and the cost of health care. Also, the cost of fighting the "war on drugs" is likely to increase as organized crime develops new patterns of involvement in the illicit drug trade.

Health Costs

Some consumption of alcohol has been shown to be beneficial in that "moderate drinkers are generally healthier, live longer, and have lower death rates than abstainers" (Thio 2004, 324). However, the physical health consequences of abusing alcohol, tobacco, and other drugs are tremendous: shortened life expectancy; higher morbidity (e.g., cirrhosis of the liver, lung cancer); exposure to HIV infection, hepatitis, and other diseases through shared needles; a weakened immune system; birth defects such as fetal alcohol syndrome; drug addiction in children; and higher death rates.

Of the 2.4 million U.S. deaths annually, over 440,000 are attributable to cigarette smoking alone. The Surgeon General calls cigarette smoking "the most important of the known modifiable risk factors for coronary heart disease in the United States." Smoking also increases the risk of high blood pressure, blood clots, strokes, lung cancer, chronic obstructive pulmonary disease, and atherosclerosis (AHA 2003b). Worldwide, it is estimated that by the year 2020 over 10 million tobacco-related deaths will occur annually.

Heavy alcohol and drug use are also associated with negative consequences for an individual's mental health. Data on both male and female adults have shown that drug users are more likely to suffer from serious mental disorders including anxiety disorders (e.g., phobias), depression, and antisocial personalities (HHS 2002). Marijuana, the drug most commonly used by adolescents, is also linked to short-term memory loss, learning disabilities, motivational deficits, and retarded emotional development. Twenty-eight percent of suicides by children 12 to 16 years of age are alcohol related (*Affects Child* 2003).

Focus on Technology

The Question of Drug Testing

The technology available to detect whether a person has taken drugs was used during the 1970s by crime laboratories, drug treatment centers, and the military. Today, employers in private industry have turned to chemical laboratories for help in making decisions on employment and retention, and parents and school officials use commercial testing devices to detect the presence of drugs. An individual's drug use can be assessed through the analysis of hair, blood, or urine. New technologies include portable breath (or saliva) alcohol testers, THC detection strips, passive alcohol sensors, interlock vehicle ignition systems, and fingerprint screening devices. Counter-technologies have even been developed—for example, shampoos that rid hair of toxins and "Urine Luck," a urine additive that is advertised to speed the breakdown of unwanted chemicals. New Mexico recently joined 11 other states that require that drivers be tested for alcohol consumption each time they try to start their cars by use of an ignition interlock devise (AP 2002).

In 1986, the President's Commission on Organized Crime recommended that all employees of private companies contracting with the federal government be regularly subjected to urine testing for drugs as a condition of employment. This recommendation was based on the belief that if employees such as air traffic controllers, airline pilots, and railroad operators are using drugs, human lives may be in jeopardy as a result of impaired job performance. In 1987, an Amtrak passenger train crashed outside Baltimore, killing 16 and injuring hundreds. There was evidence of drug use by those responsible for the train's safety. As a result, the Supreme Court ruled in 1989 (by a vote of 7–2) that it is constitutional for the Federal Railroad Administration to administer a drug test to railroad crews if they are involved in an accident. Testing those in "sensitive" jobs for drug use may save lives.

An alternative perspective is that drug testing may be harmful. One concern is the accuracy of the tests and the possible impact of false positives. An innocent person, for example, could lose his or her job. Concern with accuracy of drug tests has led the U.S. Department of Health and Human Services to begin an investigation of all federally certified drug testing laboratories (Brannigan 2000).

A second issue concerns the constitutionality of drug testing. At the heart of the debate is the Fourth Amendment, which states that "the right of the people to be secure in their persons . . . against unreasonable searches and seizures, shall not be violated." Specifically at issue is the definition of "special needs"—an exception to the Fourth Amendment. The special needs exception argues that when a circumstance arises (e.g., drug use among student athletes) that requires action (e.g., controlling drug use), an exception to the Fourth

The societal costs of drug-related health concerns are also extraordinary—an estimated $14 billion annually (ONDCP 2003m). Health costs include medical services for drug users, the cost of disability insurance, the effects of secondhand smoke, the spread of AIDS, and the medical costs of accident and crime victims, as well as unhealthy infants and children. For example, cocaine use in pregnant women may lead to low birth weight babies, increased risk of spontaneous abortions, and abnormal placental functioning (Klutt 2000). Additionally, women who smoke while using oral contraceptives are at a greater risk of coronary heart disease and stroke than nonsmoking women who are taking oral contraceptives (AHA 2003b).

Treatment Alternatives

Prevention is always preferable to treatment. Prevention techniques fall into one of two categories (Hanson 2002). First are what may be called "risk and protective" strategies. Here, factors known to be associated with drug use (e.g., child abuse) are targeted and those that help insulate a person from drug use (e.g., stable family) are

Focus on Technology

Amendment based on "special needs" (e.g., requiring drug testing as a condition of eligibility) may be made. Thus in 1995, the Supreme Court ruled that random drug testing of student athletes in public schools, where a pattern of drug use had been established and athletes voluntarily submitted to physical examinations, is not unconstitutional. However, a recent U.S. Supreme Court decision held that students in any competitive activity, including cheerleading and choir, may be subject to random drug testing even if no pattern of drug use has been established (Greenhouse, 2002). Ironically, student drug testing has not been found to reduce student drug use (Yamaguchi, Johnson, and O'Malley, 2003).

The issue continues to grow in complexity as drug testing spreads to other venues. Michigan, for example, became the first state in the United States to pass a law that requires that new welfare applicants submit to drug testing. Those who refuse to take the test will be denied benefits and those who fail the test will be required to participate in a drug treatment program or have their benefits cut by 25 percent (Alcoholism and Drug Abuse Weekly 2000). However, a recent class action suit filed on behalf of welfare recipients has resulted in an injunction preventing implementation of the state policy. Not surprisingly, the suit claims that such a stipulation, that is, requiring drug tests for welfare eligibility, is a Fourth Amendment violation.

Of equal significance is a case involving the drug testing of pregnant women, during routine pregnancy examinations, who exhibited certain warning signs of drug use (Kahn 2000). The case, now before the Supreme Court, was brought by 10 women who had tested positive for cocaine and were arrested after giving birth. In *Ferguson v. City of Charleston* the legal issue is whether the hospital violated the plaintiffs' Fourth Amendment rights by disclosing to nonmedical personnel the results of the urine analyses. The hospital argues that it is an effective and nonintrusive means of protecting the "special needs" of the fetus (CNN 2000). The question in a complex and increasingly technologically dependent society is how to balance the rights of an individual with the needs of society as a whole.

Sources: *Alcoholism and Drug Abuse Weekly.* 2000. September 18, 6; Associated Press. 2002. "New Year Ushers in New Laws." *Daily Reflector,* A1; Brannigan, Martha. 2000. "Labs that Test Transportation Workers for Drugs Face Inquiry over Samples." *Wall Street Journal,* October 2, A4; CNN.com. 2000. "Attorney Jennifer Granick Discusses U.S. Supreme Court Case About Drug Testing of Pregnant Women." October 11; Greenhouse, Linda. 2002. "Justices Allow Schools Wider Use of Random Drug Tests for Pupils." New York Times Online. NYTimes.com. June 28; Kahn, Jeffery. 2000. "Criminally Pregnant." CNN.com. October 30; Yamaguchi, Ryoko, Loyd Johnson, and Patrick O'Malley. 2003. "Student Drug Testing not Effective in Reducing Drug Use." News and Information Services. Ann Arbor: University of Michigan.

encouraged. The second group of prevention techniques rather than dealing with reducing the vulnerability of an individual focuses on "the dynamics of the situations, beliefs, motives, reasoning and reactions that enter into the choice to abuse or not abuse drugs" (Hanson 2002, 3).

Treatment of drug users has become increasingly important in part as a response to the greater need for treatment programs (HHS 2002). In a recent poll, two-thirds of Americans responded that they wanted treatment rather than punishment for nonviolent drug offenders. In 2003, approximately one-third of the "war on drugs" budget was spent on treatment and educational programs, the remainder on law enforcement and interdiction efforts (Drug Policy Alliance 2003c).

In a nationwide survey of recovering substance abusers, 19 percent said their main problem was drugs, 4 percent reported problems with drugs *and* alcohol but said drugs were the primary problem, and 22 percent reported that alcohol and other drugs were equally problematic. Two-thirds of the sample were between the ages of 35 and 59, 41 percent were white-collar professionals, 17 percent were retired, and 8 percent were unemployed (Willing 2002). Persons who are interested in overcoming chemical dependency have a number of treatment alternatives from which to choose. Some options include family therapy, counseling, private and

People who smoke are more likely to be heavy drinkers and current illicit drug users. Some evidence suggests that giving up smoking leads to a reduction in alcohol consumption.

state treatment facilities, community care programs, pharmacotherapy (i.e., use of treatment medications), behavior modification, drug maintenance programs, and employee assistance programs. Two commonly used techniques are inpatient/outpatient treatment and peer support groups.

Inpatient/Outpatient Treatment

Inpatient treatment refers "to the treatment of drug dependence in a hospital and includes medical supervision of detoxification" (McCaffrey 1998, 2). Most inpatient programs last between 30 and 90 days and target individuals whose withdrawal symptoms require close monitoring (e.g., alcoholics, cocaine addicts). Some drug-dependent patients, however, can be safely treated as outpatients. Outpatient treatment allows individuals to remain in their home and work environments and is often less expensive. In outpatient treatment the patient is under the care of a physician who evaluates the patient's progress regularly, prescribes needed medication, and watches for signs of a relapse.

The longer a patient stays in treatment, the greater the likelihood of a successful recovery (NIDA 2003). Variables that predict success include the user's motivation to change, support of family and friends, criminal justice or employer intervention, a positive relationship with therapeutic staff, and a program of recovery that addresses many of the needs of the patient.

Peer Support Groups

12-Step Programs. Both Alcoholics Anonymous (AA) and Narcotics Anonymous (NA) are voluntary associations whose only membership requirement is the desire to stop drinking or taking drugs. AA and NA are self-help groups in that they are operated by nonprofessionals, offer "sponsors" to each new member, and proceed along a continuum of 12 steps to recovery. Members are immediately immersed in a fellowship of caring individuals with whom they meet daily or weekly to affirm their commitment. Some have argued that AA and NA members trade their addiction to drugs for feelings of interpersonal connectedness by bonding with other group members. In a survey of recovering addicts, over 50 percent reported using a self-help program such as AA in their recovery (Willing 2002). AA boasts a membership of over 2 million (Kornblum & Julian 2004) alcoholics.

Symbolic interactionists emphasize that AA and NA provide social contexts in which people develop new meanings. Abusers are surrounded by others who offer positive labels, encouragement, and social support for sobriety. Sponsors tell the new members that they can be successful in controlling alcohol and/or drugs "one day at a time" and provide regular interpersonal reinforcement for doing so. Although thought of as a "crutch" by some, AA members may also take medications to help prevent relapses. In a study of 222 AA members, Rychtarik and colleagues (2000) found that although over half of those surveyed thought the use of relapse-preventing medication was or might be a good idea, 29 percent reported pressures from others to stop taking the medication.

Therapeutic Communities. In **therapeutic communities,** which house between 35 and 500 people for up to 15 months, participants abstain from drugs, develop marketable skills, and receive counseling. Synanon, which was established in 1958, was the first therapeutic community for alcoholics and was later expanded to include other drug users. More than 400 residential treatment centers are now in existence, including Daytop Village and Phoenix House. The longer a person stays at such a facility, the greater the chance he or she has of overcoming dependency. Symbolic interactionists argue that behavioral changes appear to be a consequence of revised self-definition and the positive expectations of others.

Strategies for Action: America Responds

Drug use is a complex social issue exacerbated by the structural and cultural forces of society that contribute to its existence. While the structure of society perpetuates a system of inequality, creating in some the need to escape, the culture of society, through the media and normative contradictions, sends mixed messages about the acceptability of drug use. Thus, developing programs, laws, or initiatives that are likely to end drug use may be unrealistic. Nevertheless, numerous social policies have been implemented or proposed to help control drug use and its negative consequences (see Table 3.2).

Government Regulations

The largest social policy attempt to control drug use in the United States was Prohibition. Although this effort was a failure by most indicators, the government continues to develop programs and initiatives designed to combat drug use (see Figure 3.2). In the 1980s the federal government declared a "war on drugs" based on the belief that controlling drug availability would limit drug use and, in turn, drug-related problems. In contrast to a **harm reduction** position that focuses

Table 3.2
Government Initiatives in the Fight Against Drugs

National Youth Anti-Drug Media Campaign—paid advertising and local initiatives to educate the nation's families, parents, and youth about the dangers of drugs.

Drug Free Communities Program—formation of local coalitions of community members to help organize existing groups and serve as a catalyst for new organizations and agencies in the fight against drugs.

Parents Drug Corp Initiative—training parents in the war against drugs including the formation of "parent drug prevention groups."

The President's Drug Treatment Initiative—a voucher program for those in the criminal justice system, health clinics, emergency rooms, schools, or the faith community, who are in need of treatment and cannot afford it.

Drug Courts Program—diverting drug offenders out of the criminal justice system and into treatment facilities, aftercare programs, and the like.

Priority Targeting Initiative—adds 329 positions to the DEA (Drug Enforcement Agency) in their fight against the manufacturing and distribution of illegal drugs in the United States.

Andean Counter-drug Initiative—support for various counter-drug activities including illicit crop reduction in Colombia, Bolivia, Peru, and the Andean region; also includes humanitarian efforts in this region—for example, support for vulnerable groups.

Expanded Support to Colombia—the additional monies "will be used to fund various programs to conduct a unified campaign against terrorism and drugs"; Colombian ground and air forces will also be trained in counter-drug techniques including coastal interdiction.

Source: Office of National Drug control Policy (ONDCP). "Exeucitve Summary: 2004 Budget" http://www.whitehousedrugpolicy.gov

Figure 3.2
Federal Drug Control Spending by Function, Fiscal Year 2004
Source: Office of National Drug Control Policy (ONDCP). 2003o. "The National Drug Control Strategy." http://www.whitehousedrugpolicy.gov

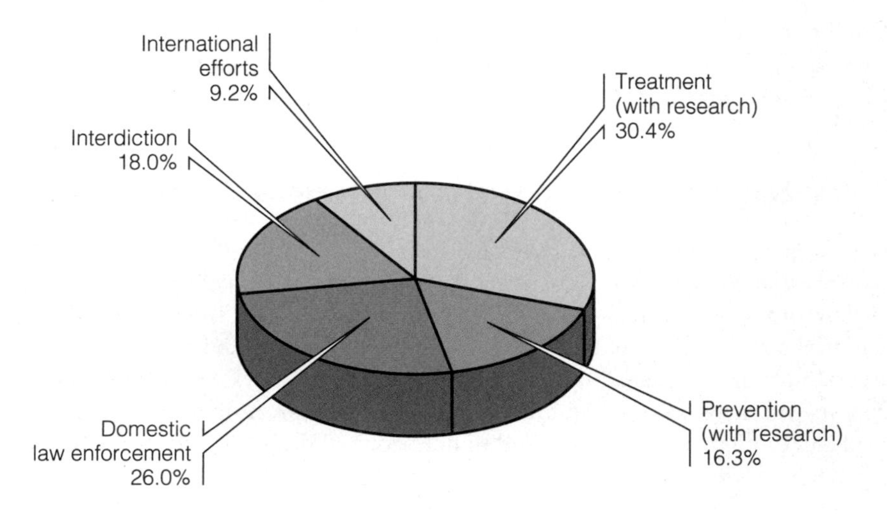

on minimizing the costs of drug use for both user and society (e.g., distributing clean syringes to decrease the risk of HIV infection), this "zero-tolerance" approach advocates get-tough law enforcement policies. However, Yale law professor Steven Duke and co-author Albert C. Gross, in their book *America's Longest War* (1994), argue that the war on drugs, much like Prohibition, has only intensified other social problems: drug-related gang violence and turf wars, the creation of syndicate-controlled black markets, unemployment, the spread of AIDS, overcrowded prisons, corrupt law enforcement officials, and the diversion of police from other serious crimes.

Consistent with conflict theory, still others argue that the war on drugs unfairly targets minorities (see Chapter 4). U.S. Department of Justice data analyzed by Human Rights Watch, an international human rights organization, indicates that, in general, black men are 13 times more likely to be incarcerated in state prisons for drug charges than white men (Fletcher 2000, A10).

Despite concerns, the war on drugs continues at an astronomical societal cost—a projected $11.7 billion in 2004. The National Control Drug Strategy has three priorities: (1) "stopping drug use before it starts," which focuses on young people, education, and communities, (2) "healing drug users," which targets the over 6 million citizens who are in need of drug treatment, and (3) "disrupting drug markets" in transit and producer countries, including the United States (ONDCP 2003o).

Many of the countries in which drug trafficking occurs are characterized by government corruption and crime, military coups, and political instability. Such is the case with Colombia, supplier of over 90 percent of cocaine in the United States. Aerial spraying of the coca plants "is a major component of Colombia's strategy for fighting the drug trade and is the program with the single greatest potential for disrupting the production of cocaine before it enters the supply train to the United States" (ONDCP 2003n, 35). To that end, the United States has channeled over $100 million in support of counter-drug activities in Colombia. When a national poll of adult Americans were asked about financial assistance to foreign countries to fight trafficking 42 percent responded that the United States was providing too much assistance (Pew 2002). Rather than foreign aid and military assistance, many argue that trade sanctions should be imposed in addition to crop eradication programs and interdiction efforts. Others, however, noting the relative failure of such programs in reducing the supply of illegal drugs entering the United States, argue that the "war on drugs" should be abandoned and that deregulation is preferable to the side effects of regulation.

Deregulation or Legalization: The Debate

Americans have mixed feelings about drugs and drug use. For example, only 34 percent believe that marijuana should be legal; however, 72 percent believe that recreational marijuana use should result in nothing more than a fine, and 47 percent have tried the drug (Time/CNN 2003). Given these results, it is not surprising that some advocate alternatives to the punitive emphasis of the last several decades.

Deregulation is the reduction of government control over certain drugs. For example, although individuals must be 21 years old to purchase alcohol and 18 to purchase cigarettes, both substances are legal and purchased freely. Further, in some states, possession of marijuana in small amounts is now a misdemeanor rather than a felony, and in other states marijuana is lawfully used for medical purposes. In 1996, both Arizona and California passed acts known as "marijuana medical bills" that made the use and cultivation of marijuana, under a physician's orders, legal. Medical use of marijuana has also been approved in Alaska, Colorado, Maine, Nevada, Oregon, and Washington. Deregulation is popular in other countries as well. Canada has legalized the medical use of marijuana and England has established a "seize and warn" policy—the marijuana is "seized" and the individual is "warned." No other sanctions are imposed (Stein 2002).

Proponents for the **legalization** of drugs affirm the right of adults to make an informed choice. They also argue that the tremendous revenues realized from drug taxes could be used to benefit all citizens, that purity and safety controls could be implemented, and that legalization would expand the number of distributors, thereby increasing competition and reducing prices. Drugs would thus be safer, drug-related crimes would be reduced, and production and distribution of previously controlled substances would be taken out of the hands of the underworld.

Those in favor of legalization also suggest that the greater availability of drugs would not increase demand, pointing to countries where some drugs have already been decriminalized. **Decriminalization,** or the removing of penalties for certain drugs, would promote a medical rather than criminal approach to drug use that would encourage users to seek treatment and adopt preventive practices. For example, making it a criminal offense to sell or possess hypodermic needles without a prescription encourages the use of nonsterile needles that spread infections such as HIV and hepatitis.

Opponents of legalization argue that it would be construed as government approval of drug use and, as a consequence, drug experimentation and abuse would increase. Further, while the legalization of drugs would result in substantial revenues for the government, because all drugs would not be decriminalized (e.g., crack), drug trafficking and black markets would still flourish. Legalization would also require an extensive and costly bureaucracy to regulate the manufacture, sale, and distribution of drugs. Finally, the position that drug use is an individual's right cannot guarantee that others will not be harmed. It is illogical to assume that a greater availability of drugs will translate into a safer society.

Collective Action

Social action groups such as Mothers Against Drunk Driving **(MADD)** have successfully lobbied legislators to raise the drinking age to 21 and to provide harsher penalties for driving while impaired. MADD, with 3.5 million members and 600 chapters, has also put pressure on alcohol establishments to stop "two for one" offers and has pushed for laws that hold the bartender personally liable if a served person is later involved in an alcohol-related accident. Even hosts in private homes can now be held liable if they allow a guest to drive who became impaired while drinking at their house.

Recently, MADD has released a plan to jump-start their fight against drunk driving. Between 1980 and 1994 drunk driving deaths decreased significantly. Since that time, however, drunk driving deaths have remained constant. The proposed eight-point plan, among other things, calls for the public to "Get MADD All Over Again," increase use of sobriety checkpoints, impose tougher sanctions for repeat offenders, and raise beer excise taxes (MADD 2003d).

"If some drunk gets out and kills your kid, you'd probably be a little crazy about it, too."

Kathy Prescott
Former MADD President

Collective action is also being taken against tobacco companies by smokers, ex-smokers, and the families of smoking victims. They charge that tobacco executives knew over 30 years ago that tobacco was addictive and concealed this fact from both the public and the government. Furthermore, they charge that tobacco companies manipulate nicotine levels in cigarettes with the intention of causing addiction. In a class action suit of over 300,000 Florida smokers, a jury ordered the top five cigarette producers to pay $145 billion—the largest settlement to date—to the plaintiffs. In 2003, Philip Morris, now also a defendant in an Illinois class action suit, may have to post a $12 billion bond. If so, the company says it will have to file bankruptcy and Florida will lose its settlement money (James 2003). In response to such legal assaults, the tobacco industry has dramatically increased their cigarette exports and foreign production.

Finally, several initiatives have resulted in statewide referendums concerning the cost-effectiveness of government policies. For example, as a result of the passage of Proposition 36, California, as well as many other states, will now require that nonviolent first- and second-time minor drug offenders receive treatment including job training, therapy, literacy education, and family counseling rather than jail time. The initiative provides over $100 million a year over the next five and a half years for community-based treatment facilities. The program is predicted to divert over 30,000 state and county prisoners to treatment programs resulting in a net savings to California taxpayers of $1.5 billion over the course of the program. Evaluation of a similar strategy in Arizona boasts a treatment success rate of 71 percent (Mann 2000; Thompson 2000). In recent years the debate over treatment versus incarceration has been on several state ballots with varying results (Drug Policy Alliance 2003d).

Understanding Alcohol and Other Drugs

In summarizing what we know about substance abuse, drugs and their use are socially defined. As the structure of society changes, the acceptability of one drug or another changes as well. As conflict theorists assert, the status of a drug as legal or illegal is intricately linked to those who have the power to define acceptable and unacceptable drug use. There is also little doubt that rapid social change, anomie, alienation, and inequality further drug use and abuse. Symbolic interactionism also plays a significant role in the process—if people are labeled "drug users" and expected to behave accordingly, drug use is likely to continue. If people experience positive reinforcement of such behaviors and/or have a biological predisposition to use drugs, the probability of their drug involvement is even higher. Thus, the theories of drug use complement rather than contradict one another.

Drug use must also be conceptualized within the social context in which it occurs. Many youths who are at high risk for drug use have been "failed by society"—they are living in poverty; victims of abuse; dependents of addicted and neglectful parents; alienated from school (Fields 2001; Siegel 2002). Despite the social origins of drug use, many treatment alternatives, emanating from a clinical model of drug use, assume that the origin of the problem lies within the individual rather than in the structure and culture of society. Although admittedly the problem may lie within the individual at the time treatment occurs, policies that address the social causes of drug abuse provide a better means of dealing with the drug problem in the United States.

"A child who reaches 21 without smoking, abusing alcohol or using drugs is virtually certain never to do so."

Joseph A. Califano, Jr.
President, Center on Addiction and Substance Abuse

As stated earlier, prevention is preferable to intervention, and given the social portrait of hard drug users—young, male, minority—prevention must entail dealing with the social conditions that foster drug use. Some data suggest that inner city adolescents are particularly vulnerable to drug involvement because of their lack of legitimate alternatives.

> Illegal drug use may be a way to escape the strains of the severe urban conditions and dealing illegal drugs may be one of the few, if not the only, way to provide for material needs. Intervention and treatment programs, therefore, should include efforts to find alternate ways to deal with the limiting circumstances of inner-city life, as well as create opportunities for youngsters to find more conventional ways of earning a living. (p. 22)

Social policies dealing with drug use have been predominantly punitive rather than preventive. Recently, however, there appears to be some movement toward educating the public and changing the culture of drugs. For example, a new national campaign, using the slogan "Honor. My Anti-Drug" features the cast members from a popular animated action series *YU-GI-OH.* The show's characters will be pictured on original art collectible stickers "each promoting the passions that keep them drug-free" (ONDCP 2003p). The stickers will be distributed at the 4,700 Blockbuster Video stores throughout the United States.

In this country and throughout the world, millions of people depend on legal drugs for the treatment of a variety of conditions, including pain, anxiety and nervousness, insomnia, overeating, and fatigue. Although drugs for these purposes are relatively harmless, the cultural message "better living through chemistry" contributes to alcohol and drug use and its consequences. But these and other drugs are embedded in a political and economic context that determines who defines what drugs, in what amounts, as licit or illicit and what programs are developed in reference to them.

Chapter Review

- **What is a drug and what is meant by drug abuse?**
  Sociologically, the term "drug" refers to any chemical substance that (1) has a direct effect on the user's physical, psychological, and/or intellectual functioning, (2) has the potential to be abused, and (3) has adverse consequences for the individual and/or society. Drug abuse occurs when acceptable social standards of drug use are violated resulting in adverse physiological, psychological, and/or social consequences.

- **How do the three sociological theories of society explain drug use?**
  Functionalists argue that drug abuse is a response to the weakening of norms in society leading to a condition known as anomie or normlessness. From a conflict perspective, drug use occurs as a response to the inequality perpetuated by a capitalist system as societal members respond to alienation from their work, family, and friends. Symbolic interactionism concentrates on the social meanings associated with drug use. If the

initial drug use experience is defined as pleasurable, it is likely to recur, and over time, the individual may earn the label of "drug user."

- **What are the most frequently used legal and illegal drugs?**
  Alcohol is the most commonly used and abused legal drug in America. The use of tobacco products is also very high with 30 percent of Americans reporting that they currently smoke cigarettes. Marijuana is the most commonly used illicit drug with 147 million marijuana users representing 3.1 percent of the world's population.

- **What are the consequences of drug use?**
  The consequences of drug use are fourfold. First is the cost to the family, often manifesting itself in higher rates of divorce, spouse abuse, child abuse, and child neglect. Second is the relationship between drugs and crime. Those arrested have disproportionately high rates of drug use. Although drug users commit more crimes, sociologists disagree as to whether drugs actually "cause" crime or whether, instead, criminal activity leads to drug involvement. Third are the economic costs, which are estimated to be over $143 billion. Finally are the health costs of abusing drugs including shortened life expectancy; higher morbidity (e.g., cirrhosis of the liver, lung cancer); exposure to HIV infection, hepatitis, and other diseases through shared needles; a weakened immune system; birth defects such as fetal alcohol syndrome; drug addiction in children; and higher death rates.

- **What treatment alternatives are available for drug users?**
  Although there are many ways to treat drug abuse, two methods stand out. The inpatient/outpatient model entails medical supervision of detoxification and may or may not include hospitalization. Twelve-step programs like Alcoholics Anonymous (AA) and Narcotics Anonymous (NA) are particularly popular as are therapeutic communities. Therapeutic communities are residential facilities where drug users learn to redefine themselves and their behavior as a response to the expectations of others and self-definition

- **What can be done about the drug problem?**
  First, there are government regulations limiting the use (law establishing 21-year-old drinking age) and distribution (prohibitions about importing drugs) of legal and illegal drugs. The government also imposes sanctions on those who violate drug regulations and provides treatment facilities for other offenders. Second, there are collective action groups—for example, Mothers Against Drunk Driving. Finally, there are local and statewide initiatives geared toward holding companies responsible for the consequences for their product—for example, class action suits against tobacco producers.

Critical Thinking

1. Are alcoholism and other drug addictions a consequence of nature or nurture? If nurture, what environmental factors contribute to such problems? Which of the three sociological theories best explains drug addiction?
2. Measuring alcohol and drug use is often very difficult. This is particularly true given the tendency for respondents to acquiesce, that is, respond in a way they believe is socially desirable. Consider this and other problems in doing research on alcohol and other drugs, and how such problems would be remedied.
3. Jeffery Reiman (2003, 37) notes that "on the basis of available scientific evidence, there is every reason to suspect that we do our bodies more damage, more irreversible damage, by smoking cigarettes and drinking liquor than by using heroin." How would a social constructionist interpret this statement?

Key Terms

anomie
chemical dependency
club drugs
crack
date-rape drugs
decriminalization
deregulation
drug
drug abuse
drug addiction
gateway drug
harm reduction
legalization
MADD
therapeutic communities

Taking a Stand

Should drugs be legalized for those 21 years old and older?

Many argue that the right of an adult to make an informed decision includes deciding to use illegal drugs. Similar to arguments about the legalization of prostitution and gambling, drug use is considered by many to be a victimless crime. Further, it is argued, any violence associated with drug use is a consequence of its being illegal, which drives its price up and forces users to turn to crime as a means of supporting their drug use. Alternatively, there is the argument that the state not only has the right but the obligation to protect its citizens. Seatbelt laws, traffic laws, and many administrative agency regulations (e.g., U.S. Food and Drug Administration) are designed to do just that. Moreover, if drugs were made legal there would be an increase in drug use and the associated costs described in this chapter.

Use Wadsworth's exclusive online resources—InfoTrac College Edition, MicroCase Online, and OVRC—to formulate a position on this topic.

The Wadsworth's Sociology Online Resources and Writing Companion will help you get started. This valuable guide will show you how to use Wadsworth's exclusive online resources when studying social problems. It will also help you to build essential research and writing skills. InfoTrac College Edition, MicroCase Online, OVRC, and an electronic copy of portions of this companion are available at http://sociology.wadsworth.com/mooney_knox_schacht/problems4e, the companion Web site for *Understanding Social Problems,* Fourth Edition.

Media Resources

The Companion Web Site for *Understanding Social Problems,* Fourth Edition

http://sociology.wadsworth.com/mooney_knox_schacht/problems4e

Supplement your review of this chapter by going to the companion Web site to take one of the Tutorial Quizzes, use the flash cards to master key terms, and check out the many other study aids you'll find there. You'll also find special features such as *Wadsworth's Sociology Online Resources and Writing Companion,* GSS Data, and Census 2000 information, data, and resources at your fingertips to help you complete that special project or do some research on your own.

Interactions CD-ROM

Go to the Interactions CD-ROM for *Understanding Social Problems,* Fourth Edition, to access additional interactive learning tools, such as in-depth review materials, corresponding practice quizzes, and other engaging resources and activities to help you study the concepts in this chapter.

“Unjust social arrangements are themselves a kind of extortion, even violence.” *John Rawls from* A Theory of Justice

Crime and Violence

Amadou Diallo was born in Liberia on September 2, 1975. When he left home to attend college in the United States, he told his mother, "You have worked so hard and struggled so long to take care of our family. I must be a success so I can take over and you may rest" (Foundation 2003a). The western African immigrant moved to the Bronx, a borough of New York City. Late one February evening, he was standing in the hallway of his apartment building when he was approached by several members of the NYPD Street Crime Unit who were investigating a rape case. Fitting the description of the rape suspect, Diallo was told to "freeze," but with limited understanding of English he continued to reach for his wallet apparently believing that the officers were asking for his identification. One police officer, thinking he saw a weapon, yelled "gun" and in seconds Diallo was dead. The four police officers fired a total of 41 shots—19 hitting the 22-year-old. Diallo was unarmed and innocent of the suspected crime (PBS 2000; Human Rights Watch 2000).

One year later, in February of 2000, all four officers were acquitted of second-degree murder after jurors found that the officers were in reasonable fear for their safety. The decision spurred countless protests and, in the wake of accusations of racial profiling, contributed to a growing distrust of the criminal justice system. In 2001, U.S. Department of Justice officials refused to file charges against the four officers stating that they could not prove, beyond a reasonable doubt, that Mr. Diallo's right to be "free from unreasonable use of force was violated" (Haughney 2001, A9). In honor of the shooting victim, the Amadou Diallo Foundation, Inc., was formed, dedicated to "promoting racial healing and cross cultural understandings" (Foundation 2003b).

Clearly what happened in the Diallo case is not representative of the millions of police–citizen interactions that take place every year. In fact, most citizens have confidence in the police. In a national survey of adult Americans 18 and older, 64.4 percent believed that the police are doing an "excellent" or "pretty good" job of solving crime (Harris Poll 2002). The police are part of a massive bureaucracy called the criminal justice system, a system that often comes under public scrutiny particularly in high profile cases such as the kidnaping of Elizabeth Smart, the murder of Lacy Peterson, and the trial of the "beltway snipers." This chapter examines the criminal justice system as well as theories, types, and demographic patterns of criminal behavior. The economic, social, and psychological costs of crime and violence are also examined. The chapter concludes with a discussion of social policies and prevention programs designed to reduce crime and violence in America.

The Global Context: International Crime and Violence

Several facts about crime are true throughout the world. First, crime is ubiquitous; that is, there is no country where crime does not exist. Second, most countries have the same components in their criminal justice systems—police, courts, and prisons. Third, worldwide, adult males comprise the largest category of crime suspects, and fourth, in all countries, theft is the most common crime committed whereas violent crime is a relatively rare event.

Dramatic differences do exist, however, in international crime and violence rates. For example, less-developed countries have higher rates of homicide than industrialized countries (Thio 2004). Among industrial nations, however, the United States has one of the highest homicide rates—six times the murder rate of Holland, five times the rate of Canada, and eight times the rate of Europe. The U.S. crime rate is also high in other categories. For example, when compared to Japan, the United States has "eighteen times more rapes, ten times more homicides, six times more burglaries, and nine times more drug related offenses" (Sullivan 2003, 320).

Recent concerns have focused on **transnational crime** defined by the U.S. Department of Justice as "organized criminal activity across one or more national borders" (USDOJ 2003a). Chinese Triads operate in large cities worldwide netting billions of dollars a year from prostitution, drugs, and other organized crime activities; Colombian cocaine cartels flourish and spread to sub-Saharan countries with needy economies; ecstasy is manufactured in the Netherlands and sold in the United States by Israeli organized crime members; and an estimated 50,000 women and children are trafficked into the United States annually for use in pornography rings and prostitution (United Nations 1997; INTERPOL 1998; Finckenauer 2000; Dobriansky 2001; USDOJ 2003a). Transnational crime is facilitated by recent trends in globalization including enhanced transportation and communication technologies. For example, the U.S. Customs Service estimates that, worldwide, there are over 100,000 Web sites involved in child pornography (ABCNews 2001). One type of transnational crime, terrorism, will be discussed in Chapter 16.

Sources of Crime Statistics

The U.S. government spends millions of dollars annually compiling and analyzing crime statistics. A **crime** is a violation of a federal, state, or local criminal law. For a violation to be a crime, however, the offender must have acted voluntarily and with intent and have no legally acceptable excuse (e.g., insanity) or justification (e.g., self-defense) for the behavior. The three major types of statistics used to measure crime are official statistics, victimization surveys, and self-report offender surveys.

"Ultimately, any crime statistic is only as useful as the reader's understanding of the processes that generated it."

Robert M. O'Brien
Sociologist, University of Oregon

Official Statistics

Local sheriffs' departments and police departments throughout the United States collect information on the number of reported crimes and arrests and voluntarily report them to the Federal Bureau of Investigation (FBI). The FBI then compiles these statistics annually and publishes them, in summary form, in the Uniform Crime Reports (UCR). The UCR lists **crime rates** or the number of crimes committed per 100,000 population, the actual number of crimes and the percentage of change over time, and clearance rates. **Clearance rates** measure the percentage of cases in which an arrest and official charge have been made and the case turned over to the courts.

These statistics have several shortcomings. For example, many incidents of crime go unreported. It is estimated that only 31 percent of rapes and sexual assaults, 55 percent of aggravated assaults, and 57 percent of robberies are actually reported to the police (USDOJ 2003b). Even if a crime is reported it may not be recorded by the police (see Figure 4.1). Alternatively, some rates may be exaggerated. Motivation for such distortions may come from the public (e.g., demanding that something be done), or from political officials (e.g., election of a sheriff), and/or organizational pressures (e.g., budget requests). For example, a police department may "crack down" on drug-related crimes in an election year. The result is an increase in the recorded number of these offenses for that year. Such an in-

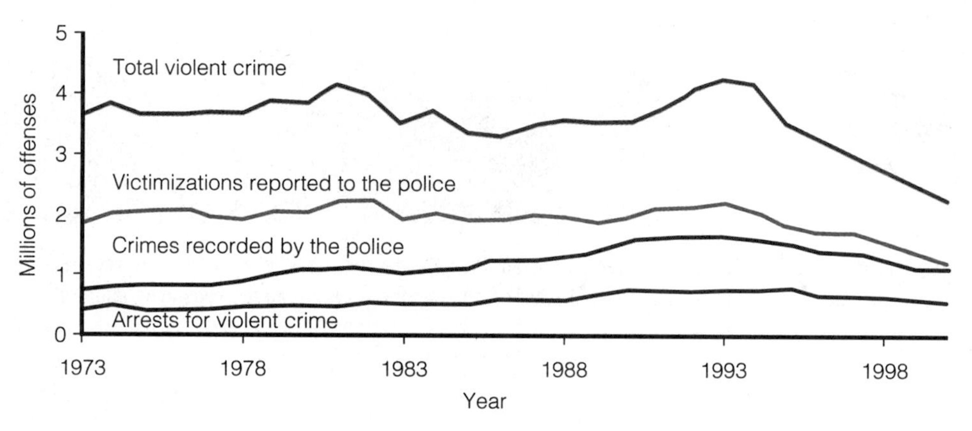

Figure 4.1
Four Measures of Serious Violent Crime

Source: Bureau of Justice Statistics.

Note: The serious violent crimes included are rape, robbery, aggravated assault, and homicide.

crease reflects a change in the behavior of law enforcement personnel, not a change in the number of drug violations. Thus, official crime statistics may be a better indicator of what police are doing than what criminals are doing.

Victimization Surveys

Victimization surveys ask people if they have been victims of crime. The Department of Justice's National Crime Victimization Survey (NCVS), conducted annually, interviews nearly 100,000 people about their experiences as victims of crime. Interviewers collect a variety of information, including the victim's background (e.g., age, race and ethnicity, sex, marital status, education, and area of residence), relationship to offender (stranger or nonstranger), and the extent to which the victim was harmed. Although victimization surveys provide detailed information about crime victims, they provide less reliable data on offenders.

Self-Report Offender Surveys

Self-report surveys ask offenders about their criminal behavior. The sample may consist of a population with known police records, such as a prison population, or it may include respondents from the general population, such as college students. Self-report data compensate for many of the problems associated with official statistics but are still subject to exaggerations and concealment. The Criminal Activities Survey in this chapter's *Self and Society* feature asks you to indicate whether you have engaged in a variety of illegal activities.

Self-report surveys reveal that virtually every adult has engaged in some type of criminal activity. Why then is only a fraction of the population labeled criminal? Like a funnel, which is large at one end and small at the other, only a small proportion of the total population of law violators is ever convicted of a crime. For an individual to be officially labeled as a criminal, his or her behavior (1) must become known to have occurred; (2) must come to the attention of the police who then file a report, conduct an investigation, and make an arrest; and finally, (3) the arrestee must go through a preliminary hearing, an arraignment, and a trial and may or may not be convicted. At every stage of the process, an offender may be "funneled" out. As Figure 4.1 indicates, the measures of crime used at various points in time lead to different results.

Criminal Activities Survey

Read each of the following questions. If, since the age of 16, you have ever engaged in the behavior described, place a "1" in the space provided. If you have not engaged in the behavior, put a "0" in the space provided. After completing the survey, read the section on interpretation to see what your answers mean.

Questions	1 (Yes)	0 (No)
1. Have you ever been in possession of drug paraphernalia?	_______	_______
2. Have you ever lied about your age, or about anything else when making application to rent an automobile?	_______	_______
3. Have you ever intentionally destroyed or erased someone else's phone messages?	_______	_______
4. Have you ever tampered with a coin-operated vending machine or parking meter?	_______	_______
5. Have you ever loaned or given away lewd or obscene materials?	_______	_______
6. Have you ever begun and/or participated in an office basketball or football pool?	_______	_______
7. Have you ever used "filthy, obscene, annoying, or offensive" language while on the telephone?	_______	_______
8. Have you ever given or sold a beer to someone under the age of 21?	_______	_______
9. Have you ever been on someone else's property (land, house, boat, structure, etc.) without that person's permission?	_______	_______
10. Have you ever forwarded a chain letter with the intent to profit from it?	_______	_______
11. Have you ever improperly gained access to someone else's e-mail or other computer account?	_______	_______
12. Have you ever written a check when you knew it was bad?	_______	_______

Interpretation

Each of the activities described in these questions represents criminal behavior that was subject to fines, imprisonment, or both under the laws of Florida in 2000. For each activity, the following table lists the maximum prison sentence and/or fine for a first-time offender. To calculate your "prison time" and/or fines, sum the numbers corresponding to each activity you have engaged in.

Maximum Prison Sentence	Maximum Fine	Offense
1. One year	$1000	Possession of drug paraphernalia
2. Five years	$5000	Fraud
3. Two months	$500	Unlawful interference with telecommunications
4. Two months	$500	Fraud
5. One year	$1000	Unlawful dissemination of obscene material
6. Two months	$500	Illegal gambling
7. Two months	$500	Harassing/obscene telecommunications
8. Two months	$500	Illegal distribution of alcohol
9. One year	$1000	Trespassing
10. One year	$1000	Illegal gambling
11. Five years	$5000	Illegal misappropriation of cybercommunication
12. One year	$1000	Worthless check

Source: Florida Criminal Code. 2000. http://www.leg.state.fl./statutes/index

Sociological Theories of Crime and Violence

Some explanations of crime and violence focus on psychological aspects of the offender such as psychopathic personalities, unhealthy relationships with parents, and mental illness. Other crime theories focus on the role of biological variables such as central nervous system malfunctioning, stress hormones, vitamin or mineral deficiencies, chromosomal abnormalities, and a genetic predisposition toward aggression. Sociological theories of crime and violence emphasize the role of social factors in criminal behavior and societal responses to it.

Structural-Functionalist Perspective

According to Durkheim and other structural-functionalists, crime is functional for society. One of the functions of crime and other deviant behavior is that it strengthens group cohesion: "The deviant individual violates rules of conduct that the rest of the community holds in high respect; and when these people come together to express their outrage over the offense . . . they develop a tighter bond of solidarity than existed earlier" (Erikson 1966, 4).

Crime may also lead to social change. For example, an episode of local violence may "achieve broad improvements in city services . . . [and] be a catalyst for making public agencies more effective and responsive, for strengthening families and social institutions, and for creating public-private partnerships" (National Research Council 1994, 9–10).

Although functionalism as a theoretical perspective deals directly with some aspects of crime and violence, it is not a theory of crime per se. Three major theories of crime and violence have developed from functionalism, however. The first, called **strain theory,** was developed by Robert Merton (1957) using Durkheim's concept of anomie, or normlessness. Merton argues that when legitimate means (e.g., a job) of acquiring culturally defined goals (e.g., money) are limited by the structure of society, the resulting strain may lead to crime.

Individuals, then, must adapt to the inconsistency between means and goals in a society that socializes everyone into wanting the same thing but provides opportunities only for some (see Table 4.1). Conformity occurs when individuals accept the culturally defined goals and the socially legitimate means of achieving them. Merton suggests that most individuals, even those who do not have easy access to the means and goals, remain conformists. Innovation occurs when an individual accepts the goals of society but rejects or lacks the socially legitimate means of achieving them. Innovation, the mode of adaptation most associated with criminal behavior, explains the high rate of crime committed by uneducated and poor individuals who do not have access to legitimate means of achieving the social goals of wealth and power.

Another adaptation is ritualism, in which the individual accepts a lifestyle of hard work, but rejects the cultural goal of monetary rewards. The ritualist goes through the motions of getting an education and working hard, yet he or she is not committed to the goal of accumulating wealth or power. Retreatism involves rejecting both the cultural goal of success and the socially legitimate means of achieving it. The retreatist withdraws or retreats from society and may become an alcoholic, drug addict, or vagrant. Finally, rebellion occurs when an individual rejects both culturally defined goals and means and substitutes new goals and means. For

Table 4.1
Merton's Strain Theory

Mode of Adaption	Seeks Culturally Defined Goals?	Uses Structurally Defined Means to Achieve Them?
1. Conformity	Yes	Yes
2. Innovation	Yes	No
3. Ritualism	No	Yes
4. Retreatism	No	No
5. Rebellion	No—seeks to replace	No—seeks to replace

Source: Adapted with permission from Robert K. Merton's *Social Theory and Social Structure* (1957). Copyright © 1957 by The Free Press; Copyright renewed 1985 by Robert K. Merton.

example, rebels may use social or political activism to replace the goal of personal wealth with the goal of social justice and equality.

Whereas strain theory explains criminal behavior as a result of blocked opportunities, **subcultural theories** argue that certain groups or subcultures in society have values and attitudes that are conducive to crime and violence. Members of these groups and subcultures, as well as other individuals who interact with them, may adopt the crime-promoting attitudes and values of the group. For example, Kubrin and Weitzer (2003) found that retaliatory homicide is a response to subcultural norms of violence that exist in some neighborhoods.

But if blocked opportunities and subcultural values are responsible for crime, why don't all members of the affected groups become criminals? **Control theory** may answer that question. Hirschi (1969), consistent with Durkheim's emphasis on social solidarity, suggests that a strong social bond between individuals and the social order constrains some individuals from violating social norms. Hirschi identified four elements of the social bond: attachment to significant others, commitment to conventional goals, involvement in conventional activities, and belief in the moral standards of society. Several empirical tests of Hirschi's theory support the notion that the higher the attachment, commitment, involvement, and belief, the higher will be the social bond and the lower the probability of criminal behavior. For example, Laub, Nagan, and Sampson (1998) found that a good marriage contributes to the cessation of a criminal career. Further, Warner and Rountree (1997) report that local community ties, although varying by neighborhood and offense, decrease the probability that crimes will occur. On the other hand, Vander Ven, Cullen, Carrozza, and Wright (2001) report that maternal employment is unrelated to behavioral problems of children.

“There are two criminal justice systems in this country. There is a whole different system for poor people. It's the same courthouse— it's not separate— but it's not equal.”

Paul Petterson
Public defender

Conflict Perspective

Conflict theories of crime suggest that deviance is inevitable whenever two groups have differing degrees of power; in addition, the more inequality in a society, the greater the crime rate in that society. Social inequality leads individuals to commit crimes such as larceny and burglary as a means of economic survival. Other individuals, who are angry and frustrated by their low position in the socioeconomic hierarchy, express their rage and frustration through crimes such as drug use, assault, and homicide. In Argentina, for example, the soaring violent crime rate is hypothesized to be "a product of the enormous imbalance in income distribution . . . between the rich and the poor" (Pertossi 2000).

According to the conflict perspective, those in power define what is criminal and what is not, and these definitions reflect the interests of the ruling class. Laws against vagrancy, for example, penalize individuals who do not contribute to the capitalist system of work and consumerism. Further, D'Alessio and Stolzenberg (2002, 178) found that "in cities with high unemployment, unemployed defendants have a substantially higher probability of pretrial detention" than employed defendants. Rather than viewing law as a mechanism that protects all members of society, conflict theorists focus on how laws are created by those in power to protect the ruling class. For example, wealthy corporations contribute money to campaigns to influence politicians to enact tax laws that serve corporate interests (Reiman 2001).

Furthermore, conflict theorists argue that law enforcement is applied differentially, penalizing those without power and benefiting those with power. For example, although race of victim should not matter, blacks are more likely to be arrested when involved in black-on-white crime than when involved in black-on-black crime (Eitle, D'Alessio, & Stolzenberg 2002). Moreover, female prostitutes are more likely to be arrested than are the men who seek their services. Unlike street criminals, corporate criminals are often punished by fines rather than by lengthy prison terms, and rape laws originated to serve the interests of husbands and fathers who wanted to protect their property—wives and unmarried daughters.

© Michael Newman/PhotoEdit

To Marxists, the cultural definition of women as property contributes to the high rates of female criminality and, specifically, involvement in prostitution, drug abuse, and petty theft. In 2003, there were 79,733 arrests for prostitution and commercial vice in the United States.

Societal beliefs also reflect power differentials. For example, "rape myths" are perpetuated by the male-dominated culture to foster the belief that women are to blame for their own victimization, thereby, in the minds of many, exonerating the offender. Such myths include the notion that when a woman says "no" she means "yes," that "good girls" don't get raped, that appearance indicates willingness, and that women secretly want to be raped. Not surprisingly, in societies where women and men have greater equality, there is less rape.

Symbolic Interactionist Perspective

Two important theories of crime and violence emanate from the symbolic interactionist perspective. The first, **labeling theory,** focuses on two questions: how do crime and deviance come to be defined as such, and what are the effects of being labeled as criminal or deviant? According to Howard Becker (1963):

> Social groups create deviance by making rules whose infractions constitute deviance, and by applying those rules to particular people and labeling them as outsiders. From this point of view, deviance is not a quality of the act a person commits, but rather a consequence of the application by others of rules and sanctions to an "offender." The deviant is one to whom the label has successfully been applied; deviant behavior is behavior that people so label. (p. 238)

Labeling theorists make a distinction between **primary deviance,** which is deviant behavior committed before a person is caught and labeled an offender, and **secondary deviance,** which is deviance that results from being caught and labeled. After a person violates the law and is apprehended, that person is stigmatized as a criminal. This deviant label often dominates the social identity of the person to whom it is applied and becomes the person's "master status," that is, the primary basis on which the person is defined by others.

Being labeled as deviant often leads to further deviant behavior because (1) the person who is labeled as deviant is often denied opportunities for engaging in nondeviant behavior, and (2) the labeled person internalizes the deviant label, adopts a deviant self-concept, and acts accordingly. For example, the teenager who is caught selling drugs at school may be expelled and thus denied opportunities to participate in nondeviant school activities (e.g., sports, clubs) and to associate with nondeviant peer groups. The labeled and stigmatized teenager may also adopt the self-concept of a "druggie" or "pusher" and continue to pursue drug-related activities and membership in the drug culture.

The assignment of meaning and definitions learned from others is also central to the second symbolic interactionist theory of crime, **differential association.** Edwin Sutherland (1939) proposed that through interaction with others, individuals learn the values and attitudes associated with crime as well as the techniques and motivations for criminal behavior. Individuals who are exposed to more definitions favorable to law violation (e.g., "crime pays") than unfavorable (e.g., "do the crime, you'll do the time") are more likely to engage in criminal behavior. Thus children who see their parents benefit from crime or who live in high-crime neighborhoods where success is associated with illegal behavior are more likely to engage in criminal behavior.

Unfavorable definitions may come from a variety of sources. Of particular concern of late is the role of video games in promoting criminal or violent behavior. One particular game, Grand Theft Auto III, has players "head bashing, looting, drug-dealing, drive-by shooting, and running over innocent bystanders with a taxi" (Richtel 2003). In response to this and other violent video games, many states now require a video rating system that differentiates between cartoon violence, fantasy violence, intense violence, and sexual violence (AP 2003).

Types of Crime

The FBI identifies eight **index offenses** as the most serious crimes in the United States. The index offenses, or street crimes as they are often called, may be against a person (called violent or personal crimes) or against property (see Table 4.2). Other types of crime include vice crime, such as drug use, gambling, and prostitution, as well as organized crime, white-collar crime, computer crime, and juvenile delinquency.

Street Crime: Violent Offenses

Data from the Uniform Crime Reports indicate that the 2002 violent crime rate decreased from the previous year by 1.4 percent (USDOJ 2003c). Remember, however, that crime statistics represent only those crimes *reported* to the police—1.4 million violent crimes in 2001. Victim surveys indicate that about one-half of all violent crimes are actually reported to the police (USDOJ 2003b).

Table 4.2
Index Crime Rates, Percent Change, and Clearance Rates, 2001

	Rate per 100,000, 2001	Percent Change in Rate, (2000–2001)	Percent Cleared, 2001
Total Index Crimes	**4,160.5**	**+2.1**	**19.6**
Murder	5.6	+3.1	62.4
Forcible Rape	31.8	+.3	44.3
Robbery	148.5	+3.7	24.9
Aggravated Assault	318.5	−.5	56.1
Violent Crime Total	504.4	+.8	46.2
Burglary	740.8	+2.9	12.7
Larceny/Theft	2,484.6	+1.5	17.6
Motor Vehicle Theft	430.6	+5.7	13.6
Arson	37.1[a]	+2.0	16.0
Property Crime Total[b]	3,656.1	+2.3	16.2

[a]Arson rates per 100,000 are calculated independently because population coverage for arson is lower than for the other index offenses—1999 rate.

[b]Property Crime totals do not include arson.

Source: Federal Bureau of Investigation. 2002. *Uniform Crime Reports, 2001.* Washington, DC: U.S. Department of Justice.

Violent crime includes homicide, assault, rape, and robbery. Homicide refers to the willful or nonnegligent killing of one human being by another individual or group of individuals. Although homicide is the most serious of the violent crimes, it is also the least common, accounting for less than 1 percent of all index crimes (FBI 2002). A typical homicide scenario includes a male killing a male with a handgun after a heated argument. The victim and offender are disproportionately young and of minority status. When a woman is murdered and the victim–offender relationship is known, she is most likely to have been killed by her husband or boyfriend; men are more likely to be killed by a stranger (FBI 2002).

Another form of violent crime, aggravated assault, involves the attacking of another with the intent to cause serious bodily injury. Like homicide, aggravated assault occurs most often between members of the same race and, as with violent crime in general, is more likely to occur in warm weather months. In 2001, the assault rate was over 50 times greater than the murder rate, assaults making up only 7.9 percent of all index crime (FBI 2002).

Rape is also classified as a violent crime and is also intra-racial, that is, the victim and offender are from the same racial group. The FBI definition of "rape" contains three elements: sexual penetration, force or the threat of force, and nonconsent of the victim. In 2001, more than 90,000 forcible rapes were reported in the United States, a slight increase from the previous year (FBI 2002). Rapes are more likely to occur in warm months in part because of the greater ease of victimization. People are outside more and later, doors are open, windows are unlocked, and so forth.

Perhaps as many as 80 percent of all rapes are **acquaintance rapes**—rapes committed by someone the victim knows. Although acquaintance rapes are the most likely to occur, they are the least likely to be reported and the most difficult to prosecute. Unless the rape is what Williams (1984) calls a **classic rape**—that is, the rapist was a stranger who used a weapon and the attack resulted in serious bodily injury—women hesitate to report the crime out of fear of not being believed. The increased use of "rape drugs" such as Rohypnol may lower reporting levels even further (see Chapter 3).

Robbery, unlike simple theft, also involves force or the threat of force, or putting a victim in fear, and is thus considered a violent crime. Officially, in 2001, over 400,000 robberies took place in the United States. Robberies are most often committed by young adults with the use of a gun (FBI 2002). In 2001, robbers stole more than $532 million in money and merchandise. As with rape, victims who resist a robbery are more likely to stop the crime but are also more likely to be physically harmed.

Street Crime: Property Offenses

Property crimes are those in which someone's property is damaged, destroyed, or stolen; they include larceny, motor vehicle theft, burglary, and arson. Property crimes have risen slightly with a 2.3 percent increase between 2000 and 2001. Larceny, or simple theft, is the most common property crime, accounting for over half of all property arrests. The average dollar value lost per larceny incident is $730. Examples of larcenies include purse-snatching, theft of a bicycle, pick-pocketing, theft from a coin-operated machine, and shoplifting.

Larcenies involving automobiles and auto accessories are the largest single category of thefts. However, because of the cost involved, motor vehicle theft is considered a separate index offense. Numbering over 1 million in 2001, motor vehicle thefts have increased 5.7 percent since 2000. Because of insurance requirements, vehicle theft is one of the most highly reported index crimes and, consequently, estimates between the UCR and the NCVS are fairly compatible. Less than 14 percent of motor vehicle thefts are cleared.

Burglary, which is the second most common index offense, entails entering a structure, usually a house, with the intent to commit a crime while inside. Official statistics indicate that in 2001, over 2.1 million burglaries occurred, a rate of 741 per 100,000 population. Most burglaries are residential rather than commercial and take place during the day when houses are unoccupied. The most common type of burglary is forcible entry, followed by unlawful entry.

Arson involves the malicious burning of the property of another. Estimating the frequency and nature of arson is difficult given the legal requirement of "maliciousness." Of the reported cases of arson, 42.2 percent involved structures (the majority of which were residential) and 32.5 percent involved movable property (e.g., boat, car), the remainder being miscellaneous (e.g., crops, timber) property (FBI 2002).

Vice Crimes

Vice crimes are illegal activities that have no complaining party and are, therefore, often called **victimless crimes.** Vice crimes include using illegal drugs, engaging in or soliciting prostitution (except for legalized prostitution, which exists in Nevada), illegal gambling, and pornography.

Most Americans view drug use as socially disruptive (see Chapter 3). Less consensus exists, nationally or internationally, that gambling and prostitution are problematic. For example, in the Netherlands, prostitution is legal. Although a country of only 16 million, the Netherlands has an estimated 25,000 to 50,000 sex workers. As other workers, sex workers in the Netherlands have access to the social security system and pay income tax. It is, however, illegal for anyone under the age of 16 to visit a prostitute, and brothels, although openly advertised, are also illegal (*The Situation* 2003).

In the United States, many states have legalized gambling, including casinos in Nevada, New Jersey, Connecticut, and others, as well as state lotteries, bingo parlors, horse and dog racing, and jai alai. Further, some have argued that there is little difference, other than societal definitions of acceptable and unacceptable behavior, between gambling and other risky ventures such as investing in the stock market. Conflict theorists are quick to note that the difference is who is making the wager.

Organized crime refers to criminal activity conducted by members of a hierarchically arranged structure devoted primarily to making money through illegal means. Although often discussed under victimless crimes because of its association with prostitution, drugs, and gambling, organized crime often uses coercive techniques. For example, organized crime groups may force legitimate businesses to pay "protection money" by threatening vandalism or violence.

The traditional notion of organized crime is the Mafia—a national band of interlocked Italian families—but members of many ethnic groups engage in organized crime.

> Chinese, Vietnamese, Korean, and Japanese gangs have been found on the East and West coasts, active in smuggling drugs and extorting money from businesses. . . . Scores of other groups can be found in various cities: Cubans running illegal gambling operations in Miami, Canadians engaging in gun smuggling and money laundering . . . and Russians carrying out extortion and contract murders in New York. (Thio 2004, 395)

Organized crime also occurs at the international level. One of the largest organized crime groups in the world is the Yakuza of Japan. It is made up of 10 crime families, the largest one composed of over 750 separate gangs each with 25 to 30 members. The young men who join the Yakuza tend to be from the lower class and must undergo a training period of five years. It is during this apprenticeship period that members learn absolute loyalty to their superiors as well as the other norms and values of the group. The Yakuza are involved in drugs, gambling, and prostitution as well as several legitimate businesses. Interestingly, the Yakuza are legal in Japan and proudly display their name at their office headquarters while recruits wear lapel pins identifying themselves as members (Thio 2004).

White-Collar Crime

White-collar crime includes both occupational crime, in which individuals commit crimes in the course of their employment, and corporate crime, in which corporations violate the law in the interest of maximizing profit. Occupational crime is motivated by individual gain. Employee theft of merchandise, or pilferage, is one of the most common types of occupational crime. Other examples include embezzlement, forgery and counterfeiting, and insurance fraud. Price fixing, antitrust violations, and security fraud are all examples of corporate crime, that is, crime that benefits the organization. In recent years several officers of major corporations, including Enron, WorldCom, Adelphia, and Imclone, have been charged with securities fraud, tax invasion, and insider trading (Eisenberg 2002). WorldCom engaged

Residents of Love Canal were victimized by corporate violence when toxic waste, dumped by Hooker Chemical Company, began to seep into the basements of homes and schools. Many residents of Love Canal moved out of the area; others complained of high rates of miscarriages, birth defects, and cancer. Although Love Canal was allegedly cleaned up, residents protested against resettling the area. After 20 years of litigation, the last suit was settled in 1998 with the City of Love Canal receiving $250,000 and the construction of a park near the spill.

in what has been called the "largest accounting fraud in history," exaggerating its worth by $9 billion. Shareholders lost over $3 billion, and 17,000 employees are in danger of losing their jobs (Ripley 2003). In part as a response to the widespread scandals of late, the U.S. Sentencing Commission has approved stiffer penalties for white-collar criminals, including a new sentencing formula (Lichtblau 2003).

Corporate violence, another form of corporate crime, refers to the production of unsafe products and the failure of corporations to provide a safe working environment for their employees. Corporate violence is the result of negligence, the pursuit of profit at any cost, and intentional violations of health, safety, and environmental regulations. For example, after more than a year of recalls in 16 countries, in August of 2000 Bridgestone/Firestone began a U.S. recall of over 6.5 million tires. The tires, many of which were standard equipment on the popular Ford Explorer, had a 10-year history of tread separation. It was only after 88 U.S. traffic deaths were linked to the defective tires, prompting a congressional investigation, that Ford and Bridgestone/Firestone acknowledged the overseas recalls and the tires' questionable safety history (Pickler 2000). More recently, Ford Motor Company has been asked by a federal judge to turn over data on their 15-passenger van. Several deaths have occurred as a result of the van rolling over and Ford is accused of hiding evidence of the problem. According to the National Highway Traffic Safety Administration, over 400 people have died in passenger van accidents since 1990 (*Chicago Tribune* 2003). Table 4.3 summarizes some of the major categories of white-collar crime.

Computer Crime

Computer crime refers to any violation of the law in which a computer is the target or means of criminal activity. It is one of the fastest growing crimes in the United States, costing an estimated $226 million dollars in 2000—over twice the previous year's estimates. Hacking, or unauthorized computer intrusion, is one type of com-

Table 4.3
Types of White-Collar Crime

Crimes Against Consumers	Crimes Against Employees
Deceptive advertising	Health and safety violations
Antitrust violations	Wage and hour violations
Dangerous products	Discriminatory hiring practices
Manufacturer kickbacks	Illegal labor practices
Physician insurance fraud	Unlawful surveillance practices
Crimes Against the Public	**Crimes Against Employers**
Toxic waste disposal	Embezzlement
Pollution violations	Pilferage
Tax fraud	Misappropriation of government funds
Security violations	Counterfeit production of goods
Police brutality	Business credit fraud

puter crime. In just one month, hackers successfully attacked the computer systems of Walt Disney World, Yahoo, eBay, and Amazon.com through "denial of service" invasions (Kong & Swartz 2000).

The increase in computer break-ins has also led to an increase in identity theft—the use of someone else's identification (e.g., social security number, birth date) to obtain credit. In 2002, the number of identity thefts doubled from the previous year becoming the most frequent complaint to the Federal Trade Commission (Lee 2003). Although mail theft is one of the most common modes of obtaining the needed information, new technologies have contributed to the increased rate of this offense. For example, weekly, thousands of stolen credit card numbers are sold online in "membership only cyberbazaars, operated largely by residents of the former Soviet Union who have become central players in credit card and identity theft" (Richtel 2002, 1).

Conklin (1998) and Reid (2003) have identified other examples of computer crime.

- Two individuals were charged with theft of 80,000 cellular phone numbers. Using a device purchased from a catalogue, the thieves picked up radio waves from passing cars, determined private cellular codes, reprogrammed computer chips with the stolen codes, and then, by inserting the new chips into their own cellular phones, charged calls to the original owners.
- A programmer made $300 a week by programming a computer to round off each employee's paycheck down to the nearest 10¢ and then to deposit the extra few pennies in the offender's account.
- A computer hacker broke into a telephone system and rigged the outcome of a radio station contest. Three hackers won a trip to Hawaii, a Porsche, and a cash prize.
- An oil company illegally tapped into another oil company's computer to get information that allowed the offending company to underbid the other company for leasing rights.

Juvenile Delinquency

In general, children under the age of 18 are handled by the juvenile court either as status offenders or as delinquent offenders. A status offense is a violation that can only be committed by a juvenile, such as running away from home, truancy, and underage drinking. A delinquent offense is an offense that would be a crime if committed by an adult, such as the eight index offenses. The most common status offenses handled in juvenile court are underage drinking, truancy, and running away. In 2001, 16.7 percent of all arrests (excluding traffic violations) were of offenders under the age of 18, and 5.4 percent were under the age of 15. Those under 18 accounted for 26 percent of all arrests for index crimes (FBI 2002).

As is the case with adults, juveniles commit more property than violent offenses and the number of violent offenses has dropped in recent years. Nonetheless, Americans are concerned about the high rate of juvenile violence including violence in schools (see Chapter 12) and gang-related violence. Gang-related crime is, in part, a function of two interrelated social forces: the increased availability of guns in the 1980s and the lucrative and expanding drug trade. It is estimated that the United States has 28,700 street gangs with over 780,200 members, three-quarters of whom are racial minorities (Bartollas 2003).

Demographic Patterns of Crime

Although virtually everyone violates a law at some time, persons with certain demographic characteristics are disproportionately represented in the crime statistics. Victims, for example, are disproportionately young, lower-class, minority males from urban areas. Similarly, the probability of being an offender varies by gender, age, race, social class, and region (see Figure 4-2).

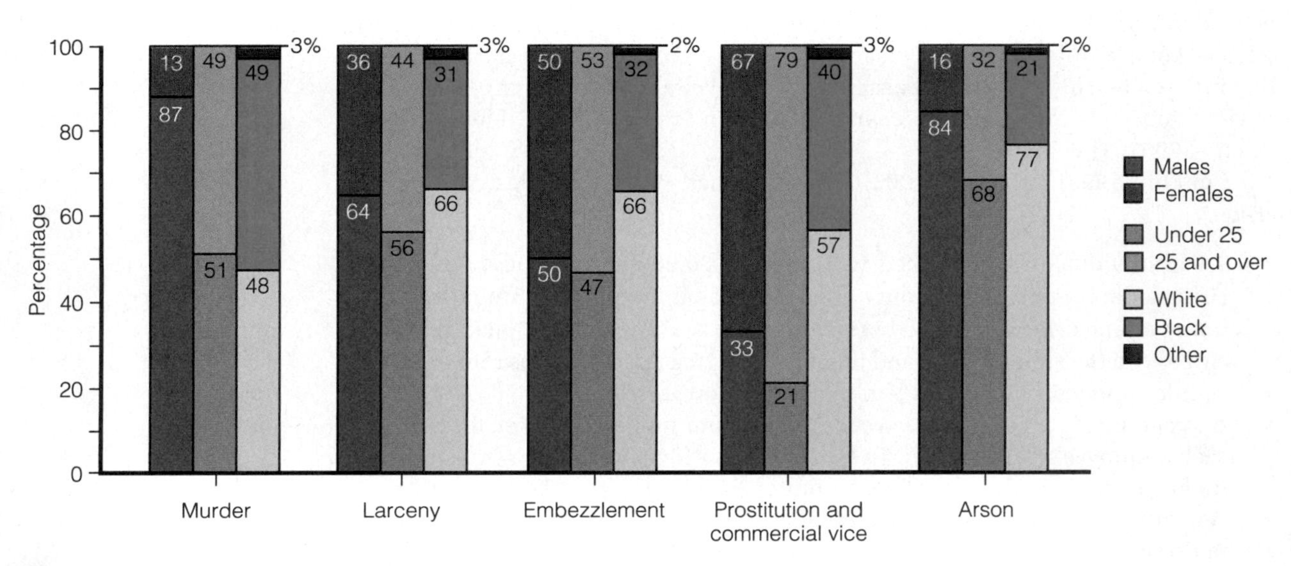

Figure 4.2
Percentage of Arrests by Sex, Age, and Race, 2001
Source: U.S. Department of Justice. 2002. *Crime in the United States, 2001.* Washington, D.C.

© A. Ramey/PhotoEdit

Females who join gangs often do so to win approval from their boyfriends who are gang members. Increasingly, however, females are forming independent "girl gangs." The most common type of female gang member remains, however, a female auxiliary to a male gang.

Gender and Crime

Both official statistics and self-report data indicate that females commit fewer violent crimes than males (see Figure 4-2). Why are females less likely to commit violent crimes than males? Some would argue that "girls are less violent than boys because they are controlled through subtle mechanisms, which include[d] learning that violence is incompatible with the meaning of their gender" (Heimer & DeCoster 1999, 305–306).

In 2001, males accounted for 83 percent of all arrests for violent crimes (FBI 2002). Females are not only less likely than males to commit serious crimes, but the monetary value of female involvement in theft, property damage, and illegal drugs is typically less than that for similar offenses committed by males. Nevertheless, a growing number of women have become involved in characteristically male criminal activities such as gang-related crime and drug use.

The recent increase in crimes committed by females has led to the development of a feminist criminology. Feminist criminology focuses on how the subordinate position of women in society affects their criminal behavior. For example, Chesney-Lind and Shelden (2004) report that arrest rates for runaway juvenile females are higher than for males not only because they are more likely to run away as a consequence of sexual abuse in the home, but also because police with paternalistic attitudes are more likely to arrest female runaways than male runaways. Feminist criminology thus adds insights into understanding crime and violence often neglected by traditional theories by concentrating on gender inequality in society.

The subordinate position of women in the United States also impacts their victimization rates. A recent report from the Harvard School of Public Health reveals that 70 percent of all female homicide victims in industrial countries are American (HSPH 2002). A female in the United States is five times more likely to be murdered than a female in Germany, eight times more likely to be murdered than a female in England, and three times more likely to be murdered than a female in Canada. When she is murdered, it is most likely by an ex-boyfriend, husband, or other intimate.

Age and Crime

In general, criminal activity is more prevalent among younger persons than older persons. The highest arrest rates are for individuals under the age of 25. Crimes committed by people in their teens or early twenties tend to be property crimes like burglary, larceny, arson, and vandalism. In 2002, 58 percent of those arrested for property index offenses were under the age of 25 (FBI 2002). The median age of people who commit more serious crimes such as aggravated assault and homicide is in the late twenties.

Why is criminal activity more prevalent among individuals in their teens and early twenties? One reason is that juveniles are insulated from many of the legal penalties for criminal behavior. Younger individuals are also more likely to be unemployed or employed in low-wage jobs. Thus, as strain theorists argue, they have less access to legitimate means for acquiring material goods.

Some research suggests, however, that high school students who have jobs become more, rather than less, involved in crime (Felson 2002). In earlier generations, teenagers who worked did so to support themselves and/or their families. Today, teenagers who work typically spend their earnings on recreation and "extras," including car payments and gasoline. The increased mobility associated with having a vehicle also increases opportunity for criminal behavior and reduces parental control.

Race, Social Class, and Crime

Race is a factor in who gets arrested. Minorities are disproportionately represented in official statistics. For example, although African-Americans represent about 13 percent of the population, they account for more than 38 percent of all violent index offenses, 31 percent of all property index offenses, and 33 percent of the crime index total (FBI 2002). Further, black males between the ages of 20 and 39 make up about a third of the prison population (BJS 2003a).

Nevertheless, it is inaccurate to conclude that race and crime are causally related. First, official statistics reflect the behaviors and policies of criminal justice actors. Thus, the high rate of arrests, conviction, and incarceration of minorities may be a consequence of individual and institutional bias against minorities (see this chapter's *Social Problems Research Up Close* feature). For example, blacks are sent to prison for drug offenses at a rate 8.2 times higher than the rate for whites. If present trends continue, by 2020 two out of every three black men between the ages of 18 and 34 will be in prison (Fletcher 2000; Dickerson 2000). These disturbing statistics have led to concerns over **racial profiling**—the practice of targeting suspects based on race status. Proponents of the practice argue that because race, as gender, is a significant predictor of who commits crime, the practice should be allowed. Opponents hold that racial profiling is little more than discrimination and should therefore be abolished.

Second, race and social class are closely related in that nonwhites are overrepresented in the lower classes. Because lower-class members lack legitimate means to acquire material goods, they may turn to instrumental, or economically motivated, crimes. Further, while the "haves" typically earn social respect through their socioeconomic status, educational achievement, and occupational role, the "have-nots" more often live in communities where respect is based on physical strength and violence, as subcultural theorists argue.

Thus the apparent relationship between race and crime may, in part, be a consequence of the relationship between these variables and social class. Philips

Social Problems Research Up Close

Race and Ethnicity in Sentencing Outcomes

Certainly one of the more pressing issues of recent times is the question of whether legally irrelevant information (e.g., race, sex of offender) affects criminal justice outcomes. Given recent accounts of alleged police brutality, accusations of racial profiling, and concern over race, class, and ethnic bias in death penalty cases, Steffensmeier and Demuth's (2000) investigation of the impact of extralegal variables in federal court sentencing is particularly timely.

Sample and Methods

Steffensmeier and Demuth (2000) investigate whether the criminal justice system discriminates on the basis of race and ethnicity. Although considerable research supports the contention that minorities compared with other members of society are more likely to receive harsher sentences (having fewer resources, perceived as "dangerous," threatening to the status quo, etc.), Steffensmeier and Demuth's (2000) analysis examines both inter- and intragroup variations. The researchers argue that court decisions take place within a powerful cultural context—a "perceptual shorthand" of sorts—where the definitions of an offender's blameworthiness, the need to protect society, and the practical limitations of the sentence imposed affect the decision-making process. Given that federal sentencing guidelines allow for some discretion and that "Hispanic defendants may seem even more culturally dissimilar and be even more disadvantaged than their black counterparts" (Steffensmeier & Demuth 2000, 709), the researchers hypothesize that (1) Hispanics will receive more severe sentences than whites or blacks, (2) in drug cases, Hispanics specifically, but minorities in general, will receive harsher sentences than whites, and (3) Hispanics identified as black will receive harsher sentences than Hispanics identified as white. The dependent variable, severity of sentence, was measured by whether the court imposed a prison sentence and, if so, the length of the sentence. The independent variables are race and ethnicity. The data include all convictions of U.S. citizens in federal courts between 1993 and 1996 (N = 89,637).

Findings and Conclusions

The authors note the "considerable consistency" of sentencing of federal criminal defendants across racial and ethnic categories for *similar cases.*

> Judges, on balance, prescribe similar sentences for similar defendants convicted of the same offense. Whether they were white, black, or Hispanic—defendants who committed more serious crimes, had more extensive criminal histories, or were convicted at trial (as opposed to a guilty plea), were much more likely to be incarcerated and receive longer prison sentences (p. 724).

They also note, however, small to moderate effects of race and ethnicity in terms of the *overall* sentencing, with Hispanics receiving harsher sentences than whites and blacks. These differences are likely the result of what is called "substantial assistance departures." When a defendant, for example, "substantially assists" the State in the prosecution or investigation of another offender, the judge may "depart" from the sentencing guidelines. Because Hispanics may be more likely to be involved in drug trafficking circles who seek retribution against those who cooperate with authorities, they may be less likely to provide "substantial assistance" and thus are less likely to have their sentences reduced.

Finally, the researchers conclude that their results "provide strong evidence for the continuing significance of race and ethnicity in the larger society in general and in organizational decision-making processes in particular" (p. 726). While acknowledging that the results are less an indication of discrimination than the unintended consequence of seemingly neutral policies, the authors call for more research in the area of race and ethnic relations.

Source: Steffensmeier, Darryl, and Stephen Demuth. 2000. "Ethnicity and Sentencing Outcomes in U.S. Federal Courts: Who Is Punished More Harshly?" *American Sociological Review* 65:705–729.

(2002, 367), in her investigation of white, black, and Latino homicide rates, concludes that:

> it nonetheless remains clear from this study that a significant portion of the racial homicide differential could be reduced by improving socioeconomic conditions for minority populations. This conclusion provides some promising policy options. For example, improving levels of education, lowering levels of poverty, and reducing the extent of male

> unemployment among minority populations might well have an impact on levels of violence and reduce the striking racial homicide differential that currently exists in the United States.

Some research indicates, however, that even when social class backgrounds of blacks and whites are comparable, blacks have higher rates of criminality (D'Alessio & Stolzenberg 2003). Further, to avoid the bias inherent in official statistics, researchers have compared race, class, and criminality by examining self-report data and victim studies. Their findings indicate that although racial and class differences in criminal offenses exist, the differences are not as great as official data would indicate.

Region and Crime

In general, crime rates, and particularly violent crime rates, are higher in urban than suburban areas, and higher in suburban areas than rural areas. Higher crime rates in urban areas result from several factors. First, social control is a function of small intimate groups socializing their members to engage in law-abiding behavior, expressing approval for their doing so and disapproval for their noncompliance. In large urban areas, people are less likely to know each other and thus are not influenced by the approval or disapproval of strangers. Demographic factors also explain why crime rates are higher in urban areas: cities have large concentrations of poor, unemployed, and minority individuals.

Crime rates also vary by region of the country. In 2001, both violent and property crimes were highest in southern states followed by western, mid-western, and northeastern states. The murder rate is particularly high in the South, with 43 percent of all murders recorded in southern states (FBI 2002). The high rate of southern lethal violence has been linked to high rates of poverty and minority populations in the South, a southern "subculture of violence," higher rates of gun ownership, and a warmer climate that facilitates victimization by increasing the frequency of social interaction.

Costs of Crime and Violence

Crime and violence often result in physical injury and loss of life. In 1997, the most recent year in which data are available, one person out of every 240 people in the United States was likely to be a victim of murder in his or her lifetime (FBI 2000). Not surprisingly, in a Gallup survey Americans ranked crime and violence the third most pressing issue in the country preceded only by education, and ethics and morals (Gallup Poll 2000b). Moreover, the U.S. Public Health Service now defines "violence" as one of the top health concerns facing Americans. In addition to death, physical injury, and loss of property, crime also has economic, social, and psychological costs.

Economic Costs of Crime and Violence

Conklin (1998, 71–72) suggests that the financial costs of crime can be classified into at least six categories. First are direct losses from crime such as the destruction of buildings through arson, of private property through vandalism, and of the environment by polluters. In 2001, the average dollar loss of destroyed or damaged property due to arson was $11,098 (FBI 2002). Second are costs associated with the transferring of property. Bank robbers, car thieves, and embezzlers have all taken property from its rightful owner at tremendous expense to the victim and society. For example, in 2001 it is estimated that $8 billion was lost as a result of motor vehicle theft; the average value per vehicle at the time of the theft was $6,646 (FBI 2002).

A third major cost of crime is that associated with criminal violence, for example, the medical cost of treating crime victims—$5 billion annually—or the loss of productivity of injured workers (Surgeon General 2002). Fourth are the costs associated with the production and sale of illegal goods and services, that is, illegal expenditures. The expenditure of money on drugs, gambling, and prostitution diverts funds away from the legitimate economy and enterprises and lowers property values in high-crime neighborhoods. Fifth is the cost of prevention and protection, the billions of dollars spent on locks and safes, surveillance cameras, guard dogs, and the like. It is estimated that Americans spend $65 billion annually on self-protection items (Surgeon General 2002).

Finally, there is the cost of the criminal justice system—law enforcement, litigative and judicial activities, corrections, and victims' assistance. The cost of the criminal justice system is estimated to be $90 billion annually and growing (Surgeon General 2002). Reasons for such growth include increases in the rates of arrest and conviction, changes in the sentencing structure, public attitudes toward criminals, the growing number of young males, and the war on drugs. Regardless of the cause, however, the staggering cost of public institutions has led to the "privatization" of prisons whereby the private sector increasingly supplies needed prison services.

What is the total economic cost of crime? One estimate suggests that the total cost of crime and violence in the United States is more than $1.7 trillion a year (Anderson 1999). Although costs from "street crimes" are staggering, the costs from "crimes in the suites" such as tax evasion, fraud, false advertising, and antitrust violations are greater than the cost of the FBI index crimes combined (Reiman 2001).

Social and Psychological Costs of Crime and Violence

Crime and violence entail social and psychological as well as economic costs. When asked, in general, "How safe do you feel in each of these locations?" 74 percent of respondents reported feeling "very safe" in their homes, 54 percent "very safe" walking in their neighborhoods after dark, and 33 percent "very safe" when downtown at night (Harris Poll 2002). Fear of crime may be fueled by media presentations that may not accurately reflect the crime picture. For example, in a content analysis of local television portrayals of crime victims and offenders, whites were overrepresented as victims and blacks as offenders when compared with official statistics (Dixon & Linz 2000).

Fear of crime and violence also affects community life.

> If frightened citizens remain locked in their homes instead of enjoying public spaces, there is a loss of public and community life, as well as a loss of "social capital"—the family and neighborhood channels that transmit positive social values from one generation to the next. (National Research Council 1994, 5–6)

This is particularly true of women and the elderly who restrict their activities, living "limited lives," as a consequence of fear (Madriz 2000).

White-collar crimes also take a social and psychological toll at both the individual and the societal level. Rosoff, Pontell, and Tillman (2002, 346) state that

"The possibility of being a victim of a crime is ever present on my mind. . . . thinking about it is as natural as . . . breathing."

40-year-old woman in New York City

white-collar crime can produce "feelings of cynicism among the public, remove an essential element of trust from everyday interaction, delegitimatize political institutions, and weaken respect for the law." Further, the authors argue, white-collar crime "encourages and facilitates" other types of crime, that is, "there is a connection, both direct and indirect, between 'crime in the suites' and 'crime in the streets.'"

Strategies for Action: Responding to Crime and Violence

> "The eight blunders that lead to violence in society:
> — Wealth without work
> — Pleasure without conscience
> — Knowledge without character
> — Commerce without morality
> — Science without humanity
> — Worship without sacrifice
> — Politics without principle
> — Rights without responsibilities"
>
> **Mohandas K. Gandhi**
> **Indian nationalist leader and peace activist**

In addition to economic policies designed to reduce unemployment and poverty, numerous social policies and programs have been initiated to alleviate the problem of crime and violence. These policies and programs include local initiatives, criminal justice policies, and legislative action.

Local Initiatives

Youth Programs. Early intervention programs acknowledge that it is better to prevent crime than to "cure" it once it has occurred. Preschool enrichment programs, such as the Perry Preschool Project, have been successful in reducing rates of aggression in young children. After random assignment of children to either a control or experimental group, experimental group members received academically oriented interventions for one to two years, frequent home visits, and weekly parent–teacher conferences. When control and experimental groups were compared, the experimental group had better grades, higher rates of high school graduation, lower rates of unemployment, and fewer arrests (Murray, Guerra, & Williams 1997).

Recognizing the link between juvenile delinquency and adult criminality, many anti-crime programs are directed toward at-risk youths. These prevention strategies, including youth programs such as Boys and Girls Clubs, are designed to keep young people "off the streets," provide a safe and supportive environment, and offer activities that promote skill development and self-esteem. According to Gest and Friedman (1994), housing projects with such clubs report 13 percent fewer juvenile crimes and a 25 percent decrease in the use of crack.

Finally, many youth programs are designed to engage juveniles in noncriminal activities and integrate them into the community. In Weed and Seed, a program under the Department of Justice, "law enforcement agencies and prosecutors cooperate in 'weeding out' criminals who participate in violent crime and drug abuse . . . and 'seeding' human services to the area, encompassing prevention, intervention, treatment and neighborhood revitalization" (Weed and Seed 2003, 1). As part of the program, Safe Havens are established in, for example, schools where multiagency services are provided for youth.

Community Programs. Neighborhood watch programs involve local residents in crime-prevention strategies. For example, MAD DADS patrol the streets in high-crime areas of the city on weekend nights, providing positive adult role models for troubled children. Members also report crime and drug sales to police, paint over gang graffiti, organize gun buy-back programs, and counsel incarcerated fathers (MAD DADS 2003). In 2002, 9,850 communities from 50 states participated in "National Night Out," a crime prevention event in which citizens, businesses, neighborhood organizations, and local officials joined together in outdoor activities to heighten awareness of neighborhood problems, promote anti-crime messages, and strengthen community ties (NNO 2003).

Mediation and victim-offender dispute resolution programs are also increasing, with over 300 such programs in the United States and Canada. The growth of these programs is a reflection of their success rate: two-thirds of cases referred result in face-to-face meetings, over 95 percent of these cases result in a written restitution agreement, and 90 percent of the written restitution agreements are completed within one year (VORP 2003).

Local initiatives also include ground-up movements to change or initiate a law. In California, for example, there is a "corporate three strikes" campaign (*Three Strikes* 2003, 1). The proposed law

> requires that the State of California take legal action to put out of business in California any corporation that has had "major violations" in a ten year period. A major violation is an intentional or grossly negligent violation of existing law that results in death, or in a fine, damages or settlement of over a million dollars.

Criminal Justice Policy

The criminal justice system is based on the principle of **deterrence,** that is, the use of harm or the threat of harm to prevent unwanted behaviors. It assumes that people rationally choose to commit crime, weighing the rewards and consequences of their actions. Thus, the recent emphasis on "get tough" measures holds that maximizing punishment will increase deterrence and cause crime rates to decrease. Research indicates, however, that the effectiveness of deterrence is a function of not only the severity of the punishment but also the certainty and swiftness of the punishment as well. Further, "get tough" policies create other criminal justice problems, including overcrowded prisons and, consequently, the need for plea bargaining and early release programs.

Law Enforcement Agencies. In 2002, the United States had 659,104 law enforcement officers with an average of 2.5 officers for every 1,000 people (FBI 2002). Ironically, despite recent increases in the number of law enforcement personnel, public opinion that "more police on the streets" will reduce violent crime has decreased (Pew Research Center 2000). Further, accusations of racial profiling, police brutality, and discriminatory arrest practices have shaken public confidence in the police. When a national sample of U.S. adults were asked, "Do you think that police brutality against blacks and Hispanics in your community happens often, occasionally, or never," over 60 percent said often or occasionally (Harris Poll 2002).

In response to such trends, the Crime Control Act of 1994 established the Office of Community Oriented Policing Services (COPS). Community-oriented policing involves collaborative efforts among the police, the citizens of a community, and local leaders. As part of community policing efforts, officers speak to citizen groups, consult with social agencies, and enlist the aid of corporate and political leaders in the fight against neighborhood crime (COPS 2003).

Officers using community policing techniques often employ "practical approaches" to crime intervention. Such solutions may include what Felson (2002) calls "situational crime prevention." Felson argues that much of crime could be prevented simply by minimizing the opportunity for its occurrence. For example, cars could be outfitted with unbreakable glass, flush-sill lock buttons, an audible reminder to remove keys, and a high-security lock for steering columns (Felson 2002).

These techniques, and community-oriented policing in general, have been fairly successful. After COPS was implemented in New York City and New Orleans, homicide rates dropped dramatically—40 percent and 18 percent, respectively (Sileo 2000).

Rehabilitation Versus Incapacitation. An important debate concerns the primary purpose of the criminal justice system: Is it to rehabilitate offenders or to incapacitate them through incarceration? Both **rehabilitation** and **incapacitation** are concerned with recidivism rates, or the extent to which criminals commit another crime. Advocates of rehabilitation believe recidivism can be reduced by changing the criminal, whereas proponents of incapacitation think it can best be reduced by placing the offender in prison so that he or she is unable to commit further crimes against the general public.

Societal fear of crime has led to a public emphasis on incapacitation and a demand for tougher mandatory sentences, a reduction in the use of probation and parole, support of a "three strikes and you're out" policy, and truth-in-sentencing laws. However, these tough measures have recently come under attack for two reasons. First is research that indicates incarceration may not deter crime. For example, a study by the Justice Department reports that 67 percent of inmates reoffend within three years of being released (Butterfield 2002a). Second is the accusation that get-tough measures, like California's "three strikes and you're out" policy, are patently unfair. Leandro Andrade stole $153 worth of children's videos from K-Mart and was sentenced to 50 years to life in prison. Is that equitable? As unjust as it sounds, those in favor of the policy would be quick to note that the man had prior convictions for burglary and shoplifting. Andrade appealed his sentence to the U.S. Supreme Court, which held that California's three strikes law did not violate the constitutional ban on grossly disproportionate sentences (*Major Rulings* 2003).

Although incapacitation is clearly enhanced by longer prison sentences, deterrence, as discussed previously, and rehabilitation may not be. Rehabilitation assumes that criminal behavior is caused by sociological, psychological, and/or biological forces rather than being solely a product of free will. If such forces can be identified, the necessary change can be instituted. Rehabilitation programs include education and job training, individual and group therapy, substance abuse counseling, and behavior modification. *Time for Freedom,* a residential facility in Florida, provides ex-inmates with jobs and a needed support system lacking in most prison release programs (Sealey 2002).

Corrections. While the debate between rehabilitation and incarceration continues, over 2 million inmates are housed in the nation's federal, state, and local institutions and their numbers continue to grow (BJS 2003a). Since 1970 the prison population has increased 500 percent (Butterfield 2002a). Such increases have usually been met with building more prisons, but recent concerns with budget deficits have made many states "close prisons, lay off guards, and consider shortening sentences" (Butterfield 2002a). Washington State, for example, is debating whether to limit the time served for those who commit nonviolent crimes, and Ohio, Michigan, and Illinois have each closed several correctional facilities in an effort to save money. Other money-saving techniques include use of "the box," a four-pound electronic device that uses the Global Positioning System (GPS) to track the location of parolees or probationers 24 hours a day. Presently, 27 states use some type of satellite surveillance system (Lee 2002).

© Billy E. Barnes/PhotoEdit

According to the U.S. Bureau of Justice Statistics, there were 1,264,000 state inmates and 161,000 federal inmates as of June 30, 2002. Typically, prisons confine persons convicted of felonies while jails confine people who have committed misdemeanors and have sentences of one year or less.

Capital Punishment. With capital punishment, the State (the federal government or a state) takes the life of a person as punishment for a crime. Thirty-eight states have capital punishment. In 2001, 66 executions took place in 15 states, with each prisoner spending, on average, 12 years on death row. There are over 3,500 inmates on death row, the youngest 19 and the oldest 86 (BJS 2003b). Over 100 countries have banned state executions including Australia, Canada, Italy, Denmark, and Ireland. In 2002, the United States, China, Iran, and Saudi Arabia were responsible for 80 percent of the world's executions (Amnesty International 2002). In this chapter's *The Human Side* feature, Robert Johnson, a professor at American University, describes his reaction to being witness to an execution.

"I feel morally and intellectually obligated to concede that the death penalty experiment has failed."

Justice Harry Blackmun
U.S. Supreme Court Justice

Proponents of capital punishment argue that executions of convicted murderers are necessary to convey public disapproval and intolerance for such heinous crimes. Those against capital punishment believe that no one, including the State, has the right to take another person's life and that putting convicted murderers behind bars for life is a "social death" that conveys the necessary societal disapproval.

Proponents of capital punishment also argue that it deters individuals from committing murder. Critics of capital punishment hold, however, that because most homicides are situational and are not planned, offenders do not consider the consequences of their actions before they commit the offense. Critics also point out that the United States has a much higher murder rate than Western European nations that do not practice capital punishment, and that death sentences are racially discriminatory. For example, a study of Maryland's death penalty practices found that "blacks who kill whites are significantly more likely to face the death penalty . . . than are blacks who kill blacks, or white killers" (Liptak 2003, 1).

Capital punishment advocates suggest that executing a convicted murderer relieves the taxpayer of the costs involved in housing, feeding, guarding, and providing medical care for inmates. Opponents of capital punishment argue that the principles that decide life and death issues should not be determined by financial considerations. In addition, taking care of convicted murderers for life may actually

The Human Side

Witness to an Execution

At 10:58 the prisoner entered the death chamber. He was, I knew from my research, a man with a checkered, tragic past. He had been grossly abused as a child, and went on to become grossly abusive of others. I was told he could not describe his life, from childhood on, without talking about confrontations in defense of a precarious sense of self—at home, in school, on the streets, in the prison yard. Belittled by life and choking with rage, he was hungry to be noticed. Paradoxically, he had found his moment in the spotlight. . . .

En route to the chair, the prisoner stumbled slightly, as if the momentum of the event had overtaken him. Were he not held securely by two officers, one at each elbow, he might have fallen. . . . Once the prisoner was seated, again with help, the officers strapped him into the chair.

Arms, legs, stomach, chest, and head were secured in a matter of seconds. Electrodes were attached to the cap holding his head and to the strap holding his exposed right leg. A leather mask was placed over his face. The last officer mopped the prisoner's brow, then touched his hand in a gesture of farewell. . . .

The strapped and masked figure sat before us, utterly alone, waiting to be killed . . . waiting for a blast of electricity that would extinguish his life. Endless seconds passed. His last act was to swallow, nervously, pathetically, with his Adam's apple bobbing. I was struck by the simple movement then, and can't forget it even now. It told me, as nothing else did, that in the prisoner's restrained body, behind that mask, lurked a fellow human being who, at some level, however primitive, knew or sensed himself to be moments from death.

. . . Finally, the electricity hit him. His body stiffened spasmodically, though only briefly. A thin swirl of smoke trailed away from his head and then dissipated quickly. The body remained taut, with the right foot raised slightly at the heel, seemingly frozen there. A brief pause, then another minute of shock. When it was over, the body was flaccid and inert.

Three minutes passed while the officials let the body cool. (Immediately after the execution, I'm told, the body would be too hot to touch and would blister anyone who did.) All eyes were riveted to the chair; I felt trapped in my witness seat, at once transfixed and yet eager for release. I can't recall any clear thoughts from that moment. One of the death watch officers later volunteered that he shared this experience of staring blankly at the execution scene. Had the prisoner's mind been mercifully blank before the end? I hope so.

The physician listened for a heartbeat. Hearing none, he turned to the warden and said, "This man has expired." The warden, speaking to the Director, solemnly intoned: "Mr. Director, the court order has been fulfilled." . . .

Source: Robert Johnson. 1989. " 'This Man Has Expired': Witness to an Execution." *Commonwealth,* January 13, 9–15. Copyright by the Commonwealth Foundation. Reprinted with permission.

be less costly than sentencing them to death, because of the lengthy and costly appeal process for capital punishment cases (Garey 1985; Myths 1998).

Nevertheless, those in favor of capital punishment argue that it protects society by preventing convicted individuals from committing another crime, including the murder of another inmate or prison official. One study of the deterrent effect of capital punishment concludes that each execution is associated with eight fewer homicides and perhaps even more (Rubin 2002). Opponents contend, however, that capital punishment may result in innocent people being sentenced to death. Since 1973, 102 death row inmates have been exonerated (*Exonerations* 2003). Further, a report by the Justice Project entitled "A Broken System" found reversible errors in two of every three death penalty cases reviewed over the study period (Herbert 2002). Stating that the system was riddled with error, in 2002 the governor of Illinois commuted 167 death penalty cases to life in prison or less (Wilgoren 2003; Napolitano 2004).

Legislative Action

Gun Control. Fueled by the Columbine school shooting and other recent images of children as both gun victims and offenders, in May 2000, tens of thousands of women descended on Washington, D.C., demanding that something be done about gun violence. Although the impact of the Million Mom March is still unknown, most Americans, and particularly women, support some restriction on handguns. When a national sample of U.S. adults were asked, "In general, would you say you favor stricter gun control, or less strict gun control," 66 percent of the respondents reported that they wanted stricter gun control (Harris Poll 2001).

Those against gun control argue that not only do citizens have a constitutional right to own guns, but that more guns may actually lead to less crime as would-be offenders retreat in self-defense (Lott 2003). Advocates of gun control, however, insist that the 200 million privately owned guns in this country, 70 million of which are handguns (Albanese 2000), significantly contribute to the violent crime rate in the United States as well as distinguishing it from other industrialized nations.

After a seven-year battle with the National Rifle Association (NRA), gun control advocates achieved a small victory in 1993 when Congress passed the **Brady Bill.** This law requires a five-day waiting period on handgun purchases so that sellers can screen buyers for criminal records or mental instability. Today, the law requires background checks of not just handgun users but also for those who purchase rifles and shotguns (Reid 2003). Presently, there is a bill before the U.S. Senate which, if passed, would give gun manufacturers immunity from any ongoing or future civil suits (Mosk 2003).

Other Legislation. Major legislative initiatives have been passed in recent years including the 1994 Crime Control Act that created community policing "three strikes and you're out" and truth-in-sentencing laws. Significant crime-related legislation presently before Congress include:

- *Victims of Trafficking and Violence Protection Act*—"the purposes of this act are to combat trafficking in persons, . . . to ensure just and effective punishment of traffickers, and to protect victims" (*State Department* 2003)
- *The Innocence Protection Act*—intended to reduce the number of wrongful executions by providing enhanced legal services and DNA testing for the accused (see this chapter's *Focus on Technology* feature)
- *Local Law Enforcement Enhancement Act*—hate crime legislation (see Chapters 7 and 9) that gives the federal government the power to help local law enforcement agencies in their investigation of hate crimes (Sieber 2003)
- *States Rights to Medical Marijuana Bill*—would allow doctors in states that have legalized marijuana for medical purposes to be free from prosecution under federal law (Bipartisan Bill 2003)

Understanding Crime and Violence

What can we conclude from the information presented in this chapter? Research on crime and violence supports the contentions of both functionalists and conflict theorists. Inequality in society, along with the emphasis on material well-being and corporate profit, produces societal strains and individual frustrations. Poverty, unemployment, urban decay, and substandard schools, the symptoms of social inequality, in turn lead to the development of criminal subcultures and conditions

Focus on Technology

DNA Evidence

In 1954, Sam Sheppard, a prominent Cleveland physician, was accused of killing his wife, Marilyn Sheppard. Over 40 years later this case was the basis for the movie *The Fugitive* and the television series of the same name. The real-life drama carries on as Sam Sheppard, Jr. continues to try to clear his father's name. This time, however, he is armed with DNA evidence—evidence that suggests that someone other than his father may have killed Marilyn Sheppard.

Increasingly, law enforcement officers both in the United States and in Europe are using DNA evidence in the identification of criminal suspects. DNA stands for deoxyribonucleic acid, which is found in the nucleus of every cell and contains an individual's complete and unique genetic makeup. Developed in the mid-1980s, DNA fingerprinting is a general term used to describe the process of analyzing and comparing DNA from different sources, including evidence found at a crime scene, for example, blood, semen, hair, saliva, fibers, and skin tissue, and the DNA of a suspect. In the United States, DNA evidence is primarily used in rape and homicide cases—there have been over 100 exonerations nationwide (Wilgorn 2002).

Concern over the number of post-conviction exonerations has led to the establishment of a National Commission on the Future of DNA Evidence, an independent panel under the National Institute of Justice. The commission provides recommendations to the U.S. Attorney General on the "use of current and future DNA methods, applications, and technologies in the operation of the criminal justice system, from the crime scene to the courtroom" (NIJ 2000b).

Some concern exists, however, about the use of DNA evidence and, specifically, questions about how donors will be selected, what methods of data collection will be used, and potential abuses of analysis results. In California, several hundred inmates brought suit against the state refusing to give up their DNA. The appellate court refused to hear the case but the governor signed a bill permitting the use of force in obtaining a sample (Kluger 2002). That is why, until recently, England had the only nationwide DNA data bank in the world. However, in 1994 the FBI initiated the Combined DNA Index System (CODIS). With almost every state contributing to it, CODIS is a national database that permits various agencies access for investigative purposes. To date, 1.8 million profiles have been registered and there have been 8,000 leads in "cold cases" (Kluger 2002; Bormann & Espinoza 2003). The extent of the collection effort has led some to ask what will happen to the samples which, among other things, reveal a donor's genetic disposition for certain diseases. One law enforcement goal is to "build a national computer system that holds the fingerprints and DNA of every felon and the ballistic signature of every gun ever used in a crime" (Kluger 2002, 40).

It is likely that use of DNA will face many legal battles but, nonetheless, the future of DNA evidence is bright. Scientists are already testing a tool that will allow eight DNA samples to be measured simultaneously followed by a determination of matches. In 2003, the president asked Congress for $1 million to help analyze the backlog of samples and to expand laboratory capabilities (Bormann & Espinoza 2003). Further, even if it doesn't survive the legal scrutiny that is likely, it remains a valuable identification technique used in biology, archeology, medical diagnosis, paleontology, and forensics.

Sources: Bormann, Dawn, and Richard Espinoza. 2003. "DNA Is Changing the Future of Law Enforcement." *Kansas City Star,* August 10. http://www.kansascity.com; Kluger, Jeffery. 2002. "How Science Solves Crimes." *Time,* October 21, 36–45; National Institute of Justice. 2000b. "Annual Report to Congress." Washington DC: U.S. Department of Justice; Wiligorn, Jodi. 2002. "Prosecutors Use DNA Test to Clear Man in '85 Rape." *New York Times,* November 14. http://www.NYTimes.com

favorable to law violation. Further, the continued weakening of social bonds between members of society and between individuals and society as a whole, the labeling of some acts and actors as "deviant," and the differential treatment of minority groups by the criminal justice system encourage criminal behavior.

Although increasing slightly in the past year, there has been a general decline in crime over the last decade making it tempting to conclude that "get-tough" criminal justice policies are responsible for the reductions. Other valid explanations exist and are likely to have contributed to the falling rates—changing demographics, community policing, stricter gun control, and a reduction in the use of crack cocaine. Nonetheless, "nail'em and jail'em" policies have been embraced by citizens and politicians alike. Get tough measures include building more prisons and imposing lengthier mandatory prison sentences on criminal offenders. Advocates of harsher prison sentences argue that "getting tough on crime" makes society safer by keeping criminals off the streets and deterring potential criminals from committing crime.

> "To do justice, to break the cycle of violence, to make Americans safer, prisons need to offer inmates a chance to heal like a human, not merely heel like a dog."
>
> **Richard Stratton**
> **Former prison inmate**

Prison sentences, however, may not always be effective in preventing crime. In fact, they may promote it by creating an environment in which prisoners learn criminal behavior, values, and attitudes from each other. Two-thirds of all parolees are rearrested within three years of release (Butterfield 2002b). With 90 percent of inmates being discharged into the community, 600,000 annually, 1,700 a day, punitive policies may be a short-sighted temporary fix (Travis & Waul 2002).

Rather than getting tough on crime after the fact, some advocate getting serious about prevention. Reemphasizing the values of honesty and, most important, taking responsibility for one's actions is a basic line of prevention with which most agree. The recent movement toward **restorative justice,** a philosophy primarily concerned with repairing the victim-offender-community relation, is in direct response to the concerns of an adversarial criminal justice system that encourages offenders to deny, justify, or otherwise avoid taking responsibility for their actions.

Restorative justice holds that the justice system should offer a "healing process rather than [be] a distributor of retribution and revenge" (Siegel 2000, 278)—that is, it should "repair the harm" (Sherman 2003, 10). Key components of restorative justice include restitution to the victim, remedying the harm to the community, and mediation. Restorative justice programs have been instituted in several states including Vermont and New York. In 2002, the United Nations Economic and Social Council endorsed the "Basic Principles on the Use of Restorative Justice Programmes in Criminal Matters" around the world (UN 2003).

Chapter Review

- **Are there any similarities between crime in the United States and crime in other countries?**
  All societies have crime and have a process by which they deal with crime and criminals, that is, they have police, courts, and correctional facilities. Worldwide, most offenders are young males and the most common offense is theft—the least common offense murder.

- **How can we measure crime?**
  There are three primary sources of crime statistics. First are official statistics—for example, the FBI's Uniform Crime Reports, which are published annually. Second are victimization surveys designed to get at the "dark figure" of crime—crime that is missed by official statistics. Finally, there are self-report studies that have all the problems of any survey research. You must be cautious who you ask and how you ask the questions.

- **What sociological theory of criminal behavior blames the schism between the culture and structure of society for crime?**
  Strain theory was developed by Robert Merton (1957) using Durkheim's concept of anomie, or normlessness. Merton argues that when legitimate means (e.g., a job) of acquiring culturally defined goals (e.g., money) are limited by the structure of society, the resulting strain may lead to crime. Individuals, then, must adapt to the inconsistency between means and goals in a society that socializes everyone into wanting the same thing but provides opportunities only for some.

- **What are index offenses?**
  Index offenses, as defined by the FBI, include two categories of crime: violent and property. Violent crimes include murder, robbery, assault, and rape; property crimes include larceny, car theft, burglary, and arson. Property crimes, although less serious than violent crimes, are the most numerous.

- **What is meant by white-collar crime?**
  White-collar crime includes two categories: occupational crime, that is, crime committed in the course of one's occupation, and corporate crime in which corporations violate the law in the interest of maximizing profits. In occupational crime, the motivation is individual gain.

- **How do social class and race impact the likelihood of criminal behavior?**
  Official statistics indicate that minorities are disproportionately represented in the offender population. Nevertheless, it is inaccurate to conclude that race and crime are causally related. First, official statistics reflect the behaviors and policies of criminal justice actors. Thus, the high rate of arrests, conviction, and incarceration of minorities may be a consequence of individual and institutional bias against minorities. Second, race and social class are closely related in that nonwhites are overrepresented in the lower classes. Because lower-class members lack legitimate means to acquire material goods, they may turn to instrumental, or economically motivated, crimes. Thus the apparent relationship between race and crime may, in part, be a consequence of the relationship between these variables and social class.
- **What are some of the economic costs of crime?**
  First are direct losses from crime such as the destruction of buildings through arson or of the environment by polluters. Second are costs associated with the transferring of property (e.g., embezzlement). A third major cost of crime is that associated with criminal violence, for example, the medical cost of treating crime victims. Fourth are the costs associated with the production and sale of illegal goods and services. Fifth is the cost of prevention and protection. Finally, there is the cost of the criminal justice system—law enforcement, litigative and judicial activities, corrections, and victims' assistance.
- **What are some of the most recent criminal justice policies in the fight against crime?**
  First, there has been an increase in the number of police on the streets in the hope that their presence will increase deterrence and, therefore, decrease crime. Second, societal fear of crime has led to a public emphasis on incapacitation and a demand for tougher mandatory sentences, a reduction in the use of probation and parole, support of a "three strikes and you're out" policy, and truth-in-sentencing laws.

Critical Thinking

1. Crime statistics are sensitive to demographic changes. Explain why crime rates in the United States began to rise in the 1960s as baby-boom teenagers entered high school, and why they may increase again as we move into the twenty-first century.
2. Some countries have high gun ownership and low crime rates. Others have low gun ownership and low crime rates. What do you think accounts for the differences between these countries?
3. The use of technology in crime-related matters is likely to increase dramatically over the next several decades. DNA testing and the use of heat sensors, blood detecting chemicals, and computer surveillance are just some of the ways science will help fight crime. As with all technological innovations, however, there is the question, "Who benefits?" Do these new technologies have gender, race, and/or class implications?

Key Terms

acquaintance rape
Brady Bill
classic rape
clearance rate
computer crime
control theory
corporate violence
crime
crime rate
deterrence
differential association
incapacitation
index offenses
labeling theory
organized crime
primary deviance
racial profiling
rehabilitation
restorative justice
secondary deviance
strain theory
subcultural theory
transnational crime
victimless crimes
white-collar crime

Taking a Stand

Should sex offenders be required to register?

Megan Kanka, an eight-year-old girl, was sexually molested and killed after she was lured into a neighbor's house with the promise of a puppy. Megan's Law requires that "sexually violent persons, persons convicted of sexually violent offenses, and those convicted of offenses against minors" register in the community where they live (Reid 2003, 521). Opponents argue that such a law leads to vigilante-style violence as community members take the law into their own hands. In 2003, the U.S. Supreme Court held that (1) states may require registration of sex offenders whose crimes took place before Megan's Law was enacted, and (2) states may publish names of offenders even if no assessment of risk has taken place (*Major Rulings* 2003).

Use Wadsworth's exclusive online resources—InfoTrac College Edition, MicroCase Online, and OVRC—to formulate a position on this topic.

The Wadsworth's Sociology Online Resources and Writing Companion will help you get started. This valuable guide will show you how to use Wadsworth's exclusive online resources when studying social problems. It will also help you to build essential research and writing skills. InfoTrac College Edition, MicroCase Online, OVRC, and an electronic copy of portions of this companion are available at http://sociology.wadsworth.com/mooney_knox_schacht/problems4e, the companion Web site for *Understanding Social Problems,* Fourth Edition.

Media Resources

The Companion Web Site for *Understanding Social Problems*, Fourth Edition

http://sociology.wadsworth.com/mooney_knox_schacht/problems4e

Supplement your review of this chapter by going to the companion Web site to take one of the Tutorial Quizzes, use the flash cards to master key terms, and check out the many other study aids you'll find there. You'll also find special features such as *Wadsworth's Sociology Online Resources and Writing Companion,* GSS Data, and Census 2000 information, data, and resources at your fingertips to help you complete that special project or do some research on your own.

Interactions CD-ROM

Go to the Interactions CD-ROM for *Understanding Social Problems,* Fourth Edition, to access additional interactive learning tools, such as in-depth review materials, corresponding practice quizzes, and other engaging resources and activities to help you study the concepts in this chapter.

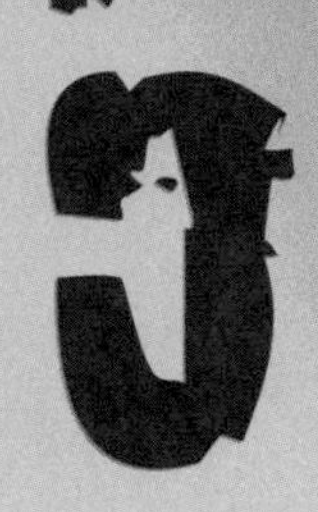

"In a democracy, we are all responsible for supporting healthy family life." *Al and Tipper Gore, from* Joined at the Heart

Family Problems

Ann has been married to Paul for 19 years. She has been unhappy in her marriage for years, but has "hung in there" because she has two young children and her marriage "wasn't that bad." When Ann first met Paul, she had just come out of a relationship with an abusive alcoholic who had spent time in prison and who was suicidal and jobless. In comparison to this nightmare, Paul offered a life of security and stability. He wasn't an alcoholic, drug user, womanizer, or abuser. Paul was a decent man with a strong work ethic and was a very involved and loving father to their two children. But during their marriage, Ann began to see that, at the core, she and Paul differed in ways that made the kind of emotional intimacy she craved an impossibility. Ann struggled silently for years before reaching the conclusion that she could no longer go on with the charade of pretending to be OK in her marriage. When Ann began to express her marital unhappiness to friends and family, she was criticized and blamed. "How can you be unhappy? Paul is such a good person and a wonderful father" her siblings told her. Ann struggled with self-doubt and uncertainty about her conviction that divorce was inevitable. Ann wondered how a divorce would affect her children, how she would survive economically, and how she would cope with being a single parent. And she wondered, "Would I have ever married Paul if I hadn't been in an abusive relationship before I met him?"

Ann's situation involves some of the problems discussed in this chapter: violence and abuse in relationships, divorce, and single-parenting. The problems discussed in every chapter of this text affect families. For example, families are affected by health problems (Chapter 2), substance abuse (Chapter 3), problems of youth and the elderly (Chapter 9), gender inequality (Chapter 7), problems with employment and poverty (Chapters 10 and 11), and so on. Issues related to same-sex couples and families are addressed in Chapter 8.

Much of the controversy surrounding family issues centers around differing views of what constitutes a family. By reviewing diversity in families worldwide and noting the changing patterns and structures of households and families in the United States, it is evident that family forms and functions vary widely.

The Global Context: Families of the World

Family is a central aspect of every society throughout the world. But family forms are diverse.

Monogamy and Polygamy

In many countries, including the United States, the only legal form of marriage is **monogamy**—a marriage between two partners. A common variation of monogamy is **serial monogamy**—a succession of marriages in which a person has more than one spouse over a lifetime but is legally married to only one person at a time. Some societies practice **polygamy**—a form of marriage in which one person may have two or more spouses. The most common form of polygamy is **polygyny**—the concurrent marriage of one man with two or more women. Polygyny is practiced in some Islamic societies, including some regions of Africa. Polygamy, which is illegal in the United States, is often referred to as **bigamy**—the criminal offense of marrying one person while still legally married to another. Some Mormon fundamentalists embrace the practice of polygyny. In Utah, a state with a high Mormon population, an estimated 30,000 residents live in polygynous marriages (Janofsky

These two Masai wives, shown with their children, share the same husband and live in two separate huts in Tanzania. Another Masai man may have sex with these women as long as he places his spear outside the hut to announce his presence inside.

© Caroline Schacht

In Utah's first trial involving bigamy charges in 50 years, Thomas A. Green of Utah, shown here with his five wives, was sentenced to five years in prison after being convicted of four counts of bigamy and one count of criminal nonsupport. Mr. Green's wives had pleaded with Judge Guy R. Burningham to keep him out of prison and were relieved that Green was not sentenced to the 25 years sought by the prosecutor in the case. Although as many as 30,000 Utahans live in polygamous marriages, Green was singled out for prosecution because of his frequent media interviews about his illegal lifestyle.

© AFP/Corbis

2001). The second form of polygamy is **polyandry**—the concurrent marriage of one woman with two or more men. Polyandry is very rare, but does occur in some groups including Tibetans and various African groups.

Same-Sex Relationships

Norms and policies concerning same-sex intimate relationships also vary around the world. In some countries, homosexuality is punishable by imprisonment or even death (see Chapter 8). In contrast, in 2001, the Netherlands became the first country in the world to offer legal marriage to same-sex couples, and in 2003, Belgium became the second (Demian 2003). The Canadian provinces of Ontario and British Columbia also ruled in 2003 to extend the option of marriage to same-sex couples. And in a number of countries (Denmark, Finland, Germany, Iceland, Greenland, Netherlands, Norway, and Sweden), same-sex couples can sign as "Registered Partners" to claim a status and benefits similar to marriage. In 2003, Quebec approved "civil unions" for same-sex couples, which confer some but not all of the rights, benefits, and responsibilities of marriage.

In the United States, the 1996 federal Defense of Marriage Act defines marriage as the union between one man and one woman and denies federal recognition of same-sex marriages; 38 states have passed laws denying recognition to any legal marriage of same-sex partners. As discussed in Chapter 8, same-sex couples in Vermont and California can apply for a "civil union" certificate that entitles them to most of the same legal rights and responsibilities of marriage. In 2003, the Massachusetts Supreme Judicial Court ruled that same- and opposite-sex couples must be given equal marriage rights under the state constitution—the first such ruling in U.S. history.

In France, couples can enter a type of marital arrangement called *Pacte civil de solidarite* (civil solidarity pact), or PAC. PACS are similar to marriage in that each party is responsible for financially supporting the other and sharing in each others' debts accumulated during the pact, and partners are eligible for the others' work benefits. But unlike traditional marriages, PACS can be broken in a short period of time and can be entered into by same-sex couples, although nearly 40 percent of couples opting for PACS in France are heterosexual couples (Daley 2000).

Family Values, Roles, and Norms

Family values, roles, and norms are also highly variable across societies. For example, unlike in Western societies, in Asian countries some marriages are arranged by the parents, who select mates for their children. Another traditional Asian practice is for the eldest son and his wife to move in with the son's parents, where his wife takes care of her husband's parents.

The roles of women and children also vary across societies. In some societies, women are considered equal partners in marriage whereas in other societies wives are expected to be subservient to their husbands. In some societies, children as young as 5 years old are expected to work full time to help support the family, whereas other societies expect children to go to school rather than contribute economically to the family.

Social norms concerning having children out of wedlock also vary widely throughout the world. Acceptance of this lifestyle ranges from 90 percent or more in parts of Western Europe to fewer than 15 percent in Singapore and India (Global Study of Family Values 1998).

Norms surrounding divorce also vary across societies. Ireland did not allow divorce under any conditions until 1995, when voters narrowly approved a public referendum allowing divorce for the first time (Thompson & Wyatt 1999). In 2000, the Egyptian Parliament voted to allow women to file for divorce on grounds of incompatibility (Eltahawy 2000). Prior to this new law, Egyptian women could file for divorce only in cases of proven physical or psychological abuse. By contrast, a man could simply say "I divorce you" three times or get a divorce by filing a paper with the marriage registrar without even notifying his wife. Under the new law, a woman who wants a divorce must return her husband's dowry and relinquish all financial claims, including alimony.

It is clear from the previous discussion that families are shaped by the social and cultural context in which they exist. As we discuss the family issues addressed in this chapter—violence and abuse, divorce, and nonmarital and teenage childbearing—we refer to social and cultural forces that shape these events and the attitudes surrounding them. Next we look at changing patterns and structures of U.S. families and households.

Changing Patterns and Structures in U.S. Families and Households

Over the last century, dramatic changes have occurred in the patterns and structures of U.S. families and households. A **family household,** as defined by the U.S. Census Bureau, consists of two or more persons related by birth, marriage, or adoption who reside together. **Nonfamily households** may consist of one person who lives alone, two or more people as roommates, or cohabiting heterosexual or homosexual couples involved in a committed relationship (see Figure 5.1). The Census Bureau considers households composed of heterosexual cohabiting couples and same-sex couples as "nonfamily" households, even though such couples function economically and emotionally as a family. Some of the significant changes in

Figure 5.1
The Many Types of U.S. Households, 2003

Source: *Statistical Abstract of the United States: 2003.* 2003. U.S. Bureau of the Census, Table 66.

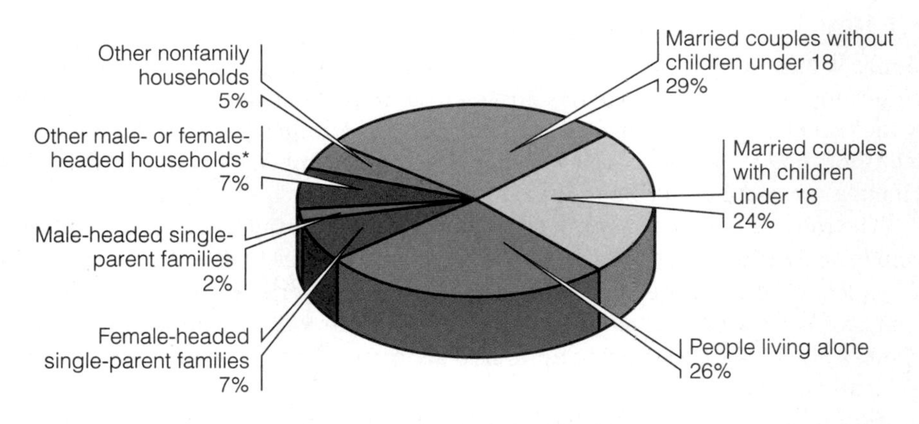

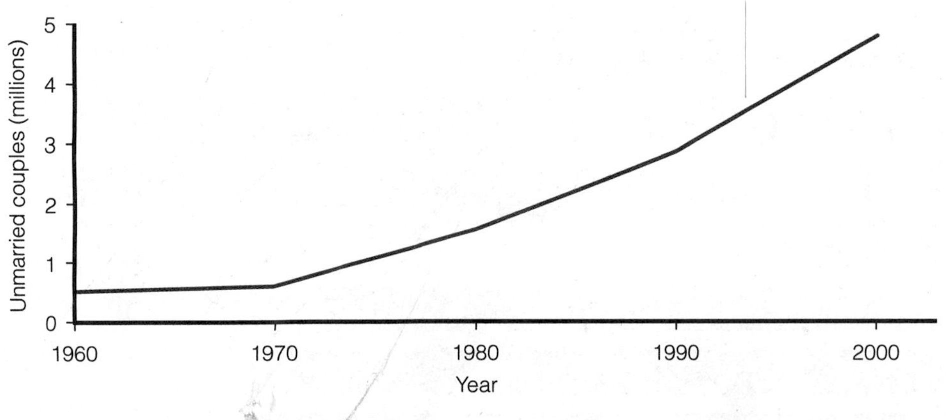

Figure 5.2
Number of Cohabiting, Unmarried Couples of the Opposite Sex, by Year, United States

Source: U.S. Bureau of the Census. *Current Population Reports,* Series P20-537; *America's Families and Living Arrangements: March 2000* and earlier reports.

U.S. families and households that have occurred over the last several decades include the following:

- *Increased singlehood and older age at first marriage.* U.S. women and men are staying single longer. In 1960, the median age at first marriage was about 20 for women and 23 for men. Now, women are about 25 and men about 27 at their first marriage (Hacker 2003). Today, 11.5 percent of women and 16.7 percent of men aged 40 to 44 have never been married—the highest figures in this nation's history (Hacker 2003; *Statistical Abstract* 2003).
- *Increased heterosexual and same-sex cohabitation.* Although women and men are staying single longer, they are not forgoing intimate relationships. From 1960 to 2000, the number of cohabiting unmarried couples skyrocketed (see Figure 5.2). Nationally, about 9 percent of coupled households are unmarried partner households (Simmons & O'Connell 2003). An estimated one-quarter of unmarried women aged 25 to 39 are currently living with a partner, and another one-quarter have lived with a partner in the past (Whitehead & Popenoe 2003). More than half of U.S. women who married in the 1990s reported that they had cohabited before marriage (Bachrach, Hindin, & Thomson 2000). For most unmarried living-together couples, cohabitation is a part of the modern courtship process; for others, cohabitation is an informal type of marriage, an alternative to marriage, or simply an alternative to living alone.

  Most of the 5.5 million cohabiting couples in 2000 were heterosexual couples, but about 1 in 9 had partners of the same sex (Simmons & O'Connell 2003). Figure 5.3 shows unmarried same-sex couples and unmarried opposite-sex couples as a percentage of all coupled households.

  Increased cohabitation among adults means that children are increasingly living in families that may function as two-parent families but do not have the social or legal recognition that married-couple families have. About four in ten (43%) opposite-sex unmarried partner households, one-fifth (22.3%) of gay male couples, and one-third (34.3%) of lesbian couples have children present in the home (Simmons & O'Connell 2003). When children are denied a legal relationship to both parents due to their unmarried status and/or sexual orientation, they may be denied Social Security survivor benefits, health care insurance, or the ability to have either parent authorize medical treatment in an emergency, among other protections. Some states, cities, counties, and employers allow unmarried partners (same-sex and/or heterosexual partners) to apply

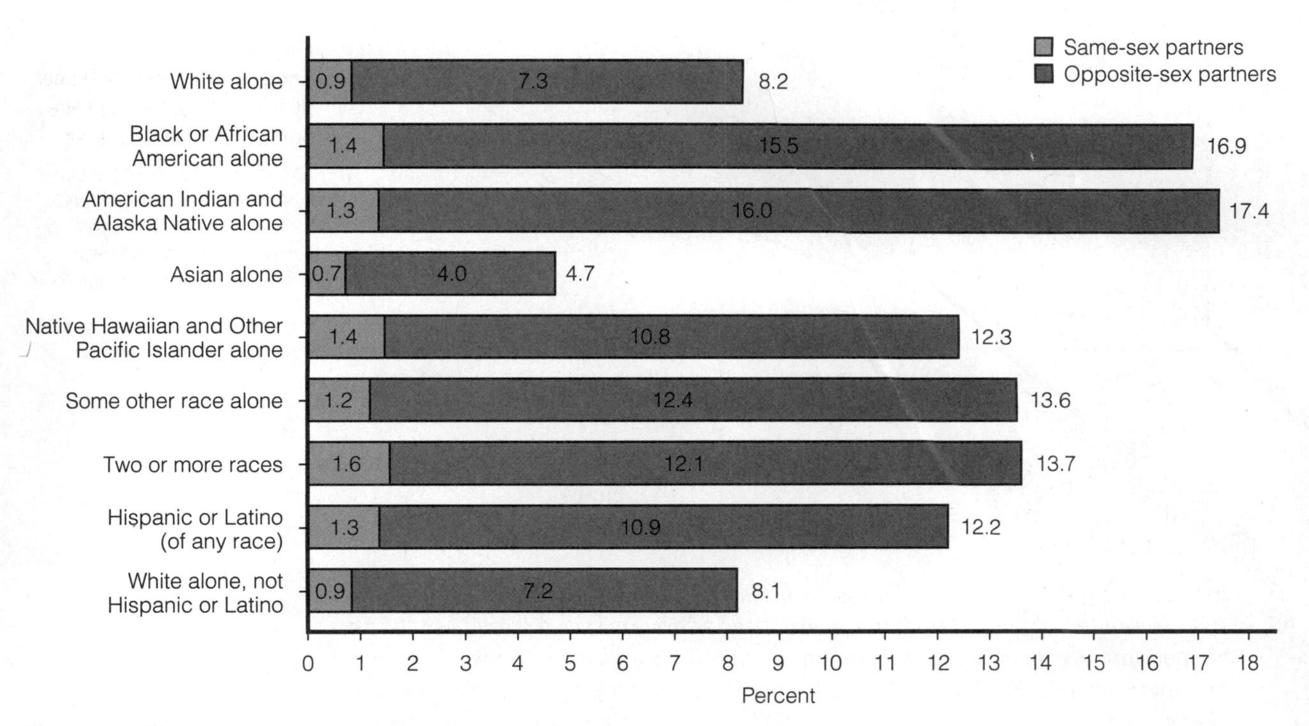

Figure 5.3
Unmarried Same-Sex and Opposite Sex Couple Households, as Percentages of All Coupled Households, by Race and Hispanic Origin, U.S.: 2000

Source: Tavia Simmons and Martin O'Connell. 2003. "Married-Couple and Unmarried-Partner Households: 2000." *Census 2000 Special Reports.* http://www.census.gov/

for a **domestic partnership** designation, which grants them some legal entitlements such as health insurance benefits and inheritance rights that have traditionally been reserved for married couples. Seven states, and certain counties in 16 other states, guarantee same-sex couples access to second-parent adoptions to establish a legal relationship with their children ("American Bar Association Supports Equal Protections for Children of Same-Sex Parents" 2003).

- *Increased births to unmarried women.* The percentage of births to unmarried women increased from 5.3 percent of total births in 1960 to 33.5 percent in 2001 (Whitehead & Popenoe 2003). Expressed another way, today about one-third of all U.S. births are to unmarried women. However, not all unwed mothers are single parents. Half of new unwed mothers are cohabiting with the fathers at the time their children are born (Sigle-Rushton & McLanahan 2002). Later in this chapter, we discuss unwed childbearing in more detail.
- *Increased single-parent families.* The rise in both divorce and nonmarital births has resulted in an increase in single-parent families, although this trend has reversed itself in recent years (see Figure 5.4). The percentage of single-parent families headed by men has also increased. In 1970, about 10 percent of children in single-parent families lived with their father (Sugarman 2003). Of the 19.8 million children under 18 living in a single-parent family in 2002, 83 percent lived with their mother and 17 percent lived with their father. As noted earlier, many children in what are considered to be "single-parent families" have two parental adults in the home who are living in an unmarried cohabiting relationship: 16 percent of children living with single fathers and

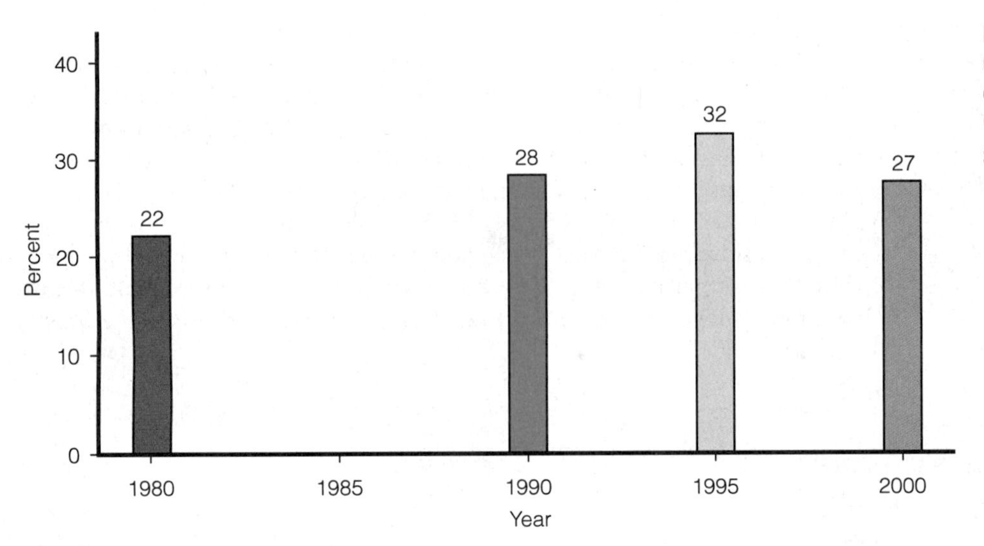

Figure 5.4
One-Parent Families, as a Percent of Family Groups with Children Under 18

Source: *Statistical Abstract of the United States: 2003,* Table 74.

9 percent of children living with single mothers also live with their parents' partners (Forum on Child and Family Statistics 2000).

- *Fewer children living in married-couple families.* The percentage of children living in married-couple families decreased from 77 percent in 1980 to 72 percent in 1990, and 68 percent in 2000. However, this downward trend shows signs of stabilizing and, perhaps, reversing in direction, as the percentage of children in married-couple families *increased* by one point (from 68% to 69%) and the percentage of black children living in married-couple families increased from 34 to 39 percent between 2000 and 2002 (Federal Interagency Forum on Child and Family Statistics 2003).
- *Increased divorce and blended families.* A century ago, divorce was highly uncommon. Whereas marriages made in 1890 had only a one in ten chance of ending in divorce, by 1930, the chances rose to one in four, and by the 1970s, the chances were one in two (Coltrane & Collins 2001). The **divorce rate**—the number of divorces per 1,000 population—doubled from 1950 to its peak in the late 1970s and early 1980s, increasing from a rate of 2.6 to 5.3. In nearly every year since the early 1980s, the divorce rate has decreased and in 2001 was 4.0 (*Statistical Abstract* 2003). Despite the decline in divorce rates over the 1990s, at today's rate an estimated half of marriages started today are expected to end in divorce (Whitehead & Popenoe 2003). About one-fourth of U.S. first-year college students have parents who are divorced (American Council on Education & University of California 2002).

  Most divorced individuals remarry and create *blended families,* traditionally referred to as *stepfamilies.* An estimated one-quarter of all children born in the United States will live with a stepparent before they reach adulthood (Mason 2003).

  Federal, state, and private-sector policies have not kept pace with the concerns of modern stepfamilies. In most states stepparents have no obligation to support their stepchildren during the marriage, nor do they have any right of custody or control. In the event of divorce, stepparents usually have no rights to custody or even visitation, and no obligation to pay child support. Stepchildren have no right of inheritance in the event of the stepparent's death (unless the stepparent has specified such inheritance in a will) (Mason 2003).

- *Increased employment of married mothers.* Employment of married women with children under age 18 rose from 24 percent in 1950 to 40 percent in 1970 to 70 percent in 1999, and nearly 60 percent of all wives with children age three or younger are in the labor force (Gilbert 2003). This increase in maternal employment means that the **traditional family** in which the husband is a breadwinner and the wife is a homemaker is no longer the norm in U.S. society. In the 1940s, about two-thirds of U.S. families fit the traditional model. In 2002, two-thirds (61%) of U.S. married-couple families with children under 18 were dual-earner marriages; less than one-third (30%) of these families involved an employed father and an unemployed mother (*Statistical Abstract* 2003).

Optimistic and Pessimistic Views of the Changing American Family

Do the recent transformations in American families signify a collapse of marriage and family in the United States? Does the trend toward diversification of family forms mean that marriage and family are disintegrating, falling apart, or even disappearing? Or, like other aspects of social life, has family simply undergone transformations in response to changes in socioeconomic conditions, gender roles, and cultural values? In coming to your own conclusions, consider that a national Gallup survey shows that Americans rank their family as *the most* important aspect of life (Moore 2003). Health was ranked second, followed by work, friends, money, and religion. In a national survey of first-year students in U.S. colleges and universities, students listed "raising a family" as one of the top two values (second only to "being well-off financially") (American Council on Education & University of California 2002).

Marriage continues to be a normative part of adult life, with about 95 percent of women and men in their early sixties having been married at least once (Hacker 2003). Although women and men are marrying later than they did in the past, and their marriages may not last as long as they once did, most people eventually do get married. And despite high rates of cohabitation, living together tends to be a precursor to marriage rather than a permanent substitute (Goldstein & Kenney 2001).

Although the high rate of marital dissolution seems to suggest a weakening of marriage, divorce may also be viewed as resulting from placing a high value on marriage, such that a less than satisfactory marriage is unacceptable. Indeed, the expectations young women and men have of marriage have changed. A survey of single women and men aged 20 to 29 found that the overwhelming majority (94%) agree that "when you marry, you want your spouse to be your soul-mate first and foremost" (Whitehead & Popenoe 2003). Whereas once the main purpose of marriage was to have and raise children, today women and men want marriage to provide adult intimacy and personal satisfaction; having and raising children is an endeavor that exists apart from marriage. A recent survey found that only 15 percent of the population agree that "when there are children in the family, parents should stay together even if they don't get along" (Whitehead & Popenoe 2003).

The high rate of out-of-wedlock childbirth and single parenting is also not necessarily indicative of a decline in the value of marriage. Edin (2000) found that in a sample of low-income single women with children, most said they would like to be married but just haven't found "Mr. Right."

Is the well-being of a family measured by the degree to which that family conforms to the idealized married, two-parent, stay-at home mom model of the 1950s? Or is family well-being measured by function rather than form? As suggested by family scholars Mason, Skolnick, and Sugarman (2003), "the important question to ask about American families . . . is not how much they conform to a particular image of the family, but rather how well do they function—what kind of love, care, and nurturance do they provide?" (p. 2). After a brief discussion of sociological theories of family problems, we look at what happens when children and adults experience violence and abuse in their intimate and family relationships.

"We recognize today that children can be effectively raised in many different family systems and that it is the emotional climate of the family, rather than its kinship structure, that primarily determines a child's emotional well-being and healthy development."

David Elkind
Child development specialist

Sociological Theories of Family Problems

Three major sociological theories—structural-functionalism, conflict theory, and symbolic interactionism—help to explain different aspects of the family institution and the problems in families today.

Structural-Functionalist Perspective

The structural-functionalist perspective views the family as a social institution that performs important functions for society, including producing new members, regulating sexual activity and procreation, socializing the young, and providing physical and emotional care for family members. According to the structural-functionalist perspective, traditional gender roles contribute to family functioning as women perform the "expressive" role of managing household tasks and providing emotional care and nurturing to family members, and men perform the "instrumental" role of earning income and making major family decisions.

According to the structural-functionalist perspective, the high rate of divorce and the rising numbers of single-parent households constitute a "breakdown" of the family that has resulted from rapid social change and social disorganization. The functionalist perspective views the breakdown of the family as a primary social problem that leads to such secondary social problems as crime, poverty, and substance abuse.

Functionalist explanations of family problems examine how changes in other social institutions contribute to family problems. For example, a structural-functionalist view of divorce examines how changes in the economy (more dual-earner marriages) and in the legal system (such as the adoption of "no-fault" divorce) contribute to high rates of divorce. Changes in the economic institution, specifically falling wages among unskilled and semiskilled men, also contribute to both intimate partner abuse and the rise in female-headed single-parent households (Edin 2000).

Conflict and Feminist Perspectives

Conflict theory focuses on how capitalism, social class, and power influence marriages and families. Feminist theory is concerned with how gender inequalities influence and are influenced by marriages and families. Feminists are critical of the traditional male domination of families—a system known as **patriarchy**—that is reflected in the tradition of wives taking their husband's last name and the view that wives are subservient to their husbands.

The overlap between conflict and feminist perspectives is evident in views on how industrialism and capitalism have contributed to gender inequality. With the onset of factory production during industrialization, workers—mainly men—left the home to earn incomes and women stayed home to do unpaid child care and domestic work. This arrangement resulted in families founded on what conflict theorist Engels calls "domestic slavery of the wife" (quoted in Carrington 2002, 32). Modern society, according to Engels, rests upon gender-based slavery, with women doing household labor for which they receive neither income nor status while men leave the home to earn an income. Times have certainly changed since Engels made his observations, with most wives today leaving the home to earn incomes. However, wives employed full-time still do the bulk of unpaid domestic labor and women are more likely than men to compromise their occupational achievement to take on child care and other domestic responsibilities. The continuing unequal distribution of wealth that favors men contributes to inequities in power and fosters economic dependence of wives on husbands (see also Chapter 7).

Economic factors have also influenced norms concerning monogamy. In societies in which women and men are expected to be monogamous within marriage, there is a double standard that grants men considerably more tolerance for being nonmonogamous. Engels explains that monogamy arose from the concentration of wealth in the hands of a single individual—a man—and from the need to bequeath this wealth to children of his own, which requires that his wife be monogamous. The "sole exclusive aims of monogamous marriage were to make the man supreme in the family and to propagate, as the future heirs to his wealth, children indisputably his own" (quoted in Carrington 2002, 32).

Feminist and conflict perspectives on domestic violence suggest that the unequal distribution of power among women and men and the historical view of women as being the property of men contribute to wife battering. When wives violate or challenge the male head-of-household's authority, the male may react by "disciplining" his wife or using anger and violence to reassert his position of power in the family.

While modern gender relations within families and within society at large are more egalitarian than in the past, male domination persists, even if less obvious. Lloyd and Emery (2000) note that "one of the primary ways that power disguises itself in courtship and marriage is through the 'myth of equality between the sexes'. . . . The widespread discourse on 'marriage between equals' serves as a cover for the presence of male domination in intimate relationships . . . and allows couples to create an illusion of equality that masks the inequities in their relationships" (pp. 25–26).

Conflict theorists emphasize that social programs and policies that affect families are largely shaped by powerful and wealthy segments of society. The interests of corporations and businesses are often in conflict with the needs of families. Corporations and businesses strenuously fought the passage of the 1993 Family and Medical Leave Act, giving people employed full-time for at least 12 months in companies with at least 50 employees up to 12 weeks of unpaid time off for parenting leave, illness or death of a family member, and elder care. Hewlett and West (1998) note that corporate interests undermine family life "by exerting enormous downward pressure on wage levels for young, child-raising adults" (p. 32). Government, which is largely influenced by corporate interests through lobbying and political financial contributions, enacts policies and laws that serve the interests of for-profit corporations, rather than families.

Symbolic Interactionist Perspective

Symbolic interactionism emphasizes that human behavior is influenced by meanings and definitions that emerge from small group interaction. Divorce, for example, was once highly stigmatized and informally sanctioned through the criticism and rejection of divorced friends and relatives. As societal definitions of divorce became less negative, however, the divorce rate increased. The social meanings surrounding single parenthood, cohabitation, and delayed childbearing and marriage have changed in similar ways. As the definitions of each of these family variations became less negative, the behaviors became more common.

The symbolic interactionist perspective is concerned with how labels affect meaning and behavior. For example, when a noncustodial divorced parent (usually a father) is awarded "visitation" rights, he may view himself as a visitor in his children's lives. The meaning attached to the visitor status can be an obstacle to the father's involvement as the label "visitor" minimizes the importance of the noncustodial parent's role and results in conflict and emotional turmoil for fathers (Pasley & Minton 2001). Fathers' rights advocates suggest replacing the term "visitation" with such terms as "parenting plan" or "time-sharing arrangement," as these terms do not minimize either parent's role.

Symbolic interactionists also point to the effects of interaction on one's self-concept, especially the self-concept of children. In a process called the "looking glass self," individuals form a self-concept based on how others interact with them. Family members, such as parents, grandparents, siblings, and spouses are significant others who have a powerful effect on our self-concepts. For example, negative self-concepts may result from verbal abuse in the family, while positive self-concepts may develop in families where interactions are supportive and loving. The importance of social interaction in children's developing self-concept suggests a compelling reason for society to accept rather than stigmatize nontraditional family forms. Imagine the effect on children who are called "illegitimate" or are teased for having two moms or dads.

The symbolic interactionist perspective is useful in understanding the dynamics of domestic violence and abuse. For example, some abusers and their victims learn to define intimate partner violence as an expression of love (Lloyd 2000). Emotional abuse often involves using negative labels (e.g., "stupid," "whore," "bad,") to define a partner or family member. Such labels negatively affect the self-concept of abuse victims, often convincing them that they deserve the abuse. Next, we discuss violence and abuse in intimate and family relationships, noting the scope, causes, and consequences of this troubling social problem.

Violence and Abuse in Intimate and Family Relationships

Although intimate and family relationships provide many individuals with a sense of intimacy and well-being, for others, these relationships involve physical violence, verbal and emotional abuse, sexual abuse, and/or neglect. Indeed, in U.S. society, people are more likely to be physically assaulted, abused and neglected, sexually assaulted and molested, and killed in their own homes and by other family members than anywhere else or by anyone else (Gelles 2000). Before reading further, you may want to take the Abusive Behavior Inventory in this chapter's *Self and Society* feature.

Abusive Behavior Inventory

Circle the number that best represents your closest estimate of how often each of the behaviors happened in your relationship with your partner or former partner during the previous six months.

1. **Never**
2. **Rarely**
3. **Occasionally**
4. **Frequently**
5. **Very frequently**

1.	Called you a name and/or criticized you.	**1 2 3 4 5**
2.	Tried to keep you from doing something you wanted to do (e.g., going out with friends, going to meetings).	**1 2 3 4 5**
3.	Gave you angry stares or looks.	**1 2 3 4 5**
4.	Prevented you from having money for your own use.	**1 2 3 4 5**
5.	Ended a discussion with you and made the decision himself/herself.	**1 2 3 4 5**
6.	Threatened to hit or throw something at you.	**1 2 3 4 5**
7.	Pushed, grabbed, or shoved you.	**1 2 3 4 5**
8.	Put down your family and friends.	**1 2 3 4 5**
9.	Accused you of paying too much attention to someone or something else.	**1 2 3 4 5**
10.	Put you on an allowance.	**1 2 3 4 5**
11.	Used your children to threaten you (e.g., told you that you would lose custody, said he/she would leave town with the children).	**1 2 3 4 5**
12.	Became very upset with you because dinner, housework, or laundry was not done when he/she wanted it done or done the way he/she thought it should be.	**1 2 3 4 5**
13.	Said things to scare you (e.g., told you something "bad" would happen, threatened to commit suicide).	**1 2 3 4 5**
14.	Slapped, hit, or punched you.	**1 2 3 4 5**
15.	Made you do something humiliating or degrading (e.g., begging for forgiveness, having to ask his/her permission to use the car or to do something).	**1 2 3 4 5**
16.	Checked up on you (e.g., listened to your phone calls, checked the mileage on your car, called you repeatedly at work).	**1 2 3 4 5**
17.	Drove recklessly when you were in the car.	**1 2 3 4 5**
18.	Pressured you to have sex in a way you didn't like or want.	**1 2 3 4 5**
19.	Refused to do housework or child care.	**1 2 3 4 5**
20.	Threatened you with a knife, gun, or other weapon.	**1 2 3 4 5**
21.	Spanked you.	**1 2 3 4 5**
22.	Told you that you were a bad parent.	**1 2 3 4 5**
23.	Stopped you or tried to stop you from going to work or school.	**1 2 3 4 5**
24.	Threw, hit, kicked, or smashed something.	**1 2 3 4 5**
25.	Kicked you.	**1 2 3 4 5**
26.	Physically forced you to have sex.	**1 2 3 4 5**
27.	Threw you around.	**1 2 3 4 5**
28.	Physically attacked the sexual parts of your body.	**1 2 3 4 5**
29.	Choked or strangled you.	**1 2 3 4 5**
30.	Used a knife, gun, or other weapon against you.	**1 2 3 4 5**

SCORING: Add the numbers you circled and divide the total by 30 points to find your score. The higher your score, the more abusive your relationship.

The inventory was given to 100 men and 78 women equally divided into groups of abusers/abused and nonabusers/nonabused. The men were members of a chemical dependency treatment program in a veterans' hospital and the women were partners of these men. Abusing or abused men earned an average score of 1.8; abusing or abused women earned an average score of 2.3. Nonabusing/abused men and women earned scores of 1.3 and 1.6, respectively.

Source: Melanie F. Shepard and James A. Campbell. The abusive behavior inventory: A measure of psychological and physical abuse. *Journal of Interpersonal Violence,* September 1992, 7, no. 3, 291–305. Inventory is on pages 303–304. Used by permission of Sage Publications, 2455 Teller Road, Newbury Park, CA 91320.

Intimate Partner Violence and Abuse

Abuse in relationships can take many forms, including emotional and psychological abuse, physical violence, and sexual abuse. **Intimate partner violence** refers to actual or threatened violent crimes committed against persons by their current or former spouses, boyfriends, or girlfriends.

Prevalence and Patterns of Intimate Partner Violence. Globally, one woman in every three has been subjected to violence in an intimate relationship (United Nations Development Programme 2000). Most (85%) acts of intimate partner violence are committed against women and slightly more than one in five (22%) U.S. women has been assaulted by an intimate partner during her lifetime (Mercy, Krug, Dahlberg, & Zwi 2003; Rennison 2003). Although women also assault their male partners, these assaults tend to be acts of retaliation or self-defense (Johnson 2001).

Factors associated with higher rates of intimate partner victimization against women include being young (ages 16 to 24), black, divorced or separated, and earning lower incomes. The rate and severity of physical violence tend to be higher among cohabiting couples compared with dating and marital partners (Johnson & Ferraro 2003; Magdol, Moffitt, Caspi, & Silva 1998; Stets & Straus 1989).

Four patterns of partner violence have been identified: common couple violence, intimate terrorism, violent resistance, and mutual violent control (Johnson & Ferraro 2003). **Common couple violence** refers to occasional acts of violence arising from arguments that get "out of hand." Common couple violence usually does not escalate into serious or life-threatening violence. **Intimate terrorism** is violence that is motivated by a wish to control one's partner and involves the systematic use of not only violence but also economic subordination, threats, isolation, verbal and emotional abuse, and other control tactics. Intimate terrorism is almost entirely perpetrated by men and is more likely to escalate over time and to involve serious injury. **Violent resistance** refers to acts of violence that are committed in self-defense. Violent resistance is almost exclusively perpetrated by women against a male partner. **Mutual violent control** is a rare pattern of abuse "that could be viewed as two intimate terrorists battling for control" (Johnson & Ferraro 2003, 169).

Intimate partner abuse also takes the form of sexual aggression, which refers to sexual interaction that occurs against one's will through use of physical force, threat of force, pressure, use of alcohol/drugs, or use of position of authority. An estimated 7 to 14 percent of married women have been raped by their husbands (Monson, Byrd, & Langhinrichsen-Rohling 1996). A national study found that about 3 percent of college women experience a completed or attempted rape during a college year (Fisher, Cullen, & Turner 2000). Nearly 90 percent of the sexually assaulted college women knew the person who assaulted them. Based on data from the National Violence Against Women survey, half of the women raped by an intimate partner and two-thirds of the women physically assaulted by an intimate partner had been victimized multiple times (Rand 2003).

Effects of Intimate Partner Violence and Abuse. Battering is the single major cause of injury to women in the United States ("Domestic Violence Fact Sheet" 1999). In 2000, at least 1,247 women and 440 men were killed by an intimate partner. In recent years, an intimate partner killed about 33 percent of female murder victims and 4 percent of male murder victims (Rennison 2003).

Many battered women are abused during pregnancy, resulting in a high rate of miscarriage and birth defects. Psychological consequences for victims of intimate partner violence can include depression, suicidal thoughts and attempts, lowered self-esteem, alcohol and other drug abuse, and post-traumatic stress disorder (National Center for Injury Prevention and Control 2000).

Battering also interferes with women's employment. Some abusers prohibit their partners from working. Other abusers "deliberately undermine women's employment by depriving them of transportation, harassing them at work, turning off alarm clocks, beating them before job interviews, and disappearing when they promise to provide child care" (Johnson & Ferraro 2003, 508). Battering also undermines employment by causing repeated absences, impairing women's ability to concentrate, and lowering their self-esteem and aspirations.

Abuse, whether physical or emotional, is no doubt a factor in many divorces. And in a survey of U.S. mayors, domestic violence was identified as a primary cause of homelessness (U.S. Conference of Mayors 2003).

About 4 in 10 female victims of intimate partner violence live in households with children under age 12 (Rennison & Welchans 2000). Many children who witness domestic violence get involved by yelling, calling for help, or intervening to try to stop the abuse (Edleson, Mbilinyi, Beeman, & Hagemeister 2003). Children who witness domestic violence are at risk for emotional, behavioral, and academic problems as well as future violence in their own adult relationships (Kitzmann, Gaylord, Hold, & Kenny 2003; Parker, Bergmark, Attridge, & Miller-Burke 2000). Children may also commit violent acts against a parent's abusing partner.

Why Do Abused Adults Stay in Abusive Relationships? Adult victims of abuse are commonly blamed for choosing to stay in their abusive relationships. But from the point of view of the victims, there are compelling reasons to stay. These reasons include love, emotional dependency, commitment to the relationship, hope that things will get better, the view that violence is legitimate because they "deserve" it, guilt, fear, economic dependency, and feeling stuck. In one study of why abused women stay in abusive relationships, the most important reason reported by women in these relationships was fear of loneliness (Hendy, Eggen, Gustitus, McLeod, & Ng 2003). Approximately one in four victims delays leaving a violent home because she fears the abuser will abuse or neglect a family pet (Fogle 2003). Victims also stay because abuse in relationships is usually not ongoing and constant, but rather occurs in cycles. The **cycle of abuse** involves a violent or abusive episode, followed by a makeup period when the abuser expresses sorrow and asks for forgiveness and "one more chance." The honeymoon period may last for days, weeks, or even months before the next outburst of violence occurs.

Child Abuse

Child abuse refers to the physical or mental injury, sexual abuse, negligent treatment, or maltreatment of a child under the age of 18 by a person who is responsible for the child's welfare. In 2001, 10 percent of victims of child maltreatment were sexually abused, 19 percent suffered physical abuse, 7 percent were psychologically maltreated, and more than half (59%) suffered **neglect**—the child caregiver's failure to provide adequate attention and supervision, food and nutrition, hygiene, medical care, and a safe and clean living environment (U.S. Dept. of Health and Human Services 2003). Parents who are alone, poor, involved with alcohol or drugs, have a large family, and are parenting one or more children with difficult characteristics are at the greatest risk of committing child abuse or neglect (Barth 2003).

Effects of Child Abuse. The effects of child abuse and neglect vary according to the frequency and intensity of the abuse or neglect. Physical injuries sustained by child abuse cause pain, disfigurement, scarring, physical disability, and death. In 2001, 1,300 U.S. children died of abuse or neglect (U.S. Dept. of Health and Human Services 2003). Head injury is the leading cause of death in abused children (Rubin et al., 2003). **Shaken baby syndrome,** whereby the caretaker, most often the father, shakes the baby to the point of causing the child to experience brain or retinal hemorrhage, most often occurs in response to a baby, who typically is younger than six months, who won't stop crying (Ricci et al. 2003; Smith 2003). Battered or shaken babies are often permanently handicapped. In addition to risk of immediate injury and death, abuse during childhood is associated with risk factors and risk-taking behaviors later in life, including depression, low academic achievement, smoking, obesity, teen pregnancy, alcohol and drug use, sexually transmitted diseases, low self-esteem, aggressive behavior, juvenile delinquency, adult criminality, and suicide (Administration for Children and Families 2003; Mercy et al. 2003).

© Carol Geddes/Sonoma Image Studios

David Pelzer, author of the autobiographical best-sellers *A Child Called "It," The Lost Boy,* and *A Man Named Dave* describes his mother holding his arm in the flame of the gas stove, requiring him to sleep in the cold garage, denying him food for days at a time, slamming his face into a soiled baby's diaper, forcing him to vomit and then eat it, forcing ammonia and dishwashing soap down his throat, throwing a knife into his stomach, and repeatedly beating him until he was black and blue. Although David's experience of child abuse was one of the most severe ever discovered in California, he has gone on to a fulfilling and successful life as a husband, father, inspirational speaker, volunteer, child abuse prevention activist, and best-selling author.

Among females, early forced sex is associated with decreased self-esteem, increased levels of depression, running away from home, alcohol and drug use, and multiple sexual partners (Jasinski, Williams, & Siegel 2000; Lanz 1995; Whiffen, Thompson, & Aube 2000). Compared with nonabused peers, sexually abused girls are also more likely to experience teenage pregnancy, to have higher numbers of sexual partners in adulthood, and to acquire sexually transmitted infections and experience forced sex (Browning & Laumann 1997; Stock, Bell, Boyer, & Connell 1997). Women who were sexually abused as children also report a higher frequency of post-traumatic stress disorder (Spiegel 2000) and suicide ideation (Thakkar et al. 2000). Spouses who were physically and sexually abused as children report lower marital satisfaction, higher stress, and lower family cohesion than spouses with no abuse history (Nelson & Wampler 2000). Adult males who were sexually abused as children tend to exhibit depression, substance abuse, and difficulty establishing intimate relationships (Krug 1989). Sexually abused males also have a higher risk of anxiety disorders, sleep and eating disorders, and sexual dysfunctions (Elliott & Briere 1992).

Effects of child sexual abuse are likely to be severe when the sexual abuse is forceful, is prolonged, and involves intercourse, and when the abuse was perpetrated by a father or stepfather (Beitchman et al. 1992). Not only has the child been violated physically, she or he has lost an important social support. One woman

The Human Side

Child Sexual Abuse—A College Student Tells Her Story

Jennifer Doyle is one of thousands of women who have experienced child sexual abuse. As a way to heal from this traumatic experience, help others who have had similar experiences, and educate the public at large, she has told her story on the Internet. The following excerpts represent bits and pieces of Jennifer's story. . . .

Remembering the Abuse

I didn't admit to the abuse for a long time. I couldn't let myself believe it. I have a tendency to "split" my father into two people. It's very hard for my mind to connect the man who came into my bed at night with the man who used to throw me up in the air and catch me until I laughed like crazy, with the man who bought me my first puppy, with the man who used to throw money in the front lawn some mornings so my sister and I could have a treasure hunt. There is a side of that man that I loathe with every cell in my body, but there is also a side of me that loves him simply because he was my father, and because he was not always "bad."

I've always known what happened. . . . I just couldn't let myself know or think about it because I could not handle the pain and couldn't deal with it. I believe this defense mechanism is called repression. I always knew something was wrong. I always knew what had happened, but I didn't want to think about it, so I didn't. I forced it to the back of my mind and, from everyone else's point of view, was a fairly normal child.

I didn't have specific memories for a while, but it's not because they weren't there—it's because I didn't allow myself access to them. I didn't want to remember. And they didn't come to me "all at once," or simply appear one day when I woke up. They started with a vague suspicion I tried to deny that became a horrible certainty I didn't know how to express. It was more of an awakening. What set it in motion, I think, was when I was in junior high, my father would brush across my knee repeatedly or rub my back incessantly, despite my constant pleas for him to stop and my attempts to move away from those hands that never seemed to go away. I felt then just so utterly dirty and used and destroyed. Once I admitted my feelings to myself, that I had felt this way before and I knew why I did—because it had happened before—it was like opening Pandora's box. I was totally unprepared for what spewed forth, and what continues to.

Sometimes when I think I've remembered all there is to know, another memory comes to mind and I think, "Oh, yes, I'd forgotten about that. Why did I forget that? Why did I remember it now? What does it mean?" Other times I'm just scared and angry and I think, "How could he do that to me? Why didn't anyone stop him? Why didn't I tell someone sooner?"

The Abuse

I think that the abuse must have started before I was five, maybe at two or three—possibly back into infancy. It started out with just fondling, and then it turned into oral sex. And after that, I'm not ready to talk about at this point. Not in detail. But I lost my virginity before I was in the second grade. I have this one memory, when I was five or six, of climbing into my father's bed, and I know he's naked, and the sheets are pale pink, and I ask him what game we're going to play this time. I know that when the abuse started, I didn't say no. I liked it; I thought it was a game. And I didn't think there was anything wrong with it. I thought

who had been sexually abused by her father described feeling that she had lost her father; he was no longer a person to love and protect her (Spiegel 2000). This chapter's *The Human Side* feature describes one young woman's experiences of being sexually abused by her father.

Elder Abuse, Parent Abuse, and Sibling Abuse

Domestic violence and abuse may involve adults abusing their elderly parents or grandparents, children abusing their parents, and siblings abusing each other.

Elder Abuse. Community-based surveys suggest that 3 percent to 6 percent of elders in the United States (over age 65) report having experienced elder abuse (Lachs, Williams, O'Brien, Hurst, & Horwitz 1997). **Elder abuse** includes physical

The Human Side

everyone played with their fathers that way.

I'm not exactly sure how long the sexual abuse went on . . . it had stopped by the time I was in sixth grade, I'm pretty sure. What I remember most is always feeling dirty and not knowing, at the time, exactly why. I remember that by seventh grade I was continually warding off my father's advances and removing his hands from my body. He also came into the bathroom a few times when I was taking a bath, assumingly by accident. I think other people (like my sister and mother) noticed but never really said anything.

In addition to the sexual abuse, my father also abused me emotionally and physically. He accused me of being a drug addict, a lesbian, and a slut. He begged me to forgive him and made me feel like it was my fault. He would pretend like he was going to hit me, and then hit the wall above my head at the last second, just to scare me. He hit me with boards and with a belt. He would sit on me for long periods of time and refuse to move until I begged him. He threw my body against walls. He grabbed my arms, pushed me into furniture, and slapped me across the face. He hit me in the back of my head and knocked me unconscious. He held a gun to my head.

Talking About the Abuse; Telling People About It

One of the hardest things for me to do is talk about my past. In the past few years, I've done a lot of talking though. I've told most of my friends that I was sexually abused by my father, and through my web page I've told a lot of strangers as well.

As for my family . . . I haven't talked to them about it much at all. I brought the subject up to my older sister one time. She says she was molested by the son of one of our parents' friends. I asked her if maybe it had been our father, and she emphatically said no and that nothing had happened to me either. I don't know if she either doesn't remember or is in denial. Since that time, she has conceded that she believes me that our father abused me, but at the same time she says she will deny it because she doesn't want to think about it. She still sees him regularly.

My mother found out over the summer of 1998, when she saw a book on sexual abuse survivors I had beside my bed. She asked me if my father had ever sexually abused me, and I said yes, and made it clear I didn't wish to discuss it further. She assumed it happened while I was in high school, and for years I didn't correct her that the abuse occurred way before then.

Once I told my sister about my web site, she told my mother, father, and stepmother—along with various other relatives. Most of them no longer talk to me. My mother asserts that sexual abuse survivors have to be careful who they tell. She also believes me, and feels that the abuse is not her fault because when she asked me if anything was going on, I would never answer her. It's true—I never did answer her, because I knew that she would do nothing to stop it and, while I was pretty sure she had to know what was going on, to admit that she knew concretely and chose not to help me would have been unbearable.

I read that most survivors go through a "tell everyone" stage, and I know I did. It's starting to come to a close now, because I've learned there's more to me than the abuse.

Source: http://www.geocities.com/midsecret78. Used by permission of Jennifer Doyle.

abuse, psychological abuse, financial abuse (such as improper use of the elder's financial resources), and neglect. Elder neglect includes failure to provide basic health and hygiene needs such as clean clothes, doctor visits, medication, and adequate nutrition. Neglect also involves unreasonable confinement, isolation of elderly family members, lack of supervision, and abandonment. Most abuses of the elderly are committed by adult children, followed by spouses (Lachs et al. 1997).

Parent Abuse. Some parents are victimized by their children's violence, ranging from hitting, kicking, and biting to pushing a parent down the stairs and using a weapon to inflict serious injury or even kill a parent. More violence is directed against mothers than against fathers, and sons tend to be more violent toward parents than are daughters (Ulman 2003). In most cases of children being violent toward their parents, parents had been violent toward the children.

Sibling Abuse. The most prevalent form of abuse in families is sibling abuse. Ninety-eight percent of the females and 89 percent of the males in one study reported having been emotionally abused by a sibling, and 88 percent of the females and 71 percent of the males reported having been physically abused by a sibling (Simonelli, Mullis, Elliott, & Pierce 2002).

Factors Contributing to Intimate Partner and Family Violence and Abuse

Research suggests that cultural, community, and individual and family factors contribute to domestic violence and abuse.

Cultural Factors. Violence in the family stems from our society's acceptance of violence as a legitimate means of enforcing compliance and solving conflicts at personal, national, and international levels (Viano 1992). Violence and abuse in the family may be linked to cultural factors such as violence in the media (see Chapter 4), acceptance of corporal punishment, gender socialization, and the view of women and children as property:

1. *Acceptance of corporal punishment.* **Corporal punishment** is the intentional infliction of pain for a perceived misbehavior (Block 2003). Many mental health professionals and child development specialists argue that it is ineffective and damaging to children. Children who experience corporal punishment display more antisocial behavior, are more violent, and have an increased incidence of depression as adults (Straus 2000). Yet many parents accept the cultural tradition of spanking as an appropriate form of child discipline. More than 90 percent of parents of toddlers reported using corporal punishment (Straus 2000). Eighty-three percent of more than 11,000 undergraduate students at the University of Iowa reported that they had experienced some form of physical punishment during their childhood (Knutson & Selner 1994). Although not everyone agrees that all instances of corporal punishment constitute abuse, undoubtedly, some episodes of parental "discipline" are abusive.
2. *Gender role socialization.* Traditional male gender roles have taught men to be aggressive and to be dominant in male-female relationships. Male abusers are likely to hold traditional attitudes toward women and male-female roles (Lloyd & Emery 2000). Traditional male gender socialization also discourages men to verbally express their feelings, which increases the potential for violence and abusive behavior (Umberson, Anderson, Williams, & Chen 2003). Anderson (1997) found that men who earn less money than their partners are more likely to be violent toward them. "Disenfranchised men then must rely on other social practices to construct a masculine image. Because it is so clearly associated with masculinity in American culture, violence is a social practice that enables men to express a masculine identity" (p. 667). Traditional female gender roles have also taught women to be submissive to their male partner's control.
3. *View of women and children as property.* Before the late nineteenth century, a married woman was considered the property of her husband. A husband had a legal right and marital obligation to discipline and control his wife through the use of physical force. The expression "rule of thumb" can be traced to an old English law that permitted a husband to beat his wife with a rod no thicker than his thumb; it was originally intended as a humane measure to limit how

"Our culture promotes sexual victimization of women and children when it encourages males to believe that they have overpowering sexual needs that must be met by whatever means available."

Edwin Schur
Sociologist

harshly men could beat their wives. This traditional view of women as property may contribute to men's doing with their "property" as they wish. A recent study of men in battering intervention programs found that about half of the men viewed battering as acceptable in certain situations (Jackson et al. 2003).

The view of women and children as property also explains marital rape and father-daughter incest. Historically, the penalties for rape were based on property rights laws designed to protect a man's property—his wife or daughter—from rape by other men; a husband or father "taking" his own property was not considered rape (Russell 1990). In the past, a married woman who was raped by her husband could not have her husband arrested because marital rape was not considered a crime. Today, marital rape is considered a crime in all 50 states.

Community Factors. Community factors that contribute to violence and abuse in the family include social isolation and inaccessible or unaffordable health care, day care, elder care, and respite care facilities:

1. *Social isolation.* Living in social isolation from extended family and community members increases a family's risk for abuse. Isolated families are removed from material benefits, care-giving assistance, and emotional support from extended family and community members.
2. *Inaccessible or unaffordable community services.* Failure to provide medical care to children and elderly family members (a form of neglect) is sometimes a result of the lack of accessible or affordable health care services in the community. Failure to provide supervision for children and adults may result from inaccessible day care and elder care services. Without elder care and respite care facilities, socially isolated families may not have any help with the stresses of caring for elderly family members and children with special needs.

Individual and Family Factors. Individual and family factors associated with intimate partner and family violence and abuse include a history of family violence, drug and alcohol abuse, and poverty.

1. *History of abuse.* Men who witnessed their fathers abusing their mothers and women who witnessed their mothers abusing their fathers are more likely to become abusive partners themselves (Babcock, Miller, & Slard 2003; Heyman & Slep 2002). Individuals who were abused as children are more likely to report being abused in an adult domestic relationship (Heyman & Slep 2002). Mothers who have been sexually abused as children are more likely to physically abuse their own children (DiLillo, Tremblay, & Peterson 2000). Although a history of abuse is associated with an *increased likelihood* of being abusive as an adult, most adults who were abused as children do not continue the pattern of abuse in their own relationships (Gelles 2000).
2. *Drug and alcohol abuse.* More than half of prison and jail inmates serving time for violence against an intimate had been using alcohol, drugs, or both at the time of the incident for which they were incarcerated (U.S. Department of Justice 1998). Alcohol use is reported as a factor in 50 percent to 75 percent of incidents of physical and sexual aggression in intimate relationships (Lloyd & Emery 2000). Alcohol and other drugs increase aggression in some individuals and enable the offender to avoid responsibility by blaming his or her violent behavior on drugs/alcohol.

3. *Poverty.* Abuse in adult relationships occurs among all socioeconomic groups. However, Kaufman and Zigler (1992) point to a relationship between poverty and child abuse:

> Although most poor people do not maltreat their children, and poverty, per se, does not cause abuse and neglect, the correlates of poverty, including stress, drug abuse, and inadequate resources for food and medical care, increase the likelihood of maltreatment. (p. 284)

Strategies for Action: Preventing and Responding to Violence and Abuse in Intimate and Family Relationships

Strategies to prevent family violence and abuse include **primary prevention** strategies that target the general population, **secondary prevention** strategies that target groups at high risk for family violence and abuse, and **tertiary prevention** strategies that target families who have experienced abuse (Gelles 1993; Harrington & Dubowitz 1993).

Primary Prevention Strategies

"Ameliorating and preventing intimate violence against women will ultimately require a deep commitment to ending both the glorification of violence and the many facets of patriarchy that are still embedded in our culture, our laws, and our intimate lives."

Sally A. Lloyd
Director of Women's Studies, Miami University

Preventing violence and abuse may require broad, sweeping social changes such as eliminating the norms that legitimize and glorify violence in society and changing the sexist character of society (Gelles 2000). Specific abuse prevention strategies include public education and media campaigns that may help reduce domestic violence by conveying the criminal nature of domestic assault and offering ways to prevent abuse. Other prevention efforts focus on parent education to teach parents realistic expectations about child behavior and methods of child discipline that do not involve corporal punishment. For example, the National Mental Health Association (2003) distributes a fact sheet on alternatives to spanking (see Table 5.1). In 13 countries, laws or court rulings prohibit corporal punishment of children (see Table 5.2).

Another strategy involves reducing violence-provoking stress by reducing poverty and unemployment and providing adequate housing, child care programs and facilities, nutrition, medical care, and educational opportunities. However, rather than strengthening the supports for poor families with children, welfare reform legislation enacted in 1996 limits cash assistance to poor single parents to two consecutive years with a five-year lifetime limit (some exceptions are granted) and forces women into the labor force. Many women going from welfare to work will experience greater hardships as a result of a loss of food stamp benefits, increases in federal housing rent, loss of Medicaid benefits, cost of transportation to work, and child care costs (Edin & Lein 1997). Some women forced to go to work and unable to afford child care will leave their children unattended, increasing child neglect. The cumulative stresses and hardships that the welfare-to-work legislation will have on single parents may very well contribute to increases in child neglect and abuse.

Secondary Prevention Strategies

Families at risk of experiencing violence and abuse include low-income families, parents with a history of depression or psychiatric care, single parents, teenage mothers, parents with few social and family contacts, individuals who experienced

Table 5.1
Effective Discipline Techniques for Parents: Alternatives to Spanking

Punishment is a "penalty" for misbehavior, but discipline is a method of teaching a child right from wrong. Alternatives to physical discipline include the following:

1. *Be a positive role model.* Children learn behaviors by observing their parents' actions, so parents must model the ways in which they want their children to behave. If a parent yells or hits, the child is likely to do the same.
2. *Set rules and consequences.* Make rules that are fair, realistic, and appropriate to a child's level of development. Explain the rules to children along with the consequences of not following them. If children are old enough, they can be included in establishing the rules and consequences for breaking them.
3. *Encourage and reward good behavior.* When children are behaving appropriately, give them verbal praise and occasionally reward them with tangible objects, privileges, or increased responsibility.
4. *Create charts.* Charts to monitor and reward behavior can help children learn appropriate behavior. Charts should be simple and focus on one behavior at a time, for a certain length of time.
5. *Give time-outs.* "Time-outs" involve removing children from a situation following a negative behavior. This can help children calm down, end the inappropriate behavior, and reenter the situation in a positive way. Explain what the inappropriate behavior is, why the time-out is needed, when the time-out will begin, and how long it will last. Set an appropriate length of time for the time-out based on age and level of development, usually just a few minutes.

Source: Based on National Mental Health Association, 2003. "Effective Discipline Techniques for Parents: Alternatives to Spanking." Strengthening Families Fact Sheet. http://www.nmha.org. Reprinted with permission.

Table 5.2
13 Countries that Prohibit Corporal Punishment in All Settings*

Austria	Israel
Croatia	Italy
Cyprus	Latvia
Denmark	Norway
Finland	Sweden
Germany	Switzerland
Iceland	

*The ban against corporal punishment in Italy and Switzerland is based on high court rulings. In the other countries, the ban is based on legislation.

Source: The Center for Effective Discipline. 2003. "Kids Are 'Unbeatable' in 13 Countries." http://www.stophitting.com

abuse in their own childhood, and parents or spouses who abuse drugs or alcohol. Secondary prevention strategies, designed to prevent abuse from occurring in high-risk families, include parent education programs, parent support groups, individual counseling, substance abuse treatment, and home visiting programs.

Tertiary Prevention Strategies

The National Domestic Violence Hotline (1-800-799-SAFE) is a 24-hour, toll-free service that provides crisis assistance and local domestic violence shelter and safe house referrals for callers across the country. Shelters provide abused women and their children with housing, food, and counseling services. Safe houses are private homes of individuals who volunteer to provide temporary housing to abused persons who decide to leave their violent homes. Some communities have abuse shelters for victims of elder abuse. Some programs offer a safe shelter for pets of domestic violence victims. Because one in four victims reports a delay in leaving dangerous domestic situations due to concerns over the safety of a pet, some domestic abuse agencies have paired with veterinary schools, humane societies, and community organizations to help victims and their pets escape violent homes (Fogle 2003).

In 2001, 275,000 children were removed from homes as a result of child abuse investigations or assessments (U.S. Dept. of Health and Human Services 2003). Such children may be placed in foster care or in the care of another family member, such as a grandparent. However, federal law requires that states have programs to prevent family breakup when desirable and possible without jeopardizing the welfare of children in the home. **Family preservation programs** are in-home interventions for families who are at risk of having a child removed from the home because of abuse or neglect.

Alternatively, a court may order an abusing spouse or parent to leave the home. Abused spouses or cohabiting partners may obtain a restraining order prohibiting the perpetrator from going near the abused partner. About half of the states and Washington, D.C., now have mandatory arrest policies that require police to arrest abusers, even if the victim does not want to press charges. However, only about half of all victims of intimate partner violence report the violence to law enforcement authorities (Rennison & Welchans 2000). Reasons that victims do not report intimate partner violence to the police include (1) believing that such violence is a private or personal matter, (2) fear of retaliation, (3) viewing the violence as a "minor" crime, (4) to protect the offender, and (5) the belief that the police will not help or will be ineffective (Rennison & Welchans 2000).

Treatment for abusers—which may be voluntary or mandated by the court—typically involves group and/or individual counseling; substance abuse counseling; and/or training in communication, conflict resolution, and anger management. Men who stop abusing their partners learn to take responsibility for their abusive behavior, develop empathy for their partner's victimization, reduce their dependency on their partners, and improve their communication skills (Scott & Wolfe 2000). However, recent evaluations of batterer intervention programs found no significant differences between treatment and control groups on re-offense rates or men's attitudes toward domestic violence (Jackson et al. 2003).

Problems Associated with Divorce

The United States has the highest divorce rate among Western nations. Despite the decline in divorce rates over the 1990s, at today's rate an estimated half of marriages started today are expected to end in divorce (Whitehead & Popenoe 2003). Divorce is considered problematic primarily because of its effects on children. More than one-quarter (26%) of U.S. children have experienced their parents' separation or divorce (Wallerstein 2003).

Social Causes of Divorce

When we think of why a particular couple gets divorced, we typically think of a number of individual and relationship factors that might have contributed to the marital breakup: incompatibility in values or goals, poor communication, lack of conflict resolution skills, sexual incompatibility, extramarital relationships, substance abuse, emotional or physical abuse or neglect, boredom, jealousy, and difficulty coping with change or stress related to parenting, employment, finances, in-laws, and illness. But understanding the high rate of divorce in U.S. society requires awareness of how the following social and cultural factors contribute to marital breakup.

1. *Changing family functions.* Before the Industrial Revolution, the family constituted a unit of economic production and consumption, provided care and protection to its members, and was responsible for socializing and educating children. During industrialization, other institutions took over these functions. For example, the educational institution has virtually taken over the systematic teaching and socialization of children. Today, the primary function of marriage and the family is the provision of emotional support, intimacy, affection, and love. When marital partners no longer derive these emotional benefits from their marriage, they may consider divorce with the hope of finding a new marriage partner to fulfill these affectional needs.
2. *Increased economic autonomy of women.* As noted earlier, before 1940 most wives were not employed outside the home and depended on their husband's income. Today, nearly two-thirds of married women are in the labor force (*Statistical Abstract* 2000). Wives who are unhappy in their marriage are more likely to leave the marriage if they have the economic means to support themselves. Unhappy husbands may also be more likely to leave a marriage if their wives are self-sufficient and can contribute to the support of the children.
3. *Increased work demands and dissatisfaction with marital division of labor.* Another factor influencing divorce is increased work demands and the stresses of balancing work and family roles. Workers are putting in longer hours, often working overtime or taking second jobs. Many employed parents, particularly mothers, come home to work a **second shift**—the work involved in caring for children and household chores (Hochschild 1989). Wives are more likely than husbands to perceive the marital division of labor—household chores and child care—as unfair (Nock 1995). This perception of unfairness can lead to marital tension and resentment, as reflected in the following excerpt:

   > My husband's a great help watching our baby. But as far as doing housework or even taking the baby when I'm at home, no. He figures he works five days a week; he's not going to come home and clean. But he doesn't stop to think that I work seven days a week. Why should I have to come home and do the housework without help from anybody else? My husband and I have been through this over and over again. Even if he would just pick up from the kitchen table and stack the dishes for me, that would make a big difference. He does nothing. On his weekends off, I have to provide a sitter for the baby so he can go fishing. When I have a day off, I have the baby all day long without a break. He'll help out if I'm not here, but the minute I am, all the work at home is mine. (quoted in Hochschild 1997, 37–38)

"It is now widely accepted that men and women have the right to expect a happy marriage, and that if a marriage does not work out, no one has to stay trapped."

Sylvia Ann Hewlett
Family advocate

Women are increasingly looking for egalitarianism in relationships. Women want to be equal partners in their marriages, not just in earning income but in sharing the work of household chores, childrearing, and marital communication and in making decisions for the family. Frustrated by men's lack of participation in marital work, women desiring relationship egalitarianism may see divorce as the lesser of two evils (Hackstaff 2003).

4. *Liberalized divorce laws.* Before 1970, the law required a couple who wanted a divorce to prove that one of the spouses was at fault and had committed an act defined by the state as grounds for divorce—adultery, cruelty, or desertion. In 1969, California became the first state to initiate **no-fault divorce,** which permitted a divorce based on the claim that there were "irreconcilable differences" in the marriage. No-fault divorce law has contributed to the U.S. divorce rate by making divorce easier to obtain. Although U.S. divorce rates started climbing before California instituted the first no-fault divorce law, the widespread adoption of such laws has probably contributed to their continued escalation. Today, all 50 states recognize some form of no-fault divorce.
5. *Increased individualism.* U.S. society is characterized by **individualism**—the tendency to focus on one's individual self-interests rather than on the interests of one's family and community. The high divorce rate has been linked to the rise of individualism. "For many, concerns with self-fulfillment and careerism diminished their commitment to family, rendering marriage and other intimate relationships vulnerable" (Demo, Fine, & Ganong 2000, 281). **Familism**—the view that the family unit is more important than individual interests—is still prevalent among Asian-Americans and Mexican-Americans, which helps to explain why the divorce rate is lower among these groups than among whites and African-Americans (Mindel, Habenstein, & Wright 1998).
6. *Increased life expectancy.* Finally, more marriages today end in divorce, in part, because people live longer than they did in previous generations. Because people live longer today than in previous generations, "till death do us part" involves a longer commitment than it once did. Indeed, one can argue that "marriage once was as unstable as it is today, but it was cut short by death not divorce" (Emery 1999, 7).

Consequences of Divorce

Divorce is considered a social problem because of the distress and difficulties associated with it. When parents have bitter and unresolvable conflict, and/or if one parent is abusing the child or the other parent, divorce may offer a solution to family problems. But divorce often has negative effects for ex-spouses and their children and also contributes to problems that affect society as a whole.

Physical and Mental Health Consequences. In a review of research on the consequences of divorce for adults, Amato (2003) cites numerous studies that found that divorced individuals, compared to married individuals, have more health problems and a higher risk of mortality; and they experience lower levels of psychological well-being, including more unhappiness, depression, anxiety, and poorer self-concepts. Both divorced and never-married individuals are, on average, more distressed than married people because unmarried people are more likely than married people to have low social attachment, low emotional support, and increased economic hardship (Walker 2001).

However, Amato's (2003) review also cites studies in which divorced individuals report higher levels of autonomy and personal growth than do married individuals. For example, many divorced mothers report improvements in career opportunities, social lives, and happiness following divorce; some divorced women report more self-confidence and some men report more interpersonal skills and a greater willingness to self-disclose.

Economic Consequences. Compared to married individuals, divorced individuals have a lower standard of living, have less wealth, and experience greater economic hardship, although this difference is considerably greater for women than men (Amato 2003). In some families at the lowest income levels, divorce can actually improve women's income because men who earn little income can be a drain on the family's finances (Demo et al. 2000). More typically, the economic costs of divorce are greater for women and children because women tend to earn less than men (see Chapter 8) and mothers devote substantially more time to household and child care tasks than fathers do. The time women invest in this unpaid labor restricts their educational and job opportunities as well as their income. Men are less likely than women to be economically disadvantaged after divorce, as they continue to profit from earlier investments in education and career (Peterson 1996). However, men with low and unstable earnings can experience financial strain following divorce, which often underlies failure to pay child support. When custodial mothers who did not receive child support were asked the reasons they did not receive it, 66 percent responded "father unable to pay" (Henry 1999). The Fragile Families Coalition estimates that over 3 million noncustodial fathers are eligible for food stamps (Henry 1999). If these fathers are so poor that they need assistance simply to put food on the table for themselves, is it fair to characterize them as "deadbeats" when they do not pay child support? Instead of being deadbeats, some dads are simply dead broke (Henry 1999).

In some cases, a man's economic support of children following divorce may end if the man can prove that he is not the biological father of the child(ren). This chapter's *Focus on Technology* feature examines issues surrounding DNA testing and child support.

Effects on Children. Reviews of recent research on the consequences of divorce for children found that children with divorced parents score lower on measures of academic success, psychological adjustment, self-concept, social competence, and long-term health; they also have higher levels of aggressive behavior and depression (Amato 2003; Wallerstein 2003). With only one parent in the home, children of divorce as well as children of never-married mothers tend to have less adult supervision than children in two-parent homes. Lack of adult supervision is related to higher rates of juvenile delinquency, school failure, and teenage pregnancy (Popenoe 1993). A survey of 90,000 teenagers found that the mere physical presence of a parent in the home after school, at dinner, and at bedtime significantly reduces the risk of teenage suicide, violence, and drug use (Resnick et al. 1997).

Many of the negative effects of divorce on children are related to the economic hardship associated with divorce. Economic hardship is associated with less effective and less supportive parenting, inconsistent and harsh discipline, and emotional distress in children (Demo et al. 2000).

Despite the adverse effects of divorce on children, some family scholars observe that in most circumstances, children adapt to divorce, "showing resiliency, not dysfunction" (Thompson & Amato 1999, xix). As with studies of adults, some

Focus on Technology

DNA Paternity Testing and Child Support

DNA paternity testing involves using samples of blood, body tissue, fluid, sperm, bone, or most commonly, buccal swabs (from inside the cheek) to determine if the genetic makeup of a particular child matches that of the man tested. With more than 99.9 percent accuracy, DNA testing is the most advanced and accurate test available to determine a child's paternity identity.

The American Association of Blood Banks reports that the 310,490 paternity tests conducted on men in 2001 ruled out 29 percent as the biological father. Courts and legislatures have had to face the question: Should a man have to continue paying child support when DNA test results prove that he is not the biological parent?

When men voluntarily acknowledge paternity, federal law gives them 60 days to rescind it. After 60 days, states set their own limits. For example, Florida allows a year after a child support order, California two years after a birth. Only four states—Maryland, Alabama, Ohio, and Georgia—have laws that release men from child support obligations for children proven by DNA testing not to be theirs biologically, with no time constraint on when relief can be obtained.

Many unwed fathers paying child support never admitted paternity. A 1996 federal welfare law requires an unwed mother to identify a father when she applies for public assistance. A court summons can be mailed to his last known address, but many men don't receive the notice. Indeed, they may know nothing of a paternity claim until they discover that their paycheck is being docked under "default" judgments of paternity that can't be contested after six months.

If a man believes that he has been falsely accused of fathering a child, he should act promptly to protect himself. The federal government requires states to provide genetic DNA testing in all contested paternity proceedings, at no cost to the man if he is indigent. However, the courts may or may not grant relief from child support based on DNA paternity testing, and some courts will not even consider DNA testing in child support hearings. California judges, for example, would not consider tests showing that Air Force Master Sgt. Raymond Jackson's three children of his former marriage were fathered by other men. Jackson's resentment is reflected in comments he made in an interview with *USA Today:* "It's like they are saying, 'Let your wife cheat on you, have children by other men, divorce you, and now you have to pay for it all' " (Kasindorf 2002, A3). After DNA testing revealed that a Texas man was not the biological father of a child he was financially supporting, the man went to court to seek an end to his financial obligation. The court did not relieve him of the economic obligation to the child, but took away all visitation rights based on the DNA evidence. In contrast, an Oregon judge excused an ex-husband from child support payments after DNA tests showed he was not the father of his former wife's son. The judge nevertheless granted him visitation rights.

Various men's groups are lobbying state legislatures to change paternity laws as a result of the new DNA technology. Bert Riddick, a teacher in California paying $1,400 a month for a teenage girl born out of wedlock whom he's never met, says, "Think of it. I can get out of jail for murder based on DNA evidence, but I can't get out of child support payments" (quoted in Kasindorf 2002, A3). Groups such as *U.S. Citizens Against Paternity Fraud* (*http://www.paternity-fraud.com*) and *New Jersey Citizens Against Paternity Fraud* applaud states (Alabama, Arkansas, Georgia, Iowa, Ohio, and Virginia) that have reshaped paternity laws to permit ex-husbands and out-of-wedlock alleged fathers to end child support on the basis of DNA evidence. Maryland has made the same change through court decisions. But some women's groups and child advocates have concerns about how such laws affect children. Valerie Ackerman of the National Center for Youth Law says, "Families are more complicated than who's biologically related to whom. . . . If there has been a relationship between a father and child, the man can't just abdicate the responsibility that he's taken on" (quoted in Kasindorf 2002, A3). Family law professor Carol Sanger (Columbia University) suggests the law should be more lenient to men who may not even know a child than to dads who have been living with the children they didn't father. Finally, Geraldine Jensen, president of the Association for Children for Enforcement of Support, suggests that all children should be DNA-tested at birth or at the time of divorce, and that maternity wards should distribute pamphlets telling men, "Get tested now if you have any questions, because doing it later will disrupt this child's life" (quoted in Kasindorf 2002, A3).

Sources: American Association of Blood Banks. 2002 (October). "Annual Report Summary for Testing in 2001." 8101 Glenbrook Rd., Bethesda MD; Kasindorf, Martin. 2002. "Men Wage Battle on 'Paternity Fraud.' " *USA Today,* December 3, A3; Myricks, Noel. 2003. "DNA Testing and Child Support: Can the Truth Really Set You Free?" *American Journal of Family Law* 17(1):31–36. *US Newswire.* 2003 (March 10). "Father Who Took DNA Paternity Fraud Case to U.S. Supreme Court Wins at State Level." COMTEX. http://www.comtexnews.com

research suggests that divorce has positive consequences for some children. For example, in highly conflictual marriages, divorce may actually improve the emotional well-being of children relative to staying in a conflicted home environment (Jekielek 1998).

Effects on College-Age Students. In a review of the literature on the effects of parental divorce on college-age students, Nielson (1999) found that when their parents divorce and their mother remarries within a few years, most children do not suffer serious long-term consequences in terms of self-confidence, mental health, or academic achievements. However, children of divorced women who do not remarry tend to experience a reduced standard of living and a lack of parental discipline. These children "fail to develop as much self-control, self-motivation, self-reliance, and self-direction as people their own age whose mothers have remarried" (Nielsen 1999, 547).

Effects on Father/Child Relationships. Children who live with their mothers may suffer from a damaged relationship with their nonresidential father, especially if he becomes disengaged from their lives. On the other hand, children may benefit from having more quality time with their fathers after parental divorce. Some fathers report that they became more active in the role of father after divorce. One father commented:

> In the last 4 1/2 years, I have developed an incredibly strong and loving bond with my two sons. I am actively involved in all aspects of their lives. I have even coached their soccer and basketball teams. . . . The time I spend with them is very quality time—if anything, the divorce has made me a better and more caring father . . . not to say this would not have happened if my marriage had worked out. (quoted in Pasley & Minton 2001, 248)

The mother's attitude toward the father's continued contact with the child affects the father/child relationship. "If the mother approves of the close contact between her child and its father, the child will benefit both from the continued affection of the father and from the parental harmony. If the mother disapproves of the father's influence, the child, feeling torn by conflicting loyalties, may fail to benefit" (Wallerstein 2003, 76–77). Some divorced mothers not only fail to encourage their children's relationships with their fathers, but they actively attempt to alienate the children from their father. (We note that some divorced fathers do likewise.) Thus, some children of divorce suffer from **parental alienation syndrome** (PAS), defined as an emotional and psychological disturbance in which children engage in exaggerated and unjustified denigration and criticism of a parent (Gardner 1998). "Children of PAS show negative parental reactions and perceptions which can be grossly exaggerated. . . . Put simply, they profess rejection and hatred of a previously loved parent, most often in the context of divorce and child custody conflicts" (Family Court Reform Council of America 2000). Parental alienation syndrome has been described as a form of "psychological kidnapping" whereby one parent manipulates children's psyches to make them hate and reject the other parent. Children who suffer from PAS are victims of a form of child abuse in which one parent essentially brainwashes the child to hate the other parent. A parent may alienate his or her child from the other parent by engaging in the following behaviors (Schacht 2000):

- Minimizing the importance of contact and relationship with the other parent
- Being rude to the other parent; refusing to speak to or tolerate the presence of the other parent, even at events important to the child; refusing to allow the other parent near the home for drop-off or pick-up visitations

- Failing to express concern for missed visits with the other parent
- Failing to display any positive interest in the child's activities or experiences during visits
- Expressing disapproval or dislike of the child's spending time with the other parent and refusing to discuss anything about the other parent ("I don't want to hear about . . .") or selective willingness to discuss only negative matters
- Making innuendoes and accusations against the other parent, including statements that are false
- Demanding that the child keep secrets from the other parent
- Destruction of gifts or memorabilia from the other parent
- Promoting loyalty conflicts (for example, offering an opportunity for a desired activity that conflicts with scheduled visitation)

Long-term effects of PAS on children are extremely serious and can include long-term depression, inability to function, guilt, hostility, alcoholism and other drug abuse, and other symptoms of internal distress (Family Court Reform Council of America 2000). Indeed, the effects on the rejected parent are equally devastating.

Some noncustodial divorced fathers discontinue contact with their children as a coping strategy for managing emotional pain (Pasley & Minton 2001). Many divorced fathers are overwhelmed with feelings of failure, guilt, anger, and sadness over the separation from their children (Knox 1998). Hewlett and West (1998) explain that "visiting their children only serves to remind these men of their painful loss, and they respond to this feeling by withdrawing completely" (p. 69). Divorced fathers commonly experience the legal system as favoring the mother in child-related matters. One divorced father commented:

> I believe that the system [judges, attorneys, etc.] have [sic] little or no consideration for the father. At some point the system creates an environment where the father loses any natural desire to see his children because it becomes so difficult, both financially and emotionally. At that point, he convinces himself that the best thing to do is wait until they are older. (quoted in Pasley & Minton 2001, 242)

As we have seen, the effects of divorce on adults and children are mixed and variable. In a review of research on the consequences of divorce for children and adults, Amato (2003) concludes that "divorce benefits some individuals, leads others to experience temporary decrements in well-being that improve over time, and forces others on a downward cycle from which they might never fully recover" (p. 206).

Strategies for Action: Alleviating Problems of Divorce

Two general strategies for responding to the problems of divorce are (1) strategies to prevent divorce and strengthen marriages and (2) strategies to strengthen post-divorce families.

Strategies to Strengthen Marriage and Prevent Divorce

A growing "marriage movement" involves efforts by religion and government to strengthen marriage and prevent divorce. Workplace and economic supports are also important in strengthening marriages and preventing divorce.

Religion. The Community Marriage Policy strategy, developed by journalist and Presbyterian layman Michael McManus, asks religious officials (who perform 75 percent of all U.S. weddings) to follow the Common Marriage Policy of the American Roman Catholic Church (Browning 2003). The Roman Catholic Common Marriage Policy includes five components: (1) a six-month minimum marriage preparation period; (2) the use of a premarital questionnaire to identify problems or potential problems the couple may have; (3) the practice of mentoring engaged and newlywed couples; (4) the use of marriage education (such as workshops, classes, and encounter groups) for engaged and married couples for the purpose of exploring the relationship, identifying problems, and learning effective communication and conflict resolutions techniques; and (5) engagement ceremonies held before the entire congregation (Browning 2003).

The Community Marriage Policy seeks to establish a common marriage policy across religious denominations to provide a united front on standards of marriage preparation. Ministers from over 147 cities have adopted the Community Marriage Policy (Browning 2003). Of course, couples who are not required by their minister, priest, or rabbi to participate in premarital or marital education can voluntarily choose to participate in such programs offered by their religious organization, local mental health organization, or private marriage and family counselors. Researchers have found that couples who participated in a widely used couples' education program called PREP (the Prevention and Relationship Enhancement Program) had a lower divorce/separation rate five years after completing the program compared with couples who did not participate (Stanley, Markman, St. Peters, & Leber 1995). Marriage education programs for newlyweds have also been found to be effective in improving communication and conflict resolution and overall marital satisfaction (Cole, Cole, & Gandolfo 2000).

Government. In recent years, state governments have begun to enact laws designed to strengthen marriage and discourage divorce. With the passing of the 1996 Covenant Marriage Act, Louisiana became the first in the nation to offer two types of marriage contracts. Under the new law, couples can voluntarily choose the standard marriage contract that allows a no-fault divorce (after a six-month separation) or a **covenant marriage** that permits divorce only under condition of fault (e.g., abuse, adultery, felony conviction) or after a two-year separation. Couples who choose a covenant marriage must also get premarital counseling. During the first two years after the law went into effect, only 3 percent of Louisiana couples applying for a marriage license applied for a covenant marriage license (Hacker 2003). Variations of the covenant marriage have also been adopted in Arizona and Arkansas.

Advocates of the covenant marriage believe that such marriages will strengthen marriages and decrease divorce. However, critics argue that covenant marriage may increase family problems by making it more emotionally and financially difficult to terminate a problematic marriage, and by prolonging the exposure of children to parental conflict (Applewhite 2003).

Another example of government strategies to strengthen marriage and prevent divorce is a 1998 Florida law that requires all ninth- or tenth-grade public school students to take a course in marriage and relationship education (Browning 2003). Florida also cuts the cost of a marriage license by half if the couple can show that they have taken a four-hour marriage education course. Since Florida passed its marriage education law, similar bills have been introduced in Minnesota, Maryland, Arizona, Connecticut, Kansas, and Michigan. Several states are considering legislation that is intended to decrease the number of divorces by extending the waiting period required before a divorce is granted or requiring proof of fault (e.g., adultery, abuse). Opponents argue that divorce law reform measures would increase acrimony between divorcing spouses (which harms the children as well as the adults involved), increase the legal costs of getting a divorce (which leaves less money to support any children), and delay court decisions on child support and custody and distribution of assets. Efforts in many state legislatures to repeal no-fault divorce laws have largely failed.

Workplace and Economic Supports. Strengthening marriages may be achieved through policies and services that provide greater resources to couples whose marriages are at risk. These supports include marriage counseling, flexible workplace policies that decrease work/family conflict, affordable child care, and economic support such as the Earned Income Tax Credit. In a nationwide poll of parents' political priorities (National Parenting Association 1996) parents indicated that they want economic help in raising their children and work policies that help them balance their work and family roles.

Strengthening Post-Divorce Families

Negative consequences of divorce for children may be minimized if both parents continue to spend time with their children on a regular and consistent basis and communicate to their children that they love them and are interested in their lives. Parental conflict, either in intact families or divorced families, negatively influences the psychological well-being of children (Demo 1992, 1993). Ongoing conflict between divorced parents also tends to result in decreased involvement of nonresidential fathers with their children (Leite & McKenry 2000). By maintaining a civil coparenting relationship during a separation and after divorce, parents can minimize the negative effects of divorce on their children.

What can society do to promote cooperative parenting by ex-spouses? One answer is to encourage, or even mandate, divorcing couples to participate in divorce mediation. In **divorce mediation** divorcing couples meet with a neutral third party, a mediator, who helps them resolve issues of property division, child custody, child support, and spousal support in a way that minimizes conflict and encourages cooperation. Children of mediated divorces adjust better to the divorce than children of litigated divorces (Marlow & Sauber 1990). An increasing number of jurisdictions and states have mandatory child custody mediation programs, whereby parents in a custody or visitation dispute must attempt to resolve their dispute through mediation before a court will hear the case.

"Although divorce ends a marriage, it does not end the family."

Ross A. Thompson and Paul R. Amato
The Postdivorce Family

Another trend aimed at strengthening post-divorce families is the establishment of parenting programs for divorcing parents. Programs such as "Sensible Approach to Divorce" (Wyandotte County, Kansas), "Parenting after Divorce" (Orange County, North Carolina), and "Focus on Children in Separation" (Jackson County, Missouri) emphasize the importance of cooperative co-parenting for the well-being of children. Parents are taught about children's reactions to divorce, nonconflictual co-parenting skills, and how to avoid negative behavior toward their ex-spouse. In some programs (such as "Focus on Children in Separation"), children participate and are taught that they are not the cause of the divorce, how to deal with grief reactions to divorce, and techniques for talking to parents about their concerns. A few states require all divorcing parents to attend a parenting program, as do more than 100 courts throughout the United States.

Nonmarital and Teenage Childbearing

As noted earlier in this chapter, about one-third of all U.S. births are to unmarried women. Over the last several decades, the rates of nonmarital childbearing has increased substantially, and while the majority of nonmarital births (64% in 2000) are among whites, blacks have a much higher *rate* of nonmarital childbearing compared to whites. However, as shown in Figure 5.5, between 1995 and 2001, the rate of nonmarital births *decreased* slightly among blacks, while continuing an upward trend among whites.

Compared with some countries, the United States has a high proportion of nonmarital births. In Germany, Italy, Greece, and Japan, less than 15 percent of births occur out of wedlock. Some countries, however, such as Iceland, Norway, and Sweden, have rates of out-of-wedlock childbearing that are much higher than that of the United States (Ventura & Bachrach 2000).

In recent years, we have witnessed a substantial decrease in rates (per 1,000) of teenage pregnancy, abortion, and births. For example, for females aged 15 to 19, the birthrate fell from 62.1 in 1991 to 45.8 in 2001—a historic low for this age group (Ozer et al., 2003). Although the largest number of teenage births are to non-Hispanic whites (42.6%), the birth rate per 1,000 for non-Hispanic whites aged 15 to 19 is less than half that of Hispanic, Black, and American Indian/Alaskan Native groups. The lowest teenage birthrates are among Asian/Pacific Islander (see Figure 5.6).

In the 1950s, when the majority of teen mothers were married and were expected to be stay-at-home wives, teenage childbearing was not a public concern. Today, most teenage births (80%) occur outside of wedlock (Mauldon 2003). Teenage births today are considered problematic because early parenthood interferes with the acquisition of education and job-related skills, and "it seems to guarantee a lifetime of poverty and hardship for a teenage mother and her baby" (Mauldon 2003, 41).

In 1970, half of all nonmarital births were to teens. However, in 1999, less than 3 in 10 (29 percent) nonmarital births were to teenagers; the remaining 70 percent of unmarried births were to women aged 20 and older (Ventura & Bachrach 2000). About half of pregnant teens carry their babies to term, 35 percent have an abortion,

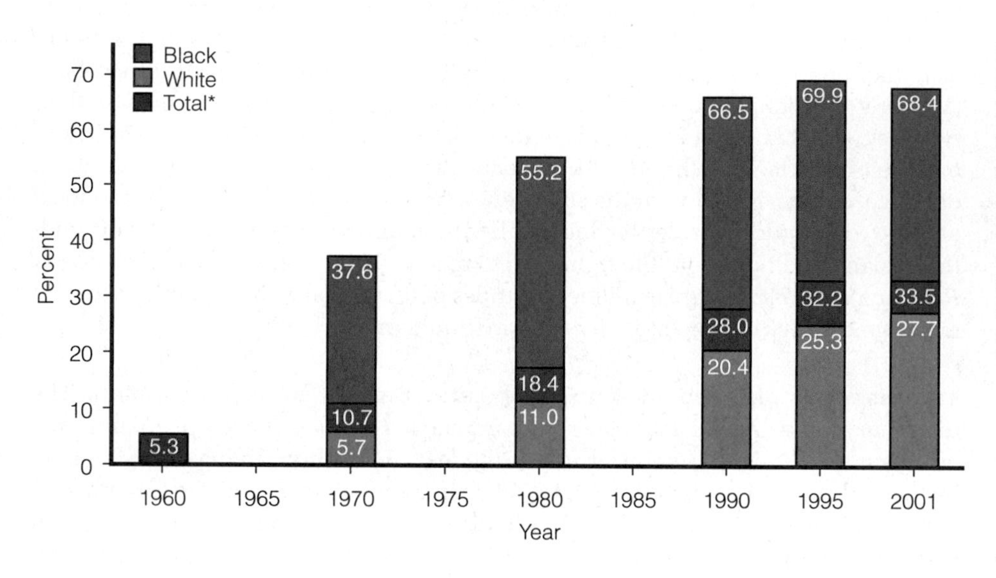

Figure 5.5
Percentage of All Live Births to Unmarried Women, by Race and Year, United States

Source: *Statistical Abstract of the United States: 1995,* Table 94; *Statistical Abstract of the United States, 1999,* Table 99; *Statistical Abstract of the United States, 2000,* Table 85; *Statistical Abstract of the United States, 2003,* Table 91.

Figure 5.6
Birthrates (per 1,000) of U.S. Teenage Females Ages 15–19, by Race and Hispanic Origin, 2001

Source: Based on Ozer, E. M., M. J. Park, T. Paul, C. D. Brindis, and C. E. Irwin, Jr. 2003. *America's Adolescents: Are They Healthy?* San Francisco: University of California, San Francisco, National Adolescent Health Information Center.

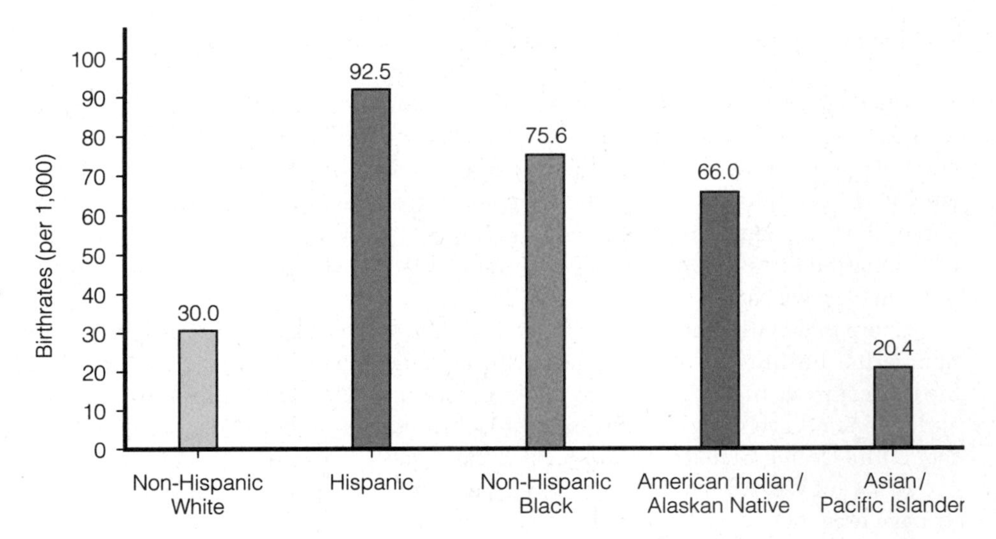

and the remainder miscarry. Most teens today who carry their babies to term (95%) keep the baby rather than place it for adoption (Jorgensen 2000).

Teen pregnancy, birth, and abortion rates in the United States are higher than in any other industrialized country. In much of Western Europe, low teen pregnancy and birth rates are attributed to the widespread availability and use of effective contraception among sexually active teens.

Social Factors Related to Nonmarital and Teenage Childbearing

High rates of childbearing among teenage and unmarried women are related to several factors, including increased social acceptance of unwed childbearing; increased singlehood, cohabitation, and same-sex relationships; and poverty.

Increased Social Acceptance of Unwed Childbearing. Having a baby outside of marriage has become more socially acceptable and does not carry the stigma it once did. In the 1950s and 1960s, more than half of U.S. women who gave birth to a baby conceived out of wedlock married before the birth of the baby. In the 1990s, less than one in four (23%) of such women married before the birth of the baby (Ventura & Bachrach 2000). In a 2001 Gallup survey of women and men in their twenties, less than half (44%) of respondents agreed that "it is wrong to have a child outside of marriage" (Whitehead & Popenoe 2003). In the same survey, 40 percent of single women agreed with the statement that "although it might not be the ideal option, you would consider having a child on your own if you reached your mid-thirties and had not found the right man to marry." This chapter's *Social Problems Research Up Close* feature examines attitudes of low-income single mothers toward marriage that help to explain why they are not married.

Increased Singlehood, Cohabitation, and Same-Sex Relationships. The dictionary once defined a *spinster* as an unmarried woman above age 30. In previous generations, being a spinster meant the fear of isolation, living alone, and being something of a social outcast. Today, women are "more confident, more self-sufficient, and more choosy than ever [and] no longer see marriage as a matter of survival and acceptance" (Edwards 2000, 48). The women's movement created both

Actress Susan Sarandon is considered, legally, a single parent even though she is cohabiting with actor Tim Robbins, the father of her two sons.

new opportunities for women and expectations of egalitarian relationships with men. A Time/CNN poll of single women found that 61 percent felt that men were "too controlling" (Edwards 2000, 50). Increasingly, college-educated women in their thirties and forties are making "the conscious decision to have a child on their own because they haven't found Mr. Adequate, let alone Mr. Right" (Drummond 2000, 54).

Increased acceptance of cohabitation and same-sex relationships has also contributed to the high rate of nonmarital births. In the 1990s, about 40 percent of out-of-wedlock births were to women who were cohabiting, presumably with the child's father (Bachu 1999), but in some cases with the mother's same-sex partner.

Poverty. Teenage pregnancy has been related to a variety of factors, including low self-esteem and hopelessness, low parental supervision, and perceived lack of future occupational opportunities (Aassve 2003; Jorgensen 2000; Luker 1996). Although lack of information about and access to contraceptives contributes to unintended teenage pregnancy, 30 percent to 40 percent of adolescent pregnancies are intended (Jorgensen 2000). Teenage females who do poorly in school may have little hope of success and achievement in pursuing educational and occupational goals. They may think that their only remaining option for a meaningful role in life is to become a parent. In addition, some teenagers feel lonely and unloved and have a baby to create a sense of feeling needed and wanted.

Social Problems Research Up Close

Perceptions of Marriage Among Low-Income Single Mothers

Single mothers with low incomes face both economic hardships and parenting challenges. These hardships and challenges could potentially be alleviated by marriage to a partner who contributes income to the family and who shares the responsibilities of housework and child care and supervision. Yet low-income single women have low rates of marriage and remarriage. Sociologist Kathryn Edin (2000) conducted research on the perceptions of marriage among low-income single mothers that helps explain why these mothers are not married. Here we summarize her research and its findings.

Sample and Methods

This study focused on transcripts and field notes collected from multiple qualitative interviews with 292 low-income single mothers. Initially, interviewers utilized the resources of community groups and local institutions to recruit eligible mothers. They then relied on references from the mothers themselves to gain access to others who might also participate in the interviews. The sample was evenly divided between African-Americans and whites in three U.S. cities. Approximately half of the participants in each city depended on welfare, and about half worked at low-wage jobs earning less than $7.50/hour.

Researchers scheduled a minimum of two interviews with each mother to maximize rapport and to encourage them to expand and clarify responses that may have been initially unclear. Interviewers asked respondents to describe the circumstances surrounding the births of their children, their current family situations and relationship with each child's father, their views about how their family situations might change over time, and their views of marriage in general. Interviewers focused on two issues in their conversations with mothers: why women with few economic resources choose to have children and why these same women fail to marry or remarry.

Findings and Conclusions

Edin found four primary reasons that low-income single mothers do not marry. These include the low economic status of available men; the desire for parental, economic, and relationship control; mistrust; and domestic violence.

1. *Low economic status of men.* Edin found that men's income is a significant factor in poor single mothers' willingness to marry, as "they simply could not afford to keep an unproductive man around the house" (p. 118). Nearly every mother in Edin's sample said that it was important for any man they would marry to have a "good job." Although total earnings is the most important aspect for single mothers, also important are the regularity of those earnings, the effort men make to find and keep employment, and the source of the income. "Women whose male partners couldn't or wouldn't find work, often lost respect for them and 'just couldn't stand' to keep them around" (p. 119). Mothers did not consider earnings from crime (e.g., drug dealing) as legitimate earnings and reported that they wouldn't marry such a man no matter how much money he made from crime.

Respectability was also an issue for poor single mothers:

> Mothers said that they could not achieve respectability by marrying someone who was frequently out of work, otherwise underemployed, supplemented his income through criminal activity, and had little chance of improving his situation over time. Mothers believed that marriage to such a man would diminish their respectability, rather than enhance it. (p. 120)

Mothers also believed that marriage to a lower-class man would be unlikely to last because the economic pressures on the relationship would be too great. Mothers talked about the "sacred" nature of marriage and believed that no "respectable" woman would marry a man whose economic situation would likely lead to a future divorce.

Edin notes that:

> In interview after interview, mothers stressed the seriousness of the marriage commitment and their belief that "it should last forever." Thus, it is not that mothers held marriage in low esteem, but rather the fact that they held it in such high esteem that convinced them to forego marriage, at least until their prospective marriage partner could prove himself worthy economically or they could find another partner who could. (pp. 120–121)

2. *Desire for parental, economic, and relationship control.* When asked what they liked best about being a single parent, the most common response was "I am in charge," or "I am in control." Most mothers felt that the presence of fathers often interfered with their parental control. Single mothers in Edin's study felt that a husband might be "too demanding" of them and thus impede their efforts to spend time with their children.

Mothers were also concerned about losing control of the family's economic situation. One mother said, "[I won't marry because] the men take over the money. I'm too afraid to lose control of my money again" (p. 121).

Regardless of whether the prospective wife worked, mothers feared that prospective husbands would expect to be the "head of the house," who, being "in charge," would want to make the "final" decisions about child rearing, finances, and other matters. Mothers also expressed the view that if they married, their husbands would expect them to do all of the household chores. Some women described their relationships with their ex-partners as "like having one more kid to take care of" (p. 122).

Most single mothers in Edin's sample wanted to marry eventually and thought that the best way to maximize their control in a marriage was by working and earning an income. One African-American mother explained,

> One thing my mom did teach me is that you must work some and bring some money into the household so you can have a say in what happens. If you completely live off a man, you are helpless. That is why I don't want to get married until I get my own [career] and get off of welfare. (p. 122)

Mothers also wanted to develop their labor market skills before marriage to ensure against destitution in the event of a divorce and to increase their bargaining power in their relationships. Mothers believed that women who were economically dependent on men had to "put up with all kinds of behavior" because they could not legitimately threaten to leave.

> Mothers felt that if they became more economically independent . . . they could legitimately threaten to leave their husbands if certain conditions (i.e., sexual fidelity) weren't met. These threats would, in turn they believed, keep a husband on his best behavior. (p. 122)

Many of the single mothers in Edin's study had held traditional gender role attitudes when they were younger and still in a relationship with their children's fathers. When the men for whom they sacrificed so much gave them nothing but pain and anguish, they felt they had been "duped" and were no longer willing to be dependent or subservient to men.

3. *Mistrust.* Many of the single mothers in Edin's sample did not marry because they did not trust their partners or men in general, and/or because their partners did not trust them. Some fathers claimed that the child was not theirs because the mother was "a whore." One partner of a pregnant woman in the sample told the interviewer, "how do I know the baby's mine? Who knows if she hasn't been stepping out on me with some other man and now she wants me to support another man's child!" (p. 124).

Mothers tended to mistrust men because their previous boyfriends and partners had been unfaithful. This experience was so common among respondents that many simply did not believe men could be faithful to only one woman. Most said they would rather never marry than to "let him make a fool out of me" (p. 124). "Nonetheless, many of these same women often held out hope of finding a man who was 'different,' one who could be trusted" (p. 125).

4. *Domestic violence.* For some mothers in Edin's sample, domestic violence in either their childhood or adult lives played a role in their negative attitudes toward marriage. Many mothers reported physical abuse during pregnancy and several mothers had miscarriages because of such abuse. One woman who had been in the hospital three times because of abuse said, "I was terrified to leave because I knew it would mean going on welfare. . . . But that is okay. I can handle that. The thing I couldn't deal with is being beat up" (p. 126). Edin commented,

> The fact that women tended to experience repeated abuse from their children's fathers before they decided to leave attests to their strong desire to make things work with their children's fathers. Many women finally left when they saw the abuse beginning to affect their children's well-being. (p. 126)

In conclusion, low-income single mothers in Edin's study were reluctant to marry the father of their children because these men had low economic status, traditional notions of male domination in household and parental decisions, and patterns of untrustworthy and even violent behavior. Given the low level of trust these mothers have of men and given their view that husbands want more control than the women are willing to give them, women realize that a marriage that is also economically strained is likely to be conflictual and short-lived. "Interestingly, mothers say they reject entering into economically risky marital unions out of respect for the institution of marriage, rather than because of a rejection of the marriage norm" (Edin 2000, 130). Low-income single mothers in Edin's study "say they are willing, even eager, to marry if the marriage represents an increase in their class standing and if . . . their prospective husband's behavior indicates he won't beat them, abuse their children, refuse to share household tasks, insist on making all decisions, be sexually unfaithful, or abuse alcohol or drugs" (p. 113):

> In short, the mothers interviewed here believe that marriage will probably make their lives more difficult than they are currently. . . . If they are to marry, they want to get something out of it. If they cannot enjoy economic stability and gain upward mobility from marriage, they see little reason to risk the loss of control and other costs they fear marriage might exact from them. Unless low-skilled men's economic situations improve and they begin to change their behaviors toward women, it is quite likely that most low-income women will continue to resist marriage. (p. 130)

Source: Edin, Kathryn. 2000. "What Do Low-Income Single Mothers Say About Marriage?" *Social Problems* 47(1):112–133. Used by permission of University of California Press.

Social Problems Related to Nonmarital and Teenage Childbearing

"For many disadvantaged teenagers, childbearing reflects— rather than causes— the limitations of their lives."

Ellen W. Freeman and Karl Rickels
Researchers

Teenage and unmarried childbirth are considered social problems because of the adverse consequences for women and children that are associated with such births.

1. *Poverty for single mothers and children.* Many unmarried mothers, especially teenagers, have no means of economic support or have limited earning capacity. Single mothers and their children often live in substandard housing and have inadequate nutrition and medical care. Even with public assistance, many unwed parents struggle to survive economically. Of the 17 percent of U.S. children living in poverty in 2002, more than half (52%) lived in single-mother homes (Fields 2003).
2. *Poor health outcomes.* Compared with older pregnant women, pregnant teenagers are less likely to receive timely prenatal care and to gain adequate weight and are more likely to smoke and use alcohol and drugs during pregnancy (Jorgensen 2000; Ventura, Curtin, & Mathews 2000). As a consequence of these and other factors, infants born to teenagers are at higher risk of low birth weight, of premature birth, and of dying in the first year of life.
3. *Low academic achievement.* Low academic achievement is both a contributing factor and a potential outcome of teenage parenthood. Teens whose parents have not graduated from high school are at higher risk for becoming pregnant (Hogan, Sun, & Cornwell 2000). Three-fifths of teenage mothers drop out of school and, as a consequence, have a much higher probability of remaining poor throughout their life. Because poverty is linked to unmarried parenthood, a cycle of successive generations of teenage pregnancy may develop.

   Some research has found that children of single mothers are more prone to academic problems. However, Henry Ricciuti, professor of human development, found that the mother's educational level and ability, rather than the absence of a father, have the most influence on a child's school readiness (Drummond 2000).
4. *Children without fathers.* About one-third of U.S. children who live in households without their fathers are the products of unmarried childbirth (Hewlett & West 1998). Shapiro and Schrof (1995) report that children who grow up without fathers are more likely to drop out of school, be unemployed, abuse drugs, experience mental illness, and be a target of child sexual abuse. They also note that "a missing father is a better predictor of criminal activity than race or poverty" (p. 39). Popenoe (1996) believes that fatherlessness is a major cause of the degenerating conditions of our young.

   However, others argue that the absence of a father is not, in and of itself, damaging to children. Rather, other conditions associated with female-headed single-parent families, such as low educational attainment of the mother, poverty, and lack of child supervision contribute to negative outcomes for children.

Strategies for Action: Interventions in Teenage and Nonmarital Childbearing

Interventions in teenage and nonmarital childbearing include efforts to prevent such births and strategies to minimize the negative effects of such births to women and children. Although most nonmarital births are to women who are not on welfare, the 1996 welfare reform law contained measures designed to discourage nonmarital childbearing. The welfare reform law of 1996 included an "illegitimacy bonus," which rewards states that reduce out-of-wedlock births and also decrease abortions. From 1999 to 2003, the federal government provided generous financial bonuses to up to five states each year that achieved the greatest declines in out-of-wedlock births and reduced their abortion rate to below its 1995 level.

The European approach to teenage sexual activity is to provide widespread confidential and accessible contraceptive services to adolescents. The provision of contraceptive services to European teens is believed to be a central factor in explaining the more rapid declines in teenage childbearing in northern and western European countries, in contrast to slower declines in the United States (Singh & Darroch 2000). Although sex education is provided in schools throughout the United States, most programs emphasize abstinence and do not provide students with access to contraception.

Other programs aim at both preventing teenage and unmarried childbearing and minimizing its negative effects by increasing the life options of teenagers and unmarried mothers. Such programs include educational programs, job training, and skill-building programs. Other programs designed to help teenage and unwed mothers and their children include prenatal programs to help ensure the health of the mother and baby, and public welfare such as WIC (Women, Infants, and Children) and TANF (Temporary Assistance to Needy Families). However, public assistance to low-income single mothers has been blamed for contributing to the problem by discouraging the poor from marrying. Up until 1996, single mothers with no or little income could receive welfare as long as they had a dependent child living with them. If single mothers on welfare married, they could lose welfare benefits. In 1996, the Personal Responsibility and Work Opportunity Reconciliation Act (PRWORA) reformed the welfare program (see Chapter 10) by limiting cash benefits to adults. For example, adults can receive welfare benefits for no more than two consecutive years (unless they work at least 20 hours per week) and are limited to five years of welfare cash benefits over their lifetime. One goal of welfare reform is to encourage marriage by decreasing the benefits an unmarried mother can claim. Despite the new time limitations on welfare benefits, marriage rates among low-income single mothers continue to be low.

Strategies to increase and support fathers' involvement with their children are relevant to both children of unwed mothers and children of divorce. At the federal and state levels, Fatherhood Initiative programs encourage fathers' involvement with children through a variety of means (U.S. Department of Health and Human Services 2000). These include promoting responsible fatherhood by improving work opportunities for low-income fathers, increasing child support collections, providing parent education training for men, supporting access and visitation by noncustodial parents, and involving boys and young men in preventing teenage pregnancy and early parenting. Because teenage parents are less likely than older parents to use positive and effective childrearing techniques, parent education programs for teen mothers and fathers are an important component of improving the lives of young parents and their children.

Understanding Family Problems

Family problems can best be understood within the context of the society and culture in which they occur. Although domestic violence, divorce, and teenage pregnancy and unmarried parenthood may appear to result from individual decisions, these decisions are influenced by a myriad of social and cultural forces.

> “It is time to boldly address the urgent need to provide every appropriate support to every kind of family— and so make it possible for all of us to discover the lasting joy and deep sense of connection experienced by those who are joined at the heart.”
>
> **Al and Tipper Gore**

The impact of family problems, including divorce, abuse, and nonmarital childbearing, is felt not only by family members but by the larger society as well. Family members experience such life difficulties as poverty, school failure, low self-esteem, and mental and physical health problems. Each of these difficulties contributes to a cycle of family problems in the next generation. The impact on society includes public expenditures to assist single-parent families and victims of domestic violence and neglect, increased rates of juvenile delinquency, and lower worker productivity.

For some, the solution to family problems implies encouraging marriage and discouraging other family forms, such as single parenting, cohabitation, and same-sex unions. But many family scholars argue that the fundamental issue is making sure that children are well cared for, regardless of whether their parents are married. Some even suggest that marriage is part of the problem, not part of the solution. Professor Martha Fineman of Cornell Law School says, “This obsession with marriage prevents us from looking at our social problems and addressing them. . . . Marriage is nothing more than a piece of paper, and yet we rely on marriage to do a lot of work in this society: It becomes our family policy, our police in regard to welfare and children, the cure for poverty” (quoted in Lewin 2000, 2). Strengthening marriage is a worthy goal because strong marriages offer many benefits to individuals and their children. However, “strengthening marriage does not have to mean a return to the patriarchal family of an earlier era. . . . Indeed, greater marital stability will only come about when men are willing to share power, as well as housework and childcare, equally with women” (Amato 1999, 184). And strengthening marriage does not mean that other family forms should not also be supported. In their book *Joined at the Heart,* Al and Tipper Gore (2002) suggest that the first and most important step to helping families is to change our way of *thinking* about families so that our view of family encompasses those who are connected emotionally and committed to one another as family—those who are “joined at the heart” (p. 327). The reality is that the postmodern family comes in many forms, each with its strengths, needs, and challenges. Given the diversity of families today, social historian Stephanie Coontz (2000) suggests, “The only way forward at this point in history is to find better ways to make *both* marriage and its alternatives work” (p. 15).

Chapter Review

- **What are some examples of diversity in families around the world?**
  Some societies recognize monogamy as the only legal form of marriage, whereas other societies permit polygamy. Societies also vary in their policies regarding same-sex couples and their norms regarding the roles of women, men, and children in the family.

- **What are some of the major changes in U.S. households and families that have occurred in the last several decades?**
  Some of the major changes in U.S. households and families that have occurred in recent decades include increased singlehood and older age at first marriage, increased heterosexual and same-sex cohabitation, increased births to unmarried women, increased single-parent families, fewer children living in married-couple families, increased divorce and blended families, and increased employment of married mothers. While some view these changes as signaling a breakdown of marriage and family, others emphasize that the forms and patterns of families are less important than the quality of love, care, and nurturance families provide for their members.

- **Feminist theories of family are most similar to which of the three main sociological theories: structural-functionalism, conflict, or symbolic interactionism?**
  Feminist theories of family are most aligned with conflict theory. Both feminist and conflict theories are concerned with how gender inequality influences and results from family patterns.

- **What are four patterns of partner violence identified by Johnson and Ferraro?**
Four patterns of partner violence are (1) common couple violence (occasional acts of violence arising from arguments that get "out of hand"); (2) intimate terrorism (violence that is motivated by a wish to control one's partner); (3) violent resistance (acts of violence that are committed in self-defense); and (4) mutual violent control (both partners battling for control).

- **What are the differences between primary prevention, secondary prevention, and tertiary prevention strategies in regard to preventing and responding to domestic violence and abuse?**
Primary prevention strategies target the general population, secondary prevention strategies target groups at high risk for family violence and abuse, and tertiary prevention strategies target families who have experienced abuse.

- **Why do many abused adults stay in abusive relationships?**
Adult victims of abuse are commonly blamed for choosing to stay in their abusive relationships. From the point of view of the victim, reasons to stay in the relationship include love, emotional dependency, commitment to the relationship, hope that things will get better, the view that violence is legitimate because they "deserve" it, guilt, fear, economic dependency, feeling stuck, and fear of loneliness. Some victims stay because they fear the abuser will abuse or neglect a family pet. Victims also stay because abuse in relationships is usually not ongoing and constant but rather occurs in cycles in which a violent or abusive episode is followed by a makeup period when the abuser expresses sorrow and asks for forgiveness and "one more chance."

- **What are some of the effects of divorce on children?**
Reviews of recent research on the consequences of divorce for children find that children with divorced parents score lower on measures of academic success, psychological adjustment, self-concept, social competence, and long-term health, and have higher levels of aggressive behavior and depression. Such effects are related to the economic hardship associated with divorce, the reduced parental supervision resulting from divorce, and parental conflict during and following divorce. In highly conflictual marriages, divorce may actually improve the emotional well-being of children relative to staying in a conflicted home environment.

- **What is divorce mediation?**
In divorce mediation divorcing couples meet with a neutral third party, a mediator, who helps them resolve issues of property division, child custody, child support, and spousal support in a way that minimizes conflict and encourages cooperation. In some states, counties, and jurisdictions, divorcing couples who are disputing child custody issues are required to participate in divorce mediation before their case can be heard in court.

- **What are some of the social problems related to nonmarital and teenage childbearing?**
Teenage and unmarried childbirth are considered social problems because of the adverse consequences for women and children that are associated with such births, which include (1) increased risk of poverty for single mothers and their children; (2) risk of poor health outcomes for babies born to teenage women; (3) risk of dropping out of school for teenage mothers and for low academic achievement of their children; and (4) increased risks for children who grow up without fathers.

- **How does the European approach to teenage sexuality compare with the U.S. approach?**
The European approach to teenage sexual activity involves providing widespread confidential and accessible contraceptive services to adolescents. Although sex education is provided in schools throughout the United States, most programs emphasize abstinence and do not provide students with access to contraception.

Critical Thinking

1. Some scholars and politicians argue that "stable families are the bedrock of stable communities." Others argue that "stable communities and economies are the bedrock of stable families." Which of these two positions would you take and why?
2. Research has suggested that secondhand smoke from cigarettes represents a health hazard for those who are exposed to it. Consequently, smoking is now banned in many public places and workplaces. Do you think that parents should be banned from smoking in enclosed areas (home or car) to protect their children from secondhand smoke? Do you think that smoking in enclosed areas with one's children present should be considered a form of child abuse? Why or why not?
3. In the United States, women are more likely than men to initiate divorce. Why do you think this is so?

Key Terms

bigamy
child abuse
common couple violence
corporal punishment
covenant marriage
cycle of abuse
divorce mediation
divorce rate
domestic partnership
elder abuse
familism
family household
family preservation program
individualism
intimate partner violence
intimate terrorism
monogamy
mutual violent control

neglect
no-fault divorce
nonfamily household
parental alienation syndrome (PAS)
patriarchy
polyandry
polygamy
polygyny
primary prevention
second shift
secondary prevention
serial monogamy
shaken baby syndrome
tertiary prevention
traditional family
violent resistance

Taking A Stand

Should corporal punishment of children be prohibited by law?

In 1979, Sweden became the first country in the world to ban corporal punishment. Eight European countries (Austria, Croatia, Cyprus, Denmark, Finland, Italy, Latvia, and Norway) have followed Sweden's lead and have banned corporal punishment in all settings, including the home. Do you think that U.S. law should ban corporal punishment of children?

Use Wadsworth's exclusive online resources—InfoTrac College Edition, MicroCase Online, and OVRC—to formulate a position on this topic.

The Wadsworth's Sociology Online Resources and Writing Companion will help you get started. This valuable guide will show you how to use Wadsworth's exclusive online resources when studying social problems. It will also help you to build essential research and writing skills. InfoTrac College Edition, MicroCase Online, OVRC, and an electronic copy of portions of this companion are available at http://sociology.wadsworth.com/mooney_knox_schacht/problems4e, the companion Web site for *Understanding Social Problems,* Fourth Edition.

Media Resources

The Companion Web Site for *Understanding Social Problems,* Fourth Edition

http://sociology.wadsworth.com/mooney_knox_schacht/problems4e

Supplement your review of this chapter by going to the companion Web site to take one of the Tutorial Quizzes, use the flash cards to master key terms, and check out the many other study aids you'll find there. You'll also find special features such as *Wadsworth's Sociology Online Resources and Writing Companion,* GSS Data, and Census 2000 information, data, and resources at your fingertips to help you complete that special project or do some research on your own.

Interactions CD-ROM

Go to the Interactions CD-ROM for *Understanding Social Problems,* Fourth Edition, to access additional interactive learning tools, such as in-depth review materials, corresponding practice quizzes, and other engaging resources and activities to help you study the concepts in this chapter.

"The 21st century will be the century in which we redefine ourselves as the first country in world history which is literally made up of every part of the world." *Kenneth Prewitt, Census Bureau director*

Race and Ethnic Relations

Dana Marie Fecho-Al Hilali, a U.S.- born Muslim journalism student from Missouri, was in New York, attending a conference of the College Media Association. Dana Marie follows the practice of hijab, or wearing a head scarf, as part of her religious practice; the rest of her clothing was contemporary Western dress. While she and a non-Muslim female friend were walking in Central Park, she noticed a police car driving back and forth, very slowly. "I felt relieved," Dana later told a reporter, "thinking the officer was keeping an eye on us, two young women alone, and maybe also in a protective way on me. Wearing hijab, I know I stand out, and sometimes my scarf invites some negative comments from people. So it was kind of good to see the police around" (Morrison 2002, 9). Imagine her shock when the patrol car pulled up alongside her and the police officer yelled out the window "You (expletive) American Taliban. Go back where you came from!" Dana Marie recalled, "It pretty much turned my sense of security upside down. . . . Here I was, an American, and I had always trusted the police to protect me. I thought I could always go to them if there was a problem. And now here was a New York City policeman calling me a terrorist, and telling me to get out of my own country" (p. 9).

Following the terrorist attacks of September 11, 2001, there was backlash of prejudice and discrimination against individuals who are (or who are perceived to be) Muslim, Middle Eastern, and/or who have names that are similar to the terrorists involved in the September 11 attack. In the United States, Muslims and people of Middle-Eastern descent are minorities. A **minority group** is a category of people who have unequal access to positions of power, prestige, and wealth in a society and who tend to be targets of prejudice and discrimination. Minority status is not based on numerical representation in society but rather on social status. For example, before Nelson Mandela was elected president of South Africa, South African blacks suffered the disadvantages of a minority, even though they were a numerical majority of the population.

In this chapter, we discuss the nature and origins of prejudice and examine the extent of discrimination and its consequences for racial and ethnic minorities. We also discuss strategies designed to reduce prejudice and discrimination. We begin by examining racial and ethnic diversity worldwide and in the United States, emphasizing first that the concept of race is based on social rather than biological definitions.

The Global Context: Diversity Worldwide

A first-grade teacher asked the class, "What is the color of apples?" Most of the children answered red. A few said green. One boy raised his hand and said "white." The teacher tried to explain that apples could be red, green, or sometimes golden, but never white. The boy insisted his answer was right and finally said, "Look inside" (Goldstein 1999). Like apples, human beings may be similar on the "inside," but are often classified into categories according to external appearance. After examining the social construction of racial categories, we review patterns of interaction among racial and ethnic groups and examine racial and ethnic diversity in the United States.

The Social Construction of Race

The concept **race** refers to a category of people who are believed to share distinct physical characteristics that are deemed socially significant. Racial groups are sometimes distinguished on the basis of such physical characteristics as skin color, hair texture, facial features, and body shape and size. Some physical variations among people are the result of living for thousands of years in different geographical regions (Molnar 1983). For example, humans living in regions with hotter climates developed darker skin from the natural skin pigment, melanin, which protects the skin from the sun's rays. In regions with moderate or colder climates, people had no need for protection from the sun and thus developed lighter skin.

Cultural definitions of race have taught us to view race as a scientific categorization of people based on biological differences between groups of individuals. Yet racial categories are based more on social definitions than on biological differences.

"Race and ethnicity are in the eye of the beholder."

Harold L. Hodgkinson
Director of the Center for Demographic Policy

Anthropologist Mark Cohen (1998) explains that distinctions among human populations are graded, not abrupt. Skin color is not black or white, but rather ranges from dark to light with many gradations of shades. Noses are not either broad or narrow, but come in a range of shapes. Physical traits such as these, as well as hair color and other characteristics, come in an infinite number of combinations. For example, a person with dark skin can have any blood type and can have a broad nose (a common combination in West Africa), a narrow nose (a common trait in East Africa), or even blond hair (a combination found in Australia and New Guinea) (Cohen 1998).

The science of genetics also challenges the notion of race. Geneticists have discovered that "the genes of black and white Americans probably are 99.9 percent alike" (Cohen 1998, B4). Furthermore, genetic studies indicate that genetic variation is greater *within* racially classified populations than *between* racial groups (Keita & Kittles 1997). Classifying people into different races fails to recognize that over the course of human history, migration and intermarriage have resulted in the blending of genetically transmitted traits. Thus, there are no "pure" races; people in virtually all societies have genetically mixed backgrounds.

The American Anthropological Association has passed a resolution stating that "differentiating species into biologically defined 'race' has proven meaningless and unscientific" (Etzioni 1997, 39). Scientists who reject the race concept now speak of *populations* when referring to groups that most people would call races (Zack 1998).

As clear evidence that race is a social rather than a biological concept, different societies construct different systems of racial classification, and these systems change over time. For example, "at one time in the not too distant past in the United States, Italians, Greeks, Jews, the Irish, and other 'white' ethnic groups were not considered to be white. Over time . . . the category of 'white' was reshaped to include them" (Rothenberg 2002, 3).

The significance of race is not biological but social and political, as race is used to separate "we" from "they" and becomes a basis for unequal treatment of one group by another. Despite the increasing acceptance that race is not a valid biological categorization, its social significance continues to be evident throughout the world.

Patterns of Racial and Ethnic Group Interaction

When two or more racial or ethnic groups come into contact, one of several patterns of interaction may occur, including genocide, expulsion or population transfer, slavery, colonialism, segregation, acculturation, assimilation, pluralism, and amalgamation. These patterns of interaction may occur when two or more groups exist in the same society or when different groups from different societies come into contact. Although not all patterns of interaction between racial and ethnic groups are

"I have a dream that my four little children will one day live in a nation where they will not be judged by the color of their skin, but by the content of their character."

Martin Luther King, Jr.
Civil rights leader

The most cited twentieth-century example of genocide is the "extermination" of over 12 million Jews and others considered by Hitler not to be members of his "superior race."

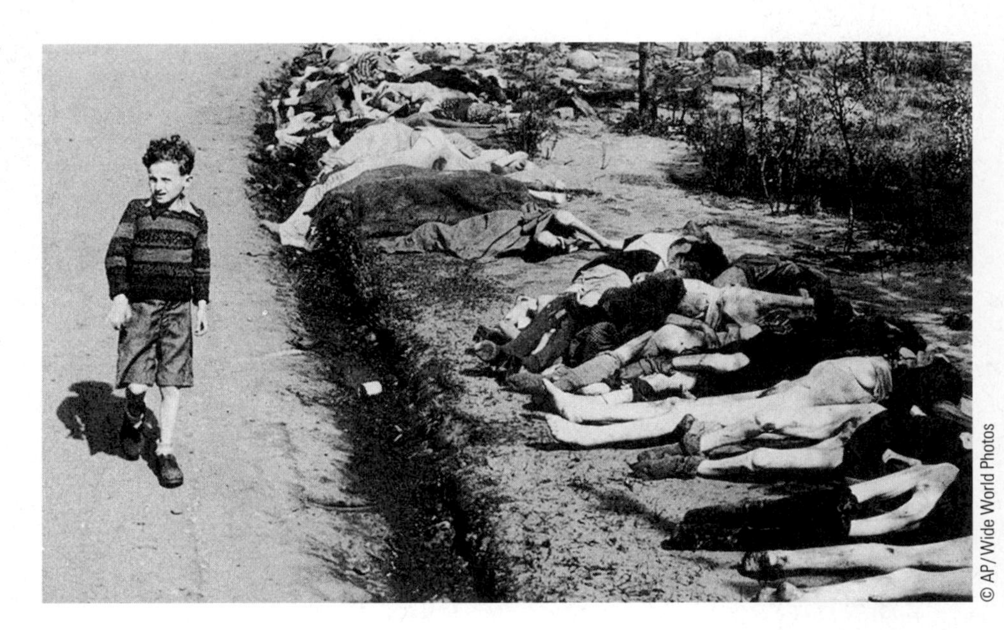

© AP/Wide World Photos

destructive, author and Mayan shaman Martin Prechtel reminds us, "Every human on this earth, whether from Africa, Asia, Europe, or the Americas, has ancestors whose stories, rituals, ingenuity, language, and life ways were taken away, enslaved, banned, exploited, twisted, or destroyed" (quoted in Jensen 2001, 13).

- **Genocide** refers to the deliberate, systematic annihilation of an entire nation or people. The European invasion of the Americas, beginning in the sixteenth century, resulted in the decimation of most of the original inhabitants of North and South America. Some native groups were intentionally killed, whereas others fell victim to diseases brought by the Europeans. In the twentieth century, Hitler led the Nazi extermination of more than 12 million people, including over 6 million Jews, in what has become known as the Holocaust. More recently, ethnic Serbs attempted to eliminate Muslims from parts of Bosnia—a process they call "ethnic cleansing."
- **Expulsion** or **population transfer** occurs when a dominant group forces a subordinate group to leave the country or to live only in designated areas of the country. The 1830 Indian Removal Act called for the relocation of eastern tribes to land west of the Mississippi River. The movement, lasting more than a decade, has been called the Trail of Tears because tribes were forced to leave their ancestral lands and endure harsh conditions of inadequate supplies and epidemics that caused illness and death. After Japan's attack on Pearl Harbor in 1941, President Franklin Roosevelt authorized the removal of any people considered threats to national security. All people on the West Coast of at least one-eighth Japanese ancestry were transferred to evacuation camps surrounded by barbed wire, where 120,000 Japanese Americans experienced economic and psychological devastation. In 1979, Vietnam expelled nearly 1 million Chinese from the country as a result of long-standing hostilities between China and Vietnam.
- **Colonialism** occurs when a racial or ethnic group from one society takes over and dominates the racial or ethnic group(s) of another society. The European invasion of North America, the British occupation of India, and the Dutch presence in

South Africa before the end of apartheid are examples of outsiders taking over a country and controlling the native population. As a territory of the United States, Puerto Rico is essentially a colony whose residents are U.S. citizens but cannot vote in presidential elections unless they move to the mainland.

- **Segregation** refers to the physical separation of two groups in residence, workplace, and social functions. Segregation may be ***de jure*** (Latin meaning "by law") or ***de facto*** ("in fact"). Between 1890 and 1910, a series of U.S. laws that came to be known as **Jim Crow laws** were enacted to separate blacks from whites by prohibiting blacks from using "white" buses, hotels, restaurants, and drinking fountains. In 1896, the U.S. Supreme Court (in *Plessy v. Ferguson*) supported de jure segregation of blacks and whites by declaring that "separate but equal" facilities were constitutional. Blacks were forced to live in separate neighborhoods and attend separate schools. Beginning in the 1950s, various rulings overturned these Jim Crow laws, making it illegal to enforce racial segregation. Although de jure segregation is illegal in the United States, de facto segregation still exists in the tendency for racial and ethnic groups to live and go to school in segregated neighborhoods.
- **Acculturation** refers to adopting the culture of a group different from the one in which a person was originally raised. Acculturation may involve learning the dominant language, adopting new values and behaviors, and changing the spelling of the family name. In some instances, acculturation may be forced, as in the California decision to discontinue bilingual education and force students to learn English in school.
- **Pluralism** refers to a state in which racial and ethnic groups maintain their distinctness but respect each other and have equal access to social resources. In Switzerland, for example, four ethnic groups—French, Italians, Germans, and Swiss Germans—maintain their distinct cultural heritage and group identity in an atmosphere of mutual respect and social equality. In the United States, the political and educational recognition of multiculturalism reflects efforts to promote pluralism.
- **Assimilation** is the process by which formerly distinct and separate groups merge and become integrated as one. Assimilation is sometimes referred to as the "melting pot," whereby different groups come together and contribute equally to a new, common culture. Although the United States has been referred to as a "melting pot," in reality, many minorities have been excluded or limited in their cultural contributions to the predominant white Anglo-Saxon Protestant tradition.

  Assimilation may be of two types: secondary and primary. **Secondary assimilation** occurs when different groups become integrated in public areas and in social institutions, such as neighborhoods, schools, the workplace, and in government. **Primary assimilation** occurs when members of different groups are integrated in personal, intimate associations, as with friends, family, and spouses.
- **Amalgamation,** also known as **marital assimilation,** occurs when different ethnic or racial groups become married or pair-bonded and produce children. Nineteen states had **antimiscegenation laws** banning interracial marriage until 1967 when the Supreme Court (in *Loving v. Virginia*) declared these laws unconstitutional. In 2000, the citizens of Alabama voted to remove the state constitution's ban on interracial marriage (even though the ban had been deemed unconstitutional by the 1967 Supreme Court ruling and could not be enforced). However, 41 percent of Alabama voters voted *against* removing the ban from the state constitution in the 2000 election, reflecting continuing disapproval of interracial marriage among some voters. Despite the lack of complete social acceptance of interracial marriage, the share of all U.S. mixed-race married couples grew from 4.4 percent in 1990 to

6.7 percent in 2000. One in 15 U.S. marriages today is interracial—up from one in 23 in 1990 (Frey 2003).

Racial and Ethnic Diversity in the United States

The United States is becoming increasingly diversified in the racial and ethnic characteristics of its population. Indeed, 2000 census data revealed that for the first time in modern U.S. history, non-Hispanic whites make up less than half of the population of California (Purdum 2001). The majority of Californians are minorities, with Hispanic residents making up nearly one-third of the state's population. Racial and ethnic group relations continue to be a major social concern in the United States and throughout the world. Despite significant improvements in U.S. race and ethnic relations over the last two centuries, a great deal of work remains to be done. And much pessimism about the future of race relations in the United States is evident. In response to a question asking whether relations between blacks and whites will always be a problem for the United States or whether a solution will eventually be worked out, 51 percent of whites and 59 percent of blacks say that race relations will always be a problem (Ludwig 2000).

Racial Diversity in the United States

The first census in 1790 divided the U.S. population into four groups: free white males, free white females, slaves, and other persons (including free blacks and Indians). To increase the size of the slave population, the **one drop of blood rule** appeared, which specified that even one drop of Negroid blood defined a person as black and, therefore, eligible for slavery. In 1960, the census recognized only two categories: white and nonwhite. In 1970, the census categories consisted of white, black, and "other" (Hodgkinson 1995). In 1990, the U.S. Bureau of the Census recognized four racial classifications: (1) white, (2) black, (3) American Indian, Aleut, or Eskimo, and (4) Asian or Pacific Islander. The 1990 census also included the category of "other." Beginning with the 2000 census, the Office of Management and Budget requires federal agencies to use a minimum of five race categories: (1) white, (2) black or African American, (3) American Indian or Alaska Native, (4) Asian,

Golf pro Tiger Woods has referred to himself as "Cablinasian"—reflecting his mixed heritage, which includes Caucasian, Black, Asian, and Indian.

→NOTE: Please answer BOTH questions 5 and 6.
5. **Is this person Spanish/Hispanic/Latino?** Mark ☒ the "No" box if **not** Spanish/Hispanic/Latino.

☐ **No,** not Spanish/Hispanic/Latino ☐ Yes, Puerto Rican
☐ Yes, Mexican, Mexican Am., Chicano ☐ Yes, Cuban
☐ Yes, other Spanish/Hispanic/Latino — *Print group.*

6. **What is this person's race?** Mark ☒ **one or more races** to indicate what this person considers himself/herself to be.

☐ White
☐ Black, African Am., or Negro
☐ American Indian or Alaska Native — *Print name of enrolled or principal tribe.*

☐ Asian Indian ☐ Japanese ☐ Native Hawaiian
☐ Chinese ☐ Korean ☐ Guamanian or Chamorro
☐ Filipino ☐ Vietnamese ☐ Samoan
☐ Other Asian — *Print race.* ☐ Other Pacific Islander — *Print race.*

☐ Some other race — *Print race.*

Figure 6.1
Reproduction of Questions on Race and Hispanic Origin from Census 2000

Source: U.S. Census Bureau, Census 2000 questionnaire.

and (5) Native Hawaiian or other Pacific Islander (Grieco & Cassidy 2001). In addition, respondents to federal surveys and the census now have the option of officially identifying themselves as being more than one race, rather than checking only one racial category (see Figure 6.1). Figure 6.2 presents 2000 Census data on the racial composition of the United States.

A mixed-race option for self-identification avoids putting children of mixed-race parents in the difficult position of choosing the race of one parent over the other when filling out race data on school and other forms. It also avoids impairment of children's self-esteem and social functioning that comes from choosing the racial category of "other." Such a category implies that the society does not recognize and respect mixed-race individuals, and thus "children growing up within mixed families may feel ashamed of their 'irregular' racial makeup and may experience rejection and alienation in the wider social community" (Zack 1998, 23).

Some critics of the new mixed-race option are concerned that the wide-scale recognition of mixed-race identity will decrease the numbers within minority groups and disrupt the solidarity and loyalty based on racial identification. What will happen, for example, to organizations and movements devoted to equal rights for blacks if much of the "black" population acquires a new mixed-racial identity? However, 2000 census data suggest that the mixed-race option will not have the large national impact that critics fear. As shown in Figure 6.2 only 2.4 percent of the U.S. population identified themselves as being of more than one race in the

Figure 6.2
Race Composition of the United States, 2000

Source: Grieco, Elizabeth M., & Rachel C. Cassidy. 2001 (March). "Overview of Race and Hispanic Origin: Census 2000 Brief." U.S. Census Bureau.

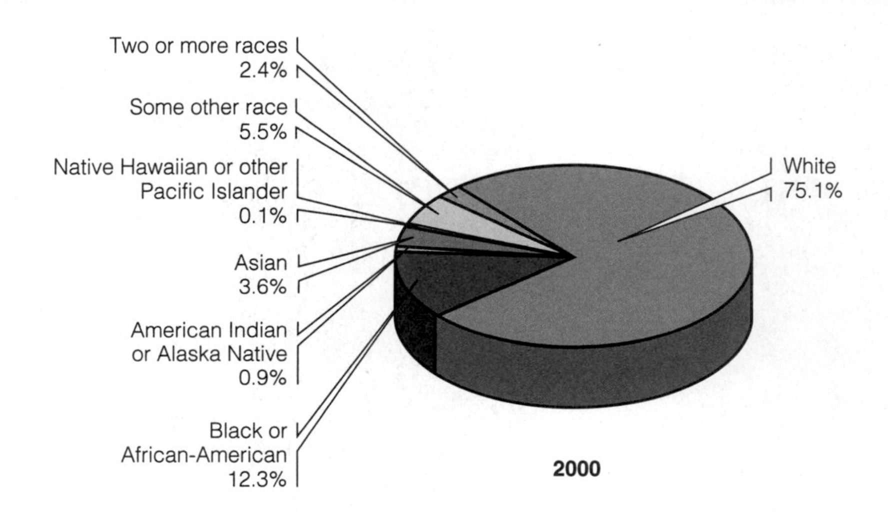

2000 Census. About 5 percent of U.S. black people said they were multiracial (Schmitt 2001).

Ethnic Diversity in the United States

Ethnicity refers to a shared cultural heritage or nationality. Ethnic groups may be distinguished on the basis of language, forms of family structures and roles of family members, religious beliefs and practices, dietary customs, forms of artistic expression such as music and dance, and national origin.

Two individuals with the same racial identity may have different ethnicities. For example, a black American and a black Jamaican have different cultural, or ethnic, backgrounds. Conversely, two individuals with the same ethnic background may identify with different races. While most Hispanics/Latinos are white, in the 2000 Census 2 percent of Latinos identified themselves as black (Navarro 2003).

The current Census Bureau classification system does not allow people of mixed Hispanic/Latino ethnicity to identify themselves as such. Individuals with one Hispanic and one non-Hispanic parent still must say they are either Hispanic or not Hispanic. And Hispanics must select one country of origin, even if their parents are from different countries.

U.S. citizens come from a variety of ethnic backgrounds. The largest ethnic population in the United States is of Hispanic origin. (The terms "Hispanic" and "Latino" are used interchangeably.) More than one in eight (13.3%) people in the United States are Hispanic or Latino and two-thirds (66.9%) of all U.S. Hispanics/Latinos are of Mexican origin (Ramirez & de la Cruz 2003). Other Hispanic Americans have origins in Central and South America (14.3%), Puerto Rico (8.6%), and Cuba (3.7%).

The fastest growing racial/ethnic population in the United States is Asian/Pacific Islanders, followed by Hispanics. Estimates project that the percentage of Asian/Pacific Islanders in the United States will increase 36.1 percent from 2000 to 2010; the percentage of Hispanics will increase 31.2 percent over the same time period (Gardyn & Fetto 2000).

The use of racial and ethnic labels is often misleading. The ethnic classification of "Hispanic/Latino," for example, lumps together such disparate groups as Puerto

Ricans, Mexicans, Cubans, Venezuelans, Colombians, and others from Latin American countries. The racial term "American Indian" includes more than 300 separate tribal groups that differ enormously in language, tradition, and social structure. The racial label "Asian American" includes individuals with ancestors from China, Japan, Korea, India, the Philippines, or one of the countries of Southeast Asia.

U.S. Immigration

The growing racial and ethnic diversity of the United States is largely the result of immigration (as well as the higher average birth rates among many minority groups). The many hardships experienced by poor people throughout the world "push" them to leave their home countries, while the economic opportunities that exist in more affluent countries "pull" them to those countries.

For the first 100 years of U.S. history, all immigrants were allowed to enter and become permanent residents. The continuing influx of immigrants, especially those coming from nonwhite, non-European countries, created fear and resentment among native-born Americans who competed with immigrants for jobs and who held racist views toward some racial and ethnic immigrant populations. Increasing pressures from U.S. citizens to restrict, or halt entirely, the immigration of various national groups led to legislation that did just that. America's open door policy on immigration ended in 1882 with the Chinese Exclusion Act, which suspended for 10 years the entrance of the Chinese to the United States and declared them ineligible for U.S. citizenship. The Immigration Act of 1917 required all immigrants to pass a literacy test before entering the United States. And in 1921, the Johnson Act for the first time introduced a limit on the number of immigrants who could enter the country in a single year, with stricter limitations for certain countries (including Africa and the Near East). The 1924 Immigration Act further limited the number of immigrants allowed into the United States and completely excluded the Japanese. Other federal immigration laws include the 1943 Repeal of the Chinese Exclusion Act, the 1948 Displaced Persons Act (which permitted refugees from Europe), and the 1952 Immigration and Naturalization Act (which permitted a quota of Japanese immigrants).

In the 1960s, most immigrants were from Europe, but now they are from Central America (predominantly Mexico) or Asia (see Figure 6.3). In 2002, more than one in 10 U.S. residents (11.5%) was born in a foreign country (Schmidley 2003)

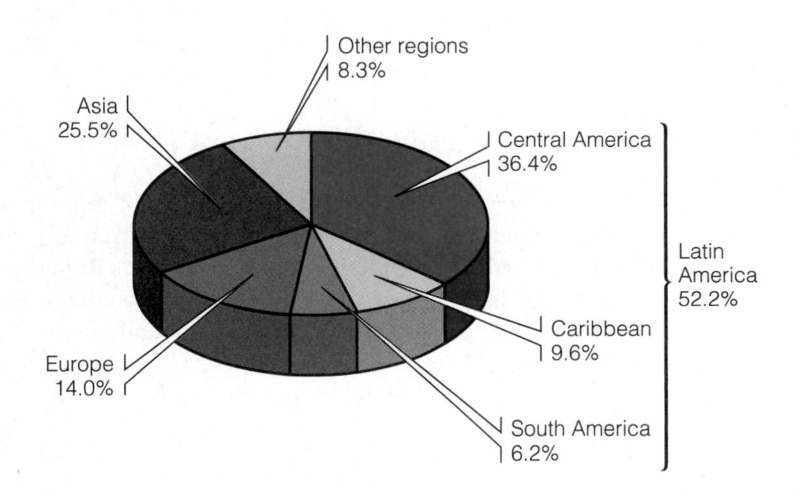

Figure 6.3
U.S. Foreign Born by Region of Birth: 2002

Source: Schmidley, Dianne. 2003 (February). "The Foreign-Born Population in the United States: March 2002." *Current Population Reports,* P20-539. Washington, DC: U.S. Census Bureau.

Table 6.1

Foreign-Born Population and Percentage of Total Population for the United States: 1890 to 2002

Year	Number (in millions)	Percent of total
2002	32.5	11.5
1990	19.8	7.9
1970	9.6	4.7
1950	10.3	6.9
1930	14.2	11.6
1890	9.2	14.8

Source: Lisa Lollock. 2001. "The Foreign-Born Population in the United States: March 2000." *Current Population Reports,* P20-534. Washington, DC: U.S. Bureau of the Census; Schmidley, Dianne. 2003. "The Foreign-Born Population in the United States: March 2002." *Current Population Reports,* P20-539. Washington, DC: U.S. Bureau of the Census.

(see Table 6.1), and one in five U.S. children 18 and under is the child of an immigrant (Fix & Capps 2002). In 2000, more than half of the nation's foreign-born population lived in four states: California (28%), New York (12%), Texas (9%), and Florida (9%). States with the fastest growing immigrant populations include North Carolina, followed by Georgia, Nevada, Arkansas, Utah, Tennessee, Nebraska, Colorado, Arizona, and Kentucky (Fix & Capps 2002).

Foreign-born residents of the United States may or may not apply for, and be granted, U.S. citizenship. In 2000, of all foreign-born U.S. residents, 35 percent were **naturalized citizens** (immigrants who applied and met the requirements for U.S. citizenship); 65 percent were not U.S. citizens (Lollock 2001). Requirements for becoming a U.S. citizen are discussed in this chapter's *Self and Society* feature.

Illegal immigration is an ongoing concern in the United States. In 1986, Congress approved the Immigration Reform and Control Act, which made hiring illegal aliens an illegal act punishable by fines and even prison sentences. This act also prohibits employers from discriminating against legal aliens who are not U.S. citizens. The more recent 1996 Illegal Immigrant Reform and Immigrant Responsibility Act increased border enforcement and penalties for illegal entry into the United States and placed restrictions on federal public benefits for immigrants (McLemore, Romo, & Baker 2001).

Many immigrants struggle to adjust to life in the United States. Immigrants are less likely than U.S. natives to graduate from high school and are more likely to be unemployed and to live in poverty (Schmidley 2003). Most legal immigrants admitted to the United States after 1996 are ineligible for welfare, public health insurance, and other benefits (Reardon-Anderson, Capps, & Fix 2002). But despite the prejudice, discrimination, and lack of social support they experience, many foreign-born U.S. residents work hard to succeed educationally and occupationally. The percentage of the foreign born with a bachelor's degree or more education (26.5%) is nearly equivalent to that of the native population (26.8%). Unemployment rates in 2002 were similar between foreign-born men (6.8%) and native men

(6.7%), but they differed between foreign-born women (7.0%) and native women (5.4%) (Schmidley 2003).

According to a study by the National Academy of Sciences, immigration produces economic benefits for the United States as a whole ("Study Finds Benefits from Immigration" 1997). Economist James Smith explained: "It's true that some Americans are now paying more taxes because of immigration, and native-born Americans without a high school education have seen their wages fall slightly because of the competition sparked by lower-skilled, newly arrived immigrants. But the vast majority of Americans are enjoying a healthier economy as a result of the increased supply of labor and lower prices that result from immigration" (p. 4A). In North Carolina, the state with the highest growth of immigrants in the 1990s, one county commissioner noted that without immigration, "our economy would shut down" (Foust, Grow, & Pascual 2002, 82). Some of immigration's biggest supporters are business leaders who want to keep wages low. Sociology professor Stephen Klineberg notes that "the accelerating immigration of Hispanics has meant massively downward pressure on their wages, negotiating power, and ability to confront their employers over violations of their rights" (quoted in Greenhouse 2003, n.p). While some negative public attitudes toward immigration stem from racism, job displacement is also a major concern. In the Los Angeles hotel industry, for example, unionized native-born black janitors making $12 an hour have largely been replaced by non-union immigrants from Mexico and El Salvador, who are paid much less (Motavalli & Zarrella 2004).

Sociological Theories of Race and Ethnic Relations

Some theories of race and ethnic relations suggest that individuals with certain personality types are more likely to be prejudiced or to direct hostility toward minority group members. Sociologists, however, concentrate on the impact of the structure and culture of society on race and ethnic relations. Three major sociological theories lend insight into the continued subordination of minorities.

Structural-Functionalist Perspective

Functionalists emphasize that each component of society contributes to the stability of the whole. In the past, inequality between majority and minority groups was functional for some groups in society. For example, the belief in the superiority of one group over another provided moral justification for slavery, supplying the South with the means to develop an agricultural economy based on cotton. Further, southern whites perpetuated the belief that emancipation would be detrimental for blacks, who were highly dependent upon their "white masters" for survival (Nash 1962).

Functionalists recognize, however, that racial and ethnic inequality is also dysfunctional for society (Schaefer 1998; Williams & Morris 1993). A society that practices discrimination fails to develop and utilize the resources of minority members. Prejudice and discrimination aggravate social problems such as crime and violence, war, poverty, health problems, family problems, urban decay, and drug use—problems that cause human suffering as well as financial burdens on individuals and society.

Self and Society

Becoming an American Citizen: Could You Pass the Test?

To become a U.S. citizen, immigrants must have been lawfully admitted for permanent residence; have resided continuously as a lawful permanent U.S. resident for at least five years; be able to read, write, speak, and understand basic English (certain exemptions apply); and must show that they have "good moral character" (Immigration and Naturalization Service 2001). Applicants who have been convicted of murder or an aggravated felony are permanently denied U.S. citizenship. In addition, applicants are denied if in the last five years they have engaged in any one of a variety of offenses, including prostitution, illegal gambling, controlled substance law violation (except for a single offense of possession of 30 grams or less of marijuana), habitual drunkenness, willful failure or refusal to support dependents, and criminal behavior involving "moral turpitude."

To become a U.S. citizen, one must take the oath of allegiance and swear to support the Constitution and obey U.S. laws, renounce any foreign allegiance, and bear arms for the U.S. military or perform services for the U.S. government when required.

Finally, applicants for U.S. citizenship must pass an examination administered by the U.S. Immigration and Naturalization Service. The following questions are typical of those on the examination given to immigrants seeking U.S. citizenship. Applicants may choose between an oral and a written test. On the oral test, they must answer all the questions correctly. On the written test, they must correctly answer 12 of 20 questions. Based on your answers to these questions, would you be granted U.S. citizenship if you were an immigrant (that is, could you correctly answer 6 out of the following 10 questions)? After selecting an answer for each of the following questions, check your answers using the answer key provided.

Sample Questions for Becoming a U.S. Citizen

1. Who becomes the president of the United States if the president and vice president should die?
   a. The speaker of the House of Representatives
   b. The Senate majority leader
   c. The chairman of the Joint Chiefs of Staff
   d. The chief justice of the Supreme Court
2. Who said, "Give me liberty or give me death"?
   a. George Washington
   b. Benjamin Franklin
   c. Patrick Henry
   d. Thomas Jefferson
3. How many branches are there in our government?
   a. 2
   b. 3
   c. 4
   d. 6
4. Which countries were our enemies during World War II?
   a. Iraq, Libya, and Turkey
   b. Germany, Japan, and the Soviet Union

The structural-functionalist analysis of manifest and latent functions also sheds light on issues of race and ethnic relations. For example, the manifest function of the civil rights legislation in the 1960s was to improve conditions for racial minorities. However, civil rights legislation produced an unexpected consequence, or latent function. Because civil rights legislation supposedly ended racial discrimination, whites were more likely to blame blacks for their social disadvantages and thus perpetuate negative stereotypes such as "blacks lack motivation" and "blacks have less ability" (Schuman & Krysan 1999).

Conflict Perspective

The conflict perspective examines how competition over wealth, power, and prestige contribute to racial and ethnic group tensions. Consistent with this perspective, the "racial threat" hypothesis views white racism as a response to perceived

c. Japan, Italy, and Germany
d. Italy, Germany, and France

5. What are the duties of Congress?
   a. To execute laws
   b. To naturalize citizens
   c. To sign bills into law
   d. To make laws
6. Which list contains three rights or freedoms guaranteed by the Bill of Rights?
   a. Right to life, right to liberty, right to the pursuit of happiness
   b. Freedom of speech, freedom of press, freedom of religion
   c. Right to protest, right to protection under the law, freedom of religion
   d. Freedom of religion, right to elect representatives, human rights
7. How many times may a senator be reelected?
   a. There is no limit
   b. Once
   c. Twice
   d. 4 times
8. Who signs bills into law?
   a. The Supreme Court
   b. The president
   c. Congress
   d. The Senate
9. How many changes or amendments are there to the Constitution?
   a. 5
   b. 9
   c. 13
   d. 27
10. Who has the power to declare war?
    a. Congress
    b. The president
    c. Chief justice of the Supreme Court
    d. Chairman of the Joint Chiefs of Staff

Answer Key

1 = a; 2 = c; 3 = b; 4 = c; 5 = d; 6 = b; 7 = a; 8 = b; 9 = d; 10 = a

Sources: Immigration and Naturalization Service, U.S. Department of Justice. 1989. By the People . . . U.S. Government Structure. Washington, DC: U.S. Government Printing Office. Immigration and Naturalization Service, 2001. "General Naturalization Requirements." http://www.ins.usdoj.gov/graphics/services/natz/general.htm

or actual threats to their economic well-being or cultural dominance by minorities. For example, between 1840 and 1870, large numbers of Chinese immigrants came to the United States to work in mining (California Gold Rush of 1848), railroads (transcontinental railroad completed in 1860), and construction. As Chinese workers displaced whites, anti-Chinese sentiment rose, resulting in increased prejudice and discrimination and the eventual passage of the Chinese Exclusion Act of 1882, which restricted Chinese immigration until 1924. More recently, white support for Proposition 209—a 1996 resolution passed in California that ended state affirmative action programs—was higher in areas with larger Latino, African-American or Asian-American populations, even after controlling for other factors (Tolbert & Grummel 2003). In other words, opposition to affirmative action programs that help minorities was higher in areas with greater racial and ethnic diversity, suggesting that whites living in diverse areas feel more threatened by the minorities. In another study, researchers interviewed individuals in white racist Internet chat rooms to examine the extent to which people would advocate interracial violence in response to alleged economic and cultural threats (Glaser, Dixit, & Green 2002). The researchers posed three scenarios that might be perceived as threatening: interracial marriage, minority in-migration (i.e., blacks moving into one's neighborhood), and job competition (i.e., competing with a black person for a job). Respondents' reactions to interracial marriage were the most volatile, followed by in-migration. The researchers concluded that violent ideation among white racists stems from perceived threats to white cultural dominance and separateness rather than from perceived economic threats.

Further, conflict theorists suggest that capitalists profit by maintaining a surplus labor force, that is, having more workers than are needed. A surplus labor force assures that wages remain low because someone is always available to take a disgruntled worker's place. Minorities who are disproportionately unemployed serve the interests of the business owners by providing surplus labor, keeping wages low, and, consequently, enabling them to maximize profits.

Conflict theorists also argue that the wealthy and powerful elite foster negative attitudes toward minorities in order to maintain racial and ethnic tensions among workers. As long as workers are divided along racial and ethnic lines, they are less likely to join forces to advance their own interests at the expense of the capitalists. In addition, the "haves" perpetuate racial and ethnic tensions among the "have-nots" to deflect attention away from their own greed and exploitation of workers.

Symbolic Interactionist Perspective

The symbolic interactionist perspective focuses on the social construction of race and ethnicity—how we learn conceptions and meanings of racial and ethnic distinctions through interaction with others—and how meanings, labels, and definitions affect racial and ethnic groups. The different connotations of the colors white and black, for example, may contribute to negative attitudes toward people of color. The white knight is good, and the black knight is evil; angel food cake is white, devil's food cake is black. Other negative terms associated with black include black sheep, black plague, black magic, black mass, blackballed, and blacklisted. The continued use of such derogatory terms as Jap, Gook, Spic, Frog, Kraut, Coon, Chink, Wop, and Mick also confirms the power of language in perpetuating negative attitudes toward minority group members.

The labeling perspective directs us to consider the role that negative stereotypes play in race and ethnicity. **Stereotypes** are exaggerations or generalizations about the characteristics and behavior of a particular group. When Americans in a 1990 National Opinion Research Center poll were asked to evaluate various racial and ethnic groups, blacks were rated least favorably (Shipler 1998). Most of the respondents labeled blacks as less intelligent than whites (53%), lazier than whites (62%), and more likely than whites to prefer being on welfare to being self-supporting (78%). Negative stereotyping of minorities leads to a self-fulfilling prophecy. As Schaefer (1998, 17) explains:

> Self-fulfilling prophecies can be devastating for minority groups. Such groups often find that they are allowed to hold only low-paying jobs with little prestige or opportunity for advancement. The rationale of the dominant society is that these minority individuals lack the ability to perform in more important and lucrative positions. Training to become scientists, executives, or physicians is denied to many subordinate group individuals, who are then locked into society's inferior jobs. As a result, the false definition

becomes real. The subordinate group has become inferior because it was defined at the start as inferior and was therefore prevented from achieving the levels attained by the majority.

The symbolic interactionist perspective is concerned with how individuals learn negative stereotypes and prejudicial attitudes. In the next section, we explore the concepts of prejudice and racism in more depth and discuss ways in which socialization and media perpetuate negative stereotypes and prejudicial attitudes toward racial and ethnic groups.

Prejudice and Racism

Prejudice refers to negative attitudes and feelings toward or about an entire category of people. Prejudice may be directed toward individuals of a particular religion, sexual orientation, political affiliation, age, social class, sex, race, or ethnicity. **Racism** is the belief that race accounts for differences in human character and ability and that a particular race is superior to others.

"I'm not against Blacks. I'm against all nonwhites."

Ku Klux Klan Member

Aversive and Modern Racism

Compared with traditional, "old-fashioned" prejudice which is blatant, direct, and conscious, contemporary forms of prejudice are often subtle, indirect, and unconscious. Two variants of these more subtle forms of prejudice include aversive racism and modern racism.

Aversive Racism. **Aversive racism** represents a subtle, often unintentional form of prejudice exhibited by many well-intentioned white Americans who possess strong egalitarian values and who view themselves as nonprejudiced. The negative feelings that aversive racists have toward blacks and other minority groups are not feelings of hostility or hate, but rather, feelings of discomfort, uneasiness, disgust, and sometimes fear (Gaertner & Dovidio 2000). Aversive racists may not be fully aware that they harbor these negative racial feelings; indeed, they disapprove of individuals who are prejudiced and would feel falsely accused if they were labeled prejudiced. "Aversive racists find Blacks 'aversive,' while at the same time find any suggestion that they might be prejudiced 'aversive' as well" (Gaertner & Dovidio 2000, 14).

Another aspect of aversive racism is the presence of pro-white attitudes, as opposed to anti-black attitudes. In several studies, respondents did not indicate that blacks were worse than whites, only that whites were better than blacks (Gaertner & Dovidio 2000). For example, blacks were not rated as being lazier than whites, but whites were rated as being more ambitious than blacks. Gaertner and Dovidio (2000) explain that "aversive racists would not characterize blacks more negatively than whites because that response could readily be interpreted by others or oneself to reflect racial prejudice" (p. 27). Compared with anti-black attitudes, pro-white attitudes reflect a more subtle prejudice that, although less overtly negative, is still racial bias.

Modern Racism. Like aversive racism, **modern racism** involves the rejection of traditional racist beliefs, but a modern racist displaces negative racial feelings onto more abstract social and political issues. The modern racist believes that serious discrimination in America no longer exists, that any continuing racial inequality is the fault of minority group members, and that demands for affirmative action for minorities are unfair and unjustified. "Modern racism tends to 'blame the victim' and place the responsibility for change and improvements on the minority groups, not on the larger society" (Healey 1997, 55). Like the aversive racist, modern racists tend to be unaware of their negative racial feelings and do not view themselves as prejudiced.

Learning to Be Prejudiced: The Role of Socialization and the Media

Psychological theories of prejudice focus on forces within the individual that give rise to prejudice. For example, the frustration-aggression theory of prejudice (also known as the scapegoating theory) suggests that prejudice is a form of hostility that results from frustration. According to this theory, minority groups serve as convenient targets of displaced aggression. The authoritarian-personality theory of prejudice suggests that prejudice arises in people with a certain personality type. According to this theory, people with an authoritarian personality—who are highly conformist, intolerant, cynical, and preoccupied with power—are prone to being prejudiced.

Rather than focus on the individual, sociologists focus on social forces that contribute to prejudice. Earlier we explained how intergroup conflict over wealth, power, and prestige give rise to negative feelings and attitudes that serve to protect and enhance dominant group interests. In the following discussion, we explain how prejudice is learned through socialization and the media.

Learning Prejudice Through Socialization. Although most researchers agree that the majority of children learn conceptions of racial and ethnic distinctions by the time they are about 6, some suggest that children as young as 3 years old have acquired prejudicial attitudes (Van Ausdale & Feagin 2001).

> Well before they can speak clearly, children are exposed to racial and ethnic ideas through their immersion in and observation of the large social world. Since racism exists at all levels of society and is interwoven in all aspects of American social life, it is virtually impossible for alert young children either to miss or ignore it. . . . Children are inundated with it from the moment they enter society. (Van Ausdale & Feagin 2001, 189–190)

In the socialization process, individuals adopt the values, beliefs, and perceptions of their family, peers, culture, and social groups. Prejudice is taught and learned through socialization, although it need not be taught directly and intentionally. Parents who teach their children to not be prejudiced, yet live in an all-white neighborhood, attend an all-white church, and have only white friends may be indirectly teaching negative racial attitudes to their children. Socialization may also be direct, as in the case of a parent who uses racial slurs in the presence of her children or forbids her children from playing with children from a certain racial or ethnic background. Children may also learn prejudicial attitudes from their peers. The telling of racial and ethnic jokes among friends, for example, perpetuates stereotypes that foster negative racial and ethnic attitudes.

Prejudice and the Media. The media contribute to prejudice by portraying minorities in negative and stereotypical ways, or by not portraying them at all. In the

2001–2002 television season, 80 percent of primary recurring characters on prime-time television were white, and 15 percent were African-American (Miller, Parker, Espejo, & Grossman-Swenson 2002). Only 2 percent were Latino/Hispanics, and Arab/Middle Easterners, Asians, and Native Americans comprised less than 1 percent of primary recurring characters. An analysis of character portrayals in the 2001–2002 prime-time television season revealed that Latinos were overrepresented as service workers, unskilled laborers, and criminals; Native Americans were typecast as spiritual advisers, and Arab and Middle Eastern characters were primarily shown in connection to terrorism in both fictional and nonfictional programming. Some television programming has attempted to counter negative stereotypes of minorities. For example, in the wake of 9/11, episodes of *The West Wing* (NBC) and *7th Heaven* (WB) called upon their audiences to distinguish between Islamic extremists and average Muslims in America (Miller et al., 2002).

Another media form that contributes to hatred of minority groups is "white power music" that contains racist lyrics such as the following:

> Niggers just hit this side of town, watch my property values go down. Bang, bang, watch them die, watch those niggers drop like flies . . . Berserkr.

"Skinhead" music, which contains anti-Semitic, racist, and homophobic lyrics, has become a leading recruitment tool for white supremacist groups ("Intelligence Briefs" 2000). The Southern Poverty Law Center has identified 123 domestic and 227 international white power bands that promote hate and intolerance through their music ("White Power Bands" 2002).

The Internet also spreads messages of hate toward minority groups through the Web sites of various white supremacist and hate group organizations (see this chapter's *Focus on Technology* feature).

Discrimination Against Racial and Ethnic Minorities

Whereas prejudice refers to attitudes, **discrimination** refers to actions or practices that result in differential treatment of categories of individuals. Although prejudicial attitudes often accompany discriminatory behavior or practices, one may be evident without the other.

Individual Versus Institutional Discrimination

Individual discrimination occurs when individuals treat persons unfairly or unequally because of their group membership. Individual discrimination may be overt or adaptive. In **overt discrimination** the individual discriminates because of his or her own prejudicial attitudes. For example, a white landlord may refuse to rent to a Mexican-American family because of her own prejudice against Mexican-Americans. Or a Taiwanese-American college student who shares a dorm room with an African-American student may request a roommate reassignment from the student housing office because he is prejudiced against blacks.

Suppose a Cuban-American family wants to rent an apartment in a predominantly non-Hispanic neighborhood. If the landlord is prejudiced against Cubans and does not allow the family to rent the apartment, that landlord has engaged in overt discrimination. But what if the landlord is not prejudiced against Cubans but still refuses to rent to a Cuban family? Perhaps that landlord is engaging in

Focus on Technology

Hate on the Web

Many Web sites promote hate and intolerance toward various minority groups. In Chapter 9, for example, we refer to a Web site that promotes antigay sentiments: http://www.godhatesfags.com. Here, we focus on the use of Internet technology to promote hatred toward racial and ethnic groups.

The Southern Poverty Law Center, a nonprofit organization that combats hate, intolerance, and discrimination, recorded an increase in hate Web sites—from 405 in 2001 to 443 in 2002 (*SPLC Report* 2003). Hate Web sites use sophisticated graphics, music, and entertaining games to lure children, teenagers, and adults (Nemes 2002). Once individuals are hooked on racist ideology, they utilize Internet discussion groups or chat rooms where they can connect with other like-minded individuals with whom they can share their racist views and have those views reinforced.

Although hate groups use the Internet as a tool for communication, organization, and recruitment, the Internet can also be used to fight hate. Now offline, for six years the Web site of "HateWatch" (http://www.hatewatch.org) was devoted to educating the public about the proliferation of hate on the Internet. In addition, it is possible that the prevalence of racist groups on the Internet may reduce hate crimes by providing outlets for hate that are nonphysical. Furthermore, the presence of hate groups on the Internet has made them more visible to the public, which in turn facilitates monitoring by government and watchdog groups (Glaser, Dixit, & Green 2002).

A number of countries, including Germany and France, have passed laws prohibiting hate speech on the Internet and have brought civil and criminal penalties against defendants. In 2001, Yahoo announced it would actively try to keep hateful material out of its auctions, classified sections, and shopping areas after a French court ordered Yahoo to pay fines of about $13,000 a day if the company did not install technology that would shield French Web users from seeing Nazi-related memorabilia in its auction site (French law prohibits the display of such material). However, in most cases, the defendants continued to publish their hate material from the United States, where free speech is protected under the First Amendment. For example, many German neo-Nazi white supremacist groups use U.S.-based Internet servers to avoid prosecution under German law (Kaplan & Kim 2000). One U.S.-based Web page posted in German offered a $7,500 reward for the murder of a young, left-wing activist, giving his home address, job, and phone number.

A number of international efforts have attempted to criminalize hate material online and establish an international legal framework for prosecuting cross-jurisdictional hate speech on the Internet. But unless there is universal ratification, such measures are not likely to be effective. The United States, holding firmly to the value of freedom of speech, is unlikely to participate in international efforts to criminalize hate speech in cyberspace. Even in the unlikely event of international agreement to prohibit online hate speech, how would laws be enforced against Web sites transmitted from satellites in outer space?

adaptive discrimination, or discrimination that is based on the prejudice of others. In this example, the landlord may fear that if he or she rents to a Cuban-American family, other renters who are prejudiced against Cubans may move out of the building or neighborhood and leave the landlord with unrented apartments. Overt and adaptive individual discrimination may coexist. For example, a landlord may not rent an apartment to a Cuban family because of his or her own prejudices and the fear that other tenants may move out.

Institutional discrimination occurs when normal operations and procedures of social institutions result in unequal treatment of minorities. Institutional discrimination is covert and insidious and maintains the subordinate position of minorities in society. When businesses move out of inner-city areas, they are removing employment opportunities for America's highly urbanized minority groups. When schools use standard intelligence tests to decide which children will be placed in

Focus on Technology

Before September 11, 2001, hate material on the Internet mostly targeted minorities such as blacks, Jews, and Asians as well as gays and lesbians. However, in the wake of 9/11, a surge of anti-Americanism has surfaced throughout the world and on the Internet. "Some websites openly praise the events of 11 September 2001, and express disappointment that only 3,000 Americans were killed. They spout slogans such as 'Osama should have killed more,' 'Americans are vermin' and 'Americans conspire to control the world'" (Nemes 2002, 194). When civil liberties groups have challenged legislation against hate speech on the Web, the Supreme Court has sided with them, ruling that such legislation violates free speech. Whether the Supreme Court continues this position in light of anti-American hate speech on the Internet remains to be seen.

© Gamma Presse Images/Liaison Agency

White supremacist groups such as Stormfront spread their message of racial hate through their Web site.

Sources: Glaser, Jack, Jay Dixit, and Donald P. Green. 2002. "Studying Hate Crime with the Internet: What Makes Racists Advocate Racial Violence?" *Journal of Social Issues* 58(1): 177–193; Kaplan, David E., and Lucian Kim. 2000 (September 25). "Nazism's New Global Threat." *U.S. News Online.* http://www.usnews.com/usnews/issue/000925/nazi.htm; Nemes, Irene. 2002. "Regulating Hate Speech in Cyberspace: Issues of Desirability and Efficacy." *Information & Communication Technology Law* 11(3): 193–220; *SPLC Report* 2003 (March). "Hate Group Numbers Rise, but New-Nazis in Disarray." 33(1):3. Southern Poverty Law Center. 400 Washington Ave., Montgomery, AL 36104.

college preparatory tracks, they are limiting the educational advancement of minorities whose intelligence is not fairly measured by culturally biased tests developed from white middle-class experiences. Institutional discrimination is also found in the criminal justice system, which more heavily penalizes crimes that are more likely to be committed by minorities. For example, the penalties for crack cocaine, more often used by minorities, have traditionally been higher than those for other forms of cocaine use even though the same prohibited chemical substance is involved. As conflict theorists emphasize, majority group members make rules that favor their own group.

In a national survey of college freshmen, more than one in five (21.8%) agreed with the statement, "Racial discrimination is no longer a major problem in the United States" (American Council on Education and University of California 2003). Yet, racial and ethnic minorities experience discrimination and its effects in almost every sphere of social life. Next, we look at discrimination in employment, housing, education, and politics. Finally, we expose the extent and brutality of physical and verbal violence against minorities.

Employment Discrimination

Despite laws against it, discrimination against minorities occurs today in all phases of the employment process, from recruitment to interview, job offer, salary, promotion, and firing decisions. In an investigation of employment agencies in the San Francisco Bay area, "testers" were sent to 17 employment agencies and found that 7 of the firms favored white applicants over black applicants for entry-level positions (*Race Relations Reporter* 1999).

> At one employment agency, a black applicant was told that interviews were not being conducted and that he should come back a week later. Hours later, a white applicant with identical educational and employment qualifications was interviewed at the agency and was offered a job within 30 minutes of completing the interview. (p. 1)

More recently, a sociologist at Northwestern University studied employers' treatment of job applicants in Milwaukee, Wisconsin, by dividing job applicant "testers" into four groups: blacks with a criminal record, blacks without a criminal record, whites with a criminal record, and whites without a criminal record (Pager 2003). Applicant testers, none of whom actually had a criminal record, were trained to behave similarly in the application process, and were sent with comparable resumes, to the same set of employers. The study found that white applicants with no criminal record were the most likely to be called back for an interview (34%), and black applicants with a criminal record were the least likely to be called back (5%). But surprisingly, white applicants *with* a criminal record (17%) were more likely to be called back for an interview than were black applicants *without* a criminal record (14%)! The researcher concluded that "the powerful effects of race thus continue to direct employment decisions in ways that contribute to persisting racial inequality" (Pager 2003, 960).

Discrimination in hiring may be unintended. For example, many businesses rely on their existing employees to refer new recruits when a position opens up. Word-of-mouth recruitment is inexpensive and efficient; some companies offer bonuses to employees who bring in new recruits. But this traditional recruitment practice tends to exclude minority workers, as they often do not have a network of friends and family members in higher positions of employment who can recruit them (Schiller 2004).

Employment discrimination contributes to the higher rates of unemployment and lower incomes of blacks and Hispanics compared with whites (see Chapters 10 and 11). Lower levels of educational attainment among minority groups account for some, but not all, of the disadvantages they experience in employment and income. As shown in Figure 6.4, average lifetime earnings of whites is higher than for blacks and Hispanics at the same level of educational attainment (Day & Newburger 2002). While the lifetime earnings of Asian/Pacific Islanders with advanced degrees are equivalent to those of whites, at every other level of educational attainment, whites earn more than Asian/Pacific Islanders.

Workplace discrimination also includes unfair treatment. In one workplace with a large Hispanic workforce, Hispanic workers were selected each week to clean the lunchroom without being paid for that work (Greenhouse 2003). One

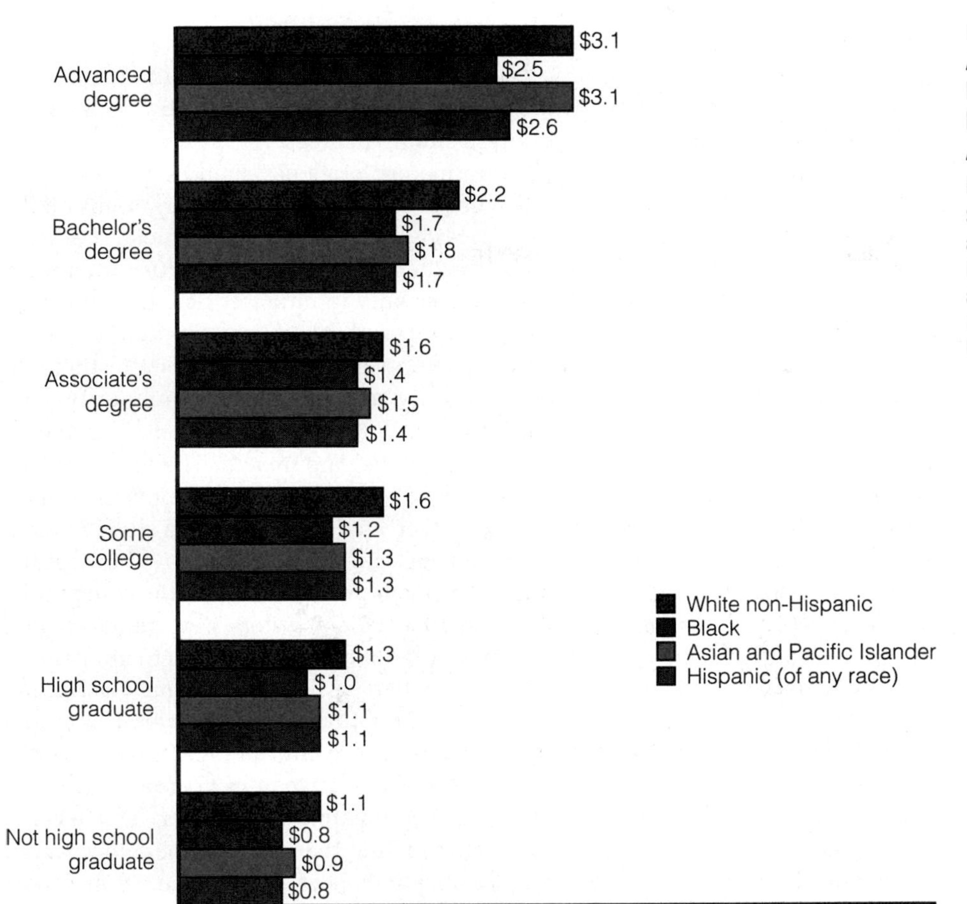

Figure 6.4
Average Work-Life Earnings Estimates for Full-time, Year-Round Workers by Educational Attainment, Race, and Hispanic Origin

Source: Day, Jennifer Cheeseman, and Eric Newburger. 2002. "The Big Payoff: Educational Attainment and Synthetic Estimates of Work-Life Earnings." Washington, DC: U.S. Census Bureau.

Hispanic worker said that this treatment was a matter of dignity and that it made the workers feel humiliated.

Since the terrorist attacks of September 11, 2001, workplace discrimination against individuals who are (or who are perceived to be) Muslim, Arabic, Middle Eastern, South Asian, or Sikh has increased. The most common complaints of post-9/11 backlash reported to the U.S. Equal Employment Opportunity Commission are alleged harassment and firing (EEOC Press Release 2003).

Housing Discrimination and Segregation

Analysis of 2000 census data revealed that although blacks and whites lived in neighborhoods that were slightly more integrated than they were in 1990, people still lived in largely segregated neighborhoods. In 2000, an average white person living in a metropolitan area (which includes city dwellers and suburban residents) lived in a neighborhood that is about 80 percent white and 7 percent black (Schmitt 2001). In contrast, the average black person lived in a neighborhood that is 33 percent white and 51 percent black.

U.S. minorities, who are disproportionately represented among the poor, tend to be segregated in concentrated areas of low-income housing, often in inner-city areas of concentrated poverty (Massey & Denton 1993). Zoning regulations in affluent suburbs restrict development of affordable housing in order to keep out "undesirables" and maintain high property values. Suburban zoning regulations that require large lot sizes, minimum room sizes, and single-family dwellings serve as barriers to low-income development in suburban areas.

Segregation also results from discriminatory practices such as redlining and geographic steering. **Redlining** occurs when mortgage companies deny loans for the purchase of houses in minority neighborhoods, arguing that the financial risk is too great. **Geographic steering** occurs when realtors discourage minorities from moving into certain areas by showing them homes only in minority neighborhoods.

Although housing discrimination is illegal, it is not uncommon. To assess discrimination in housing, researchers use a method called "paired testing." In a paired test, two individuals—one minority and the other nonminority—are trained to pose as home seekers and interact with real estate agents, landlords, rental agents, and mortgage lenders to see how they are treated. The testers are assigned comparable or identical income, assets, and debt as well as housing preferences, family circumstances, education, and job characteristics. In a study in Jacksonville, Florida, white and African-American volunteer "testers" tried to rent the same apartments. More than half of the apartment owners tested in the study broke laws against racial discrimination (Halton 1998). Black testers were quoted higher rents and security deposits, told that units were not available, or were not granted appointments or applications. Another paired testing study of housing discrimination in 23 metropolitan areas found that in the rental market, whites were more likely to receive information about available housing units and had more opportunities to inspect available units than did blacks and Hispanics (Turner, Ross, Galster, & Yinger 2002). The incidence of discrimination was greater for Hispanic renters than for black renters. The same study found that in the home sales market, white homebuyers were more likely to be able to inspect available homes and to be shown homes in more predominantly non-Hispanic white neighborhoods than were comparable blacks and Hispanics. Whites were also more likely to receive information and assistance with financing. In other research on mortgage lending discrimination, minorities were less likely to receive information about loan products, they received less time and information from loan officers, they were often quoted higher interest rates, and they had higher loan denial rates than whites, other things being equal (Turner & Skidmore 1999). This chapter's *Social Problems Research Up Close* feature presents research on housing discrimination conducted by college students.

Educational Discrimination and Segregation

Both institutional and individual discrimination in education negatively affect racial and ethnic minorities and help to explain why minorities (with the exception of Asian-Americans) tend to achieve lower levels of academic attainment and success (see also Chapter 12). Institutional discrimination is evidenced by inequalities in school funding—a practice that disproportionately hurts minority students (Kozol 1991). Nearly half of school funding comes from local taxes. In 2002, the federal government supplied 8 percent of educational expenditures and state government contributed 48 percent; local governments provided the remainder (Schiller 2004). Because minorities are more likely than whites to live in economically disadvantaged areas, they are more likely to go to schools that receive inadequate funding. Inner-city schools, which primarily serve minority students, receive less funding per student than do schools in more affluent, primarily white areas.

For example, inequalities in school funding resulted in New York City receiving $2,000 less per pupil than Buffalo, Rochester, Syracuse, and Yonkers. New York State Supreme Court Judge Leland DeGrasse found that New York City's system of school funding violated federal civil rights laws because it disproportionately hurt minority students (more than 70% of the state's Asian, black, and Hispanic students live in New York City) (Goodnough 2001).

Another institutional education policy that is advantageous to whites is the policy that gives preference to college applicants whose parents or grandparents are alumni. The overwhelming majority of alumni at the highest ranked universities and colleges are white. Thus, white college applicants are the primary beneficiaries of legacy admissions policies. The *Journal of Blacks in Higher Education* asked the nation's 50 highest-ranked universities and colleges to provide information on their legacy acceptance rates. At 24 of the 25 colleges/universities that responded, the legacy admittance rate was higher—often substantially higher—than the overall acceptance rate. For example, at Princeton and Harvard, one in ten applicants was accepted for admission, but at least one in three legacy applicants was admitted ("Details on the Huge Advantage for Legacy Applicants" 2003).

Minorities also experience individual discrimination in the schools, as a result of continuing prejudice among teachers. One college student completing a teaching practicum reported that some of the teachers in her school often spoke about children of color as "wild kids who slam doors in your face" (Lawrence 1997, 111). In a survey conducted by the Southern Poverty Law Center, 1,100 educators were asked if they had heard racist comments from their colleagues in the past year. More than one-quarter of survey respondents answered "Yes" (*Teaching Tolerance* 2000). It is likely that teachers who are prejudiced against minorities discriminate against them, giving them less teaching attention and less encouragement.

Racial and ethnic minorities are also treated unfairly in educational materials, such as textbooks, which often distort the history and heritages of people of color (King 2000). For example, Zinn (1993) observes, "To emphasize the heroism of Columbus and his successors as navigators and discoverers, and to de-emphasize their genocide, is not a technical necessity but an ideological choice. It serves, unwittingly—to justify what was done" (p. 355).

Finally, racial and ethnic minorities are largely isolated from whites in an increasingly segregated school system. A study by the Civil Rights Project at Harvard University finds that U.S. schools in the 2000–2001 year were more segregated than they were in 1970 (Orfield 2001). The upward trend in school segregation is due to large increases in minority student enrollment, continuing white flight from urban areas, the persistence of housing segregation, and the termination of court-ordered desegregation plans. Court-mandated busing became a means to achieve equality of education and school integration in the early 1970s after the Supreme Court (in *Swann v. Charlotte-Mecklenberg*) endorsed busing to desegregate schools. But in the 1990s, lower courts lifted desegregation orders in dozens of school districts (Winter 2003a). In the words of Bradley Schiller (2004), "Racial isolation in the schools is still the hallmark of the American educational system" (p. 178).

Political Discrimination

Historically, African-Americans have been discouraged from political involvement by segregated primaries, poll taxes, literacy tests, and threats of violence. However, tremendous strides have been made since the passage of the 1965 Voting Rights Act,

Social Problems Research Up Close

An Undergraduate Sociology Class Uncovers Racial Discrimination in Housing

Previous research indicates that Americans can infer the race of a speaker through the speaker's accent, grammar, and diction, thus offering rental agents an opportunity to discriminate over the phone. Under the guidance of their sociology professor (Douglas Massey) and a postdoctoral fellow (Garvey Lundy), students in an undergraduate sociology research methods class at the University of Pennsylvania designed a study to determine whether rental agents discriminated over the phone.

Sample and Methods

Students involved in this study developed a data collection instrument that consisted of a standard script that auditors (students posing as renters) followed in their telephone interactions with rental agents. For example, if a machine answered the call, the auditor said, "Hello. My name is ________. I'm interested in the apartment you advertised in ________. Please call me back at ________." If a rental agent answered the phone, the auditor said, "Hello. My name is ________. I'm interested in the apartment you advertised in ________. Are any apartments still available?" Other questions auditors asked included "How much do I have to put down?" and "Are there any other fees?"

After devising the data collection instrument, students chose 79 rental listings from three sources: *Apartments for Rent* magazine, *The Apartment Hunter,* published by the *Philadelphia Inquirer,* and the Sunday Real Estate Section of the *Philadelphia Inquirer* itself. The listings covered all zones of the Philadelphia metropolitan area. Four male and nine female auditors telephoned the selected rental agents, following the script and presenting a profile of a recent college graduate in his or her early to mid-twenties with an annual income of $25,000 to $30,000. The auditors spoke White Middle-Class English (WME), Black Accented English (BAE), or Black English Vernacular (BEV). Black Accented English and Black English Vernacular are widely spoken by African-Americans in the United States. Massey and Lundy (2001) explain that "although BEV and BAE may both be identified as 'black sounding,' we suspect that most listeners can tell the difference between the two dialects and that they attach different class labels to each style of speech" (p. 456). Specifically, when an African-American speaks Standard English with a black pronunciation of certain words (BAE), listeners conclude that the speaker is a middle-class black person, whereas the combination of nonstandard grammar with a black accent (BEV) signals lower-class status.

Based on this assumption, the researchers were able to employ six independent variables to test for a three-way interaction between race (black-white), gender (male-female), and class (lower-middle). The six independent variables were (1) male BEV, (2) female BEV, (3) male BAE, (4) female BAE, (5) male WME, and (6) female WME. The dependent variables included the various responses of the rental agents, such as whether the agent returned the auditor's phone call, whether the rental agent indicated that a housing unit was still available, and whether the agent indicated

which prohibited literacy tests and provided for poll observers. Blacks have won mayoral elections in Atlanta, Cleveland, Washington, D.C., and New York City, and the governorship in Virginia. However, racial minorities and Hispanics continue to be underrepresented in political positions and voting participation.

Discrimination against racial and ethnic minorities in the U.S. political process persists. After "voting irregularities" in the 2000 national elections, the National Association for the Advancement of Colored People (NAACP) and several civil rights groups filed a lawsuit in Florida to eliminate unfair voting practices (NAACP Press Release 2001). In the 2000 election, thousands of black voters complained that they were wrongfully turned away from the polls or had trouble casting their ballots. Complaints were not limited to Florida; black voters in about a dozen other states reported similar unfair treatment (NAACP Press Release 2000).

that an application fee was required (and if so, how much).

Findings and Conclusions

The researchers found "clear and often dramatic evidence of phone-based racial discrimination" (p. 466). Compared with whites, African-Americans were less likely to speak to a rental agent (their calls were less likely to be returned). African-Americans were also less likely to be told that a unit was available, more likely to pay application fees, and more likely to have credit mentioned as a potential problem in qualifying for a lease.

These racial effects were exacerbated by gender and class. Lower-class blacks experienced less access to rental housing than middle-class blacks, and black females experienced less access than black males. Lower-class black females were the most disadvantaged group. They experienced the lowest probability of contacting and speaking to a rental agent and, even if they did make contact, they faced the lowest probability of being told of a housing unit's availability. Lower-class black females also faced the highest chance of paying an application fee. On average, lower-class black females were assessed $32 more per application than white middle-class males.

The share of auditors reaching an agent and the share being told a unit was available was combined to indicate an overall measure of access to rental units in the Philadelphia housing market. Whereas more than three-quarters (76 percent) of white middle-class males gained access to a potential rental unit, the figure dropped to 63 percent for middle-class black men (those speaking BAE), 60 percent for white middle-class females (those speaking WME), 57 percent for black middle-class females (those speaking BAE), 44 percent for lower-class black men (those speaking BEV), and only 38 percent for lower-class black women (those speaking BEV). "In other words, for every call a white male makes to find out about a rental unit in the Philadelphia housing market, a poor black female must make two calls to achieve the same level of access, roughly doubling her time and effort compared with his" (p. 461). In sum, "being identified as black on the basis of one's speech pattern clearly reduces access to rental housing, but being black and female lowers it further, and being black, female, and poor lowers it further still" (p. 467).

These findings suggest that much housing discrimination probably occurs over the phone. "Through technology, a racist landlord may discriminate without actually having to experience the inconvenience or discomfort of personal contact with his or her victim" (Massey & Lundy 2001, 454–455). The authors also conclude that "telephone audit studies offer social scientists a cheap, effective, and timely way to measure the incidence and severity of racial discrimination in urban housing markets" (p. 455).

Source: Massey, Douglas S., and Garvey Lundy. 2001. "Use of Black English and Racial Discrimination in Urban Housing Markets: New Methods and Findings." *Urban Affairs Review* 36(4):252–469.

Hate Crimes

In June 1998, James Byrd, Jr., a 49-year-old father of three, was walking home from a niece's bridal shower in the small town of Jasper, Texas. According to police reports, three white men riding in a gray pickup truck saw Byrd, a black man, walking down the road and offered him a ride. The men reportedly drove down a dirt lane and, after beating Byrd, chained him to the back of the pickup truck and dragged him for two miles down a winding, narrow road. The next day, police found Byrd's mangled and dismembered body. The three men who were arrested had ties to white supremacist groups. James Byrd had been brutally murdered simply because he was black.

The murder of James Byrd exemplifies a **hate crime**—an unlawful act of violence motivated by prejudice or bias. Examples of hate crimes, also known as

Posters such as the one depicted in this photo are part of the efforts to increase political participation of minorities.

© Mark Richards/PhotoEdit

"bias-motivated crimes" and "ethnoviolence" include intimidation (e.g., threats), destruction/damage of property, physical assault, and murder. In 2001, 10 hate crime murders were reported to the FBI, 5 of which were motivated by ethnicity/national origin bias, 4 by racism, and 1 by sexual-orientation bias (Federal Bureau of Investigation 2003). However, in the same year, the Southern Poverty Law Center identified 21 victims of hate murders in the United States, 6 of whom were perceived as being from the same ethnic or religious group as the September 11 terrorists; 11 were either gay, transgender, or perceived as such; 1 was black; 1 was Hispanic; and 2 were Asian-American ("The Forgotten" 2002). FBI hate crime data undercounts the actual number of hate crimes, including hate crime murders, because (1) not all U.S. jurisdictions report hate crimes to the FBI (reporting is voluntary); (2) victims are often reluctant to report hate crimes to the authorities; (3) law enforcement agencies shy away from classifying crimes as hate crimes because it makes their community "look bad"; and (4) it is difficult to prove that crimes are motivated by hate or prejudice. As a result, fewer than one-sixth of U.S. hate crimes are reported to the FBI.

From the first year that FBI hate crime data were published in 1992, the majority of hate crimes have been based on racial bias (see Figure 6.5). From 2000 to 2001, religious-bias hate crimes, especially anti-Islamic religion incidents, increased significantly, although more than half of religious-bias incidents were anti-Jewish (FBI 2003). The number of hate crimes motivated by ethnicity/national origin more than doubled from 2000 to 2001, presumably as a result of the terrorist attacks of September 11, 2001. This chapter's *Human Side* feature describes some of the victims of the post-9/11 backlash.

"The vast majority of the world's 1.2 billion Muslims are as offended by a violent act carried out in the name of Islam as most Christians are horrified by atrocities perpetrated by Serbian Christians or the Real Irish Republican Army."

Charles Kimball
Wake Forest University

Levin and McDevitt (1995) found that the motivations for hate crimes were of three distinct types: thrill, defensive, and mission. Thrill hate crimes are committed by offenders who are looking for excitement and attack victims for the "fun of it." Defensive hate crimes involve offenders who view their attacks as necessary to protect their community, workplace, or college campus from "outsiders" or to pro-

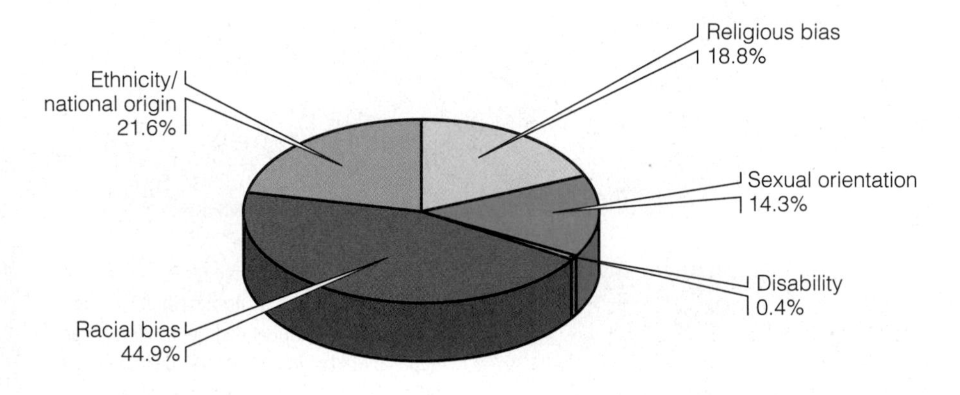

Figure 6.5
Hate Crime Incidences, by Category of Bias: 2001

Source: Federal Bureau of Investigation 2003. *Hate Crime Statistics 2001.* http://www.fbi.gov

tect their racial and cultural purity from being "contaminated" by interracial marriage and childbearing. A study of white racists in Internet chat rooms found that the topic of interracial marriage was more likely than other topics (such as blacks moving into one's neighborhood and competing for one's job) to elicit advocacy of violence (Glaser, Dixit, & Green 2002). For example, one respondent said, "better kill her. kill him and her. pull a oj . . . im not kidding. i would do it if it was my sister. i would gladly go to prison then live a free life knowing some mud babies were calling me uncle whitey" (p. 184).

Mission hate crimes are perpetrated by white supremacist group members or other offenders who have dedicated their lives to bigotry. The Ku Klux Klan, the first major racist, white supremacist group in the United States, began in Tennessee shortly after the Civil War. Klansmen have threatened, beaten, mutilated, and lynched blacks as well as whites who dared to oppose them. Other racist groups known to engage in hate crimes are the Identity Church Movement, neo-Nazis, and the skinheads. The number of hate groups identified by the Southern Poverty Law Center rose from 457 in 1999 to 602 in 2000, 676 in 2001, and 708 in 2002 (*SPLC Report* 2003; "The Year in Hate" 2001). Some of this increase is due to improved counting techniques rather than the appearance of new groups during the year.

Hate on Campus. Karl Nichols, a white residence hall director at the University of Mississippi, learned about racism in college, but not in the classroom and not from a textbook. Two chunks of asphalt were hurled through Karl's dormitory room window along with a note warning, "You're going to get it, you Godforsaken nigger-lover" ("Hate on Campus" 2000, 10). The next night, someone attempted to set Karl's door on fire. According to the university's investigation of these incidents, Karl Nichols "may have violated racist taboos . . . by openly displaying his affinity for African-American individuals and black culture, by dating black women, by playing [black] music . . . and by promoting diversity" in his dormitory (p. 10).

Karl Nichols is not alone. At Brown University in Rhode Island, a black student is beaten by three white students who tell her she is a "quota" who doesn't belong at a university. At the State University of New York at Binghamton, an Asian-American student is left with a fractured skull after a racially motivated assault by three students. Two students at the University of Kentucky—one white,

The Human Side

Post 9/11 Backlash Hate Crimes

A photograph of Egyptian-American Adel Karas, 48, is placed along with an American flag and flowers at the door to his shop, International Market in San Gabriel, California, after he was shot and killed just days following the September 11, 2001, terrorist attacks. Mr. Karas is one of many victims of post-9/11 hate crime.

The innocent people who died in the terrorist attacks of September 11, 2001, and the families and friends who still grieve, are not the only victims of that fateful day. Following 9/11, a backlash of hate has been directed toward Muslims and people of Middle-Eastern decent (or those perceived as such), as well as those whose names are similar to the names of the terrorists.

Abdo Ali Ahmed, 51, Reedley, California On September 29, 2001, Abdo Ali Ahmed, a 51-year-old Muslim from Yemen crossed the dusty lot from his family's modest home to his convenience store to begin a 14-hour workday. Two days before, Ahmed had found a death-threat note on his car's windshield: "We're going to kill all you [expletive] Arabs." When taunted or threatened in person, Ahmed would calmly reply, "I am a citi-

one black—were crossing the street just off campus when they were attacked by 10 white men. The attackers yelled racist slurs at the black student and choked him until he couldn't speak or move. The assailants called the white student a "nigger-lover" as they broke his hand and nose. "I definitely thought I was going to lose my life," the black student said later. The white student was shocked by the incident, commenting, "I didn't know that much hate existed" ("Hate on Campus" 2000, 7).

zen." When he found the note, he threw it in the trash, telling one of his eight children not to bother calling the police. "God will take care of it," he said. But around 4 P.M. that afternoon, the threats were made real. Ahmed was shot three times in the torso, a few feet from the American flag he had taped in the window of his store. Ahmed crumpled on the floor and died in his wife's arms while waiting for medical attention. The Fresno County Sheriff's Department investigated Ahmed's murder as a botched robbery. A local resident summed up sentiments of the local Muslim community: "Mr. Ahmed is one more victim of the tragedy of September 11."

Adel Karas, 48, San Gabriel, California
Adel Karas was tolerance personified. For 20 years, the Egyptian ran an international market that lived up to its name, selling tortillas and African drums alongside the Budweiser and beach balls. Karas greeted customers in Spanish, Arabic, and English. On September 15, 2001, the grocer was shot in his store by two young men. Karas crawled outside where he was found bleeding on the sidewalk. He died soon after being taken to a hospital. Apparently, Karas's attackers mistook Karas for a Muslim and targeted him for revenge just four days after the September 11 terrorist attacks. In fact, the 48-year-old father of three was an Orthodox Christian who came to the United States to escape persecution by the Muslim majority in Cairo.

Vasudev Patel, 49, Mesquite, Texas
On October 4, 2001, an armed man walked into Vasudev Patel's Shell station and convenience store in a suburb of Dallas and shot Patel. The next afternoon, police used the store's video surveillance camera evidence to arrest 32-year-old Mark Anthony Stroman, a two-time convicted felon. At first, Stroman, who had left the store empty-handed after the shooting, said his motive was robbery. But then he changed his story, telling investigators that he killed Patel out of revenge: His sister, he claimed, had died in the World Trade Center on September 11. If Stroman had come to Patel's store looking for revenge on Arabs or Muslims, he chose the wrong target: Patel was a Hindu and native of India.

Balbir Singh Sodhi, 49, Mesa, Arizona
Balbir Singh Sodhi, a Sikh who immigrated from India in 1988, ran a Chevron station in Mesa, Arizona. After September 11, he shopped around for a flag he could fly to show his support for the United States. With the Stars and Stripes in record demand, he still hadn't found one on Saturday, September 15. That afternoon, as Sodhi tended to the landscaping in front of his station, he was shot and killed with a gun fired from a passing pickup truck. The shooter, 42-year-old Frank Roque, then drove down the road and shot (but did not kill) a Lebanese-American gas-station clerk, then opened fire on the home of a family with Afghan roots. After the shooting spree, he went to a nearby sports bar, announcing loudly, "They're investigating the murder of a turban-head down the street."

While these individuals represent victims of extreme hate, a poll by the Council on American-Islamic Relations showed that more than half of America's seven million Muslims (57%) say they have experienced bias or discrimination since 9/11 (Morrison 2002). On the positive side, the same poll found that most American Muslims (79%) also experienced special kindness or support from friends or colleagues of other faiths. Acts of kindness include verbal assurances, support, and even offers to help guard local mosques and Islamic schools.

Sources: Adapted from "The Forgotten." 2002 (Spring). *Intelligence Report,* No. 105. Montgomery, AL: Southern Poverty Law Center. Morrison, Pat. 2002 (September 6). "September 11: A Year Later—American Muslims Are Determined Not to Let Hostility Win." *National Catholic Reporter* 38(38):9–10.

According to the FBI, one in 10 hate crimes occurs at schools or colleges (FBI 2003). The FBI listed 286 hate crimes on college campuses in 2001, but the U.S. Department of Education recorded 487 such crimes in the same year (Willoughby 2003). Far more common than hate crimes are "bias incidents," which are events that aren't crimes but still can have the same negative and divisive effects. Howard J. Ehrilich, director of the Prejudice Institute in Baltimore, estimates that each year one-quarter of racial and ethnic minority college students and up to 5 percent of white college students are targets of bias-motivated name-calling, e-mails, telephone calls, verbal aggression, and other forms of psychological intimidation (Willoughby 2003).

Strategies for Action: Responding to Prejudice, Racism, and Discrimination

"The problem is . . . we don't believe we are as much alike as we are. Whites and blacks, Catholics and Protestants, men and women. If we saw each other as more alike, we might be very eager to join in one big human family in this world, and to care about that family the way we care about our own."

Morrie Schwartz
Sociologist
(from *Tuesdays with Morrie*)

Because racial and ethnic tensions exist worldwide, strategies for combating prejudice, racism, and discrimination globally require international cooperation and commitment. The World Conference Against Racism, Racial Discrimination, Xenophobia, and Related Intolerance, held in Durban, South Africa, in 2001 exemplifies international efforts to reduce racial and ethnic tensions and inequalities and increase harmony among the various racial and ethnic populations of the world. Unfortunately, the U.S. delegation to this conference withdrew because of the expectation that hateful language would be used against Israel (due to Israel's treatment of Palestinians). While international strategies to combat racism and discrimination are vital, in this section we focus on legal/political and educational strategies within the United States.

Legal/Political Strategies

Legal and political strategies to reduce prejudice, racism, and discrimination include increasing minority participation and representation in government, eliminating inequalities in school funding and creating equal opportunities through affirmative action. (Because affirmative action includes voluntary programs and policies in addition to legally mandated ones, we consider affirmative action under a separate heading.) Although it is readily apparent that such strategies can prevent or reduce discriminatory practices, the effects of legal and political strategies on prejudice are more complex. Legal policies that prohibit discrimination can actually increase modern forms of prejudice, as in the case of individuals who conclude that because laws and policies prohibit discrimination, any social disadvantages of minorities must be their own fault. On the other hand, any improvement in the socioeconomic status of minorities that results from legal/political policies may help to replace negative images of minorities with positive images.

Increasing Minority Representation and Participation in Government. Increasing minority representation and participation in government promises to increase minorities' voices in influencing public policy that addresses their interests. In addition, minority representation in government affects race relations. One study found a high degree of mistrust of U.S. government among African-Americans; however, those who believed that blacks could influence the political process were less likely to distrust the government (Parsons, Simmons, Shinhoster, & Kilburn 1999). The researchers concluded that African-Americans' "distrust of government will not be reduced until African Americans perceive that they have more of a role to play in their government" (p. 218).

Various national, state, and local minority groups and organizations encourage minorities to register to vote and to vote in governmental elections. Such efforts contributed to an upsurge in black voter participation in the 2000 presidential election (National Coalition on Black Civic Participation 2000). Efforts to increase voting participation have also targeted Asian-Americans and Hispanic populations.

Representation of racial and ethnic minorities in elected government positions has also increased. Members of the 108th U.S. Congress include a record 25 Hispanic members, as well as 39 African-Americans, 7 Asians/Pacific Islanders, and 3 Native Americans; 8 members were foreign born (Arner 2003).

Reducing Disparities in Education. Legal remedies have also sought to address institutional discrimination in education by reducing or eliminating disparities in school funding. As noted earlier, schools in poor districts—which predominantly serve minority students—have traditionally received less funding per pupil than do schools in middle- and upper-class districts (which predominantly serve white students). In recent years, more than two dozen states have been forced by the courts to come up with a new system of financing schools to increase inadequate funding of schools in poor districts (Goodnough 2001).

Affirmative Action

Affirmative action refers to a broad range of policies and practices in the workplace and educational institutions to promote equal opportunity as well as diversity in the workplace and on campuses. Affirmative action represents an attempt to compensate for the effects of past discrimination and prevent current discrimination against women and racial and ethnic minorities. Vietnam veterans and people with disabilities may also qualify under affirmative action policies. Although the largest category of affirmative action beneficiaries is women, the majority of students in two sociology classes did not know that women were covered by affirmative action (Beeman, Chowdhry, & Todd 2000).

Federal Affirmative Action. Affirmative action policies developed in the 1960s from federal legislation requiring that any employer (universities as well as businesses) receiving contracts from the federal government must make "good faith efforts" to increase the pool of qualified minorities and women (U.S. Department of Labor 2002). Such efforts can be made through expanding recruitment and training programs. Hiring decisions are to be made on a nondiscriminatory basis.

Affirmative Action in Higher Education. The Supreme court's 1974 ruling in *University of California Board of Regents v. Bakke* marks the beginning of the decline of affirmative action. Alan Bakke, a white male, had applied to the University of California at Davis medical school and was rejected, even though his grade point average and score on the medical school admissions test were higher than those of several minority applicants who had been admitted. The medical school had established fixed racial quotas, guaranteeing admission to 16 minority applicants regardless of their qualifications. Bakke claimed that such quotas discriminated against him as a white male and that the University of California had violated his Fourteenth Amendment right to equal protection under the law. The Supreme Court ruled by a 5 to 4 vote (showing how split the court was) in Bakke's favor, concluding that the University of California unwittingly engaged in "reverse discrimination" that was unconstitutional. Affirmative action programs, the court ruled, could not use fixed quotas in admission, hiring, or promotion policies. However, the court affirmed the right for universities and employers to consider race as a factor in admission, hiring, and promotion in order to achieve diversity.

Since the Bakke case, numerous legal battles have challenged affirmative action (Olson 2003). In *Hopwood v. Texas,* Cheryl Hopwood and three other white individuals who had been denied admission to the University of Texas Law School claimed that their Fourteenth Amendment rights to equal protection under the law had been violated by the university's affirmative action admission policies. The fifth circuit court of appeals decided in 1997 that the University of Texas could no longer use race as a factor in awarding financial aid, admitting students, and hiring and promoting faculty. Higher courts refused to review the decision. But in 2003, the U.S. Supreme Court, after hearing the appeals from two white applicants who applied but were not accepted to the University of Michigan, affirmed the right of colleges to consider race in admissions, but rejected Michigan's use of a point system to do so. In the University of Michigan's undergraduate admissions procedure, minority status provided 20 points on a 150-point scale for admission. The court found that the problem with a point system is that for some applicants it can turn race into the decisive factor instead of just one of many (Winter 2003b). In response, the University of Michigan abandoned its point system and created an undergraduate admissions policy similar to its law school admissions policy that may serve as a model for how other universities can achieve a diverse student body while following court guidelines. In the new "holistic review" approach, the university will consider the unique circumstances of each student, prioritizing academics and treating all other factors, including race, equally. Applicants are now required to write more essays, including one on cultural diversity. Other colleges and universities that are committed to having a diverse student body but want to avoid charges of "reverse discrimination" ask students to write essays about "cultural traditions," ask if they speak English as a second language, and increase efforts to recruit at high schools with large minority populations.

Attitudes Toward Affirmative Action. A number of public opinion polls suggest that a slight majority of Americans support affirmative action (Pew Research Center 2003; Plous 2003). But public opinion on the issue depends on how the survey question is worded and framed. Using terms such as "affirmative action," "equal," and "opportunity" in survey questions yields more support for affirmative action policies, while the use of "special preferences," "preferential treatment," and "quotas" tends to lessen support. And survey questions that ask whether respondents favor "affirmative action programs for women and minorities" elicit more favorable responses than questions that ask about affirmative action for minorities only (Paul 2003).

Surveys about affirmative action in higher education yield contradictory results. In a 2003 Time/CNN Poll that asked, "Do you approve or disapprove of affirmative action admissions programs at colleges and law schools that give racial preferences to minority applicants?" 54 percent disapproved (Polling Report 2003). Among first-year college students, more than half (52.7%) agreed that "affirmative action in college admissions should be abolished" (American Council on Education and University of California 2002). However, a Pew Research Center (2003) survey asked, "Do you think affirmative action programs designed to increase the number of black and minority students on college campuses is a good thing or a bad thing?" Sixty percent said it was a good thing.

Women are more likely than men to support affirmative action (Pew Research Center 2003)—not a surprising finding given that women are the largest category targeted to benefit from affirmative action. Blacks, Hispanics, and Asian-Americans are more likely than whites to support affirmative action (Paul 2003; Pew Research Center 2003; Tolbert & Grummel 2003).

Supporters of affirmative action suggest that such policies have many social benefits. In a review of over 200 scientific studies of affirmative action, Holzer and

Neumark (2000) concluded that these policies produce benefits for women, minorities, and the overall economy. Holzer and Neumark (2000) found that employers adopting affirmative action increase the relative number of women and minorities in the workplace by an average of 10 percent to 15 percent. Since the early 1960s, affirmative action in education has contributed to an increase in the percentage of blacks attending college by a factor of three and the percentage of blacks enrolled in medical school by a factor of four. Black doctors choose more often than their white medical school classmates to practice medicine in inner cities and rural areas serving poor or minority patients (Holzer & Neumark 2000). Increasing the numbers of minorities in educational and professional positions also provides positive role models for other, especially younger, minorities "who can identify with them and form realistic goals to occupy the same roles themselves" (Zack 1998, 51).

Opponents of affirmative action suggest that such programs constitute "reverse discrimination," which hurts whites. However, using 2000 data showing there are 1.3 million unemployed blacks and 112 million employed whites, if every unemployed black worker in the United States were to displace a white worker, only 1 percent of white workers would be affected. Because affirmative action pertains only to job-qualified applicants, the actual percentage of affected whites would be a fraction of 1 percent (Plous 2003). The main causes of unemployment among the white population are corporate downsizing, computerization and automation, and factory relocations outside the United States, not affirmative action.

Some critics of affirmative action argue that it undermines the self-esteem of women and minorities. While affirmative action may have this effect in rare cases, in many cases affirmative action may raise the self-esteem of women and minorities by providing them with opportunities for educational advancement and employment (Plous 2003). Another critique of affirmative action is that it fails to help the most impoverished of minorities—those whose deep and persistent poverty impairs their ability to compete not only with whites but with other more advantaged minorities (Wilson 1987).

Opposition to affirmative action threatens the future of such policies and programs and the future educational and occupational opportunities of minorities. After California passed Proposition 209—an initiative to abolish affirmative action—black students enrolling at the UCLA law school declined from 10.3 percent in 1996 to 1.4 percent in 2000 (Greenberg 2003). Black and Latino enrollment has also significantly declined in the UC Berkeley undergraduate program and law school, the UC San Diego School of Medicine, and the University of Texas Law School.

It is ironic that President George W. Bush, who does not support affirmative action, was himself a beneficiary of a long-standing policy that gives preferential treatment to college applicants. When President Bush applied to highly selective Yale University in 1964, he had a mediocre academic record, but he was nevertheless admitted into Yale because he was a "legacy" applicant—that is, the son and grandson of distinguished alumni.

Educational Strategies

Educational strategies to reduce prejudice, racism, and discrimination have been implemented in schools (primary, secondary, and higher education levels), in communities and community organizations, and in the workplace.

"Teach tolerance. Because open minds open doors for all our children."

Radio Public Service Announcement

Multicultural Education in Schools and Communities. In schools across the nation, **multicultural education,** which encompasses a broad range of programs and strategies, works to dispel myths, stereotypes, and ignorance about minorities,

More than 600,000 teachers receive *Teaching Tolerance* magazine—a free resource for teaching tolerance education in the classroom (Dees 2000). Other materials available free from the Southern Poverty Law Center (http://www.spl-center.org) include *101 Tools for Tolerance, Responding to Hate at School,* and *Ten Ways to Fight Hate.*

promotes tolerance and appreciation of diversity, and includes minority groups in the school curriculum (see also Chapter 12). With multicultural education, the school curriculum reflects the diversity of American society and fosters an awareness and appreciation of the contributions of different racial and ethnic groups to American culture. The Southern Poverty Law Center's program "Teaching Tolerance" has published and distributed materials and videos designed to promote better human relations among diverse groups to schools, colleges, religious organizations, and a variety of community groups across the nation.

Many colleges and universities have made efforts to promote awareness and appreciation of diversity by offering courses and degree programs in racial and ethnic studies, and multicultural events and student organizations. A national survey by the Association of American Colleges and Universities found that 54 percent of colleges and universities required students to take at least one course that emphasizes diversity and another 8 percent were in the process of developing such a requirement (Humphreys 2000). Evidence suggests a number of positive outcomes for both minority and majority students who take college diversity courses, including increased racial understanding and cultural awareness, increased social interaction with students who have backgrounds different from their own, improved cognitive development, increased support for efforts to achieve educational equity, and higher satisfaction with their college experience (Humphreys 1999). Courses in "whiteness studies" are also emerging in college curricula. Sociologist Dalton Conley (2002) explains that whiteness studies "serve to rectify something wrong with the way we study race in America: By traditionally focusing on minority groups, the implicit message that scholarship projects is that nonwhites are 'deviant,' that's why they are studied" (n.p).

Diversification of College Student Populations. Efforts to recruit and admit racial and ethnic minorities in institutions of higher education have also been found to foster positive relationships among diverse groups and enrich the educational experience of all students. In the United States, about one in five undergraduates at four-year colleges is a minority (American Council on Education and the American Association of University Professors 2000). Research indicates that racial and ethnic diversity on campus provides educational benefits for all students—minority and nonminority alike (American Council on Education and the American Association of University Professors 2000). Gurin (1999) found that students with the most exposure to diverse populations during college had the most cross-racial interactions five years after leaving college. A poll of law students at Harvard Law School and the University of Michigan found that nearly 90 percent of the students said that diversity in the classroom provided them with a better educational experience (*Race Relations Reporter* 1999). Nearly 90 percent of the law students said that the contact they had with students of different racial or ethnic backgrounds influenced them to change their view on some aspect of civil rights.

Diversity Training in the Workplace. Increasingly, corporations have begun to implement efforts to reduce prejudice and discrimination in the workplace through an educational approach known as diversity training. Broadly defined, diversity training involves "raising personal awareness about individual 'differences' in the workplace and how those differences inhibit or enhance the way people work together and get work done" (Wheeler 1994, 10). Diversity training may address such issues as stereotyping and cross-cultural insensitivity, as well as provide workers with specific information on cultural norms of different groups and how these norms affect work behavior and social interactions.

In a survey of 45 organizations that provide diversity training, Wheeler (1994) found that for 85 percent of the respondents, the primary motive for offering diversity training was to enhance productivity and profits. In the words of one survey respondent, "The company's philosophy is that a diverse work force that recognizes and respects differing opinions and ideas adds to the creativity, productivity, and profitability of the company" (p. 12). Only 4 percent of respondents said they offered diversity training out of a sense of social responsibility.

Understanding Race and Ethnic Relations

After considering the material presented in this chapter, what understanding about race and ethnic relations are we left with? First, we have seen that racial and ethnic categories are socially constructed; they are largely arbitrary, imprecise, and misleading. Although some scholars suggest we abandon racial and ethnic labels, others advocate adding new categories—multiethnic and multiracial—to reflect the identities of a growing segment of the U.S. and world population.

Conflict theorists and functionalists agree that prejudice, discrimination, and racism have benefited certain groups in society. But racial and ethnic disharmony has created tensions that disrupt social equilibrium. Symbolic interactionists note that negative labeling of minority group members, which is learned through interaction with others, contributes to the subordinate position of minorities.

Prejudice, racism, and discrimination are debilitating forces in the lives of minorities. In spite of these negative forces, many minority group members succeed in living productive, meaningful, and prosperous lives. But many others cannot overcome the social disadvantages associated with their minority status and become victims of a cycle of poverty (see Chapter 10). Minorities are disproportionately poor, receive inferior education and health care, and with continued discrimination in the workplace, have difficulty improving their standard of living.

Achieving racial and ethnic equality requires alterations in the structure of society that increase opportunities for minorities—in education, employment and income, and political participation. In addition, policy makers concerned with racial and ethnic equality must find ways to reduce the racial/ethnic wealth gap and foster wealth accumulation among minorities (Conley 1999). Social class is a central issue in race and ethnic relations. Bell Hooks (2000) warns that focusing on issues of race and gender can deflect attention away from the larger issue of class division that increasingly separates the "haves" from the "have-nots." Addressing class inequality must, suggests Hooks, be part of any meaningful strategy to reduce inequalities suffered by minority groups. Civil rights activist Lani Guinier (1998) suggests that "the real challenge is to . . . use race as a window on issues of class, issues of gender, and issues of fundamental fairness, not just to talk about race as if it's a question of individual bigotry or individual prejudice. The issue is more than about making friends—it's about making change." But, as Shipler (1998) notes, making change requires members of society to recognize that change is necessary, that there is a problem that needs rectifying.

> One has to perceive the problem to embrace the solutions. If you think racism isn't harmful unless it wears sheets or burns crosses or bars blacks from motels and restaurants, you will support only the crudest anti-discrimination laws and not the more refined methods of affirmative action and diversity training. (p. 2)

Chapter Review

- **What is meant by the idea that race is socially constructed?**
  Racial categories are based more on social definitions than on biological differences. Genetically, the genes of black and white Americans are 99.9 percent alike and there are no "pure" races; people in virtually all societies have genetically mixed backgrounds. Different societies construct different systems of racial classification, and these systems change over time. The significance of race is not biological but social and political, as race is used to separate "we" from "they" and becomes a basis for unequal treatment of one group by another.

- **What are two types of assimilation?**
  Assimilation is the process by which formerly distinct and separate groups merge and become integrated as one. Secondary assimilation occurs when different groups become integrated in public areas and in social institutions, such as neighborhoods, schools, the workplace, and in government. Primary assimilation occurs when members of different groups are integrated in personal, intimate associations, as with friends, family, and spouses.

- **Beginning with the 2000 Census, what are the five race categories used to identify the race composition of the United States?**
  Beginning with the 2000 census, the five race categories are (1) white, (2) black or African-American, (3) American Indian or Alaska Native, (4) Asian, and (5) Native Hawaiian or other Pacific Islander. In addition, respondents to federal surveys and the census now have the option of officially identifying themselves as being more than one race, rather than checking only one racial category.

- **What is an ethnic group?**
  An ethnic group is a population that has a shared cultural heritage or nationality. Ethnic groups may be distinguished on the basis of language, forms of family structures and roles of family members, religious beliefs and practices, dietary customs, forms of artistic expression such as music and dance, and national origin. The largest ethnic population in the United States is Hispanics/Latinos.

- **What percentage of the U.S. population (in 2002) was born outside the United States?**
  In 2002, more than one in ten U.S. residents (11.5%) was born in a foreign country and one in five U.S. children 18 and under was the child of an immigrant.

- **What is a manifest function and latent function of the civil rights movement?**
  The manifest function of the civil rights legislation in the 1960s was to improve conditions for racial minorities. However, civil rights legislation produced an unexpected consequence, or latent function. Because civil rights legislation supposedly ended racial discrimination, whites were more likely to blame blacks for their social disadvantages and thus perpetuate negative stereotypes such as "blacks lack motivation" and "blacks have less ability."

- **How does contemporary prejudice differ from more traditional, "old-fashioned" prejudice?**
  Traditional, old-fashioned prejudice is easy to recognize, as it is blatant, direct, and conscious. More contemporary forms of prejudice are often subtle, indirect, and unconscious.

- **Is it possible for an individual to discriminate without being prejudiced?**
  Yes. In overt discrimination, an individual discriminates because of his or her own prejudicial attitudes. But sometimes individuals who are not prejudiced discriminate because of someone else's prejudice. For example, a store clerk may watch black customers more closely because the store manager is prejudiced against blacks and has instructed the employee to follow black customers in the store closely. Discrimination based on someone else's prejudice is called "adaptive discrimination."

- **Are U.S. schools segregated?**
  Racial and ethnic minorities are largely isolated from whites in an increasingly segregated school system. A study by the Civil Rights Project at Harvard University finds that U.S. schools in 2000–2001 were more segregated than they were in 1970. The upward trend in school segregation is due to large increases in minority student enrollment, continuing white flight from urban areas, the persistence of housing segregation, and the termination of court-ordered desegregation plans.

- **According to FBI data, the majority of hate crimes are committed against what group?**
  Since the FBI began publishing hate crime data in 1992, the majority of hate crimes have been based on racial bias. From 2000 to 2001, religious-bias hate crimes, especially anti-Islamic religion incidents, increased significantly, although more than half of religious-bias incidents were anti-Jewish. The number of hate crimes motivated by ethnicity/national origin more than doubled from 2000 to 2001, presumably as a result of the terrorist attacks of September 11, 2001.

- **What group constitutes the largest beneficiary of affirmative action policies?**
  Affirmative action policies are designed to benefit racial and ethnic minorities, women, and in some cases, Vietnam veterans and people with disabilities. The largest category of affirmative action beneficiaries is women.

Critical Thinking

1. At colleges and universities around North America, a number of professors are endorsing Holocaust denial, race-based theories of intelligence, and other racist ideas. For example, Associate Professor Arthur Butz of Northwestern University publicly rejects the claim that millions of Jews were exterminated in the Holocaust ("Hate on Campus" 2000). Professor Edward M. Miller of the University of New Orleans has concluded that blacks are "small-headed, over-equipped in genitalia, oversexed, hyper-violent and . . . unintelligent" (p. 9). Professor Glayde Whitney of Florida State University and Professor J. Philippe Rushton of the University of Western Ontario have both described blacks as having smaller brains. How should institutions of higher learning respond to such racist claims made by faculty members? What role does the right to free speech and academic freedom play?
2. Women, most of whom are white, are the largest category designated to benefit from affirmative action. Yet a survey of 35 introductory sociology texts published in the 1990s found that nearly 90 percent of the texts did not mention affirmative action in their sections on gender inequality, and only 20 percent of texts included women in their definitions of affirmative action (Beeman, Chowdhry, & Todd 2000). Why do you think many textbooks overlook or minimize the benefits women may receive from affirmative action?
3. Do you think the time will ever come when a racial classification system will no longer be used? Why or why not? What arguments can be made for discontinuing racial classification? What arguments can be made for continuing it?

Key Terms

acculturation
adaptive discrimination
affirmative action
amalgamation
antimiscegenation laws
assimilation (primary and secondary)
aversive racism
colonialism
de facto segregation
de jure segregation
discrimination
ethnicity
expulsion
genocide
geographic steering
hate crime
individual discrimination
institutional discrimination
Jim Crow laws
marital assimilation
minority group
modern racism
multicultural education
naturalized citizen
one drop of blood rule
overt discrimination
pluralism
population transfer
prejudice
primary assimilation
race
racism
redlining
secondary assimilation
segregation
stereotype

Taking a Stand

Should race be a factor in adoption placements? Should people be discouraged from adopting a child that is of a different race from that of the adoptive parents? Why or why not?

Transracial adoption means placing a child who is of one race with adoptive parents of another race. In the United States transracial adoptions usually involve placing a nonwhite child with Caucasian adoptive parents. People choose to adopt transracially for a number of reasons, including: (1) there are fewer young Caucasian children available for adoption than in the past; (2) some adoptive parents feel connected to a particular race through ancestry or through personal experiences such as travel or military service; and (3) some adoptive parents want to care for a child in need, no matter what the race of the child. Some adoption experts believe that children placed for adoption should always be placed with a family with at least one parent of the same race as the child so that the child can develop a strong racial identity. Other experts believe that race is not an important factor in adoption and that having a loving family who can meet the needs of the child is all that matters.

Use Wadsworth's exclusive online resources—InfoTrac College Edition, MicroCase Online, and OVRC—to formulate a position on these topics.

The Wadsworth's Sociology Online Resources and Writing Companion will help you get started. This valuable guide will show you how to use Wadsworth's exclusive online resources when studying social problems. It will also help you to build essential research and writing skills. InfoTrac College Edition, MicroCase Online, OVRC, and an electronic copy of portions of this companion are available at http://sociology.wadsworth.com/mooney_knox_schacht/problems4e, the companion Web site for *Understanding Social Problems,* Fourth Edition.

Media Resources

The Companion Web Site for *Understanding Social Problems,* Fourth Edition

http://sociology.wadsworth.com/mooney_knox_schacht/problems4e

Supplement your review of this chapter by going to the companion Web site to take one of the Tutorial Quizzes, use the flash cards to master key terms, and check out the many other study aids you'll find there. You'll also find special features such as *Wadsworth's Sociology Online Resources and Writing Companion,* GSS Data, and Census 2000 information, data, and resources at your fingertips to help you complete that special project or do some research on your own.

Interactions CD-ROM

Go to the Interactions CD-ROM for *Understanding Social Problems,* Fourth Edition, to access additional interactive learning tools, such as in-depth review materials, corresponding practice quizzes, and other engaging resources and activities to help you study the concepts in this chapter.

"Only a radical transformation of the relationship between women and men to one of full and equal partnership will enable the world to meet the challenges of the 21st century." *Beijing Declaration and Platform for Action*

Gender Inequality

Since she was little, all Sharon Fullilove wanted to do was to fly like a bird. For her, graduation from high school meant she was headed off to the Air Force Academy, the only college she had ever wanted to attend, indeed the only one she had applied to. She had "the right stuff"—a cheerleader, a dance champion, a track star, she sang and did comedy, and was a straight A student. Her mother, an Air Force colonel stationed at the Academy, couldn't have been prouder (Thomas 2003; CNN.com 2003).

One evening after watching a movie Sharon and two friends accepted the offer of a ride home by a male upperclassman they all knew. She was the last to be dropped off and it was then that the senior cadet forced himself on her. Knowing full well of the "good-ole boy" culture she did not report the rape. However, unable to keep her secret and on the advice of her mother, four months after the crime had occurred she reported it. The case was closed with no arrests or punishment and Fullilove, disillusioned, quit the Academy (Thomas 2003; CNN.com 2003).

Over the years there had been several accusations of sexual assault against Air Force cadets but few could have anticipated the results of a Department of Defense study. Of the 579 female cadets surveyed, one in ten seniors reported being the victim of rape or attempted rape; more than 80 percent of the attacks went unreported (Smith 2003). Female cadets spoke of sexual harassment, accusations of promiscuity, and allegations of "violating rules against drinking, fraternization with upperclassmen, and having sex in the dormitories" (CBSNews.com 2003b, 1). First-year Air Force Academy cadets are now required to take a course called "Street Smarts" whereby cadets learn to protect themselves and to avoid potentially dangerous situations. Said one Academy psychologist, "Are you the one girl going to Denver for an overnight in a hotel with five guys? Probably not a good idea" (CBSNews.com 2003b). In the last decade, only one Air Force Academy cadet has been court-martialed for rape—he was acquitted (Thomas 2003).

Some have argued that sexual assault is a reflection of gender inequality. The term "gender inequality," however, begs the question. Unequal in what way? Depending on the issue, both women and men are victims of inequality. When income, career advancement, and sexual harassment are the focus, women are most often disadvantaged. But when life expectancy, mental and physical illness, and access to one's children following divorce are considered, it is often men who are disadvantaged. In this chapter, we seek to understand inequalities for both genders.

This chapter looks at **sexism**—the belief that innate psychological, behavioral, and/or intellectual differences exist between women and men and that these differences connote the superiority of one group and the inferiority of the other. As with race and ethnicity, such attitudes often result in prejudice and discrimination at both the individual and institutional levels. Individual discrimination is reflected by the physician who will not hire a male nurse because he or she believes that women are more nurturing and empathetic and are, therefore, better nurses. Institutional discrimination, that is, discrimination built into the fabric of society, is exemplified by the difficulty many women experience in finding employment; they may have no work history and few job skills as a consequence of living in traditionally defined marriage roles.

Discerning the basis for discrimination is often difficult because the different types of minority status may intersect. For example, elderly African-American and Hispanic women are more likely to receive lower wages and work in fewer prestigious jobs than younger white women. They may also experience discrimination if they are "out" as homosexuals. Such **double or triple jeopardy** occurs when a person is a member of two or more minority groups. In this chapter, however, we emphasize the impact of gender inequality. **Gender** refers to the social definitions and expectations associated with being female or male and should be distinguished from **sex**, which refers to one's biological identity.

The Global Context: The Status of Women and Men

Research indicates that although some progress has been made, millions of women around the world remain victims of violence, discrimination, and abuse. See, for example, the following worldwide statistics (Austin 2000; Leeman 2000; World Bank 2003; Belkachla 2003):

- Over 60 million young girls, predominantly in Asia, are listed as "missing" and are likely the victims of infanticide or neglect.
- 2 million girls between the ages of 5 and 15 are forced into the sex trade each year.
- Less than 10 percent of national parliament seats are held by women.
- 500,000 women each year die of complications from childbirth.
- Two-thirds of the world's 862 million women are illiterate.
- One in three women has been abused, beaten, or coerced into sex.

One specific type of violence suffered by millions of women is female genital mutilation (FGM). Clitoridectomy and infibulation are two forms of FGM. In a clitoridectomy, the entire glans and shaft of the clitoris and the labia minora are removed or excised. With infibulation the two sides of the vulva are stitched together shortly after birth, leaving only a small opening for the passage of urine and menstrual blood. After marriage, the sealed opening is reopened to permit intercourse and delivery. After childbirth, the woman is often reinfibulated. Although some progress is being made, worldwide about 100 to 140 million women and girls have undergone genital mutilation (WHO 2001; FGC 2003).

© David Turnley/CORBIS

In traditional Muslim societies, women are forbidden to show their faces or other parts of their bodies when in public. As pictured, Muslim women wear a veil to cover their faces and a *chador,* a floor-length loose-fitting garment, to cover themselves from head to toe. Although some women adhere to this norm out of fear of repercussions, many others believe veiling was first imposed on Muhammad's wives out of respect for women and the desire to protect them from unwanted advances. There are more than half a million Muslims living in the United States.

Self and Society

The Beliefs About Women Scale (BAWS)

INSTRUCTIONS: The statements listed below describe different attitudes toward men and women. There are no right or wrong answers, only opinions. Indicate how much you agree or disagree with each statement, using the following scale: A = Strongly disagree; B = Slightly disagree; C = Neither agree nor disagree; D = Slightly agree; E = Strongly agree

_______ 1. Women are more passive than men.
_______ 2. Women are less career-motivated than men.
_______ 3. Women don't generally like to be active in their sexual relationships.
_______ 4. Women are more concerned about their physical appearance than men are.
_______ 5. Women comply more often than men do.
_______ 6. Women care as much as men do about developing a job/career.
_______ 7. Most women don't like to express their sexuality.
_______ 8. Men are as conceited about their appearance as women are.
_______ 9. Men are as submissive as women are.
_______ 10. Women are as skillful in business-related activities as men are.
_______ 11. Most women want their partner to take the initiative in their sexual relationships.
_______ 12. Women spend more time attending to their physical appearance than men do.
_______ 13. Women tend to give up more easily than men do.
_______ 14. Women dislike being in leadership positions more than men do.
_______ 15. Women are as interested in sex as men are.
_______ 16. Women pay more attention to their looks than most men do.
_______ 17. Women are more easily influenced than men are.
_______ 18. Women don't like responsibility as much as men do.
_______ 19. Women's sexual desires are less intense than men's.
_______ 20. Women gain more status from their physical appearance than men do.

The Beliefs About Women Scale (BAWS) consists of 15 separate subscales; only four are used here. The items for these four subscales and coding instructions are as follows:

Subscales:

_______ 1. Women are more passive than men. (Items 1, 5, 9, 13, 17)
_______ 2. Women are interested in careers less than men. (Items 2, 6, 10, 14, 18)
_______ 3. Women are less sexual than men. (Items 3, 7, 11, 15, 19)
_______ 4. Women are more appearance conscious than men. (Items 4, 8, 12, 16, 20)

Items 1–5, 7, 11–14, and 16–20 should be scored as follows: Strongly agree = +2, Slightly agree = +1, Neither agree nor disagree = 0, Slightly disagree = −1, and Strongly disagree = +2.

Items 6, 8, 9, 10, and 15 are scored so that: Strongly agree = −2, Slightly agree = −1, Neither agree nor disagree = 0, Slightly disagree = +1, and Strongly disagree = +2. Scores range from −40 to +40; sub-scale scores range from −10 to +10. The higher your score, the more traditional your gender beliefs about men and women.

Source: William E. Snell, Jr., Ph.D. (1997). College of Liberal Arts, Department of Psychology, Southeast Missouri State University. Reprinted with permission.

The societies that practice clitoridectomy and infibulation do so for a variety of economic, social, and religious reasons. A virgin bride can inherit from her father, thus making her an economic asset to her husband. A clitoridectomy increases a woman's worth because a woman whose clitoris is removed is thought to have less sexual desire and therefore to be less likely to be tempted to have sex before marriage. Older women in the community also generate income by performing the surgery so its perpetuation has an economic function (Kopelman 1994, 62). Various cultural beliefs also justify FGM. In Muslim cultures, for example, female circumcision is justified on both social and religious grounds. Muslim women are regarded as inferior to men: they cannot divorce their husbands, but their husbands can divorce them; they are restricted from buying and inheriting property; and they are not allowed to have custody of their children in the event of divorce. Female circumcision is merely an expression of the inequality and low social status women have in Muslim society.

Inequality in the United States

Although attitudes toward gender equality are becoming increasingly liberal, the United States has a long history of gender inequality. (You can assess your own beliefs about gender equality in this chapter's *Self and Society* feature.) Women have had to fight for equality: the right to vote, equal pay for comparable work, quality education, entrance into male-dominated occupations, and legal equality. Even today most U.S. citizens agree that American society does not treat women and men equally—women have lower incomes, hold fewer prestigious jobs, earn fewer academic degrees, and are more likely than men to live in poverty.

Increasingly, however, society recognizes that men are also the victims of gender inequality. When U.S. college students were asked to list the best and worst things about being the opposite sex, the same qualities, although in opposite categories, emerged (Cohen 2001). For example, what males list as the best thing about being female (e.g., free to be emotional), females list as the worst thing about being male (e.g., not free to be emotional). Similarly, what females list as the best thing about being male (e.g., higher pay), males listed as the worst thing about being female (e.g., lower pay). As Cohen notes (2001, 3), although "some differences are exaggerated or oversimplified, . . . we identif[ied] a host of ways in which we 'win' or 'lose' simply because we are male or female."

Sociological Theories of Gender Inequality

Both structural-functionalism and conflict theory concentrate on how the structure of society and, specifically, its institutions contribute to gender inequality. However, these two theoretical perspectives offer opposing views of the development and maintenance of gender inequality. Symbolic interactionism, on the other hand, focuses on the culture of society and how gender roles are learned through the socialization process.

"The history of mankind is a history of repeated injuries and usurpations on the part of man toward woman, having in direct object the establishment of a tyranny over her."

Manifesto, Seneca Falls, New York, 1848

Structural-Functionalist Perspective

Structural-functionalists argue that pre-industrial society required a division of labor based on gender. Women, out of biological necessity, remained in the home performing such functions as bearing, nursing, and caring for children. Men, who were physically stronger and could be away from home for long periods of time, were responsible for providing food, clothing, and shelter for their families. This division of labor was functional for society and, over time, became defined as both normal and natural.

Industrialization rendered the traditional division of labor less functional, although remnants of the supporting belief system still persist. Today, because of day care facilities, lower fertility rates, and the less physically demanding and dangerous nature of jobs, the traditional division of labor is no longer as functional. Thus modern conceptions of the family have, to some extent, replaced traditional ones—families have evolved from extended to nuclear, authority is more egalitarian, more women work outside the home, and greater role variation exists in the division of labor. Functionalists argue, therefore, that as the needs of society change, the associated institutional arrangements also change.

Conflict Perspective

Many conflict theorists hold that male dominance and female subordination are shaped by the relationship men and women have to the production process. During the hunting and gathering stage of development, males and females were economic equals, each controlling their own labor and producing needed subsistence. As society evolved to agricultural and industrial modes of production, private property developed and men gained control of the modes of production, while women remained in the home to bear and care for children. Male domination was furthered by inheritance laws that ensured that ownership would remain in their hands. Laws that regarded women as property ensured that women would remain confined to the home.

As industrialization continued and the production of goods and services moved away from the home, the male-female gaps continued to grow—women had less education, lower incomes, and fewer occupational skills and were rarely owners. World War II necessitated the entry of large numbers of women into the labor force, but in contrast to previous periods, many did not return home at the end of the war. They had established their own place in the workforce and, facilitated by the changing nature of work and technological advances, now competed directly with men for jobs and wages.

Conflict theorists also argue that continued domination by males requires a belief system that supports gender inequality. Two such beliefs are that (1) women are inferior outside the home (e.g., they are less intelligent, less reliable, and less rational) and (2) women are more valuable in the home (e.g., they have maternal instincts and are naturally nurturing). Thus, unlike functionalists, conflict theorists hold that the subordinate position of women in society is a consequence of social inducement rather than biological differences that led to the traditional division of labor.

Symbolic Interactionist Perspective

Although some scientists argue that gender differences are innate, symbolic interactionists emphasize that through the socialization process both females and males are taught the meanings associated with being feminine and masculine. Gender assignment begins at birth as a child is classified as either female or male. However, the learning of gender roles is a lifelong process whereby individuals acquire society's definitions of appropriate and inappropriate gender behavior.

© Marjorie Farrell/The Image Works

Women as well as girls are often portrayed provocatively as a means of selling a product or service. This billboard is a good example of the cultural emphasis placed on women's physical appearance.

Gender roles are taught in the family, the school, and in peer groups, and by media presentations of girls and boys, and women and men (see the section "The Social Construction of Gender Roles" later in this chapter). Most important, however, gender roles are learned through symbolic interaction as the messages others send us reaffirm or challenge our gender performances. As Lorber (1998, 213) notes:

> Gender is so pervasive that in our society we assume it is bred into our genes. Most people find it hard to believe that gender is constantly created and recreated out of human interaction, out of social life, and is the texture and order of social life. Yet gender, like culture, is a human production that depends on everyone constantly "doing gender."

Feminist theory, although also consistent with a conflict perspective, incorporates many aspects of symbolic interactionism. Feminists argue that conceptions of gender are socially constructed as societal expectations dictate what it means to be female or what it means to be male. Thus, for example, women are generally socialized into **expressive** or nurturing and emotionally supportive roles, and males are more often socialized into **instrumental** or task-oriented roles. These roles are then acted out in countless daily interactions as boss and secretary, doctor and nurse, football player and cheerleader "do gender."

Feminists also hold that gender "is a central organizing factor in the social world and so must be included as a fundamental category of analysis in sociological research" (Renzetti & Curran 2003, 8). Feminists, noting that the impact of the structure and culture of society is not the same for different groups of women and men, encourage research on gender that takes into consideration the differential effects of age, race and ethnicity, and sexual orientation.

Gender Stratification: Structural Sexism

As structural-functionalists and conflict theorists agree, the social structure underlies and perpetuates much of the sexism in society. **Structural sexism,** also known as "institutional sexism," refers to the ways the organization of society, and specifically its institutions, subordinate individuals and groups based on their sex classification. Structural sexism has resulted in significant differences between the education and income levels, occupational and political involvement, and civil rights of women and men.

Education and Structural Sexism

Literacy rates worldwide indicate that women are less likely to be able to read and write than men, with millions of women being denied access to even the most basic education. For example, on the average, women in South Asia have only half as many years of education as their male counterparts. Further, of the 150 million children aged 6 to 11 who are not in school, 90 million are girls (World Bank 2003).

In 2002, few differences existed between U.S. men and women in their completion rates of high school and college degrees (U.S. Census 2003). However, as Figure 7.1 indicates, dramatic differences appear in the types of advanced degrees men and women earned. For example, although women received 67 percent of all psychology doctorates, they earned only 17 percent of all doctorates in engineering. Note that with the exception of psychology, men are more likely to earn a doctorate than women in every degree listed. Although the general trend is for men to have higher levels of education than women, in recent years African-American women's educational levels have increased at a faster rate than African-American men's educational levels. Black women receive twice as many college degrees and twice as many master's degrees as black men. They also earn 50 percent more Ph.D.s, J.D.s, M.D.s, and D.D.S.s (Urban Institute 2003).

One explanation for why women earn fewer advanced degrees than men is that women are socialized to choose marriage and motherhood over long-term career preparation. From an early age, women are exposed to images and models of femininity that stress the importance of domestic family life. When 821 undergraduate women were asked to identify their lifestyle preference, less than 1 percent selected being unmarried and working full-time. In contrast, 53 percent selected "gradua-

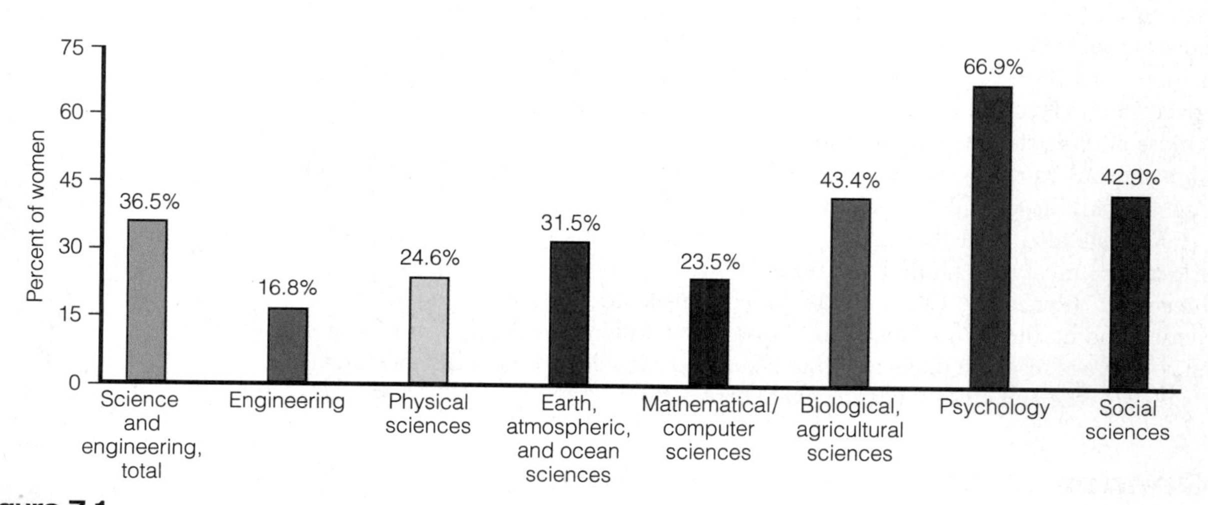

Figure 7.1
Science and Engineering Doctorates Awarded to Women, 2001
Source: National Science Foundation, Division of Science Resource Studies. *Science and Engineering Doctorate Awards:* 2001.

tion, full-time work, marriage, children, stop working at least until youngest child is in school, then pursue a full-time job" as their preferred lifestyle sequence (Schroeder, Blood, & Maluso 1993, 243). Only 6 percent of 535 undergraduate men selected this same pattern.

Structural limitations also discourage women from advancing in the education profession. For example, women seeking academic careers may find that securing a tenure track position is more difficult than for men. McBrier (2003), in examining sex inequality in law schools, found that women progress through the professorial ranks at a slower pace than men. She attributes the differences to a variety of variables including family and geographic constraints, social capital differences, degree prestige, and prior work history.

Long, Allison, and McGinnis (1993) examined the promotions of 556 men and 450 women with Ph.D.s in biochemistry. They found that women were less likely to be promoted to associate or full professor, were held to a higher standard than men, and were particularly disadvantaged in more prestigious departments. Even in public schools, where women make up 66 percent of all classroom teachers, only 45 percent of principals and assistant principals are women (*Statistical Abstract* 2002).

Income and Structural Sexism

In general, the higher one's education, the higher one's income. Yet even when men and women have identical levels of educational achievement and both work full-time, women, on the average, earn significantly less than men (see Table 7.1). Racial differences also exist. Although white women earned 75 percent as much as white men, African-American and Hispanic-American women earned just 54 and 41 percent, respectively, of white men's salaries (*Statistical Abstract* 2002).

Table 7.1
Effects of Education and Sex on Income, 2000

	Average Annual Income	
Educational Attainment	**Men**	**Women**
Total, 25 years and over	45,595	25,483
Less than 9th grade	18,312	11,213
9th to 12th grade (no diploma)	23,960	14,031
High School Graduate (Includes Equivalency)	33,790	19,348
Some college, no degree	40,842	24,685
Associate Degree	45,067	26,641
Bachelor's Degree or More	74,027	40,247
Bachelor's Degree	64,912	35,680
Master's Degree	80,672	46,929
Professional Degree	115,289	60,875
Doctorate Degree	96,330	62,567

Source: *Statistical Abstract 2002,* 122nd ed. 2002. U.S. Bureau of the Census. Washington, DC: U.S. Government Printing Office.

Investigating the gender income gap, a team of researchers (Kilbourne, Farkas, Beron, Weir, & England 1994) analyzed data from the National Longitudinal Survey that included more than 5,000 women and 5,000 men. They concluded that occupational pay is gendered and that "occupations lose pay if they have a higher percentage of female workers or require nurturant skills" (p. 708). Cohen and Huffman (2003) report similar findings but also found that in occupations with high female representation, men performing the same job as women earn higher salaries. Budig (2003) reports that regardless of the male-female distribution in the job, men are paid more than women.

Two hypotheses are frequently cited in the literature for why the income gender gap continues to exist. One is called the **devaluation hypothesis.** It argues that women are paid less because the work they perform is socially defined as less valuable than the work performed by men. The other hypothesis, the **human capital hypothesis,** argues that female-male pay differences are a function of differences in women's and men's levels of education, skills, training, and work experience.

Tam (1997), in testing these hypotheses, concludes that human capital differences are more important in explaining the income gender gap than the devaluation hypothesis. Marini and Fan (1997) also found support for the human capital hypothesis, although their research supports a third category of variables as well. They found that organizational variables (characteristics of the business, corporation, or industry) explain, in part, the gender income gap. For example, women and men upon career entry are channeled by employers into gender-specific jobs that carry different wage rates.

Work and Structural Sexism

Women now make up one-third of the world's labor force. Worldwide, women tend to work in jobs that have little prestige and low or no pay, where no product is produced, and where women are the facilitators for others. Women are also more likely to hold positions of little or no authority within the work environment and to have more frequent and longer periods of unemployment (United Nations 2000a; Athreya 2003). Although improvements are being made, women of color are even less likely to hold positions of power (EEOC 2003). In an investigation of female and male African-American and white firefighters, black women were the most subordinated group, as black males and white females relied on their superordinate gender and race statuses, respectively (Yoder & Aniakudo 1997).

No matter what the job, if a woman does it, it is likely to be valued less than if a man does it (Barko 2003). For example, in the early 1800s, 90 percent of all clerks were men, and being a clerk was a very prestigious profession. As the job became more routine, in part because of the advent of the typewriter, the pay and prestige of the job declined and the number of female clerks increased. Today, female clerks predominate, and the position is one of relatively low pay and prestige.

The concentration of women in certain occupations and men in other occupations is referred to as **occupational sex segregation** (see Table 7.2). For example, women are overrepresented in semiskilled and unskilled occupations, and men are disproportionately concentrated in professional, administrative, and managerial positions. Although the pace is slow, increasingly men are applying for jobs traditionally held by women, leading to such terms as "mannies" (male nannies) and "murses" (male nurses) (Cullen 2003).

In some occupations, sex segregation has decreased in recent years. For example, the percentage of female physicians increased from 16 percent to 29 percent between 1983 and 2001, female dentists increased from 7 percent to 20 percent, fe-

Table 7.2
Highly Sex-Segregated Occupations, 2001

Female-Dominated Occupations	Percentage of Female Workers
Child care workers	97
Cleaners and servants	96
Dental hygienists	97
Dietitians	86
Elementary school teachers	83
Librarians	85
Prekindergarten and kindergarten teachers	98
Receptionists	97
Registered nurses	93
Secretaries	98
Speech therapists	92
Teachers' aides	94
Typists	95
Male-Dominated Occupations	**Percentage of Male Workers**
Announcers	79
Airplane pilots and navigators	97
Architects	77
Automobile mechanics	95
Clergy	85
Construction workers	97
Dentists	80
Engineers	90
Firefighters	97
Lawyers	71
Mechanics and repairers	95
Physicians	71
Police and detectives	86

Source: *Statistical Abstract 2002,* 122nd ed. 2002. U.S. Bureau of the Census. Washington, DC: U.S. Government Printing Office.

male engineers increased from 6 to 11 percent, and female clergy increased from 6 to 15 percent (*Statistical Abstract* 2002).

Nevertheless, despite these and other changes, women are still heavily represented in low-prestige, low-wage **"pink-collar" jobs** that offer few benefits. Even those women in higher-paying jobs are often victimized by a **glass ceiling**—an invisible barrier that prevents women and other minorities from moving into top corporate positions. A study of the Fortune 500 companies revealed that 393 had no women in their top five executive positions (Jones 2003).

Sex segregation in occupations continues for several reasons (Martin 1992; Williams 1995; Renzetti & Curan 2003). First, cultural beliefs about what is an "appropriate" job for a man or a woman still exist. Cejka and Eagly (1999) report that the more college students believed that an occupation was male- or female-dominated, the more they attributed success in that occupation to masculine or feminine characteristics. Further, males and females continue to be socialized to learn different skills and acquire different aspirations. A government report documents that work at age 12 is sex-segregated: girls baby-sit and boys do lawn work (BLS 2000).

Opportunity structures differ as well. Women have fewer opportunities in the more prestigious and higher-paying male-dominated professions. The result? Women are twice as likely as men to be minimum-wage earners and, when paid hourly, have a median average earning of $9.57 compared to $11.36 for men (BLS 2002; BLS 2003). Women may also be excluded by male employers and employees who fear the prestige of their profession will be lessened with the entrance of women, or who simply believe that "the ideal worker is normatively masculine" (Martin 1992, 220).

Finally, because family responsibilities primarily remain with women, working mothers may feel pressure to choose professions that permit flexible hours and career paths, sometimes known as "mommy tracks" (Moen & Yu 2000). Thus, for example, women dominate the field of elementary education, which permits them to be home when their children are not in school. Nursing, also dominated by women, often offers flexible hours.

Politics and Structural Sexism

Women received the right to vote in 1920 with the passage of the Nineteenth Amendment. Even though this amendment went into effect almost 80 years ago, women still play a rather minor role in the political arena. In general, the more important the political office, the lower is the probability that a woman will hold it. Although women are 52 percent of the population, the United States has never had a woman president or vice president and, until 1993 when a second woman was appointed, had only one female U.S. Supreme Court justice. The highest-ranking woman ever to serve in U.S. government was Madeleine Albright, who became the U.S. Secretary of State in 1997. In 2003, women represented only 10 percent of all governors and held only 13.6 percent of all U.S. congressional seats (see Table 7.3). Worldwide, the percentage of national and local legislative seats held by women is less than 10 percent (World Bank 2003).

Table 7.3
Percentage of Women Elected by Level and Type of Government Position, 2003

Level of Government/Position	No. Seats	No. Women	Percentage Held by Women
U.S. President	1	0	0.0
U.S. Vice President	1	0	0.0
U.S. Congress	535	73	13.6
House	435	59	13.5
Senate	100	14	14.0
Governors	50	6	12.0
State Legislators	7,382	1,648	22.3

Source: Center for American Women in Politics. "Women in Elected Offices, 2003." http://rci.rutgers.edu/cawp

In response to the underrepresentation of women in the political arena, some countries have instituted quotas. In India, a 1993 amendment held one-third of all seats in local contests for women. Eight thousand women were elected. A 1996 law in Brazil requires that a minimum of 20 percent of each party's candidates be women. Countries with similar policies include Finland, Germany, Mexico, South Africa, and Spain (Sheehan 2000).

The relative absence of women in politics, as in higher education and in high-paying, high-prestige jobs in general, is a consequence of structural limitations. Running for office requires large sums of money, the political backing of powerful individuals and interest groups, and a willingness of the voting public to elect women. Thus, minority women have even greater structural barriers to election and, not surprisingly, represent an even smaller percentage of elected officials. Nonetheless, 80 percent of U.S. women believe that by 2024 a woman will be in the White House (Thomas 1999).

Civil Rights, the Law, and Structural Sexism

The 1963 Equal Pay Act and Title VII of the 1964 Civil Rights Act made it illegal for employers to discriminate in wages or employment on the basis of sex. Nevertheless, such discrimination still occurs as evidenced by the thousands of grievances filed each year with the Equal Employment Opportunity Commission (EEOC). One technique used to justify differences in pay is the use of different job titles for the same work. The courts have ruled, however, that jobs that are "substantially equal," regardless of title, must result in equal pay.

"The price of inequality is just too high."

Nasfis Sadik
United Nations Population Fund

Women are also discriminated against in employment. Discrimination, although illegal, takes place at both the institutional and the individual level (see Chapter 6). Institutional discrimination includes screening devices designed for men, hiring preferences for veterans, the practice of promoting from within an organization based on seniority, and male-dominated recruiting networks (Reskin & McBrier 2000). For example, the Augusta National Golf Club, home of the Masters Golf Tournament and a virtual "who's who of the corporate world," prohibits women from joining (Gandy 2003). One of the most blatant forms of individual discrimination is sexual harassment (see the Sexual Harassment section).

Discrimination takes place in other forms as well. In the United States, having lower incomes, shorter work histories, and less collateral, women often have difficulty obtaining home mortgages or rental property. Until fairly recently, husbands who raped their wives were exempt from prosecution. Even today, some states require a legal separation agreement and/or separate residences for a raped wife to receive full protection under the law. Women in the military have traditionally been restricted in the duties they can perform, and finally, since the U.S. Supreme Court's 1973 *Roe v. Wade* decision, which made abortion legal, the right of a woman to obtain an abortion has steadily been limited and narrowed by subsequent legislative acts and judicial decisions. The debate continues with several recent court decisions (see Chapter 15).

The Social Construction of Gender Roles: Cultural Sexism

As social constructionists note, structural sexism is supported by a system of cultural sexism that perpetuates beliefs about the differences between women and men. **Cultural sexism** refers to the ways the culture of society—its norms, values, beliefs, and symbols—perpetuates the subordination of an individual or group because of the sex classification of that individual or group.

For example, the *belief* that females are less valuable than males has serious consequences. China has 20 percent fewer girls than boys in the birth to 4 years old age group. China's "missing girl" phenomenon, a growing problem, is a consequence of selective abortions of female fetuses and premature death of female infants due to a withholding of nourishment and health care (Banister 2003). Cultural sexism takes place in a variety of settings including the family, the school, and media as well as in everyday interactions.

Family Relations and Cultural Sexism

From birth, males and females are treated differently. The toys male and female children receive convey different messages about appropriate gender behavior. For example, 90 percent of girls aged 3 to 11 have a Barbie doll (Dittrich 2002b). Recently, retail giant Toys-R-Us, after much criticism, removed store directories labeled "Boy's World" and "Girl's World." Similarly, toy manufacturer Mattel came under fire after producing a pink, flowered Barbie computer for girls, and a blue Hot Wheels computer for boys. The social significance of the gender-specific computers and the public criticism came after it was revealed that the accompanying software packages were different—the boys' package had more educational titles (Bannon 2000). This chapter's *Focus on Technology* feature documents the negative consequences of such seemingly harmless differences.

Household Division of Labor. Little girls and boys work within the home in approximately equal amounts until the age of 18 when the female-to-male ratio begins to change (Robinson & Bianchi 1997; UNFPA 2003). In a study of household labor in 10 Western countries, Bittman and Wajcman (2000) report that "women continue to be responsible for the majority of hours of unpaid labor" ranging from a low of 70 percent in Sweden to a high of 88 percent in Italy (p. 173). The fact that women, even when working full-time, contribute significantly more hours to home care than men is known as the "second shift" (Hochschild 1989). Not surprisingly, women compared to men, have less free time, are solely responsible for children during their free time, and are more likely to feel "rushed" (Mattingly & Bianchi 2003).

Three explanations for the continued traditional division of labor emerge from the literature. The first explanation is the "time-availability approach." Consistent with the structural-functionalist perspective, this position claims that role performance is a function of who has the time to accomplish certain tasks. Because women are more likely to be at home, they are more likely to perform domestic chores.

A second explanation is the "relative resources approach." This explanation, consistent with a conflict perspective, suggests that the spouse with the least power is relegated the most unrewarding tasks. Because men have more education, higher incomes, and more prestigious occupations, they are less responsible for domestic labor.

"Gender role ideology," the final explanation, is consistent with a symbolic interactionist perspective. It argues that the division of labor is a consequence of traditional socialization and the accompanying attitudes and beliefs. Females and males have been socialized to perform various roles and to expect their partners to perform other complementary roles. Women typically take care of the house, men the yard. This division of labor is learned in the socialization process through the

Focus on Technology

Women, Men, and Computers

Technology has changed the world in which we live. The technological revolution has brought the possibility of greater gender equality for, unlike tasks dominating industrialization, sex differences in size, weight, and strength are less relevant. Although feminists have long decried the gendering of technology (see Chapter 15), surely computers and other information technologies are gender neutral—or are they?

According to a study by the National Institute on Media and the Family, 92 percent of children and adolescents between the ages of 2 and 17 play video games. Girls, however, spend significantly less time playing video games than boys. On any given day, 44 percent of boys compared to 17 percent of girls report playing video games (Kaiser Family Foundation 2003). In a *Children Now* (2001) study of 1,716 video characters in over 70 games, 64 percent of the characters were human males and 17 percent human females, the remainder being nonhuman characters. Further, of the 874 player-controlled characters in the study, 73 percent were males and 12 percent females (*Children Now* 2001).

Half of all female characters were props or bystanders, that is, part of the background. Of the female characters displayed, 11 percent were hyper-sexualized having very large breasts and very small waists. Another 7 percent had disproportionate or extremely thin bodies. Twenty percent had exposed breasts, 13 percent exposed buttocks, and 30 percent exposed midriffs.. Males were also stereotyped. Over one in three males, 35 percent, were portrayed as overly muscular, and nearly as many appeared unaffected by violence (*Children Now* 2001).

In adolescence and beyond, girls are not as interested in computers as boys. When they are interested, unlike boys who consider a computer a toy to explore, something that's fun, girls define computers as a tool to accomplish a task, a kind of homework helper. Says Jane Margolis, a researcher at Carnegie Mellon (quoted in Breidenbach 1997, 69):

> Girls want to do something constructive with computers, while boys get into hacking and using computers for their own sake. . . . Computers are just one interest of many for girls, while they become an object of love and fascination for boys.

If women do not pursue computer-based information technologies, they will have an "intellectual and workplace handicap that can only get worse as technology grows more prevalent" (Currid 1996, 114). For example, whereas social work positions, held primarily by women, are projected to grow by 36 percent, systems analyst and computer engineer positions—two of the four fastest-growing occupations between 1998 and 2008—are projected to grow 94 and 108 percent, respectively (*Statistical Abstract* 2002).

The culture and structure of high-tech occupations, however, are often not conducive to female employment. Rapidly changing knowledge bases make taking time off for motherhood almost impossible; long hours make child care arrangements and time away from home difficult; and lingering stereotypes such as "women can't handle stress" and women don't "understand technology as well as men" (*Informationweek* 1996) make advancement difficult. A Massachusetts Institute of Technology report on the gender gap in computing found that "women are often judged as less qualified than men even when their performance is identical" (Breidenbach 1997, 69).

Women earn less than a third of the bachelor's degrees awarded in computer and information sciences, and only 18 percent of advanced degrees in these areas (Stabiner 2003). Nonetheless, there are some signs of optimism. The number of women getting computer science degrees, although small, has increased over the last several decades (*Statistical Abstract* 2002). Further, some companies such as Sun Microsystems and Hewlett-Packard have begun aggressive recruitment and hiring programs for women and people of color. However, for significant changes to take place in reference to women, men, and computers we must recognize that computers specifically, and technology in general, are not gender neutral and, in fact, have emerged and flourished within the context of a male-dominated industry.

Sources: Susan Breidenbach. 1997. "Where Are All the Women?" *Network World* 14(41):68–69; Cheryl Currid. 1996. "Bridging the Gender Gap: Women Will Lose Out Unless They Catch Up with Men in Technology Use." *Informationweek,* April 1, 114; Children Now. 2001. *Fair Play: Violence, Gender, and Race in Video Games. Children Now:* Oakland, CA: http://www.childrennow.org; *Informationweek.* 1996. "Women Gain in IS Ranks," September 16, 158; *Kaiser Family Foundation.* 2003. "Video Game Key Facts." http://www.kff.org; Karen Stabiner. 2003. "Where the Girls Aren't." *New York Times,* January 12. http://www. NYTimes.com; *Statistical Abstract of the United States: 2002,* 122nd ed. 2002. U.S. Bureau of the Census. Washington, DC: U.S. Government Printing Office; Janese Swanson. 2003. "Guidelines for Improving Video Games for Girls." *Media Awareness Network Tip Sheet.* http://media-awareness.ca/english /resources; Candee Wilde. 1997. "Women Cut Through IT's Glass Ceiling." *Informationweek,* January 20, 83–86.

media, schools, books, and toys. A test of the three positions found that although all three had some support, gender role ideology was the weakest of the three in predicting work allocation (Bianchi, Milkie, Sayer, & Robinson 2000).

The School Experience and Cultural Sexism

Sexism is also evident in the schools. It can be found in the books students read, the curricula and tests they are exposed to, and the different ways teachers interact with students.

Textbooks. The bulk of research on gender images in textbooks and other instructional materials documents the way males and females are portrayed stereotypically. For example, Purcell and Stewart (1990) analyzed 1,883 storybooks used in schools and found that they tended to depict males as clever, brave, adventurous, and income-producing and females as passive and as victims. Females were more likely to be in need of rescue and were also depicted in fewer occupational roles than males. Witt (1996), in a study of third-grade textbooks from six publishers, reports that little girls were more likely to be portrayed as having both traditionally masculine *and* feminine traits whereas little boys were more likely pictured as having masculine characteristics only. These results are consistent with research that suggests that boys are much less free to explore gender differences than females, and with Purcell and Stewart's conclusion that boys are often depicted as having "to deny their feelings to show their manhood" (1990, 184). Although some recent evidence suggests that the frequency of male and female textbook characters is increasingly equal, portrayals of girls and boys largely remain stereotypical (Evans & Davies 2000).

"In the school room, more than any other place, does the difference of sex, if there is any, need to be forgotten."

Susan B. Anthony
Feminist

Curricula and Testing. Encouragement to participate in sports, academic programs, and extracurricular activities is gender-biased despite Title IX of the 1972 Educational Amendments Act, which prohibits officials from "tracking" students by sex (Orecklin 2003). Although women's and girls' participation is now at an all-time high with, for example, over 150,000 women participating in NCAA sporting events, differences remain in the sports males and females play (*Statistical Abstract* 2002). Males are more likely to participate in competitive sports that emphasize traditional male characteristics such as winning, aggression, physical strength, and dominance. Women are more likely to participate in sports that emphasize individual achievement (e.g., figure skating) or cooperation (e.g., synchronized swimming).

The differing expectations and/or encouragement that females and males receive also contribute to their varying abilities, as measured by standardized tests, in such disciplines as math and science. Boys and girls have the same mathematics and science proficiency at age 9; by age 13, males outperform females in science, although not math. By age 17, males outperform females in both math and science. Are such differences a matter of aptitude? In an experiment at the University of Waterloo, male and female college students, all of whom said they were good in math, were shown either gender-stereotyped or gender-neutral advertisements. When, subsequently, female students who had seen the female stereotyped advertisements took a math test, they performed not only lower than women who had seen the gender-neutral advertisements but lower than their male counterparts (Begley 2000). Research also indicates that standardized tests themselves are biased—almost exclusively being timed, multiple-choice tests—a format favoring males according to some advocates (Smolken 2000).

Teacher-Student Interactions. Sexism is also reflected in the way teachers treat their students. Millions of young girls are subjected to sexual harassment by male teachers who then fail them when they refuse the teachers' sexual advances (Quist-Areton 2003). After interviewing 800 adolescents, parents, and teachers in three school districts in Kenya, Mensch and Lloyd (1997) report that teachers were more likely to describe girls as lazy and unintelligent. "And when the girls do badly," the researchers remark, "it undoubtedly reinforces teachers' prejudices, becoming a vicious cycle." Similarly, in the United States, Sadker and Sadker (1990) observed that elementary and secondary school teachers pay more attention to boys than to girls. Teachers talk to boys more, ask them more questions, listen to them more, counsel them more, give them more extended directions, and criticize and reward them more frequently. However, a book by philosopher Christina Sommers (2000), entitled *The War Against Boys,* argues that it is "boys, not girls, on the weak side of the educational gender gap" (p. 14). Noting that boys are at a higher risk for learning disabilities and lag behind in reading and writing scores, Sommers argues that the belief that females are educationally shortchanged is untrue (see Chapter 12). Further, as one researcher notes, the "problem with boys" is not just in the United States but exists throughout the developed world (Poe 2004).

Media, Language, and Cultural Sexism

One concern voiced by social scientists in reference to cultural sexism is the extent to which the media portrays females and males in a limited and stereotypical fashion, and the impact of such portrayals. A study of 1,200 children between the ages of 10 and 17 produced these results (Dittrich 2002a):

- Seven out of 10 girls said they wanted to look like a television character.
- A majority of boys and girls said women on TV are thinner than in real life
- 16 percent of girls said they had dieted or exercised to look like a character on television.
- When asked to name their most admired TV characters, boys' and girls' top five slots were filled by men.
- The children associated "worrying about appearance and weight, crying, whining and weakness" with female TV characters; playing sports, wanting to be kissed or to have sex, and being a leader were associated with being a male character.

Men are also victimized by media images. A recent study of 1,000 adults found that two-thirds of the respondents thought that women in television advertisements were pictured as "intelligent, assertive and caring" while men were portrayed as "pathetic and silly" (Abernathy 2003).

Like media images, both the words we use and the way we use them can reflect gender inequality. The term "nurse" carries the meaning of "a woman who . . ." and the term "engineer" suggests "a man who . . ." Terms like "broad," "old maid," and "spinster" have no male counterpart. Language is so gender-stereotyped that the placement of male or female before titles is sometimes necessary as in the case of "female police officer" or "male prostitute." Further, as symbolic interactionists note, the embedded meanings of words carry expectations of behavior.

Virginia Sapiro (1994) has shown how male-female differences in communication style reflect the structure of power and authority relations between men and women. For example, women are more likely to use disclaimers ("I could be wrong but . . .") and self-qualifying tags ("That was a good movie, wasn't it?"), reflecting less certainty about their opinions. Communication differences between women and men also reflect different socialization experiences. Women are more often passive and polite in conversation; men are less polite, interrupt more often, and talk more (Renzetti & Curran 2003).

Social Problems and Traditional Gender Role Socialization

Cultural sexism, transmitted through the family, school, media, and language, perpetuates traditional gender role socialization. Gender roles, however slowly, are changing. As one commentator observed (Fitzpatrick 2000, 1), "The hard lines that once helped to define masculine [and feminine] identity are blurring. Women serve in the military, play pro basketball, run corporations and govern. Men diet, undergo cosmetic surgery, bare their souls in support groups and cook."

Despite this **"gender tourism"** (Fitzpatrick 2000), most research indicates that traditional gender roles remain dominant, particularly for males who, in general, have less freedom to explore the gender continuum. Social problems that result from traditional gender socialization include the feminization of poverty, social-psychological and health costs, and conflict in relationships.

The Feminization of Poverty

Globally, the percentage of female households is increasing dramatically, with one-third of households in developing nations headed by a woman (WGI 2003). Often at the poverty level, many of these households are headed by young women with dependent children and older women who have outlived their spouses. Women perform two-thirds of the world's work yet earn only one-tenth of the world's income, and own less than 1 percent of the world's property (UNICEF 2003).

A "report card" of U.S. efforts to reduce poverty among women was released by U.S. Women Connect, a nonprofit activist group. Although the United States received a "B" for placing women in decision-making positions, it received an "F" for efforts to reduce female poverty. Citing federal statistics, the group reports that although the overall poverty rate in the United States has decreased, female poverty has increased over the last five years (Winfield 2000). As noted earlier, both individual and institutional discrimination contribute to the economic plight of women.

Traditional gender role socialization also contributes to poverty among women. Women are often socialized to put family ahead of their education and careers. Women are expected to take primary responsibility for child care, which contributes to the alarming rate of single-parent poor families in the United States. Hispanic and black female-headed households are the poorest of all families headed by a single woman (BLS 2002). Further, a study of the relationship between marital status, gender, and poverty in the United States, Australia, Canada, and France indicates that never-married women compared with ever-married women in all four countries are more likely to live in poverty (Nichols-Casebolt & Krysik 1997).

Social-Psychological and Other Health Costs

Many of the costs of traditional gender socialization are social-psychological in nature. Reid and Comas-Diaz (1990) noted that the cultural subordination of women results in women having low self-esteem and being dissatisfied with their roles as spouses, homemakers/workers, mothers, and friends. In a study of self-esteem among more than 1,160 students in grades 6 through 10, girls were significantly more likely to have "steadily decreasing self-esteem," whereas boys were more likely to fall into the "moderate and rising" self-esteem group (Zimmerman, Copeland, Shope, & Dielman 1997).

Not all researchers have found that women have a more negative self-concept than men. Summarizing their research on the self-concepts of women and men in the United States, Williams and Best (1990) found "no evidence of an appreciable difference" (p. 153). They also found no consistency in the self-concepts of women and men in 14 countries: "[I]n some of the countries the men's perceived self was noticeably more favorable than the women's, whereas in others the reverse was found" (p. 152). More recent research also reports that women are becoming more assertive and desirous of controlling their own lives rather than merely responding to the wishes of others or the limitations of the social structure (Burger & Solano 1994).

Men also suffer from traditional gender socialization (Gupta 2003). Men experience enormous cultural pressure to be successful in their work and earn a high income. Research indicates that men who have higher incomes feel more "masculine" than those with lower incomes (Rubenstein 1990). Not surprisingly, males are more likely than females to value materialism and competition over compassion and self-actualization (Beutel & Marini 1995; McCammon, Knox, & Schacht 1998; Cohen 2001). Traditional male socialization also discourages males from expressing emotion—part of what Pollack (2000a) calls the "boy code." This chapter's *The Human Side* feature describes the problems and pressures of being male in American society.

"I learned from my father how to work. I learned from him that work is life and life is work, and work is hard."

Philip Roth
Author

On the average, men in the United States die about six years earlier than women, although gender differences in life expectancy have been shrinking (PRB 2003). Traditional male gender socialization is linked to males' higher rates of cirrhosis of the liver, most cancers, homicide, drug- and alcohol-induced deaths, suicide, and firearm and motor vehicle accidents (*Statistical Abstract* 2002). At every stage of life "American males have poorer health and a higher risk of mortality than females" (Gupta 2003, 84). Men engage in risky behaviors more often than women—from smoking, drinking, and abusing drugs to not wearing a seatbelt and working in dangerous environments (Gupta 2003). However, 15 percent of males compared to 24 percent of females are clinically depressed at some point in their life (Mazure, Keita & Blehar 2002). Females are also much more likely to be anorexic or bulimic—seven million girls suffer from eating disorders (Renzetti & Curran 2003).

Although men have higher rates of HIV/AIDS worldwide, the disease disproportionately affects women in many areas of the world. For example, in sub-Saharan Africa, 58 percent of those infected are women. Women's inequality contributes to the spread of the disease (Heyzer 2003). First, in many of these societies "women lack the power in relationships to refuse sex or negotiate protected sex" (Heyzer 2003, 1). Second, women are often the victims of rape and sexual assault with little social or legal recourse. Finally, some women turn to prostitution as a means of supporting themselves and their children. With little education and no training or work history, they have few options. Ironically, it is women who care for those who are ill, often dropping out of school or quitting jobs to do so, furthering their subordinate role (McGregor 2003).

Are gender differences in morbidity and mortality a consequence of socialization differentials or physiological differences? Although both nature and nurture

The Human Side

Real Boys' Voices

Psychologist and author William Pollack traveled from coast to coast talking to boys about the "boy code."

Brad, 14, from a suburb in the Northwest
Guys aren't supposed to be weak or vulnerable. Guys aren't supposed to be sweet. A friend of mine died in the hospital. . . . I knew that, as a guy, I was supposed to be strong and I wasn't supposed to show any emotion. . . . I was supposed to be tough . . . when I went home, I just sat by myself and let myself cry. (p.17)

Sam, 16, from a city in New England
I think most of the macho stuff guys do is stupid . . . like the kids who do wrestling moves in the hall. At the same time, there are things that I wouldn't do because I'm a guy. I've never gone to a guy friend, for example, and said, "I'm feeling hurt right now and let's talk about it." (p. 31)

Gordon, 18, from a small town in the South
. . . all the men in my family . . . have been the epitome of negativity. Some have become wrapped up in infidelity, some abuse, some alcoholism. I don't want to become a man, because I don't want to become this. (p. 53)

Jeff, 16, from a small town in New England
Your virginity is what determines whether you're a man or a boy in the eyes of every teenage male. Teenage men see sex as a race: the first one to the finish line wins. (p. 69)

Brett, 17, from a city in the South
I think most guys are kind of isolated because it's thought of as weird if you have any really close guy friends. To get around it, guys will go fishing or hunting or bowling or something else "masculine," and then talk about personal or serious things while they're doing that activity. (p. 116)

Jesse, 17, from a suburb in New England
From the girls I've spoken to about relationships, one of their biggest complaints is that they're doing all the giving and the guy is doing all the taking. Girls also tend to be better able to understand social situations. Girls can look at someone and tell what they're feeling. They have more social intuitiveness, more than we clueless guys do. I think that makes them more aware of what's happening in a relationship than we are. (p. 253)

Graham, 17, from a suburb in the West
We would live in a better society if guys could share their feelings more easily. But guys still hear mixed messages from our society. On the one hand they hear that it's OK now to talk about their feelings, but on the other hand they still hear that they have to be tough and that only girls get emotional. My friend who talked to me and cried about his girlfriend was on the football team. His teammates would laugh at him if he tried to talk to them about that sort of stuff. (p. 272)

Jake, 16, from a suburb in southern New England
Ever since I've played Little League the word "win" has been forced into my mind. When I was eight years old, the coach would tell us at the beginning of the season that we were just out there for fun, but I knew that it wasn't true. Every day that there was a game, my day would be ruined. (pp. 280–281)

Dylan, 17, from a suburb of Chicago
If I get in shape, if I develop a more attractive body, I'd be more popular. It's like the way life is around here, what society shows you. It's a problem to be naturally skinny like me. You're not as athletic or muscular or attractive; you're not as good as the other kids are. (p. 302)

Kirk, 18, from a suburb in the Northwest
I think it's hard growing up in the year 2000. It's definitely hard for a guy. Going through high school is tough. I have pressures in sports, school, life all rolled into one. My parents pressure me to do well in school, do well in sports, and I pressure myself to do well in life. . . . I worry about life a lot. I feel that everything is going to work out for everybody except me, that I'll be left in the dust. (p. 341)

are likely to be involved, social rather than biological factors may be dominant. As part of the "masculine mystique," men tend to engage in self-destructive behaviors—heavy drinking and smoking, poor diets, lack of exercise, stress-related activities, higher drug use, and a refusal to ask for help. Men are also more likely to work in hazardous occupations than women. Women's higher rates of depression are also likely to be rooted in traditional gender roles. The heavy burden of child care and household responsibilities, the gender pay and occupation gap, and fewer socially acceptable reactions to stress (e.g., it is more acceptable for males than females to drink alcohol) contribute to gender differences in depression.

Conflict in Relationships

Worldwide, gender inequality influences relationships. Research indicates that the distribution of resources impacts the decision-making process. For example, a study by the International Food Policy Research Institute found that "assets brought to marriage have an impact on bargaining power within the marriage": the more assets the more power (UNFPA 2003). In five of the developing countries studied—Bangladesh, Ethiopia, Ghana, the Philippines, and South Africa—men brought more land and other assets to the marriage and thus had more power in the decision-making process. It should be noted, however, that over half of women's work is unpaid and, thus, is not viewed as household income. When the value of household work is included in income totals, women contribute between 40 and 60 percent of the household income (UNFPA 2003).

Gender inequality also has an impact on relationships in the United States. For example, negotiating work and home life can be a source of relationship problems. Whereas men in traditional versus dual-income relationships are more likely to report being satisfied with household task arrangements, women in dual-income families are the most likely to be dissatisfied with household task arrangements (Baker, Kriger, & Riley 1996). Further, the belief that one's partner is not performing an equitable portion of the housework is associated with a reduction in the perception of spousal social support (Van Willigen & Drentea 1997).

We must consider also, of course, the practical difficulties of raising a family, having a career, and maintaining a happy and healthy relationship with a significant other. In one survey, over 80 percent of both men and women responded that changing gender roles make it more difficult to have a successful marriage (Morin & Rosenfeld 2000). Successfully balancing work, marriage, and children may require a number of strategies, including (1) a mutually satisfying distribution of household labor, (2) rejection of such stereotypical roles as "super-mom" and "breadwinner dad" (see this chapter's *Social Problems Research Up Close* feature), (3) seeking outside help from others (e.g., child care providers, domestic workers), and (4) a strong commitment to the family unit. Finally, violence in relationships is gender-specific (see Chapters 4 and 5). Although men are more likely to be victims of violent crime, women are more likely to be victims of rape and domestic violence. Violence against women reflects male socialization that emphasizes aggression and dominance over women. Male violence is a consequence of gender socialization and a definition of masculinity which holds that "as long as nobody is seriously hurt, no lethal weapons are employed, and especially within the framework of sports and games—football, soccer, boxing, wrestling—aggression and violence are widely accepted and even encouraged in boys" (Pollack 2000b, 40).

Social Problems Research Up Close

Family, Gender Ideology, and Social Change

One of the most important questions concerning gender is the extent to which gender ideologies affect family roles. An investigation by Zuo and Tang (2000) addresses this issue by focusing on two research questions: (1) Are men less likely than women to believe in the equality of roles? and (2) Is the male "breadwinner" status predictive of beliefs about gender ideology?

Sample and Methods

Data for this investigation came from a randomly selected national sample of married persons collected by the Bureau of Sociological Research at the University of Nebraska. As part of a larger longitudinal study, respondents (N = 400 married men and 640 married women) were interviewed in 1980, 1983, and 1992. All were between the ages of 18 and 55, 95 percent were white, and 67 percent had 1992 annual incomes between $25,000 and $45,000. The independent variable, *breadwinner status,* was measured by the proportion of a husband's income to the total family income. For example, a husband who provided 90 percent of the total family income received a higher score than a husband who provided 50 percent of the total family income. The higher a respondent's score, the higher their breadwinner status and the lower the breadwinner status of their spouse. *Gender ideology,* the dependent variable, was measured by the extent to which a respondent agreed or disagreed with statements concerning (1) the wife's economic role (e.g., ". . . a woman should not be employed if jobs are scarce"), (2) the provider role (e.g., ". . . a husband should be the main breadwinner even if his wife works"), and (3) the women's maternal role (e.g., ". . . a woman's most important task in life is being a mother"). In combination, these three variables indicated the extent to which respondents adhered to a traditional or egalitarian (i.e., equal partners) gender ideology.

Findings and Conclusions

The results signify that over the years studied, both men and women have shifted toward a more egalitarian gender role ideology. The shift, however, is greater for women than for men. One notable exception is in reference to beliefs about a woman's maternal role. Here, men held more egalitarian beliefs than women. The authors caution, however, that this result does not necessarily indicate that women hold more traditional beliefs about motherhood than men. It may be, for example, that women's stress over the lack of child care facilities outside of the home is responsible for gender differences on this indicator.

Results also indicate that the higher a husband's breadwinner status, that is, the more he contributes to household finances, the more likely he is to hold traditional gender beliefs. Conversely, the lower a husband's breadwinner status, the more likely he is to hold egalitarian gender beliefs. Similarly, the higher a wife's breadwinner status, that is, the more she contributes to household finances, the more likely she is to hold egalitarian beliefs, and the lower her breadwinner status, the more likely she is to hold traditional beliefs.

The authors conclude that the results of their study support what is called the *benefits hypothesis.* The benefits hypothesis holds that men whose wives earn high wages, that is, men who have lower breadwinner statuses, are likely to embrace rather than be threatened by role equality. Given the empirical support for this hypothesis, the authors predict a continued narrowing in male-female differences in gender role ideology.

> The present trend is that men's breadwinner status continues to decline; more and more individuals perform non-gendered family roles. Based on these facts, it may be predicted that the movement toward egalitarianism for both men and women will continue and a further decrease in the gender gap in gender ideology is down the road. (2000, 36)

Source: Jiping Zuo and Shengming Tang. 2000. "Breadwinner Status and Gender Ideologies of Men and Women Regarding Family Roles." *Sociological Perspectives* 43:29–44.

Strategies for Action: Toward Gender Equality

Measuring gender inequality for comparative purposes is a difficult task. The United Nations, however, has created a gender empowerment index. This index combines gender inequality in three areas—political power, division of labor, and income—into one measure. Using this measure, the country that is the most "gender equal" is Norway, followed by Iceland, Sweden, Denmark, and Finland. The United States is ranked eleventh in gender equality (GEM 2003).

Grassroots Movements

Efforts to achieve gender equality in the United States have been largely fueled by the feminist movement. Despite a conservative backlash, feminists, and to a lesser extent men's activists groups, have made some gains in reducing structural and cultural sexism in the workplace and in the political arena.

Feminism and the Women's Movement. **Feminism** is the belief that women and men should have equal rights and responsibilities. The American feminist movement began in Seneca Falls, New York, in 1848 when a group of women wrote and adopted a women's rights manifesto modeled after the Declaration of Independence. Although many of the early feminists were primarily concerned with suffrage, feminism has its "political origins . . . in the abolitionist movement of the 1830s," when women learned to question the assumption of "natural superiority" (Anderson 1997, 305). Early feminists were also involved in the temperance movement, which advocated restricting the sale and consumption of alcohol, although their greatest success was the passing of the Nineteenth Amendment in 1920, which recognized women's right to vote.

The rebirth of feminism almost 50 years later was facilitated by a number of interacting forces: an increase in the number of women in the labor force, the publication of Betty Friedan's book *The Feminine Mystique,* an escalating divorce rate, the socially and politically liberal climate of the 1960s, student activism, and the establishment of the Commission on the Status of Women by John F. Kennedy. The National Organization for Women (NOW) was established in 1966 and remains the largest feminist organization in the United States with more than 100,000 members. One of NOW's hardest-fought battles was the struggle to win ratification of the Equal Rights Amendment (ERA), which states that "equality of rights under the law shall not be denied or abridged by the United States, or by any state, on account of sex." The proposed amendment passed both the House of Representatives and the Senate in 1972 but failed to be ratified by the required 38 states by the 1978 deadline, later extended to 1982. Thirty-five states have ratified the ERA, and it is presently in several state legislatures awaiting action.

Supporters of the ERA argue that its opponents used scare tactics—saying the ERA would lead to unisex bathrooms, mothers losing custody of their children, and mandatory military service for women—to create a conservative backlash. Susan Faludi in *Backlash: The Undeclared War Against American Women* (1991) contends that contemporary arguments against feminism are the same as those levied against the movement a hundred years ago and that the negative consequences predicted by opponents of feminism (e.g., women unfulfilled and children suffering) have no empirical support.

Today, a new wave of feminism is being led by young women and men who grew up with the benefits won by their mothers but shocked by the realities of the Air Force scandal, Paula Jones's accusations of sexual harassment against a sitting president, and packs of men roaming Central Park openly assaulting women. These young feminists are more inclusive than their predecessors, welcoming all who champion the cause of global equality. Not surprisingly, the new feminists are likely to attract a more diverse group of supporters than their predecessors as future feminist efforts focus on "gender equality" over "gender sameness" (Parker 2000). Some observers, however, note that the new diversity of the women's movement may contribute to "a tension within feminism between the felt urgency to present concerns and grievances as a single unified group of women, and the need to give voice to the variations in concerns and grievances that exist among feminists on the basis of race and ethnicity, social class, sexual orientation, age, physical ability/disability, and a host of other factors" (Renzetti & Curran 2003, 25).

The Men's Movement. As a consequence of the women's rights movement, men began to reevaluate their own gender status. In *Unlocking the Iron Cage: The Men's Movement, Gender Politics, and American Culture,* Michael Schwalbe (1996) examines the men's movement as both participant and researcher. For three years, he attended meetings and interviewed active members. His research indicates that participants, in general, are white middle-class men who feel they have little emotional support, question relationships with their fathers and sons, and are overburdened by responsibilities, unsatisfactory careers, and what is perceived as an overly competitive society.

As with any grassroots movement, the men's movement has a variety of factions. Some men's organizations advocate gender equality, that is, they are profeminists; others developed to oppose "feminism" and what was perceived as male bashing. For example, the Promise Keepers, part of a Christian men's movement, and Louis Farrakhan's Nation of Islam, have often been criticized as patriarchal and antifeminist (Renzetti & Curran 2003).

"As I talked to boys across America, I'm struck by how trapped they feel. Our culture puts boys in a gender straightjacket."

William Pollack
Psychologist

Today, issues of custody and fathers' rights, led by such groups as Dads against Discrimination, Texas Father's Alliance, and the National Coalition of Free Men (NCFM), headline the men's rights movement and have led to increased visibility. Many members of such groups argue that society portrays men as "disposable" and that as fathers and husbands, workers and soldiers, they feel that they can simply be replaced by other men willing to do the "job." They also hold that nothing in society is male-affirming and that the social reform of the last 30 years has "been the deliberate degradation and dis-empowerment of men economically, legally, and socially" (NCFM 1998, 7). Still other men's advocates concentrate less on men's rights and more on personal growth, advocating "the restoration of earlier versions of masculinity" (Cohen 2001, 393).

Public Policy

A number of statutes have been passed to help reduce gender inequality. They include the 1963 Equal Pay Act, Title VII of the Civil Rights Act of 1964, Title IX of the Educational Amendments Act of 1972, the Family Leave Act of 1993, the 1994 Violence Against Women Act, and the Victims of Trafficking and Violence Protection Act of 2000. The National Organization for Women encourages women to be politically active, to run for political office, and to participate in the decision-making processes of the nation. Recently, public policy has focused on two issues—sexual harassment and affirmative action.

Sexual Harassment. During the 1980s and 1990s, the courts held that Title VII of the 1964 Civil Rights Act prohibited **sexual harassment** involving members of the opposite sex. In 1998, the U.S. Supreme Court extended protection to victims of same-sex harassment. Studies on sexual harassment indicate that it is a widespread problem occurring in a variety of settings including college campuses, public schools, military academies, and the workplace (Renzetti & Curran 2003).

Sexual harassment can be of two types: (1) *quid pro quo,* in which an employer requires sexual favors in exchange for a promotion, salary increase, or any other employee benefit, and (2) the existence of a hostile environment that unreasonably interferes with job performance, as in the case of sexually explicit comments or insults being made to an employee. According to a 1993 Supreme Court decision, a person no longer has to demonstrate "severe psychological damage" to win damages. Later guidelines "stress that in order for a behavior to constitute sexual harassment it must be severe and repetitive; a single, inappropriate act is not considered sexual harassment" (Renzetti & Curran 2003, 129).

Some research suggests that the number of incidents of sexual harassment is inversely proportional to the number of women in an occupational category (Fitzgerald & Shullman 1993; *Civil Rights Monitor* 2000). For example, female doctors (Schneider & Phillips 1997) and lawyers (Rosenberg, Perlstadt, & Phillips 1997) report high rates of sexual harassment—in the first case by male patients and in the second by male colleagues. Sexual harassment is a worldwide phenomenon. Seventy percent of female government employees in Japan report being sexually harassed at work (Yamaguchi 2000).

Affirmative Action. The 1964 Civil Rights Act provided for **affirmative action** to end employment discrimination based on sex and race (see Chapter 6). Such programs require employers to make a "good faith effort" to provide equal opportunity to women and other minorities. However, in response to the growing sentiment that affirmative action programs constitute "reverse discrimination," recent court decisions have begun to dismantle affirmative action programs. In 1996, a California ballot initiative abolished racial and sexual preferences in government programs that included state colleges and universities (Chavez 2000). Washington state voters passed a similar initiative in 1998, and in 1999 the governor of Florida signed an executive order ending that state's affirmative action program. However, in 2003, the U.S. Supreme Court held that universities have a "compelling interest" in a diverse student population and, therefore, may take minority status into consideration when making admissions decisions (*Major Rulings* 2003) (see Chapter 6).

International Efforts

According to a World Bank report, gender inequality in the transition countries of Europe and Central Asia is growing as "health, pension, and education systems have crumbled" (*Popline* 2003, p. 4). Interestingly, in the transition countries of South Asia, the "burden of transformation" has disproportionately fallen on women; in Russia, Ukraine, and Belarus, the greater burden has fallen on men, while in Central and Eastern Europe men and women equally share the disparities.

The Convention to Eliminate All Forms of Discrimination Against Women (CEDAW), also known as the International Women's Bill of Rights, was adopted by the United Nations in 1979. CEDAW establishes rights for women not previously recognized internationally in a variety of areas, including education, politics, work, law, and family life. The United States signed the document on July 17, 1980, although it has yet to be ratified by the required two-thirds vote of the U.S. Senate. Over 170 countries have ratified the treaty, including every country in Europe and South and Central America. The United States is the only industrialized country that has not ratified the document (Mitchell 2002). Contrary to what some critics argue, provisions in CEDAW would not supersede existing U.S. law (United Nations 2000b; Rabin 2000).

© Kim Kulish/Corbis

© Bob Sacha

Definitions of appropriate gender roles change over time. Fifty years ago, women playing hockey and winning the Olympics would have been unimaginable. Men's roles, although more slowly, are also changing.

In addition to the CEDAW and many other global efforts, individual countries have instituted programs or policies designed to combat sexism and gender inequality. For example, Japan has implemented the Basic Law for a Gender-Equal Society, a "blueprint for gender equality in the home and workplace" (Yumiko 2000, 41). Wage differences remain high, however, with women earning just 65 percent of men's salaries (CEDAW 2003). A new South African Bill of Rights prohibits discrimination on the basis of, among other things, gender, pregnancy, and marital status (IWRP 2000), and China has recently established the Programme for the Development of Chinese Women, which focuses on empowering women in the areas of education, human rights, health, child care, employment, and political power (WIN News 2000). Further, in 2003, a bill was introduced into Norway's Parliament that, if passed, would require that 40 percent or more of corporation board members be women. The law, which would impact over 600 companies, is expected to pass. The 40 percent rule is to be in effect by 2005 and companies who have not yet met the quota will be in jeopardy of losing their board certification (Alvarez 2003).

"Give to every human being every right that you claim for yourself."

Thomas Paine
Political and Social Activist

Understanding Gender Inequality

Gender roles and the social inequality they create are ingrained in our social and cultural ideologies and institutions and are, therefore, difficult to alter. For example, in almost all societies women are primarily responsible for child care and men for military service and national defense (World Bank 2003). Nevertheless, as we have seen in this chapter, growing attention to gender issues in social life has spurred some change. Women who have traditionally been expected to give domestic life first priority are now finding it acceptable to be more ambitious in seeking a career outside the home. In a recent survey of college freshmen, only 21 percent agreed with the statement: "The activities of a married woman are best confined to the home and family" (Sax, Lindholm, Astin, Korn, & Mahoney 2002). Men's roles are also changing. Men who have traditionally been expected to be aggressive and task-oriented are now expected to be more caring and nurturing.

The fact that gender lines are becoming blurred confirms that gender is not an either-or phenomenon but rather exists on a continuum from femininity to masculinity. Defining people as *either* feminine *or* masculine is an oversimplification of gender roles. Men and women, in varying degrees, have both feminine and masculine characteristics, that is, are **androgynous.** In many cultures androgyny was and is embraced. For example, a "berdache" is a person who adopts the gender role of the opposite sex; this status was held in high esteem in Asian, South Pacific, and North American Indian cultures (Renzetti & Curran 2003).

Although traditionally it has been women who have fought for gender equality, men also have much to gain. They too are victims of discrimination and gender stereotyping. Eliminating gender stereotypes and redefining gender in terms of equality does not mean simply liberating women, but liberating men and our society as well. "What we have been talking about is allowing people to be more fully human and creating a society that will reflect that humanity. Surely that is a goal worth striving for" (Basow 1992, 359). Regardless of whether traditional gender roles emerged out of biological necessity as the functionalists argue or economic oppression as the conflict theorists hold, or both, it is clear today that gender inequality carries a high price: poverty, loss of human capital, feelings of worthlessness, violence, physical and mental illness, and death. Surely, the costs are too high to continue to pay.

Chapter Review

- **Does gender inequality exist worldwide?**
There is no country in the world where men and women are treated equally. While women suffer in terms of income, education, and occupational prestige, men are more likely to suffer in terms of mental and physical health, mortality, and the quality of their relationships.

- **How do the three major sociological theories view gender inequality?**
Structural-functionalists argue that the traditional division of labor was functional for pre-industrial society and, over time, has become defined as both normal and natural. Today, however, modern conceptions of the family have, to some extent, replaced traditional ones. Conflict theorists hold that male dominance and female subordination evolved in relation to the means of production—from hunting and gathering societies where females and males were economic equals to industrial societies where females were subordinate to males. Symbolic interactionists emphasize that through the socialization process both females and males are taught the meanings associated with being feminine and masculine.

- **What is meant by structural and cultural sexism?**
Structural sexism refers to the ways in which the organization of society, and specifically its institutions, subordinate individuals and groups based on their sex classification. Structural sexism has resulted in significant differences between education and income levels, occupational and political involvement, and civil rights of women and men. Structural sexism is supported by a system of cultural sexism that perpetuates beliefs about the differences between women and men. Cultural sexism refers to the ways the culture of society—its norms, values, beliefs, and symbols—perpetuates the subordination of an individual or group because of the sex classification of that individual or group.

- **What are some of the problems caused by traditional gender roles?**
First is the feminization of poverty. Women are socialized to put family ahead of education and careers, a belief that is reflected in their less prestigious occupations and lower incomes. Second are social-psychological and health costs. Women tend to have lower self-esteem and higher rates of depression and eating disorders than men. Men, on the average, die about six years earlier than women and have higher rates of cirrhosis of the liver, most cancers, homicide, drug- and alcohol-induced deaths, suicide, and firearm and motor vehicle accidents (*Statistical Abstract* 2002). Last, gender inequality influences the nature of relationships. For example, although men are more likely to be victims of violent crime, women are more likely to be victims of domestic violence.

- **What strategies can be employed to end gender inequality?**
Grassroots movements, such as feminism and the women's rights movement and the men's rights movement, have made significant inroads in the fight against gender inequality. Their accomplishments, in part, have been the result of successful lobbying for passage of laws concerning sex discrimination, sexual harassment, and affirmative action. Besides these national efforts, international efforts continue as well. One of the most important is the Convention to Eliminate All Forms of Discrimination Against Women (CEDAW), also known as the International Women's Bill of Rights, which was adopted by the United Nations in 1979.

Critical Thinking

1. Some research suggests that "men and women with more androgynous gender orientations—that is to say, those having a balance of masculine and feminine personality characteristics—show signs of greater mental health and more positive self-images" (Anderson 1997, 34). Do you agree or disagree? Why or why not?
2. Recent evidence suggests that a "gender gap" exists in the number of men and women entering college, particularly among African-Americans, with women attending at higher rates than men. Although the number of females in the population is slightly higher, the difference does not explain the projected gap in enrollments. Why are black women entering college at a higher rate than black men?
3. Why are women more likely to work in traditionally male occupations than men are to work in traditionally female occupations? Are the barriers that prevent men from doing "women's work" cultural, structural, or both? Explain.

Key Terms

affirmative action
androgynous
cultural sexism
devaluation hypothesis
double or triple (multiple) jeopardy
expressive roles
feminism
gender
gender tourism
glass ceiling
human capital hypothesis
instrumental roles
occupational sex segregation
pink-collar jobs
sex
sexism
sexual harassment
structural sexism

Taking a Stand

Are gender roles a matter of biology or environment?

The argument is an old one—nature versus nurture. Physically, men and women are different from birth. Men, in general, are stronger, taller, heavier, and have more facial hair. Women develop breasts, have higher-pitched voices, menstruate, and bear children. But are these physical charactersitics related to behavioral differences? Are women *innately* nurturers? Are men *innately* aggressive? Many, noting for example that most societies are (were) patriarchal, would answer yes. Others, however, would be quick to note the role of socialization in traditional gender role assignment. Cross-cultural evidence is mixed. Some societies are characterized by traditional gender roles; in others, androgyny dominates, and in still others, traditional gender roles are reversed.

Use Wadsworth's exclusive online resources—InfoTrac College Edition, MicroCase Online, and OVRC—to formulate a position on this topic.

The Wadsworth's Sociology Online Resources and Writing Companion will help you get started. This valuable guide will show you how to use Wadsworth's exclusive online resources when studying social problems. It will also help you to build essential research and writing skills. InfoTrac College Edition, MicroCase Online, OVRC, and an electronic copy of portions of this companion are available at http://sociology.wadsworth.com/mooney_knox_schacht/problems4e, the companion Web site for *Understanding Social Problems,* Fourth Edition.

Media Resources

The Companion Web Site for *Understanding Social Problems,* Fourth Edition

http://sociology.wadsworth.com/mooney_knox_schacht/problems4e

Supplement your review of this chapter by going to the companion Web site to take one of the Tutorial Quizzes, use the flash cards to master key terms, and check out the many other study aids you'll find there. You'll also find special features such as *Wadsworth's Sociology Online Resources and Writing Companion,* GSS Data, and Census 2000 information, data, and resources at your fingertips to help you complete that special project or do some research on your own.

Interactions CD-ROM

Go to the Interactions CD-ROM for *Understanding Social Problems,* Fourth Edition, to access additional interactive learning tools, such as in-depth review materials, corresponding practice quizzes, and other engaging resources and activities to help you study the concepts in this chapter.

"Homophobia alienates mothers and fathers from sons and daughters, friend from friend, neighbor from neighbor, Americans from one another. So long as it is legitimated by society, religion, and politics, homophobia will spawn hatred, contempt, and violence, and it will remain our last acceptable prejudice."

Byrne Fone

Issues in Sexual Orientation

Sheila Hein, 51, a civilian Army employee who worked as a management analyst, died on September 11, 2001, when a hijacked American Airlines jet crashed into the Pentagon. When rescue workers found Sheila's remains, she was wearing the gold band that her partner of 18 years, Peggy Neff, had given her. Assisted by the Human Rights Campaign, a Washington organization that lobbies for gay rights, and the Lambda Legal Defense and Education Fund, Neff filed to receive compensation from a September 11 Victim Compensation Fund established by the Justice Department. Congress created the compensation fund 11 days after the terrorist attacks. Families that apply to it are barred from suing anyone—except the terrorists themselves and their accomplices—for damages as a result of the terrorist attacks. "Words cannot express what I have lost," Neff wrote in an affidavit filed with her federal claim. "She was my entire world and my soulmate, my closest confidante and my best friend" (Vogel 2003, B01). Neff had been denied state aid from Virginia, which limited victim compensation benefits to spouses, parents, grandparents, siblings, and children. When the Justice Department decided to award victim compensation to Peggy Neff, gay rights advocates hailed the decision, saying it was the first known time the federal government gave recognition to a same-sex relationship.

The horrendous events of 9/11 revealed how vulnerable gay and lesbian individuals are in times of crisis—experiencing discriminatory treatment and lack of social and legal recognition is commonplace in the lives of nonheterosexual individuals, couples, and their families. In this chapter we examine prejudice and discrimination toward nonheterosexual individuals: homosexual (or gay) women (also known as lesbians), homosexual (or gay) men, and bisexual individuals.

It is beyond the scope of this chapter to explore how sexual diversity and its cultural meanings vary throughout the world. Rather, this chapter focuses on Western conceptions of diversity in sexual orientation. The term **sexual orientation** refers to the classification of individuals as heterosexual, bisexual, or homosexual, based on their emotional and sexual attractions, relationships, self-identity, and lifestyle. **Heterosexuality** refers to the predominance of emotional and sexual attraction to persons of the other sex. **Homosexuality** refers to the predominance of emotional and sexual attraction to persons of the same sex, and **bisexuality** is emotional and sexual attraction to members of both sexes. Lesbians, gays, and bisexuals, sometimes referred to collectively as **the lesbigay population,** are considered to be part of a larger population referred to as the transgendered community. **Transgendered individuals** include "a range of people whose gender identities do not conform to traditional notions of masculinity and femininity" (Cahill, Ellen, & Tobias 2002, 10). Transgendered individuals include not only homosexuals and bisexuals, but also cross-dressers, transvestites, and transsexuals. Cross-dressers are individuals, usually heterosexual men, who occasionally dress in the clothing of the other sex; transvestites are homosexual men who occasionally dress as women; and transsexuals are individuals who have undergone hormone treatment and sex reassignment surgery to change their gender identity (McCammon, Knox, & Schacht 2004). Much of the current literature on the treatment and political and social agendas of the lesbigay population includes other members of the transgendered community; hence the term **LGBT** or **GLBT** is often used to refer collectively to lesbians, gays, bisexuals, and transgendered individuals.

In the Life, the gay and lesbian television newsmagazine, reported Sheila Hein (left) and Peggy Neff's (right) story in its April 2002 episode on PBS. The story "Victims of 9/11" received a 2003 Emmy nomination for "Single News Feature: 9/11."

After summarizing the legal status of lesbians and gay men

around the world, we discuss the prevalence of homosexuality, heterosexuality, and bisexuality in the United States, review explanations for sexual orientation diversity, and apply sociological theories to better understand societal reactions to sexual diversity. Then, after detailing the ways in which nonheterosexuals are victimized by prejudice and discrimination, we end the chapter with a discussion of strategies to reduce antigay prejudice and discrimination.

The Global Context: A World View of Laws Pertaining to Homosexuality

Homosexual behavior has existed throughout human history and in most, perhaps all, human societies (Kirkpatrick 2000). A global perspective on laws and social attitudes regarding homosexuality reveals that countries vary tremendously in their treatment of same-sex sexual behavior—from intolerance and criminalization to acceptance and legal protection. In 84 countries, sexual activity between consenting adults of the same sex is illegal (International Gay and Lesbian Human Rights Commission 2003a). In 55 of these countries, laws criminalizing same-sex sexual behavior apply to both female and male homosexuality; in 29 of these countries, the laws apply to male homosexuality only. Legal penalties vary for violating laws that prohibit homosexual sexual acts. In nine countries, individuals found guilty of engaging in same-sex sexual behavior may receive the death penalty (see Table 8.1).

In general, countries throughout the world are moving toward increased legal protection of sexual orientation minorities. According to the International Gay and Lesbian Human Rights Commission (1999), 22 countries have national laws that ban various forms of discrimination against gays, lesbians, and bisexuals. In 1996 South Africa became the first country in the world to include in its constitution a clause banning discrimination based on sexual orientation. Fiji, Canada, and Ecuador also have constitutions that ban discrimination based on sexual orientation ("Constitutional Protection" 1999).

In recent years legal recognition of same-sex relationships has become more widespread. In 2001, the Netherlands became the first country in the world to offer full, legal marriage to same-sex couples. Same-sex married couples and opposite-sex married couples in the Netherlands will be treated identically, with two exceptions. Unlike opposite-sex marriages, same-sex couples married in the Netherlands are unlikely to have their marriages recognized as fully legal abroad. Regarding children, parental rights will not automatically be granted to the nonbiological

Table 8.1
Countries in Which Homosexual Acts Are Subject to the Death Penalty

Mauritania	Pakistan	United Arab Emirates
Sudan	Iran	Yemen
Afghanistan	Saudi Arabia	Somalia

Sources: "Jail, Death Sentences in Africa." 2001 (February 21). PlanetOut.com. http://www.planetout.com; Mackay, J. 2001. "Global Sex: Sexuality and Sexual Practices Around the World." *Sexuality and Relationship Therapy,* 16:71–82.

© AP/Wide World Photos

Gert Kasteel, left, and Dolf Pasker were among the world's first same-sex couples to marry legally under Dutch law after the Netherlands became the first country to allow same-sex marriages.

spouse in gay couples. To become a fully legal parent, the spouse of the biological parent must adopt the child. In 2003, Belgium passed a law allowing same-sex marriages but disallowing any adoptions. Also in 2003, the provinces of Ontario and British Columbia, Canada, granted full equal marriage rights to same-sex couples (International Gay and Lesbian Human Rights Commission 2003b).

Other countries recognize same-sex **registered partnerships,** which are federally recognized relationships that convey most but not all of the rights of marriage (some countries also offer registered partnerships to opposite-sex couples). Federally recognized registered partnerships for same-sex couples are available in Australia, Belgium, Brazil, Canada, Denmark, Finland, France, Germany, Greenland, Hungary, Iceland, Israel, Italy, New Zealand, Norway, Portugal, Spain, and Sweden (International Gay and Lesbian Human Rights Commission 2003b). Two U.S. states—Vermont and California—give legal recognition to same-sex couples. As we discuss later in this chapter, same-sex couples in Vermont and California are granted state-level rights, not the more than 1,000 federal benefits that go along with civil marriage. In 2003, the Massachusetts Supreme Court ruled that same-sex marriages are allowable under the State Constitution.

Human rights treaties and transnational social movement organizations have increasingly asserted the rights of people to engage in same-sex relations. International organizations such as Amnesty International, which resolved in 1991 to defend those imprisoned for homosexuality, the International Lesbian and Gay Association (founded in 1978), and the International Gay and Lesbian Human Rights Commission (founded in 1990) continue to fight antigay prejudice and discrimination. As this book goes to press, the United Nations Human Rights Commission is considering a Resolution on Sexual Orientation and Human Rights. This landmark resolution recognizes the existence of sexual orientation–based discrimination around the world; affirms that such discrimination is a violation of human rights; and calls all governments to promote and protect the human rights of all people, regardless of their sexual orientation (International Gay and Lesbian Human Rights Commission 2004). Despite the worldwide movement toward increased acceptance and protection of homosexual individuals, the status and rights of lesbians and gays in the United States continues to be one of the most divisive issues in American society.

Homosexuality and Bisexuality in the United States: A Demographic Overview

Before looking at demographic data concerning homosexuality and bisexuality in the United States, it is important to understand the ways in which identifying or classifying individuals as "homosexual," "gay," "lesbian," and "bisexual" is problematic.

Sexual Orientation: Problems Associated with Identification and Classification

The classification of individuals into sexual orientation categories (e.g., "gay," "straight," "bisexual," "lesbian," "homosexual," "heterosexual") is problematic for a number of reasons. First, due to the social stigma associated with nonheterosexual identities, many individuals conceal or falsely portray their sexual orientation identities to protect themselves against prejudice and discrimination. But more important, distinctions among sexual orientation categories are simply not as clear-cut as many people would believe.

Consider the early research on sexual behavior by Kinsey and his colleagues (1948, 1953), which found that although 37 percent of men and 13 percent of women had had at least one same-sex sexual experience since adolescence, very few of the individuals reported exclusive homosexual behavior. These data led Kinsey to conclude that most people are not exclusively heterosexual or homosexual. Rather, Kinsey suggested an individual's sexual orientation may have both heterosexual and homosexual elements. In other words, Kinsey suggested that heterosexuality and homosexuality represent two ends of a sexual orientation continuum, and that most individuals are neither entirely homosexual nor entirely heterosexual, but fall somewhere within this continuum. Most people are, in other words, to some degree, bisexual.

Sexual orientation classification is also complicated by the fact that sexual behavior, attraction, love, desire, and sexual orientation identity do not always match. For example, "research conducted across different cultures and historical periods (including present-day Western culture) has found that many individuals develop passionate infatuations with same-gender partners in the absence of same-gender sexual desires . . . whereas others experience same-gender sexual desires that never manifest themselves in romantic passion or attachment" (Diamond 2003, 173).

> "Although it is typically presumed that heterosexual individuals only fall in love with other-gender partners and gay-lesbian individuals only fall in love with same-gender partners, this is not always so."
>
> **Lisa M. Diamond**
> **Psychologist**

Consider the findings of a national study of U.S. adults that investigated (1) sexual attraction to persons of the same sex, (2) sexual behavior with people of the same sex, and (3) homosexual self-identification (Michael, Gagnon, Laumann, & Kolata 1994). This survey found that 4 percent of women and 6 percent of men said they are sexually attracted to individuals of the same sex, and 4 percent of women and 5 percent of men reported that they had had sexual relations with a same-sex partner after age 18. Yet less than 3 percent of the men and less than 2 percent of women in this study identified themselves as homosexual or bisexual (Michael et al. 1994). What these data tell us is that first, "those who acknowledge homosexual desires may be far more numerous than those who actually act on those desires" (Black, Gates, Sanders, & Taylor 2000, 140). Second, not all people who are sexually attracted to or have had sexual relations with individuals of the same sex view themselves as homosexual or bisexual. A final difficulty in labeling a person's sexual orientation is that an individual's sexual attractions, behavior, and identity may change across time.

The Prevalence of Nonheterosexual Adults and Same-Sex Cohabiting Couples in the United States

Despite the difficulties inherent in categorizing individuals into sexual orientation categories, recent data reveal the prevalence of individuals who identify as lesbian, gay, or bisexual and of cohabiting same-sex couples living in the United States. As noted earlier, in the national survey by Michael et al. (1994), less than 3 percent of the men and less than 2 percent of women in this study identified themselves as homosexual or bisexual. In the 2000 presidential election, self-identified gays, lesbians, and bisexuals represented 4 percent of voters nationwide ("Post-Election Analysis" 2000). More recently, Smith and Gates (2001) estimated that there are more than 10 million gay and lesbian adults in the United States, which represents between 4 percent and 5 percent of the total U.S. adult population.

The 2000 Census found that about one in nine (594,000) unmarried-partner households in the United States involve partners of the same sex (Simmons & O'Connell 2003). Nationally, 51 percent of same-sex couples had male partners. Census 2000 data also revealed that 99.3 percent of U.S. counties reported same-sex cohabiting partners, compared to 52 percent of counties in 1990 (Bradford, Barrett, & Honnold 2002).

Why are data on the numbers of U.S. GLBT adults and couples relevant? Primarily because census numbers on the prevalence of GLBT adults and couples can influence laws and policies that affect gay individuals and their families. In anticipation of the 2000 Census, the National Gay and Lesbian Task Force Policy Institute and the Institute for Gay and Lesbian Strategic Studies conducted a public education campaign urging people to "out" themselves on the 2000 Census. The slogan was, "The more we are counted, the more we count" (Bradford et al. 2002, 3). "The fact that the Census documents the actual presence of same-sex couples in nearly every state legislative and U.S. Congressional district means anti-gay legislators can no longer assert that they have no gay and lesbian constituents" (Bradford et al. 2002, 8).

Other Demographic Data on the Gay Population

Census data from 1990 and 2000 provide a variety of demographic data on the gay population (Bradford et al. 2002; Simmons & O'Connell 2003). Same-sex couples are more likely to live in metropolitan areas than in rural areas. However, the largest proportional increases in the number of same-sex couples self-reporting in 2000 compared to 1990 came in rural, sparsely populated states. States with the highest

Palms of Manasota

Residents of America's first gay and lesbian retirement community, Palms of Manasota, in Plametto, Florida.

percentage of same-sex unmarried partners of all coupled households include California, Massachusetts, Vermont, and New York.

In the 1990 Census, 31 percent of the lesbian and bisexual women in same-sex relationships and 19 percent of the gay and bisexual men in same-sex relationships reported that they had been previously married—or were currently married—to a person of the opposite sex. The 2000 Census revealed that one-third (33%) of female same-sex householders were living with their children (under 18 years). Of male same-sex householders, 22 percent had their children present in the household.

The best data on age and sexual orientation comes from the 1990 Census, which found that in 11 percent of female same-sex couples, at least one member was 60 years of age or older, and in 8 percent, a member was 65 years or older. Data on the elderly gay and lesbian population are necessary to address economic support, social support, health care, and housing needs of this population.

The Origins of Sexual Orientation Diversity

One of the prevailing questions raised regarding sexual orientation centers on its origin or "cause." Questions about the "causes" of sexual orientation are typically concerned with the origins of homosexuality and bisexuality. Because heterosexuality is considered normative and "natural," causes of heterosexuality are rarely considered.

Much of the biomedical and psychological research on sexual orientation attempts to identify one or more "causes" of sexual orientation diversity. The driving question behind this research is, Is sexual orientation inborn? Or is it learned or acquired from environmental influences? While a number of factors have been correlated with sexual orientation, including genetic factors, gender role behavior in childhood, and fraternal birth order, there is no single theory that can explain diversity in sexual orientation (McCammon et al. 2004).

Beliefs About What "Causes" Homosexuality

Aside from what "causes" homosexuality, sociologists are interested in what people *believe* about the "causes" of homosexuality. Most gays believe that homosexuality is an inherited, inborn trait. In a national study of homosexual men, 90 percent believe that they were born with their homosexual orientation; only 4 percent believe that environmental factors are the sole cause (Lever 1994). The percentage of Americans who believe that homosexuality is something a person is born with increased from 13 percent in 1977 to 40 percent in 2002 (Newport 2002).

Individuals who believe that homosexuality is biologically based tend to be more accepting of homosexuality. In contrast, "those who believe homosexuals choose their sexual orientation are far less tolerant of gays and lesbians and more likely to conclude homosexuality should be illegal than those who think sexual orientation is not a matter of personal choice" (Rosin & Morin 1999, 8).

Can Homosexuals Change Their Sexual Orientation?

Individuals who believe that homosexuals choose their sexual orientation tend to think that homosexuals can and should change their sexual orientation. Various forms of **reparative therapy** or **conversion therapy** are dedicated to changing

The Human Side

Can Homosexuals Change Their Sexual Orientation? A Case Study

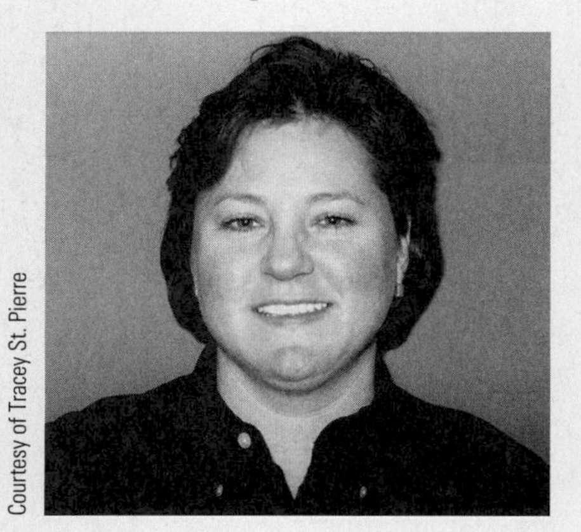
Courtesy of Tracey St. Pierre

I sat peacefully in the prayer circle, nodding in agreement or whispering my "amen," as different women prayed. As I peeked up, my eyes met those of my church counselor. She was watching me. With a nod and a quick wave of her hand, she instructed me to sit in a more "feminine manner." This is something we had talked about before, my needing to become more feminine.

I knew I was different at an early age. When my hormones started raging in high school, I didn't go boy crazy. I fell in love with another girl. When we met, sparks flew. We were both in love for the first time. But this was a small town in the buckle of the Bible Belt during the late 70s. Society, organized religion, and Anita Bryant, all with uncompromising certainty, declared that our kind of love was wrong. As a result, we kept our 2-year relationship a secret. We lived in constant fear that someone would find out and label us as lesbians.

By the time I left for college, I was desperate to talk to someone about my relationship and my sexual orientation. As it happened, that someone came in the form of the church counselor. When I told her about my relationship, she said that God did not create me gay and the love I had shared was sinful. She assured me that God could heal me and make me "whole again"—a real woman. She prayed for me, laid her hands on me and rebuked the demonic "spirit of homosexuality" that I had "allowed to control me." She gave me Bible verses to memorize, told me to avoid temptation and to break off my 2-year relationship. She encouraged me to develop "godly" relationships with women—but not too close.

We scheduled time together to work on my femininity. The first afternoon we spent in front of a mirror. I learned how to apply make-up—eyeliner, mascara, eye shadow, lipstick, the works. Another day, she criticized my short, popular Dorothy Hamill haircut that all the girls had. "Let it grow longer," she said. She told me to rid my closet of my old jeans, sweat pants, and gym shorts and replace them with skirts and dresses. My family, particularly my mother and tomboy sister (heterosexual and married now for 20 years), noticed the changes in me. I was zealous in my pursuit of heterosexuality. God would give me a husband, the counselor said.

My church counselor explained that my same-sex attraction was because of either sexual abuse, a flaw in my upbringing, or a deficient parent-child relationship. But try as

homosexuals' sexual orientation. Some religious organizations sponsor "ex-gay ministries," which claim to "cure" homosexuals and transform them into heterosexuals through prayer and other forms of "therapy" (see this chapter's *The Human Side* feature). Ex-gay ministries attract new recruits by claiming that there is a "cure" for people who are unhappy being gay. Critics of ex-gay ministries take a different approach:

> The cure for unhappiness is not the "ex-gay" ministries—but coming out with dignity and self-respect. It is not gay men and lesbians who need to change . . . but negative attitudes and discrimination against gay people that need to be abolished. (Besen 2000, 7)

The American Psychiatric Association, American Psychological Association, American Academy of Pediatrics, and the American Medical Association agree that sexual orientation cannot be changed and that efforts to change sexual orientation do not work and may, in fact, be harmful (Human Rights Campaign 2000). According to the American Psychological Association, close scrutiny of reports of "suc-

The Human Side

I might, I could never pinpoint what "made me gay." I have wonderful, loving parents (married for 45 years now), four brothers and a sister (all heterosexual), and no history of abuse. In the meantime, I prayed, fasted (sometimes for days), and literally begged God to give me an attraction to men.

For me, the journey of self-acceptance began after almost 15 years of celibacy and intense personal struggle. After years of trying to suppress my sexual orientation, I could not shake a crush on a fellow female congressional staffer. So I decided to seek the help of a mental health professional. It didn't take long for me to realize that I am a lesbian, and happily so. My therapist said that the church did no more than brainwash me, and I now know that "reparative therapy"—which purports to be able to change one's sexual orientation from homosexual to heterosexual—is a lie.

My story is not unique. The reality is that societal prejudice and religious intolerance can drive people to take drastic actions, often with devastating consequences. In my work at the Human Rights Campaign, I have heard horror stories about parents who coerce their children to try to change. One 21-year-old man struggling with his orientation and his church's "gay deprogramming" recently wrote me. He was raised in a strict religious family who, upon learning of his orientation, threatened to disown him and have him excommunicated unless he changed. He went into the program voluntarily because he couldn't bear to lose his family, but other young men, he said, were forced against their will after being kidnapped. They strapped him to a chair with electrodes and sensors and showed him pictures of nude men, shocking him when he became aroused, he said. They continued this until he didn't respond. He finally fled the program after being sexually abused by a male orderly. The experience left him traumatized with incredible feelings of self-hatred, fear and thoughts of suicide. He knows that he is still gay.

I hope that the American people will see that these groups who call gay people sinners, telling us we can and should change when we cannot, are simply masking their message of prejudice in religious terms for political and monetary gain. It is unconscionable to me that people use God, religion or "Christian love" as an excuse for namecalling and discrimination. It is simply wrong. As for me, I have never felt as complete or peaceful as I have since coming out. And, I have never felt closer to God.

—Tracey St. Pierre

Source: Human Rights Campaign. 2000. *Finally Free: Personal Stories: How Love and Self-Acceptance Saved Us from "Ex-Gay" Ministries.* Human Rights Campaign Foundation. 919 18th St. N.W., Suite 800. Washington, DC 20006. Used by permission.

cessful" reparative therapy reveal that (1) many claims come from organizations with an ideological perspective on sexual orientation rather than from unbiased researchers; (2) the treatments and their outcomes are poorly documented; and (3) the length of time that clients are followed after treatment is too short for definitive claims to be made about treatment success (Human Rights Campaign 2000). In addition, at least 13 ministries of one reparative therapy group, Exodus, have closed because their directors reverted to homosexuality (Fone 2000).

Sociological Theories of Sexual Orientation

Sociological theories do not explain the origin or "cause" of sexual orientation diversity; they help explain societal reactions to homosexuality and bisexuality, and ways in which sexual identities are socially constructed.

Structural-Functionalist Perspective

Structural-functionalists, consistent with their emphasis on institutions and the functions they fulfill, emphasize the importance of monogamous heterosexual relationships for the reproduction, nurturance, and socialization of children. From a functionalist perspective, homosexual relations, as well as heterosexual nonmarital relations, are defined as "deviant" because they do not fulfill the family institution's main function of producing and rearing children. Clearly, however, this argument is less salient in a society in which (1) other institutions, most notably schools, have supplemented the traditional functions of the family; (2) reducing (rather than increasing) population is a societal goal; and (3) same-sex couples can and do raise children.

Some functionalists argue that antagonisms between heterosexuals and homosexuals may disrupt the natural state, or equilibrium, of society. Durkheim, however, recognized that deviation from society's norms may also be functional. As Durkheim observed, deviation "may be useful as a prelude to reforms which daily become more necessary" (Durkheim [1938] 1993, 66). Specifically, the gay rights movement has motivated many people to reexamine their treatment of sexual orientation minorities and has produced a sense of cohesion and solidarity among members of the gay population (although bisexuals have often been excluded from gay and lesbian communities and organizations). Gay activism has been instrumental in advocating more research on HIV and AIDS, more and better health services for HIV and AIDS patients, protection of the rights of HIV-infected individuals, and HIV/AIDS public education. Such HIV/AIDS prevention strategies and health services benefit the society as a whole.

Finally, the structural-functionalist perspective is concerned with how changes in one part of society affect other aspects. With this focus on the interconnectedness of society, we note that urbanization has contributed to the formation of strong social networks of gays and bisexuals. Cities "acted as magnets, drawing in gay migrants who felt isolated and threatened in smaller towns and rural areas" (Button, Rienzo, & Wald 1997, 15). Given the formation of gay communities in large cities, it is not surprising that the gay rights movement first emerged in large urban centers.

Other research has demonstrated that the worldwide rise in liberalized national policies on same-sex relations and the lesbian and gay rights social movement has been influenced by three cultural changes: the rise of individualism, increasing gender equality, and the emergence of a global society in which nations are influenced by international pressures (Frank & McEneaney 1999). Individualism "appears to loosen the tie between sex and procreation, allowing more personal modes of sexual expression" (p. 930).

> Whereas once sex was approved strictly for the purpose of family reproduction, sex increasingly serves to pleasure individualized men and women in society. This shift has involved the casting off of many traditional regulations on sexual behavior, including prohibitions of male-male and female-female sex. (Frank & McEneaney 1999, 936)

Gender equality involves the breakdown of sharply differentiated sex roles, thereby supporting the varied expressions of male and female sexuality. Globalization permits the international community to influence individual nations. For example, when Zimbabwe president Robert Mugabe pursued antihomosexual policies in 1995, 70 members of the U.S. Congress signed a letter asking him to halt his antihomosexual campaign. Many international organizations and human rights associations joined the protest. The pressure of international opinion led Zimbabwe's Supreme Court to rule in favor of lesbian and gay groups' right to organize.

Conflict Perspective

Conflict theorists, particularly those who do not emphasize a purely economic perspective, note that the antagonisms between heterosexuals and nonheterosexuals represent a basic division in society between those with power and those without power. When one group has control of society's institutions and resources, as in the case of heterosexuals, they have the authority to dominate other groups. The recent battle over gay rights is just one example of the political struggle between those with power and those without it.

A classic example of the power struggle between gays and straights took place in 1973 when the American Psychiatric Association (APA) met to revise its classification scheme of mental disorders. Homosexual activists had been appealing to the APA for years to remove homosexuality from its list of mental illnesses but with little success. The view of homosexuals as mentally sick contributed to their low social prestige in the eyes of the heterosexual majority. In 1973, the APA's board of directors voted to remove homosexuality from its official list of mental disorders. The board's move encountered a great deal of resistance from conservative APA members and was put to a referendum, which reaffirmed the board's decision (Bayer 1987).

More currently, gays and lesbians are waging a political battle to win civil rights protections in the form of laws prohibiting discrimination on the basis of sexual orientation (discussed later in this chapter). Conflict theory helps to explain why many business owners and corporate leaders oppose civil rights protection for gays and lesbians. Employers fear that such protection would result in costly lawsuits if they refused to hire homosexuals, regardless of the reason for their decision. Business owners also fear that granting civil rights protections to homosexual employees would undermine the economic health of a community by discouraging the development of new businesses and even driving out some established firms (Button et al. 1997).

However, some companies are recognizing that implementing antidiscrimination policies that include sexual orientation is good for the "bottom line." The majority (61%) of Fortune 500 companies have included sexual orientation in their nondiscrimination policies, and employers are increasingly offering benefits to domestic partners of LGBT employees (Human Rights Campaign 2003). Gay-friendly work policies help employers maintain a competitive edge in recruiting and maintaining a talented and productive work force.

In summary, conflict theory frames the gay rights movement and the opposition to it as a struggle over power, prestige, and economic resources. Recent trends toward increased social acceptance of homosexuality may, in part, reflect the corporate world's competition over the gay and lesbian consumer dollar.

Symbolic Interactionist Perspective

Symbolic interactionism focuses on the meanings of heterosexuality, homosexuality, and bisexuality, how these meanings are socially constructed, and how they influence the social status and self-concepts of nonheterosexual individuals. The meanings we associate with same-sex relations are learned from society—from family, peers, religion, and the media. The negative meanings associated with homosexuality are reflected in the current slang use of the phrase "that's so gay," or "you're so gay," which is meant to convey something that is considered bad or valueless, as a synonym for "dumb" or "stupid" (Kosciw & Cullen 2002).

Historical and cross-cultural research on homosexuality reveals the socially constructed nature of homosexuality and its meaning. Although many Americans assume that same-sex romantic relationships have always been taboo in our society, during the nineteenth century "romantic friendships" between women were encouraged and regarded as preparation for a successful marriage. The nature of these friendships bordered on lesbianism. President Grover Cleveland's sister Rose wrote to her friend Evangeline Whipple in 1890: "It makes me heavy with emotion . . . all my whole being leans out to you. . . . I dare not think of your arms" (Goode & Wagner 1993, 49).

The symbolic interactionist perspective also points to the effects of labeling on individuals. Once individuals become identified or labeled as lesbian, gay, or bisexual, that label tends to become their **master status.** In other words, the dominant heterosexual community tends to view "gay," "lesbian," and "bisexual" as the most socially significant statuses of individuals who are identified as such. Esterberg (1997) notes that "unlike heterosexuals, who are defined by their family structures, communities, occupations, or other aspects of their lives, lesbians, gay men, and bisexuals are often defined primarily by what they do in bed. Many lesbians, gay men, and bisexuals, however, view their identity as social and political as well as sexual" (p. 377).

Negative social meanings associated with homosexuality also affect the self-concepts of nonheterosexual individuals. **Internalized homophobia**—a sense of personal failure and self-hatred among lesbians and gay men resulting from social rejection and stigmatization—has been linked to increased risk for depression, substance abuse and addiction, anxiety, and suicidal thoughts (Bobbe 2002; Gilman et al. 2001). The negative meanings associated with homosexuality are explored in the following section.

Heterosexism, Homophobia, and Biphobia

The United States, along with many other countries throughout the world, is predominantly heterosexist. **Heterosexism** refers to "the institutional and societal reinforcement of heterosexuality as the privileged and powerful norm" (SIECUS 2000). Heterosexism is based on the belief that heterosexuality is superior to homosexuality; it results in prejudice and discrimination against homosexuals and bisexuals. Prejudice refers to negative attitudes, whereas discrimination refers to behavior that denies individuals or groups equality of treatment. Before reading further, you may wish to complete this chapter's *Self and Society* feature, which assesses your behaviors toward individuals you perceive to be homosexual.

Homophobia

SIECUS, the Sex Information and Education Council of the United States, "strongly supports the right of each individual to accept, acknowledge, and live in accordance with his or her orientation . . . and deplores all forms of prejudice and discrimination against people based on sexual orientation" (SIECUS 2000, 1). Nevertheless, negative attitudes toward homosexuality are reflected in the high percentage of the U.S. population who disapprove of homosexuality. The Gallup Organization reports that nearly half (46%) of U.S. adults view homosexuality as a sin and 44 percent say that homosexuality should *not* be considered an acceptable alternative lifestyle (Newport 2002).

The term **homophobia** is commonly used to refer to negative attitudes and emotions toward homosexuality and those who engage in it. Homophobia is not necessarily a clinical phobia (that is, one involving a compelling desire to avoid the feared object in spite of recognizing that the fear is unreasonable). Other terms that refer to negative attitudes and emotions toward homosexuality include "homonegativity" and "antigay bias."

In general, persons who are more likely to have negative attitudes toward homosexuality and to oppose gay rights are older, attend religious services, are less educated, live in the South or Midwest, and reside in small rural towns (Curtis 2003; Loftus 2001; Page 2003). Having positive contact with homosexuals or having homosexuals as friends is associated with less homophobia (Simon 1995). Public opinion surveys also indicate that men are more likely than women to have negative attitudes toward gays (Moore 1993). But many studies on attitudes toward homosexuality do not distinguish between attitudes toward gay men and attitudes toward lesbians (Kite & Whitley 1996). Research that has assessed attitudes toward male versus female homosexuality has found that heterosexual women and men hold similar attitudes toward lesbians, but men are more negative toward gay men (Louderback & Whitley 1997; Price & Dalecki 1998).

A study of undergraduates (110 men, 98 women) attending a Canadian university found that attitudes toward gay men were more negative than toward lesbians (Schellenberg, Hirt, & Sears 1999). When compared with science or business majors, students in the arts and social sciences had more positive attitudes toward gay men, and women were more positive than men.

Cultural Origins of Homophobia. Why do many Americans disapprove of homosexuality? Antigay bias has its roots in various aspects of U.S. culture.

1. *Religion.* Most Americans who view homosexuality as unacceptable say they object on religious grounds (Rosin & Morin 1999). Indeed, conservative Christian ideology has been identified as the best predictor of homophobia (Plugge-Fouse 2001). Many religious leaders teach that homosexuality is sinful and prohibited by God. The Roman Catholic Church rejects all homosexual expression and resists any attempt to validate or sanction the homosexual orientation. Some fundamentalist churches have endorsed the death penalty for homosexual people and teach the view that AIDS is God's punishment for engaging in homosexual sex. The Westboro Baptist Church (Topeka, Kansas), headed by the antigay Reverend Fred Phelps, maintains a Web site called godhatesfags.com. Members of this church have held antigay demonstrations near the funerals of people who have died from AIDS, carrying signs reading "Gays Deserve to Die."

Some religious groups, such as the Quakers, are accepting of homosexuality and other groups have made reforms toward increased acceptance of lesbians and gays. Some Episcopal priests perform "ceremonies of union" between same-sex couples; some Reform Jewish groups sponsor gay synagogues; and the United Church of Christ allows homosexuals to be ordained (Fone 2000). In June 2003, the Reverend V. Gene Robinson was elected bishop of the Episcopal diocese of New Hampshire, becoming the first openly gay bishop in the church's history. Although the official position of the United Methodist Church is one that condemns homosexuality, some Methodist ministers advocate acceptance of and equal rights for lesbians and gay men (The United Methodist Church and Homosexuality 1999).

When the Reverend V. Gene Robinson was elected bishop of the Episcopal diocese in New Hampshire in 2003, he became the first openly gay bishop in the church's history.

Self and Society

The Self-Report of Behavior Scale (Revised)

This questionnaire is designed to examine which of the following statements most closely describes your behavior during past encounters with people you thought were homosexuals. Rate each of the following self-statements as honestly as possible by choosing the frequency that best describes your behavior.

A = Never **B = Rarely** **C = Occasionally** **D = Frequently** **E = Always**

_______ 1. I have spread negative talk about someone because I suspected that he or she was gay.
_______ 2. I have participated in playing jokes on someone because I suspected that he or she was gay.
_______ 3. I have changed roommates and/or rooms because I suspected my roommate to be gay.
_______ 4. I have warned people whom I thought were gay and who were a little too friendly with me to keep away from me.
_______ 5. I have attended anti-gay protests.
_______ 6. I have been rude to someone because I thought that he or she was gay.
_______ 7. I have changed seat locations because I suspected the person sitting next to me to be gay.
_______ 8. I have had to force myself to stop from hitting someone because he or she was gay and very near me.
_______ 9. When someone I thought to be gay has walked toward me as if to start a conversation, I have deliberately changed directions and walked away to avoid him or her.
_______ 10. I have stared at a gay person in such a manner as to convey to him or her my disapproval of his or her being too close to me.
_______ 11. I have been with a group in which one (or more) person(s) yelled insulting comments to a gay person or group of gay people.
_______ 12. I have changed my normal behavior in a restroom because a person I believed to be gay was in there at the same time.
_______ 13. When a gay person has "checked" me out, I have verbally threatened him or her.
_______ 14. I have participated in damaging someone's property because he or she was gay.
_______ 15. I have physically hit or pushed someone I thought was gay because he or she brushed his or her body against me when passing by.
_______ 16. Within the past few months, I have told a joke that made fun of gay people.
_______ 17. I have gotten into a physical fight with a gay person because I thought he or she had been making moves on me.
_______ 18. I have refused to work on school and/or work projects with a partner I thought was gay.
_______ 19. I have written graffiti about gay people or homosexuality.
_______ 20. When a gay person has been near me, I have moved away to put more distance between us.

2. *Marital and procreative bias.* Many societies have traditionally condoned sex only when it occurs in a marital context that provides for the possibility of producing and rearing children. Although Vermont, Massachusetts, and California have granted legal recognition of same-sex couples (which we discuss later in this chapter), as of this writing, same-sex marriages are not legally recognized in the United States. Further, even though assisted reproductive technologies make it possible for gay individuals and couples to have children, many people believe that these advances should be used only by heterosexual married couples.

3. *Concern about HIV and AIDS.* Although most cases of HIV and AIDS worldwide are attributed to heterosexual transmission, the rates of HIV and AIDS in the United States are much higher among gay and bisexual men than among other groups. Because of this, many people associate HIV and AIDS with homosexuality and bisexuality. This association between AIDS and homosexuality has fueled anti-

Scoring: The SBS-R is scored by totaling the number of points endorsed on all items (Never = 1; Rarely = 2; Occasionally = 3; Frequently = 4; Always = 5), yielding a range from 20 to 100 total points. The higher the score, the more negative the attitudes toward homosexuals.

Comparison data: The SBS was originally developed by Sunita Patel (1989) in her psychology master's thesis research at East Carolina University. College men (from a university campus and from a military base) were the original participants (Patel, Long, McCammon, & Wuensch 1995). The scale was revised by Shartra Sylivant (1992), who used it with a coed high school student population, and by Tristan Roderick (1994), who involved college students to assess its psychometric properties. The scale was found to have high internal consistency. Two factors were identified: a passive avoidance toward homosexuals, and active or aggressive reactions.

In Roderick's study (Roderick, McCammon, Long, & Allred 1998) the mean score for 182 college women was 24.76. The mean score for 84 men was significantly higher, at 31.60. In Sylivant's (1992) high school sample, the mean scores were 33.74 for young women and 44.40 for young men.

Sources:
Patel, S. 1989. *Homophobia: Personality, Emotional, and Behavioral Correlates.* Unpublished master's thesis, East Carolina University, Greenville, NC.
Patel, S., T. E. Long, S. L. McCammon, and K. L. Wuensch. 1995. "Personality and Emotional Correlates of Self-reported Antigay Behaviors." *Journal of Interpersonal Violence,* 10, 354–366.
Roderick, T. 1994. *Homonegativity: An Analysis of the SBS-R.* Unpublished master's thesis, East Carolina University, Greenville, NC.
Roderick, T., S. L. McCammon, T. E. Long, and L. J. Allred. 1998. "Behavioral Aspects of Homonegativity." *Journal of Homosexuality,* 36, 79–88.
Sylivant, S. 1992. *The Cognitive, Affective, and Behavioral Components of Adolescent Homonegativity.* Unpublished master's thesis. East Carolina University, Greenville, NC.

gay sentiments. During a 1985 mayoral campaign in Houston, the challenging candidate was asked what he would do about the AIDS problem. His answer: he would "shoot the queers" (Button et al. 1997, 70). Lesbians, incidentally, have a very low risk for sexually transmitted HIV—a lower risk than heterosexual women.

4. *Rigid gender roles.* Antigay sentiments also stem from rigid gender roles. When Cooper Thompson (1995) was asked to give a guest presentation on male roles at a suburban high school, male students told him that the most humiliating put-down was being called a "fag." The boys in this school gave Thompson the impression they were expected to conform to rigid, narrow standards of masculinity in order to avoid being labeled in this way.

From a conflict perspective, heterosexual men's subordination and devaluation of gay men reinforces gender inequality. "By devaluing gay men . . . heterosexual men devalue the feminine and anything associated with it" (Price & Dalecki 1998,

155–156). Negative views toward lesbians also reinforce the patriarchal system of male dominance. Social disapproval of lesbians is a form of punishment for women who relinquish traditional female sexual and economic dependence on men. Not surprisingly, research findings suggest that individuals with traditional gender role attitudes tend to hold more negative views toward homosexuality (Louderback & Whitley 1997).

5. *Myths and negative stereotypes.* Prejudice toward homosexuals may also stem from some of the myths and negative stereotypes regarding homosexuality. One negative myth about homosexuals is that they are sexually promiscuous and lack "family values" such as monogamy and commitment to relationships. Although some homosexuals do engage in casual sex, as do some heterosexuals, many homosexual couples develop and maintain long-term committed relationships. Among lesbians, between 64 and 80 percent report that they are in a committed relationship at any given time. And between 46 and 60 percent of gay men report being in a committed relationship (Cahill, Ellen, & Tobias 2002).

Another myth that is not supported by data is that homosexuals, as a group, are child molesters. In fact, 95 percent of all reported incidents of child sexual abuse are committed by heterosexual men (SIECUS 2000). Most often the abuser is a father, stepfather, or heterosexual relative of the family. Research cited by Cahill and Jones (2003) finds that a child's risk of being molested by his or her relative's heterosexual partner is over 100 times greater than by someone who is homosexual or bisexual. Furthermore, "when a man abuses a young girl, the problem is not heterosexuality. . . . Similarly, when a priest sexually abuses a boy or under-age teen, the problem is not homosexuality. The problem is child abuse" (Cahill & Jones 2003, 1).

Changing Attitudes Toward Homosexuality. Over the last decade, attitudes toward the morality of homosexuality have become more liberal, and support for protecting civil rights of gays and lesbians has increased (Loftus 2001). Part of the explanation for these changing attitudes is the increasing levels of education in the U.S. population, as individuals with more education tend to be more liberal in their attitudes toward homosexuality. Another explanation for these changing attitudes is the positive depiction of gays and lesbians in the popular media. In 1998 Ellen DeGeneres came out on her sitcom *Ellen,* and by 2004 many television viewers had seen gay and lesbian characters in television shows such as *Dawson's Creek, Will & Grace, Buffy the Vampire Slayer,* and *All My Children.* The 2001–2002 prime time television season saw an increased visibility of gay and lesbian characters (Miller, Parker, Espejo, & Grossman-Swenson 2002). "Honest, non-stereotyped and diverse portrayals of gays and lesbians in prime time can offer youth a realistic representation of the gay community . . . [and] can offer positive role models for gay and lesbian youth" (p. 21).

Gays on television entered a new era with the debut of *Queer Eye for the Straight Guy* featuring five gay men (the "Fab Five") whose mission is to transform a style-deficient and culture-deprived straight man from "drab to fab" in fashion, food and wine, interior design, grooming, and culture. Scott Seomin, entertainment media director for the Gay and Lesbian Alliance Against Defamation, suggests that "What [*Queer Eye for the Straight Guy*] tells viewers is we can be helpful and they can learn from us" (quoted in Hallett 2003, 38). In *Boy Meets Boy,* a television dating show, a gay man gets a chance to pick a mate from among 15 contestants. After narrowing his choices, the leading man finds out that some of the contestants are heterosexual. If the leading man chooses a gay man, he wins a cash prize and a va-

© AP/Wide World Photos

The "Fab Five" of *Queer Eye for the Straight Guy* provide positive portrayals of gay men as being sophisticated and helpful to straight men.

cation with his guy of choice. If he chooses a straight guy, the straight guy wins a cash prize. This show may help dispel stereotypes of gay, and straight, men. Scott Seomin comments, "Viewers are going to watch this show and think they can pick out the gay guys from the straight guys, and they'll be wrong" (quoted in Hallett 2003, 38).

Greater acceptance of homosexuality may also be due to increased personal contact between heterosexuals and openly gay individuals as more gay and lesbian Americans are coming out to their family and friends. Polls of U.S. adults find that more than half of respondents (56%) say they have a friend or acquaintance who is gay or lesbian, nearly one-third (32%) say they work with someone who is gay or lesbian, and nearly one-quarter (23%) say that someone in their family is gay or lesbian (Newport 2002). Contact with openly gay individuals reduces negative stereotypes and ignorance, and increases support for gay and lesbian equality (Wilcox & Wolpert 2000).

Biphobia

Just as the term "homophobia" is used to refer to negative attitudes toward gay men and lesbians, **biphobia** refers to "the parallel set of negative beliefs about and stigmatization of bisexuality and those identified as bisexual" (Paul 1996, 449). Although both homosexual- and bisexual-identified individuals are often rejected by heterosexuals, bisexual-identified women and men also face rejection from many homosexual individuals. Thus, bisexuals experience "double discrimination."

Biphobia includes negative stereotyping of bisexuals; the exclusion of bisexuals from social and political organizations of lesbians and gay men; and fear and distrust of, as well as anger and hostility toward, people who identify themselves as bisexual (Firestein 1996). Individuals who are biphobic often believe that bisexuals are actually homosexuals afraid to acknowledge their real identity or homosexuals maintaining heterosexual relationships to avoid rejection by the heterosexual mainstream. Bisexual individuals are sometimes viewed as heterosexuals who are looking for exotic sexual experiences. One negative stereotype that encourages biphobia is the belief that bisexuals are, by definition, nonmonogamous. However, many bisexual women and men prefer and have long-term committed monogamous relationships.

Lesbians seem to exhibit greater levels of biphobia than do gay men. This is because many lesbian women associate their identity with a political stance against sexism and patriarchy. Some lesbians view heterosexual and bisexual women who "sleep with the enemy" as traitors to the feminist movement.

Effects of Homophobia and Heterosexism on Heterosexuals

The homophobic and heterosexist social climate of our society is often viewed in terms of how it victimizes the gay population. However, heterosexuals are also victimized by homophobia and heterosexism. "Hatred, fear, and ignorance are bad for the bigot as well as the victim" (*Homophobia 101* 2000).

"Injustice anywhere is a threat to justice everywhere. We are caught in an inescapable network of mutuality, tied in a single garment of destiny. Whatever affects one directly affects all indirectly."

Martin Luther King, Jr.

Because of the antigay climate, heterosexuals, especially males, are hindered in their own self-expression and intimacy in same-sex relationships. "The threat of victimization (i.e., antigay violence) probably also causes many heterosexuals to conform to gender roles and to restrict their expressions of (nonsexual) physical affection for members of their own sex" (Garnets, Herek, & Levy 1990, 380). Homophobic epithets frighten youth who do not conform to gender role expectations, leading some youth to avoid activities that they might otherwise enjoy and benefit from (arts for boys, athletics for girls, for example) (*Homophobia 101,* 2000).

Some cases of rape and sexual assault are related to homophobia and compulsory heterosexuality. For example, college men who participate in gang rape, also known as "pulling train," entice each other into the act "by implying that those who do not participate are unmanly or homosexual" (Sanday 1995, 399). Homonegativity also encourages early sexual activity among adolescent men. Adolescent male virgins are often teased by their male peers, who say things like "You mean you don't do it with girls yet? What are you, a fag or something?" Not wanting to be labeled and stigmatized as a "fag," some adolescent boys "prove" their heterosexuality by having sex with girls.

Antigay cultural attitudes also affect family members and friends of homosexuals, who live with the fear that their lesbian or gay friend or family member will be victimized by antigay prejudice and discrimination. Youth with gay and lesbian family members are often taunted by their peers.

As we discuss later in this chapter, extreme homophobia contributes to instances of physical violence against homosexuals—acts known as hate crimes. But hate crimes are crimes of perception, meaning that victims of antigay hate crimes may not be homosexual; they may just be perceived as being homosexual. Many heterosexuals have been victims of antigay physical violence because the attacker(s) perceived the victim to be gay.

Antigay harassment has also been a factor in many of the school shootings in recent years. In March 2001, 15-year-old Charles Andrew Williams fired more than 30 rounds in a San Diego suburban high school, killing 2 and injuring 13 others. A woman who knew Williams reported that the students had teased Williams and called him gay (Dozetos 2001). According to the Gay, Lesbian, and Straight Education Network (GLSEN), Williams's story is not unusual. Referring to a study of harassment of U.S. students that was commissioned by the American Association of University Women, a GLSEN report concluded, "For boys, no other type of harassment provoked as strong a reaction on average; boys in this study would be less upset about physical abuse than they would be if someone called them gay" (Dozetos 2001).

Discrimination Against Sexual Orientation Minorities

In June 2003, a Supreme Court decision in *Lawrence v. Texas* invalidated state laws that criminalize sodomy—oral and anal sexual acts. This historic decision overruled a 1986 Supreme court case (*Bowers v. Hardwick*), which upheld a Georgia **sodomy law** as constitutional. The 2003 ruling, which found sodomy laws to be discriminatory and unconstitutional, removes the stigma and criminal branding that sodomy laws have long placed on GLBT individuals. Prior to this historic ruling, sodomy was illegal in 13 states: Alabama, Florida, Idaho, Kansas, Louisiana, Mississippi, Missouri, North Carolina, Oklahoma, South Carolina, Texas, Utah, and Virginia, and four states (Kansas, Missouri, Oklahoma, and Texas) targeted only same-sex acts. Penalties for engaging in sodomy ranged from a $200 fine to 20 years' imprisonment. In states that criminalized both same- and opposite-sex sodomy, sodomy laws were usually not used against heterosexuals but were used primarily against gay men and lesbians.

Like other minority groups in American society, homosexuals and bisexuals experience various forms of discrimination. Next, we look at sexual orientation discrimination in the workplace, in family matters, and in violent expressions of hate. This chapter's *Focus on Technology* feature discusses the discriminatory effects of Internet filtering and monitoring technology on sexual orientation minorities.

Discrimination in the Workplace

The majority of Americans agree that homosexuals should have equal rights in the workplace. Gallup polls reveal that the percentage of U.S. adults who believe that homosexuals should have equal rights in terms of job opportunities has increased from 56 percent in 1977 to 88 percent in 2003 (Newport 2002; Page 2003). Yet as of February 2004, it was legal in 36 states to fire, decline to hire or promote, or otherwise discriminate against an employee because of his or her sexual orientation (National Gay and Lesbian Task Force 2003).

A majority of Americans support the idea of gays serving in the military (Newport 2002). Nevertheless, sexual orientation discrimination occurs in the military. In 1992, a gay Navy serviceman was fired after revealing his homosexuality but was reinstated by orders of a federal judge. In 1993, President Clinton instituted a "Don't ask, don't tell" policy in which recruiting officers are not allowed to ask about sexual orientation, and homosexuals are encouraged not to volunteer such information. Although the "Don't ask, don't tell" policy was intended to protect homosexuals from discrimination in the military, the number of service members discharged from the armed forces under the policy doubled between 1994 to 1998 (Human Rights Watch 2001). More than 5,400 service

After the Supreme court invalidated anti-sodomy laws in June 2003, lesbians, gays, bisexuals, and their allies gathered throughout the country to celebrate the landmark ruling.

Focus on Technology

The Impact of Internet Filtering and Monitoring on Sexual Orientation Minorities

Joan Garry is the executive director of the Gay & Lesbian Alliance Against Defamation (GLAAD) and a mother of three children. When her 10-year-old daughter Sarah approached her for information about COLAGE (Children of Lesbians and Gays Everywhere), a support organization for the children of lesbian and gay parents, they signed on to America Online to look up the Web site. The COLAGE site was "Web Restricted": Sarah could not access the site because her computer was equipped with AOL's filtering software called *Kids Only.* In fact, when Joan tried to look up various family, youth, and national organization Web sites with lesbian and gay content, most sites came up "Web Restricted" as well (Garry 1999). For example, the Kids Only software blocked access to several youth-oriented gay and lesbian resource sites, including PFLAG (Parents, Family and Friends of Lesbians and Gays), Family Pride, !OutProud!, GLSEN (Gay, Lesbian & Straight Education Network), and Oasis Magazine, a gay and lesbian youth Webzine (Javier 1999).

An explosion of filtering technologies to help maintain "decency" and "community standards" on the Internet has occurred. Filtering technologies are promoted as tools to help parents, schools, libraries, and communities prevent children's access to sexually explicit and pornographic material on the Internet. However, some filtering software also denies users access to several lesbian, gay, and bisexual youth resource sites as well as to health (especially sexual health) sites. Filters block Web sites by screening the keywords that index the sites and blocking any sites with keywords in "objectionable" categories. Filtering products can be customized by selecting topics or categories for blocking or can be used at a default setting.

The words "gay" and "lesbian" and their slang synonyms are often among those categories that are blocked by Internet filtering products. For example, the filtering software "CyberSitter" automatically filters out words and phrases like "gay," "lesbian," "gay rights," and "gay community" (Javier 1999). In a study of seven filtering products, nearly one in three "safe sex" sites were blocked by at least one of the filters at the *least restrictive* setting, and at the most restrictive setting, one-quarter (24%) of 3,053 health Web sites were blocked (Rideout, Richardson, & Resnick 2002).

Given the impact of Internet filtering on the gay community, the Gay & Lesbian Alliance Against Defamation does not support the use of filtering software. Instead, they advocate parental oversight, school supervision, and training of young Internet users (Bowes 1999). However, the Children's Internet Protection Act (CIPA), passed by Congress in 2000, requires schools and libraries that receive certain types of federal funding to use Internet filtering devices. The CIPA requirement for libraries was struck down in Spring 2002 by a circuit court on the grounds that it violates the First Amendment. But in June 2003, the U.S. Supreme court in *United States v. American Library Association* reversed the lower court's decision and upheld the CIPA requirement, stipulating that public libraries must turn off the filter upon request by an adult patron.

Another concern of sexual orientation minorities is the use of monitoring software that allows parents, teachers, and other authority figures to track sites a Web surfer tries to access. The monitoring software industry markets its products by claiming that they allow for parental awareness without censorship, and that parental use of monitoring software encourages open family communication. But for youth who are not ready to reveal their sexual orientation to their family, "such software could potentially 'out' them before they are ready, leading to strained family relations and deeper isolation" (Javier 1999, 8).

The best-known story of a gay person affected by an "online outing" is that of a naval officer who was outed in 1998 after an America Online employee divulged his personal data and online history to the Navy. This invasion of privacy cost this officer his career.

members were discharged under the policy from 1994 to 1999. Women were discharged at a disproportionately high rate. The "Don't ask, don't tell" policy also provided an additional means for servicemen to harass lesbian service members by threatening to "out" those who refused their advances or threatened to report them, thus ending their careers (Human Rights Watch 2001). Discrimination toward gays in the military is not surprising; in the largest survey ever conducted on military attitudes toward homosexuality, 80 percent of those questioned had heard offensive comments about gays within the previous year (Ricks 2000).

Focus on Technology

Web sites also record information about your Internet usage. When you visit a free news Web site, you may be required to "subscribe" by giving them your name and e-mail address. After subscribing, you read a few articles. But unknown to you, the news Web site has put a "cookie," or piece of computer code, on the hard drive of your computer. This cookie contains a unique identifier permitting the news site to recognize you when you return to that site. Cookies can do helpful things like remember your password for accessing that site and tailor Web pages for preset preferences. But they can also contain a record of what you did on that site, including what searches you made and what articles you read (Aravosis 1999).

Although some Web sites have strict privacy policies, others sell information about site visitors to corporate America. One company boasts on its Web site that it is "the world's oldest and largest mailing list manager and broker for Gay, Lesbian, and HIV-related names, currently managing almost two million names, which we estimate to be about 65 percent of all those commercially available in this segment" (Aravosis 1999, 33). Databases containing the names of gay consumers are valuable because gays are perceived as a "wealthy and wired market." A 1998 study found that the average household income of lesbians and gay men on the Internet was $57,300, slightly higher than the $52,000 for the general Internet population. Another study by Computer Economics predicts that within the next few years, the worldwide gay and lesbian Internet population will increase from 9.2 million to 17.1 million (Aravosis 1999).

The Internet has been a useful tool for gays, lesbians, and bisexuals. Going online has allowed the gay community to create safe places for support and information. However, Internet filtering and monitoring software that has been installed on computers in homes, schools, libraries, and workplaces throughout the country represents a threat to the gay community. Filtering and monitoring technologies make it impossible or dangerous for closeted gays or lesbians to seek out support and information about their community.

> We live in an age in which the Internet has become an extremely important part of the coming out process for many gay and lesbian youth. In many cases, it can be a lifeline to those in geographically isolated areas. To deny basic educational and support resources to lesbian and gay youth could seriously endanger their physical and emotional well-being. (Appendix A 1999, 46)

Sources: Appendix A. 1999. "Frequently Asked Questions." In *Access Denied Version 2.0, The Continuing Threat Against Internet Access and Privacy and Its Impact on the Lesbian, Gay, Bisexual and Transgender Community,* pp. 45–48. New York: Gay & Lesbian Alliance Against Defamation; Aravosis, John. 1999. "Privacy: The Impact on Lesbian, Gay, Bisexual and Transgender Community. In *Access Denied Version 2.0, The Continuing Threat Against Internet Access and Privacy and Its Impact on the Lesbian, Gay, Bisexual and Trangender Community,* pp. 30–33. New York: Gay & Lesbian Alliance Against Defamation; Bowes, John. 1999. "Conclusions." In *Access Denied Version 2.0, The Continuing Threat Against Internet Access and Privacy and Its Impact on the Lesbian, Gay, Bisexual and Trangender Community,* pp. 38–44. New York: Gay & Lesbian Alliance Against Defamation; Dozetos, Barbara. 2001 (March 20). "ACLU Sues over Censorship of GLBT Sites." PlanetOut.com. http://www.planetout.com; Garry, Joan M. 1999. "Introduction: How Access and Privacy Impact the Lesbian, Gay, Bisexual and Transgender Community." In *Access Denied Version 2.0, The Continuing Threat Against Internet Access and Privacy and its Impact on the Lesbian, Gay, Bisexual and Trangender Community,* pp. 3–5. New York: Gay & Lesbian Alliance Against Defamation; Javier, Loren. 1999. "The World Since Access Denied." In *Access Denied Version 2.0 The Continuing Threat Against Internet Access and Privacy and Its Impact on the Lesbian, Gay, Bisexual and Trangender Community,* pp. 6–9. New York: Gay & Lesbian Alliance Against Defamation; Rideout, Victoria, Caroline Richardson, & Paul Resnick. 2002. *See No Evil: How Internet Filters Affect the Search for Online Health Information.* Washington, DC: Kaiser Family Foundation. http://www.kff.org; Schneider, Karen G. 1999. "Access: The Impact on the Lesbian, Gay, Bisexual and Transgender Community." In *Access Denied Version 2.0, The Continuing Threat Against Internet Access and Privacy and Its Impact on the Lesbian, Gay, Bisexual and Trangender Community,* pp. 10–15. New York: Gay & Lesbian Alliance Against Defamation.

Discrimination in Family Relationships

In addition to discrimination in the workplace, sexual orientation minorities experience discrimination in policies concerning marriage. States and judges have also used a person's sexual orientation to deny custody and visitation, adoption, and foster care.

Same-Sex Marriage. Before the 2003 Massachusetts Supreme Court ruling in *Goodridge v. Department of Public Health,* no state had declared that same-sex couples have a constitutional right to be legally married. In response to growing efforts to secure legal recognition of same-sex couples, opponents of same-sex marriage have prompted antigay marriage legislation. In 1996, Congress passed and President Clinton signed the **Defense of Marriage Act,** which states that marriage is a "legal union between one man and one woman" and denies federal recognition of same-sex marriage. In effect, this law allows states to either recognize or not recognize same-sex marriages performed in other states. In early 2004, Ohio was poised to become the 38th state to pass anti-marriage legislation denying recognition of same-sex marriages (Rostow 2004).

A new threat to gay and lesbian families is legislation known as "Super DOMAs." While federal and state Defense of Marriage Acts ("DOMAs") prohibit recognition of same-sex marriages, Super DOMAs aim to prohibit any kind of recognition of same-sex relationships. Super DOMAs potentially endanger employer-provided domestic partner benefits, joint and second-parent adoptions, health care decision-making proxies, or any policy or document that recognizes the existence of a same-sex partnership. As of this writing, one state—Nebraska—has passed "Super DOMA legislation" (Cahill & Slater 2004).

In 2003, President G. W. Bush signed a proclamation declaring October 12–18 "Marriage Protection Week" in an effort to define marriage as "a union between a man and a woman." And a group of conservative Republicans are pushing for a Federal Marriage Amendment that would amend the U.S. Constitution to define marriage as being between a man and a woman.

Advocates of same-sex marriage argue that banning same-sex marriages or refusing to recognize same-sex marriages granted in other states is a violation of civil rights that denies same-sex couples the many legal and financial benefits that are granted to heterosexual married couples (Sullivan 2003). For example, married couples have the right to inherit from a spouse who dies without a will; to avoid inheritance taxes between spouses; to make crucial medical decisions for a partner and take family leave to care for a partner in the event of the partner's critical injury or illness; to receive Social Security survivor benefits; and to include a partner in his or her health insurance coverage. Other rights bestowed on married (or once-married) partners include assumption of spouse's pension, bereavement leave, burial determination, domestic violence protection, reduced-rate memberships, divorce protections (such as equitable division of assets and visitation of partner's children), automatic housing lease transfer, and immunity from testifying against a spouse. Another argument for same-sex marriage is that it would promote monogamy and stability among all couples and therefore strengthen the institution of marriage (Sullivan 2003).

Without legal recognition of same-sex families, children living in gay- and lesbian-headed households are denied a range of securities that protect children of heterosexual married couples. These include the right to get health insurance coverage and Social Security survivor benefits from a nonbiological parent. In some cases, children in same-sex households lack the automatic right to continue living with their nonbiological parent should their biological mother or father die (Tobias & Cahill 2003).

Whereas advocates of same-sex marriage argue that as long as same-sex couples cannot be legally married, they will not be regarded as legitimate families by the larger society, opponents do not want to legitimize homosexuality as a socially acceptable lifestyle. Opponents of same-sex marriage who view homosexuality as unnatural, sick, and/or immoral do not want their children to learn that homosexuality is an accepted "normal" lifestyle. The most common argument against same-sex marriage is that it subverts the stability and integrity of the heterosexual family.

© AP/Wide World Photos

© Kim Kulish/CORBIS

In February 2004, the city and county of San Francisco became the first government in the United States to grant marriage licenses to same-sex couples. By the end of March 2004, 3,955 same-sex couples had been granted marriage licenses in San Francisco. Same-sex marriage licenses had also been granted in New Mexico, New Jersey, Oregon, and New York.

However, Sullivan (1997) suggests that homosexuals are already part of heterosexual families:

> [Homosexuals] are sons and daughters, brothers and sisters, even mothers and fathers, of heterosexuals. The distinction between "families" and "homosexuals" is, to begin with, empirically false; and the stability of existing families is closely linked to how homosexuals are treated within them. (p. 147)

Child Custody and Visitation. Several respected national organizations—including the Child Welfare League of America, the American Psychological Association, the American Psychiatric Association, and the National Association of Social Workers—have taken the position that a parent's sexual orientation is irrelevant in determining child custody (Landis 1999). In a review of research on family relationships of lesbians and gay men, Patterson (2001) concluded that "the greater majority of children with lesbian or gay parents grow up to identify themselves as heterosexual" and that "concerns about possible difficulties in personal development among children of lesbian and gay parents have not been sustained by the results of research" (p. 279). Patterson (2001) notes that the "home environments provided by lesbian and gay parents are just as likely as those provided by heterosexual parents to enable psychosocial growth among family members" (p. 283). Nevertheless, some court judges are biased against lesbian and gay parents in custody and visitation disputes. For example, in 1999, the Mississippi Supreme Court denied custody of a teenage boy to his gay father and instead awarded custody to his heterosexual mother who remarried into a home "wracked with domestic violence and excessive drinking" ("Custody and Visitation" 2000, 1).

Adoption and Foster Care. A 2003 poll of U.S. adults reveals that Americans are almost evenly split on the issue of the right of gay couples to adopt children: 49 percent support allowing gay couples to adopt and 48 percent oppose it (Page 2003). Gay and lesbian people have adopted in at least 22 states and the District of Columbia; however, Florida and Mississippi forbid adoption by gay and lesbian people, Utah forbids adoption by any unmarried couple (which includes all same-sex couples), and in Arkansas, lesbians and gay men are prohibited from serving as foster parents ("Adoption/Foster Care Laws in the U.S." 2002).

Most adoptions by gay people have been by individual gay men or lesbians who adopt the biological children of their partners. Such second-parent or stepparent adoptions ensure that the children can enjoy the benefits of having two legal parents, especially if one parent dies or becomes incapacitated. However, in four states (Colorado, Nebraska, Ohio, and Wisconsin) court rulings have decided that the state adoption law does not allow for second-parent or stepparent adoption by members of same-sex couples ("Second-Parent/Stepparent Adoption in the U.S." 2002).

Hate Crimes Against Sexual Orientation Minorities

On October 6, 1998, Matthew Shepard, a 21-year-old student at the University of Wyoming, was abducted and brutally beaten. He was found tied to a wooden ranch fence by two motorcyclists who had initially thought that he was a scarecrow. His skull had been smashed, and his head and face had been slashed. The only apparent reason for the attack: Matthew Shepard was gay. On October 12, Matthew died of his injuries. Media coverage of his brutal attack and subsequent death focused nationwide attention on hate crimes against sexual orientation minorities.

"The prejudice against gays and lesbians today is meaner, nastier— a vindictiveness that's rooted in hatred, not ignorance."

Betty DeGeneres
Comedienne Ellen DeGeneres's mom

Other incidents of brutal hate crimes against gays include the murder of Billy Jack Gaither in Alabama in February 1999. Two men who claimed to be angry over a sexual advance made by Gaither plotted his murder, beat Gaither to death with an ax handle, and then burned his body on a pyre of old tires. In July 1999 at Fort Campbell, Kentucky, Private First Class Barry Winchell was beaten to death by another soldier with a baseball bat because Winchell was perceived to be homosexual.

In eighteenth-century America, where laws against homosexuality often included the death penalty, violence against gays and lesbians was widespread and included beatings, burnings, various kinds of torture, and execution (Button et al. 1997). Although such treatment of sexual orientation minorities is no longer legally condoned, gays, lesbians, and bisexuals continue to be victimized by hate crimes. Anti-LGBT hate crimes are crimes against individuals or their property that are based on bias against the victim because of his or her perceived sexual orientation or gender identity. Such crimes include verbal threats and intimidation, vandalism, sexual assault and rape, physical assault, and murder.

According to the FBI Uniform Crime Reports, 1,244 incidents of sexual orientation hate crimes were reported in 2002, representing 16.7 percent of all hate crimes reported to the FBI (Federal Bureau of Investigation 2003). But as discussed in Chapter 6, FBI hate crime statistics underestimate the incidence of hate crimes. The National Coalition of Antiviolence Programs reported 1,968 incidents of anti-gay hate crimes in 2002—a 1 percent increase from 2001—and 12 anti-LGBT murders (Patton 2003). Fifty-nine percent of victims of anti-LGBT violence in 2002 were male. And because it is not uncommon for heterosexual men and women to be mistaken for gay men and lesbians, one in ten victims of anti-LGBT violence in 2002 identified as being heterosexual.

Antigay Hate in Schools and on Campuses

America's schools are not safe places for gay, lesbian, and bisexual youth. More than two-thirds (69%) of gay and lesbian students have been verbally, physically, or sexually harassed at school (Chase 2000). A survey of 904 LGBT youth (grades 6–12) from 48 states found that homophobic language is pervasive in U.S. schools, and that harassment is a common experience for LGBT youth. Findings from this study include the following (Kosciw & Cullen 2002):

- More than 80 percent of students had heard antigay words like "faggot" or "dyke" frequently or often, and over 90 percent had heard the phrase "that's so gay" or "you're so gay." Almost one-fourth of the youth (23.6%) reported hearing homophobic remarks from faculty or school staff. The frequency of homophobic remarks was higher than that of sexist and racist remarks.
- More than one-third of the youth reported experiencing physical harassment in the past year (being shoved, pushed) because of their sexual orientation; nearly 10 percent reported that physical harassment occurred frequently.
- More than 20 percent of the youth reported having been physically assaulted (being punched, kicked, injured with a weapon) in the past year due to their sexual orientation.
- Nearly one-third of youth (30.8%) had missed a day of school in the past month because they felt unsafe.

Given the harsh treatment of LGBT youth in school settings, it is not surprising that 40 percent of gay youth report schoolwork being negatively affected by conflicts around sexual orientation (*Homophobia 101* 2000). Over one-fourth of gay youth drop out of school—usually to escape the harassment, violence, and alienation they endure there (Chase 2000). Lesbian, gay, and bisexual youth reporting high levels of victimization at school also have higher levels of substance use, suicidal thoughts, and sexual risk behaviors than heterosexual peers reporting high levels of at-school victimization (Bontempo & D'Augelli 2002). A survey of youths' risk behavior conducted by the Massachusetts Department of Education in 1999 found that 30 percent of gay teens attempted suicide in the previous year, compared with 7 percent of their straight peers (reported in Platt 2001).

Antigay hate is also common among college students (see this chapter's *Social Problems Research Up Close* feature). In a survey of 484 young adults at six community colleges in California, 10 percent reported physically assaulting or threatening people whom they believed to be homosexual and 24 percent reported calling homosexuals insulting names (Franklin 2000). The researcher concluded that overall, findings suggest that many young adults believe that antigay harassment and violence is socially acceptable.

Strategies for Action: Reducing Antigay Prejudice and Discrimination

Many of the efforts to change policies and attitudes regarding sexual orientation minorities have been spearheaded by organizations such as the Human Rights Campaign (HRC); the National Gay and Lesbian Task Force (NGLTF); Gay and Lesbian Alliance Against Defamation (GLAAD); the International Lesbian and Gay Association (ILGA); Lambda Legal Defense and Education Fund; the National Center for Lesbian Rights; and the Gay, Lesbian, and Straight Education Network (GLSEN).

Social Problems Research Up Close

Campus Climate for GLBT People: A National Perspective

Since the mid-1980s, numerous studies have documented the hostile climate that GLBT students, staff, faculty, and administrators experience on college campuses. A recent survey by the National Gay and Lesbian Task Force Policy Institute represents a significant contribution to the research literature in this area.

Sample and Methods

Due to the difficulty in identifying lesbian, gay, bisexual, and transgender individuals, purposeful sampling of GLBT individuals and snow-ball sampling were used. Contacts were made with "out" GLBT individuals on campus who were asked to share the survey with other members of the GLBT community who were not so open about their sexual/gender identity.

Surveys, both paper-and-pencil and online versions, were administered to students, faculty, staff, and administrators at 14 colleges and universities from around the country. The surveys, which contained 35 questions and an additional space for respondents to provide commentary, were designed to collect data about (1) respondents' personal campus experiences as members of the GLBT community; (2) their perception of the climate for GLBT members of the academic community; and (3) their perceptions of institutional actions, including administrative policies and academic initiatives regarding GLBT issues and concerns on campus. A total of 1,669 usable surveys were returned, representing the following:

- 1,000 students (undergraduate and graduate), 150 faculty, and 467 staff/administrators
- 720 men, 848 women
- 326 people of color
- 572 gay people (mostly male), 458 lesbians, 334 bisexuals, and 68 transgender
- 825 "closeted" people

Findings and Discussion

Some of the findings of the Campus Climate Assessment include the following:

Lived Oppressive Experiences

Nineteen percent of the respondents reported that, within the past year, they had feared for their physical safety because of their sexual orientation/gender identity, and 51 percent concealed their sexual/gender identity to avoid intimidation. GLBT people of color were more likely than white GLBT people to conceal their sexual/gender identity to avoid harassment. Many nonwhite respondents reported feeling out of place at predominantly white GLBT settings. In the words of one student, "As a chicana, I felt ostracized even more. Forget about feeling a sense of community when you're a member of two minority groups" (Rankin 2003, 25). The study also found that 27 percent of faculty/staff/administrators and 40 percent of students indicated that they had concealed their sexual identity to avoid discrimination in the past year.

Respondents were also asked if they had experienced harassment in the past year. Harassment was defined as "conduct that has interfered unreasonably with your ability to work or learn on this campus or has created an offensive, hostile, intimidating working or learning environment."

Undergraduate students were the most

But the effort to reduce antigay prejudice and discrimination is not just a "gay agenda"; many heterosexuals and mainstream organizations (such as the National Education Association) have worked on this agenda as well.

Many of the advancements in gay rights have been the result of political action and legislation. Barney Frank (1997), an openly gay U.S. representative, emphasizes the importance of political participation in influencing social outcomes. He notes that demonstrative and cultural expressions of gay activism, such as "gay pride" celebrations, marches, demonstrations, or other cultural activities promoting gay rights are important in organizing gay activists. However, he points out:

> Too many people have seen the cultural activity as a substitute for democratic political participation. In too many cases over the past decades we have left the political arena to our most dedicated opponents [of gay rights], whose letter writing, phone calling, and lobbying have easily triumphed over our marching, demonstrating, and dancing. The most important lesson . . . for people who want to make America a fairer place is that pol-

Social Problems Research Up Close

likely to have experienced harassment (36%), followed by faculty (27%), graduate students (23%), and staff (19%). Derogatory remarks were the most common form of harassment (89%). Other types of harassment included verbal harassment or threats, anti-GLBT graffiti, threats of physical violence, denial of services, and physical assault.

Perceptions of Anti-GLBT Oppression on Campus

About three-quarters of faculty, students, administrators, and staff rated the campus climate as homophobic. In contrast, most respondents rated the campus generally (not specific to GLBT people) as friendly, concerned, and respectful. Thus, "even though respondents feel that the overall campus climate is hospitable, heterosexism and homophobia are still prevalent" (p. 31).

Institutional Actions

Participants were asked to respond to several questions about institutional actions regarding GLBT concerns on campus. Fewer than half (37%) agreed that "the college/university thoroughly addresses campus issues related to sexual orientation/gender identity" and only 22 percent agreed that "the curriculum adequately represents the contributions of GLBT persons." However, the majority of respondents agreed that their classrooms or their job sites were accepting of GLBT persons (63%) and that the college/university provided resources on GLBT issues and concerns (71%). These positive responses may be due in part to the inclusion of sexual orientation in the nondiscrimination policies of all but one of the participating colleges/universities. Several of the campuses also provided domestic partner benefits, had GLBT resource centers, and offered safe-space programs.

These findings indicate that intolerance and harassment of GLBT students, staff, faculty, and administrators continue to be prevalent on U.S. campuses. It is important to note that the colleges and universities that participated in this study are not representative of most institutions of higher learning in the United States and "may be among the most gay-friendly campuses in the country" (Rankin 2003, 3). Rankin explains that "all of the institutions who participated in this survey had a visible GLBT presence on campus, including, in most cases, a GLBT campus center. Most had sexual orientation nondiscrimination policies. As only 100 of the 5,500 U.S. colleges and universities have GLBT student centers, the 14 universities surveyed here are not representative of most institutions of higher education in the U.S." (p. 3). Thus, Rankin suggests that the findings of this study "may significantly understate the problems facing GLBT students and staff at U.S. colleges and universities" (p. 3).

Source: Susan R. Rankin. 2003. *Campus Climate for Gay, Lesbian, Bisexual, and Transgendered People: A National Perspective.* New York: The National Gay and Lesbian Task Force Policy Institute. http://www.ngltf.org

itics—conventional, boring, but essential politics—will ultimately have a major impact on the extent to which we can rid our lives of prejudice. (Frank 1997, xi)

Next, we look at efforts to reduce employment discrimination against sexual orientation minorities, provide recognition and support to lesbian and gay families, and include sexual orientation in hate crime legislation. We also overview educational policies and programs designed to reduce intolerance and support nonheterosexual students in schools and on campuses.

Reducing Employment Discrimination Against Sexual Orientation Minorities

The majority of Americans agree that no one should be denied employment, fired, passed over for promotion, or otherwise discriminated against at work solely because of his or her sexual orientation. Yet, as of this writing, federal law protects individuals from discrimination only on the basis of race, religion, national origin, sex, age, and disability. A bill called the **Employment Nondiscrimination Act (ENDA)** would make it illegal to discriminate based on sexual orientation. The ENDA would not apply to religious organizations, businesses with fewer than 15 employees, or the military. ENDA was introduced in Congress in 1994. In 1996, the bill came within one vote of passing the Senate. It was reintroduced in Congress in 1999 and 2002. At the time of this writing, the Senate has reintroduced ENDA and a vote by Congress is pending. In its current form, ENDA, if passed, would *not* protect transgender employees against discrimination based on gender expression or identity.

With the absence of federal legislation prohibiting antigay discrimination, some state and local governments as well as private employers have taken measures to prohibit employment discrimination based on sexual orientation. The scope of these measures varies, from prohibiting discrimination in public employment only to comprehensive protection against discrimination in public and private employment, education, housing, public accommodations, credit, and union practices.

Local and State Bans on Antigay Employment Discrimination. In 1974, Minneapolis became the first municipality to ban antigay job discrimination. By the end of 2002, 119 cities and 23 counties prohibited sexual orientation discrimination in the private workplace (Human Rights Campaign 2003). In 1982, Wisconsin became the first state to prohibit antigay employment discrimination. By the end of 2003, 14 states (and the District of Columbia) had laws banning discrimination based on sexual orientation (see Table 8.2).

Nondiscrimination Policies in the Workplace. In 1975, AT&T became the first employer to add sexual orientation to its nondiscrimination policy. By the end of 2002, 2,221 employers, including 61 percent of Fortune 500 companies, had nondiscrimination policies that included sexual orientation (see Table 8.3). The Cracker Barrel restaurant chain, which fired at least 11 openly gay and lesbian employees in 1991 stating that it would refuse employment to people "whose sexual

Table 8.2
States with Laws* Banning Discrimination Based on Sexual Orientation

Wisconsin (1982)	New Hampshire (1997)
Massachusetts (1989)	Nevada (1999)
Connecticut (1991)	Rhode Island** (2001)
Hawaii (1991)	Maryland (2001)
New Jersey (1992)	New York (2002)
Vermont (1992)	California** (2003)
Minnesota** (1993)	New Mexico** (2003)

*Laws covering both public and private sector workplaces
**Includes banning discrimination against gender identity

Source: National Gay & Lesbian Task Force. 2003 (August). "GLBT Civil Rights in the U.S." http://www.ngltf.org

Table 8.3
Number of Employers with Nondiscrimination Policies that Include Sexual Orientation

Fortune 500 Companies	304
Other Private Sector Employers	1,208
Colleges and Universities	382
Federal Departments and Agencies	38

Source: Human Rights Campaign. 2003. *The State of the Workplace for Lesbian, Gay, Bisexual and Transgendered Americans, 2002.* Washington DC: Human Rights Campaign. http://www.hrc.org

preferences fail to demonstrate normal heterosexual values" voted in 2002 to add sexual orientation to its nondiscrimination policy (Button et al. 1997, 126; *Law-Briefs* 2003).

A list of employers that prohibit sexual orientation discrimination is available at the Human Rights Campaign WorkNet Web site at http://www.hrc.org/worknet. Whereas gay rights advocates promote doing business with companies that are "gay friendly," groups that oppose gay rights advocate boycotting such businesses. For example, the Web site godhatesfags.com urges its visitors to avoid doing business with any company that is listed on the Human Rights Campaign WorkNet Web site.

Effects of Nondiscrimination Laws and Policies. A survey of 126 U.S. cities and counties that had implemented laws or policies prohibiting discrimination on the basis of sexual orientation found that such laws and policies sometimes had negative effects and "resulted in divisions in the community, greater controversy or tension, or the increased mobilization of those opposed to gay rights" (Button et al. 1997, 120). However, the more commonly cited effects were positive and included the following:

- Reduced discrimination against lesbians and gays in public employment and other institutions covered by the law.
- Increased feelings of security, comfort, and acceptance by lesbians and gay men, resulting in lesbians and gay men being more likely to "come out of the closet"—to reveal their sexual orientation to others (or to at least stop hiding it).
- An increased sense of legitimacy among gays and lesbians that helped them overcome "internalized homophobia."

Providing Recognition and Support to Lesbian and Gay Couples and Families

In 2003, the Massachusetts Supreme Court ruled (in *Goodridge v. Department of Public Health*), that denying same-sex couples the protections, benefits, and obligations of civil marriage violated the basic premises of individual liberty and equality in the Massachusetts Constitution (Cahill & Slater 2004). As of this writing, no other state has affirmed the right of same-sex couples to be legally married. However, in recent years, the gay rights movement has witnessed significant progress in the provision of benefits to lesbian and gay couples, parents, and co-parents.

Recognition of Same-Sex Couples. Some states, counties, cities, and workplaces allow unmarried couples, including gay couples, to register as **domestic partners.** The rights and responsibilities granted to domestic partners vary from place to place but may include coverage under a partner's health and pension plan, rights of inheritance and community property, tax benefits, access to married student housing, child custody and child and spousal support obligations, and mutual responsibility for debts.

In 1991 the Lotus Development Corporation became the first major American firm to extend domestic partner recognition to gay and lesbian employees. By the end of 2002, the Human Rights Campaign (2003) identified 5,698 employers that provided domestic partner health insurance benefits to their employees—an increase of 16 percent from the previous year. Before ending his duties as governor of California, Gray Davis signed landmark legislation mandating that any company doing business with the state offer domestic partner benefits to its employees. To allow companies time to change their policies, the law, known as "AB 17," does not go into effect until January 2007. However, even when companies offer domestic partner benefits to same-sex partners of employees, these are usually taxed as income by the federal government, whereas spousal benefits are not.

Under Vermont's civil union system, passed in 2000, same-sex couples may apply to a town clerk for a civil union license; have that license "certified" by a judge, justice of the peace, or member of the clergy; and then receive a civil union certificate. A **civil union** is a legal status parallel to civil marriage under Vermont state law. Civil union status entitles same-sex couples to all the rights and responsibilities available under state law to married couples.

"If the love of two people, committing themselves to each other exclusively for the rest of their lives, is not worthy of respect, then what is?"

Andrew Sullivan

Unlike marriage for heterosexual couples, the rights of partners in same-sex civil unions in Vermont and domestic partners in California are not recognized by federal law, so they do not have the more than 1,000 federal protections that go along with civil marriage. Nor is the legal status granted to same-sex couples in Vermont and California recognized in other states.

In ten states and the District of Colombia, as well as in several dozen municipalities, same-sex partners of public employees are granted domestic partner benefits. Three states—Hawaii, New Jersey, and California—have enacted laws that provide varying degrees of protection for domestic partners. The California law (the Domestic Partner Rights and Responsibilities Act of 2003) is the most comprehensive, granting same-sex domestic partners nearly all of the rights and responsibilities granted to married couples in the state (Cahill & Slater 2004).

Gay and Lesbian Parental Rights. As noted earlier, 2000 Census data revealed that one-third (33%) of female same-sex householders and 22 percent of male same-sex householders were living with their children (under 18 years) (Simmons, & O'Connell 2003). A number of policies and rulings reflect the increasing provision of parental rights and responsibilities to gay and lesbian parents and co-parents.

- In an unprecedented ruling, the California Board of Equalization granted head-of-household tax status to a nonbiological lesbian co-parent ("NCLR Wins Equal Tax Benefits for Non-biological Lesbian Mother" 2000). This is the first ruling by any state tax board that provides equitable tax status for gay and lesbian families.
- In 2000, a New Jersey appeals court awarded visitation to a nonbiological lesbian co-parent, reasoning that the woman had a close "parent-type relationship" with the children ("Custody and Visitation" 2000).

- A Massachusetts court ruled that two women may be listed as "mother" on a birth certificate, when one of the women donated the egg for the child and the other carried the child (*LAWBriefs* 2000).
- A Pennsylvania court ordered a nonbiological, nonadoptive parent to pay child support for the five children she jointly parented with her former partner (*LAWBriefs* 2003).

Antigay Hate Crimes Legislation

Hate crime laws call for tougher sentencing when prosecutors can prove that the crime committed was a hate crime. As of February 2004, 29 states and the District of Columbia have hate crime laws that include sexual orientation; 15 states have hate crime laws that do not include sexual orientation; 2 states have hate crime laws that address crimes motivated by bias or prejudice but do not list categories; and 5 states have no hate crime laws ("Hate Crime Laws in the U.S." 2004). As of this writing, gay rights advocates are lobbying Congress to pass federal hate crime legislation covering sexual orientation, gender, and disability. Formerly known as the Hate Crimes Prevention Act, the renamed Local Law Enforcement Enhancement Act of 2000 would make hate crimes based on sexual orientation, gender, and disability a federal offense.

Educational Strategies: Policies and Programs in the Public Schools

If schools are to promote the health and well-being of all students, they must address the needs of gay, lesbian, and bisexual youth and promote acceptance of sexual orientation diversity within the school setting. One strategy for promoting tolerance for diversity among students involves establishing and enforcing a school policy prohibiting antigay behavior. In 1994, Massachusetts state legislators passed such a policy prohibiting discrimination against gay and lesbian students. This law permits students who have suffered antigay discrimination, and who were not protected by the school administration, to bring lawsuits against their schools. Harassment and/or discrimination based on sexual orientation is prohibited by law, regulation, or policy in at least 13 states and the District of Columbia. But the majority of schools do not have any policies banning antigay harassment. Schools that do not protect students against harassment may face legal challenges. Between 1996 and 2002, a number of court rulings have held school districts responsible for failing to protect GLBT students from discrimination, violence, and harassment, and have ordered school districts to pay between $40,000 and $1 million in damages (Cianciotto & Cahill 2003).

One resource for creating a "harassment-free" climate is the Gay, Lesbian, and Straight Education Network (GLSEN)—the leading national organization fighting against harassment and discrimination in K-12 schools. GLSEN has conducted training for nearly 400 school staffs around the country and developed the faculty training program of the Massachusetts Department of Education's "Safe Schools for Gay and Lesbian Students" program—the first statewide effort aimed at ending homophobia in schools (*Homophobia 101* 2000). The National Education Association has also implemented national training programs to educate teachers in every state about the role they can and must play to stop antigay harassment in their schools (Chase 2000).

Some schools have established school-based support groups for LGBT students. Such groups can help students increase their self-esteem and overcome their sense of isolation, provide information and resources, and serve as a resource for parents. School counselors may be trained to work with gay, lesbian, and bisexual youth and their parents. Education about sexual orientation can be implemented in sex education or health education classes or in conflict resolution or diversity curricula. In-service training for teachers and other staff is important and may include examining the effects of antigay bias, dispelling myths about homosexuality, and brainstorming ways to create a more inclusive environment (Mathison 1998).

Gay-straight alliances (GSAs) are school-sponsored clubs for gay teens and their straight peers. First established in a Los Angeles high school in 1984, there are more than 1,000 GSAs in U.S. schools today (Kosciw & Cullen 2002). However, most public schools offer little to no support and education regarding sexual orientation diversity. Most schools have no support groups or special counseling services for gay and lesbian youth.

Campus Policies and Programs Dealing with Sexual Orientation

D'Emilio (1990) suggests that colleges and universities have the ability and the responsibility to promote gay rights and social acceptance of homosexual people:

> For reasons that I cannot quite fathom, I still expect the academy to embrace higher standards of civility, decency, and justice than the society around it. Having been granted the extraordinary privilege of thinking critically as a way of life, we should be astute enough to recognize when a group of people is being systematically mistreated. We have the intelligence to devise solutions to problems that appear in our community. I expect us also to have the courage to lead rather than follow. (p. 18)

Student groups have been active in the gay liberation movement since the 1960s. Because of the activism of students and the faculty and administrators who support them, nearly 400 U.S. colleges and universities have nondiscrimination policies that include sexual orientation (Singh & Wathington 2003). Other measures to support the LGBT college student population include gay and lesbian studies programs, social centers, and support groups, as well as campus events and activities that celebrate diversity. Many campuses have "Safe Zone" or "Ally" programs designed to visibly identify students, staff, and faculty who support the LGBT population. Safe Zone or Ally programs may require a training session that provides a foundation of knowledge needed to be an effective ally to LGBT students and those questioning their sexuality. Participants in Safe Zone or Ally programs display some type of sign or placard outside their office or residence hall room that signifies them as individuals who are willing to provide a safe haven, a listening ear, and support for LGBT people and those struggling with sexual orientation issues (Safe Zone Resources 2003).

Strategies for reducing antigay prejudice and discrimination are influenced largely by politicians, religious leaders, courts, and educators who will continue to make decisions that either promote or hinder the well-being of sexual orientation minorities. According to the Gay and Lesbian Victory Fund, a political organization that aims to help LGBT individuals elected to public office, there are 249 openly gay LGBT officeholders in the United States, a tiny fraction of the 511,039 total elected officials at the local, state, and federal levels (Johnston 2003a).

Ultimately, however, each individual must decide to embrace either an inclusive or exclusive ideology; collectively, those individual decisions will determine

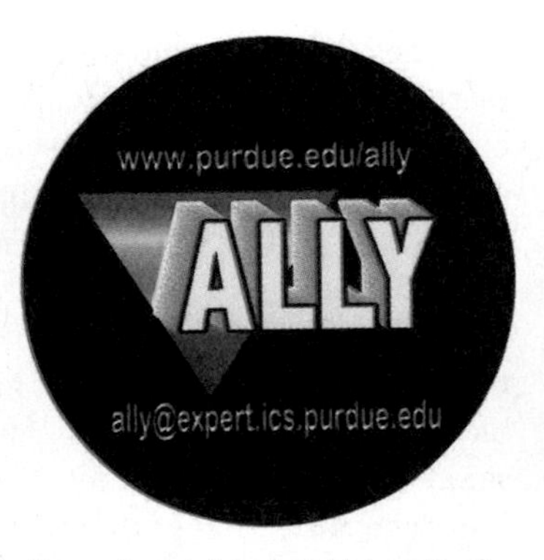

Pictured above are logos of safe zone programs at three universities/colleges, from left to right: University of Pennsylvania, University of North Carolina at Chapel Hill, Purdue University, and Cornell College.

the future treatment of sexual orientation minorities. In addition, lesbigay individuals must find their own strategies for living in a homophobic and biphobic society. Representative Tammy Baldwin (D–Wisconsin)—the first out lesbian and the first openly gay nonincumbent elected to Congress—offered the following advice in a speech entitled "Never Doubt," which she delivered at the Millennium March on Washington in April 2000:

> If you dream of a world in which you can put your partner's picture on your desk, then put his picture on your desk . . . and you will live in such a world.
>
> If you dream of a world in which there are more openly gay elected officials, then run for office . . . and you will live in such a world.
>
> And if you dream of a world in which you can take your partner to the office party, even if your office is the U.S. House of Representatives, then take her to the party. I do, and now I live in such a world.
>
> Remember, there are two things that keep us oppressed—them and us. We are half of the equation.

"Resistance begins with people confronting pain, whether it's theirs or somebody else's, and wanting to do something to change it."

bell hooks
Author

Understanding Issues in Sexual Orientation

Recent years have witnessed a growing acceptance of lesbians, gay men, and bisexuals as well as increased legal protection and recognition of these marginalized populations. The advancements in gay rights are notable and include the Supreme Court's 2003 decriminalization of sodomy, the growing adoption of nondiscrimination policies covering sexual orientation, the increased recognition of domestic partnerships, the state marriage rights and responsibilities offered to same-sex couples in Vermont and California, the Massachusetts ruling that same-sex marriage is allowable under the state constitution, and the election of the first openly gay bishop in the New Hampshire Episcopal diocese. But the winning of these battles in no way signifies that the war is over. Gay, lesbian, and bisexual individuals continue to be frequent victims of discrimination, harassment, and violence. In some countries, homosexuality is formally condemned with penalties ranging from fines to imprisonment and even death.

As both functionalists and conflict theorists note, nonheterosexuality challenges traditional definitions of family, childrearing, and gender roles. Every victory in achieving legal protection and social recognition for sexual orientation minorities fuels the backlash against them by groups who are determined to maintain traditional notions of family and gender. Often, this determination is rooted in and derives its strength from uncompromising religious ideology.

As symbolic interactionists note, meanings associated with homosexuality are learned. Powerful individuals and groups opposed to gay rights focus their efforts on maintaining the meanings of homosexuality as negative to keep the gay, lesbian, and bisexual population marginalized. Since taking office, President George W. Bush issued over 250 proclamations, including those honoring National School Lunch Week and Wright Brothers Day, although he has refused to recognize Gay Pride Month with an official proclamation (Tobias & Cahill 2003). Bush did issue a presidential proclamation for Marriage Protection Week as part of an effort to define marriage as being between a woman and a man, thus denying marriage rights to same-sex couples. Another direct attack on gay rights is the proposal of a Federal Marriage Amendment that would amend the Constitution to define marriage as being between one woman and one man. New York Senator Hillary Rodham Clinton notes that the push for the amendment represents the first time an attempt has been made to amend the Constitution to specifically deny rights to any group of individuals. She also suggests that the current campaign against gay rights is a political strategy to drive wedges between Americans and to divert the public's attention away from social problems. "They'd rather talk about taking away rights and undermining the ability of Americans to live their own lives, to have their own families," said Clinton, "than to talk about the miserable economy, to talk about their miserable foreign policy, to talk about their rollback of environmental laws and workers' rights, education and health care!" (quoted in Johnston 2003b).

Political efforts to undermine gay rights and recognition must contend with the fact that "gay is everywhere": 2000 Census data reveal that 99.3 percent of U.S. counties reported same-sex cohabiting partners, compared to 52 percent of counties in 1990. With the gay population being more "out" than ever before, more heterosexual individuals are reporting having family, work-related, or friendship/acquaintance ties to lesbian and gay individuals. This increased contact, as well as gay-straight alliances in schools and more frequent portrayals of lesbians and gays in the media, have inspired many heterosexually identified individuals to support the gay rights movement.

But, as one scholar noted, "The new confidence and social visibility of homosexuals in American life have by no means conquered homophobia. Indeed it stands as the last acceptable prejudice" (Fone 2000, 411). True, the American public is becoming increasingly supportive of gay rights. But as Yang (1999) points out, as the antigay minority diminishes in size, "it often becomes more dedicated and impassioned" (p. ii).

> Our task in the coming years is to get the heterosexual Americans who support our cause to feel as passionately outraged by the injustices we face and to be as strongly motivated to act in support of our rights as our adversaries are in their opposition to our rights. (p. iii)

Chapter Review

- **Are there any countries in the world that have national laws that ban discrimination against gays, lesbians, and bisexuals?**
  Yes, more than 20 countries have national laws that ban discrimination based on sexual orientation.

- **Is there any country where same-sex couples can be legally married?**
  Yes. In 2001 the Netherlands became the first country in the world to offer legal marriage to same-sex couples. Belgium and the provinces of Ontario and British Columbia, Canada, also grant legal marriage to same-sex couples.

- **In what ways is the classification of individuals into sexual orientation categories problematic?**
  Classifying individuals into sexual orientation categories is problematic for a number of reasons: (1) some people conceal or falsely portray their sexual orienta-

tion identity; (2) attractions, love, behavior, and self-identity do not always match; and (3) sexual orientation can change over time.

- **What is the relationship between beliefs about what "causes" homosexuality and attitudes toward homosexuality?**
Individuals who believe that homosexuality is biologically based or inborn tend to be more accepting of homosexuality. In contrast, individuals who believe homosexuals choose their sexual orientation are less tolerant of gays and lesbians.

- **What is the position of the American Psychiatric Association, American Psychological Association, American Academy of Pediatrics, and American Medical Association on conversion or reparative therapy for gays and lesbians?**
These organizations agree that sexual orientation cannot be changed and that efforts to change sexual orientation (conversion or reparative therapy) do not work and may, in fact, be harmful.

- **What is internalized homophobia? What are the effects of internalized homophobia on lesbian and gay individuals?**
Internalized homophobia is a sense of personal failure and self-hatred among lesbian and gay individuals resulting from social rejection and stigmatization. Internalized homophobia has been linked to increased risk for depression, substance abuse and addiction, anxiety, and suicidal thoughts.

- **What factor has been identified as the best predictor of homophobia?**
Conservative Christian ideology has been identified as the best predictor of homophobia.

- **What factors help to explain the increased acceptance of homosexuality and support for gay rights in the last decade?**
Factors that help explain the increased acceptance of homosexuality and support of gay rights include the increasing levels of education among U.S. adults; the positive depiction of gays and lesbians in the popular media; and increased contact between heterosexuals and openly gay and lesbian individuals.

- **Is employment discrimination based on sexual orientation illegal in all 50 states?**
As of this writing (Spring 2004), it is legal in 36 states to discriminate against someone on the basis of sexual orientation.

- **What is the Federal Marriage Amendment?**
The Federal Marriage Amendment is a proposal to amend the U.S. Constitution to define marriage as being between a man and a woman.

- **What two pieces of important legislation regarding gay rights did Governor Gray Davis sign into law shortly before he was recalled from office?**
Before leaving office, California's Governor Gray Davis passed landmark legislation (known as AB 17) mandating that any company doing business with the state offer domestic partner benefits to its employees. Governor Davis also signed into law the Domestic Partner Rights and Responsibilities Act of 2003, which gives same-sex domestic partners in California nearly all of the rights and responsibilities granted to married couples in the state.

- **What are "Safe Zone" or "Ally" programs?**
Safe Zone or Ally programs are designed to visibly identify students, staff, and faculty who support the LGBT population. Participants in Safe Zone or Ally programs display a sign or placard outside their office or residence hall room that signifies them as individuals who are willing to provide a safe haven and support for LGBT people and those struggling with sexual orientation issues.

Critical Thinking

1. How is the homosexual population similar to and different from other minority groups?
2. Do you think that social acceptance of homosexuality leads to the creation of laws that protect lesbians and gays? Or does the enactment of laws that protect lesbians and gays help to create more social acceptance of gays?
3. A Virginia court ruling denied Sharon Bottoms (a lesbian) custody of her son and awarded custody to Sharon's mother. Some people support this court decision on the belief that lesbian mothers are more likely to influence their children to be homosexual. Do you see any contradictions in the court's ruling in the Sharon Bottoms case?

Key Terms

biphobia
bisexuality
civil union
conversion therapy
Defense of Marriage Act
domestic partners
Employment Nondiscrimination Act (ENDA)
GLBT
heterosexism
heterosexuality
homophobia
homosexuality
internalized homophobia
lesbigay population
LGBT
master status
registered partnerships
reparative therapy
sexual orientation
sodomy laws
transgendered individuals

Taking a Stand

Should sex education in the public schools include information on sexual orientation issues?

Most public school students will take sex education by the time they graduate from high school. In a national survey by the Kaiser Family Foundation (2000), 76 percent of parents of youths grades 7–12 indicated that sex education should include discussion of homosexuality, yet only 41 percent of students said that "homosexuality" was covered in sex education class. What should sex education classes teach students about sexual orientation?

Use Wadsworth's exclusive online resources—InfoTrac College Edition, MicroCase Online, and OVRC—to formulate a position on this topic.

The Wadsworth's Sociology Online Resources and Writing Companion will help you get started. This valuable guide will show you how to use Wadsworth's exclusive online resources when studying social problems. It will also help you to build essential research and writing skills. InfoTrac College Edition, MicroCase Online, OVRC, and an electronic copy of portions of this companion are available at http://sociology.wadsworth.com/mooney_knox_schacht/problems4e, the companion Web site for *Understanding Social Problems,* Fourth Edition.

Media Resources

The Companion Web Site for *Understanding Social Problems,* Fourth Edition

http://sociology.wadsworth.com/mooney_knox_schacht/problems4e

Supplement your review of this chapter by going to the companion Web site to take one of the Tutorial Quizzes, use the flash cards to master key terms, and check out the many other study aids you'll find there. You'll also find special features such as *Wadsworth's Sociology Online Resources and Writing Companion,* GSS Data, and Census 2000 information, data, and resources at your fingertips to help you complete that special project or do some research on your own.

Interactions CD-ROM

Go to the Interactions CD-ROM for *Understanding Social Problems,* Fourth Edition, to access additional interactive learning tools, such as in-depth review materials, corresponding practice quizzes, and other engaging resources and activities to help you study the concepts in this chapter.

"The quality of a nation is reflected in the way it recognizes that its strength lies in its ability to integrate the wisdom of elders with the spirit and vitality of its children and youth." *Margaret Mead, Anthropologist*

Problems of Youth and Aging

"Who am I?" asked Nathan Grieco. "That's a question I haven't asked myself for quite some time. Many things (mostly bad) have happened in my life. There are so many it would take me two whole lifetimes to type about it" (quoted in Carpenter & Kopas 1999, 1).

There is no doubt that Nathan Grieco was depressed. Things had not been easy for him. His girlfriend had broken up with him, he was socially awkward and not very popular at school, and although his grades were good, he had been diagnosed with attention deficit disorder and was on medication.

The greatest "torture" in his life, however, was the ongoing custody battle between his feuding parents. After eight years there was no end in sight for him and his two younger brothers. After several occasions of being forced to see their father, a man they described as abusive and overbearing, the most recent court order held that failure to visit their father would result in contempt of court charges against their mother and her possible incarceration. Having no legal rights of their own, they had few choices. Within a year of the order, at age 16, Nathan Grieco was dead—found by his mother, kneeling next to his bed, with a leather belt around his neck. The coroner ruled that there was "insufficient evidence of either a suicide or an accident" (Carpenter & Kopas 1999, 1).

Like too many other youths, Nathan Grieco defined his life as one of "endless torment." Interestingly, some research suggests that depression is "curvilinear" with age, that is, highest at the extremes of the age continuum (DeAngelis 1997). Depression is not the only characteristic shared by the young and the old. Both groups are often the victims of stereotyping, physical abuse, age discrimination, and poverty; both groups are also major population segments of American society. In this chapter, we examine the problems and potential solutions associated with youth and aging. We begin by looking at age in a cross-cultural context.

The Global Context: Youth and Aging Around the World

The young and the old receive different treatment in different societies. Differences in the treatment of the dependent young and old have traditionally been associated with whether the country is developed or less developed. Although proportionately more elderly live in developed countries than in less-developed ones, these societies have fewer statuses for the elderly to occupy. Their positions as caretakers, homeowners, employees, and producers are often usurped by those aged 18 to 64. Paradoxically, the more primitive the society, the more likely that society is to practice senilicide—the killing of the elderly. In some societies the elderly are considered a burden and left to die or, in some cases, actually killed. For example, in Malawi, the elderly are killed for their body parts, which bring a high price on the black market (*African Eye* 2003).

Not all societies treat the elderly as a burden. Scandinavian countries provide government support for in-home care workers for elderly who can no longer perform such tasks as cooking and cleaning. Eastern cultures such as Japan revere the elderly, in part, because of their presumed proximity to honored ancestors. Japan has the highest proportion of elderly of any country in the world—18.5 percent of their population (*Zhong Xin Net* 2003).

Societies also differ in the way they treat children. In less-developed societies, children work as adults, marry at a young age, and pass from childhood directly to adulthood with no recognized period of adolescence. In contrast, in industrialized nations, children are often expected to attend school for 12 to 16 years and during this time to remain financially and emotionally dependent on their families.

Because of this extended period of dependence, the United States treats "minors" and adults differently. There is a separate justice system for juveniles and age limits for driving, drinking alcohol, joining the military, entering into a contract, marrying, dropping out of school, and voting. These limitations would not be tolerated if placed on individuals on the basis of sex or race. Hence **ageism,** the belief that age is associated with certain psychological, behavioral, and/or intellectual traits, at least in reference to children, is significantly more tolerated than sexism or racism in the United States.

"I'm a little older now. I've moved from the sandwich generation to the club sandwich generation."

Ellen Goodman
Syndicated columnist

Despite this differential treatment, people in the United States are fascinated with youth and being young. This was not always the case. The elderly were once highly valued in the United States—particularly older men who headed families and businesses. Younger men even powdered their hair, wore wigs, and dressed in a way that made them look older. It should be remembered, however, that in 1900 the average life expectancy in the United States was 47 and over half the population was under the age of 16. Being old was rare and respected; to some it was a sign that God looked upon the individual favorably.

One theory argues that the shift from valuing the old to valuing the young took place during the transition from an agriculturally based society to an industrial one. Land, which was often owned by elders, became less important as did their knowledge and skills about land-based economies. With industrialization, technological skills, training, and education became more important than land ownership. Called **modernization theory,** this position argues that as a society becomes more technologically advanced, the position of the elderly declines (Tirrito 2003).

Youth and Aging

Age is largely socially defined. Cultural definitions of "old" and "young" vary from society to society, from time to time, and from person to person. For example, in ancient Greece or Rome where the average life expectancy was 20 years, one was old at 18; similarly, one was old at 30 in medieval Europe and at 40 in the United States in 1850.

Age is also a variable that has a dramatic impact on one's life (items 1–4 in the following list are from Matras 1990):

1. Age determines one's life experiences because the date of birth determines the historical period in which a person lives. Twenty years ago cell phones and palm-sized computers couldn't have been imagined.
2. Different ages are associated with different developmental stages (physiological, psychological, and social) and abilities. Ben Franklin observed, "At 20 years of age the will reigns; at 30 the wit; at 40 judgment."
3. Age defines roles and expectations of behavior. The expression "act your age" implies that some behaviors are not considered appropriate for people of certain ages.
4. Age influences the social groups to which one belongs. Whether one is part of a sixth-grade class, a labor union, or a senior's bridge club depends on one's age.
5. Age defines one's legal status. Sixteen-year-olds can get a driver's license, 18-year-olds can vote and get married without their parents' permission, and 65-year-olds are eligible for Social Security benefits.

Table 9.1
Percentage of Population in Three Age Groups: United States, 1950, 2000, and 2050

Year	All ages	Under 18 years	18–64 years	65 years and over
		Percentage		
1950	100.0	31.3	60.6	8.2
2000	100.0	25.7	61.9	12.4
2050	100.0	23.7	56.0	20.3

Notes:Data are for the resident population. Data for 1950 exclude Alaska and Hawaii. See Appendix II, Population.

Sources: U.S. Census Bureau, 1980 Census of Population, General Population Characteristics, United States Summary (PC80-1-B1) [includes data for 1950]; 2000 Census of Population, Profiles of General Demographic Characteristics, United States, http://www.census.gov/prod/cen2000/dp1/2kh00.pdf accessed on September 27, 2001; Projections of the Total Resident Population by 5-Year Age Groups, and Sex with Special Age Categories: Middle Series, 2050 to 2070, http://www.census.gov/population/projections/nation/summary/np-t3-g.txt accessed on September 27, 2001.

Childhood, Adulthood, and Elderhood

Every society assigns different social roles to different age groups. **Age grading** is the assignment of social roles to given chronological ages (Matras 1990). Although the number of age grades varies by society, most societies make at least three distinctions—childhood, adulthood, and elderhood (See Table 9.1).

Childhood. The period of childhood in our society is from birth through age 17 and is often subdivided into infancy, childhood, and adolescence. Infancy has always been recognized as a stage of life, but the social category of childhood developed only after industrialization, urbanization, and modernization took place. Before industrialization, infant mortality was high because of the lack of adequate health care and proper nutrition. Once infants could be expected to survive infancy, the concept of childhood emerged, and society began to develop norms in reference to children. In the United States, child labor laws prohibit children from being used as cheap labor, educational mandates require that children attend school until the age of 16, and federal child pornography laws impose severe penalties for the sexual exploitation of children.

Adulthood. The period from age 18 through 64 is generally subdivided into young adulthood, adulthood, and middle age. Each of these statuses involves dramatic role changes related to entering the workforce, getting married, and having children. The concept of "middle age" is a relatively recent one that has developed as life expectancy has been extended. Some people in this phase are known as members of the **sandwich generation** because they are often emotionally and economically responsible for both their young children and their aging parents.

Elderhood. At age 65 one is likely to be considered elderly, a category that is often subdivided into the young-old, old, and old-old. Membership in one of these categories does not necessarily depend on chronological age. The healthy, active, independent elderly—increasing in number—are often considered to be the young-old, whereas the old-old are less healthy, less active, and more dependent.

Sociological Theories of Age Inequality

Three sociological theories help explain age inequality and the continued existence of ageism in the United States. These theories—structural-functionalism, conflict theory, and symbolic interactionism—are discussed in the following sections.

Structural-Functionalist Perspective

Structural-functionalism emphasizes the interdependence of society—how one part of a social system interacts with other parts to benefit the whole. From a functionalist perspective, the elderly must gradually relinquish their roles to younger members of society. This transition is viewed as natural and necessary to maintain the integrity of the social system. The elderly gradually withdraw as they prepare for death, and society withdraws from the elderly by segregating them in housing such as retirement villages and nursing homes. In the interim, the young have learned through the educational institution how to function in the roles surrendered by the elderly. In essence, a balance in society is achieved whereby the various age groups perform their respective functions: the young go to school, adults fill occupational roles, and the elderly, with obsolete skills and knowledge, disengage. As this process continues, each new group moves up and replaces another, benefiting society and all its members.

This theory is known as **disengagement theory** (Tirrito 2003). Some researchers no longer accept this position as valid, however, given the increased number of elderly who remain active throughout life (Riley 1987). In contrast to disengagement theory, **activity theory** emphasizes that the elderly disengage in part because they are structurally segregated and isolated, not because they have a natural tendency to do so (Tirrito 2003). For those elderly who remain active, role loss may be minimal. In studying 1,720 respondents who reported using a senior center in the previous year, Miner, Logan, and Spitze (1993) found that those who attended were less disengaged and more socially active than those who did not.

© Reuters NewMedia Inc./CORBIS

Despite age grading, many singers such as Stevie Nicks, Mick Jagger, and Cher continue to perform and are commercially successful. Tina Turner, almost 60 in this picture, defies age stereotypes.

Conflict Perspective

The conflict perspective focuses on age grading as another form of inequality as both the young and the old occupy subordinate statuses. Some conflict theorists emphasize that individuals at both ends of the age continuum are superfluous to a capitalist economy. Children are untrained, inexperienced, and neither actively producing nor consuming in an economy that requires both. Similarly, the elderly, although once working, are no longer productive and often lack required skills and levels of education. Both young and old are considered part of what is called the dependent population; that is, they are an economic drain on society. Hence, children are required to go to school in preparation for entry into a capitalist economy, and the elderly are forced to retire.

Other conflict theorists focus on how different age strata represent different interest groups that compete with one another for scarce resources. Debates about funding for public schools, child health programs, Social Security, and Medicare largely represent conflicting interests of the young versus the old.

Symbolic Interactionist Perspective

The symbolic interactionist perspective emphasizes the importance of examining the social meaning and definitions associated with age. Teenagers are often portrayed as lazy, aimless, and awkward. The elderly are also defined in a number of stereotypical ways contributing to a host of myths surrounding the inevitability of physical and mental decline. Table 9.2 identifies some of these myths.

Media portrayals of the elderly contribute to their negative image. The young are typically portrayed in active, vital roles and are often overrepresented in commercials. In contrast, the elderly are portrayed as difficult, complaining, and burdensome and are often underrepresented in commercials. A study of the elderly in popular 1940s through 1980s films concluded that "older individuals of both genders were portrayed as less friendly, having less romantic activity, and enjoying fewer positive outcomes than younger characters at a movie's conclusion" (Brazzini, McIntosh, Smith, Cook, & Harris 1997, 541). Media images are powerful—a recent study found that children as young as 5 years old have already developed negative stereotypes of the elderly (*EurekAlert* 2003).

"Since it is the Other within us who is old, it is natural that the revelation of our age should come to us from outside— from others."

Simone de Beauvoir
Feminist author

The elderly are also portrayed as childlike in terms of clothes, facial expressions, temperament, and activities—a phenomenon known as **infantilizing elders.** For example, young and old are often paired together. A promotional advertisement for the movie *Just You and Me, Kid* with Brooke Shields and George Burns described it as "the story of two juvenile delinquents." Jack Lemmon and Walter Matthau in *Grumpy Old Men* get "cranky" when they get tired, and the media focus on images of Santa visiting nursing homes and local elementary school children teaching residents arts and crafts. Finally, the elderly are often depicted in role reversal, cared for by their adult children as in the situation comedies *Golden Girls, Frasier,* and *King of Queens.*

Negative stereotypes and media images of the elderly engender **gerontophobia**—a shared fear or dread of the elderly, which may create a self-fulfilling prophecy. For example, in a 20-year study of 600 respondents, researchers found that elderly people with positive perceptions of aging (e.g., seeing the elderly as wise) lived, on the average, seven and a half years longer than those who had a negative image of aging (e.g., considering most elderly to be senile) (Ramirez 2002).

Table 9.2
Myths and Facts About the Elderly

Health
Myth The elderly are always sick; most are in nursing homes.
Fact Over 85 percent of the elderly are healthy enough to engage in their normal activities. Only 6 percent are confined to a nursing home.

Mental Status
Myth The elderly are senile.
Fact Although some of the elderly learn more slowly and forget more quickly, most remain oriented and mentally intact. Only 20–25 percent develop Alzheimer's disease or some other incurable form of brain disease. Senility is not inevitable as one ages.

Employment
Myth The elderly are inefficient employees.
Fact Although only about 17 percent of men and 9 percent of women 65 years old and over are still employed, those who continue to work are efficient workers. When compared with younger workers, the elderly have lower job turnover, fewer accidents, and less absenteeism. Older workers also report higher satisfaction in their work (AOA 2003).

Politics
Myth The elderly are not politically active.
Fact In 1994, 1996, 1998, 1999, 2000, and 2002 individuals age 65 and older were more likely to be registered to vote and/or to vote than any other age group.

Sexuality
Myth Sexual satisfaction disappears with age.
Fact Many elderly people report active and satisfying sex lives. For example, of couples 75 years old and older, over 25 percent report having sexual intercourse once a week (Toner 1999).

Adaptability
Myth The elderly cannot adapt to new working conditions.
Fact A high proportion of the elderly are flexible in accepting change in their occupations and earnings. Adaptability depends on the individual: many young are set in their ways, and many older people adapt to change readily.

Source: Administration on Aging. 2000. "Profile of Older Americans: 1999." http://www.aoa.dhhs.gov/aoa/stats; Administration on Aging. 2003. "Profile of Older Americans: 2002." http://www/aoa.gov; Robert H. Binstock. 1986. "Public Policy and the Elderly." *Journal of Geriatric Psychiatry* 19:115–143; Federal Elections Commission. "Voter Registration and Turnout by Age, Gender, and Race." http://www.fec.gov; Erdman B. Palmore. 1984. "The Retired." In *Handbook on the Aged in the United States,* ed. Erdman B. Palmore, pp. 63–75. Westport, CT: Greenwood Press; Teresa Seeman and Nancy Adler. 1998. "Older Americans: Who Will They Be?" National Forum, Spring, 22–25; *Statistical Abstract 2002.* 2002, 122nd ed. U.S. Bureau of the Census. Washington, DC: U.S. Government Printing Office. Robin Toner. 1999. "A Majority over 45 Say Sex Lives Are Just Fine." *New York Times,* August 4, A10.

Problems of Youth

The number of people under the age of 18—72 million in 2003—is the largest in history and will grow to 80 million by the year 2020. Although recent evidence indicates that conditions for children are improving, numerous problems remain. Indeed, some of our most pressing social problems can be traced to early childhood experiences and adolescent behavioral problems. Table 9.3 provides a summary of problems associated with youth in America.

Child Labor

Child labor involves children performing work that is hazardous; that interferes with a child's education; or that harms a child's health or physical, mental, spiritual, or moral development (U.S. Department of Labor 1995). Even though virtually every country in the world has laws that limit or prohibit the extent to which children can be employed, child labor persists throughout the world. An estimated

Table 9.3
Every Day in America

1	young person under 25 dies from HIV infection.
5	children or youth under 20 commit suicide.
9	children or youth under 20 are homicide victims.
9	children or youth under 20 die from firearms.
34	children and youth under 20 die from accidents.
77	babies die.
157	babies are born at very low birthweight (less than 3 lbs., 4 oz.).
180	children are arrested for violent crimes.
367	children are arrested for drug abuse.
401	babies are born to mothers who had late or no prenatal care.
825	babies are born at low birthweight (less than 5 lbs. 8 oz.).
1,310	babies are born without health insurance.
1,329	babies are born to teen mothers.
2,019	babies are born into poverty.
2,319	babies are born to mothers who are not high school graduates.

Source: Children's Defense Fund. 2003. "Every Day in America." http://www.childrensdefense.org/everyday.html

250 million school-age children are child laborers (BBC 2003). To grasp the scale of child labor, imagine a country as populous as the United States, in which the entire population consists of child laborers.

Child laborers work in factories, workshops, construction sites, mines, quarries, and fields, on deep-sea fishing boats, at home, on the street, and on the battlefield where, globally, 300,000 child soldiers endure the rigors of armed conflict (Becker 2004). Child laborers make bricks, shoes, soccer balls, fireworks and matches, furniture, toys, rugs, and clothing. They work in manufacturing of brass, leather goods, and glass. They tend livestock and pick crops. In Egypt, over 1 million children ages 7 to 12 work each year in cotton pest management. They endure routine beatings by their foremen as well as exposure to heat and pesticides ("Underage and Unprotected" 2001). Children typically earned the equivalent of about $1 per day and worked from 7:00 A.M. to 6:00 P.M. daily, with one midday break, seven days a week.

Illegal and oppressive employment of children also occurs in the United States in restaurants, grocery stores, meat-packing plants, sweatshops in urban garment districts, and in agriculture. Somewhere between 300,000 and 800,000 U.S. child workers labor on commercial farms alone, frequently under dangerous and grueling conditions (HRW 2003a). Child farmworkers in the United States often work 12-hour days, sometimes beginning at 3:00 or 4:00 A.M. Many are exposed to dangerous pesticides that cause cancer and brain damage, with short-term symptoms including rashes, headaches, dizziness, nausea, and vomiting. They often work in 100°F temperatures without adequate access to drinking water and are sometimes forced to work without access to toilets or handwashing facilities (HRW 2003a). Agriculture is also one of the most dangerous occupations, and children (as well as adults) sustain high rates of injury from work with knives, other sharp tools, and heavy equipment.

Juvenile farmworkers are less protected under U.S. law than are juveniles working in safer occupations. Under the federal Fair Labor Standards Act (FLSA):

(1) children working on U.S. farms may be employed at age 12, whereas the minimum age for employment in other occupations is 14; (2) the number of hours child farmworkers may work on school days is not limited, whereas in other occupations, children under age 16 are limited to three hours of work per day when school is in session; (3) FLSA does not require overtime pay for agricultural workers as it does for other occupations; and (4) juveniles working in agriculture may engage in hazardous work at age 16, whereas for other occupations, the minimum age for hazardous work is 18 (HRW 2000). Because 85 percent of migrant and seasonal farm workers nationwide are racial and ethnic minorities, the FLSA's bias against farmworker children is a form of institutional discrimination.

Child Prostitution and Trafficking. One of the worst forms of child labor is child prostitution and child trafficking. Worldwide, it is estimated that there are about 1 million child prostitutes; in the U.S., 300,000 (UNICEF 2003b; Dorman 2001). In poor countries, the sexual services of children are often sold by their families in an attempt to get money. Some children are kidnapped or lured by traffickers with promises of employment, only to end up in a brothel. The U.S. Immigration and Naturalization Service identified 250 brothels in 26 American cities where forced prostitutes, including children, are taken. Americans also engage in child-sex tourism abroad, particularly in Southeast Asia. Of 240 identified cases in which legal action was taken against foreigners for sexually abusing children in this region, researchers found that about one-quarter of the violators were from the United States (Dorman 2001).

Orphaned and Street Children

Worldwide, there are millions of orphans, many kept in institutions, some living on the streets, and still others, with no means of supporting themselves, turning to sex work, crime, or drugs. In many of these institutions children are simply warehoused with little food, clean water, or medical care (HRW 2003b). Of late, the number of orphans has increased dramatically for at least two reasons. First is the devastation of the HIV/AIDS pandemic. Globally, the number of children orphaned by the disease, predominantly in developing nations, is over 13 million estimated to grow to 25 million by 2010 (*CBS News* 2003). Over 50 percent of children 0 to 14 years old in Zimbabwe, Botswana, Zambia, Kenya, and Uganda have lost one or both parents to HIV/AIDS (UNICEF 2003c). Second, children are increasingly orphaned from armed conflicts. For example, the civil war in Rwanda is responsible for hundreds of thousands of orphans, many of whom are now street children.

It is estimated that over 100 million children live on the streets, 40 million in Latin America alone. Many of the street children are "addicted to inhalants, such as cobbler's glue, which offers them an escape from reality in exchange for a host of physical and psychological problems" (CASA 2003). Often resorting to prostitution, many of the street children have HIV/AIDS. For example, of Mexico's 2 million street children, at least 7 percent are HIV/AIDS positive (CASA 2003). The number of orphaned and street children is predicted to grow dramatically over the next decade.

Children's Rights

Historically, children have had little control over their lives. They have been "double dependent" on both their parents and the state. Indeed, colonists in America regarded children as property. Beginning in the 1950s, however, the view that children should have greater autonomy became popular and was codified in several legal decisions and international treaties. In 1959, the United Nations General Assembly approved the *Declaration on the Rights of the Child,* which held that health care, housing, and education, as well as freedom from abuse, neglect, and exploitation, are fundamental children's rights.

A second measure, the *Convention on the Rights of the Child,* further articulated the rights of children and was adopted by the United Nations in 1989. Countries ratifying the document have made significant improvements in the lives of children. For example, the Democratic Republic of the Congo's draft constitution now prohibits conscription into the army before age 18. Only two countries, the United States and Somalia, have failed to ratify the *Convention.* One reason the United States has not signed the pact may be Article 11, which requires that children be assured "the highest attainable standard of health." Some people are concerned that accepting such a position would result in cases being brought to court that they are simply unwilling, at present, to hear. The two most recent provisions of the *Convention* deal with the involvement of children in armed conflicts, the sale of children, child prostitution, and child pornography (UNICEF 2003a).

Children are both discriminated against and granted special protections under the law. Although legal mandates require that children go to school until age 16, other laws provide a separate justice system whereby children have limited legal responsibility based on their age status. Concern with youth violence has many Americans questioning the wisdom of a separate legal structure for minors. The laws are changing: most states have lowered the age of accountability, capital punishment of 16-year-olds is allowed in a number of states, and the number of parent-liability laws has increased. Although children's rights may expand in some areas such as self-determination, recent legal changes may cost minors their protected status in the courts. In 2001, 13-year-old Lionel Tate was convicted of first-degree murder and sentenced to life in prison without parole. In January of 2004, Tate's sentence was reduced to time served, and after only three years he was released (CNN 2004).

Poverty and Economic Discrimination

Based on a recent worldwide study, the U.S. poverty rate for children is more than double that of every other major industrialized country. In 2001, 16 percent of U.S. children lived below the poverty line and 7 percent, a 17 percent increase from 2000, live in extreme poverty, that is, live in families who earn half the federal established poverty level (NCCP 2003). In 2001, 30.2 percent of black children were poor, 28 percent of Hispanic children, 12 percent of Asian children, and 9 percent of non-Hispanic white children (CDF 2003a). Childhood poverty is related to school failure (CDF 2003a), negative involvement with parents (Harris & Marmer 1996), stunted growth, reduced cognitive abilities, limited emotional development (Brooks-Gunn & Duncan 1997), and a higher likelihood of dropping out of school (Duncan, Yeung, Brooks-Gunn, & Smith 1998).

Three decades ago the elderly were the poorest age group in the United States; today it is children ("Snapshots" 2000). Although Social Security, Medicare, housing subsidies, and Supplemental Security Income (SSI) keep millions of elderly out of poverty, the United States has no universal government policy to protect children. Further, recent welfare reform has led to cutbacks in what few programs do

The Human Side

Growing Up in the Other America

Author Alex Kotlowitz conducted a two-year participant observation research study of children in a Chicago housing project. The following description captures the horrific living conditions endured by Lafeyette, Pharoah, and Dede—three children living in the "jects."

The children called home "Hornets" or, more frequently, "the projects" or, simply, the "jects" (pronounced jets). Pharoah called it "the graveyard." But they never referred to it by its full name: the Governor Henry Horner Homes.

Nothing here, the children would tell you, was as it should be. Lafeyette and Pharoah lived at 1920 West Washington Boulevard, even though their high-rise sat on Lake Street. Their building had no enclosed lobby; a dark tunnel cut through the middle of the building, and the wind and strangers passed freely along it. Those tenants who received public aid had their checks sent to the local currency exchange, since the building's first-floor mailboxes had all been broken into. And since darkness engulfed the building's corridors, even in the daytime, the residents always carried flashlights, some of which had been handed out by a local politician during her campaign.

Summer, too, was never as it should be. It had become a season of duplicity.

On June 13, a couple of weeks after their peaceful afternoon on the railroad tracks, Lafeyette celebrated his twelfth birthday. Under the gentle afternoon sun, yellow daisies poked through the cracks in the sidewalk as children's bright faces peered out from behind their windows. Green leaves clothed the cottonwoods, and pastel cotton shirts and shorts, which had sat for months in layaway, clothed the children. And like the fresh buds on the crabapple trees, the children's spirits blossomed with the onset of summer.

Lafeyette and his nine-year-old cousin Dede danced across the worn lawn outside their building, singing the lyrics of an L.L. Cool J rap, their small hips and spindly legs moving in rhythm. The boy and girl were on their way to a nearby shopping strip, where Lafeyette planned to buy radio headphones with $8.00 he had received as a birthday gift.

Suddenly, gunfire erupted. The frightened children fell to the ground. "Hold your head down!" Lafeyette snapped, as he covered Dede's head with her pink nylon jacket. If he hadn't physically restrained her, she might have sprinted for home, a dangerous action when the gangs started warring. "Stay down," he ordered the trembling girl.

The two lay pressed to the beaten grass for half a minute, until the shooting subsided. Lafeyette held Dede's hand as they cautiously crawled through the dirt toward home. When they finally made it inside, all but fifty cents of Lafeyette's birthday money had trickled from his pockets.

Source: Alec Kotlowitz. 1991. *There Are No Children Here.*

benefit children. For example, more than half of food stamp recipients are under the age of 18 (CDF 2003a).

Children are also the victims of discrimination in terms of employment, age restrictions, wages, training programs, and health benefits. Traditionally, children worked on farms and in factories but were displaced during the Industrial Revolution. In 1938, Congress passed the Fair Standards Labor Act, which required factory workers to be at least 16. Although the law was designed to protect children, it was also discriminatory in that it prohibited minors from having free access to jobs and economic independence. Today, 14- and 15-year-olds are restricted in the number of hours per day and hours per week they can work; those under 14 are, in general, prohibited from working. Of those youths who are old enough to work, the 2001 summer unemployment rate was 59 percent, the highest it has been in 55 years (CDF 2003b).

Children, Violence, and the Media

Gang violence, suicide, child abuse, and crime are all-too-common childhood experiences (see this chapter's *The Human Side* feature). When a sample of 12- to 17-year-olds were asked about the most important problems facing their age group, "drugs" was the number one response; "crime and violence in school" ranked third, and "other crime and violence" ranked fifth (CASA 2000).

According to the Children's Defense Fund, compared with children in 25 other industrialized nations *combined,* U.S. children are 12 times more likely to die from gunfire, 16 times more likely to be murdered by a gun, 11 times more likely to commit suicide with a firearm, and 9 times more likely to die from a firearm accident (CDF 2003c). Child abuse and neglect remain at epidemic levels, with an estimated 3 million children victimized annually; children under 18 account for 26 percent of all arrests (FBI 2002).

School violence, and particularly the tragic deaths of 12 students and one teacher at Columbine High School, has focused attention on the relationship between youth violence, guns, and the media. Forty-three percent of respondents in a Gallup survey reported fearing for their child's safety at school—almost twice the number than in 1977. When asked about the causes of violence, the majority of respondents reported that movies; video or computer games; song lyrics on CDs, tapes, or the radio; and television were each an "extremely serious" or "serious" problem (Gallup Poll 2000). The typical child watches 28 hours of television a week. By the time the child is 18 he or she will have seen 16,000 murders and 200,000 other acts of violence on television. Ironically, the Center for Media and Public Affairs reports that commercial television for children is 50 to 60 times more violent than prime-time TV with, for example, cartoons averaging 80 violent acts an hour (CDF 2003d).

Children's Health

In addition to the problems already discussed, children suffer from a variety of health problems, both mental and physical. One in five children, for example, has a diagnosable mental disorder such as depression or schizophrenia (see Chapter 2). Many children, particularly girls, suffer from eating disorders such as anorexia nervosa or bulimia. It is estimated that 1 percent of young girls will develop anorexia and, of that number, 10 percent will die from the disease. Suicide is the third most common cause of death of 15- to 24-year-olds, the sixth most common cause of death of 5- to 15-year-olds, and the third most common cause of death of college students. Suicide rates among the young have tripled since 1960. More youth die from suicide than die from heart disease, cancer, HIV/AIDS, stroke, birth defects, and chronic lung disease combined (NMHA 2003).

Kids in Crisis

Childhood is a stage of life that is socially constructed by structural and cultural forces of the past and present. The old roles for children as laborers and farm helpers are disappearing, yet no new roles have emerged. While being bombarded by the media, children must face the challenges of an uncertain future, peer culture, music videos, divorce, poverty, and crime. Parents and public alike fear children are becoming increasingly involved with sex, drugs, alcohol, and violence. Some even argue that childhood as a stage of life is disappearing.

Despite the many problems associated with childhood, the condition of children in the United States is better than the public's perception of it. For example, 12 percent of children lack health insurance but 93 percent of a national sample of adults believed that 20 to 30 percent of children lack health insurance. Further, almost half of the adults sampled believed that 30 percent of children live in poverty when in fact 17 percent live in poverty. The significance of these discrepancies lies in the public's influence on social policy. If the public has an inaccurate picture of the state of children in the United States they may incorrectly impact policymakers (Peterson 2003).

Despite some gains, conditions remain intolerable for millions of children in the United States and around the world—homelessness, sexual exploitation, low birth weight, hunger, childhood depression, absent or inadequate health care, and dangerous living conditions (e.g., foster care, child abuse). A report from the National Research Council recently called for a "new national dialogue focused on rethinking the meaning of both shared responsibility for children and a strategic investment in their future" (Jacobson 2000, 1). Such an investment, particularly in preventive initiatives, is not only humane but would save millions of dollars spent on solving child-related problems after they occur.

Demographics: The "Graying of America"

In recent years, the ratio of children to old people has changed dramatically (see Table 9.1 and Figure 9.1). In 2001, one in every eight Americans was over the age of 65; in 2030 one in four people will be over the age of 65 (AOA 2003a).

These statistics reflect three significant demographic changes. First, 76 million baby boomers born between 1946 and 1964 are getting older. Second, life expectancy has increased as a result of a general trend toward modernization including better medical care, sanitation, and nutrition. Finally, lowered birth rates mean fewer children and a higher percentage of the elderly. For example, in Japan, low birth rates and rising life expectancies have contributed to the highest proportion of elderly in the world (*Zhong Xin Net* 2003).

Age and Race and Ethnicity

In 2001, about 16 percent of the elderly were racial/ethnic minorities. Minority populations are projected to represent 25 percent of the U.S. elderly population by 2030, up from 13 percent in 1990. Although the elderly white population is projected to increase 81 percent between 1999 and 2030, the elderly minority population is expected to increase 219 percent in the same time period (AOA 2003b). The growth of the elderly minority population is a consequence of higher fertility rates and immigration patterns, particularly among Hispanics. Given their higher rates of diabetes, arthritis, and cardiovascular disease, the increased numbers of older minorities presents a unique challenge to health care providers (AOA 2000a; Newman 2000).

Age and Gender

In the United States, elderly women outnumber elderly men. The sex ratio for persons 60 and older is 77 men for every 100 women (UN 2003). Men die at an earlier age than women for both biological and sociological reasons—heart disease, stress,

Figure 9.1
U.S. Population Pyramids
Source: U.S. Bureau of the Census, International Data Base. Washington, D.C.: U.S. Department of Commerce.

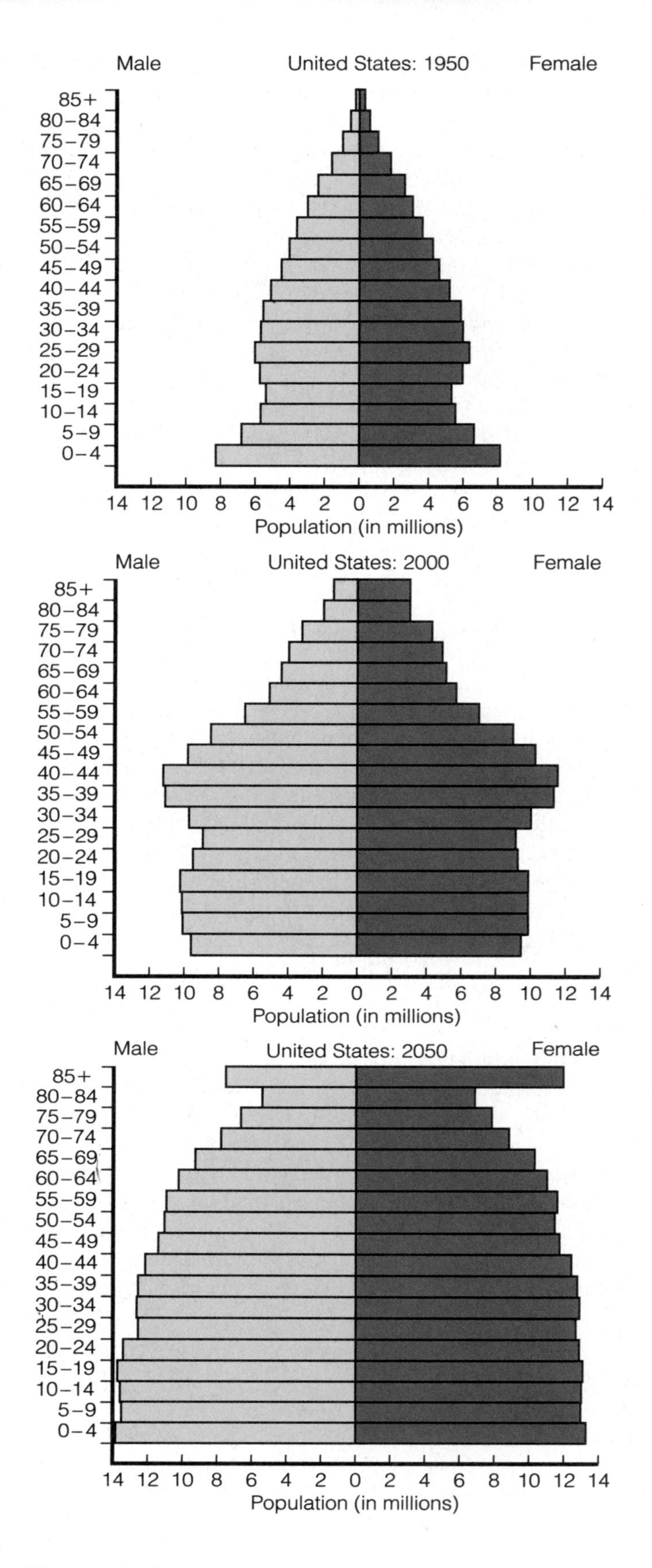

© Tony Freeman/Photo Edit

Women, as they get older, are often portrayed as sexually unattractive, something men are rarely subjected to. Advertising contributes to this image by idealizing youth in advertisements that advocate the use of wrinkle creams, hair dyes, and getting one's "girlish figure" back.

and occupational risk (see Chapter 2 and Chapter 7). The fact that women live longer results in a sizable number of elderly women who are poor. Not only do women, in general, make less money than men, but older women may have spent their savings on their husband's illness, and as homemakers, they are not eligible for Social Security benefits. Further, retirement benefits and other major sources of income may be lost with a husband's death. Seventy percent of all elderly poor are women, half of whom were not poor before the death of their husbands.

Age and Social Class

How long a person lives is influenced by his or her social class. In general, the higher the social class, the longer one lives, the fewer the debilitating illnesses, the greater the number of social contacts and friends, the less likely one is to define oneself as "old," and the greater the likelihood of success in adapting to retirement. Higher social class is also related to fewer residential moves, higher life satisfaction, more leisure time, and more positive self-rated health. Of those 65 and older, 26 percent of those with annual incomes of $35,000 or higher reported that their health was "excellent" whereas only 10 percent of those with incomes under $10,000 rated their health as such. Functional limitations such as problems with walking, dressing, and bathing were also less common among higher income groups (Seeman & Adler 1998). In short, the higher one's socioeconomic status, the longer, happier, and healthier one's life.

Problems of the Elderly

The increase in the number of the elderly, worldwide, presents a number of institutional problems. The **dependency ratio**—the number of societal members who are under 18 or are 65 and over compared with the number of people who are between 18 and 64—is increasing. In 2000, there were 62 "dependents" for every

Self and Society

Facts on Aging Quiz

Answer the following questions about the elderly and assess your knowledge of the world's fastest-growing age group.

	True	False
1. Lung capacity tends to decline in old age.	______	______
2. The majority of old people say they are seldom bored.	______	______
3. Old people tend to become more religious as they age.	______	______
4. The health and economic status of old people will be about the same or worse in the year 2010 (compared with younger people).	______	______
5. A person's height tends to decline in old age.	______	______
6. The aged are more fearful of crime than are younger persons.	______	______
7. The proportion of blacks among the aged is growing.	______	______
8. The majority of old people live alone.	______	______
9. The five senses all tend to weaken in old age.	______	______
10. Medicare pays over half of the medical expenses of the elderly.	______	______
11. Older persons who reduce their activity tend to be happier than those who do not.	______	______
12. Older persons have more injuries in the home than younger persons.	______	______
13. Physical strength tends to decline with age.	______	______
14. The aged are the most law abiding of all adult age groups.	______	______
15. Older persons have more acute illnesses than do younger persons.	______	______

Answers: 1, 2, 5–7, 9, 13, and 14 are true. The remainder are false.

Source: Erdman B. Palmore. 1999. *Ageism: Negative and Positive.* New York: Springer Publishing Co. Used by permission.

100 persons between 18 and 64. By 2050, the estimated ratio will be 80 to 100 (AOA 2003c). This dramatic increase, and the general movement toward global aging, may lead to a shortage of workers and military personnel, foundering pension plans, and declining consumer markets. It may also lead to increased taxes as governments struggle to finance elder care programs and services, heightening intergenerational tensions as societal members compete for scarce resources. In addition to these macro-level concerns, the elderly face a number of challenges of their own. This chapter's *Self and Society* feature tests your knowledge of the aged and some of these concerns.

Work and Retirement

What one does (occupation), for how long (work history), and for how much (wages) are important determinants of retirement income. Indeed, employment is important because it provides the foundation for economic resources later in life. Yet, for the elderly who want to work, entering and remaining in the labor force may be difficult because of negative stereotypes, lower levels of education, reduced geographic mobility, fewer employable skills, and discrimination. Nonetheless, older

Americans are working longer than ever before. According to a recent study, 42 percent of people over the age of 65 are working full- or part-time during their retirement years (Burtner 2003).

In 1967, Congress passed the Age Discrimination in Employment Act (ADEA), which was designed to ensure continued employment for people between the ages of 40 and 65. In 1986, the upper limit was removed, making mandatory retirement illegal in most occupations. Nevertheless, thousands of cases of age discrimination occur annually. Although employers cannot advertise a position by age, they can state that the position is an "entry level" one or that "2 to 3 years' experience" is required. Despite strong evidence to the contrary, a study conducted by the National Council of Aging revealed that "50 percent of employers surveyed believed that older workers cannot perform as well as younger workers" (Reio & Sanders-Reto 1999). Further, if displaced, older workers remain unemployed longer than younger workers, often have to accept lower salaries than they earned earlier, and may be more likely to give up looking for work (Goldberg 2000).

Retirement is a relatively recent phenomenon. Before passage of the Social Security laws, individuals continued to work into old age. Today, Social Security payments are limited to those over 65 years of age—age 67 by the year 2022. Most people, however, retire before age 65, with 60 percent of the labor force retiring at age 62 (Goldberg 2000).

Retirement is difficult in the United States as "work" is often equated with "worth." A job structures one's life and provides an identity; retirement often culturally signifies the end of one's productivity. Retirement may also involve a dramatic decrease in personal income. The desire to remain financially independent, a lack of confidence in the Social Security system, increased educational levels and technological skills, and the desire to continue working have led to a recent reduction in early retirements (AOA 2000b). Additionally, the number of people who stop working once they retire is also decreasing as retirees open their own businesses, work part-time, continue their educations, volunteer in the community, and begin second careers (Ennis 2000; Costa 2000; Goldberg 2000; AARP 2000).

Poverty

In 2002, the median income for persons 65 years old and older was $14,152, and 38 percent of the elderly reported incomes of $10,000 or less (AOA 2003b). Poverty among the elderly varies dramatically by gender, race, ethnicity, marital status, and age: women, minorities, those who are single or widowed, and the "old-old" are most likely to be poor (NCSC 2000a; Willson & Hardy 2002). Nine percent of elderly whites are poor compared to 22 percent of elderly blacks and elderly Hispanics. Elderly poor are more likely to live in central cities, in rural areas, and in southern states (AOA 2003d).

Elderly women are more likely to be poor than elderly men. Nearly half of elderly Hispanic women, and 4 of 10 elderly black women are living in poverty. The comparable percentage for white women is half that of minorities. Minority women are more likely to have been working in low-paying jobs with no retirement plan, to have little or no savings, and to have fewer resources to fall back on. For example, older black households have an estimated median net worth of $13,000 compared to older white households with an estimated median net worth of $181,000 (Hounsell & Humphlett 2003). For many minority women Social Security is the sole source of support. The highest poverty rate, over 50 percent, is among elderly Hispanic women who live alone or with nonrelatives (AOA 2003b).

Actually titled "Old Age, Survivors, Disability, and Health Insurance," Social Security is a major source of income for many elderly. When Social Security was established in 1935, it was never intended to be a person's sole economic support in old age; rather, it was to supplement other savings and assets. However, a Social Security Administration survey revealed that when asked "major sources of income," 90 percent of the elderly reported Social Security, 59 percent income from assets, 41 percent public and private pensions, and 22 percent earnings (AOA 2003d). Spending on the elderly has increased over time as have Social Security benefits. Today, 92 percent of Americans aged 65 and over receive Social Security payments, and another 3 percent will be eligible once they retire.

Although Social Security may keep as many as 4 out of 10 elderly from being poor (NCSC 2000a), the Social Security system has been criticized for being based on number of years of paid work and pre-retirement earnings. Hence women and minorities, who often earn less during their employment years, receive less in retirement benefits. Another concern is whether funding for the Social Security system will be adequate to provide benefits for the increased numbers of aged in the next decades. Fear of its demise has led some economists and government officials to recommend privatization. Such a system would allow workers to put the money they now have deducted from their payroll for Social Security (FICA) into a personally owned and invested retirement account.

Health Issues

The biology of aging is called **senescence.** It follows a universal pattern but does not have universal consequences. "Massive research evidence demonstrates that the aging process is neither fixed nor immutable. Biologists are now showing that many symptoms that were formerly attributed to aging—for example, certain disturbances in cardiac function or in glucose metabolism in the brain—are instead produced by disease" (Riley & Riley 1992, 221). Biological functioning is also intricately related to social variables. Altering lifestyles, activities, and social contacts affects mortality and morbidity. For example, a longitudinal study of men and women between ages 70 and 79 found that regular physical activity, higher levels of ongoing positive social relationships, and a sense of self-efficacy enhanced physical and cognitive functioning (Seeman & Adler 1998).

"If exercise could be put in a pill it would be the number one anti-aging medicine."

Robert Butler
Gerontologist

Biological changes are consequences of either **primary aging** caused by physiological variables such as cellular and/or molecular variation (e.g., gray hair) or secondary aging. **Secondary aging** entails changes attributable to poor diet, lack of exercise, increased stress, and the like. Secondary aging exacerbates and accelerates primary aging.

Alzheimer's disease received national attention when former president Ronald Reagan announced that he had the disease. Named for German neurologist Alois Alzheimer, the debilitating disease affects both the mental and physical condition of some 4.5 million Americans—a projected 13 million by 2050 (NIH 2003).

Over a quarter (27%) of the elderly rate their health as fair or poor; 41.6 percent of blacks and 35.1 percent of Hispanics rate their health as fair or poor compared to 26 percent of older whites (AOA 2003b). Although elderly Americans are in better health than in previous generations, health often declines with age. Older people account for 36 percent of all hospital stays, averaging 6.4 days per year compared to 4.6 days for people under 65 (AOA 2003b). Health is a major quality-of-life issue for the elderly, especially because they face higher medical bills with reduced incomes. Older Americans spend three times as much on health care as their younger counterparts, over half of which is spent on insurance. The poor elderly, often women and/or minorities, spend an even higher proportion of their resources on health care.

Medicare was established in 1966 to provide medical coverage for those over the age of 65 and today insures approximately 40 million people (NCHS 2003). Although it is widely assumed that the medical bills of the elderly are paid by the government, the elderly are responsible for as much as 25 percent of their total health costs. Medicare, for example, pays about half the cost of a visit to the doctor, but does not pay for prescriptions, most long-term care, dental care, glasses, and hearing aids. The difference between Medicare benefits and the actual cost of medical care is called the **medigap.** In 2003, President Bush proposed a $400 billion legislative package designed to improve and modernize Medicare including adding coverage of many of the items listed above (AOA 2003e).

Because health is associated with income, the poorest old are often the most ill: they receive less preventive medicine, have less knowledge about health care issues, and have limited access to health care delivery systems. Medicaid is a federally and state-funded program for those who cannot afford to pay for medical care. However, eligibility requirements often disqualify many of the aged poor, often minorities and women.

Living Arrangements

The elderly live in a variety of contexts, depending on their health and financial status. Most elderly do not want to be institutionalized but prefer to remain in their own homes or in other private households with friends and relatives. Of the noninstitutionalized elderly population, 69 percent live in a family setting although that number decreases with age (AOA 2003b). Homes of the elderly, however, are usually older, located in inner-city neighborhoods, in need of repair, and often too large to be cared for easily. Six percent of the elderly, nearly one-and-a-half million households, live in homes that need serious repairs or modifications (NCSC 2000b).

Although many of the elderly poor live in government housing or apartments with subsidized monthly payments, the wealthier aged often live in retirement communities. These are often planned communities, located in states with warmer climates, and are often very expensive. These communities offer various amenities and activities, have special security, and are restricted by age. One criticism of these communities is that they segregate the elderly from the young and discriminate against younger people by prohibiting them from living in certain areas.

Those who cannot afford retirement communities or may not be eligible for subsidized housing often live with relatives in their own home or in the homes of others. It is estimated that more than 22 million people provide care for aging family members (Jackson 2003). Almost three-quarters are women, many of whom are daughters who care not only for their elderly parents but for their own children as well (see this chapter's *Social Problems Research Up Close* feature).

Other living arrangements include shared housing, modified independent living arrangements, and nursing homes. With shared housing, people of different ages live together in the same house or apartment; they have separate bedrooms but share a common kitchen and dining area. They share chores and financial responsibilities. In modified independent living arrangements, the elderly live in their own house, apartment, or condominium within a planned community where special services such as meals, transportation, and home repairs are provided. Skilled or semiskilled health care professionals are available on the premises, and the residences have call buttons so help can be summoned in case of an emergency.

Social Problems
Research Up Close

Children and Grandchildren as Primary Caregivers

The single most significant demographic change of the next several decades will be the dramatic increase in the number of elderly worldwide. The increase in the elderly is associated with a variety of concerns including who will care for the growing number of the old. The present study by Dellman-Jenkins, Blankemeyer, and Pinkard (2000) examines a recent trend in caregiving—young adult children and grandchildren as primary caregivers.

Sample and Methods

The present research focuses on three areas: (1) characteristics of young caregivers and the assistance they provide, (2) the consequences of being a primary caregiver for the individual, and (3) caregivers' social support. Specifically, the researchers "compared the caregiving motivations, experiences and support needs of young adult grandchildren providing assistance to grandparents with those of young adult children caring for their older parents" (p. 178). Participants were recruited from a variety of sources including social service agencies, hospitals, and assisted-living facilities. An open-ended interview was used in conjunction with a 65-item questionnaire. The final sample was composed of 43 caregivers—20 daughters, 2 sons, 19 granddaughters, and 2 grandsons (N = 43). Eighty-one percent of the sample respondents were white, 53 percent were married, and 54 percent lived with the care-recipient; the remainder living in a separate residence.

Findings and Conclusions

Caregiver Role and Assistance Given. Children, compared with grandchildren, were significantly more likely to respond that they were caring for their parents because no one else would. Grandchildren were most likely to respond that they were caring for their grandparent(s) out of a sense of duty and the desire to avoid nursing home care. Grandchildren also reported volunteering to care for grandparent(s) to help their parents.

> When Alzheimer's became apparent, my parents weren't coping well with it and were mean to her (not physically). I couldn't stand it . . . I told them I would take her to my house. My dad told me it would be the biggest mistake I ever made. Three years later . . . I still disagree.

The modal category for length of caregiving for both grandchildren and children was one to five years with approximately equal proportions of both providing a round-the-clock (47%) versus daily assistance (53%). Both sets of caregivers provided transportation, companionship and emotional support, personal care, and household support (e.g., cleaning house, paying bills, making appointments).

CONSEQUENCES OF ROLE

Caregivers in general, and single caregivers in particular, reported that caregiving activities interfered with their social life. All caregivers reported spending at least three hours a day in the caregiving role. One 20-year-old granddaughter commented, "I sometimes get mad because I think I'm doing too much for my grandpa and I don't have a life of my own . . . then I feel selfish for having such thoughts" (p. 184). Caregivers also reported strained family relations, particularly with spouse and children.

> I would like to go camping with my husband once in a while, but I can't just get up and go away, because of taking care of my grandparents. Even though they have the medical alert, I'm afraid they won't use it if something goes wrong.

Further, careers were negatively affected as relocating, job performance, and achievement of long-term goals became difficult because of the demands of caregiving. Increases in stress were also reported by both groups although significantly higher for grandchildren (81%) than children (59%). Positive outcomes were also expressed with over 97 percent of respondents identifying some benefit or reward. Grandchildren were most likely to note maintaining a strong relationship with the care-recipient, whereas children more often listed prevention of nursing home placement as the primary benefit.

SOURCES OF SUPPORT

All respondents reported seeking informal assistance from some source. Children most often turned to their siblings for help, whereas grandchildren most often turned to their spouse or dating partner. Formal support, although seldom sought, most often came from nursing/home health care providers or community support groups.

The two generations of caregivers, children and grandchildren, differ little in their caregiving activities, behaviors, and strains. They do, however, differ in their motivation to assist, grandchildren more often noting attachment rather than need or obligation as the primary reason for becoming a caregiver. Further, the authors conclude, grandchildren are "more likely to report personal rewards (e.g., greater closeness, positive memories), while children were more apt to report instrumental rewards (e.g., providing quality home-care and avoiding nursing home placement)" (p. 185).

Source: Dellman-Jenkins, Mary, Maureen Blankemeyer, and Odessa Pinkard. 2000. "Young Adult Children and Grandchildren in Primary Caregiver Roles to Older Relatives and Their Service Needs." *Family Relations* 49:177–187.

© Tom Miner/The Image Works

Little did members of this age cohort know that they were to become the middle-aged "sandwich generation," emotionally and economically responsible for both their children and their aging parents. The stress and pressures from these caregiver roles are often part of what's called the "mid-life crisis."

Nursing homes are residential facilities that provide full-time nursing care for residents. Nursing homes may be private or public. Private facilities are very expensive and are operated for profit by an individual or a corporation (see Figure 9.2). Public facilities are nonprofit and are operated by a government agency, religious organization, or the like. The probability of being in such an extended care facility is associated with race, age, and sex: whites, the old-old, and women are more likely to be in residence. The elderly with chronic health problems are also more likely to be admitted to nursing homes. Nursing homes vary dramatically in cost, services provided, and quality of care. A recent study found that of the 17,000 nursing homes in the United States, 90 percent were understaffed (Pear 2002).

The fastest growing facilities for the elderly are assisted-living quarters—36,399 in 2002, a 48 percent increase from 1998 (McCoy 2003). Assisted living facilities, although not as "full service" as nursing homes, offer private living units with the confidence of an around-the-clock staff. In 2003, the government proposed federal guidelines that require such facilities to reveal all costs, services, and policies and to have an adequately trained staff on duty at all times. Further, the proposal suggests the creation of a National Center for Excellence in Assisted Living and increases state and federal funding for the Long-Term Care Ombudsman Program (McCoy 2003).

Victimization and Abuse

Elder abuse refers to physical or psychological abuse, financial exploitation, or medical abuse or neglect of the elderly and is a global problem (Hooyman & Kiyak 1999; Green 2003). Even in Japan, where the elderly are traditionally revered, incidents of elder abuse are on the rise. Although it can take place in private homes by family members, the elderly, like children, are particularly vulnerable to abuse when they are institutionalized. In 1990 the Nursing Home Reform Act was passed, establishing various rights for nursing care residents: the right to be free of mental and physical abuse, the right not to be restrained unless necessary as a

Figure 9.2
Nursing Home and Home Care Costs

Source: The MetLife Mature Market Institute Market Survey of Nursing Home and Home Care Costs: 2003.

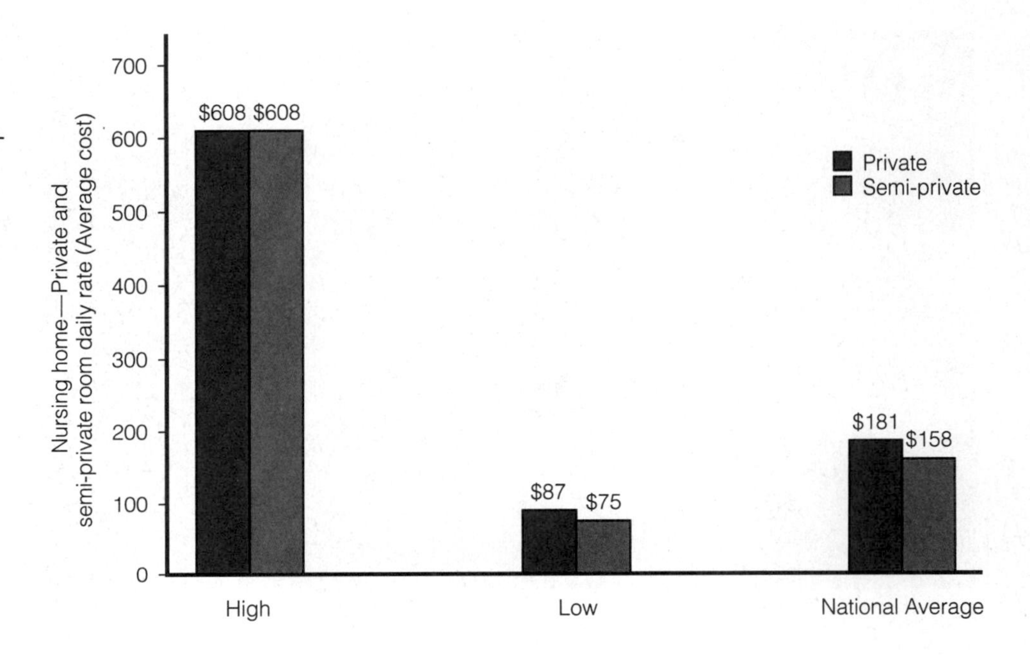

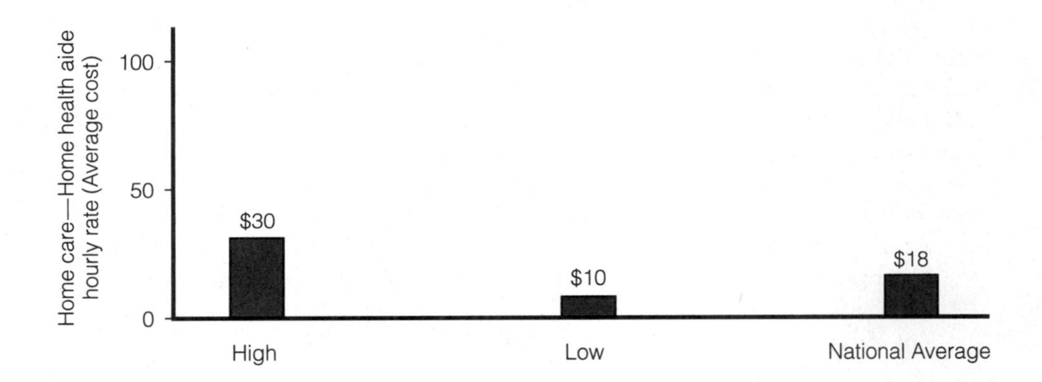

safety precaution, the right to choose one's physician, and the right to receive mail and telephone communication (Harris 1990, 362).

Whether the abuse occurs within the home or in an institution, the victim is most likely to be female, widowed, white, on a limited income, and in her mid-70s. The abuser tends to be an adult child or spouse of the victim who misuses alcohol. Some research suggests that the perpetrator of the abuse is more often an adult child who is financially dependent on the elderly victim (Mack & Jones 2003). Whether the abuser is an adult child or a spouse may simply depend on whom the elder victim lives with.

Many of the problems of the elderly are compounded by their lack of interaction with others, loneliness, and inactivity. This is particularly true for the old-old. The elderly are also segregated in nursing homes and retirement communities, separated from family and friends, and isolated from the flow of work and school. As with most problems of the elderly, the problems of isolation, loneliness, and inactivity are not randomly distributed. They are higher among the elderly poor, women, and minorities. A cycle is perpetuated—being poor and old results in being isolated and engaging in fewer activities. Such withdrawal affects health, which makes the individual less able to establish relationships or participate in activities.

Quality of Life

Although some elderly do suffer from declining mental and physical functioning, many others do not. Being old does not mean being depressed, poor, and sick. Less than 6 percent of those 65 and over suffer from any type of depression (NIMH 2003). Interestingly, Blumenthal and colleagues report that depression in the elderly is better treated by exercise than medication alone, or by medication and exercise combined (Livni 2000).

Among the elderly who are depressed, two social factors tend to be in operation. One is society's negative attitude toward the elderly. Words and phrases such as "old," "useless," and "a has-been" reflect cultural connotations of the aged that influence feelings of self-worth. The roles of the elderly also lose their clarity. How is a retiree supposed to feel or act? What does a retiree do? As a result, the elderly become dependent on external validation that may be weak or absent.

The second factor contributing to depression among the elderly is the process of "growing old." This process carries with it a barrage of stressful life events all converging in a relatively short time period. These include health concerns, retirement, economic instability, loss of significant other(s), physical isolation, job displacement, and increased awareness of the inevitability of death as a result of physiological decline. For example, 73.6 percent of those over the age of 80 report having at least one disability (AOA 2003b). All of these events converge on the elderly and increase the incidence of depression and anxiety, and may affect the decision not to prolong life (see the *Focus on Technology* feature in this chapter). The elderly commit suicide at a higher rate than any other age group.

Strategies for Action: Growing Up and Growing Old

Activism by or on behalf of children or the elderly has been increasing in recent years and, as their numbers grow, such activism is likely to escalate and to be increasingly successful. For example, global attention to the elderly led to 1999 being declared the "International Year of Older Persons," and "the first nearly universally ratified human rights treaty in history" deals with children's rights (UNICEF 1998). Such activism takes several forms, including collective action through established organizations and the exercise of political and economic power.

> "For age is opportunity,
> no less than youth itself;
> although in another dress.
> And as the evening twilight fades away
> The sky is filled with stars
> Invisible by day."
>
> **Henry Wadsworth Longfellow**
> **Poet**

Collective Action

Countless organizations are working on behalf of children, including the Children's Defense Fund, UNICEF, Children's Partnership, Children Now, and the Children's Action Network. Many successes take place at the local level where parents, teachers, corporate officials, politicians, and citizens join together in the interest of children. In Kentucky, AmeriCorp volunteers raised the reading competency of underachieving youths 116 percent in just six months; in Dayton, Ohio, behavioral problems of students at an elementary school were significantly reduced once "character education" was introduced; and in the Los Angeles Crenshaw High School, student entrepreneurs established "Food from the Hood," a garden project that is expected to earn $50,000 in profit this year.

> "No man or woman stands as tall as one who stoops to help a child."
>
> **Teddy Roosevelt**
> **Former U.S. President**

Some programs combine the interests of both the young and the old. The Adopt a Grandparent Program arranges for children to visit nursing home residents, and the Foster Grandparent Program pairs children with special needs with low-income

Focus on Technology

Physician-Assisted Suicide and Social Policy

Given the dramatic increase in the number of elderly and the technological ability to extend life, the debate over physician-assisted suicide (PAS) is likely to continue. When 2,000 doctors of terminally ill patients were surveyed, 6 percent said they had assisted in patient suicides and 33 percent said they would prescribe lethal amounts of drugs if permitted by law (Finsterbusch 2001). Further, in a study of U.S. physicians, of those who responded, 20 percent reported requests to hasten death and, of that number, 59 percent said they would honor a request to hasten death under some circumstances (Meier, Emmons, Litke, Wallenstein, and Morrison 2003).

Americans, in general, support PAS. In a national survey of U.S. adults, 65 percent agreed with the statement that "the law should allow doctors to comply with the wishes of a dying patient in severe distress who asks to have his or her life ended." Furthermore, 63 percent disagreed with the 1997 U.S. Supreme Court decision that individuals do not have a constitutional right to doctor-assisted suicide (Taylor 2002). This decision, however, did validate the concept of double effect. **Double effect** refers to the use of medical interventions to relieve pain and suffering but which may also hasten death.

One of the most important factors influencing a physician's decision to restrict technological interventions is "family preference" (Randolph, Zollo, Wigton, & Yeh 1997). The family's decision, however, is most often based on the physician's recommendations to limit care whether it be withdrawal of life support (food, water, or mechanical ventilation), administering medications to end life (intravenous vasopressors), or withholding certain procedures that would prolong life (cardiopulmonary resuscitation) (Luce 1997).

As of 2003, only Oregon recognizes the right of PAS with its Death with Dignity Act. Two physicians must agree that the patient is terminally ill and is expected to die within six months, the patient must ask three times for death both orally and in writing, and the patient must swallow the barbiturates him- or herself rather than be injected with a drug by the physician. In 2002, 38 physician-assisted suicides occurred in Oregon—71 percent were male, 97 percent were white, 53 percent were married, and 84 percent had cancer. The median age was 69 (Vollmar 2003). Loss of autonomy and a "diminished ability to participate in activities that make life enjoyable" were the two most commonly cited "end-of-life concerns" (Annual Report 2003, 20).

In 2001, John Ashcroft, the U.S. Attorney General, legally challenged Oregon's Death with Dignity Act (DWD) by instructing the Drug Enforcement Administration to "prosecute physicians and pharmacists who prescribe and dispense controlled substances under Oregon law" (Annual Report 2003). Oregon then brought suit against the Attorney General resulting in a finding favorable to the state: the federal government could

elderly. Further, the Children's Defense Fund, the Child Welfare League of America, the AARP, and the National Council on Aging joined together in 1986 to create Generations United, a "national organization that focuses solely on promoting inter-generational strategies, programs and policies" (*Generations United* 2003, 1). It is the only national organization acting as an advocate on behalf of the young and the old.

More than a thousand organizations are directed toward realizing political power, economic security, and better living conditions for the elderly. One of the earliest and most radical groups is the Gray Panthers, founded in 1970 by Margaret Kuhn. The Gray Panthers were responsible for revealing the unscrupulous practices of the hearing aid industry, persuading the National Association of Broadcasters to add "age" to "sex" and "race" in the Television Code of Ethics statement on media images, and eliminating the mandatory retirement age. In view of these successes, it is interesting that the Gray Panthers, with only 40,000 members, is a relatively small organization when compared with the AARP.

The AARP has more than 35 million members age 50 and above. It has an annual revenue of $636 million and a staff of over 2,000 (Welch 2003). Services of the

Focus on Technology

not overturn Oregon's DWD law. The case has been appealed by the U.S. Department of Justice (Vollmar 2003).

The national debate over PAS was fueled by images of Dr. Jack Kevorkian administering a deadly dose of drugs, at the request of a terminally ill patient, on the CBS evening show *60 Minutes.* Although Kevorkian was convicted of second-degree murder for the *60 Minutes* death, advocates of PAS have tried to get Oregon-like provisions passed in other states. As of 2003, Arizona, Hawaii, and North Carolina all have pending PAS legislation.

Despite what some say is a movement toward a greater acceptance of PAS, the official position of the American Medical Association is that physicians must respect the patient's decision to forgo life-sustaining treatment but should not participate in physician-assisted suicide. One argument against PAS is that the practice is subject to abuses—a spouse with self-serving interests, a depressed patient making a hasty decision, or an overburdened family pressuring a vulnerable loved one. Concern also exists that legalizing PAS may disproportionately end the lives of minority, ethnic, or psychiatrically disturbed individuals (Allen 1998).

Some would argue, however, that ultimately the decision should reside with the patient. As one elderly person said:

> I came into this world as a human-being and I wish to leave in the same manner. Being able to walk, to communicate, to take care of my own needs, to think, to feel. . . . There is no need for me to reexperience my first few months of life though my last months in this world. . . . I do not wish to be once again in a diaper. Just the thought of me losing control over my body frightens me (Leichtenritt & Rettig 2000, 3).

Sources: C. L. Allen. 1998. "Euthanasia: Why Torture Dying People When We Have Sick Animals Put Down?" *Australian Psychologist* 33:12–15; Annual Report. 2003. "Oregon's Death with Dignity Act: The Second Year Experience." Oregon's Death with Dignity Act Annual Report, 2002. Center for Heath Statistics. Oregon Health Division; Kurt Finsterbusch, 2001. *Clashing Views on Controversial Social Issues.* Guilford, CN: Dushkin Publishing; R. D. Leichtentritt and K. D. Rettig. 2000. "Conflicting Value Considerations for End-of-Life Decisions." Poster session at 62nd Annual Meeting of National Council on Family Relations, Minneapolis, Minnesota, November 12; J. M. Luce. 1997. "Withholding and Withdrawal of Life Support: Ethical, Legal, and Clinical Aspects." *New Horizons* 5:30–37; Diane Meier, Carol-Ann Emmons, Ann Litke, Sylvan Wallenstein, and Sean Morrison. 2003. "Characteristics of Patients Requesting and Receiving Physician Assisted Suicide." *Archives of Internal Medicine* 163:1537–1542; A. G. Randolph, M. B. Zollo, R. S. Wigton, and T. S. Yeh. 1997. "Factors Explaining Variability Among Caregivers in the Intent to Restrict Life-Support Interventions in a Pediatric Intensive Care Unit." *Critical Care Medicine* 25:435–439; Humphrey Taylor. 2002. "2-to-1 Majorities Continue to Support Rights to Both Enthanasia and Doctor Assisted Suicide." Harris Poll #2. January 9. http://www.harrisinteractive. com; Valeria Vollmar. 2003. "Recent Developments in Physician-Assisted Suicide." http://www. willamette.edu/ wucl/pas

AARP include discounted mail-order drugs, investment opportunities, travel information, volunteer opportunities, and health insurance. The AARP is the largest volunteer organization in the United States with the exception of the Roman Catholic Church. Not surprisingly, it is one of the most powerful lobbying groups in Washington.

Political Power

Children are unable to hold office, to vote, or to lobby political leaders. Child advocates, however, acting on behalf of children, have wielded considerable political influence in such areas as child care, education, health care reform, and crime prevention. Further, funding of such programs is supported by most Americans. A recent Children's Defense Fund study found that the majority of those surveyed favored federally funded after-school programs and subsidized child care even if it meant raising taxes (CDF 2000).

As conflict theorists emphasize, the elderly compete with the young for limited resources. They have more political power than the young and more political power

Protests are not limited to the young. The elderly have already been active in protesting changes to Medicare and other government policies that they believe are detrimental. As the number of elderly grows, social activism is likely to increase.

in some states than in others. In Florida there is concern that the elderly may eventually wield too much political power and act as a voting bloc, demanding excessive services at the cost of other needy groups. For example, if the elderly were concentrated in a particular district, they could block tax increases for local schools. To the extent that future political issues are age-based and the elderly are able to band together, their political power may increase as their numbers grow over time. By 2030, almost half of all adults in developed countries and two-thirds of all voters will be near or at retirement age. The growing political power of the aged is already becoming evident; the Netherlands has a political party called the Pension Party.

Economic Power

Although children have little economic power, the economic power of the elderly has grown considerably in recent years, leading one economist to refer to the elderly as a "revolutionary class" (Thurow 1996). The 2001 median income for males 65 and over was $19,688; for females 65 and over, $11,3133. Households headed by persons 65 and over had a median income of $33,938 (AOA 2003d). Although these incomes are significantly lower than for men and women between 45 and 54 years of age, income is only one source of economic power. The fact that many elderly own their homes, have substantial savings and investments, enjoy high levels of disposable income, and are growing in number contribute to their increased importance as consumers.

Finally, in many parts of the country the elderly have become a major economic power as part of what is called the "**mailbox economy.**" The mailbox economy refers to the tendency for a substantial portion of local economies to be dependent on pension and Social Security checks received in the mail by older residents (Atchley 2000). States with the highest proportion of the elderly—as a percentage of each state's population—include Florida (17.6%), Pennsylvania (15.6%), West Virginia (15.3%), Iowa (14.9%), and North Dakota (14.7%) (AOA 2003b).

Fighting Discrimination

The number of age-based complaints to the U.S. Equal Employment Opportunity Commission (EEOC) has risen as the population ages and the economy worsens. In 1999, there were 14,141 complaints; in 2002, 19,921—a 41 percent increase. Of those filing complaints, 64 percent were between the ages of 40 and 59 (Nicholson 2003).

Modeled after Title VII (see Chapter 11), the Age Discrimination in Employment Act was passed in 1967 and protects workers between the ages of 40 and 65. In 1978, the upper limit was extended to 70. In 1996, the U.S. Supreme Court heard *O'Connor v. Consolidated Coin Corp.*, which held that workers do not have to prove that they were replaced by someone under 40 to have a finding of age discrimination. One of the most significant cases was heard in 2003—*EEOC and Arnett et al. v. CalPERS* (California Public Employees' Retirement System). In this case the EEOC recovered $250 million dollars for California public safety officers who had been discriminated against on the basis of age. Age discrimination suits have the lowest success rate of any of the eight protected classes including sex, race, and religious discrimination (AARP 2003).

Government Policy

When registered voters were asked what they thought was the single most important issue facing the government, three of the four top answers dealt with issues confronting the elderly—health care (discussed in Chapter 2), Social Security, and Medicare/Medicaid (CBS/New York Times Poll 2000). As mentioned earlier, Medicaid provides health care for the poor; Medicare is a national health care insurance program designed for people over the age of 65. In 1988, Congress passed the Medicare Catastrophic Coverage Act (MCCA), which was the most significant change in Medicare since its establishment in 1966. The new benefits included unlimited hospitalization, an upper limit on the amount of money recipients would pay for physicians' services, home health care and nursing home services, and unlimited hospice care. These changes were particularly significant because many of the illnesses of the elderly are chronic in nature.

The reforms were financed by increasing monthly medical premiums $4 a month and imposing an annual fee based on a person's federal income tax bracket. The maximum premium paid was $800 per person and $1,600 per couple (Harris 1990; Torres-Gil 1990). The AARP initially supported the reforms but later withdrew its support as did many other organizations. The additional monies paid were simply not worth the new benefits, they contended. The AARP also argued that the elderly should not have to bear the burden of reforms necessary for the general public. In 1989, under pressure from the AARP and other organizations of the elderly, Congress repealed the MCCA. Pressured by public demands and fear of bankruptcy, Congress is currently in the process of reforming health care policy (see Chapter 2) including Medicare. For example, Congress is considering several drug prescription benefit packages for Medicare recipients (Welch 2003). Congress is also considering the Elder Justice Act, which, if passed, will oversee services and prevention related to elder abuse including monitoring long-term care facilities (National Center for Victims of Crime 2003).

Other government initiatives for the elderly include the Elderly Nutrition Program and the Older Americans Act Amendments of 2000 (AOA 2001). The Elderly Nutrition Program provides delivered meals to locations where the elderly congregate—homes, senior centers, and schools, and provides nutritional training, counseling, and education. The Older Americans Act Amendments 2000 created the National Family Caregiver Support Program. This program will "help hundreds of thousands of family caregivers of older loved ones who are ill or who have disabilities" (AOA 2003f). The National Family Caregiver Support Program has been funded at $125 million a year through fiscal year 2005. Among other things, the program will provide respite care, supplemental services, and individual counseling and training for caregivers (AOA 2003f).

In 1997 Congress passed the State Children's Health Care Insurance Program (SCHIP), which "is the largest single expansion of health insurance coverage for children in more than 30 years" (HHS 1999, 1). The SCHIP was designed for families who cannot afford health insurance but whose incomes are too high to qualify for Medicaid. The program, at a cost of $24 billion, will provide health care to some of the 10 million children who do not have health care coverage (see Chapter 2). It should be noted, however, that revenues lost due to President Bush's tax cut initiative are more than enough funds to provide full health coverage and Head Start's comprehensive preschool services for all the children in America who need it—9.2 million children without health insurance and 1.8 million who need Head Start but do not receive it (CDF 2003e).

Finally, in 2003, President Bush signed the Trafficking Victims Protection Reauthorization Act, which provides over $200 million in federal funds to "combat the practice of human trafficking—including women and children forced into prostitution" (Thompson 2003, 1). It is estimated that, annually, 18,000–20,000 victims of human trafficking are brought into the United States.

Understanding Youth and Aging

What can we conclude about youth and aging in American society? Age is an ascribed status and, as such, is culturally defined by role expectations and implied personality traits. Society regards both the young and the old as dependent and in need of the care and protection of others. Society also defines the young and the old as physically, emotionally, and intellectually inferior. As a consequence of these and other attributions, both age groups are sociologically a minority with limited opportunity to obtain some or all of society's resources.

Although both the young and the old are treated as minority groups, different meanings are assigned to each group. In the United States, in general, the young are more highly valued than the old. Functionalists argue that this priority on youth reflects the fact that the young are preparing to take over important statuses while the elderly are relinquishing them. Conflict theorists emphasize that in a capitalist society, both the young and the old are less valued than more productive members of society. Conflict theorists also point out the importance of propagation, that is, the reproduction of workers, which may account for the greater value placed on the young than the old. Finally, symbolic interactionists describe the way images of the young and the old intersect and are socially constructed.

The collective concerns for the elderly and the significance of defining ageism as a social problem have resulted in improved economic conditions for the elderly. Currently, they are one of society's more powerful minorities. Research indicates, however, that despite their increased economic status, the elderly are still subject to discrimination in such areas as housing, employment, and medical care and are victimized by systematic patterns of stereotyping, abuse, and prejudice.

In contrast, the position of children in the United States, although improving, remains tragic with one in five children living in poverty. Wherever there are poor families, there are poor children who are educated in inner-city schools, live in dangerous environments, and lack basic nutrition and medical care. Further, age-based restrictions limit the entry of these children into certain roles (e.g., employee) and demand others (e.g., student). Although most of society's members would agree that children require special protections, concerns regarding quality-of-life issues and rights of self-determination are only recently being debated.

Age-based decisions are potentially harmful. If budget allocations were based on indigence rather than age, more resources would be available for those truly in need. Further, age-based decisions may encourage intergenerational conflict. Government assistance is a zero-sum relationship—the more resources one group gets, the fewer resources another group receives.

Social policies that allocate resources on the basis of need rather than age would shift the attention of policy makers to remedying social problems rather than serving the needs of special interest groups. Age should not be used to cause negative effects on an individual's life any more than race, ethnicity, gender, or sexual orientation. Although eliminating all age barriers or requirements is unrealistic, a movement toward assessing the needs of individuals and their abilities would be more consistent with the American ideal of equal opportunity for all.

Chapter Review

- **What problems do the young and old have in common?**
  Among others, both the young and the old are victims of stereotypes, physical abuse, age discrimination, and poverty.
- **What age distinctions are commonly made in most societies?**
  Most societies make a distinction between childhood, adulthood, and elderhood. Childhood, often subdivided into infancy, childhood, and adolescence, is usually thought of as from birth to 17 years old, adulthood from 18 to 64 years old, and elderhood 65 years of age and over.
- **What is disengagement theory?**
  According to disengagement theory, in order to achieve a balanced society, the elderly must relinquish their roles to younger members. Thus, the young go to school, adults fill occupational roles, and the elderly, with obsolete skills and knowledge, disengage.
- **What are some of the problems of youth?**
  Problems of the young include (1) child labor, (2) orphaned children, (3) street children, (4) limited civil rights, (5) poverty, (6) economic discrimination, (7) violence by and against children (e.g., gang violence, suicide, child abuse, and crime), and (8) health concerns.
- **What three independent variables impact the consequences of aging?**
  Race, gender, and social class impact the consequences of aging. Racial minorities, women, and the lower classes are more likely to suffer adversely from the aging process.
- **What are some of the problems of the elderly?**
  For the elderly who want to work, entering and remaining in the labor force may be difficult because of negative stereotypes, lower levels of education, reduced geographic mobility, fewer employable skills, and discrimination. Retirement may also be difficult due to role transition and lowered income. Poverty is a problem as well with 38 percent of the elderly reporting incomes of $10,000 or less. Some elderly also suffer from chronic health problems, including depression. Finally, elder abuse, although also in private homes, is particularly problematic in institutionalized settings.
- **In terms of strategies for action, what are some of the similarities between the young and the old?**
  Both the young and the old have organizations working on their behalf—for example, the Children's Defense Fund and UNICEF, and the AARP and the Gray Panthers. Both groups have advocates working for their political and economic interests although the elderly, according to conflict theorists, already wield considerable political and economic power. Finally, both the young and the old have benefited from government policies such as the Older Americans Act, Medicare, and the State Children's Health Care Insurance Program (SCHIP).

Critical Thinking

1. In many ways, American society discriminates against children. Children are segregated in schools, in a separate justice system, and in the workplace. Identify everyday examples of the ways in which children are treated like "second- class" citizens in the United States.
2. **Age pyramids** pictorially display the distribution of people by age (see Figure 9.1). How do different age pyramids influence the treatment of the elderly?
3. Regarding children and the elderly, what public policies or programs from other countries might be beneficial if they were adopted in the United States? Do you think policies from other countries would necessarily be successful here?

Key Terms

activity theory
age grading
age pyramids
ageism
dependency ratio
disengagement theory
double effect
elder abuse
gerontophobia
infantilizing elders
mailbox economy
medigap
modernization theory
primary aging
sandwich generation
secondary aging
senescence

Taking a Stand

Should Social Security be privatized?

Social Security is one of the many government programs known as entitlements. Funded by a payroll tax, Social Security in fiscal 2003 cost the government 400 billion dollars. The growing number of baby boomers now nearing retirement means that in the coming decades the Social Security system may become bankrupt. Currently, various plans have been proposed to fix the system. One of the most radical plans would in effect privatize Social Security.

Use Wadsworth's exclusive online resources—InfoTrac College Edition, MicroCase Online, and OVRC—to formulate a position on this topic.

The Wadsworth's Sociology Online Resources and Writing Companion will help you get started. This valuable guide will show you how to use Wadsworth's exclusive online resources when studying social problems. It will also help you to build essential research and writing skills. InfoTrac College Edition, MicroCase Online, OVRC, and an electronic copy of portions of this companion are available at http://sociology.wadsworth.com/mooney_knox_schacht/problems4e, the companion Web site for *Understanding Social Problems,* Fourth Edition.

Media Resources

The Companion Web Site for *Understanding Social Problems,* Fourth Edition

http://sociology.wadsworth.com/mooney_knox_schacht/problems4e

Supplement your review of this chapter by going to the companion Web site to take one of the Tutorial Quizzes, use the flash cards to master key terms, and check out the many other study aids you'll find there. You'll also find special features such as *Wadsworth's Sociology Online Resources and Writing Companion,* GSS Data, and Census 2000 information, data, and resources at your fingertips to help you complete that special project or do some research on your own.

Interactions CD-ROM

Go to the Interactions CD-ROM for *Understanding Social Problems,* Fourth Edition, to access additional interactive learning tools, such as in-depth review materials, corresponding practice quizzes, and other engaging resources and activities to help you study the concepts in this chapter.

"If all of the afflictions of the world were assembled on one side of the scale and poverty on the other, poverty would outweigh them all." *Rabba, Mishpatim 31:14*

Poverty

The Global Context: Poverty and Economic Inequality Around the World

Sociological Theories of Poverty and Economic Inequality

Economic Inequality and Poverty in the United States

Consequences of Poverty and Economic Inequality

Strategies for Action: Antipoverty Programs, Policies, and Proposals

Understanding Poverty

Chapter Review

After the terrorist attacks on September 11, 2001, the following e-mail circulated:

On September 11 2001, 36,000 children worldwide died of hunger.

Where: Poor countries.

News Stories: none.

Newspaper articles: none.

Military alerts: none.

Presidential proclamations: none.

Papal messages: none.

Messages of solidarity: none.

Minutes of silence: none.

Homage to the innocent children: none.

As tragic as poverty is for millions of people throughout the world, poverty rarely makes headline news. This chapter examines the extent of poverty globally and in the United States, focusing on the consequences of poverty for individuals, families, and societies. Theories of poverty and economic inequality are presented and strategies for rectifying economic inequality and poverty are considered.

The Global Context: Poverty and Economic Inequality Around the World

Who are the poor? Are rates of world poverty increasing, decreasing, or remaining stable? The answers depend on how we define and measure poverty.

Defining and Measuring Poverty

Poverty has traditionally been defined as the lack of resources necessary for material well-being—most importantly food and water, but also housing, land, and health care. This lack of resources that leads to hunger and physical deprivation is known as **absolute poverty.** In contrast, **relative poverty** refers to a deficiency in material and economic resources compared with some other population. Although many lower-income Americans, for example, have resources and a level of material well-being that millions of people living in absolute poverty can only dream of, they are relatively poor compared with the American middle and upper classes.

Various measures of poverty are used by governments, researchers, and organizations. Next, we describe international and U.S. measures of poverty.

International Measures of Poverty. The World Bank sets a "poverty threshold" of $1 per day to compare poverty in most of the developing world, $2 per day in Latin America, $4 per day in Eastern Europe and the Commonwealth of Independent States (CIS), and $14.40 per day in industrial countries (which corresponds to the income poverty line in the United States). Another poverty measure is based on whether individuals are experiencing hunger, which is defined as consuming less than 1,960 calories a day.

Table 10.1
Measures of Human Poverty in Developing and Industrialized Countries

	Longevity	Knowledge	Decent Standard of Living
For developing countries	Probability at birth of not surviving to age 40	Adult illiteracy	A composite measure based on 1. Percentage of people without access to safe water 2. Percentage of people without access to health services 3. Percentage of children under five who are underweight
For industrialized countries	Probability at birth of not surviving to age 60	Adult functional illiteracy rate	Percentage of people living below the income poverty line, which is set at 50% of median disposable income

Source: Adapted from the United Nations Development Programme 2000 *Human Development Report 2000.* New York: Oxford University Press.

In industrial countries, national poverty lines are sometimes based on the median household income of a country's population. According to this relative poverty measure, members of a household are considered poor if their household income is less than 50 percent of the median household income in that country.

Recent poverty research concludes that poverty is multidimensional and includes such dimensions as food insecurity; poor housing; unemployment; psychological distress; powerlessness; hopelessness; lack of access to health care, education, and transportation; and vulnerability (Narayan 2000). To capture the multidimensional nature of poverty, the United Nations Development Programme (1997) developed a composite measure of poverty: the **Human Poverty Index (HPI).** Rather than measure poverty by income, three measures of deprivation are combined to yield the Human Poverty Index: (1) deprivation of a long, healthy life, (2) deprivation of knowledge, and (3) deprivation in decent living standards. As shown in Table 10.1, the Human Poverty Index for developing countries **(HPI-1)** is measured differently than for industrialized countries **(HPI-2).** Among the 17 industrialized countries for which the HPI-2 was calculated, Sweden has the lowest level of human poverty (6.5%), followed by Norway (7.2%), and the Netherlands (8.4%) (*Human Development Report* 2003). The highest rate of human poverty is in the United States (15.8%), followed by Ireland (15.3%), and the United Kingdom (14.8%). The Human Poverty Index is a useful complement to income measures of poverty and "will serve as a strong reminder that eradicating poverty will always require more than increasing the income of the poorest" (UNDP 1997, 19).

"Human poverty is more than income poverty— it is the denial of choices and opportunities for living a tolerable life."

Human Development Report 1997

U.S. Measures of Poverty. In 1964 the Social Security Administration devised a poverty index based on a 1955 Agriculture Department survey that estimated the cost of an economy food plan for a family of four. Because families with three or more members spent one-third of their income on food at the time, the poverty line was set at three times the minimum cost of an adequate diet. Poverty thresholds differ by the number of adults and children in a family and by the age of the family head of household (see Table 10.2). Poverty thresholds are adjusted each year for inflation. Anyone living in a household with pre-tax income below the official poverty line is considered "poor." Individuals living in households that are above the

Table 10.2
Poverty Thresholds: 2003 (householder under 65 years old)

One adult	$9,573
Two adults	$12,321
One adult, one child	$12,682
Two adults, one child	$14,810
Two adults, two children	$18,660

Source: U.S. Census Bureau. 2004 (January 30). "Poverty 2003." http://www.census.gov

poverty line, but not very much above it, are classified as "near poor," and those living below 50 percent of the poverty line live in "deep poverty." A common working definition of "low-income" households are those with incomes that are between 100 and 200 percent of the federal poverty line, or up to twice the poverty level.

The U.S. poverty line has been criticized on several grounds. First, the poverty line is based on the assumption that low-income families spend one-third of their household income on food. That was true in 1955, but because other living costs (e.g., housing, medical care, and child care) have risen more rapidly than food costs, low-income families today spend less than one-fifth (rather than one-third) of their income on food (Schiller 2004). So current poverty lines should be based on multiplying food costs by five rather than three. This would raise the official poverty line by two-thirds, making the poverty level consistent with public opinion regarding what a family needs to escape poverty.

Another shortcoming of the official poverty line is that it is based solely on money income and does not take into consideration noncash benefits received by many low-income persons, such as food stamps, Medicaid, and public housing. Family assets, such as savings and property, are also excluded in official poverty calculations. In addition, the poverty index fails to account for tax burdens that affect the amount of disposable income available to meet basic needs. The U.S. poverty line also disregards regional differences in the cost of living, and because poverty rates are based on surveys of households, the homeless—the most destitute of the poor—are not counted among the poor.

The Extent of Global Poverty and Economic Inequality

More than 1.2 billion people—one in five people on this planet—survive on less than $1 a day. In sub-Saharan Africa, half the population lives on less than $1 a day (*Human Development Report* 2003). Nearly half of the world's population lives on less than $2 per day (World Health Organization 2002). South Asia has the greatest number of people affected by poverty, and sub-Saharan Africa has the highest proportion of people in poverty. Every day, nearly one in five (18%) of the world's population goes hungry. In South Asia, one in four goes hungry, and in sub-Saharan Africa, as many as one in three goes hungry (*Human Development Report* 2003).

Global economic inequality has reached unprecedented levels. In 2000, the average income of the richest 20 countries was 37 times that of the poorest 20 countries—a gap that doubled in the past 40 years (World Bank 2001). Although inequality between nations accounts for most of the inequality in global distribution of income, within-nation income differences are growing (Goesling 2001).

Sociological Theories of Poverty and Economic Inequality

The three main theoretical perspectives in sociology—structural-functionalism, conflict theory, and symbolic interactionism—offer insights into the nature, causes, and consequences of poverty and economic inequality. Before reading further, you may want to take the "Attitudes Toward Economic Opportunity in the United States" survey in this chapter's *Self and Society* feature.

Structural-Functionalist Perspective

According to the structural-functionalist perspective, poverty and economic inequality serve a number of positive functions for society. Decades ago, Davis and Moore (1945) argued that because the various occupational roles in society require different levels of ability, expertise, and knowledge, an unequal economic reward system helps to assure that the person who performs a particular role is the most qualified. As people acquire certain levels of expertise (e.g., B.A., M.A., Ph.D., M.D.), they are progressively rewarded. Such a system, argued Davis and Moore, motivates people to achieve by offering higher rewards for higher achievements. If physicians were not offered high salaries, for example, who would want to endure the arduous years of medical training and long, stressful hours at a hospital?

The structural-functionalist view of poverty suggests that a certain amount of poverty has positive functions for society. Although poor people are often viewed as a burden to society, having a pool of low-paid, impoverished workers ensures that someone will be willing to do dirty, dangerous, and difficult work that others refuse to do. Poverty also provides employment for those who work in the "poverty industry" (e.g., welfare workers) and supplies a market for inferior goods such as older, dilapidated homes and automobiles (Gans 1972).

The structural-functionalist view of poverty and economic inequality has received a great deal of criticism from contemporary sociologists, who point out that those in many important occupational roles such as child care workers are poorly paid (the average salary of a child care worker is less than $17,000 per year) (Bureau of Labor Statistics 2002) whereas many individuals in nonessential roles (e.g., professional sports stars and entertainers) earn astronomical sums of money. Functionalism also ignores the role of inheritance in the distribution of wealth.

Conflict Perspective

Conflict theorists regard economic inequality as resulting from the domination of the **bourgeoisie** (owners of the means of production) over the **proletariat** (workers). The bourgeoisie accumulate wealth as they profit from the labor of the proletariat, who earn wages far below the earnings of the bourgeoisie. The proletariat, dependent on the capitalist system, continue to be exploited by the wealthy and accept the belief that poverty is a consequence of personal failure rather than a flawed economic structure.

Conflict theorists note how laws and policies benefit the wealthy and contribute to the gap between the haves and the have-nots. Wealthy corporations use financial political contributions to influence politicians to enact policies that

Self and Society

Attitudes Toward Economic Opportunity in the United States

In a 1998 Gallup poll of 5,001 U.S. adults, respondents were asked questions to assess their attitudes toward economic opportunity in the United States. After responding to the questions below, compare your answers with the results from a national sample.

1. Using a 1 to 5 scale, where "1" means not at all important, and "5" means extremely important, indicate how important each of the following is as a reason for a person's success.

Item	Ranking (1 = not at all important; 5 = extremely important) 1 2 3 4 5
a. Hard work and initiative	______
b. Member of a particular race/ethnic group	______
c. Getting right education and training	______
d. Dishonesty and willingness to take whatever one can get	______
e. Parents and family	______
f. Willingness to take risks	______
g. Gender (whether one is male or female)	______
h. Connections/knowing the right people	______
i. Money inherited from family	______
j. Ability or talent one is born with	______
k. Good luck/in right place at right time	______
l. Physical appearance/good looks	______

2. For the following two questions, indicate your answer from the choices provided:
   a. Why are some people poor?
      Answer choices: ______ lack of effort
      ______ circumstances beyond their control
      ______ both ______ don't know
   b. Why are some people rich?
      Answer choices: ______ strong effort
      ______ circumstances beyond their control
      ______ both ______ don't know

benefit the wealthy. Laws and policies that favor the rich—sometimes referred to as **wealthfare** or **corporate welfare**—include low-interest government loans to failing businesses, special subsidies and tax breaks to corporations, and other laws and policies that benefit corporations and the wealthy. In 2003, corporate tax breaks cost American taxpayers more than $170 billion (Citizens for Tax Justice 2002). A study of 250 large companies found that 41 companies paid no federal income tax

3. Complete the following sentence with one of the two choices provided:
   a. The economic system in the United States:
      _______ is basically fair, since all Americans have an equal opportunity to succeed.
      _______ is basically unfair, since all Americans do not have an equal opportunity to succeed.

How Do Your Answers Compare with a National Sample of U.S. Adults?

1. This figure reveals the percentages of U.S. adults who rated the items in question #1 as important for success.

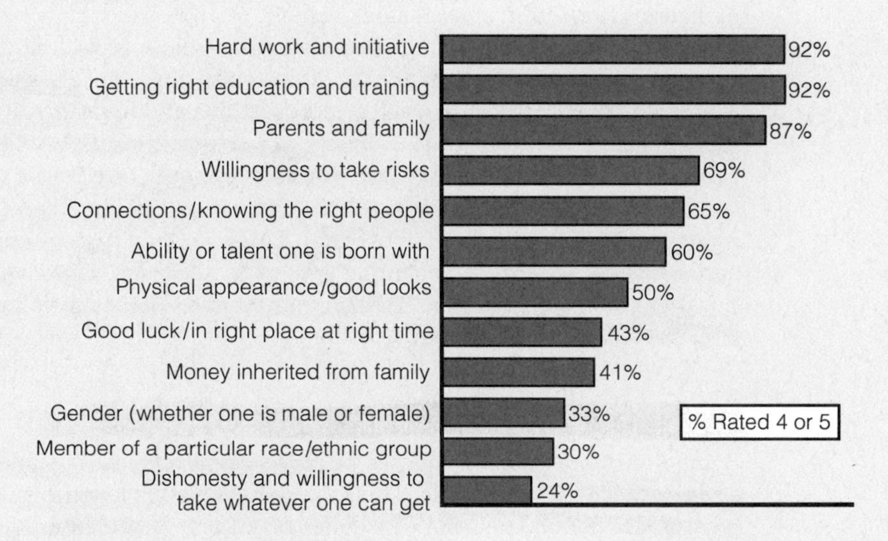

2a. In explaining why some people are poor, 43 percent of respondents indicated "lack of effort," 41 percent indicated "circumstances beyond their control," and 16 percent indicated "both" or "don't know."

2b. In explaining why some people are rich, 53 percent of respondents indicated "strong effort," 12 percent indicated "circumstances beyond their control," and 15 percent indicated "both" or "don't know."

3. Sixty-eight percent of respondents indicated that they believe the nation's economic system is basically fair; 29 percent believe it is basically unfair; and 3 percent had no opinion.

Source: Adapted from Gallup News Service Social Audit. 1998. (April 23 to May 31). http://www.gallup.com/poll/socialaudits/have_havenot.asp (Used by permission).

in at least one year from 1996 to 1998 (McIntyre & Nguyen 2000). In those tax-free years, the 41 companies reported $25.8 billion in profits. But instead of paying $9 billion in federal income tax at the 35 percent rate, these companies received $3.2 billion in rebate checks from the U.S. Treasury. The study found that 71 of the 250 companies paid taxes at less than half the official 35 percent corporate rate during 1996 to 1998. Corporate income taxes in the United States have fallen so much in the last few decades that they are nearly the lowest among the world's developed countries. U.S. corporate taxes were 4.1 percent of gross domestic in 1965. This figure fell to 1.5 percent of GDP in 2002 (McIntyre 2003).

Corporate welfare is provided by government, but it is taxpayers and communities who pay the price. In 2003, corporate tax breaks cost American taxpayers more than $170 billion (Citizens for Tax Justice 2002). Consider the case of Seaboard Corporation, an agribusiness corporate giant that received at least $150 million in economic incentives from federal, state, and local governments between 1990 and 1997 to build and staff poultry- and hog-processing plants in the United States, support its operations in foreign countries, and sell its products (Barlett & Steele 1998). Taxpayers picked up the tab not just for the corporate welfare but also for the costs of new classrooms and teachers (for schooling the children of Seaboard's employees, many of whom are immigrants), homelessness (because Seaboard's low-paid employees are unable to afford housing), and dwindling property values resulting from smells of hog waste and rotting hog carcasses in areas surrounding Seaboard's hog plants. Meanwhile, wealthy investors in Seaboard have earned millions in increased stock values.

Conflict theorists also note that throughout the world "free-market" economic reform policies have been hailed as a solution to poverty. Yet, while such economic reform has benefited many wealthy corporations and investors, it has also resulted in increasing levels of global poverty. As companies relocate to countries with abundant supplies of cheap labor, wages decline. Lower wages lead to decreased consumer spending, which leads to more industries closing plants, going bankrupt, and/or laying off workers (downsizing). This results in higher unemployment rates and a surplus of workers, enabling employers to lower wages even more. Chossudovsky (1998) suggests that "this new international economic order feeds on human poverty and cheap labor" (p. 299).

Symbolic Interactionist Perspective

Symbolic interactionism focuses on how meanings, labels, and definitions affect and are affected by social life. This view calls attention to ways in which wealth and poverty are defined and the consequences of being labeled as "poor." Individuals who are viewed as poor—especially those receiving public assistance (i.e., welfare)—are often stigmatized as lazy, irresponsible, and lacking in abilities, motivation, and moral values. Wealthy individuals, on the other hand, tend to be viewed as capable, motivated, hard working, and deserving of their wealth.

The symbolic interaction perspective also focuses on the meanings of being poor. A qualitative study of over 40,000 poor women and men in 50 countries around the world explored the meanings of poverty from the perspective of those who live in poverty (Narayan 2000). Among the study's findings is that the experience of poverty involves psychological dimensions such as powerlessness, voicelessness, dependency, shame, and humiliation.

Meanings and definitions of wealth and poverty vary across societies and across time. For example, the Dinka are the largest ethnic group in the sub-Saharan African country of Sudan. By global standards, the Dinka are among the poorest of the poor, being among the least modernized peoples of the world. In the Dinka culture, wealth is measured in large part by how many cattle a person owns. But to the Dinka, cattle have a social, moral, and spiritual value as well as an economic value. In Dinka culture, a man pays an average "bridewealth" of 50 cows to the family of his bride. Thus, men use cattle to obtain a wife to beget children, especially sons, to ensure continuity of their ancestral lineage, and according to Dinka religious beliefs, their linkage with God. Although modernized populations might label the Dinka as poor, the Dinka view themselves as wealthy. As one Dinka elder explained, "It is for cattle that we are admired, we, the Dinka. . . . All over the world, people look to us because of cattle . . . because of our great wealth; and our wealth is cattle" (Deng 1998, 107). Deng (1998) notes that many African peoples who are poor by U.S. standards resist being labeled as poor.

Definitions of poverty also vary within societies. For example, in Ghana men associate poverty with a lack of material assets, whereas for women, poverty is defined as food insecurity (Narayan 2000).

The symbolic interactionist perspective emphasizes that norms, values, and beliefs are learned through social interaction and that social interaction influences the development of one's self-concept. Lewis (1966) argued that, over time, the poor develop norms, values, beliefs, and self-concepts that contribute to their own plight. According to Lewis, the **culture of poverty** is characterized by female-centered households, an emphasis on gratification in the present rather than in the future, and a relative lack of participation in society's major institutions. "The people in the culture of poverty have a strong feeling of marginality, of helplessness, of dependency, of not belonging. . . . Along with this feeling of powerlessness is a widespread feeling of inferiority, of personal unworthiness" (Lewis 1998, 7). Early sexual activity, unmarried parenthood, joblessness, reliance on public assistance, illegitimate income-producing activities (e.g., selling drugs), and substance use are common among the **underclass**—people living in persistent poverty. The culture of poverty view emphasizes that the behaviors, values, and attitudes exhibited by the chronically poor are transmitted from one generation to the next, perpetuating the cycle of poverty. Critics of the culture of poverty approach point out that behaviors, values, and attitudes of the underclass emerge from the constraints and blocked opportunities that have resulted largely from the disappearance of work as jobs have moved out of inner-city areas to the suburbs (Jargowsky 1997; Van Kempen 1997; Wilson 1996).

> Where jobs are scarce . . . and where there is a disruptive or degraded school life purporting to prepare youngsters for eventual participation in the workforce, many people eventually lose their feeling of connectedness to work in the formal economy; they no longer expect work to be a regular, and regulating, force in their lives. . . . These circumstances also increase the likelihood that the residents will rely on illegitimate sources of income, thereby further weakening their attachment to the legitimate labor market. (Wilson 1996, 52–53)

"Many of today's problems in the inner-city ghetto neighborhoods— crime, family dissolution, welfare, low levels of social organization, and so on— are fundamentally a consequence of the disappearance of work."

William Julius Wilson
Sociologist

From the late 1960s through the 1980s, poverty became more and more concentrated in inner-city neighborhoods, and conditions in those neighborhoods steadily deteriorated. But during the economic boom of the late 1990s, when unemployment was low, poverty became less concentrated (Kingsley & Pettit 2003). This decrease in the concentration of poverty during strong economic times supports the view that it is economic conditions, rather than a "culture of poverty," that explains chronic, intergenerational poverty.

Economic Inequality and Poverty in the United States

The United States is a nation of tremendous economic variation ranging from the very rich to the very poor. Signs of this disparity are visible everywhere, from opulent mansions perched high above the ocean in California to shantytowns in the rural South where people live with no running water or electricity.

Economic Inequality in the United States

The 1990s was a decade of U.S. economic growth: interest rates were down, unemployment low, and stock market averages reached record levels before declining at the end of 1999. At the close of the twentieth century, the United States had experienced the longest period of peacetime economic expansion in history. But contrary to the adage that "a rising tide lifts all boats," economic prosperity has not been equally distributed in the United States.

For example, from 1995 to 1999, the income of the top 1 percent of taxpayers grew by 59 percent, while that of the bottom half grew by only 9 percent. In 1999, the average CEO pay—$12.4 million—was 475 times the pay of the average worker (Anderson, Cavanagh, Collins, Hartman, & Yeskel 2000). By 2002, average CEO pay had declined to $7.4 million while worker pay rose slightly, lowering the CEO-worker pay ratio from 475:1 to 282:1. However, this 282-to-1 ratio is nearly seven times as large as the 1982 ratio of 42-to-1 (Anderson, Cavanagh, Collins, Hartman, & Klinger 2003). If the average annual pay of production workers had risen at the same rate since 1990 as it has for CEOs, their 2002 annual earnings would have been $68,057 instead of $26,272. If the federal minimum wage had grown at the same rate as CEO pay, it would have been $14.40 in 2002, instead of $5.15.

The distribution of wealth is much more unequal than the distribution of wages or income. **Wealth** refers to the total assets of an individual or household, minus liabilities (mortgages, loans, and debts). Wealth includes the value of a home, investment real estate, the value of cars, unincorporated business, life insurance (cash value), stocks/bonds/mutual funds/trusts, checking and savings accounts, individual retirement accounts (IRAs), and valuable collectibles. In 2002, the wealthiest 1 percent of households owned 38 percent of all national wealth, while the bottom 80 percent of households owned only 17 percent (Mishel, Bernstein, & Boushey 2003). For another illustration of inequality in wealth, consider that in 2001 the net worth of families in the top 10 percent of incomes was $833,600 compared to $7,900 net worth of families in the lowest fifth of income earners (Andrews 2003).

Patterns of Poverty in the United States

Poverty is not as widespread or severe in the United States as it is in many less-developed countries. Nevertheless, poverty represents a significant social problem in the United States. In 2002, the official U.S. poverty rate was 12.1 percent, up from 11.7 percent in 2001 (Proctor & Dalaker 2003). This translates into 34.6 million Americans living in poverty in 2002. Poverty rates (2 year average, 2000–2001) vary considerably among the states, from 6.1 percent in New Hampshire to 18.9 percent in Mississippi.

Poverty rates also vary according to age, education, sex, family structure, race/ethnicity, and labor force participation. As discussed in this chapter's *Social Problems Research Up Close* feature, media portrayals of the poor do not accurately reflect the demographic characteristics of the poor.

Age and Poverty. Children are more likely than adults to live in poverty (see Table 10.3). More than a third (35.1%) of the U.S. poor population are children (Proctor & Dalaker 2003). Child poverty rates are much higher in the United States than in Canada or any other Western European industrialized country (Vleminckx & Smeeding 2001).

Since the late 1950s the poverty rate among the elderly has experienced a downward trend, largely as a result of more Social Security benefits and the growth

Table 10.3
U.S. Poverty Rates, by Age, 2002

Age	Poverty Rate
Under 18 years	16.7
Under age 5	19.8
18 to 64	10.6
65 years and over	10.4

Source: Proctor, Bernadette, and Joseph Dalaker. 2003. *Poverty in the United States: 2002.* Current Population Reports P60-222. U.S. Census Bureau. Washington, DC: U.S. Government Printing Office.

of private pensions (see also Chapter 9). In 1959, the poverty rate among U.S. elderly was 35.2 percent; this rate fell to 25.3 in 1969, 15.2 in 1979, 11.4 in 1989, and reached a record low of 9.7 in 1999. The elderly poverty rate increased slightly to 10.4 percent in 2002 (Proctor & Dalaker 2003).

Education and Poverty. Education is one of the best insurance policies for protecting an individual against living in poverty. In general, the higher a person's level of educational attainment, the less likely that person is to be poor (see also Chapter 12). Adults without a high school diploma are the most vulnerable to poverty, followed by those with a high school diploma but no college degree. Nearly two-thirds (63%) of children in low-income families (100 to 200 percent of the poverty level) have parents without any college degree (National Center for Children in Poverty 2003).

Sex and Poverty. Women are more likely than men to live below the poverty line—a phenomenon referred to as the **feminization of poverty.** The 2002 poverty rate of U.S. females was 13.3 percent, compared to 10.9 percent for males (U.S. Census Bureau 2003). As discussed in Chapter 8, women are less likely than men to pursue advanced educational degrees and tend to be concentrated in low-paying jobs, such as service and clerical jobs. However, even with the same level of education and the same occupational role, women still earn significantly less than men. Women who are minorities and/or who are single mothers are at increased risk of being poor.

Family Structure and Poverty. Poverty is much more prevalent among female-headed, single-parent households than among other types of family structures (see Figure 10.1). The relationship between family structure and poverty helps to explain why women and children have higher poverty rates than men (see also Chapter 5). As shown in Figure 10.2, the majority of children living under the poverty line live with their single mothers.

In other countries, poverty rates of female-headed families are lower than those in the United States. For example, poverty rates of female-headed households are less than 10 percent in Belgium, France, Great Britain, Ireland, Luxembourg, the Netherlands, Norway, and Poland (Pressman 1998). Unlike the United States, these countries offer a variety of supports for single mothers, such as income supplements, tax breaks, universal child care, national health care, and higher wages for female-dominated occupations.

Social Problems Research Up Close

Media Portrayals of the Poor

In the 1990s, intense political activity surrounding welfare reform placed poverty and welfare high on the nation's agenda. Throughout this period, the media focused significant attention on poverty and welfare reform issues. Researchers Clawson and Trice (2000) examined photographs of the poor found in newsmagazines during this period to determine whether the media perpetuate inaccurate and stereotypical images of the poor.

Sample and Methods

The sample consisted of every story on the topics of poverty, welfare, and the poor that appeared between January 1, 1993, and December 31, 1998, in five newsmagazines (*Business Week, Newsweek, New York Times Magazine, Time,* and *U.S. News & World Report*). A total of 74 stories were included in the sample, with a total of 149 photographs of 357 poor people.

In analyzing the photographs, researchers noted the race/ethnicity (white, black, Hispanic, Asian American, or undeterminable), sex (male or female), age (young: under 18; middle-aged: 18–64; or old: 65 and over); residence (urban or rural), and employment status (working/job training or not working). The researchers also analyzed whether each poor individual was portrayed in stereotypical ways, such as pregnant, engaging in criminal behavior, taking or selling drugs, drinking alcohol, smoking cigarettes, or wearing expensive clothing or jewelry. After coding all the photographs according to the aforementioned variables, the researchers compared the portrayal of poverty in their sample of photographs to poverty statistics reported by the U.S. Census Bureau or the U.S. House of Representatives Committee on Ways and Means.

Findings and Conclusions

Clawson and Trice found that the newsmagazine photographs overestimated the percentage of the poor who are black. U.S. Census data (from 1996) showed that African-Americans made up 27 percent of the poor, but in the magazine portrayals, they made up nearly half (49%) of the poor. Whites, who according to Census data made up 45 percent of the poor, were depicted in the magazine portrayals as 33 percent of the poor. There were no magazine portrayals of Asian-Americans in poverty, and Hispanics were underrepresented by 5 percent. The researchers suggest that "this underrepresentation of poor Hispanics and Asian-Americans may be part of a larger phenomenon in which these groups are ignored by the media in general" (pp. 56–57).

The elderly were also underrepresented in the magazine portrayals of the poor. According to Census data, the elderly made up 9 percent of the true poor, yet they were only 4 percent of the magazine poor. Magazine portrayals of the poor exaggerated the percentage of the poor who are women, depicting 76 percent of the poor as women, compared to the Census figure of 62 percent.

Figure 10.1
U.S. Poverty Rates, by Family Structure: 2002

Source: Proctor, Bernadette, and Joseph Dalaker. 2003. *Poverty in the United States: 2002.* Current Population Reports P60–222. U.S. Census Bureau. Washington, DC: U.S. Government Printing Office.

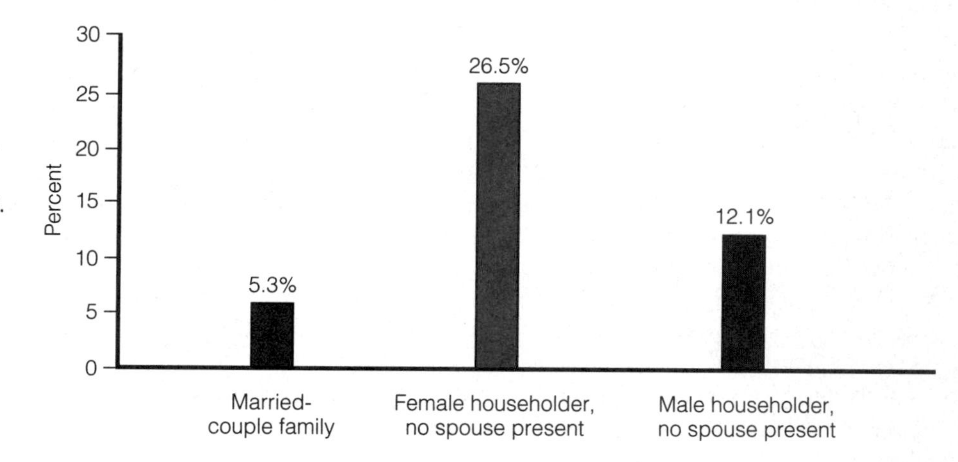

Social Problems Research Up Close

Magazine depictions implied that poverty is primarily an urban problem. Ninety-six percent of the poor were shown in urban settings, compared to Census data showing that 77 percent of the poor resided in urban areas. The authors note that "the urban underclass is often linked with various pathologies and antisocial behavior. Thus, this emphasis on the urban poor does not promote a positive image of those in poverty" (p. 60). And the media portrayals of the poor in this study created the impression that most poor people do not work: less than one-third (30%) of poor adults were shown working or participating in job training programs. In reality, half the poor worked in full- or part-time jobs, according to Census Bureau data.

Finally, the researchers analyzed the extent to which the newsmagazines portrayed the poor as having stereotypical traits. They found that media portrayals did not show poor mothers as having large numbers of children. Also, the researchers noted that the media did not overly emphasize other stereotypical characteristics associated with the poor. Of the 357 people in the sample of photographs, only three were shown engaging in criminal behavior and another three were shown with drugs. No alcoholics were presented, and only one person was smoking a cigarette. "However, of those seven stereotypical portrayals, only the person smoking was white—the others were either black or Hispanic" (p. 61). Only one poor woman was pregnant, so the media were not suggesting that poor women have babies to obtain larger welfare checks. However, the one pregnant woman shown was Hispanic. The researchers also noted whether the portrayals of the poor supported the image of the "welfare queen" stereotype (welfare recipients who do not really need assistance and who spend their welfare checks on luxuries). Of the 39 individuals who were shown with flashy jewelry or fancy clothes, "blacks and Hispanics were somewhat more likely to be portrayed this way than whites" (p. 61).

Clawson and Trice found that, overall, the portrayals of poor people in the five newsmagazines they analyzed did not reflect the reality of poverty; instead, they provided an inaccurate and stereotypical picture of poverty. The portrayals of poverty are important because they affect public opinion, which in turn, affects public policy. "Thus, if attitudes on poverty-related issues are driven by inaccurate and stereotypical portrayals of the poor, then the policies favored by the public (and political elites) may not adequately address the true problems of poverty" (p. 61).

Source: Clawson, Rosalee A., and Rakuya Trice. 2000. "Poverty As We Know It: Media Portrayals of the Poor." *Public Opinion Quarterly* 64:53–64.

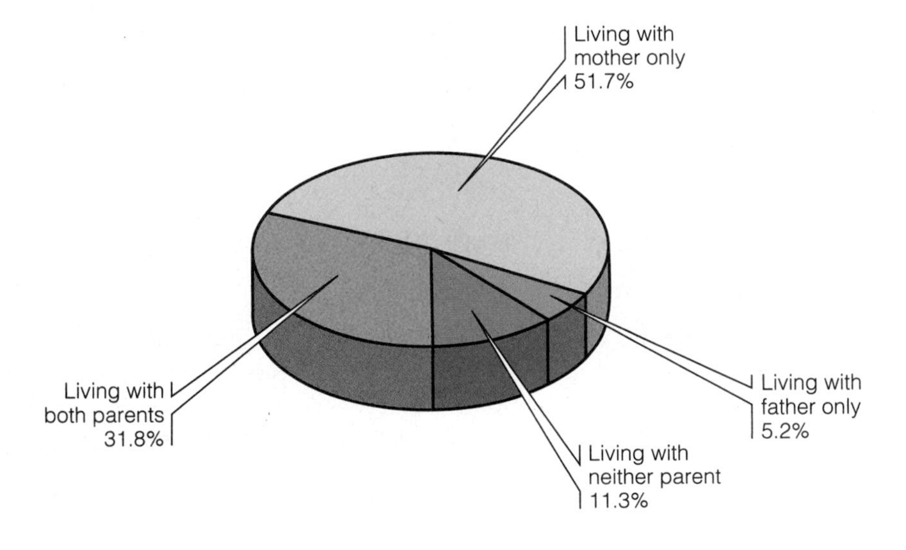

Figure 10.2
Percentage of U.S. Children Living Under the Poverty Line, by Family Type: 2002

Source: Fields, Jason. 2003. "Children's Living Arrangements and Characteristics: March 2002." *Current Population Reports,* P20–547. Washington, DC: U.S. Census Bureau.

Figure 10.3
U.S. Poverty Rates, by Race and Ethnicity: 2002

Source: Proctor, Bernadette and Joseph Dalaker. 2003. *Poverty in the United States: 2002.* Current Population Reports P60–222. U.S. Census Bureau. Washington, DC: U.S. Government Printing Office; Bernstein, Robert. 2002 (September 24). "Poverty Rate Rises, Household Income Declines." U.S. Census Bureau. http://www.census.gov

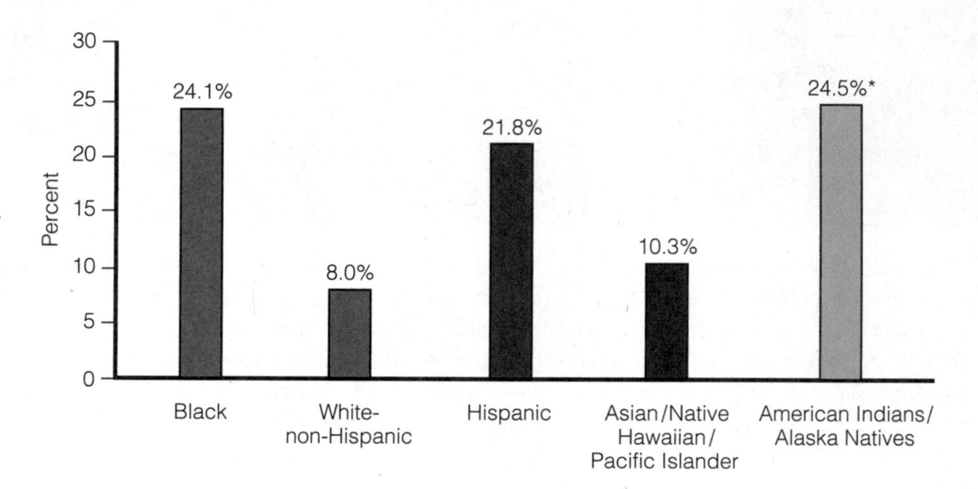

Race/Ethnicity and Poverty. Nearly half (45%) of the poor in the United States are non-Hispanic whites (Proctor & Dalaker 2003). However, as displayed in Figure 10.3, poverty rates are higher among blacks, Hispanics, and Native American/Alaska Natives than among non-Hispanic whites. As discussed in Chapter 6, past and present discrimination has contributed to the persistence of poverty among minorities. Other contributing factors include the loss of manufacturing jobs from the inner city, the movement of whites and middle-class blacks out of the inner city, and the resulting concentration of poverty in predominantly minority inner-city neighborhoods (Massey 1991; Wilson 1987, 1996). Finally, blacks and Hispanics are more likely to live in female-headed households with no spouse present—a family structure that is associated with high rates of poverty.

Labor Force Participation and Poverty. A common image of the poor is that they are jobless and unable or unwilling to work. Although the poor in the United States are primarily children and adults who are not in the labor force, many U.S. poor are classified as working poor. The **working poor** are individuals who spend at least 27 weeks per year in the labor force (working or looking for work), but whose income falls below the official poverty level. In 2002, 37.9 percent of all U.S. poor (ages 16 and over) worked; 11.2 percent worked year-round full time (U.S. Census Bureau 2003).

Consequences of Poverty and Economic Inequality

Poverty is associated with health problems, problems in education, problems in families and parenting, and housing problems. These various problems are interrelated and contribute to the perpetuation of poverty across generations, feeding a cycle of intergenerational poverty. In addition, poverty and economic inequality breed social conflict and war.

Health Problems and Poverty

In Chapter 2, we noted that poverty has been identified as the world's leading health problem. In developing countries, absolute poverty is associated with unsafe water and sanitation and indoor air pollution from heating and cooking fumes (World Health Organization 2002). Persistent poverty is associated with higher rates of infant mortality and childhood deaths and lower life expectancies among adults. Poverty often causes chronic malnutrition, which can result in permanent brain damage, learning disabilities, and mental retardation in infants and children (Hill 1998). In the United States, low socioeconomic status is associated with higher incidence and prevalence of health problems, disease, and death (Malatu & Schooler 2002). Poor children and adults also receive inadequate and inferior health care, which exacerbates their health problems. Finally, poverty is linked to higher levels of mental health problems, including stress, depression, and anxiety (Leventhal & Brooks-Gunn 2003).

Economic inequality also affects psychological and physical health. Streeten (1998) cited research suggesting that "perceptions of inequality translate into psychological feelings of lack of security, lower self-esteem, envy, and unhappiness, which, either directly or through their effects on life-styles, cause illness" (p. 5). Poor and middle-income adults who live in states with the greatest gap between the rich and the poor are much more likely to rate their own health as poor or fair than people who live in states where income is more equitably distributed (Kennedy, Kawachi, Glass, & Prothrow-Stith 1998).

Educational Problems and Poverty

Research indicates that children living in poverty are more likely to suffer academically than are children who are not poor. "Overall, poor children receive lower grades, receive lower scores on standardized tests, are less likely to finish high school, and are less likely to attend or graduate from college than are nonpoor youth" (Seccombe 2001, 323). The various health problems associated with childhood poverty contribute to poor academic performance. The poor often attend schools that are characterized by lower-quality facilities, overcrowded classrooms, and a higher teacher turnover rate (see also Chapter 12). Children living in poor inner-city ghettos "have to contend with public schools plagued by unimaginative curricula, overcrowded classrooms, inadequate plant and facilities, and only a small proportion of teachers who have confidence in their students and expect them to learn" (Wilson 1996, xv). Because poor parents have less schooling, on average, than do nonpoor parents, they may be less able to encourage and help their children succeed in school. However, research suggests that family income is a stronger predictor of ability and achievement outcomes than are measures of parental schooling or family structure (Duncan & Brooks-Gunn 1997). Poor parents have fewer resources to provide educational experiences (e.g., travel), private tutoring, books, and computers for their children. And with the skyrocketing cost of tuition and other fees, many poor parents cannot afford to send their children to college. The cost for low-income families of sending a child to a four-year public college or university rose from 13 percent of family income in 1980 to 25 percent in 2000 (Washburn 2004).

Poverty also presents obstacles to educational advancement among poor adults. Women and men who want to further their education in order to escape poverty may have to work while attending school or may be unable to attend school because of unaffordable child care, transportation, and/or tuition/fees/books.

Family Stress and Parenting Problems Associated with Poverty

In some cases, family problems contribute to poverty. For example, domestic violence causes some women to flee from their homes and live in poverty without the economic support of their husbands. Poverty also contributes to family problems. The stresses associated with poverty contribute to substance abuse, domestic violence, child abuse and neglect, divorce, and questionable parenting practices. For example, poor parents unable to afford child care expenses are more likely to leave children home without adult supervision. Poor parents are more likely than other parents to use harsh physical disciplinary techniques, and they are less likely to be nurturing and supportive of their children (Mayer 1997; Seccombe 2001).

Another family problem associated with poverty is teenage pregnancy. Poor adolescent girls are more likely to have babies as teenagers or become young single mothers. Early childbearing is associated with numerous problems, such as increased risk of premature or low birth weight babies, dropping out of school, and lower future earning potential as a result of lack of academic achievement. Luker (1996) notes that "the high rate of early childbearing is a measure of how bleak life is for young people who are living in poor communities and who have no obvious arenas for success" (p. 189). For poor teenage women who have been excluded from the American dream and disillusioned with education, "childbearing . . . is one of the few ways . . . such women feel they can make a change in their lives" (p. 182).

> Having a baby is a lottery ticket for many teenagers: it brings with it at least the dream of something better, and if the dream fails, not much is lost. . . . In a few cases it leads to marriage or a stable relationship; in many others it motivates a woman to push herself for her baby's sake; and in still other cases it enhances the woman's self-esteem, since it enables her to do something productive, something nurturing and socially responsible. . . . To the extent that babies can be ill or impaired, mothers can be unhelpful or unavailable, and boyfriends can be unreliable or punitive, childbearing can be just another risk gone wrong in a life that is filled with failures and losses. (Luker 1996, 182)

Housing Problems and Homelessness

The following description of housing in a low-income inner city neighborhood is not atypical of U.S. housing conditions for the poor.

> From the outside, Jamal's building looks like an ordinary house that has seen better days. . . . But once you walk through the front door, all resemblance to a real home disappears. . . . The building has been broken up into separate living quarters, a rooming house with whole families squeezed into spaces that would not even qualify as bedrooms in most homes. . . . Six families take turns cooking their meals in the only kitchen. . . .
>
> The plumbing breaks down without warning. . . . Windows . . . are cracked and broken, pieced together by duct tape that barely blocks the steady, freezing draft blowing through on a winter evening. Jamal is of the opinion that for the princely sum of $300 per month, he ought to be able to get more heat. (Newman 1999, 3–9)

Many poor families and individuals live in housing units that lack central heating and air conditioning, sewer or septic systems, and electric outlets in one or more rooms; many have no telephones. Housing units of the poor are also more

© Elena Rooraid/Photo Edit

Some cities and communities have laws or ordinances that prohibit the homeless from sleeping on public benches and soliciting money.

Table 10.4
Characteristics of the Homeless in U.S. Cities

41% are single men	14% are single women
40% are families with children	5% are unaccompanied minors
23% are mentally ill	30% are substance abusers
17% are employed	
10% are veterans	

Source: The United States Conference of Mayors. 2003. *Hunger and Homelessness Survey: A Status Report on Hunger and Homelessness in America's Cities.* Washington, DC: United States Conference of Mayors.

likely to have holes in the floor, a leaky roof, and open cracks in the walls or ceiling. In addition, poor individuals are more likely than the nonpoor to live in high-crime neighborhoods.

Even substandard housing would be a blessing to the hundreds of thousands of men, women, and children in the United States who are homeless on a given day. Homelessness, a growing problem in the United States, affects men, women, and children (see Table 10.4). In an annual survey of 25 U.S. cities, officials reported that requests for emergency shelter increased in the past year in 20 of these 25 cities (United States Conference of Mayors 2003). City officials cited lack of affordable housing as the major cause of homelessness, followed by mental illness and the lack of mental health services, low-paying jobs, substance abuse and the lack of substance abuse services, unemployment, domestic violence, poverty, and prison release. This chapter's *The Human Side* feature presents glimpses of what it is like to be homeless.

The Human Side

Life on the Streets: New York's Homeless

While a sociology graduate student at Columbia University, Gwendolyn Dordick undertook a study of homeless people living in New York City. She spent 15 months with four groups of homeless people: inhabitants of a large bus terminal, residents of a shantytown, occupants of a large public shelter, and clients of a small, church-run private shelter. In this *Human Side* feature, we present some of Dordick's observations, as well as excerpts from her conversations with homeless individuals she encountered at the Station and the Shanty.

The Station

The Station, located in Manhattan's West Side, encompasses bus terminals and depots, ticketing windows, shops, and fast-food restaurants. Scattered among the commuters and visitors are homeless women and men, and young adolescent boys and girls who have either run away or were kicked out by their guardians. One homeless man described living on the edge:

> Living on the streets makes you do a lot of things that you wouldn't normally do. . . . Comin' into this environment I've done a lot of things I said I wouldn't do. . . . There was some people that came along in a van and just threw sandwiches on the street and I picked them up and ate them. . . . The guilt almost killed me . . . but my stomach said, "Hey, listen, you better eat this food." (pp. 5–6)

Homeless individuals often rely on one another, offering each other companionship, friendship, and protection. Although they have little to share, they often share what they have with their fellow homeless friends.

> These people, when I got down here, these people reached out to me because they knew, they already knew what it was like. They're not afraid to help their fellow man. As soon as I got down here I met Ron and the fellows and they didn't push me away. I mean I didn't know where to go, I didn't know where to eat, I didn't know where to sleep. They just invited me right in. And ever since then at least I've been healthy, and I've been clean since I've met them. (p. 13)

Homeless individuals are often treated harshly by police. One homeless man lamented:

> You may have an invalid laying down here. He's got problems and the [Station] cops will come up and kick him. Like he's an animal with no rights. (p. 11)

Pregnancy is common among homeless women who do not have access to or cannot afford contraception. Pregnancy can have disastrous consequences for these women, who fear having a baby and having to care for a baby. According to one man in the Station:

> It's one thing being homeless, but pregnant and homeless? Some women have their babies right out here; others get rid of them. . . . Some of them abort theirself by sticking hangers up their vaginas. I've seen that myself. This young girl didn't want a baby and she stuck a hanger up her vagina. She had to go to the hospital. (p. 25)

The Shanty

A barricaded makeshift community of 20 or so residents sits on a formerly vacant lot visible from the nearby streets and a bridge that crosses the East River. The Shanty consists of 15 makeshift dwellings, or "huts" as they are called by the residents, which are made of a variety of discarded materials such as pieces of wood and boards, cardboard, mattresses, fabric, and plastic tarps.

Intergenerational Poverty

As we have seen, problems associated with poverty, such as health and educational problems, create a cycle of poverty from one generation to the next. Poverty that is transmitted from one generation to the next is called **intergenerational poverty.** In a study of intergenerational poverty using a national longitudinal survey of families, researchers found considerable mobility out of childhood poverty: three-quarters of white poor children and over half of black poor children escaped poverty in early adulthood (Corcoran & Adams 1997). However, both white and black children in poor families were still much more likely to be poor in early adulthood than were children raised in nonpoor families.

The Human Side

The materials are fastened with nails, twine, or fabric. One of the residents has tapped into a source of electricity by running a wire from a lamppost into several huts, providing electricity for light and heat. The researcher notes that as in the case of the residents of the Station, welfare plays a minimal role in the lives of residents of the Shanty. "So difficult is negotiating the system that most forgo their entitlements" (p. 58). One resident explains:

> I don't get welfare. I just can't . . . do it. I hate those people in there. They make you . . . sit and ask you questions that don't make any sense. . . . You're homeless but you have to have an address. What kind of shit is that? Give me a break. They want you to get so . . . upset that you do get up and walk out. (p. 58)

Although drug use is common among residents of the Shanty, using drugs in public is a violation of norms of "etiquette." One resident explained that using drugs in the presence of a nonuser is disrespectful:

> For me to just take my works out and shoot, I would feel uncomfortable in front of you. It's not right . . . very disrespectful. God forbid, I could be an influence. I could cause you to do it. (pp. 71–72)

Although survival among the homeless requires hustling, buying, trading, and selling, not everything is for sale. Some belongings have sentimental value that outweighs their economic value. One resident of the Shanty treasures a small gold key:

> There's a golden rule about gifts. You treasure them. You don't give them away, you don't sell them. I have right here a little key, a skeleton key. A little kid handed it to me four years ago. And every time somebody see that and say, "what is that?" I say it's a key to the world. I wouldn't give it to anyone if it was given to me. I treasure it. (p. 78)

Friendships and love relationships among the homeless suffer from the stresses of drug addictions and impoverished conditions. Nevertheless, Dordick explains that the homeless "survive through their personal relationships" (p. 193). Relationships are critical to securing the material resources needed to survive and to creating—to the extent that it is possible—a safe and secure environment. One resident of the Shanty, Richie, conveys the importance of a love relationship:

> Regardless of what people might think and say, most of us that might have a woman it's all we want. We don't look for anyone else. We really don't. . . . I'm happy just to take care of my woman. . . . I happen to love my girl. . . . Really, I know it sounds corny, but that's the truth. . . . As a matter of fact, we're gonna be married soon.

A Final Note

Virtually all the homeless individuals Dordick encountered expressed the desire to escape homelessness and be self-reliant—and they want to be understood. In the words of one homeless man at the Station:

> You never see me sleep in the street. I worked 32 years of my life. Went to prison in '85. . . . I was brought up with a certain degree of independence. . . . And now all I need is two dollars to go and sit in a movie all night long. My pride is too good to beg. I don't want you to help me, Miss. I want you to understand me. (p. 5)

Source: Excerpted and reprinted by permission from Dordick, Gwendolyn. 1997. *Something Left to Lose: Personal Relations and Survival Among New York's Homeless.* Philadelphia: Temple University Press.

Intergenerational poverty creates a persistently poor and socially disadvantaged population sometimes referred to as the underclass. The term "underclass" usually refers to impoverished individuals, often those who live in economically distressed neighborhoods (ghettos, slums, or barrios) with low educational attainment, chronic unemployment or underemployment, criminal involvement, unstable family structures, and welfare dependency. Although the underclass is stereotyped as being composed of minorities living in inner-city or ghetto communities, the underclass is a heterogeneous population that includes poor whites living in urban and nonurban communities (Alex-Assensoh 1995).

William Julius Wilson attributes intergenerational poverty and the underclass to a variety of social factors, including the decline in well-paid jobs and their movement out of urban areas, the resultant decline in the availability of marriageable males able to support a family, declining marriage rates and an increase in out-of-wedlock births, the migration of the middle-class to the suburbs, and the impact of deteriorating neighborhoods on children and youth (Wilson 1987, 1996).

War and Social Conflict

Poverty is often the root cause of conflict and war within and between nations, as "the desperation of the poor is never quiet for long" (Speth 1998, 281). In the developing world, most of the people recruited for armed conflict are unemployed. "They don't have education opportunities and they don't really see what the future holds for them other than war and misery" ("Reducing Poverty Is Key to Global Stability" 2003, 4). Not only does poverty breed conflict and war, but war also contributes to poverty. War devastates infrastructures, homes, and businesses and leaves widows and orphans to fend for themselves. Military spending associated with war diverts resources away from economic development and social spending on health and education. Among the 21 countries with extreme food emergencies in 2002, in 15, these emergencies were sparked by war, civil unrest, and the lingering effects of past conflicts (*Human Development Report* 2003).

In the United States, the widening gap between the rich and poor may lead to class warfare (Hooks 2000). Briggs (1998) asks how long the United States can maintain social order "when increasing numbers of persons are left out of the banquet while a few are allowed to gorge?" (p. 474). Although Karl Marx predicted that the have-nots would revolt against the haves, Briggs does not foresee a revival of Marxism; "the means of surveillance and the methods of suppression by the governments of industrialized states are far too great to offer any prospect of success for such endeavors" (p. 476). Instead, Briggs predicts that American capitalism and its resulting economic inequalities will lead to social anarchy—a state of political disorder and weakening of political authority.

Strategies for Action: Antipoverty Programs, Policies, and Proposals

In the United States, federal, state, and local governments have devoted considerable attention and resources to antipoverty programs for the last 50 years. Here we describe some of these programs and proposals, discuss international responses to poverty, and note the role of charity and the nonprofit sector in poverty alleviation.

"People on welfare are just like you and me. They have the same basic hopes and fears. They want a job that brings self-worth and validation. They want to support their families and contribute to their communities. They want pride and dignity, just like those of us who have been lucky enough never to need public assistance."

Alexis M. Herman
Secretary, U.S. Department of Labor

Government Public Assistance and Welfare Programs in the United States

Many public assistance programs stipulate that households are not eligible for benefits unless their income and/or assets fall below a specified guideline. Programs that have eligibility requirements based on income are called **means-tested programs.** Government public assistance programs designed to help the poor include cash support, food programs, housing assistance, medical care, educational assistance, job training programs, child care, child support enforcement, and the earned income tax credit (EITC).

Cash Support. Publicly funded cash support programs include Supplemental Security Income (SSI) and Temporary Assistance to Needy Families (TANF). Supplemental Security Income Federal SSI, administered by the Social Security Administration, provides a minimum income to poor people who are aged 65 or older, blind, or disabled. Under the 1996 welfare reforms, the definition of disability has been sharply restricted and the eligibility standards tightened.

Temporary Assistance to Needy Families. Before 1996, a cash assistance program called **Aid to Families with Dependent Children (AFDC)** provided single parents (primarily women) and their children with a minimum monthly income. In 1996, Congress passed the **Personal Responsibility and Work Opportunity Reconciliation Act (PRWORA),** commonly referred to as "welfare reform," that ended AFDC and replaced it with a program called **Temporary Assistance to Needy Families (TANF).** In 2001, TANF families received an average monthly amount of $351 (U.S. Department of Health and Human Services 2003a). Within two years of receiving benefits, adult TANF recipients must be either employed or involved in work-related activities, such as on-the-job training, job search, and vocational education. A lifetime limit of five years is set for families receiving benefits, and able-bodied recipients aged 18 to 50 and without dependents have a two-year lifetime limit. Some exceptions to these rules may be made for individuals with disabilities, victims of domestic violence, residents of high unemployment areas, and those caring for young children. To qualify for TANF benefits, unwed mothers under the age of 18 are required to live in an adult-supervised environment (e.g., with their parents) and to receive education and job training. Legal immigrants who entered the country before August 22, 1996, may receive TANF, but those who entered after this date may only receive services after they have been in the United States for five years.

Although TANF was scheduled to be reauthorized in 2002, Congress was unable to come to agreement on a reauthorization bill, so the law was extended. TANF reauthorization is currently a priority for the 108th Congress.

Food Assistance. According to a report by the Agriculture Department, nearly 3.8 million families were hungry in 2002 to the point that someone in the household skipped meals because the family could not afford them—an increase of 13 percent from 2000 (Associated Press 2003). Food assistance, including food stamps; school lunch and breakfast programs; the Special Supplemental Food Program for Women, Infants, and Children (WIC); and nutrition programs for the elderly is designed to help individuals and families who cannot afford an adequate diet. The largest food assistance program is the food stamp program, which issues monthly benefits through coupons or Electronic Benefits Transfer (EBT), using a plastic card similar to a credit card and a personal identification number (PIN). In

This poster is a public service message designed to encourage economically distressed families to apply for food stamps.

USDA Food Stamp Program

2002 the typical food stamp household had a gross income of $633 per month and received a monthly food stamp benefit of $173 (USDA 2003). The majority of food stamp households do not receive cash welfare benefits. In 2002, about one in five (21%) food stamp households received TANF benefits (USDA 2003). The Farm Bill of 2002 restored food stamp benefits for all immigrant children no matter what their date of entry to the United States.

Housing Assistance. Housing costs represent a major burden for the poor. In 25 U.S. cities, low-income households spend, on average, nearly half (46%) of their income in housing (U.S. Conference of Mayors 2003). Requests for assisted housing by low-income individuals and families increased in 83 percent of these cities in the last year (U.S. Conference of Mayors 2003). Federal housing programs include public housing, Section 8 housing, and other private project-based housing.

The **public housing** program, initiated in 1937, provides federally **subsidized housing** owned and operated by local public housing authorities (PHAs). Public housing has been plagued by problems. To save costs and avoid public opposition, high-rise public housing units were built in inner-city projects. The concentration of poor families in deteriorating neighborhoods led to increases in crime, drugs, vandalism, and violence. One survey found that one in five residents living in public housing reported feeling unsafe in his or her neighborhood (HUD 2000). The 2.6 million residents of public housing are more than twice as likely to suffer from firearm-related crimes than other U.S. residents (HUD 2000). The Hope VI Urban Demonstration Program was established in 1992 to transform the nation's most distressed public housing projects by rebuilding the physical structure of public housing developments, expanding the opportunities of its residents, and building a sense of community among residents.

Rather than build new housing units for low-income families, **Section 8 housing** relies on existing housing. With Section 8 housing, federal rent subsidies are provided either to tenants (in the form of certificates and vouchers) or to private landlords. Other private project-based housing includes privately owned housing units that do not receive rent subsidies but receive other federal subsidies such as interest rate reductions. Unlike public housing that confines low-income families to high-poverty neighborhoods, Section 8 and other private project-based housing attempt to disperse low-income families. However, because of opposition by residents in middle-class neighborhoods, most Section 8 housing units remain in low-income areas.

The level of housing assistance available is sorely inadequate to meet the housing needs of low-income Americans. In 25 U.S. cities, applicants for housing assistance must wait an average of 24 to 27 months, and nearly half of the survey cities have stopped accepting applications for at least one assisted housing program due to the excessive length of the waiting list (U.S. Conference of Mayors 2003).

Medical Care. Medical care assistance programs include Indian Health Services, maternal and child health services, and Medicaid, which provides medical services and hospital care for the poor through reimbursements to physicians and hospitals. However, many low-income individuals and families do not qualify for Medicaid and either cannot afford health insurance or cannot pay the deductible and co-payments under their insurance plan. In the earlier AFDC welfare program, all recipients were automatically entitled to Medicaid. Under the TANF program, states decide who is eligible for Medicaid; eligibility for cash assistance does not automatically convey eligibility for Medicaid. A provision of the 1996 welfare reform legislation guarantees welfare recipients at least one year of transitional Medicaid when leaving welfare for work (see also Chapter 2).

Educational Assistance. Educational assistance includes Head Start and Early Head Start programs and college assistance programs (see also Chapter 12). Head Start and Early Head Start programs provide educational services for disadvantaged infants, toddlers, and preschool-age children and their parents. Evaluations of Head Start and Early Head Start programs indicate that they improve children's cognitive, language, and social-emotional development and strengthen parenting skills (Administration for Children and Families 2002). According to the Children's Defense Fund (2003), every $1 invested in high-quality early childhood care and education saves as much as $7 by increasing the likelihood that children will be literate, go to college, and be employed, and by decreasing the likelihood that they will drop out of schools, be dependent on welfare, or be arrested for criminal activity.

To alleviate economic barriers for low-income persons wanting to attend college, the federal government offers grants, loans, and work opportunities. The Pell grant program aids students from low-income families. The guaranteed student loan program enables college students and their families to obtain low-interest loans with deferred interest payments. The federal college-work-study program provides jobs for students with "demonstrated need."

Job Training Programs. Various employment and job training programs are available to help individuals out of poverty (see also Chapter 11). These include summer youth employment programs, Job Corps, and training for disadvantaged adults and youth. These programs fall under the Job Training and Partnership Act (JTPA), a federally funded program passed in 1982 and amended in 1992. A primary shortcoming of job training programs has been that "they spread too little money among too many trainees, with the result that few are in training long enough for it to make a sufficient impact on their posttraining wages" (Levitan, Mangum, & Mangum, 1998, 29).

Child Care Assistance. In the United States, lack of affordable, good child care is a major obstacle to employment for single parents and a tremendous burden on dual-income families and employed single parents. The cost of full day care in a day care center ranges from $4,000 to $10,000 per year (Children's Defense Fund 2002). In many cases, low-income families have placed their children in low-cost, often lower-quality, and unstimulating care and nearly 7 million children are left home alone each week.

Some public and private sector programs and policies provide limited assistance with child care. The Dependent Care Assistance Plan provisions of the 1981 Economic Recovery Tax Act permits individuals to exclude the value of employer-provided child care services from their gross income. However, few employers provide on-site child care or subsidies for child care. At the same time, Congress increased the amount of the child care tax credit and modified the federal tax code to allow taxpayers to shelter pretax dollars for child care in "flexible spending plans." The Family Support Act of 1988 offered additional funding for child care services for the poor (in conjunction with mandatory work requirements). The Child Care and Development Block Grant, which became law in 1990, targeted child care funds to low-income groups and the Personal Responsibility and Work Opportunity Reconciliation Act of 1996 appropriated funds for child care. But child care assistance is inadequate; only 14 percent of the nearly 16 million children under age 13 who are eligible for child care assistance receive any help (Children's Defense Fund 2002). With recent budget shortfalls, states have made drastic cuts in child care services for low-income employed parents. According to Sonya Michel (1998), "the reluctance to make adequate provision for childcare is . . . symptomatic of a deeper aversion on the part of many legislators and public officials to helping poor and low-income women become truly economically independent, a status which is, in turn, essential to their ability to form autonomous households" (pp. 47–48).

Child Support Enforcement. Half of all U.S. children living below the poverty line in 2001 lived with their mothers and had fathers living elsewhere, making them eligible to receive child support. In 1996, less than one-third of poor children (31%) living with single mothers received child support. To encourage child support from absent parents, the Personal Responsibility and Work Opportunity Act of 1996 requires states to set up child support enforcement programs. The welfare reform law established a Federal Case Registry and National Directory of New Hires to track delinquent parents across state lines, increased the use of wage withholding to collect child support, and allowed states to seize assets and to revoke driving licenses, professional licenses, and recreational licenses of parents who fall behind in their child support. These efforts to improve child support compliance have been modestly successful: the percentage of poor children in single-mother households receiving child support increased from 31 percent in 1996 to 36 percent in 2001 (Sorensen 2003). Among poor families receiving some support, child support as a share of a poor family's income was 30 percent in 2001 (Sorensen 2003).

While gains in child support are good news for poor children and their single parents, over 60 percent of poor children in single-mother households do not receive child support. One reason is that the fathers of these tend to be unemployed or have low incomes themselves, limiting their ability to pay child support. Another reason that child support receipt is low among poor families is that most of the support paid to children receiving public assistance goes to the government rather than to the children, to recoup part of the cost of assistance already paid to the family; this reduces the incentive of fathers to pay child support.

Earned Income Tax Credit. The federal **earned income tax credit (EITC),** created in 1975, is a refundable tax credit based on a working family's income and number of children. The EITC is designed to offset Social Security and Medicare payroll taxes on working poor families and to strengthen work incentives. In 2003, an eligible family of four with two children could receive a credit of up to $4,204. Almost one out of every seven families who file federal income tax returns claim the federal EITC, which lifts more children out of poverty than any other program (Johnson, Llobrera, & Zahradnik 2003). In 2003, 17 states offered state EITCs and two local governments offered local EITCs.

Welfare in the United States: Myths and Realities

Public attitudes toward welfare assistance and welfare recipients are generally negative. Rather than view poverty as the problem, many Americans view welfare as the problem. What are some of the common myths about welfare that perpetuate negative images of welfare and welfare recipients?

MYTH 1. People receiving welfare are lazy, have no work ethic, and prefer to have a "free ride" on welfare rather than work.

Reality. Most recipients of TANF cash benefits and food stamps are children and therefore are not expected to work. Unemployed adult welfare recipients experience a number of barriers that prevent them from working, including poor health, job scarcity, lack of transportation, lack of education, and/or the desire to stay home and care for their children (which often stems from the inability to pay for child care or the lack of trust in child care providers) (Zedlewski 2003). Welfare recipients who stay home to care for children *are* doing very important work: parenting. "Raising children is work. It requires time, skills, and commitment. While we as a society don't place a monetary value on it, it is work that is invaluable—and indeed, essential to the survival of our society" (Albelda & Tilly 1997, 111).

It is also important to note that many adults receiving public assistance are either employed or in the labor force looking for work. In 2001, more than a quarter of adult TANF recipients were employed, earning an average monthly income of $686 (U.S. Dept. of Health and Human Services 2003a). More than a quarter of food stamp recipients in 2002 had earnings, typically $633 per month (USDA 2003).

Finally, most adult welfare recipients would rather be able to support themselves and their families than rely on public assistance. The image of a welfare "free loader" lounging around enjoying life is far from the reality of the day-to-day struggles and challenges of supporting a household on a monthly TANF check of $351 (which was the average monthly cash assistance to TANF families in 2001).

MYTH 2. Most welfare mothers have large families with many children.

Reality. Mothers receiving welfare have no more children, on average, than mothers in the general population. In FY 2001, the average number of persons in TANF families was 2.6. The TANF families averaged two recipient children; two in five families had only one child and one in 10 families had more than three children (U.S. Department of Health and Human Services 2003a).

MYTH 3. Welfare benefits are granted to many people who are not really poor or eligible to receive them.

Reality. Although some people obtain welfare benefits through fraudulent means, it is much more common for people who are eligible to receive welfare not to receive benefits. Only about half of families poor enough to qualify for TANF receive monthly cash assistance, and six out of 10 of those eligible for the Food Stamp Program receive benefits (Food Research and Action Center 2004; Fremstad 2004). In 25 U.S. cities, only 33 percent (on average) of eligible low-income households are receiving public housing assistance (U.S. Conference of Mayors 2003). As surmised from the data in Table 10.5, one-third of persons living below the poverty level in 2000 lived in households that did not receive any form of means-tested assistance.

A main reason for not receiving benefits is lack of information; people don't know they are eligible. Many people who are eligible for public assistance do not apply for it because they do not want to be stigmatized as lazy people who just want a "free ride" at the taxpayers' expense—their sense of personal pride prevents them from receiving public assistance. Others want to avoid the administrative and transportation hassles involved in obtaining it (Zedlewski, Nelson, Edin, Koball, & Roberts 2003). Finally, some individuals who are eligible for public assistance do not

Table 10.5
Percentage of Persons Below Poverty Level in Households Receiving Means-Tested Assistance: 2000

Total:	66.2%
Receiving Cash Assistance:	25.2%
Receiving Food Stamps:	33.8%
One or More Persons in Household Covered by Medicaid:	50.1%
Live in Public or Subsidized Housing:	18.5%

Source: "Annual Demographic Survey, March Supplement." 2001. U.S. Census Bureau. *Current Population Survey, March 2001.* http://ferret.bls.census.gov

receive it because it is not available. As noted earlier, in 25 U.S. cities, applicants for housing assistance must wait an average of 24 to 27 months before housing becomes available. Nearly half of the survey cities have stopped accepting applications for at least one assisted housing program due to the excessive length of the waiting list (U.S. Conference of Mayors 2003).

Minimum Wage Increase and "Living Wage" Laws

In 2002, 2.2 million workers (3% of all hourly-paid workers) earned wages at or below the minimum wage of $5.15. About half of minimum wage workers were under age 25, and twice as many women as men reported earning $5.15 or less (4% versus 2%) (Bureau of Labor Statistics 2003). A full-time worker earning the $5.15 an hour would earn $10,712 per year, well below the 2003 federal poverty line of $14,810 for a family of three. Further, the purchasing power of the minimum wage has declined because increases in the minimum wage have not kept up with inflation. The result is that the minimum wage, when adjusted for inflation, is worth less today than it was in 1979 (Mishel et al. 2003).

Clearly, raising the minimum wage would benefit low-wage workers and reduce poverty. Some states have established a minimum wage higher than the federal minimum wage. In 2004, states with higher minimum wages (ranging from $6.15 to $7.15) included Alaska, California, Connecticut, Delaware, Hawaii, Maine, Massachusetts, Oregon, Rhode Island, Vermont, and Washington, as well as the District of Columbia (U.S. Department of Labor 2004).

Those opposed to increasing the minimum wage argue that such an increase would result in higher unemployment, as businesses would reduce wage costs by hiring fewer employees. However, research has failed to find any systematic, significant job loss associated with minimum wage increases (Economic Policy Institute 2000).

In January 2004, 116 cities and counties had living wage laws (Living Wage Resource Center 2004). **Living wage laws** require state or municipal contractors, recipients of public subsidies or tax breaks, or, in some cases, all businesses to pay employees wages significantly above the federal minimum, enabling families to live above the poverty line. Living wage laws are not only good for individuals and families; they are also good for business. Research findings show that businesses that pay their employees a living wage have lower worker turnover and absenteeism, reduced training costs, higher morale and higher productivity, and a stronger consumer market (Kraut, Klinger, & Collins 2000). Over 50 business owners have signed a Living Wage Covenant, pledging to pay their employees over $8 an hour, as well as publicly advocate higher wages for all low-income workers.

Charity, Nonprofit Organizations, and Nongovernmental Organizations

Various types of aid to the poor are provided through individual and corporate donations to charities and nonprofit organizations, including faith-based organizations. In 2000, 89 percent of U.S. households gave charitable contributions averaging $1,620 per household (Independent Sector 2001).

Charity involves giving not only money and valuable goods but also time and effort in the form of volunteering. A national survey found that 44 percent of U.S. adults (age 21 and over) volunteered with a formal organization in 2000. When asked why they gave their time, respondents cited compassion as the most common reason. Other reasons cited by survey participants include (1) the belief that those who have more should help those with less; (2) they knew someone who would benefit from their volunteering; and (3) volunteering was a good way to meet people (Independent Sector 2001).

> "All too many of those who live in affluent America ignore those who exist in poor America; in doing so, the affluent American will eventually have to face themselves with the question. . . . : How responsible am I for the well-being of my fellows? To ignore evil is to become an accomplice to it."
>
> **Martin Luther King, Jr.**
> **Civil rights activist**

Nongovernmental organizations (NGOs) address many issues related to human rights, social justice, and environmental concerns. The number of international NGOs grew from fewer than 400 in 1900 to 26,000 in the year 2000—more than four times as many as existed just 10 years earlier (Knickerbocker 2000; Paul 2000). At the Millennium Forum meeting in 2000, representatives from over 1,000 NGOs called for a UN Global Poverty Eradication Fund to ensure that poor people have

The National Student Campaign Against Hunger and Homelessness (NSCAHH), started two decades ago by state Public Interest Research Groups (PIRGs) and USA for Africa, is the largest student network fighting hunger and homelessness in the country, with more than 600 participating campuses in 45 states. In this photo, two students from the University of Connecticut are participating in a food drive.

access to credit (Deen 2000). The NGOs declared that poverty is the most widespread violation of human rights and called upon the United Nations and governments around the world to make poverty alleviation a priority.

International Responses to Poverty

In the 1990s, the share of people living in extreme poverty fell from 30 percent to 23 percent (*Human Development Report* 2003). Two main approaches for achieving poverty reduction throughout the world include promoting economic growth and investing in "human capital."

Promoting Economic Growth. Economic growth, over the long term, generally reduces poverty (United Nations 1997). An expanding economy creates new employment opportunities and increased goods and services. As employment prospects improve, individuals are able to buy more goods and services. The increased demand for goods and services, in turn, stimulates economic growth. As emphasized in Chapter 13, economic development requires controlling population growth and protecting the environment and natural resources, which are often destroyed and depleted in the process of economic growth.

> "Trying to eradicate hunger while population continues to grow rapidly is like trying to walk up a down escalator."
>
> **Lester R. Brown**
> **World Watch Institute**

However, economic growth does not always reduce poverty; in some cases it increases it. For example, growth resulting from technological progress may reduce demand for unskilled workers. Growth does not help poverty reduction when public spending is diverted away from meeting the needs of the poor and instead is used to pay international debt, finance military operations, and support corporations that do not pay workers fair wages. The World Bank lends about $30 billion a year to developing nations to pay primarily for roads, bridges, and industrialized agriculture that mostly benefit corporations. "Relatively little attention or money has been given to developing basic social services, building schools and clinics, and building decent public sanitation and clean water systems in some of the world's poorest countries" (Mann 2000, 2). Thus, "economic growth, though essential for poverty reduction, is not enough. Growth must be pro-poor, expanding the opportunities and life choices of poor people" (*Human Development Report* 1997, 72–73). Because three-fourths of poor people in most developing countries depend on agriculture for their livelihoods, economic growth to reduce poverty must include raising the productivity of small-scale agriculture. Not only does improving the productivity of small-scale agriculture create employment, it also reduces food prices. This chapter's *Focus on Technology* feature examines agricultural biotechnology as a strategy for alleviating global hunger.

Investing in Human Capital. Promoting economic development in a society requires having a productive workforce. Yet, in many poor countries, large segments of the population are illiterate and without job skills, and/or are malnourished and in poor health. Thus a key feature of poverty reduction strategies involves investing in human capital. The term **human capital** refers to the skills, knowledge, and capabilities of the individual. Investments in human capital involve programs and policies that enhance the individual's health, skills, knowledge, and capabilities. Such programs and policies include those that provide adequate nutrition, sanitation, housing, health care (including reproductive health care and family planning), and educational and job training.

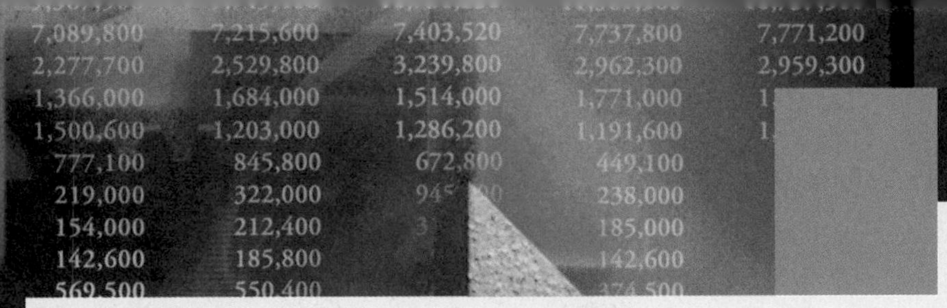

Focus on Technology

Is Agricultural Biotechnology the Solution for Global Hunger?

Biotechnology is any technique that uses living organisms or substances from those organisms to make or modify a product or develop microorganisms for specific uses (see also Chapter 15). Through **agricultural biotechnology**—the application of biotechnology to agricultural crops—biotech companies have developed products known as **genetically modified organisms (GMOs),** also referred to as genetically engineered foods (GE foods), and transgenic crops. The most common GE trait is herbicide tolerance (produces crops that tolerate weed-killing chemicals), followed by insect resistance (James 2002).

Between 1996 and 2002, the global area of transgenic crops increased by 35-fold, from 4.3 million to 145 million acres. In 2002, 16 countries planted transgenic crops, but four countries grew 99 percent of the global GE crop area: the United States grew 66 percent of the global total followed by Argentina (23%), Canada (6%), and China (4%) (James 2002).

In the United States, at least 10 GE foods have been government approved for human consumption: corn, soy, flax, tomato, beets, canola, potato, papaya, cotton, and squash (Center for Food Safety 2003; Genetic ID 2003). An estimated 60 to 70 percent of processed foods in U.S. markets contain some form of genetically modified ingredient, most often corn or soy, followed by canola, and cotton (cottonseed oil) (Public Issues Education Project 2003). Yet a national survey of U.S. adults found that only half were aware that foods containing GM ingredients are currently sold in stores, and while most Americans are likely to consume foods with GM ingredients every day, only one-fourth of the survey sample said they had consumed food containing GE ingredients (Hallman, Hebden, Aquino, Cuite, & Lang 2003).

Scientists, academics, environmentalists, public health officials, policy makers, corporations, farmers, and citizens throughout the world are deeply divided over the use of agricultural biotechnology. Not surprisingly, supporters of agricultural technology emphasize its potential benefits, while critics focus on the potential risks.

Can Biotechnology Alleviate Hunger and Malnutrition?

Supporters of agricultural biotechnology commonly cite the alleviation of hunger and malnutrition as a main benefit, claiming that agricultural biotechnology can enable farmers to produce crops with higher yields. However, research findings suggest that GM crop yields are not significantly higher than conventional crop yields and in some cases are *lower than* conventional crop yields (Mendelson 2002).

Biotech companies promote the use of genetic engineering to enhance the nutritional value of foods as a strategy for alleviating nutritional deficiencies in the diets of poor populations. "Golden Rice," genetically modified to contain vitamin A, has been touted as a remedy for vitamin A deficiencies among poor children in developing countries. However, "Golden Rice" has never entered the market because to get an adequate level of vitamin A from this rice, a four-year-old child would have to eat 27 cups of rice per day (Mendelson 2002).

Critics of GM foods argue that the world already produces enough food for all people to have a healthy diet. According to the United Nations Development Programme, if all the food produced worldwide were distributed equally, every person would be able to consume 2,760 calories a day (*Human Development Report* 2003). Biotechnology, critics argue, will not alter the fundamental causes of hunger, which are poverty and lack of access to food and to land on which to grow food.

Critics argue that biotechnology companies use the issue of poverty and hunger in the developing world to justify GM crops. Such claims are promoted by the biotech industry-created consortium, the "Council for Biotechnology Information," which has a $250 million public relations budget to tout the benefits of GE foods (Altieri 2003). In fact, most GE food products and research dollars for the development of GE foods target the more affluent nations' agriculture and consumers.

Skeptics suggest that transgenic crop technology can actually *increase* hunger and poverty. The corporate control of GM seeds, protected by patents and intellectual property rights, threatens the age-old farming practice of saving seeds from one crop to use for the next season. Instead, farmers who use GM seeds must purchase new seeds each season—an expense that many subsistence farmers cannot afford.

Are GMOs Safe for Human and Environmental Health?

Even if agricultural biotechnology has the potential to alleviate hunger and malnutrition through increased crop yields and nutritionally enhanced foods, a number of human and environmental health and safety concerns must be considered. Human health concerns include possible toxicity, carcinogenicity, food intolerance, antibiotic resistance buildup, decreased nutritional value, and food allergens in GM foods. Biotech companies claim that GM foods that have been approved by the Food and Drug Administration are safe for human consumption and even cite potential health benefits, such as the use of biotechnology

continued

continued

to remove allergens that naturally occur in foods such as nuts, making these foods safer to eat for individuals who have allergies to these foods (Bailey 2004). But critics claim that research on the effects of GM crops and foods on human health is inadequate, especially concerning long-term effects.

Biotech skeptics are also concerned about the environmental effects of transgenic crops. Biotech companies claim that crops that are genetically designed to repel insects negate the need for chemical (pesticide) control and thus reduce pesticide poisoning of land, water, animals, foods, and farmworkers. However, critics are concerned that insect populations can build up resistance to GM plants with insect-repelling traits, which would necessitate increased rather than decreased use of pesticides. Indeed, a recent study found that while GE crops substantially reduce pesticide use in the first few years of planting, GE crops have increased the overall volume of pesticides applied to corn, soybeans, and cotton over an eight-year period (Benbrook 2003).

Another health and environmental risk is the spread of traits from GM plants to non-GM plants, the effects of which are unknown. In 2003, an analysis of corn grown in nine Mexican states found that 24 percent of the samples tested positive for contamination by several varieties of GM corn, including *Starlink,* produced as cattle feed and deemed unfit for human consumption due to the presence a bacterial protein that is not broken down by the human digestive system, and is therefore a potential allergen (ETC Group 2003b; Ruiz-Marrero 2002). Mexican farmers and community members view the contamination of Mexican corn as an attack on Mexican culture. In the words of one Mexican citizen, "Our seeds, our corn, is the basis of the food sovereignty of our communities. It's much more than a food, it's part of what we consider sacred, of our history, our present and future" (quoted in ETC Group 2003b, 2).

GM seed contamination is of particular concern with regard to seed sterility technology. To maintain control over their products, biotech companies have developed "terminator" seeds which cause the plant to produce sterile seeds. Due to public opposition to "terminator" seeds, in 1999 Monsanto agreed not to market its terminator technology. However, Monsanto has recently adopted a positive stance on genetic seed sterilization, suggesting that the commercialization of terminator technology may occur in the future (ETC Group 2003a). Could the seed sterility trait in terminator crops inadvertently contaminate both traditional crops and wild plant life? The possible ramifications of widespread plant sterility could be devastating to life on earth.

Biotech critics also raise concerns about insufficient safeguards and regulatory mechanisms. In 2000, Taco Bell taco shells, made by Kraft Foods, were recalled after traces of *Starlink* corn were found in the taco shells. No one—from farmers to grain dealers to Kraft—could explain how it got mixed into corn meant for taco shells. The traces of unapproved corn were not found by the United States Department of Agriculture's Food Safety and Inspection Service, nor by the Department of Health and Human Service's Food and Drug Administration. Rather, the traces were discovered by Genetically Engineered Food Alert—a coalition of biotech skeptics. In addition to taco shells, *Starlink* contamination caused a recall of more than 300 corn-based foods. Another example of contamination occurred in 2002, when traces of corn genetically engineered to produce an "edible vaccine" to prevent piglets from diarrhea were found mixed in with soybeans that would be processed into dozens of food items (Hickey & Mittal 2003). These incidents of contamination raise disturbing questions about the regulatory oversight of GE foods to govern food safety, assess risks, monitor compliance, and enforce regulations. But such safeguards are nonexistent in some countries and, as the aforementioned contamination incidents suggest, even when regulatory systems are in place, they are not foolproof.

The Labeling of Foods with GM Ingredients

A 2003 national survey of U.S. adults as well as a compilation of 18 public opinion polls on genetically modified foods reveal that the overwhelming majority of U.S. adults says that GE foods should be labeled as such (Center for Food Safety 2002; Hallman et al., 2003). Without such labeling, consumers cannot exercise their right to make their own choices regarding the purchase and consumption of GE foods. More than 30 countries require labeling of products containing GE ingredients, as well as GE whole foods. Yet no labeling requirement for GE foods exists in the United States (although many food products that do not contain GE foods label their products as "GMO-free"). The Genetically Engineered Food Right to Know Act of 2002 calls for food companies to label all foods containing GE ingredients. Opponents of this bill argue that such a labeling requirement would be too expensive to implement and that consumers would not use the information. Yet when asked how a GM label would affect their purchasing decision, 52 percent of a national sample of U.S. adults said it would make them less willing to purchase the product, compared to 38 percent who said it would make no difference (Hallman et al., 2003).

Worldwide Government Reactions to Biotechnology

In 2000, worldwide concern about the safety of GMOs resulted in 130 nations signing the landmark Biosafety Protocol, which requires that producers of a GMO must demonstrate it is safe before it is widely used. The Biosafety Protocol also allows countries to ban the import of GM crops based on suspected health, ecological, or social risks. As of May 2004, the 25 member countries of the European Union and 6 other countries have banned the importation of GE crops; 36 countries have banned the commercial planting of GE crops (Center for Food Safety and International Forum on Globalization 2003).

Zambia, a country facing widespread famine, refused GM food aid from the United States after its scientists concluded there was insufficient evidence to demonstrate its safety. Other African countries, although plagued by hunger and malnutrition, also reject GM foods. In 1998, all African delegates (except those from South Africa) to the U.N. Food and Agriculture Organization (FAO) released the following statement against GM foods:

> European citizens have been exposed to an aggressive publicity campaign . . . to convince [us] that the world needs genetic engineering to feed the hungry. Organized and financed by Monsanto, one of the world's biggest chemical companies . . . this campaign gives a totally distorted and misleading picture of the potential of genetic engineering to feed developing countries. We, the undersigned delegates of African countries . . . strongly object that the image of the poor and hungry from our countries is being used by giant multinational corporations to push a technology that is neither safe, environmentally friendly, nor economically beneficial to us. . . . We think it will destroy the diversity, the local knowledge and the sustainable agricultural systems that our farmers have developed for millenia and that it will thus undermine our capacity to feed ourselves. (Hickey & Mittal 2003, 4)

Concluding Remarks

Many citizens have clearly taken a stand either for or against agricultural biotechnology. However, many more are uncertain and struggle to make sense out of the competing claims of the benefits and safety versus the potential hazards of GMOs, and the complex ethical and sociopolitical implications of using these technologies.

Lester Brown (2001) of the World Watch Institute suggests that "perhaps the largest question hanging over the future of biotechnology is the lack of knowledge about the possible environmental and human health effects of using genetically modified crops on a large scale over the long term" (p. 52). This lack of knowledge calls for more research to answer questions about the potential risks of GM crops. In the meantime, providing food for hungry populations can be achieved through promoting sustainable agricultural practices that have already been shown to be effective in increasing food crops for small farmers in developing countries (Altieri 2003).

Sources: Altieri, Miguel. 2003 (June 10). "The Case Against Agricultural Biotechnology: Why Are Transgenic Crops Incompatible with Sustainable Agriculture in the Third World?" Corpwatch. www.corpwatch.org; Bailey, Ronald. 2004. "Scientific Arguments Against Biotechnology Are Fallacious." In *Genetically Engineered Foods,* ed. Nancy Harris, pp. 80–93. Farmington Hills, MI: Greenhaven Press; Benbrook, Charles M. 2003 (November). *Impacts of Genetically Engineered Crops on Pesticide Use in the United States: The First Eight Years.* BioTech InfoNet, Technical Paper Number 6. www.biotech-info.net; Brown, Lester R. 2001. "Eradicating Hunger: A Growing Challenge." In *State of the World 2001,* eds. Lester R. Brown, Christopher Flavin, and Hilary French, pp. 43–62. New York: W.W. Norton; Center for Food Safety and International Forum on Globalization. 2004. "Worldwide Regulation, Prohibition, and Production of Genetically Modified Crops and Foods." www.centerforfoodsafety.org; ETC Group. 2003a. "Broken Promise? Monsanto Promotes Terminator Seed Technology." News Release (April 23). www.etcgroup.org; ETC Group. 2003b. "Contamination by Genetically Modified Maize in Mexico Much Worse than Feared." www.etcgroup.org; Genetic ID. 2003. "Field Guide to the GMOs. www.genetic-id.com; Hallman, William K., W. Carl Hebden, Helen L. Aquino, Cara L. Cuite, and John T. Lang. 2003. *Public Perceptions of Genetically Modified Foods: A National Study of American Knowledge and Opinion.* Food Policy Institute. Rutgers University, New Brunswick, NJ. www.foodpolicyinstitute.org; Hickey, Ellen, and Anuradha Mittal. 2003. *Voices from the South: The Third World Debunks Corporate Myths on Genetically Engineered Crops.* Food First Institute for Food and Development Policy and Pesticide Action Network North America. www.foodfirst.org; *Human Development Report. 2003.* 2003. United Nations Development Programme. New York: Oxford University Press; James, Clive. 2002. "Global Status of Commercialized Transgenic Crops: 2002." International Service for the Acquisition of Agri-Biotech Applications." www.isaaa.org; Mendelson, Joseph. 2002. "Why Biotechnology Will Not Feed the World." Center for Food Safety. www.centerforfoodsafety.org; Public Issues Education Project. 2003. "GE Foods in the Market." www.geo-pie.cornell.edu; Ruiz-Marrero, Carmelo. 2002. "Genetic Pollution: Starlink Corn Invades Mexico." CorpWatch. www.corpwatch.org.

Poor health is both a consequence and a cause of poverty; improving the health status of a population is a significant step toward breaking the cycle of poverty. A cross-country comparison of children living in households that survive on a dollar a day found that these very poor children have radically different chances of dying in childhood and being malnourished depending on the country in which they live. In countries with higher levels of per capita public spending on health, children living on a dollar a day had significantly lower levels of mortality and malnutrition (Wagstaff 2003).

Investments in education are also critical for poverty reduction. Increasing the educational levels of a population better prepares individuals for paid employment and for participation in political affairs that affect poverty and other economic and political issues. Improving the educational level and overall status of women in developing countries is also associated with lower birth rates, which in turn fosters economic development.

One way to help poor countries invest in human capital and reduce poverty is to provide debt relief. The Heavily Indebted Poor Countries Initiative, launched in 1996 by the International Monetary Fund and the World Bank and endorsed by 180 governments, relieves low-income countries of their debt to donors, enabling countries to redirect their debt savings to health and education (*Human Development Report* 2003).

Understanding Poverty

As we have seen in this chapter, economic prosperity has not been evenly distributed; the rich have become richer, while the poor have become poorer. Meanwhile, the United States has implemented welfare reform measures that essentially weaken the safety net for the impoverished segment of the population—largely children. Welfare reform legislation of 1996 has achieved its goal of reducing welfare rolls across the country. From 1996 to 2003, the welfare caseload was cut by 59.5 percent (U.S. Department of Health and Human Services 2003b). Advocates of welfare reform argue that transitions from welfare to work benefit children by creating positive role models in their working mothers, promoting maternal self-esteem, and fostering career advancement and higher family earnings. Critics of welfare reform argue that reforms increase stress on parents, force young children into inadequate child care, reduce parents' abilities to monitor the behavior of their adolescents, and deepen the poverty of many families. Of those who leave TANF, only 60 percent are able to find employment, and of those, only a fraction is able to earn a living wage (Parisi, Grice, & Taquino 2003). Although the long-term effects of welfare reform are not yet known, one study of the impact of welfare reform on children concluded that reforms will help some children and hurt others (Duncan & Chase-Lansdale 2001). As we discuss in the next chapter (Work and Unemployment), many of the jobs available to those leaving welfare for work are low paying, have little security, and offer few or no benefits. Without decent wages, and without adequate assistance in child care, housing, health care, and transportation, many families who leave welfare for work find their situation becomes worse, not better.

A common belief among U.S. adults is that the rich are deserving and the poor are failures. Blaming poverty on individual rather than structural and cultural factors implies not only that poor individuals are responsible for their plight, but also that they are responsible for improving their condition. If we hold individuals accountable for their poverty, we fail to make society accountable for making investments in human capital that are necessary to alleviate poverty. Such human capital investments include providing health care, adequate food and housing, education, child care, and job training. Economist Lewis Hill (1998) believes that "the fundamental cause of perpetual poverty is the failure of the American people to invest adequately in the human capital represented by impoverished children" (p. 299). Blaming the poor for their plight also fails to recognize that there are not enough jobs for those who want to work and that many jobs fail to pay wages that enable families to escape poverty. And last, blaming the poor for their condition diverts attention away from the recognition that the wealthy—individuals and corporations—receive far more benefits in the form of wealthfare or corporate welfare, without the stigma of welfare.

Ending or reducing poverty begins with the recognition that doing so is a worthy ideal and an attainable goal. Imagine a world where everyone had comfortable shelter, plentiful food, adequate medical care, and education. If this imaginary world were achieved, and absolute poverty were effectively eliminated, what would the effects be on such social problems as crime, drug abuse, family problems (e.g., domestic violence, child abuse, divorce, and unwed parenthood), health problems, prejudice and racism, and international conflict? But it would be too costly to eliminate poverty—or would it? According to one source, the cost of eradicating poverty worldwide would be only about 1 percent of global income—and no more than 2 to 3 percent of national income in all but the poorest countries (*Human Development Report* 1997). Certainly the costs of allowing poverty to continue are much greater than that.

Chapter Review

- **What is the difference between "absolute poverty" and "relative poverty"?**
  Absolute poverty refers to a lack of basic necessities for life, such as food, clean water, shelter, and medical care. In contrast, relative poverty refers to a deficiency in material and economic resources compared with some other population.

- **What share of the world's population lives on less than $2 per day?**
  According to the World Health Organization, nearly half of the world's population lives on less than $2 per day.

- **Which sociological perspective criticizes wealthy corporations for using financial political contributions to influence politicians to enact policies that benefit corporations and the wealthy?**
  The conflict perspective is critical of wealthy corporations using financial political contributions to influence laws and policies that favor corporations and the rich. Such laws and policies, sometimes referred to as wealthfare or corporate welfare, include low-interest government loans to failing businesses, and special subsidies and tax breaks to corporations.

- **In the United States, what age group has the highest rate of poverty?**
  U.S. children are more likely than adults to live in poverty. More than a third of the U.S. poor population are children. Child poverty rates are much higher in the United States than in Canada or any other Western European industrialized country.

- **According to officials in 25 U.S. cities, what is the main cause of homelessness?**
  City officials in 25 U.S. cities cited lack of affordable housing as the major cause of homelessness, followed by mental illness and the lack of mental health services, substance abuse and the lack of substance abuse services, low-paying jobs, domestic violence, unemployment, poverty, prison release, downturn in the economy, limited life skills, and change and cuts in public assistance programs.

- **Which federal program lifts more children out of poverty than any other program?**
  The federal earned income tax credit (EITC), created in 1975, is a refundable tax credit based on a working family's income and number of children. The EITC lifts more children out of poverty than any other program.

Critical Thinking

1. Should someone receiving welfare benefits be entitled to spend some of his or her money on "nonessentials" such as cosmetics, eating out, lottery tickets, and cable TV? Why or why not?
2. Parenti (1998) points out that reports of income inequality based on U.S. census data are misleading because they do not take into account the super rich—the top 1 percent of income earners. For years, the Census Bureau never interviewed anyone who had an income higher than $300,000. The reportable upper limit of $300,000 was the top figure allowed by the bureau's computer program. In 1994, the bureau raised the upper limit to $1 million. But this figure still excludes the richest 1 percent—the hundreds of billionaires and thousands of multimillionaires who make many times more than $1 million a year. "The super rich simply have been computerized out of the Census Bureau's picture" (Parenti 1998, 36). How does the exclusion of the super rich from census data distort reports of economic inequality? Who benefits from this distortion?
3. The poor in the United States have low rates of voting and thus have minimal influence on elected government officials and the policies they advocate. What strategies might be effective in increasing voter participation among the poor?

Key Terms

absolute poverty
agricultural biotechnology
Aid to Families with Dependent Children (AFDC)
bourgeoisie
corporate welfare
culture of poverty
earned income tax credit (EITC)
feminization of poverty
genetically modified organisms (GMOs)
HPI-1
HPI-2
human capital
Human Poverty Index (HPI)
intergenerational poverty
living wage laws
means-tested programs
Personal Responsibility and Work Opportunity Reconciliation Act (PRWORA)
poverty
proletariat
public housing
relative poverty
Section 8 housing
subsidized housing
Temporary Assistance to Needy Families (TANF)
underclass
wealth
wealthfare
working poor

Taking A Stand

Do laws that prohibit homeless people from begging as well as sleeping in public unfairly punish the homeless?

According to the National Law Center on Homelessness and Poverty (2002), there are more than 750,000 homeless people on any given night and only 250,000 spaces available in shelters, which means that hundreds of thousands of homeless people have no place to be, except in public. Many cities have passed laws that prohibit homeless people from begging as well as sleeping and even sitting in public. Do you think that such laws unfairly punish the homeless? Or are these laws necessary to protect the public?

Use Wadsworth's exclusive online resources—InfoTrac College Edition, MicroCase Online, and OVRC—to formulate a position on this topic.

The Wadsworth's Sociology Online Resources and Writing Companion will help you get started. This valuable guide will show you how to use Wadsworth's exclusive online resources when studying social problems. It will also help you to build essential research and writing skills. InfoTrac College Edition, MicroCase Online, OVRC, and an electronic copy of portions of this companion are available at http://sociology.wadsworth.com/mooney_knox_schacht/problems4e, the companion Web site for *Understanding Social Problems,* Fourth Edition.

Media Resources

The Companion Web Site for *Understanding Social Problems,* Fourth Edition

http://sociology.wadsworth.com/mooney_knox_schacht/problems4e

Supplement your review of this chapter by going to the companion Web site to take one of the Tutorial Quizzes, use the flash cards to master key terms, and check out the many other study aids you'll find there. You'll also find special features such as *Wadsworth's Sociology Online Resources and Writing Companion,* GSS Data, and Census 2000 information, data, and resources at your fingertips to help you complete that special project or do some research on your own.

Interactions CD-ROM

Go to the Interactions CD-ROM for *Understanding Social Problems,* Fourth Edition, to access additional interactive learning tools, such as in-depth review materials, corresponding practice quizzes, and other engaging resources and activities to help you study the concepts in this chapter.

"When a man tells you that he got rich through hard work, ask him whose." *Don Marquis, Journalist*

Work and Unemployment

In 1998, a fifth-grade class in Denver was in the middle of a history unit on American slavery when they learned that black people were still slaves in the Sudan. The students were shocked and decided to take action. They began saving their lunch money toward purchasing the freedom of slaves, and some students contributed cash given to them as birthday presents. In the end, they raised enough money to free two Sudanese slaves. With the help of their teacher, Barbara Vogel, this fifth-grade class founded S.T.O.P—a student group dedicated to ending slavery, which has grown into a nationwide campaign to free slaves.

The persistence of modern day slavery is just one of many concerns regarding the well-being of workers throughout the world. In this chapter, we examine problems of work and unemployment, including slavery, sweatshop labor, health and safety hazards in the workplace, job dissatisfaction and alienation, work/family concerns, and declining labor strength and representation. We begin by looking at the global economy.

Before reading further, you may want to complete the "Attitudes Toward Corporations" survey in the *Self and Society* feature of this chapter. It might be interesting to retake this survey after reading this chapter and see how your attitudes may have changed.

The Global Context: The Economy in the Twenty-First Century

In May 2004, the European Union (EU) accepted 10 new member nations, forming the largest single trading bloc in the world and representing a quarter of the world's wealth. Residents of EU countries can buy and sell goods and services in any of the 25 member countries without tariff barriers, and most of the EU countries share a common currency, the euro. The European Union reflects the increasing globalization of economic institutions. The **economic institution** refers to the structure and means by which a society produces, distributes, and consumes goods and services.

In recent decades, innovations in communication and information technology have spawned the emergence of a **global economy**—an interconnected network of economic activity that transcends national borders and spans the world. The globalization of economic activity means that increasingly our jobs, the products and services we buy, and our nation's political policies and agendas influence and are influenced by economic activities occurring around the world. After summarizing the two main economic systems in the world—capitalism and socialism—we describe how industrialization and post-industrialization have changed the nature of work, and we look at the emergence of free trade agreements and transnational corporations.

Socialism and Capitalism

Socialism is an economic system in which the means of producing goods and services are collectively owned. In a socialist economy, the government controls income-producing property. Theoretically, goods and services are equitably distributed according to the needs of the citizens. Socialist economic systems emphasize collective well-being rather than individualistic pursuit of profit.

Under **capitalism,** private individuals or groups invest capital (money, technology, machines) to produce goods and services to sell for a profit in a competitive market. Whereas socialism emphasizes social equality, capitalism emphasizes individual freedom. Capitalism is characterized by economic motivation through profit, the determination of prices and wages primarily through supply and demand, and the absence of government intervention in the economy. More people are working today in a capitalist economy than ever before in history (Went 2000). Critics of capitalism argue that it creates too many social evils, including alienated workers, poor working conditions, near-poverty wages, unemployment, a polluted and depleted environment, and world conflict over resources.

Both capitalism and socialism claim that they result in economic well-being for society and its members. In reality, capitalist and socialist countries have been unable to fulfill their promises. Although the overall standard of living is higher in capitalist countries, so is economic inequality. Some theorists have suggested that capitalist countries will adopt elements of socialism, and socialist countries will adopt elements of capitalism. This idea, known as the **convergence hypothesis,** is reflected in the economies of Germany, France, and Sweden, which are sometimes called "integrated economies" because they have elements of both capitalism and socialism.

Industrialization, Post-Industrialization, and the Changing Nature of Work

The nature of work has been shaped by the Industrial Revolution, the period between the mid-eighteenth century and the early nineteenth century when the factory system was introduced in England. **Industrialization** dramatically altered the nature of work: machines replaced hand tools; steam, gasoline, and electric power replaced human or animal power. Industrialization also led to the development of the assembly line and an increased division of labor as goods began to be mass produced. The development of factories contributed to the emergence of large cities where the earlier informal social interactions dominated by primary relationships were replaced by formal interactions centered around secondary groups. Instead of the family-centered economy characteristic of an agricultural society, people began to work outside the home for wages.

Post-industrialization refers to the shift from an industrial economy dominated by manufacturing jobs to an economy dominated by service-oriented, information-intensive occupations. Post-industrialization is characterized by a highly educated workforce, automated and computerized production methods, increased government involvement in economic issues, and a higher standard of living.

The three fundamental work sectors (primary, secondary, and tertiary) reflect the major economic transformations in society—the Industrial Revolution and the Post-Industrial Revolution. The primary work sector involves the production of raw materials and food goods. In developing countries with little industrialization, about 60 percent of the labor force works in agricultural activities; in the United States less than 2 percent of the workforce is in farming (*Report on the World Social Situation* 1997; *Statistical Abstract* 2002). The secondary work sector involves the production of manufactured goods from raw materials (e.g., paper from wood). The tertiary work sector includes professional, managerial, technical-support, and service jobs. The transition to a post-industrialized society is marked by a decrease in manufacturing jobs and an increase in service and information-technology jobs in the tertiary work sector. For example, between January 1998 and August 2003, U.S. manufacturing employment dropped from 17.6 million to 14.6 million and its share of gross domestic product (GDP) fell from 16.3 percent to 13.9 percent (Bivens, Scott, & Weller 2003).

Attitudes Toward Corporations

Part One

How good a job do you think corporations are doing these days? Using letter grades like those in school, give corporations an A, B, C, D, or F in the following:

	Letter Grade
1. Paying their employees good wages	______
2. Being loyal to employees	______
3. Making profits	______
4. Keeping jobs in America	______

Part Two

Large corporations are doing things that some people think are serious problems, whereas others think they are not serious problems. For each of the following practices, indicate whether you think this is a serious problem.

	Serious Problem	Not a Serious Problem	Don't Know
5. Not providing health care and pensions to employees	______	______	______
6. Not paying employees enough so that they and their families can keep up with the cost of living	______	______	______
7. Paying CEOs 200 times what their employees make	______	______	______
8. Laying off large numbers of workers even when they are profitable	______	______	______

Part Three

Which of the following statements comes closer to your view? (check one).

9A. A major problem with the economy today is government waste and inefficiency. Excessive government spending and high taxes burden middle-class families and slow economic growth. Our government debt drives up interest rates, making it much harder for businesses to invest and create jobs.

OR

9B. A major problem with the economy today is politicians' catering to the interests of powerful corporations and wealthy campaign contributors at the expense of working families. That is why politicians are not doing anything to stop large corporations from laying off large numbers of employees, denying health benefits, moving jobs overseas, and raiding pension funds.

9A. ______ 9B. ______

10A. Wasteful and inefficient government is preventing the middle class from getting ahead and doing better. Excessive government spending and high taxes burden working families and slow economic growth. The federal debt drives up interest rates and taxes, hurts consumers and business, and reduces job-creating investments. Red tape and excessive regulation are hurting business.

OR

10B. Corporate greed is preventing the middle class from getting ahead and doing better. In the past when people did their jobs well they could earn a decent wage and provide a better life for their children. Now, corporate America is squeezing their employees—cutting wages, downsizing jobs, and eliminating pensions and health benefits. Companies say they can't afford to treat employees better, but many have growing profits, record stock prices, and huge salaries for their executives.

10A. ________ 10B. ________

11A. Large corporations are laying people off, cutting benefits, and moving jobs overseas mainly because they have gotten greedy and are squeezing employees to maximize profits.

OR

11B. Large corporations are laying people off, cutting benefits, and moving jobs overseas mainly because they have to in order to stay in business and provide jobs.

11A. ________ 11B. ________

Results of a National Sample

You may want to compare your answers to this survey with responses from a national sample of U.S. adults.

Part One Responses

Percentages do not total 100 because some individuals responded "Don't know."

	A	B	C	D	F
1.	11%	26%	36%	12%	7%
2.	10%	16%	28%	23%	19%
3.	52%	26%	9%	3%	2%
4.	12%	16%	31%	20%	18%

Part Two Responses

	Serious	Not Serious	Don't Know
5.	82%	15%	3%
6.	76%	19%	5%
7.	79%	14%	7%
8.	81%	14%	5%

Part Three Responses

9A. 33%; 9B. 40%; (21% answered "Both," and 6% answered "Don't know")
10A. 28%; 10B. 46%; (22% answered "Both," and 4% answered "Don't know")
11A. 70%; 11B. 22%; (7% answered "Don't know")

Source: Adapted from "Corporate Irresponsibility: There Ought to Be Laws." 1996. EDK Poll, Washington, DC: Preamble Center for Public Policy. http://www.preamble.org/polledk.html (December 12, 1998). Used by permission.

But a large number of U.S. workers are not educated and skilled enough for many of these tertiary-level positions (Koch 1998). In developing countries, many individuals with the highest level of skill and education leave the country in search of work abroad, leading to the phenomenon known as the **brain drain.** Although U.S. employers benefit as they pay lower wages to foreign workers, U.S. workers are displaced and developing countries lose valuable labor. In recent years, highly skilled foreign workers are finding employment with U.S. companies in their own countries, as work sent abroad includes not only manufacturing jobs but also the highly skilled jobs of aeronautical engineer, software designer, and stock analyst as examples (Uchitelle 2003).

The Globalization of Trade and Free Trade Agreements

Just as industrialization and post-industrialization changed the nature of economic life, so has the globalization of trade—the expansion of trade of raw materials, manufactured goods, and agricultural products across national and hemispheric borders. The first set of global trade rules were adopted through the General Agreement on Tariffs and Trade (GATT) in 1947. Members of GATT met periodically to revise trade agreements in negotiations called "rounds." In1995, GATT was renamed the World Trade Organization (WTO).

In the 1980s and early 1990s, U.S. officials began negotiating regional free trade agreements that would open doors to U.S. goods in neighboring countries and reduce the massive U.S. trade deficit, which had grown from $25.3 billion in 1980 to $122 billion in 1985 (Schaeffer 2003). **Free trade agreements** are pacts between two countries, or a group of countries, that make it easier to trade goods across national boundaries. Free trade agreements reduce or eliminate foreign restrictions on exports, reduce or eliminate tariffs (or taxes) on imported goods, and prevent U.S. technology from being copied and used by competitors through protection of "intellectual property rights." Treaties such as the Canada-U.S. Free Trade Agreement, the North American Free Trade Agreement (NAFTA), the Free Trade Area of the Americas (FTAA), and the U.S.-Central American Free Trade Agreement (CAFTA) are designed to accomplish these trade goals.

U.S. officials have also used Section 301 of the Trade Acts of 1984 and 1988 to force trade negotiations with individual countries. If U.S. trade officials determine that other countries have denied U.S. corporations "reasonable" access to domestic markets, sold their goods in the United States at below-market prices, or failed to protect the patents and copyrights of U.S. companies, Section 301 allows the United States to impose retaliatory sanctions and tariffs on goods from these countries (Schaeffer 2003).

Through GATT and the WTO, free trade agreements, and Section 301, U.S. trade officials have expanded trading opportunities, benefiting large export manufacturing and service industries in the North, specifically aircraft, auto, computer, pharmaceutical, and entertainment industries in Western Europe, the United States, and Japan. But trade globalization also hurt the U.S. steel and textile-apparel industries and the workers employed in them; small businesses who cannot compete with large retail chain stores, supermarkets, and franchises; and small farmers (Schaeffer 2003). Foreign workers have also been hurt by trade agreements. Consider the effects of NAFTA, which requires Mexico to allow free entry and exit of investment and lifts trade barriers for Mexican exports to the United States, making production there for export to the United States more profitable. In the 10 years after NAFTA went into effect, U.S. workers lost 879,280 jobs, and real wages in Mexico fell. NAFTA has also contributed to rising income inequality, suppressed real wages for production workers, weakened workers' collective bargaining powers and ability to organize unions, and reduced their benefits (Scott 2003).

Free trade agreements also undermine the ability of national, state, and local governments to implement environmental and food or product safety policies.

> Any country that decides, for example, to ban the export of raw logs as a means of conserving its forests, or ban the use of carcinogenic pesticides, can be charged under the WTO by member states on behalf of their corporations for obstructing the free flow of trade and investment. A secret tribunal of trade officials would then decide whether these laws were "trade restrictive" under the WTO rules and should therefore be struck down. Once the secret tribunal issues its edict, no appeal is possible. The country convicted is obligated to change its laws or face the prospect of perpetual trade sanctions. (Clarke 2002, 44)

Transnational Corporations

While free trade agreements have increased business competition around the world, resulting in lower prices for consumers for some goods, they also opened markets to monopolies (and higher prices) because they facilitated the development of large-scale transnational corporations. **Transnational corporations** (TNCs), also known as *multinational corporations,* are corporations that have their home base in one country and branches, or affiliates, in other countries. In less than 20 years, the number of transnational corporations has increased from seven to over 45,000. The top 100 economies around the world are transnational corporations rather than nations, and the combined yearly revenues of the largest corporations are greater than those of 182 nations, which are home to over four-fifths of the world's population (Clarke 2002). Three to six TNCs control 85 to 90 percent of world wheat, corn, coffee, cotton, and tobacco exports, 90 percent of forest product exports, and about 90 percent of iron ore exports (Schaeffer 2003).

Transnational corporations provide jobs for U.S. managers, secure profits for U.S. investors, and help the United States compete in the global economy. Transnational corporations benefit from increased access to raw materials, cheap foreign labor, and the avoidance of government regulations. "By moving production plants abroad, business managers may be able to work foreign employees for long hours under dangerous conditions at low pay, pollute the environment with impunity, and pretty much have their way with local communities. Then the business may be able to ship its goods back to its home country at lower costs and bigger profits" (Caston 1998, 274–275). Transnational corporations contribute to the trade deficit in that more goods are produced and exported from outside the United States than from within. Transnational corporations also contribute to the budget deficit, as the United States does not get tax income from U.S. corporations abroad, yet TNCs pressure the government to protect their foreign interests; as a result, military spending increases. Third, TNCs contribute to U.S. unemployment by letting workers in other countries perform labor that could be performed by U.S. employees. An estimated 15 percent of the 2.81 million jobs lost in America have gone overseas (Uchitelle 2003). Finally, transnational corporations are implicated in an array of other social problems such as poverty resulting from fewer jobs, urban decline resulting as factories move away, and racial and ethnic tensions resulting from competition for jobs.

Sociological Theories of Work and the Economy

Numerous theories in economics, political science, and history address the nature of work and the economy. In sociology, structural-functionalism, conflict theory, and symbolic interactionism serve as theoretical lenses through which we may better understand work and economic issues and activities.

Structural-Functionalist Perspective

According to the structural-functionalist perspective, the economic institution is one of the most important of all social institutions. It provides the basic necessities common to all human societies, including food, clothing, and shelter. By providing for the basic survival needs of members of society, the economic institution contributes to social stability. After the basic survival needs of a society are met, surplus materials and wealth may be allocated to other social uses, such as maintaining military protection from enemies, supporting political and religious leaders, providing formal education, supporting an expanding population, and providing entertainment and recreational activities. Societal development is dependent on an economic surplus in a society (Lenski & Lenski 1987).

Although the economic institution is functional for society, elements of it may be dysfunctional. For example, before industrialization, agrarian societies had a low degree of division of labor in which few work roles were available to members of society. Limited work roles meant that society's members shared similar roles and thus developed similar norms and values (Durkheim [1893] 1966). In contrast, industrial societies are characterized by many work roles, or a high degree of division of labor, and cohesion is based not on the similarity of people and their roles but on their interdependence. People in industrial societies need the skills and services that others provide. The lack of common norms and values in industrialized societies may result in ***anomie***—a state of normlessness—which is linked to a variety of social problems including crime, drug addiction, and violence (see Chapters 3 and 4).

The structural-functionalist perspective is also concerned with how changes in one aspect of society affect other aspects. How, for example, does the level of economic development of a country affect the subjective life satisfaction and core values of its population? Data from the World Value Survey indicate that as one moves from subsistence-level economies in developing countries to advanced industrialized societies, there is a large increase in the percentage of the population who consider themselves very happy or satisfied with their lives as a whole. But above a certain economic point, the curve levels off. In other words, "moving from a starvation level to a reasonably comfortable existence makes a big difference," but once this level is reached, further economic development does not increase subjective well-being (Inglehart 2000, 219). Economic development level also affects core values, as economic insecurity breeds xenophobia and deference to authority, while a sense of basic security fosters values such as self-expression (rather than deference to authority) and not only a tolerance of cultural diversity, but a sense that cultural differences are stimulating and exotic (Inglehart 2000).

"If one grows up with a feeling that survival can be taken for granted, instead of feeling that survival is uncertain, it influences almost every aspect on one's worldview."

Ronald Inglehart
Institute for Social Research
University of Michigan

Conflict Perspective

According to Karl Marx, the ruling class controls the economic system for its own benefit and exploits and oppresses the working masses. Whereas structural-functionalism views the economic institution as benefiting society as a whole, conflict theory holds that capitalism benefits an elite class that controls not only the economy but other aspects of society as well—the media, politics and law, education, and religion.

As an indication of the ties between business and government, consider that both George W. Bush and Dick Cheney come from the oil industry. Also, the Bush cabinet included as many or more corporate executives than any previous administration, and Bush's transition teams for the Department of Energy, Department of Health and Human Services, and the Department of Labor were made up almost entirely of people affiliated with or working for corporate interests (Mokhiber & Weissman 2001).

Corporate interests also find their way into politics through large political contributions. During the 2000 national election cycle, the Democratic and Republican parties raised a record $463.1 million in soft money contributions—nearly double the amount raised in the 1996 election. And soft money contributions to 2000 congressional campaigns were triple that of 1996 (Common Cause 2001). **Soft money** is money that flows through a loophole to provide political parties, candidates, and contributors a means to evade federal limits on political contributions. Critics of this system of campaign financing argue that corporations and interest groups purchase political influence through financial contributions. Due to the high cost of political campaigns, many candidates rely on funds from special interests, who then expect (and often get) special treatment in the law. The special treatment can be in the form of business taxes, environmental loopholes, subsidies, or lower standards of consumer and worker protection, for example. Although the top soft money donors in the 2000 elections included two labor unions, most large contributors were corporations, including Philip Morris, AT&T, Amway, American Financial Group Insurance Company, and Microsoft (Common Cause 2000).

A survey of business leaders' views on political fund-raising found that the main reasons U.S. corporations make political contributions is fear of retribution and to buy access to lawmakers ("Big Business for Reform" 2000). Although 75 percent say political donations give them an advantage in shaping legislation, nearly three-quarters (74%) say business leaders are pressured to make large political donations. Half of the executives said their colleagues "fear adverse consequences for themselves or their industry if they turn down requests" for contributions.

Penalties for violating health and safety laws in the workplace provide an example of legal policy that favors corporate interests. Suppose a corporation is guilty of a serious violation of health and safety laws, in which "serious violation" is defined as one that poses a substantial probability of death or serious physical harm to workers. What penalty do you think that corporation should pay for such a violation? According to a report by the AFL-CIO (2002), serious violations of workplace health and safety laws carry an average penalty of only $910.00. As discussed in Chapter 4, penalties for corporate crimes tend to be much less severe than those applied to individuals who violate the law. For example, under federal law, causing the death of a worker by willfully violating safety rules—a misdemeanor with a six-month maximum prison term—is a less serious crime than harassing a wild burro on federal lands, which is punishable by a year in prison (Barstow & Bergman 2003).

Corporate power is also reflected in the policies of the International Monetary Fund (IMF) and the World Bank, which pressure developing countries to open their economies to foreign corporations, promoting export production at the expense of local consumption, encouraging the exploitation of labor as a means of attracting foreign investment, and hastening the degradation of natural resources as countries sell their forests and minerals to earn money to pay back loans. Ambrose (1998) asserts that "for some time now, the IMF has been the chief architect of the global economy, using debt leverage to force governments around the world to give big corporations and billionaires everything they want—low taxes, cheap labor, loose regulations—so they will locate in their countries" (p. 5).

Symbolic Interactionist Perspective

According to symbolic interactionism, the work role is a central part of a person's identity. When making a new social acquaintance, one of the first questions we usually ask is "What do you do?" The answer largely defines for us who that person is. For example, identifying a person as a truck driver provides a different social meaning than identifying someone as a physician. In addition, the title of one's work status—maintenance supervisor or university professor—also gives meaning and self-worth to the individual. An individual's job is one of his or her most important statuses; for many, it represents a "master status," that is, the most significant status in a person's social identity.

"No race can prosper 'til it learns there is as much dignity in tilling a field as in writing a poem."

Booker T. Washington
Address to the Atlanta Exposition, September 18, 1895

Symbolic interactionism emphasizes that attitudes and behavior are influenced by interaction with others. The applications of symbolic interactionism in the workplace are numerous—employers and managers are concerned with using interpersonal interaction techniques that achieve the attitudes and behaviors they want from their employees; union organizers are concerned with using interpersonal interaction techniques that persuade workers to unionize; and job-training programs are concerned with using interpersonal interaction techniques that are effective in motivating participants.

Problems of Work and Unemployment

Next, we examine unemployment and other problems associated with work. Problems of workplace discrimination based on gender, race and ethnicity, and sexual orientation are addressed in other chapters. Minimum wage and living wage issues are discussed in Chapter 10. Here we discuss problems concerning slavery, sweatshop labor, health and safety hazards in the workplace, job dissatisfaction and alienation, work/family concerns, unemployment and underemployment, and labor unions and the struggle for workers' rights.

Slavery

To most Americans, slavery is a thing of the past. But today, an estimated 27 million people worldwide are bought and sold, held captive against their will, brutalized, and exploited for profit (Bales 1999). Slavery is defined as the loss of free will, a condition in which a person is forced through violence or the threat of violence to give up the ability to sell freely his or her labor power (Bales & Robbins 2001). Slavery expert Kevin Bales (1999) explains that the resurgence of slavery around the world is linked to three main factors: (1) rapid growth in population, especially in the developing world; (2) social and economic changes that have displaced many rural dwellers to urban centers and their outskirts, where people are powerless and jobless and are vulnerable to exploitation and slavery; and (3) government corruption that allows slavery to go unpunished, even though it is illegal in every country.

Slavery exists all over the world but is most prevalent in India, Pakistan, Bangladesh, and Nepal. Most slaves are forced to work in agriculture, mining, prostitution, and factories. Slave labor produces goods we use every day, including sugar from the Dominican Republic, chocolate from Ivory Coast, paper clips from China, carpets from Nepal, and cigarettes from India. Many of the slaves throughout the world are children. (Child Labor is discussed in Chapter 9).

Forms of Slavery. Modern slavery is different from the old form most people know, which is **chattel slavery.** In chattel slavery, slaves were considered property that could be bought and sold. In the past, the high cost of purchasing a slave (around $40,000 in today's money) gave the master incentive to provide a minimum standard of care to ensure that the slave would be healthy enough to work and generate profit for the long term. Today, slaves are cheap, costing an average of around $100 (Cernasky 2002). Because they are so cheap and abundant, slaves are no longer a major investment worth maintaining. If slaves become ill or injured, too old to work, or troublesome to the slaveholder, they are dumped or killed and replaced with another slave (Cernasky 2002).

Although chattel slavery still exists in some areas, most slaves today are not "owned" but are rather controlled by violence or the threat of violence. The most common form of slavery today is called debt bondage or bonded labor. Debtors are usually illiterate, landless, rural poor who take out a loan simply to survive or to pay for a wedding, funeral, medicines, fertilizer, or other necessities. Debtors must work for the creditor to pay back the loan, but often they are unable to repay it. Creditors can keep debtors in bondage indefinitely in two main ways. They can charge the debtors illegal fines (for workplace "violations" or for poorly performed work) and can also charge laborers for food, tools, and transportation to the work site while keeping wages too low for the debt to ever be repaid. Alternatively, creditors can claim that all the labor performed by the debtor is collateral for the debt and cannot be used to reduce it (Miers 2003).

Another common form of slavery is forced labor, where individuals are lured by the promise of a good job and instead find themselves enslaved. Migrant workers are particularly vulnerable. Organized crime rings are sometimes involved in the international trafficking in human beings, which often flows from developing nations to the West. A form of forced labor most common in South Asia is sex slavery, where girls are forced into prostitution by their own husbands, fathers, and brothers to earn money to pay family debts. Other girls are lured by offers of good jobs and then forced to work in brothels under the threat of violence.

Slavery in the United States. Kevin Bales estimates that there are between 100,000 and 150,000 slaves in the United States today, mostly due to human trafficking for domestic work, migrant farm labor, or work in the sex industry (Cockburn 2003). Migrant workers are tricked into working for little or no pay as means of repayment for debts from their transport across the U.S. border, similar to debt bondage in South Asia. Traffickers posing as employment agents lure women into the United States with the promise of good jobs and education, but then place them in "jobs" where they are forced to do domestic or sex work.

Sweatshop Labor

A U.S. Department of Labor investigation of the Daewoosa Samoa garment factory in American Samoa—a factory that produces men's sportswear for J.C. Penney—found that garment factory workers lived and worked under conditions of poor sanitation, malnutrition, electrical hazards, fire hazards, machinery hazards, illegally low wages, sexual harassment and invasion of privacy, workplace violence and corporal punishment, and overcrowded barracks where two workers were forced to share each bed (National Labor Committee 2001). Female workers reported that the company owner routinely enters their barracks to watch them shower and dress. Workers reported incidents in which security guards slapped and kicked workers. The food provided to the workers at the Daewoosa Samoa garment factory consisted of a watery broth of rice and cabbage. The factory owner, Mr. Kil Soo Lee, ignores court orders to improve worker conditions at his factory and when investigators show up to inspect the work site, Mr. Lee "happens to have $60,000 in a paper bag to flaunt in front of investigators" as a bribe (U.S. Department of Labor, quoted in National Labor Committee 2001).

The workers at the Daewoosa Samoa garment factory are among the millions of people worldwide who work in **sweatshops**—work environments that are characterized by less-than-minimum wage pay, excessively long hours of work (often without overtime pay), unsafe or inhumane working conditions, abusive treatment of workers by employers, and/or the lack of worker organizations aimed at negotiating better work conditions. Sweatshop labor conditions occur in a wide variety of industries, including garment production, manufacturing, mining, and agriculture. The dangerous conditions of sweatshops result in high rates of illness, injury, and death. In one tragic example of death resulting from sweatshop conditions, at least 53 workers, including 10 children, were burned to death in a fire at a Sagar Chowdury garment factory in Bangladesh (Hargis 2001). The fire, caused by an electrical short circuit, engulfed the entire factory with 900 workers who were *locked inside.* Local residents and firefighters broke open the locked gates of the building and rescued survivors.

Sweatshop Labor in the United States. Sweatshop conditions in overseas garment and footwear industries have been widely publicized. However, many Americans do not realize the extent to which sweatshops exist in the United States. The Department of Labor estimates that over half of the country's 22,000 sewing shops violate minimum wage and overtime laws and 75 percent violate safety and health laws ("The Garment Industry" 2001). The majority of garment workers in the United States are immigrant women who typically work 60 to 80 hours a week, of-

Farmworkers in Immokalee, Florida, typically earn 40 cents for every 32-pound bucket of tomatoes they pick and haul. This means that in order to earn 50 dollars, a farmworker must pick two tons of tomatoes.

ten earning less than minimum wage, with no overtime, and many face verbal and physical abuse.

Migrant farm workers, who process more than 85 percent of the fruits and vegetables grown in the United States, also work under sweatshop conditions. Many live in substandard and crowded housing provided by their employer and lack access to safe drinking water as well as bathing and sanitary toilet facilities. Farmworkers commonly suffer from heat exhaustion, back and muscle strains, injuries due to the use of sharp and heavy farm equipment, and illness resulting from pesticide exposure (Austin 2002). Working 12-hour days under hazardous conditions, farmworkers have the lowest annual family incomes of any U.S. wage and salary workers, and more than 60 percent of them live in poverty (Thompson 2002).

Health and Safety Hazards in the U.S. Workplace

Many workplaces are safer today than in generations past. Nevertheless, fatal and disabling occupational injuries and illnesses still occur in troubling numbers. The incidence of illnesses due to hazardous working conditions is probably much higher than reported statistics show. The Bureau of Labor Statistics (2002) explains, "Some conditions (for example, long-term latent illnesses caused by exposure to carcinogens) often are difficult to relate to the workplace and are not adequately recognized and reported" (p. 4).

Workplace Fatalities. The International Labor Organization estimates that 1.1 million workers worldwide die on the job or from occupational disease each year (*Multinational Monitor* 2000). In 2002, 5,524 U.S. workers—92 percent of whom were men—died of fatal work-related injuries (Bureau of Labor Statistics 2003b). The most common type of job-related fatality involves transportation accidents (see Figure 11.1). Industries with the highest rates (per 100,000 workers) of fatal injuries

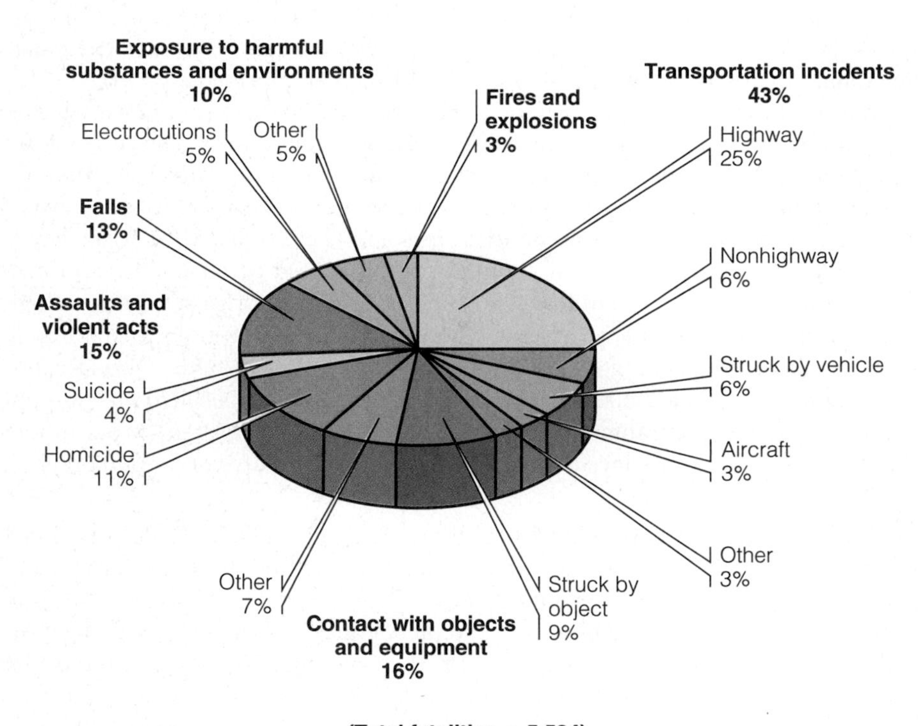

Figure 11.1
The Manner in Which Workplace Fatalities Occurred: 2002

Source: Bureau of Labor Statistics. 2003. "National Census of Fatal Occupational Injuries, 2002." U.S. Department of Labor. Washington, DC.

Figure 11.2
Occupational Injuries and Illnesses Involving Days Away from Work: 2001

Source: Bureau of Labor Statistics. 2002. "Survey of Occupational Injuries and Illnesses." U.S. Department of Labor. Washington, DC.

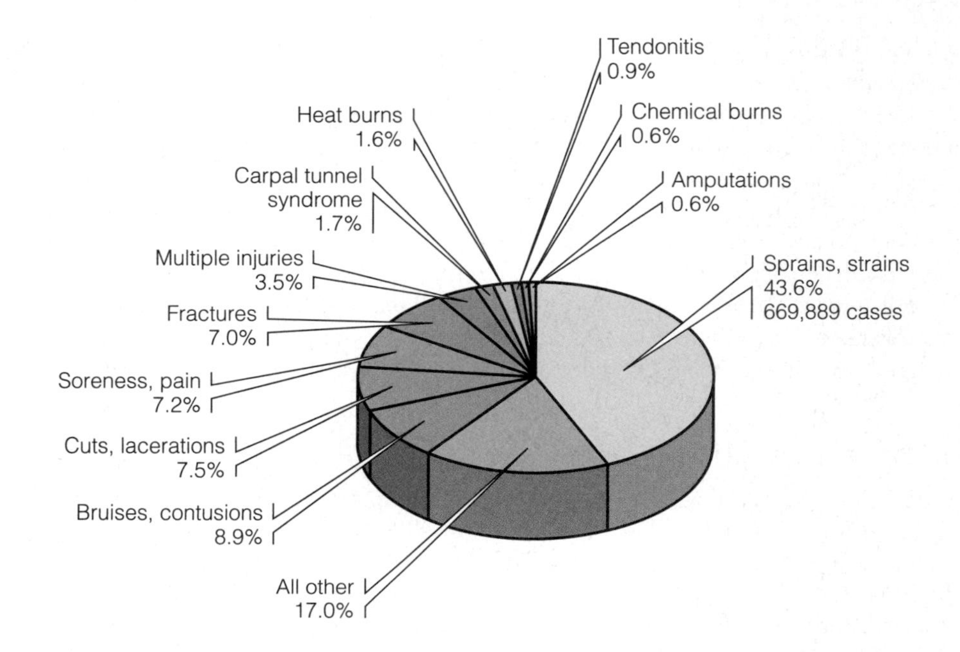

include agriculture, forestry, fishing, mining, construction, and transportation and public utilities.

Occupational Illnesses and Nonfatal Injuries. The Bureau of Labor Statistics (2002) reported a total of 5.2 million nonfatal occupational injuries and illnesses in private industry in 2001—a rate of 5.7 cases per 100 full-time workers. This was the lowest rate since the Bureau began reporting this information in the early 1970s. As shown in Figure 11.2, sprains and strains are the most common nonfatal occupational injury/illness involving days away from work. Truck drivers suffer the most injuries and illnesses involving days away from work, followed by nursing aides, who commonly experience back strains from lifting patients (see Figure 11.3).

The most common types of workplace illness are disorders associated with repeated motion or trauma, such as carpal tunnel syndrome (a wrist disorder that can cause numbness, tingling, and severe pain), tendonitis (inflammation of the tendons), and noise-induced hearing loss. Such disorders—referred to by a number of terms, including **cumulative trauma disorders** and **repetitive motion disorders**—are muscle, tendon, vascular, and nerve injuries that result from repeated or sustained actions or exertions of different body parts. Jobs that are associated with high rates of upper-body cumulative trauma disorders include computer programming, manufacturing, meatpacking, poultry processing, and clerical/office work. Repetitive motion disorders are classified as illness, not as injury, because they are not sudden, instantaneous traumatic events. As shown in Figure 11.4, carpal tunnel syndrome results in more days absent from work than fractures or amputations.

Job Stress. Another work-related health problem is job stress. When a national sample of U.S. workers was asked, "In general, how stressed do you feel at work?" nearly one in five (19%) responded "quite a bit" or "extremely"; a third (33%) responded "somewhat," and 47 percent responded either "a little" or "not at all" ("Attitudes in the American Workplace IX" 2003). A national Gallup poll found that nearly one-third (31%) of workers are "somewhat dissatisfied" or "completely dissatisfied" with the amount of on-the-job stress they experience (Newport 2002). In

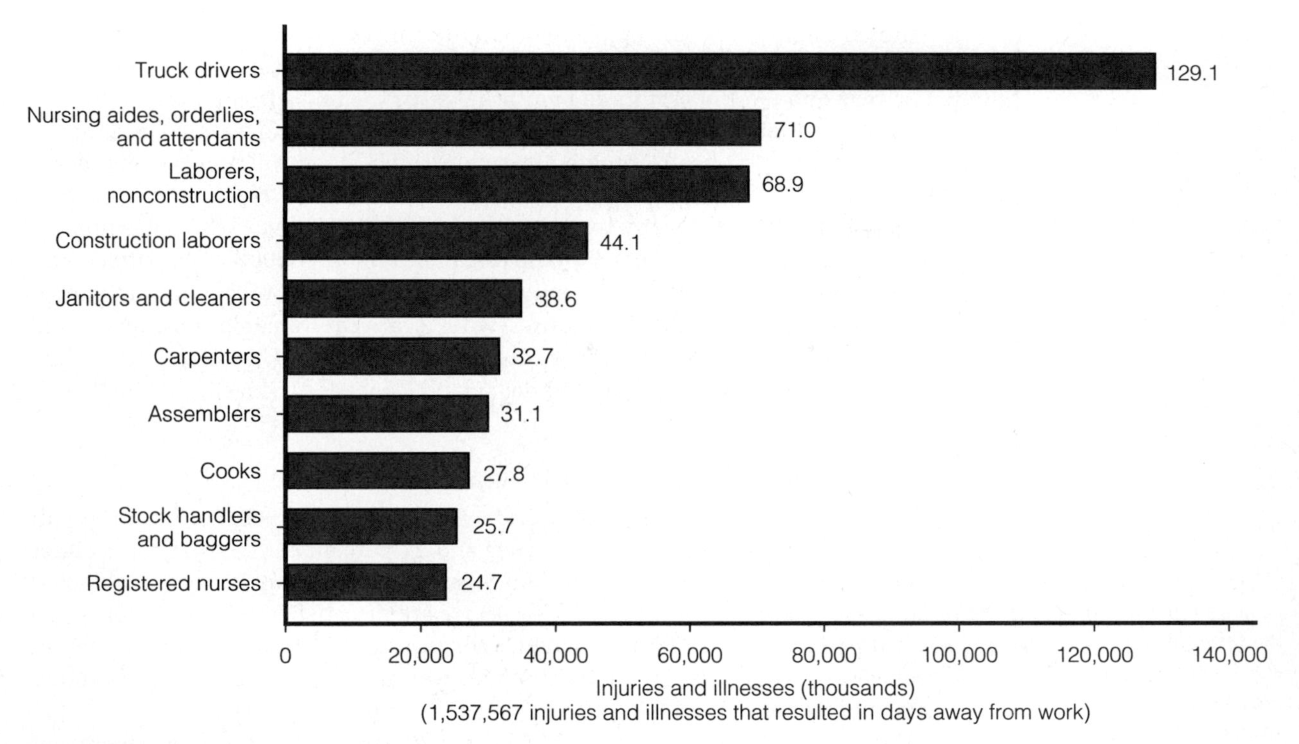

Figure 11.3
Occupations with the Most Injuries and Illnesses Involving Days Away from Work: 2001
Source: Bureau of Labor Statistics. 2002. "Survey of Occupational Injuries and Illnesses." U.S. Department of Labor. Washington, DC.

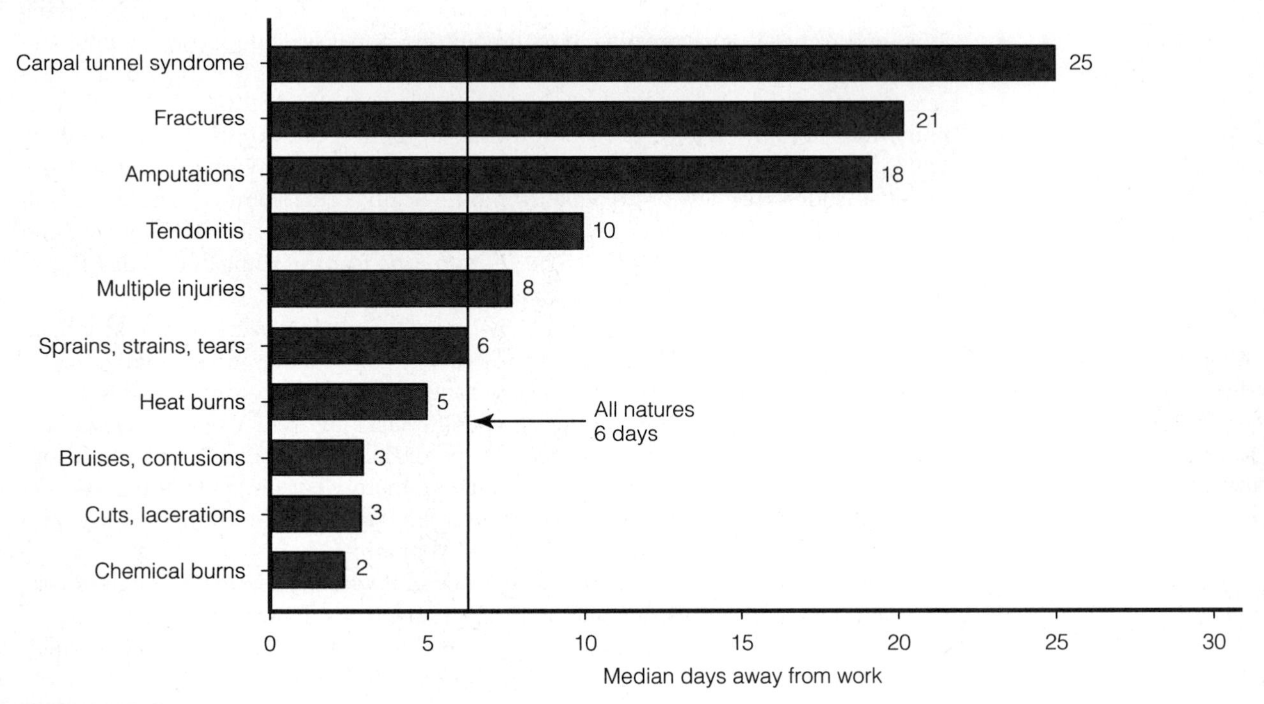

Figure 11.4
Median Days Away from Work Due to Nonfatal Occupational Injury or Illness by Nature: 2001
Source: Bureau of Labor Statistics. 2002. "Survey of Occupational Injuries and Illnesses." U.S. Department of Labor. Washington, DC.

a national sample of U.S. employees, more than one-quarter (28%) "felt overworked" and 28 percent "felt overwhelmed" by how much work they had to do often or very often in the past three months (Galinsky, Kim, & Bond 2001).

Prolonged job stress, also known as **job burnout,** can cause or contribute to physical and mental health problems, such as high blood pressure, ulcers, headaches, anxiety, and depression. Taking time off to heal and "recharge one's batteries" is not an option for many workers: one-half of the U.S. workforce has no paid sick leave and one-quarter has no paid vacation (Watkins 2002). In contrast, employers in France and Spain are legally required to provide 30 days of paid annual leave to their workers. Other countries with legal requirements for annual paid leave include Finland (24 days); Norway (21 days); the United Kingdom, the Netherlands, Italy, and Germany (20 days); and Canada (10 days) (Watkins 2002).

Job Dissatisfaction and Alienation

How prevalent is job dissatisfaction? Surveys yield different results. A Gallup poll (2002) found that 9 percent of U.S. whites and 24 percent of U.S. blacks are either somewhat or very dissatisfied with their jobs. Another national survey conducted in 2003 found that 51 percent of U.S. workers are dissatisfied with their jobs—up from 40 percent in 1995 (Franco 2003). This survey found that only about one in three workers was content with his or her wages, and, not surprisingly, satisfaction levels tend to rise with earnings.

"Clearly the most unfortunate people are those who must do the same thing over and over again, every minute, or perhaps twenty to the minute. They deserve the shortest hours and the highest pay."

John Kenneth Galbraith
American economist

Factors that contribute to job satisfaction include income, prestige, a feeling of accomplishment, autonomy, a sense of being challenged by the job, opportunities to be creative, congenial coworkers, the feeling that one is making a contribution, fair rewards (pay and benefits), promotion opportunities, and job security (Bavendam 2000; Gordon 1996). These factors often overlap—for example, high-paying jobs tend to have more prestige, be more autonomous, provide more benefits, and permit greater creativity. Yet many jobs lack these qualities, leaving workers dissatisfied. Many workers are worried about losing their job and feel that their company does not have a strong sense of loyalty to them (Bond, Galinsky, & Swanberg 1997). Workers are also dissatisfied with declining wages. Workers in low-wage, low-status jobs with few or no benefits and little job security are vulnerable to feeling dissatisfaction with not only their jobs but also with their limited housing and lifestyle options. In this chapter's *The Human Side* feature, some of the dissatisfactions associated with low-wage work are expressed.

"Without work, all life goes rotten, but when work is soulless, life stifles and dies."

Albert Camus
Philosopher

One form of job dissatisfaction is a feeling of **alienation.** Work in industrialized societies is characterized by a high degree of division of labor and specialization of work roles. As a result, workers' tasks are repetitive and monotonous and often involve little or no creativity. Limited to specific tasks by their work roles, workers are unable to express and utilize their full potential—intellectual, emotional, and physical. According to Marx, when workers are merely cogs in a machine, they become estranged from their work, the product they create, other human beings, and themselves. Marx called this estrangement "alienation."

Alienation usually has four components: powerlessness, meaninglessness, normlessness, and self-estrangement. Powerlessness results from working in an environment in which one has little or no control over the decisions that affect one's work. Meaninglessness results when workers do not find fulfillment in their work. Workers may experience normlessness if workplace norms are unclear or conflicting. For example, many companies that have family leave policies informally discourage workers from using them. Or workplaces that officially promote nondiscrimination in reality practice discrimination. Alienation also involves a feeling of self-estrangement, which stems from the workers' inability to realize their full human potential in their work roles and lack of connections to others.

Work/Family Concerns

Spouses, parents, and adult children caring for elderly parents increasingly struggle to balance their work and family responsibilities. When Hochschild (1997) asked a sample of employed parents, "Overall, how well do you feel you can balance the demands of your work and family?" only 9 percent said "very well" (pp. 199–200).

In nearly two-thirds (61%) of married couples with children under 18, and in more than half (54%) of married couples with children under age 6, both parents are employed. And 77 percent of women in female-headed single-parent households and 85 percent of men in male-headed single-parent households are employed (Bureau of Labor Statistics 2003c). A major concern of employed parents is arranging, and paying for, child care. About 3.3 million children under age 13 (15% of 6- to 12-year-olds) are left without adult supervision for some period of time each week (Vandivere, Tout, Zaslow, Calkins, & Capizzano 2003).

Another work/family concern involves caring for elderly family members. More than one-third (35%) of workers say they provided care for a relative or in-law 65 or older in the past year (Bond, Thompson, Galinsky & Prottas 2002). In the Strategies for Action section later in this chapter, we discuss policies and programs that address the work/family concerns of Americans today.

Unemployment and Underemployment

The International Labor Organization (2004) reported that in 2003 an estimated 185.9 million people worldwide—6.2 percent of the labor force—were unemployed. This is the highest level of global unemployment ever recorded.

Measures of **unemployment** in the United States consider an individual to be unemployed if he or she is currently without employment, is actively seeking employment, and is available for employment. Rates of unemployment are higher among racial and ethnic minorities (see Chapter 6), and among those with lower levels of education (see Table 11.1).

In 2000, the U.S. unemployment rate dipped to a 31-year low of 4 percent (Mishel, Bernstein, & Boushey 2003). But unemployment rose rapidly following the

Table 11.1
Unemployment Rates of U.S. Adults 25 Years and Over, by Education: October 2003

Level of Educational Attainment	Unemployment Rate
Less than high school diploma	8.9%
High school graduate	5.5%
College graduate	3.0%

Source: "The Employment Situation: October 2003." Bureau of Labor Statistics. Washington, DC: U.S. Department of Labor.

The Human Side

Excerpts from an Interview with Barbara Ehrenreich, Author of *Nickel and Dimed: On (Not) Getting By in America*

Barbara Ehrenrich's bestselling book *Nickel and Dimed* describes the day-to-day struggles of low-wage work and surviving on a low-wage income.

Barbara Ehrenreich, a successful journalist with a Ph.D, wondered how anyone could survive on low-income wages: six to seven dollars an hour. To find out, she lived for one year on wages she earned doing low-wage work, moving from Florida to Maine to Minnesota taking jobs as a waitress, hotel maid, house cleaner, nursing home aide, and Wal-Mart sales person, and living in the cheapest lodging she could find that offered an acceptable level of safety. Ehrenreich chronicled her experiences in *Nickel and Dimed: On (Not) Getting By in America*—a book that became a *New York Times* bestseller. In this *Human Side* feature, we present excerpts from an interview of Barbara Ehrenreich by Jamie Passaro in which Barbara talks about her experiences as a low-wage earner and her viewpoints concerning low-wage work in America.

Passaro: What surprised you most during your months of low-wage work?

Ehrenreich: It was a surprise to me how challenging these jobs were. I was expecting that I would be doing dull, repetitive work, that I would be bored out of my mind. Instead I was struggling all the time, physically and mentally, to master these jobs. At Wal-Mart I had to memorize the locations of hundreds of clothing items so I could put everything back in its exact place. In the nursing home I had about fifteen minutes to learn the names and dietary requirements of thirty patients. It took all the concentration I had. So I no longer use the word *unskilled* to describe any job.

Passaro: How do you think your experience would have been different if you were a man?

Ehrenreich: A lot of low-wage jobs are really for either sex now because, as heavy industry declines, the "masculine" jobs of the past are not there anymore. There are men working at Wal-Mart and in restaurants and in nursing homes. The only difference for me is that a man probably would not have been as fearful as I was about living in a creepy residential motel with no privacy or security. . . .

Passaro: When you went back to your middle-class life after working low-wage jobs, how were you different?

Ehrenreich: I was more impatient with affluent people who don't see these problems or who aren't particularly interested and brush them off . . .

Passaro: Why do you think class inequality is such a taboo subject in the mainstream media?

Ehrenreich: It undercuts the American myth that anybody can become rich, that it's just a matter of personal ability and determination. . . . We like to tell ourselves that everybody is equal. To admit that large numbers of people are systematically held back is hard, because it means upward mobility is not an option for everybody. But

events of September 11, 2001, and in 2003, the unemployment rate was about 6 percent, slightly lower than the 6.8 percent average of all the industrialized countries (International Labor Organization 2004). However, other industrialized countries have higher wages than the United States and provide more social supports, such as universal health care and unemployment insurance, for their citizens.

One cause of unemployment is **corporate downsizing**—the corporate practice of discharging large numbers of employees. Simply put, the term *downsizing* is a euphemism for mass firing of employees (Caston 1998). The Labor Department used to keep monthly records of mass layoffs by U.S. companies, but the Bush adminis-

The Human Side

that's the way it is. There are just too many things pressing poor people down, keeping them where they are. . . . The poor have become "invisiblized" in our society. They're given very little mention in the news and entertainment media. You just don't hear about them. The media system is fed by corporate advertising, and advertisers want "good demographics"—that is, they want to reach mostly the upper middle-class. . . .

Passaro: Many editors claim the middle class isn't interested in reading about poverty or the working poor. So why did *Nickel and Dimed* grab the attention of the media and the middle-class people who are presumably reading it?

Ehrenreich: One reason is that *Nickel and Dimed* is very personal and subjective, not preachy. It's not about the poor in general. It's just about me trying to survive. So people who are completely unfamiliar with the world of low-wage work can see it through the eyes of someone who is somewhat like them. . . .

Passaro: Do you think that we should boycott chain stores and restaurants that don't pay a living wage?

Ehrenreich: And then where are you going to shop or eat? At an upscale restaurant where the busboys and the dishwashers still earn little above minimum wage and the coffee beans have been picked by children in Central America? A lot of people come up to me and say, "I'll never go to Wal-Mart again." Well, terrific. So you go to a nice little boutique, which also pays its retail clerks seven dollars an hour and maybe gets its very expensive clothes from sweatshops, too. You could pay two hundred dollars for a dress that some poor seamstress made five dollars for sewing. These problems are so widespread, it's hard for me to see how boycotting a single business would help much.

That said, if a boycott were called on some particular business, and there were a focused campaign surrounding it, I would respect it. . . .

Passaro: I feel guilty wherever I shop.

Ehrenreich: There's no avoiding that guilt. What are you going to do? Weave your own cloth like Gandhi tried to do?

What we can do to help hardworking people trapped in poverty is fight for increasing social benefits, universal health insurance, and a universal child-care subsidy. We can demand that cities build affordable housing. . . . Another possibility would be to tell the courts to get serious about enforcing the law against firing people for union activity. That's the law, but it's not enforced. You could also join the living-wage movement, which is using whatever leverage it has to convince individual cities to raise wages. . . .

Passaro: Education is often seen as the best way to move people out of poverty, yet menial jobs are always going to exist. Does the nature of these jobs need to change?

Ehrenreich: I get a little annoyed when someone says, "What's wrong with these people? Why don't they get an education?" Well, great, but then who's going to take care of your elderly grandmother in the nursing home? Who's going to wait on you when you go to a restaurant or a discount store? These are important jobs, jobs that need to be done, jobs that take intelligence and concentration and sometimes a great deal of compassion. Why don't we just pay people decently for doing them?

Source: Jamie Passaro. 2003 (January). "Fingers to the Bone: Barbara Ehrenreich on the Plight of the Working Poor." *The Sun,* pp. 4-10. Used by permission.

tration quietly discontinued this program, citing lack of funding as the reason (Lazarus 2003). Another cause of U.S. unemployment is **job exportation,** the relocation of jobs to other countries where products can be produced more cheaply. **Automation,** or the replacement of human labor with machinery and equipment, also contributes to unemployment.

Unemployment figures do not include "discouraged" workers, who have given up on finding a job and are no longer looking for employment. **Underemployment** is a broader term that includes unemployed workers as well as (1) those working part-time but who wish to work full-time ("involuntary" part-timers); (2) those who want to work but have been discouraged from searching by their lack of success ("discouraged" workers), and (3) others who are neither working nor seeking work but who indicate that they want and are available to work and have looked for employment in the last 12 months. The underemployment rate tends to be higher than the unemployment rate.

Labor Unions and the Struggle for Workers' Rights

"The Labor Movement: the folks who brought you the weekend."

From a bumper sticker, 1995

Labor unions originally developed to protect workers and represent them at negotiations between management and labor. Labor unions have played an important role in fighting for fair wages and benefits, healthy and safe work environments, and other forms of worker advocacy. In 2001, the average wage of union workers was $21.40, compared to $16.67 for non-union workers, and unionized workers received insurance and pension benefits worth more than double those of non-union employees (Mishel et al. 2003). Union workers also get more paid time off than non-union workers.

In the *Justice for Janitors 2000* campaign, approximately 100,000 janitors who are members of the Service Employees International Union (SEIU) participated in strikes, rallies, and/or protests in cities across the country in an effort improve employment conditions (Wright 2001). The janitors won increased wages, expanded health care benefits, and the restoration of full-time jobs. The wage raises won will help janitors lift their families out of poverty, which was a major goal of the national campaign.

Labor unions are also influential in achieving better working conditions. For example, the United Food and Commercial Workers (UFCW), the country's largest union representing poultry processing workers, was instrumental in the formation of an Occupational Safety and Health Administration (OSHA) rule that established a federal workplace "potty" policy governing when employees can use the bathroom while on the job. According to UFCW international president Doug H. Dority, "For years workers in food processing industries have had to suffer the indignity of being denied the right to go to the bathroom when needed, just to maintain ever-increasing assembly-line speeds" ("New OSHA Policy Relieves Employees" 1998, 8). Dority claims that "poultry processors often have no other choice than to relieve themselves where they stand on the assembly line because their floor boss will not let them leave their workstation" (p. 8). The new OSHA rule mandates that employers must make toilet facilities available so that employees can use them when they need to.

The increasing numbers of women in labor unions, which nearly doubled from 20 to 39 percent between 1960 and 1998, has helped to strengthen labor unions' advocacy for women ("Labor's 'Female Friendly' Agenda" 1998). For example, a number of unions have been successful in bargaining for expanded family leave benefits, subsidized child care, elder care, and pay equity.

Despite the successes of labor unions' efforts to improve wages, benefits, and working conditions, the strength and membership of unions in the United States have declined over the last several decades. **Union density**—the percentage of workers who belong to unions—grew in the 1930s and peaked in the 1940s and 1950s, when 35 percent of U.S. workers were unionized. In the 1960s and 1970s, U.S. corporations mounted an offensive attack on labor unions, "aiming to tame them or maim them" (Gordon 1996, 207). Corporations hired management consultants to help them develop and implement anti-union campaigns. They threatened unions with decertification, fired union leaders and organizers, and threatened to relocate their plants unless the unions and their members "behaved."

> One management consultant firm . . . was unusually blunt in broadcasting its methods. A late-1970s blurb promoting its manual promised: "We will show you how to screw your employees (before they screw you)—how to keep them smiling on low pay—how to maneuver them into low-pay jobs they are afraid to walk away from" (Gordon 1996, 208).

In 2000, the percentage of American workers belonging to unions had fallen to 13.5 percent, its lowest point in six decades, but rose to 15.4 percent in 2001 (Greenhouse 2001; Mishel et al. 2003). Reasons for the decline in union representation include the loss of manufacturing jobs, which tend to have higher rates of unionization than other industries (Greenhouse 2001). Job growth has been in high technology and financial services where unions have little presence. In addition, globalization has led to layoffs and plants closing at many unionized work sites, as companies move to other countries to find cheaper labor.

Although the 1935 National Labor Relations Act guarantees the right to unionize and to strike, this legislation does not cover agricultural or domestic workers, certain categories of supervisory workers, or independent contractors. About 40 percent of government employees are denied basic collective bargaining rights. At the federal level only postal workers enjoy such rights. And workers who do have the right to unionize and to strike risk their jobs by doing so. In the 1990s, more than 20,000 U.S. workers each year were fired or discriminated against because of their union-related activities (Human Rights Watch 2001). In the United States, at least one in ten union supporters campaigning to form a union is illegally fired. For every 30 people who vote for a union in elections in any one year, one will be illegally fired. These workers can seek reinstatement and back pay by reporting their case to the National Labor Relations Board (NLRB), which is estimated to have a backlog of almost 25,000 cases involving unfair labor practices committed by employers opposing trade union activity (International Confederation of Free Trade Unions 2003).

One study found that in more than half of all union-organizing drives, employers threatened to close the plant (Mokhiber & Weissman 1998). Where union organizing drives are successful, employers carry out the threat and close the plant, in whole or in part, 15 percent of the time. In today's climate of expanding trade agreements and skyrocketing levels of corporate migration, plant-closing threats continue to be among the most powerful anti-union strategies.

Labor Union Struggles Around the World. International norms established by the United Nations and the International Labor Organization declare the rights of workers to organize, negotiate with management, and strike (Human Rights Watch 2000). In European countries, labor unions are generally strong. However, in many less-developed countries and countries undergoing economic transition, workers and labor unions struggle to have a voice in matters of wages and working conditions. A survey of 133 countries by the International Confederation of Free Trade Unions (2003) found that both corporations and governments repressed union efforts. The survey found that in 2002, 213 trade union members around the world were assassinated or disappeared, 2,000 were arrested, nearly 1,000 were injured, and about 30,000 were fired due to their union activity.

Colombia is the world's most dangerous country for trade union activity, with 184 cases of murder of trade unionists in 2002. In China, long-term imprisonment, beatings, internment in psychiatric hospitals or labor camps, and harassment of families are systematically used to discourage free trade unions. Violence against trade unionists is often brutal. After a rural workers' demonstration in Haiti, two elderly members of the Batay Ouvriyéé trade union were dragged out of a house by

College student groups across the country have participated in boycotts against Coca-Cola in protest of the violence against union leaders at Colombia Coca-Cola plants.

Kirschner/United Students Against Sweatshops

company-sponsored thugs, mutilated with knives, beheaded, and dumped into a hole. In Zimbabwe, a unionist was attacked in his home in the middle of the night. While his wife was beaten, he was thrown out of the house, assaulted with chains, pipes and whips, and left for dead (International Confederation of Free Trade Unions 2003).

Strategies for Action: Responses to Workers' Concerns

Government, private business, human rights organizations, labor organizations, college student activists, and consumers play important roles in responding to the concerns of workers. Next we look at responses to sweatshop labor, health and safety concerns, work-family policies and programs, workforce development programs, efforts to strengthen labor, and challenges to corporate power and economic globalization.

Efforts to End Slavery

More than 50 years ago, the United Nations' Universal Declaration of Human Rights stated in its Article 4 that "no one shall be held in slavery or servitude; slavery and the slave trade shall be prohibited in all their forms." Yet slavery persists throughout the world. The international community has drafted treaties on slavery but many countries have yet to ratify and implement the different treaties.

One strategy to fight slavery is punishment. Slave traffickers often avoid punishment because, explains a former official of the U.S. Agency for International Development, "government officials in dozens of countries assist, overlook, or actively collude with traffickers" (quoted in Cockburn 2003, 16). In many countries, the justice system is more likely to jail or expel sex slaves than to punish traffickers ("Sex

Trade Enslaves Millions of Women, Youth" 2003). But in 25 countries, slave trafficking is actively prosecuted and treated as a serious crime.

In the United States, the Victims of Trafficking and Violence Protection Act, passed by Congress in 2000, protects slaves against deportation if they testify against their former owners. Convicted slave traffickers in the United States are subject to prison sentences, as shown in the following examples (Cockburn 2003):

- Louisa Satia and Kevin Waton Nanji each received nine years for luring a 14-year-old girl from Cameroon with promises of schooling, then isolating her in their Maryland home, raping her, and forcing her to work as their domestic servant for three years.
- Sardar and Nadira Gasanov were sentenced to five years each for recruiting women from Uzbekistan with promises of jobs, taking their passports, and forcing them to work in strip clubs and bars in Texas.
- Juan, Ramiro, and Jose Ramos each received 10 to 12 years for transporting Mexicans to Florida and forcing them to work as fruit pickers.

U.S. corporations are also being held accountable for enterprises that involve forced labor and other human rights and labor violations. In 2003, the Unocal oil company became the first corporation in history to stand trial in the United States for human rights violations abroad (George 2003). Unocal is accused of involvement in a pipeline project that used Myanmar (formerly Burma) military personnel to provide "security" for a natural gas pipeline project in the remote Yadana region near the Thai border. According to the 9th U.S. Circuit Court of Appeals, the soldiers' true role was to force villagers in the pipeline region to work without pay—a modern form of slavery. The military also forced villagers living along the pipeline route to relocate without compensation, raped and assaulted villagers, and imprisoned and/or executed those who opposed them. The court asserted that Unocal knew the military were enslaving the people and violating their human rights.

© AP/Wide World Photos

Myanmar workers help build a road. Despite the International Labor Organization's long campaign of pressuring Myanmar to eradicate forced labor, human rights groups claim that the Myanmar army continues to use the forced labor of rural villagers.

As of this writing, the court case against Unocal is pending. Even if Unocal wins, the attention brought to this case may help prevent corporations from exploiting local peoples in the name of profit.

Responses to Sweatshop Labor

A recently formed Fair Labor Association (FLA) involves six leading apparel and footwear companies who voluntarily participate in a monitoring system to inspect their overseas factories and require them to meet minimum labor standards, such as not requiring workers to work more than 60 hours a week. However, critics point out a number of problems with the Fair Labor Association, including these: (1) standards are too low (allows below-poverty wages and excessive overtime); (2) only 10 percent of companies' factories must be monitored yearly; (3) companies can influence which factories are inspected and who does the inspection; and (4) FLA does not uphold workers' right to organize (Benjamin 1998). Critics suggest that companies use their participation in FLA as a marketing tool. Once "certified" by the FLA, companies can sew a label into their products saying the products were made under fair working conditions.

Pressure from college students and other opponents of sweatshop labor and consumer boycotts of products made by sweatshop labor have resulted in some improvements in factories that make goods for companies such as Nike and Gap, which have cut back on child labor, use less dangerous chemicals, and require fewer employees to work 80-hour weeks (Greenhouse 2000). At many factories, supervisors have stopped hitting employees, have improved ventilation, and have stopped requiring workers to obtain permission to use the toilet. But improvements are not widespread, and oppressive forms of labor continue throughout the world. According to the National Labor Committee, two areas where "progress seems to grind to a halt" are efforts to form unions and efforts to achieve wage increases (Greenhouse 2000).

Responses to Worker Health and Safety Concerns

Over the last few decades, health and safety conditions in the U.S. workplace have improved as a result of media attention, demands by unions for change, more white-collar jobs, and regulations by the Occupational Safety and Health Administration (OSHA). Through OSHA, the government develops, monitors, and enforces health and safety regulations in the workplace. Since OSHA was created three decades ago, workplace fatalities have dropped by 75 percent (*Multinational Monitor* 2000). But much work remains to be done to improve worker safety and health. Inadequate funding leaves OSHA unable to do its job effectively, with only 2,238 federal and state inspectors responsible for monitoring and enforcing job safety laws at nearly 6 million workplaces (AFL-CIO 2002). At current staffing levels, it would take federal OSHA employees 119 years to inspect just once each workplace under its jurisdiction.

Because "the task of monitoring and enforcement simply cannot be effectively carried out by a government administrative agency," Kenworthy (1995) suggests that the United States follow the example of many other industrialized countries: turn over the bulk of responsibility for health and safety monitoring to the workforce (p. 114). Worker health and safety committees are a standard feature of companies in many other industrialized countries and are mandatory in most of Europe. These committees are authorized to inspect workplaces and cite employers for violations of health and safety regulations.

In developing countries, governments fear that strict enforcement of workplace regulations will discourage foreign investment (*Multinational Monitor* 2000). Investment in workplace safety in developing countries, whether by domestic firms or foreign multinationals, is far below that in the rich countries. Unless global standards of worker safety are implemented and enforced in *all* countries, millions of workers throughout the world will continue to suffer under hazardous work conditions. Low unionization rates, as well as workers' fears of losing their jobs—or their lives—if they demand health and safety protections leave most workers powerless to improve their working conditions.

Business and industry often fight against efforts to improve safety and health conditions in the workplace. For example, in 1999, after a 10-year struggle between labor and business, the Occupational Safety and Health Administration issued ergonomic standards requiring employers to implement ergonomic programs in jobs where musculoskeletal disorders occur. **Ergonomics** refers to the designing or redesigning of the workplace to prevent and reduce cumulative trauma disorders. According to OSHA, the new ergonomic standards would prevent 4.6 million workers over the next 10 years from experiencing painful, potentially debilitating work-related musculoskeletal disorders (U.S. Department of Labor 2001). But business and industry representatives pressured Congress and President George W. Bush to repeal the ergonomic standard soon after Bush took office in 2000. Now, there are no regulations requiring employers to assess ergonomic hazards in the workplace (such as excessive repetition or poor workstation design) or to take steps to reduce these hazards.

Behavior-Based Safety Programs. A controversial health and safety strategy used by business management is behavior-based safety programs. Instead of examining how work processes and conditions compromise health and safety on the job, **behavior-based safety programs** direct attention to workers themselves as the problem. Behavior-based safety programs claim that 80 to 96 percent of job injuries and illnesses are caused by workers' own carelessness and unsafe acts (Frederick & Lessin 2000). These programs focus on teaching employees and managers to identify, "discipline," and change unsafe worker behaviors that cause accidents, and to encourage a work culture that recognizes and rewards safe behaviors.

Critics contend that behavior-based safety programs divert attention away from the employer's failure to provide safe working conditions. They also say that the real goal of behavior-based safety programs is to discourage workers from reporting illness and injuries. Workers whose employers have implemented behavior-based safety programs describe an atmosphere of fear in the workplace, such that workers are reluctant to report injuries and illnesses for fear of being labeled an "unsafe worker." At one factory that had implemented a behavioral safety program, when a union representative asked workers during shift meetings to raise their hands if they were afraid to report injuries, about half of 150 workers raised their hands (Frederick & Lessin 2000). Worried that some workers feared even raising their hand in response to the question, the union representative asked a subsequent group to write "yes" on a piece of paper if they were afraid to report injuries. Seventy percent indicated they were afraid to report injuries. Asked why they would not report injuries, workers said, "we know that we will face an inquisition," "we would be humiliated," and "we might be blamed for the injury."

A new rule issued by OSHA protects workers by prohibiting discrimination against an employee for reporting a work-related fatality, injury, or illness ("Workers at Risk" 2003). This rule also prohibits discrimination against an employee for filing a safety and health complaint or asking for health and safety records.

Work/Family Policies and Programs

"Our leaders talk as though they value families, but act as though families were a last priority."

Sylvia Hewlett And Cornel West
Family advocates

Since 1950, the percentage of U.S. women in the labor force has nearly doubled (see Table 11.2). The influx of women into the workforce has been accompanied by an increase in government and company policies designed to help women and men balance their work and family roles. Such policies are referred to by a number of terms, including "work/family," "work-life," and "family-friendly" policies.

The Federal Family and Medical Leave Act. In 1993, President Clinton signed into law the **Family and Medical Leave Act** (FMLA), which requires all companies with 50 or more employees to provide eligible workers (who work at least 25 hours a week and have been working for at least a year) with up to 12 weeks of job-protected, *unpaid* leave so they can care for a seriously ill child, spouse, or parent; stay home to care for their newborn, newly adopted, or newly placed child; or take time off when they are seriously ill. Yet, 40 percent of the workforce is not covered by the FMLA (Watkins 2002). Lower wage earners are the least likely to have family and medical leave benefits and typically have few if any resources to fall back on in times of family illness or crisis.

State Family and Medical Leave Initiatives. Because of the inadequate provisions of the federal Family and Medical Leave Act provision, many states have considered initiatives for some type of paid family and medical leave. In 2002, California became the first state in the country to adopt a comprehensive family leave policy that provides workers up to six weeks of time off with about 55 percent of their regular pay while caring for a newborn or newly adopted child, or when a family member is seriously ill (Watkins 2002). Some states have extended family leave protections to additional workers. For example, Oregon law covers workers in companies with 25 or more employees who have been on the job for at least six months (compared to the 50 employees and one-year employment requirements set by the FMLA). Other states that require smaller companies to provide family leave include Massachusetts, New Hampshire, Iowa, Minnesota, and Vermont.

Table 11.2
Percentage of U.S. Women Aged 20 and Older in the Labor Force; 1950–2002

Year	Percentage of Women 20+ in Labor Force
1950	33.3
1960	37.6
1970	43.3
1980	51.3
1990	58.0
2002	60.5

Source: Bureau of Labor Statistics. 2003. "Table A-1: Employment Status of the Civilian Population by Sex and Age." http://stats.bls.gov

Employer-Provided Work/Family Policies. Aside from government-mandated work/family policies, some corporations and employers have "family-friendly" work policies and programs, including unpaid or paid family and medical leave, child care assistance (e.g., plans that allow employees to pay for child care with pretax dollars), assistance with elderly parent care, and flexible work options. Only 2 percent of U.S. workers have paid family leave provided by their employers (Lovell & Rahmanou 2000).

Offering employees more flexibility in their work hours helps parents balance their work and family demands. Flexible work arrangements, which benefit child-free workers as well as employed parents, include flextime, job sharing, a compressed workweek, and teleworking. **Flextime** allows the employee to begin and end the workday at different times as long as 40 hours per week are maintained. A **compressed workweek** allows employees to condense their work into fewer days (e.g., four 10-hour days each week). With **job sharing,** two workers share the responsibility of one job. **Telework** allows employees to work part- or full-time at home or at a satellite office (see this chapter's *Focus on Technology* feature). A study of U.S. companies found that the more women and minorities a company has in managerial positions, the more likely that company is to offer flexible work options (Galinsky & Bond 1998).

Barriers to Employees' Use of Work/Family Benefits. Even when work/family policies and programs are available, many employees do not take advantage of them. One barrier to using work/family benefits is lack of awareness: many employees do not know what benefits are available to them. In a survey of employees covered by the FMLA, only 38 percent correctly reported that the FMLA applied to them and about one-half said they did not know whether it did (Cantor et al. 2001). In a study of employees in seven organizations, more than two-thirds of employees were unaware of or mistaken about at least one work life policy or practice. These employees believed that a policy was in operation that in fact was not, or that a policy that did formally exist did not (Still & Strang 2003).

Other barriers that discourage employee use of work/family benefits are related to economic and job security. Many workers who are covered by the FMLA do not use it because they cannot afford to take leave without pay. A 2000 survey found that 88 percent of workers who needed time off but did not take it said they would have taken leave if they could have received pay during their absence; nearly one-third of workers who needed leave but did not take it cited worry about losing their job as a reason for not taking leave (Cantor et al. 2001). Professor Phyllis Moen (2003) explains, "many are afraid to take advantage of existing [work/family] options because doing so might signal a less-than-full commitment to their jobs, exacting costs in future promotions, salary increases, and even job security" (p. 335).

Corporate initiatives in work/family benefits are a step in the right direction. However, companies that innovate in the area of work life by adopting many family-friendly programs do not show particularly high levels of benefit use among their employees (Still & Strang 2003). Although several magazines and professional organizations give awards and recognition to companies that offer family-friendly benefits, Moen suggests that "a better gauge of the worker friendliness of corporations might be the proportion of their workforces actually using them" (p. 335). In addition, "corporate leaders, along with those handing out rewards, need to recognize that job security is also a key component of friendliness or work-life quality" (Moen 2003, 335).

Focus on Technology

Telework: The New Workplace of the Twenty-First Century

The ever-widening use of modern technology in the workplace, such as computers, the Internet, e-mail, fax machines, copiers, mobile phones, and personal digital assistants, makes it possible for many workers to perform their jobs at a variety of locations. The term **telework** (also known as **"telecommuting"**) refers to flexible and alternative work arrangements that involve use of information technology. There are four types of telework (Pratt 2000): (1) home-based telework; (2) satellite offices where all employees telework for one employer; (3) telework centers, which are occupied by employees from more than one organization; and (4) mobile workers. Most (89%) teleworkers are home based (Bowles 2000). Some people telework full-time, but a larger number telework one or two days a week.

The number of companies offering telecommuting increased from 19.5 percent in 1996 to 28 percent in 1999. In 2001, 15 percent of the U.S. workforce (19.8 million workers) reported teleworking from home at least once a week (U.S. Office of Employment Policy 2004). Four in 5 teleworkers are in jobs classified as managerial, profession, or sales. Workers with children under age 18 are more likely to telework than those without children.

Telework holds potential benefits for employers, workers, and the environment. After presenting some of these benefits, we discuss concerns related to telework.

Benefits of Telework for Employers

ATTRACTS AND HELPS RETAIN EMPLOYEES
Companies regard telework and other flexible work arrangements as important in recruiting and maintaining good employees. Telework can also lower turnover and thus save companies expenses associated with hiring and training replacement employees. A 1997 AT&T survey of telecommuters showed that 36 percent of employees would quit or find another work-at-home job if their employer decided they could not work at home (cited in Lovelace 2000). However, Bowles (2000) reports that companies are beginning to express dissatisfaction with telework "because they believe that it causes resentment among office-bound colleagues and weakens corporate loyalty" (p. 2).

REDUCES COSTS
AT&T saved about $550 million from 1991 to 1998 by eliminating offices that teleworkers don't need, consolidating others, and reducing related overhead costs (Lovelace 2000).

INCREASES WORKER PRODUCTIVITY
Several studies of managers and employees at large companies conclude that telework increases worker productivity (Lovelace 2000).

Benefits of Telework for Employees

INCREASES JOB AND LIFE SATISFACTION
Studies have shown that employee satisfaction among teleworkers is higher than for their non-teleworking counterparts (Lovelace 2000). Much of the job satisfaction among telecommuters is related to the job flexibility that enables them to balance work and family demands.

HELPS BALANCE WORK/FAMILY DEMANDS
Telework can provide flexibility to working parents and adults caring for aging parents, thus reducing role conflict and strengthening family life. One father of three children described his being home when his children came back from school as being "the most significant impact" of his telecommuting (Riley, Mandavilli & Heino 2000, 5). He also took time during the day to take his children to school, to the doctor's office, and to run errands. However, one national study of children whose parents work at home found that older children (grades 7 to 12) were more likely to agree that "my father does not have the energy to do things with me because of his job" and "my father has not been in a good mood with me because of his job" than the children of fathers who work in an office (Galinsky & Kim 2000). The effects of telework on parent/child relationships seems to depend then on how each parent interacts with his or her children.

EXPANDS WORK OPPORTUNITIES FOR AMERICANS OUTSIDE THE ECONOMIC MAINSTREAM
Telework may expand job opportunities for rural job seekers who lack local employment opportunities, and for low-income urban job seekers who lack access to suburban jobs (Kukreja & Neely 2000). Telework can also bring work opportunities to individuals with disabilities. Some of the technologies that have been developed for individuals with serious disabilities include "Eye Gaze" (a communication system that allows people to operate a computer with their eyes); "Magic Wand Keyboard" (for people with limited or no hand movement); and "Switched Adapted Mouse and Trackball" (that allows clicking the mouse with parts of the body other than the hand) (Bowles 2000).

AVOIDS THE COMMUTE
For many teleworkers, the primary motivation for working from home is to reduce or eliminate the long and stressful commute that so many Americans now endure. In one AT&T unit, the average teleworker gained nearly five weeks per year by eliminating a 50-minute daily commute (Lovelace 2000).

Environmental Benefits of Telework

Telework can reduce pollution by reducing the need for transportation to the workplace, thus reducing the pollution associated with vehicle emissions. The National Environmental Policy Institute says that "telecommuting presents a non-coercive way for corporations to help the nation achieve environmental goals and improve quality of life" (quoted in Lovelace 2000, 3).

Concerns About Telework

BLURRED BOUNDARIES BETWEEN HOME AND WORK
People who work at home may find themselves on call around the clock, responding to e-mail, pagers, faxes, and voice mail. Without clear boundaries between home and work, teleworkers may feel that they are unable to escape the work environment and mind-set (Pratt 2000). Questions about overtime pay may arise when work spills over into personal time. Having a separate office within the home and a routine work schedule may help create the psychological boundary between work and family/leisure. But for some teleworkers, learning to "log off" is a challenge.

ZONING REGULATIONS
Teleworkers who work at home full time must contend with zoning regulations that may prohibit residents from having an "office" in their home.

LOSING BENEFITS AS A CONTRACT EMPLOYEE
Some employers attempt to convert the teleworker into a contract worker. This type of worker lacks job protections and benefits (Bowles 2000).

SOCIAL ISOLATION
Does telework lead to social isolation for those who live and work at home? Evidence suggests that teleworkers are able to maintain personal relationships with co-workers and are included in office networks. However, for rural and disabled individuals, telework may contribute to social isolation.

EXCLUSION OF THE DISENFRANCHISED
Lower socioeconomic groups are less likely than more affluent populations to have access to and skills in the Internet and other modern forms of information technology. Bowles (2000) suggests that "the eventual success of telework programs in the future must . . . account for the masses of people left behind. . . . All must be included in the new economy; it is not a luxury, but a must" (p. 9).

Sources: Bowles, Diane O. 2000. "Growth in Telework." Paper presented at the symposium *Telework and the New Workplace of the 21st Century,* Xavier University, New Orleans, October 16, 2000. U.S. Department of Labor. http://www.dol.gov/dol/asp/public/telework/htm; Galinsky, Ellen, and Stacy S. Kim. 2000. "Navigating Work and Parenting by Working at Home: Perspectives of Workers and Children Whose Parents Work at Home." Paper presented at the symposium *Telework and the New Workplace of the 21st Century,* Xavier University, New Orleans, October 16, 2000. U.S. Department of Labor. http://www.dol.gov/dol/asp/public/telework/htm; Kukreja, Anil, and George M. Neely, Sr. 2000. "Strategies for Preventing the Digital Divide." Paper presented at the symposium *Telework and the New Workplace of the 21st Century,* Xavier University, New Orleans, October 16, 2000. U.S. Department of Labor. http://www.dol.gov/dol/asp/public/telework/htm; Lovelace, Glenn. 2000. "The Nuts and Bolts of Telework." Paper presented at the symposium *Telework and the New Workplace of the 21st Century,* Xavier University, New Orleans, October 16, 2000. U.S. Department of Labor. http://www.dol.gov/dol/asp/public/telework/htm; Pratt, Joanne H. 2000. "Telework and Society—Implications for Corporate and Societal Cultures." Paper presented at the symposium *Telework and the New Workplace of the 21st Century,* Xavier University, New Orleans, October 16, 2000. U.S. Department of Labor. http://www.dol.gov/dol/asp/public/telework/htm; Riley, Patricia, Anu Mandavilli, and Rebecca Heino. 2000. "Observing the Impact of Communication and Information Technology on 'Net-Work.'" Paper presented at the symposium *Telework and the New Workplace of the 21st Century,* Xavier University, New Orleans, October 16, 2000. U.S. Department of Labor. http://www.dol.gov/dol/asp/public/telework/htm. U.S. Office of Employment Policy. *The Balancing Act* (March 11): all.

Workforce Development and Job-Creation Programs

The International Labor Organization (2003) estimates that at least one billion new jobs are needed in the coming decade to meet the United Nations goal of halving extreme poverty by 2015. Developing a workforce and creating jobs involves far-reaching efforts, including those designed to improve health and health care, alleviate poverty and malnutrition, develop infrastructures, and provide universal education.

In the United States, workforce development programs have provided a variety of services, including assessment to evaluate skills and needs, career counseling, job search assistance, basic education, occupational training (classroom and on-the-job), public employment, job placement, and stipends or other support services for child care and transportation assistance (Levitan, Mangum, & Mangum 1998). Workforce development programs primarily assist youths, the handicapped, welfare recipients, displaced workers, the elderly, farmworkers, Native Americans, and veterans. Numerous studies have looked at the effectiveness of workforce development programs. In general, "evaluations indicate that employment and training programs enhance the earnings and employment of participants, although the effects vary by service population, are often modest because of brief training durations and the inherent difficulty of alleviating long-term deficiencies, and are not always cost effective" (Levitan, Mangum, & Mangum 1998, 199).

Efforts to prepare high school students for work include the establishment of technical and vocational high schools and high school programs and school-to-work programs. School-to-work programs involve partnerships between business, labor, government, education, and community organizations that help prepare high school students for jobs (Leonard 1996). Although school-to-work programs vary, in general, they allow high school students to explore different careers, and they provide job skill training and work-based learning experiences, often with pay (Bassi & Ludwig 2000).

In one strategy for creating jobs, local, state, and federal governments provide benefits to corporations in the form of subsidies, tax breaks, real estate, and low-interest loans with the hope that this "corporate welfare" will result in new jobs (see also Chapter 10). However, most recent job creation in the United States is with small- and medium-sized companies. Although Fortune 500 companies are the biggest beneficiaries of corporate welfare, they have eliminated more jobs than they have created in the past decade (Barlett & Steele 1998). And many of the jobs that are created are part-time or temporary jobs.

With the new limits on welfare (see Chapter 10), more adults with low levels of education and job training are going to be entering the workforce in the coming years. The cuts in welfare benefits exacerbate the need to provide not only workforce development programs, but also jobs that pay a living wage. As shown in this chapter's *Social Problems Research Up Close* feature, single mothers who work in low-wage jobs often have more hardships than those who are dependent on welfare.

Efforts to Strengthen Labor

Although efforts to strengthen labor are viewed as problematic to corporations and employers, such efforts have potential to remedy many of the problems facing workers. In an effort to strengthen their power, some labor unions have merged with one another. Labor union mergers result in higher membership numbers, thereby increasing the unions' financial resources, which are needed to recruit new members and to withstand long strikes.

Because workers must fight for labor protections within a globalized economic system, their unions must cross national boundaries to build international cooperation and solidarity. Otherwise, employers can play working and poor people in different countries against each other. An example of international union cooperation occurred in 1994 when Ford factory workers in Cuatitlan, Mexico, went on strike to protest layoffs and poor working conditions. Members of the United Auto Workers in a U.S. Ford factory sent money to support the action. "Their reasoning was: if the Mexicans win, that's good for us as well as them because Ford won't be so quick to threaten to move production to Mexico" (Went 2000, 126). More recently, leaders from 21 unions in 11 countries on five continents resolved to form a global union network at International Paper Company (IP), the largest paper company in the world. One union leader remarked, "IP crosses national borders in search of the highest profits, and the unions . . . have resolved to match that corporate globalization with a globalization of workers' solidarity" ("Unions Forge Global Network" 2002).

The National Labor Relations Board (NLBR) and the courts play an important role in upholding workers' rights to unionize and sanctioning employers who violate these rights. In 1998, the National Labor Relations Board issued nearly 24,000 reinstatement and "back-pay" orders or other remedial orders to workers wrongfully fired or demoted for participating in union-related activities (Human Rights Watch 2001). The National Labor Relations Board and the courts have held that employer threats to close the plant if the union succeeds in organizing can be unlawful under certain circumstances (Bronfenbrenner 2000). For example, *Guardian Industries Corp. v. NLRB* held that it was unlawful for a supervisor to say to an employee, "If we got a union in there, we'd be in the unemployment line." However, under the employer free speech provisions of the Taft-Hartley Act, the courts have permitted the employer to predict a plant closing in situations where it is based on an objective assessment of the economic consequences of unionization. When an employer is found guilty of making an unlawful threat of plant closure, the typical remedy is a cease and desist order coupled with the posting of a notice promising not to make such statements in the future.

Challenges to Corporate Power and Globalization

Challenges to corporate power and globalization include campaign finance reform and the antiglobalization movement.

Campaign Finance Reform. In the United States, advocates for campaign finance reform have challenged the power that corporations have in influencing laws and policies. Campaign finance reform efforts were rewarded when Congress passed the McCain-Feingold bill known as the Bipartisan Campaign Reform Act of 2002. This law helps to remove the corrupting influence of "soft money" from federal elections so that corporations, labor unions, and wealthy donors will no longer be able to buy political influence and access. The law also prohibits corporations and unions from funding broadcast ads run shortly before elections designed to influence the election or defeat of candidates. Such ads may be funded by individual contributions, but broadcast stations must keep a public record of political ads and who paid for them. Legal challenges to the Bipartisan Campaign Reform Act of

Social Problems Research Up Close

Making Ends Meet: Survival Strategies Among Low-Income and Welfare Single Mothers

As welfare recipients reach the time limit established by welfare legislation of 1996 for receiving welfare benefits, they are forced into the workforce. But as individuals leave welfare for work, they often find themselves in low-paying jobs, often with no or few benefits. How do individuals in low-income jobs compare with those dependent on welfare in terms of their well-being? And how do both low-wage earners and welfare recipients survive on income that does not meet their basic needs? Researchers Kathryn Edin and Laura Lein (1997) conducted research to answer these questions.

Sample and Methods

The sample consisted of 379 African-American, white, and Mexican-American single mothers from four cities (Chicago, San Antonio, Boston, and Charleston, South Carolina), who either received welfare cash assistance (Aid to Families with Dependent Children, or AFDC) (N = 214) or nonrecipients who held low-wage jobs earning $5 to $7 an hour between 1988 and 1992 (N = 165). Edin and Lein used a "snowball sampling" technique in which each mother who was interviewed was asked to refer researchers to one or two friends who might also participate in interviews. Nearly 90 percent of the mothers contacted agreed to be interviewed.

Researchers collected data through conducting multiple semistructured in-depth interviews with women in the sample. Interview topics included the mothers' income and job experience, types and amount of welfare benefits they received, spending behavior, housing situation, use of medical care and child care, and hardships the women and their children experienced because of lack of financial resources.

Interviewing the mothers more than once was an important research strategy in gathering accurate information. Mothers who were unclear about their expenditures in the first interview could keep careful track of what they spent between interviews and give a more precise accounting of their spending in a later interview. Also, some mothers who insisted they received no child support later revealed that the child's father "helped out" every week by providing cash. "Most mothers termed absent fathers' cash contributions as 'child support' only if it was collected by the state" (p. 13).

Findings and Conclusions

Low-wage–earning single mothers had a higher monthly reported income than welfare-reliant mothers. However, the expenses of wage-earning mothers were also higher. This is because employed mothers usually have to pay for child care, transportation to work, and additional clothing to wear to work. If newly employed mothers have a federal housing subsidy, every extra $100 in cash income raises their rent by $30 (Jencks 1997). And employed mothers are usually not eligible for Medicaid, which means that they have more out-of-pocket medical expenses and often go uninsured.

The monthly expenses of both groups of women exceeded their reported monthly income, forcing women to use various strate-

2002 resulted in a Supreme Court hearing. In December 2003, the Supreme Court upheld the main provisions of the act—another victory for the campaign to tame the power of big money in the U.S. political process.

The Antiglobalization Movement. Challenges to corporate globalization have also taken root in the United States and throughout the world. Antiglobalization activists have targeted the World Trade Organization (WTO), the International Monetary Fund (IMF), and the World Bank as forces that advance corporate-led globalization at the expense of social goals like justice, community, national sovereignty, cultural diversity, ecological sustainability, and workers' rights.

In 1999, 50,000 street protesters and Third World delegates demonstrated in Seattle in opposition to the policies of the World Trade Organization that promoted corporate-led globalization. The brutal assaults on largely peaceful demonstrators by Seattle police dressed in their Darth Vader-like uniforms in full view of television cameras has made the Seattle WTO protest the "grand symbol of the crisis of globalization" (Bello 2001).

Social Problems Research Up Close

gies to make ends meet. Cash welfare and food stamps covered only three-fifths of welfare-reliant mothers' expenses. The main job of low-wage earning mothers covered only 63 percent of their expenses. Edin and Lein found that women relied on three basic strategies to make ends meet: work in the formal, informal, or underground economy; cash assistance from absent fathers, boyfriends, relatives, and friends; and cash assistance or help from agencies, community groups, or charities in paying overdue bills. Welfare recipients had to keep their income-generating activities hidden from their welfare caseworkers and other government officials. Otherwise, their welfare checks would be reduced by nearly the same amount as their earnings. Many of the wage-earning mothers also concealed income generated "on the side" in order to maintain food stamps, housing subsidies, or other benefits that would have been reduced or eliminated if they had reported this additional income.

Most of the single mothers in the study described experiencing serious material hardship during the previous 12 months. Material hardships included not having enough food and clothes, not receiving needed medical care, not having health insurance, having the utilities or phone cut off, not having a phone, and being evicted and/or homeless. An important finding was that wage-reliant mothers experienced more hardship than welfare-reliant mothers. In addition to the increased financial pressures of child care costs, transportation, health care, and work clothing, employed mothers worried about not providing adequate supervision of their children and struggled with balancing work and parenting responsibilities, especially when their children were sick. Nevertheless, almost all of the mothers said they would rather work than rely on welfare. They believed that work provided important psychological benefits and increased self-esteem, avoided the stigma of welfare, and enabled them to be good role models for their children.

Harvard University scholar Christopher Jencks (1997) comments on the implications of Edin and Lein's (1997) research:

> As the new time limits on welfare receipt begin to take effect, more and more single mothers will have to take jobs. Most of these newly employed mothers will have more income than they had on welfare, so their official poverty rate will fall. But they will also have more expenses than they had on welfare, and they will get fewer noncash benefits. Edin and Lein's findings dramatize the likely result. Between 1988 and 1992, mothers who held low-wage jobs reported substantially more income than those who collected welfare, but they also reported more hardship. If this pattern persists in the years ahead, time limits will probably bring both a decline in the official poverty rate and an increase in material hardship. (p. x)

Source: Based on Kathryn Edin and Laura Lein. 1977. *Making Ends Meet.* New York: Russel Sage Foundation.

Another confrontation between pro-globalization and antiglobalization forces occurred at the 2000 meeting of the International Monetary Fund (IMF) and the World Bank in Washington, D.C. About 30,000 protesters descended on America's capital and found a large section of the northwest part of the city walled off by some 10,000 police. For four days, the protesters tried, unsuccessfully, to break through the police barrier to reach the IMF-World Bank complex, resulting in hundreds of arrests.

In 2003, the fifth meeting of the WTO in Cancun, Mexico, collapsed when 21 developing nations walked out of the meeting after the U.S. and European Union refused to concede on agricultural subsidies that hurt poorer nations. Two months later, an estimated 20,000 union members, environmentalists, and religious and human rights activists from North, Central, and South America marched through the streets of Miami to protest the Free Trade Area of the Americas (FTAA). Media attention to these events contributes to the growing worldwide awareness of the forces of corporate globalization and its social, environmental, and economic effects.

In November 2003, an estimated 20,000 union members, environmentalists, and religious and human rights activists from North, Central, and South America marched through the streets of Miami to protest the Free Trade Area of the Americas (FTAA).

AP/Wide World Photos

Understanding Work and Unemployment

On December 10, 1948, the General Assembly of the United Nations adopted and proclaimed the Universal Declaration of Human Rights. Among the articles of that declaration are the following:

> Article 23. Everyone has the right to work, to free choice of employment, to just and favourable conditions of work and to protection against unemployment.
>
> Everyone, without any discrimination, has the right to equal pay for equal work.
>
> Everyone who works has the right to just and favourable remuneration ensuring for himself and his family an existence worthy of human dignity, and supplemented, if necessary, by other means of social protection.
>
> Everyone has the right to form and to join trade unions for the protection of his interests.
>
> Article 24. Everyone has the right to rest and leisure, including reasonable limitation of working hours and periodic holidays with pay.

More than half a century later, workers around the world are still fighting for these basic rights as proclaimed in the Universal Declaration of Human Rights.

> "What the public wants is called 'politically unrealistic.' Translated into English, that means power and privilege are opposed to it."
>
> **Noam Chomsky**

To understand the social problems associated with work and unemployment, we must first recognize the power and influence of governments and corporations on the workplace. We must also be aware of the role that technological developments and post-industrialization have on what we produce, how we produce it, where we produce it, and who does the producing. In regard to what we produce, the United States is moving away from producing manufactured goods to producing services. In regard to production methods, the labor-intensive blue-collar assembly line is declining in importance, and information-intensive white-collar occupations are increasing. Because of increasing corporate multinationalization, U.S. jobs are being exported to foreign countries where labor and raw materials are cheap and regulations are lax. In developing countries, investment in workplace safety is far below that in the rich nations.

Decisions made by U.S. corporations about what and where to invest influence the quantity and quality of jobs available in the United States. As conflict theorists argue, such investment decisions are motivated by profit, which is part of a capitalist system. Profit is also a driving factor in deciding how and when technological devices will be used to replace workers and increase productivity. But if goods and services are produced too efficiently, workers are laid off and high unemployment results. When people have no money to buy products, sales slump, recession ensues, and social welfare programs are needed to support the unemployed. When the government increases spending to pay for its social programs, it expands the deficit and increases the national debt. Deficit spending and a large national debt make it difficult to recover from the recession, and the cycle continues.

What can be done to break the cycle? Those adhering to the classic view of capitalism argue for limited government intervention on the premise that business will regulate itself via an "invisible hand" or "market forces." For example, if corporations produce a desired product at a low price, people will buy it, which means workers will be hired to produce the product, and so on.

Ironically, those who support limited government intervention also sometimes advocate government intervention to bail out failed banks and lend money to troubled businesses. Such government help benefits the powerful segments of our society. Yet when economic policies hurt less powerful groups, such as minorities, there has been a collective hesitance to support or provide social welfare programs. It is also ironic that such bail-out programs, which contradict the ideals of capitalism, are needed because of capitalism. For example, the profit motive leads to multinationalization, which leads to unemployment, which leads to the need for government programs. The answers are as complex as the problems.

Chapter Review

- **The United States is described as a "post-industrialized" society. What does that mean?**
  Post-industrialization refers to the shift from an industrial economy dominated by manufacturing jobs to an economy dominated by service-oriented, information-intensive occupations. The U.S. post-industrialized economy is characterized by a highly educated workforce, automated and computerized production methods, increased government involvement in economic issues, and a higher standard of living.

- **What are transnational corporations?**
  Transnational corporations (TNCs) are corporations that have their home base in one country and branches, or affiliates, in other countries. Transnational corporations dominate the world economy today. In less than 20 years, the number of transnational corporations has increased from 7 to over 45,000, and the top 100 economies around the world are transnational corporations rather than nations.

- **According to data from the World Value Survey, how does the level of economic development of a country affect the subjective life satisfaction of its population?**
  Data from the World Value Survey indicate that as one moves from subsistence-level economies in developing countries to advanced industrialized societies, there is a large increase in the percentage of the population who consider themselves very happy or satisfied with their lives as a whole. But once a society moves from a starvation level to a reasonably comfortable existence, the increase in life satisfaction levels off.

- **Does slavery still exist today? If so, where?**
  Slavery exists today all over the world, including in the United States, but is most prevalent in India, Pakistan, Bangladesh, and Nepal. Most slaves are forced to work in agriculture, mining, prostitution, and factories.

- **What is the most common cause of job-related fatality?**
  The most common type of job-related fatality involves transportation accidents.

- **How does unionization benefit employees?**
  Compared to non-union workers, union workers have higher average wages, receive more insurance and pension benefits, and get more paid time off.

- **What are ergonomics?**
  Ergonomics refers to the designing or redesigning of the workplace to prevent and reduce cumulative trauma disorders. After OSHA instituted ergonomic standards in the workplace to help prevent painful, potentially debilitating work-related musculoskeletal disorders, business and industry representatives pressured Congress and President George W. Bush to repeal

the ergonomic standard soon after Bush took office in 2000.

- **What is the the Federal Family and Medical Leave Act?**
  In 1993, President Clinton signed into law the Family and Medical Leave Act (FMLA), which requires all companies with 50 or more employees to provide eligible workers (who work at least 25 hours a week and have been working for at least a year) with up to 12 weeks of job-protected, *unpaid* leave so they can care for a seriously ill child, spouse, or parent; stay home to care for their newborn, newly adopted, or newly placed child; or take time off when they are seriously ill.

Critical Thinking

1. Union membership is higher among black employees than among white employees. For example, in 2001, 15.2 percent of white workers belonged to a union, compared to 20.5 percent of black workers (Mishel et al. 2003). What might explain the higher rate of unionization among black workers?
2. Public approval of labor unions is higher among Democrats than among Republicans. A 2003 Gallup poll found that 74 percent of Democrats approved of unions, compared with only 41 percent of Republicans (Moore 2002). Why do you think this is so?
3. In the late 1990s, a glut of media attention was given to the positive economic indicators in the United States: low inflation rates, low unemployment rates, and surging stock market values. How might a conflict theorist view the media's portrayal of the U.S. economy?

Key Terms

alienation
anomie
automation
behavior-based safety programs
brain drain
capitalism
chattel slavery
compressed workweek
convergence hypothesis
corporate downsizing
cumulative trauma disorders
economic institution
ergonomics
Family and Medical Leave Act
flextime
free trade agreements
global economy
industrialization
job burnout
job exportation
job sharing
labor unions
post-industrialization
repetitive motion disorders
socialism
soft money
sweatshop
telecommuting
telework
transnational corporations
underemployment
unemployment
union density

Taking A Stand

Should employers have the right to monitor their employees' e-mails from the work site?

In one survey, three out of four workers admitted to sending personal e-mails from work (Werhane, Radin, & Bowie 2004). The courts and legislatures have determined that employers have the right to monitor employees' e-mails that are sent or received on technology equipment (i.e., computers, servers) owned or managed by the employer. More than 50 percent of all companies monitor Internet use and e-mail of employees (Werhane et al. 2004). Do you think that employers should have this right? Or is it an invasion of employees' privacy?

Use Wadsworth's exclusive online resources—InfoTrac College Edition, MicroCase Online, and OVRC—to formulate a position on this topic.

The Wadsworth's Sociology Online Resources and Writing Companion will help you get started. This valuable guide will show you how to use Wadsworth's exclusive online resources when studying social problems. It will also help you to build essential research and writing skills. InfoTrac College Edition, MicroCase Online, OVRC, and an electronic copy of portions of this companion are available at http://sociology.wadsworth.com/mooney_knox_schacht/problems4e, the companion Web site for *Understanding Social Problems,* Fourth Edition.

Media Resources

The Companion Web Site for *Understanding Social Problems,* Fourth Edition

http://sociology.wadsworth.com/mooney_knox_schacht/problems4e

Supplement your review of this chapter by going to the companion Web site to take one of the Tutorial Quizzes, use the flash cards to master key terms, and check out the many other study aids you'll find there. You'll also find special features such as *Wadsworth's Sociology Online Resources and Writing Companion,* GSS Data, and Census 2000 information, data, and resources at your fingertips to help you complete that special project or do some research on your own.

Interactions CD-ROM

Go to the Interactions CD-ROM for *Understanding Social Problems,* Fourth Edition, to access additional interactive learning tools, such as in-depth review materials, corresponding practice quizzes, and other engaging resources and activities to help you study the concepts in this chapter.

"The quality of our public schools directly affects us all—as parents, as students, and as citizens. Yet too many children in America are segregated by low expectations, illiteracy, and self-doubt. In a constantly changing world that is demanding increasingly complex skills from its workforce, children are literally being left behind." *George W. Bush, U.S. President*

Problems in Education

The Global Context: Cross-Cultural Variations in Education

Sociological Theories of Education

Who Succeeds? The Inequality of Educational Attainment

Problems in the American Educational System

Strategies for Action: Trends and Innovations in American Education

Understanding Problems in Education

Chapter Review

Sean, Dan, and Lance were eating lunch together in the school cafeteria when they decided to go outside. As they were walking up a hill, they saw two figures with guns in the distance. Must be Annihilation, Sean thought, a paintball game that seniors played. Odd. Those guns looked real—not like the plastic models he had seen before. "Pop, Pop, Pop." The guns were suddenly turned toward the school and before Sean knew it, he was the only one of his three friends standing. Sean turned to look for the paintball that had just grazed his neck. They must be frozen he thought—he was bleeding—but as he turned he was shot three times in the abdomen. He began to run. Why am I running from paintballs he thought, as he headed toward school. Then it hit—the bullet that really hurt—in the back, striking the spine, exiting through the hip.

Sean survived the attack at Columbine High School in Littleton, Colorado, as did Lance. But Dan Rohrbough did not, leaving others the grim task of trying to make some sense out of his death and the deaths of 12 others. Sean tries not to think about it, preferring to pretend it never happened, wanting to put it and his wheelchair behind him. But, embarrassingly, people keep staring at him. What are they looking at, he wonders. Then he thought, maybe that's how Dylan and Eric felt. (Pollock 2000)

Violence is just one of the many issues that must be addressed in today's schools (see this chapter's *Social Problems Research Up Close* feature). Students continue to graduate from high school being unable to read, work simple math problems, or write grammatically correct sentences. Graduates discover that they are ill prepared for corporations that demand literate, articulate, informed employees. Teachers leave the profession because of uncontrollable discipline problems, inadequate pay, and overcrowded classrooms. Students and teachers alike are "dumbing down," lowering their standards, expectations, and role performances to fit increasingly undemanding and unresponsive systems of learning.

And yet it is education that is often claimed as a panacea—the cure-all for poverty and prejudice, drugs and violence, war and hatred, and the like. Can one institution, riddled with problems, be a solution for other social problems? In this chapter we focus on this question and what is being called an educational crisis. We begin with a look at education around the world.

The Global Context: Cross-Cultural Variations in Education

Looking only at the American educational system might lead one to conclude that most societies have developed some method of formal instruction for their members. After all, the United States has more than 96,000 schools, 6.5 million teachers, and 73 million students (U.S. Census 2003). In reality, many societies have no formal mechanism for educating the masses. As a result, worldwide, over 30 percent of females and 17 percent of males are illiterate (PRB 2003). The problem of illiteracy is greater in developing rather than developed nations where, for example, the adult illiteracy rate is as high as 80 percent (UNDP 2003).

Comparing the United States to seven other developed nations (Canada, France, Germany, Italy, Japan, the Russian Federation, and the United Kingdom) reveals some interesting findings. Preschool enrollment of 3- to 5-year-olds is lower in the United States than in France, Germany, Italy, and Japan (NCES 2003a). U.S. completion rates of secondary education are higher than in six of the eight countries, and although U.S. elementary schools have a fairly low student-teacher ratio, U.S.

Social Problems Research Up Close

Guns, Kids, and Schools

Reducing school violence is consistently listed as one of the top education priorities in public opinion polls and is a focus of President Bush's educational reform package entitled "No Child Left Behind." In an effort to curb violence, schools have established "zero-tolerance policies." For example, nationwide, over 90 percent of schools have zero-tolerance for students carrying firearms or other weapons to school (*Executive Summary* 2001). Sociologist Pamela Roundtree (2000) addresses the issue of violence in schools by asking adolescents why they carry weapons.

Sample and Methods

The respondents in this study were sixth-through twelfth-grade students who had participated in a state-sponsored research project, the Kentucky Youth Survey. Because weapon carrying is likely to vary by region, data from three distinct areas of the state were collected: (1) Urban County (wealthier, north-central part of state), (2) Western County (rural, tobacco growing area), and (3) Eastern County (poorer, high unemployment, mining region). Race and sex distributions varied between county samples, with whites and females the majority of each of the three samples.

The dependent variable is possession of weapons at school, measured by whether a student reported carrying a weapon to school in the 30 days prior to the survey. In addition to the standard demographic variables of sex, race, age, and family socioeconomic background, variables thought to be predictive of carrying weapons were measured. The independent variables include (1) fear of crime indicators (prior victimization, fear of victimization), (2) criminal involvement indicators (previous arrest, drug involvement), (3) pro-weapon socialization indicators (weapon ownership or use by respondent, weapon ownership by parent, weapon carrying by peers), and (4) social isolation indicators (disattachment from school, church). Each of these categories of indicators was predicted to be directly related to the likelihood of carrying a weapon to school, that is, as fear of crime, criminal involvement, pro-weapon socialization, and social isolation increase, the likelihood of carrying weapons increases.

Findings and Conclusion

Carrying a weapon to school is a relatively rare event, with 5 percent or less of students reporting that they had carried a weapon to school in the previous 30 days. Possession was slightly lower in Urban than in the Eastern or Western County, Urban County being the most industrialized and the wealthiest of the three.

In general, age, race, and sex were unrelated to the likelihood of a student taking a weapon to school. Only in Eastern County did sex significantly predict carrying a weapon with males 700 percent more likely than females to possess a weapon in school. Surprisingly, prior victimization and fear of crime were unrelated to weapon carrying. However, drug dealing was predictive of weapon carrying in both Eastern and Western Counties, and student drug use as an indicator of criminal involvement was significantly related to weapon possession in each of the three counties.

Pro-weapon socialization had an even stronger effect than criminal involvement. Of the three measures of this variable—weapon ownership or use by respondent, ownership by parent, or carrying by peers—carrying by peers had the strongest relationship with carrying a weapon. With each "best friend" the respondent reported as having carried a gun to school, the likelihood of the respondent carrying a gun to school increased by 75 to 100 percent. Social isolation variables were not related to carrying a weapon in any consistent way.

Unlike many studies on adult weapon carrying, the results of this study indicate that peer-based socialization has a much larger impact on the probability of carrying a weapon than fear of crime variables. However, consistent with research on adults, carrying a weapon to school was significantly related to criminal involvement and, specifically, to drug involvement.

Reference: *Executive Summary.* 2001. "Indicators of School Crime and Safety." http://www.nces.ed.gov/pubs2001/crime2000

Source: Pamela Wilcox Roundtree. 2000. "Weapons at School: Are the Predictors Generalizable across Context?" *Sociological Spectrum* 20:291–324.

Because teaching is a highly respected vocation in Japan, students treat their teachers with respect and obedience. Teaching is less well regarded in the United States, and the consequences are felt by our teachers every day.

© Karen Kasmauski / Woodfin Camp & Associates

secondary schools have the second highest student-teacher ratio following Canada. Further, U.S. schoolteachers have a higher than average starting salary when compared to comparable teachers in the other countries and are disproportionately white females (Grant & Murray 2004). Finally, U.S. university graduation rates are the second highest of the eight countries, surpassed only by the United Kingdom.

Differences also exist in the everyday operations of a school. Some countries empower professionals to organize and operate their school systems. Japan, for example, hires professionals to develop and implement a national curriculum and to administer nationwide financing for its schools. In contrast, school systems in the United States are often run at the local level by school boards and PTAs composed of lay people. In effect, local communities raise their own funds and develop their own policies for operating the school system in their area; the result is a lack of uniformity from district to district. Thus, parents who move from one state to another often find that the quality of education available to their children differs radically. In countries with professionally operated and institutionally coordinated schools such as Japan, teaching has traditionally been a prestigious and respected profession. Consequently, Japanese students are attentive and obedient to their teachers. In the United States, the phrase "Those who can, do; those who can't, teach" reflects the lack of esteem for the teaching profession. The insolence and defiance among students in American classrooms are evidence of the disrespect that many American students have for their teachers.

Sociological Theories of Education

The three major sociological perspectives—structural-functionalism, conflict theory, and symbolic interactionism—are important in explaining different aspects of American education.

Structural-Functionalist Perspective

According to structural-functionalism, the educational institution serves important tasks for society, including instruction, socialization, the provision of custodial care, and the sorting of individuals into various statuses (Sadovnik 2004). Many social problems, such as unemployment, crime and delinquency, and poverty, may be linked to the failure of the educational institution to fulfill these basic functions (see Chapters 4, 10, and 11). Structural-functionalists also examine the reciprocal influences of the educational institution and other social institutions, including the family, political institution, and economic institution.

Instruction. A major function of education is to teach students the knowledge and skills that are necessary for future occupational roles, self-development, and social functioning. Although some parents teach their children basic knowledge and skills at home, most parents rely on schools to teach their children to read, spell, write, tell time, count money, and use computers. As discussed later, many U.S. students display a low level of academic achievement. The failure of schools to instruct students in basic knowledge and skills both causes and results from many other social problems.

> "The beautiful thing about learning is that nobody can take it away from you."
>
> **B. B. King, Musician**

Socialization. The socialization function of education involves teaching students to respect authority—behavior that is essential for social organization (Merton 1968). Students learn to respond to authority by asking permission to leave the classroom, sitting quietly at their desks, and raising their hands before asking a question. Students who do not learn to respect and obey teachers may later disrespect and disobey employers, police officers, and judges.

The educational institution also socializes youth into the dominant culture. Schools attempt to instill and maintain the norms, values, traditions, and symbols of the culture in a variety of ways, such as celebrating holidays (Martin Luther King Jr. Day, Thanksgiving); requiring students to speak and write in standard English; displaying the American flag; and discouraging violence, drug use, and cheating.

As the number and size of racial and ethnic minority groups have increased, American schools are faced with a dilemma: Should public schools promote only one common culture, or should they emphasize the cultural diversity reflected in the American population? Some evidence suggests that most Americans believe that schools should do both—they should promote one common culture and emphasize diverse cultural traditions (Elam, Rose, & Gallup 1994).

Multicultural education, education that includes all racial and ethnic groups in the school curriculum, promotes awareness and appreciation for cultural diversity (see Chapter 6). In 2003, 80 percent of a U.S. adult sample responded that "teach[ing] students to get along with people from different backgrounds" is a very important role for colleges to perform (Selingo 2003).

Sorting Individuals into Statuses. Schools sort individuals into statuses by providing credentials for individuals who achieve various levels of education, at various schools, within the system. These credentials sort people into different statuses—for example, "high school graduate," "Harvard alumna," and "English major." Further, schools sort individuals into professional statuses by awarding degrees in such fields as medicine, nursing, and law. The significance of such statuses lies in their association with occupational prestige and income—the higher one's

Figure 12.1
Unemployment Rate of Persons 25 Years and Over, by Highest Level of Education: 2001

Source: U.S. Department of Labor, Bureau of Labor Statistics, Office of Employment and Unemployment Statistics, Current Population Survey, 2001.

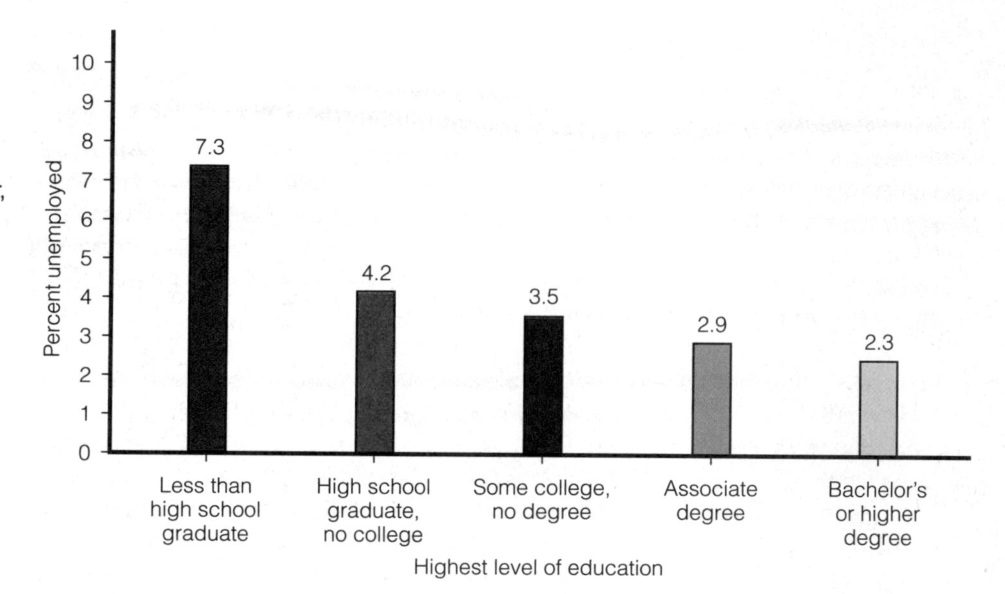

education, the higher one's income. Further, unemployment rates are tied to educational status (see Figure 12.1).

Custodial Care. The educational system also serves the function of providing custodial care (Merton 1968), which is particularly valuable to single-parent and dual-earner families, and the likely reason for the increase in enrollments of 3- to 5-year-olds. In 1970, 37 percent of 3- to 5-year-olds were enrolled in formal classes. Today, 52 percent of this age group are enrolled in either preschool or kindergarten (U.S. Census 2003).

The school system provides supervision and care for children and adolescents until they are 16–12 years of school totaling almost 13,000 hours per pupil! Yet some school districts are increasing class hours. For example, the Knowledge Is Power program in Houston, Texas, requires students to attend school several weeks in the summer, on alternate Saturdays, and from 7:25 A.M. to 5:00 P.M. during the regular school year. Working parents, the hope that increased supervision will reduce delinquency rates, and higher educational standards that require longer hours of study are some of the motivations behind the "more time" movement (Wilgoren 2001).

Conflict Perspective

Conflict theorists emphasize that the educational institution solidifies the class positions of groups and allows the elite to control the masses. Although the official goal of education in society is to provide a universal mechanism for achievement, in reality educational opportunities and the quality of education are not equally distributed.

Conflict theorists point out that the socialization function of education is really indoctrination into a capitalist ideology (Sadovnik 2004). In essence, students are socialized to value the interests of the state and to function to sustain it. Such in-

© James D. Wilson/Woodfin Camp & Associates

© Dan Habib/Impact Visuals

Per-pupil expenditure, averaging about $7,000 a year, varies dramatically between school districts and states.

doctrination begins in kindergarten. Rosabeth Moss Kanter (1972) coined the term "the organization child" to refer to the child in nursery school who is most comfortable with supervision, guidance, and adult control. Teachers cultivate the organization child by providing daily routines and rewarding those who conform. In essence, teachers train future bureaucrats to be obedient to authority.

Further, to conflict theorists, education serves as a mechanism for **cultural imperialism,** or the indoctrination into the dominant culture of a society. When cultural imperialism exists, the norms, values, traditions, and languages of minorities are systematically ignored. A Mexican-American student recalls his feelings about being required to speak English (Rodriquez 1990, 203):

> When I became a student, I was literally "remade"; neither I nor my teachers considered anything I had known before as relevant. I had to forget most of what my culture had provided, because to remember it was a disadvantage. The past and its cultural values became detachable, like a piece of clothing grown heavy on a warm day and finally put away.

Conflict theorists are also quick to note that learning is increasingly a commercial enterprise as necessary financial support for equipment, laboratories, and technological upgrades are funded by corporations anxious to bombard students with advertising and other pro-capitalist messages. For example, Snapple is the "exclusive beverage vendor" for New York's 1,200 public schools. In return for this arrangement, the school system will receive a minimum of $8 million a year from Snapple "in commissions and sponsorship money for athletic programs" (Day 2003).

Finally, the conflict perspective focuses on what Kozol (1991) calls the "savage inequalities" in education that perpetuate racial disparities. Kozol documents gross inequities in the quality of education in poorer districts, largely composed of minorities, compared with districts that serve predominantly white middle-class and upper-middle-class families. Kozol reveals that schools in poor districts tend to receive less funding and have inadequate facilities, books, materials, equipment, and personnel. For example, nationally, the richest school districts spend 56 percent more per pupil than the poorest school districts (CDF 2003).

The Human Side

They Think I'm Dumb

The following excerpt poignantly recounts a conversation between author-ethnographer Thomas Cottle and Ollie, an 11-year-old boy, labeled "dumb."

You know what, Tom?" he said, looking down at his ice cream as though it suddenly had lost its flavor, "nobody, not even you or my Dad, can fix things now. The only thing that matters in my life is school, and there they think I'm dumb and always will be. I'm starting to think they're right. . . ." "Even if I look around and know that I'm the smartest in my group, all that means is that I'm the smartest of the dumbest, so I haven't gotten anywhere at all, have I? I'm right where I always was. Every word those teachers tell me, even the ones I like most, I can hear in their voice that what they're really saying is, 'alright you dumb kids, I'll make it easy as I can, and if you don't get it then, then you'll never get it.' That's what I hear every day, man. From every one of them. Even the other kids talk to me that way too. . . ."

"I'll tell you something else," he was saying, unaware of the ice cream that was melting on his hand. "I used to think, man, that even if I wasn't so smart, that I could talk in any class in that school, if I did my studying, I mean, and have everybody in that class, all the kids and the teacher too, think I was alright. Maybe better than alright. . . ."

"Then they told me, like on a Friday, that today would be my last day in that class. That I should go to it today, you know, but that on Monday I had to switch to this other one. They just gave me a different room number, but I knew what they were doing. Like they were giving me one more day with the brains, and then I had to go to be with the dummies, where I was suppose to be. . . ."

"So I went with the brains one more day, on that Friday like I said, in the afternoon. But the teacher didn't know I was moving, so she acted like I belonged there. Wasn't her fault. All the time I was just sitting there thinking this is the last day for me. This is the last time I'm ever going to learn anything, you know what I mean? Real learning."

He had not looked up at me even once since leaving the ice cream store. . . . "From now on," he was saying, "I knew I had to go back where they made me believe I belonged . . . then the teacher called on me, and this is how I know just how not smart I am. She called on me, like she always did, like she'd call on anybody, and she asked me a question. I knew the answer, 'cause I'd read it the night before in my book. . . . So I began to speak, and suddenly I couldn't say nothing. Nothing, man. Not a word. Like my mind died in there. And everybody was looking at me, you know, like I was crazy or something. My heart was beating real fast. I knew the answer, man. And she was just waiting, and I couldn't say nothing. And you know what I did? I cried. I sat there and cried, man, 'cause I couldn't say nothing. That's how I know how smart I am. That's when I really learned at that school, how smart I was. I mean, how smart I thought I was. I had no business being there. Nobody smart's sitting in no class crying. That's the day I found out for real. That's the day that made me know for sure."

Ollie's voice had become so quiet and hoarse that I had to lean down to hear him. We were walking in silence; I was almost afraid to look at him. At last he turned toward me, and for the first time I saw the tears pouring from his eyes. His cheeks were bathed in them. Then he reached over and handed me his ice cream cone.

"I can't eat it now, man," he whispered. "I'll pay you back for it when I get some money."

Source: Excerpted from Thomas Cottle. 1976. *Barred from School.* Washington, DC: The New Republic Book Company, pp.138–140. Used by permission.

Symbolic Interactionist Perspective

Whereas structural-functionalism and conflict theory focus on macro-level issues such as institutional influences and power relations, symbolic interactionism examines education from a micro perspective. This perspective is concerned with individual and small group issues, such as teacher-student interactions and the self-fulfilling prophecy.

Teacher-Student Interactions. Symbolic interactionists have examined the ways students and teachers view and relate to each other. For example, children from economically advantaged homes may be more likely to bring to the classroom social and verbal skills that elicit approval from teachers. From the teachers' point of view, middle-class children are easy and fun to teach: they grasp the material quickly, do their homework, and are more likely to "value" the educational process. Children from economically disadvantaged homes often bring fewer social and verbal skills to those same middle-class teachers, who may, inadvertently, hold up social mirrors of disapproval. Teacher disapproval contributes to lower self-esteem among disadvantaged youth.

Self-Fulfilling Prophecy. The **self-fulfilling prophecy** occurs when people act in a manner consistent with the expectations of others. For example, a teacher who defines a student as a slow learner may be less likely to call on that student or to encourage the student to pursue difficult subjects. He or she may also be more likely to assign the student to lower ability groups or curriculum tracks (Riehl 2004). As a consequence of the teacher's behavior, the student is more likely to perform at a lower level.

A classic study by Rosenthal and Jacobson (1968) provides empirical evidence of the self-fulfilling prophecy in the public school system. Five elementary school students in a San Francisco school were selected at random and identified for their teachers as "spurters." Such a label implied that they had superior intelligence and academic ability. In reality, they were no different from the other students in their classes. At the end of the school year, however, these five students scored higher on their intelligence quotient (IQ) tests and made higher grades than their classmates who were not labeled as spurters. In addition, the teachers rated the spurters as more curious, interesting, and happy and more likely to succeed than the nonspurters. Because the teachers expected the spurters to do well, they treated the students in a way that encouraged better school performance. *The Human Side* feature in this chapter illustrates how negative labeling affects a student's self-concept and performance.

Who Succeeds? The Inequality of Educational Attainment

Figure 12.2 shows the extent of the variation in highest level of education attained by persons 25 years of age and older in the United States. As noted earlier, conflict theory focuses on such variations in discussions of education inequalities. Educational inequality is based on social class and family background, race and ethnicity, and gender. Each of these factors influences who succeeds in school.

Social Class and Family Background

One of the best predictors of educational success and attainment is socioeconomic status. Children whose families are in middle and upper socioeconomic brackets are more likely to perform better in school and to complete more years of education than children from lower socioeconomic-class families. Muller and Schiller (2000) report that students from higher socioeconomic backgrounds are more likely to enroll in advanced mathematics course credits and to graduate from high school—two indicators of future educational and occupational success. Further, when compared to low-income students, high-income students are six times more likely to graduate with a bachelor's degree in five years (Toppo 2004).

"Educational reform measures alone can have only modest success in raising the educational achievements of children from low-income families. The problems of poverty must be attacked directly."

Richard J. Murnane
Harvard University

Figure 12.2
Highest Level of Education Attained by Persons 25 Years Old: March 2001

Note: Detail may not sum to totals due to rounding.

Source: U.S. Department of Commerce, Bureau of the Census, Current Population Suevey, unpublished data.

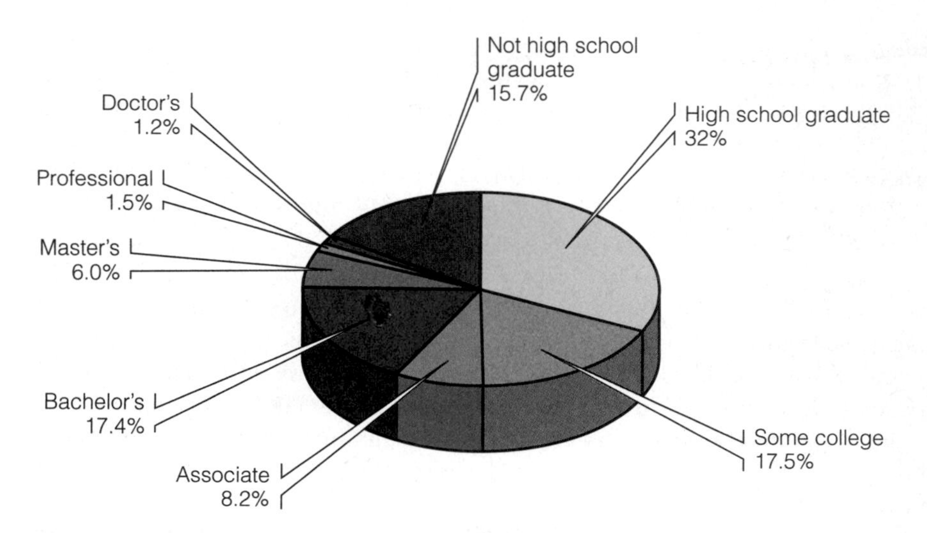

Families with low incomes have fewer resources to commit to educational purposes. Low-income families have less money to buy books, computers, tutors, and lessons in activities such as dance and music and are less likely to take their children to museums and zoos. Parents in low-income brackets are also less likely to expect their children to go to college, and their behavior may lead to a self-fulfilling prophecy. Disproportionately, children from low-income families do not go to college (Levinson 2000).

Disadvantaged parents are also less involved in their children's education. For example, 87 percent of non-poor children are read to frequently by a family member compared to 74 percent of poor children (NCES 2003b). Although working-class parents may value the education of their children, in contrast to middle- and upper-class parents, they are intimidated by their child's schools and teachers, don't attend teacher conferences, and have less access to educated people to consult about their child's education (Lareau 1989).

Because low-income parents are often themselves low academic achievers, their children are exposed to parents who have limited language and academic skills. Children learn the limited language skills of their parents, which restricts their ability to do well academically. Low-income parents may be unable to help their children with their math, science, and English homework because they often do not have the academic skills to do the assignments. Interestingly, Call, Grabowski, Mortimer, Nash, and Lee (1997) report that even among impoverished youths, parental education is one of the best predictors of a child's academic success.

Children from poor families also have more health problems and nutritional deficiencies. In 1965 Project **Head Start** began to help preschool children from disadvantaged homes. It provided an integrated program of health care, parental involvement, education, and social services. Today, over 900,000 3- to 5-year-olds are enrolled in Head Start (HHS 2003). Graduates of Head Start "score better on intelligence and achievement tests, their health status is better, and they have the socioemotional traits to help them adjust to school" (Zigler, Styfco, & Gilman 2004, 341).

Nonetheless, in 2003 President Bush mandated a nationwide assessment of all 4-year-olds in the program (CDF 2003b). The evaluation measures if and how children are learning, and facilitates comparisons between local Head Start programs.

Some fear that the assessment is a precursor to dismantling the program. The Children's Defense organization, a child advocacy group, in response to such fears, has begun airing television advertisements which state "Call Congress . . . Tell Congress Head Start's not broken, so don't break it" (CDF 2003c).

Assessments of Early Head Start, a program for infants and toddlers from low-income families, have already been conducted. One evaluation found that "after a year or more of program services, when compared with a randomly assigned control group, 2-year-old Early Head Start children performed significantly better on a range of measures of cognitive, language, and socio-emotional development" (Summary Report 2001).

Lack of adequate funding for Head Start has long been a problem as is equality of educational funding in general. Children who live in lower socioeconomic conditions receive fewer public educational resources. Schools that serve low socioeconomic districts are largely overcrowded and understaffed, lacking adequate building space and learning materials.

The U.S. tradition of decentralized funding means that local schools depend upon local taxes, usually property taxes. About 47 percent of school funding comes from local sources (Wenglinsky 2004). The amount of money available in each district varies by the socioeconomic status, or SES, of the district. For example, in New York, schools teaching the poorest students receive $2,152 per student less than schools teaching the wealthiest students (Schemo 2003). This system of depending on local communities for financing has several consequences:

- Low-SES school districts are poorer because less valuable housing means lower property values; in the inner city, houses are older and more dilapidated; less desirable neighborhoods are hurt by "white flight," with the result that the tax base for local schools is lower in deprived areas.
- Low-SES school districts are less likely to have businesses or retail outlets where revenues are generated; such businesses have closed or moved away.
- Because of their proximity to the downtown area, low-SES school districts are more likely to include hospitals, museums, and art galleries, all of which are tax-free facilities. These properties do not generate revenues.
- Low-SES neighborhoods are often in need of the greatest share of city services; fire and police protection, sanitation, and public housing consume the bulk of the available revenues. Precious little is left for education in these districts.
- In low-SES districts, a disproportionate amount of the money has to be spent on maintaining the school facilities, which are old and in need of repair, so less is available for the children themselves.

Although the state provides additional funding to supplement local taxes, it is not enough to lift schools in poorer districts to a level that even approximates the funding available to schools in wealthier districts.

Race and Ethnicity

Socioeconomic status interacts with race and ethnicity (Lareau & Horvat 2004). Because race and ethnicity are so closely tied to socioeconomic status, it appears that race or ethnicity alone can determine school success. Although race and ethnicity also have independent effects on educational achievement (Bankston & Caldas 1997; Jencks & Phillips 1998), their relationship is largely a result of the association between race/ethnicity and socioeconomic status. As Table 12.1 indicates, educational attainment has increased over time and varies by race and ethnicity. In general, the high school graduation gap between racial and ethnic groups is narrowing;

Table 12.1
Educational Attainment by Race, Ethnicity, and Sex, 1970 and 2000

	1970		2000	
	M	F	M	F
Four Years of High School or More				
White	54.0	55.0	84.2	84.0
Black	30.1	32.5	78.7	78.3
Hispanic	37.9	34.2	56.6	57.5
Asian	NA	NA	88.2	83.4
Total	51.9	52.8	84.2	84.0
Four Years of College or More				
White	14.4	8.4	28.5	23.9
Black	4.2	4.6	16.3	16.7
Hispanic	7.8	4.3	10.7	10.6
Asian	NA	NA	47.6	40.7
Total	13.5	8.1	27.8	23.6

NA = Not Available

Source: *Statistical Abstract of the United States,* 2002, 122nd edition, Tables 208, 209.

the college graduation gap, however, is getting wider—whites and Asians on one side, Hispanics and African-Americans on the other.

One reason that some minority students have academic difficulty is that they did not learn English as their native language. For example, in 2003 there were 9.8 million children in U.S. schools who speak a language other than English in their homes; for the majority of these children that language is Spanish. This number represents nearly 1 in 5 children between the ages of 5 and 17 (U.S. Census 2003), and the proportion is likely to increase. Because of high birth rates, increased immigration, and low levels of private school enrollment, the Hispanic school population continues to grow dramatically, tripling since 1968 (Frankenberg & Lee 2002).

To help American children who do not speak English as their native language, some educators advocate **bilingual education**—teaching children in both English and their non-English native language. Advocates claim that bilingual education results in better academic performance of minority students, enriches all students by exposing them to different languages and cultures, and enhances the self-esteem of minority students. Critics argue that bilingual education limits minority students and places them at a disadvantage when they compete outside the classroom, reduces the English skills of minorities, costs money, and leads to hostility with other minorities who are also competing for scarce resources.

Another factor that hurts minority students academically is that many tests used to assess academic achievement and ability are biased against minorities. Questions on standardized tests often require students to have knowledge that is specific to the white middle-class majority culture, and students for whom English is not their native language are seriously disadvantaged.

In addition to being hindered by speaking a different language and being from a different cultural background, minority students in white school systems are also

© Michael Newman/PhotoEdit

The debate over bilingual education is likely to grow. By 2040, less than half of all school-age children will be non-Hispanic whites.

disadvantaged by overt racism and discrimination. Much of the educational inequality experienced by poor children results because a high percentage of them are also nonwhite or Hispanic. Discrimination against minority students takes the form of unequal funding, as discussed earlier, as well racial profiling and school segregation.

Recent studies indicate that minority students, and specifically black students, may be the victims of what is being called "learning while black" (Morse 2002a). The allegation is:

> not unlike police who stop people on the basis of race, teachers and school officials discipline black students more often—and more harshly—than whites. The result: black students are more likely to slip behind in their studies and abandon school all together—if they're not kicked out first. (Morse 2002a, 50)

One study of 11,000 middle school students found that black students were more than twice as likely as whites to be sent to the principal's office or be suspended. Black students were also four times as likely to be expelled. While the debate is likely to continue, it must be noted that differences in discipline patterns are not necessarily a consequence of racism. They may reflect, for example, differences in behavior.

In 1954, the U.S. Supreme Court ruled in *Brown v. Board of Education* that segregated education was unconstitutional because it was inherently unequal. Despite this ruling, many schools today are racially segregated. In 1966, a landmark study entitled Equality of Educational Opportunity (Coleman et al. 1966) revealed the extent of segregation in U.S. schools. In this study of 570,000 students and 60,000 teachers in 4,000 schools, the researchers found that almost 80 percent of all schools attended by whites contained 10 percent or fewer blacks, and that whites outperformed minorities (excluding Asian-Americans) on academic tests. Coleman and his colleagues emphasized that the only way of achieving quality education for all racial groups was to desegregate the schools. This recommendation, known as the **integration hypothesis,** advocated busing to achieve racial balance.

Table 12.2
Segregation in Public Schools

	Percentage* of Black and Hispanic Students in Segregated Schools			
	School at Least 50% Minority		School at Least 90% Minority	
Year	Blacks	Hispanics	Blacks	Hispanics
1968–1969	77	55	64	23
1980–1981	63	68	33	28
1991–1992	66	73	34	34
1998–1999	70	76	37	37

*Rounded to nearest whole percent.

Source: Gary Orfield. 2001. *Schools More Separate: Consequences of a Decade of Resegregation.* Cambridge, MA: Harvard University, the Civil Rights Project.

In spite of the Coleman report, court-ordered busing, and an emphasis on the equality of education, public schools remain largely segregated. As shown in Table 12.2, most black and Hispanic U.S. students attend schools that are predominantly minority in enrollment. A study by the Civil Rights Project at Harvard found that since 1986 there has been a trend toward "resegregation" (Frankenberg & Lee 2002; Orfield 2002). Research documents the harmful effects of this continued practice. After examining the reading and mathematics achievement levels of a nationally representative sample of high school students, Roscigno (1998, 1051) concludes that "school racial composition matters . . . in the direction one would expect, even with class composition and other familial and educational attributes accounted for. Attending a black segregated school continues to have a negative influence on achievement." Nonetheless, for financial as well as political reasons, busing has essentially been abandoned.

Gender

Worldwide, as in the United States, women receive less education than men. An estimated 862 million adults in the world are illiterate, and two-thirds of them are women (UNESCO 2003). Further, "gender gaps in education have not narrowed significantly since 1990 when the world's governments committed themselves to eliminating gender disparity in education by 2000 at the Education for All Summit" (Briefing Paper 2003).

Historically, U.S. schools have discriminated against women. Prior to the 1830s, U.S. colleges accepted only male students. In 1833, Oberlin College in Ohio became the first college to admit women. Even so, in 1833, female students at Oberlin were required to wash male students' clothes, clean their rooms, and serve their meals and were forbidden to speak at public assemblies (Fletcher 1943; Flexner 1972).

In the 1960s, the women's movement sought to end sexism in education. Title IX of the Education Amendments of 1972 states that no person shall be discriminated against on the basis of sex in any educational program receiving federal funds. These guidelines were designed to end sexism in the hiring and promoting of teachers and administrators. Title IX also sought to end sex discrimination in granting admission to college and awarding financial aid. Finally, the guidelines called for an increase in opportunities for female athletes by making more funds available to their programs. Although gender inequality in education continues to be a problem worldwide, the push toward equality has had some effect. For example, in 1970, nearly twice as many men as women had four years of college or more—8.1 percent compared to 13.5 percent. By 2000, 23.6 percent of women and 27.8 percent of men had four years of college or more (see Table 12.1).

Traditional gender roles account for many of the differences in educational achievement and attainment between women and men. As noted in Chapter 7, schools, teachers, and educational materials reinforce traditional gender roles in several ways. Some evidence suggests, for example, that teachers provide less attention and encouragement to girls than to boys, and that textbooks tend to stereotype females and males in traditional roles (Evans & Davies 2000; Spade 2004).

Studies of academic performance indicate that females tend to lag behind males in math and science. One explanation is that women experience workplace discrimination in these areas, and this restricts their occupational and salary opportunities. The perception of restricted opportunities, in turn, negatively affects academic motivation and performance among girls and women (Baker & Jones 1993).

Most of the research on gender inequality in the schools focuses on how female students are disadvantaged in the educational system. But what about male students? For example, schools fail to provide boys with adequate numbers of male teachers to serve as positive role models. To remedy this, some school systems actively recruit male teachers, especially in the elementary grades where female teachers are in the majority.

The problems that boys bring to school may indeed require schools to devote more resources and attention to them. More than 70 percent of students with learning disabilities such as dyslexia are male, as are 75 percent of students identified as having serious emotional problems. Boys are also more likely than girls to have speech impairments, to be labeled as mentally retarded, to exhibit discipline problems, to drop out of school, and to feel alienated from the learning process (this chapter's *Self and Society* feature assesses student alienation) (Bushweller 1995; Goldberg 1999; Sommers 2000). As discussed in Chapter 7, the argument that girls have been educationally shortchanged has recently come under attack as some academicians charge that it is boys, not girls, who have been left behind (Sommers 2000).

Problems in the American Educational System

A 2002 survey by the National Center for Education Statistics asked respondents to "grade" the nation's schools. Based on a 4-point scale, the average grade given was 2.08—barely above a C (NCES 2003c). This is particularly troublesome given that federal funding for schools has increased dramatically since 1990 (see Figure 12.3). Consistent with the nation's "grade" is President Bush's emphasis on educational reform and the problems his administration hopes to address—low academic achievement, high dropout rates, questionable teacher training, and school violence. These and other problems contribute to the widespread concern over the quality of education in America.

Figure 12.3
Federal On-Budget Funds for Education, by Level or Other Educational Purpose: 1965–2002

Source: U.S. Office of Management and Budget, Budget of the U.S. Government, fiscal years 1967 to 2003; National Science Foudation, Federal Funds for Research and Development, fiscal years 1967 to 2002; and unpublished data.

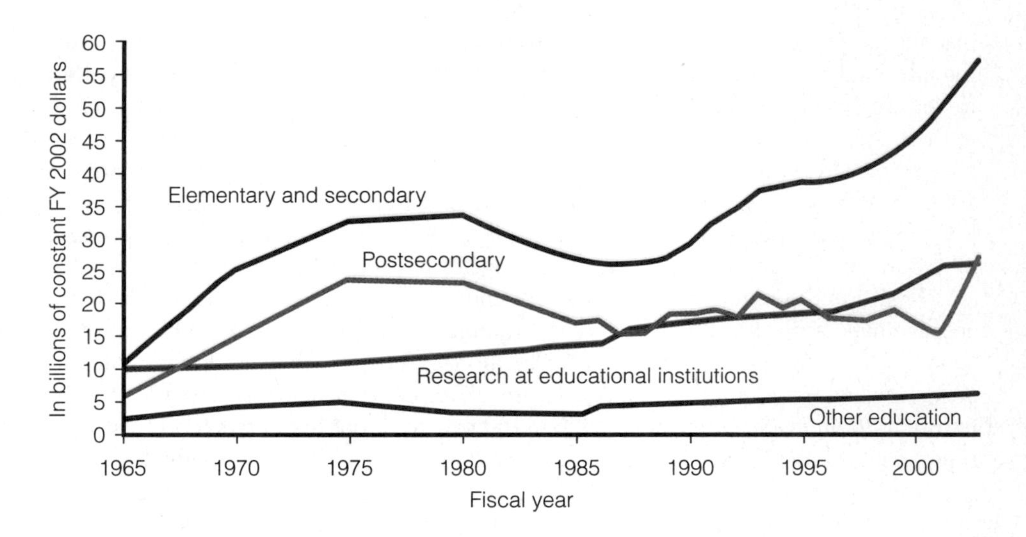

Low Levels of Academic Achievement

The most recent national data available indicate that more than 44 million adults in this country cannot "fill out an application, read a food label, or read a simple story to a child"—that is, they are **functionally illiterate** (*Literacy* 2003, 1). Functionally illiterate adults are disproportionately poor, over the age of 55, uneducated, and members of racial or ethnic minority groups (NAAL 2000).

"The more you know, the more you know you don't know."

Unknown

Among children, illiteracy tends to be highest among students who attend the poorest schools, yet apathy and ignorance may be found among students at more affluent schools as well. For example, an ABC News Special entitled "Burning Questions: America's Kids: Why They Flunk" began with the following interview with students from middle-class high schools:

Interviewer: Do you know who's running for president?

First Student: Who, run? Ooh. I don't watch the news.

Interviewer: Do you know when the Vietnam War was?

Second Student: Don't even ask me that. I don't know.

Interviewer: Which side won the Civil War?

Third Student: I have no idea.

Interviewer: Do you know when the American Civil War was?

Fourth Student: 1970.

However, results from the National Assessment of Educational Progress (NAEP), a nationwide testing effort, have improved over time. For example, mathematics proficiency of fourth, eighth, and twelfth graders has increased since 1990 (NCES 2003d). Writing scores for fourth and eighth graders have improved since 1998, although changes for twelfth graders are not statistically significant (NCES 2003e). A recent report issued by the National Commission on Writing in America's Schools and Colleges recommends that the time spent on writing, about 15 percent of the time spent on watching television, be doubled (Lewin 2003).

Although some international comparisons are improving, American students are still outperformed by many of their foreign counterparts. An international report examining student performances in 30 industrialized nations found that even though the United States is one of the leaders in educational spending, U.S. 15-year-olds scored only "average" in math, reading, and science. Other results from the report include these findings (International Education Report 2003a):

- Average class size varies dramatically—from 36 in Korea to 20 in, among other countries, Australia, Denmark, Greece, Italy, and Norway.
- On the average, a teacher works 792 hours a year; the range is from 650 hours in Japan to 950 hours or more in the United States, Scotland, and New Zealand.
- Of all foreign students in the countries studied, more attend colleges and universities in the United States than any other country.

Further, the Progress in International Reading Literacy Study indicates that in reference to reading, American youth scored above the international average. The report further concluded, "In the United States there are significant gaps in reading literacy achievement between racial/ethnic groups, between students in high poverty schools and other public schools, and also between boys and girls" (International Education Report 2003a, 1).

School Dropouts

Globally, dropout rates vary considerably between countries. In the United States approximately 11 percent (4 million) of 16- to 24-year-olds are dropouts—that is, they are not presently enrolled in and did not graduate from high school. In some U.S. cities the dropout rate is as high as 50 percent. However, in the Netherlands, the dropout rate is near zero (BBC 2003).

Why do students drop out of high school? An analysis of data from the National Educational Longitudinal Study offers some clues. Teachman, Paasch, and Carver (1997) found that staying in school between tenth and twelfth grade is a function of several variables. Children who attend Catholic schools; are from intact, two-parent families; and come from families with greater income and educational levels have lower odds of dropping out of school. The authors conclude that social capital—that is, resources that are a consequence of specific social relationships—is predictive of school continuation.

The economic and social consequences of dropping out of school are significant. Dropouts are more likely than those who complete high school to be unemployed (see Figure 12.1) and to earn less when they are employed (*Digest of Educational Statistics* 2001). Individuals who do not complete high school are also more likely to engage in criminal activity, come from a home with a low income, have poorer health and lower rates of political participation, and require more government services such as welfare and health care assistance (Rumberger 1987; Natriello 1995; NCES 2002).

Violence in the Schools

Students at school between the ages of 12 and 18 were the victims of 1.9 million crimes in 2002. Most school crime, however, is not serious, with theft accounting for 64 percent of the total crimes against students. Despite the public outrage over school violence in the wake of Columbine and other school shootings, the chance of a child's being killed at school is quite rare—about 1 in a million (NCJRS 2003).

Nonetheless, certain school characteristics are associated with an increase in the probability of a student's being the victim of a crime. For example, schools in

Self and Society

Student Alienation Scale

Indicate your agreement to each statement by selecting one of the responses provided:

1. It is hard to know what is right and wrong because the world is changing so fast.
   ______ Strongly agree ______ Agree ______ Disagree ______ Strongly disagree
2. I am pretty sure my life will work out the way I want it to.
   ______ Strongly agree ______ Agree ______ Disagree ______ Strongly disagree
3. I like the rules of my school because I know what to expect.
   ______ Strongly agree ______ Agree ______ Disagree ______ Strongly disagree
4. School is important in building social relationships.
   ______ Strongly agree ______ Agree ______ Disagree ______ Strongly disagree
5. School will get me a good job.
   ______ Strongly agree ______ Agree ______ Disagree ______ Strongly disagree
6. It is all right to break the law as long as you do not get caught.
   ______ Strongly agree ______ Agree ______ Disagree ______ Strongly disagree
7. I go to ball games and other sports activities at school.
   ______ Always ______ Most of the time ______ Some of the time ______ Never
8. School is teaching me what I want to learn.
   ______ Strongly agree ______ Agree ______ Disagree ______ Strongly disagree
9. I go to school parties, dances, and other school activities.
   ______ Always ______ Most of the time ______ Some of the time ______ Never
10. A student has the right to cheat if it will keep him or her from failing.
   ______ Strongly agree ______ Agree ______ Disagree ______ Strongly disagree
11. I feel like I do not have anyone to reach out to.
   ______ Always ______ Most of the time ______ Some of the time ______ Never
12. I feel that I am wasting my time in school.
   ______ Always ______ Most of the time ______ Some of the time ______ Never
13. I do not know anyone that I can confide in.
   ______ Strongly agree ______ Agree ______ Disagree ______ Strongly disagree
14. It is important to act and dress for the occasion.
   ______ Always ______ Most of the time ______ Some of the time ______ Never
15. It is no use to vote because one vote does not count very much.
   ______ Strongly agree ______ Agree ______ Disagree ______ Strongly disagree

neighborhoods with gang activity have higher rates of student victimization. Further, students who report knowing another student who brought a gun to school, or who actually saw a student with a gun, have higher rates of victimization. Finally, the prevalence of alcohol and other drugs in a school is indirectly related to school violence. Schools in neighborhoods with higher gang activity have students who are more likely to report alcohol and drug availability (Addington, Ruddy, Miller, Devoe, & Chandler 2004).

A study by the National Research Council investigated school shootings in rural and suburban communities. The report argues that school shootings more closely resemble the rampage killings that occur in workplaces than they do urban

16. When I am unhappy, there are people I can turn to for support.
_______ Always _______ Most of the time _______ Some of the time _______ Never
17. School is helping me get ready for what I want to do after college.
_______ Strongly agree _______ Agree _______ Disagree _______ Strongly disagree
18. When I am troubled, I keep things to myself.
_______ Always _______ Most of the time _______ Some of the time _______ Never
19. I am not interested in adjusting to American society.
_______ Strongly agree _______ Agree _______ Disagree _______ Strongly disagree
20. I feel close to my family.
_______ Always _______ Most of the time _______ Some of the time _______ Never
21. Everything is relative and there just aren't any rules to live by.
_______ Strongly agree _______ Agree _______ Disagree _______ Strongly disagree
22. The problems of life are sometimes too big for me.
_______ Always _______ Most of the time _______ Some of the time _______ Never
23. I have lots of friends.
_______ Strongly agree _______ Agree _______ Disagree _______ Strongly disagree
24. I belong to different social groups.
_______ Strongly agree _______ Agree _______ Disagree _______ Strongly disagree

Interpretation

This scale measures four aspects of alienation: powerlessness, or the sense that high goals (e.g., straight A's) are unattainable; meaninglessness, or lack of connectedness between the present (e.g., school) and the future (e.g., job); normlessness, or the feeling that socially disapproved behavior (e.g., cheating) is necessary to achieve goals (e.g., high grades); and social estrangement, or lack of connectedness to others (e.g., being a "loner"). For items 1, 6, 10, 11, 12, 13, 15, 18, 19, 21, and 22, the response indicating the greatest degree of alienation is "strongly agree" or "always." For all other items, the response indicating the greatest degree of alienation is "strongly disagree" or "never."

Source: Rosalind Y. Mau. 1992. "The Validity and Devolution of a Concept: Student Alienation." *Adolescence* 27(107):739–740. Used by permission of Libra Publishers, Inc., 3089 Clairemont Drive, Suite 383, San Diego, California 92117.

shootings (Bowman 2002, 1): "The urban [shooting] cases tended to be classic disputes that spilled into school territory, but the shooters in the rural and suburban cases consciously picked the schools as a place where a general grievance might be resolved." Between 1992 and 2002 thirteen multiple-victim shootings in U.S. rural and suburban schools were responsible for 44 deaths and 88 injuries. The National Research Council report concludes that it will be very difficult to identify assailants in advance since there is no consistent profile of a school shooter.

In response to violence, many schools throughout the country have police officers patrolling the halls, require students to pass through metal detectors before entering school, and conduct random locker searches. Video cameras set up in classrooms, cafeterias, halls, and buses purportedly deter some student violence. Seventy-five percent of new schools opened in 2002 were equipped with surveillance cameras (Dillon 2003). More than 2,000 schools nationwide conduct peer mediation and conflict resolution programs to help youth resolve conflict in nonviolent ways.

A relatively new way to deal with school "troublemakers," the alternative school, houses students who have committed a variety of offenses while allowing them to continue with their education. In the 2000–2001 school year, 39 percent of all U.S. school districts had an alternative school—66 percent of urban districts, 41 percent of suburban districts, and 35 percent of rural districts. Recently a class action suit has been filed by a group of inner-city teenagers seeking to overturn a state law. The law requires students who have committed a wide range of offenses—in or out of school—to be placed in an alternative school upon completing their sentences in juvenile detention (Steptoe 2003).

Inadequate School Facilities and Personnel

A report entitled "The Condition of America's Schools: A National Disgrace" documents the troubling condition of U.S. schools. According to the report, the American Society of Civil Engineers' Report Card on America's Infrastructure resulted in an overall grade of D-minus for U.S. schools (Crampton & Thompson 2002). State-by-state comparisons indicate dramatic differences. For example, infrastructure funding needs for Vermont equal $220 million, funding needs for New York, $47.6 billion. Further, many schools present environmental hazards.

> A significant number of schoolchildren and teachers in the United States are exposed on an almost daily basis to environmental hazards including volatile chemicals, airborne lead and asbestos, and noise pollution while they are at school. Some school hazards are linked to the aging of many of the nation's schools, to the ongoing siting of schools in close proximity to contaminated waste sites, and to the burgeoning population of school-aged children that has forced financially constrained school districts to use portable classrooms to increase their classroom space. (Wakefield 2002, 1)

School lunch programs have also come under attack. Incidents of large-scale food poisoning have increased 10 percent since 1990, with students reporting a wide range of illnesses from food poisoning to hepatitis A to salmonella (Morse 2002b). There are also concerns about the sale of "junk food" in school cafeterias in light of increasing rates of child obesity (Nash 2003). Less than 20 percent of school lunches meet the government's requirement that fewer than 10 percent of calories come from saturated fat.

School districts with inadequate funding often have difficulty attracting qualified school personnel. Unfortunately, these poorer districts are in dire need of talented teachers who can meet the needs of children from diverse backgrounds and of varying abilities. The number of minority teachers who can serve as role models, have similar life experiences, and have similar language and cultural backgrounds is far too few for the number of minority students (NCES 2000). Black and Hispanic male children may perceive that white female teachers are less understanding of what they experience than minority male teachers.

Undisciplined and violent students and inadequate teaching materials and school facilities contribute to low teacher morale. Low morale interferes with teaching effectiveness and drives some teachers out of the profession. The relatively low pay of teachers also contributes to morale problems and the questionable quality of instruction.

Deficient Teachers

In a National Press Club speech, former Department of Education Secretary Richard Riley spoke of the need for "a talented, dedicated, well-prepared teacher in every classroom." Given that over the next decade over 2 million additional teachers will be needed, and nearly 60 percent of today's teachers are over the age of 40, the challenge is not an insignificant one (PRB 2003).

Individuals who enter and remain in the teaching profession are not necessarily competent and effective teachers. There is evidence that those who go into teaching, on the average, have lower college entrance exams than the average college student (NCES 2003e). More than 25 percent of new teachers are placed in the classroom without state licensing credentials. Students at public rather than private schools and in poor rather than wealthier schools are more likely to be taught by teachers who did not major in the subject matter they are teaching (CDF 2003a). Further, a Department of Education survey found that less than 36 percent of current teachers report feeling "very well prepared" to initiate curriculum and performance standards, and less than 20 percent feel ready to meet the needs of the diverse student population (White House Fact Sheet 2002).

In response to such concerns, many states have implemented mandatory competency testing, which requires prospective teachers to pass the National Teacher's Examination or some other test of knowledge on the subject they intend to teach. In 1997, over 50 percent of prospective teachers in Massachusetts failed the state's competency exam, some unable to spell such basic words as "burned" and "abolished" (Leo 1998; Silber 1998). In response to such scandalous results, federal monies are now tied to "state 'report cards' that aggregate the test scores of would-be teachers by teacher preparation institution and thus produce a passing or failing 'grade' for each institution and state" (Cochran-Smith 2000, 163). Further, a recent report recommends that teacher pay raises be tied to students' performance (Associated Press 2004).

Additionally, more than half of the states have adopted **alternative certification programs,** whereby college graduates with degrees in fields other than education may become certified if they have "life experience" in industry, the military, or other relevant jobs. *Teach for America,* originally conceived by a Princeton University student in an honors thesis, is an alternative teacher education program aimed at recruiting liberal arts graduates into teaching positions in economically deprived and socially disadvantaged schools. After completing an eight-week training program, recruits are placed as full-time teachers in rural and inner-city schools. Critics argue that these programs may place unprepared personnel in schools.

"Education is the key to prosperity and the wisest investment we can make in our children's and our nation's future."

Richard Riley
Former U.S. Secretary of Education

Strategies for Action: Trends and Innovations in American Education

Americans rank improving education as one of their top priorities (Harris 2003). Recent attempts to improve schools include raising graduation requirements, barring students from participating in extracurricular activities if they are failing academic subjects, lengthening the school year, and prohibiting dropouts from obtaining driver's licenses. Further, there is a nationwide movement to eliminate **social promotion,** the passing of students from grade to grade even if they are failing. However, educational reformers are calling for changes that go beyond get-tough policies that maintain the status quo.

© Reuters/Kevin Lamarque/Hulton Archive

Dr. Roderick R. Page, the seventh U.S. Secretary of Education, oversees 4,700 employees and a budget of over $35 billion.

National Educational Policy

President Bush's education plan, "No Child Left Behind," was signed into law in January 2002. The plan is organized around four principles. The first is *accountability*. Each year, every state will be required to test third- through eight-grade students' math and reading abilities. Parents, administrators, and others will have access to the data, and states will issue a report card rating the performance of each school, teacher, and student.

The second principle is *flexibility*. The new law will permit federal funds to be transferred between programs, and local schools will have more input into how federal funds are being used. *Expanding options for parents* is the third principle. After a school is identified as failing, parents will be able to transfer their child to a better performing or charter school. Also, supplemental funds (between $500 to $1,000) will be provided for children in failing schools that can be used for a tutor, summer school programs, or any other school service.

The last principle is using *teaching methods* that are known to work. Money will be provided to "reading first"—a presidential plan to help children read. Teacher quality will be improved as local schools are better able to recruit and retain excellent teachers. Last, local schools will be able to use federal funds "for hiring new teachers, increasing teacher pay, improving teacher training and development and other uses" (*Fact Sheet* 2003). By 2006, according to the plan, only "highly qualified" teachers will be hired.

Recent reports on how the plan is doing, however, are troublesome. Just eight months after it was implemented, the Department of Education released a list of 9,000 "failing schools" leading many states to say the bar is simply too high.

> Utah announced it was removing some of the more difficult questions from its statewide exam. Ohio recently "refined" its criteria for calculating poor performing schools; afterward, the number receiving F's fell from 760 to 200. Michigan, California and Nevada are weighing similar actions. . . . Experts warn that the incentive to dumb down standards will only grow as the stricter provisions of the new law take effect. (Morse 2002c, 22).

In fact, so many schools were labeled as failing—57 percent in Delaware, 50 percent in Missouri, 45 percent of West Virginia—that the federal government has replaced the term "failing" with "in need of improvement" (Kronholz 2003).

> "Our society does not need to make its children first in the world in mathematics and science. It needs to care for its children— to reduce violence, to respect honest work of every kind, to reward excellence at every level, to ensure a place for every child . . ."
>
> **Nel Noddings**
> **Educational reformer**

Character Education

Seventy-five percent of over 4,500 students surveyed by Rutgers' Education Center reported cheating at least once in their academic careers, and more than half of the respondents reported plagiarizing information from the Internet (Slobogin 2002). Particularly worrisome, over half of the students surveyed didn't think that "copying questions and answers from a test" is cheating. As one student put it:

> What's important is getting ahead. The better grades you have the better school you get into, the better you're going to do in life. And if you learn to cut corners to do that, you're going to be saving yourself time and energy. In the real world, that's what's going to be going on. The better you do, that's what shows. It's not how moral you were in getting there.

To many educators and the general public, statements like this signify the need for **character education.**

Most school curricula neglect this side of education—the moral and interpersonal aspects of developing as an individual and as a member of society. President Bush's educational reform policy includes support for character education. Proponents of character education argue that "with intentional, thoughtful character education, schools can become communities in which virtues such as responsibility, hard work, honesty, and kindness are taught, expected, celebrated, and continually practiced" (FAQ 2003). For example, service learning programs are increasingly popular at universities and colleges nationwide. Service learning programs are community-based initiatives in which students volunteer in the community and receive academic credit for doing so. Studies on student outcomes have linked service learning to enhanced civic responsibility and moral reasoning, a reduction of risky behaviors, and higher levels of self-esteem (Independent Sector 2002; Jacobs 1999; Ramierz-Valles & Brown 2003).

How does the public feel about character education? In a national poll, 76 percent of respondents favored requiring schools to teach about values and morality (NPR 2003). Character education also occurs to some extent in schools that have peer mediation and conflict resolution programs. Such programs teach the value of nonviolence, collaboration, and helping others, as well as skills in interpersonal communication and conflict resolution.

Computer Technology in Education

Computers in the classroom allow students to access large amounts of information (see this chapter's *Focus on Technology* feature). The proliferation of computers both in school and at home may mean that teachers will become facilitators and coaches rather than sole providers of information. Not only do computers enable students to access enormous amounts of information including that from the World Wide Web, but they also allow students to progress at their own pace. However, computer technology is not equally accessible to all students. Students in poorer school districts are less likely to have access to computers in school or at home (NCES 2003f) (see Chapter 15). In general, however, access to computers has increased dramatically over the years. In 2000, the ratio of students to instructional computers was 5:1; in 1983, 125:1 (*Indicators* 2003).

Interestingly, the conclusion of one of the largest studies of school computers was that students who use computers often scored lower on math tests than their low-use counterparts. The Educational Testing Service's study of 14,000 fourth and eighth graders concluded that how the computers were used—repetitive math drills versus real-life simulation—was responsible for the test variations. Also of note, students in classrooms where teachers are trained in computer use do better than students in classrooms where teachers are less skilled (Weiner 2000). Minority and low-income students are less likely to have teachers highly skilled in computer technology than their middle- and upper-income counterparts.

The Enhancing Education Through Technology (ED TECH) program is part of the *No Child Left Behind Act* of 2002. The goals of the program are threefold: (1) to improve student achievement through the use of technology resources, (2) to ensure that teachers integrate technology into the curriculum in such a way as to improve student achievement, and (3) to help students become technically literate by the eighth grade. With the added funding the program provides, schools will be able to purchase additional technology in support of these goals (NCES 2003f).

Focus on Technology

Distance Learning and the New Education

Imagine never having an 8 o'clock class or walking into the lecture room late. Imagine no room and board bills, or having to eat your roommate's cooking. Imagine going to class when you want, even 3 o'clock in the morning. Imagine not worrying about parking! The future of higher education? Maybe. It's possible that the World Wide Web and other information technologies have so revolutionized education that the above scenarios are a *fait accompli.*

What is distance learning? **Distance learning** separates, by time or place, the teacher from the student. They are, however, linked by some communication technology: video-conferencing, satellite, computer, audiotape or videotape, real-time chat room, closed-circuit television, electronic mail, or the like.

Today, it is possible to earn a bachelor's degree, graduate degree, and/or professional degree via the Net. Although enrollment rates have increased dramatically since 1990, in 2000, only 8 percent of undergraduates and 12 percent of master's students were enrolled in distance education courses (NCES 2002). A higher percentage of students at two-year institutions were enrolled than at four-year institutions, and the most common medium was the Internet followed by live audio or television.

The benefits of distance learning are clear. It provides a less expensive, accessible, and often more convenient way to complete a college degree. There are even pedagogical benefits. Research suggests that "students of all ages learn better when they are actively engaged in a process, whether that process comes in the form of a sophisticated multimedia package or a low-tech classroom debate on current events" (Carvin 1997). Distance education also benefits those who have historically been disadvantaged in the classroom. A review of research on gender differences suggests that females outperform males in distance learning environments; they are also more likely to enroll in distance education courses (Koch 1998; NCES 2002).

But all that glitters is not gold. Evidence suggests that students feel more estranged from their distance learning instructors than from teachers in conventional classrooms. Among distance education users, a higher proportion of them "were less satisfied than more satisfied with the quality of instruction they received in their distance education classes compared with their regular classes" (NCES 2003).

Additionally problematic is the proliferation of "virtual degrees." London's Strassford University has an impressive brochure complete with ivy-trimmed buildings and an enviable history dating back to the reign of Queen Victoria. But in fact, it doesn't exist. It's a fake, and fake degrees in this era of electronic diplomas come with bogus transcripts, letters of recommendation, and a "backup" telephone number if anyone should want to verify the diploma's legitimacy (CBS News 2003). Most alarmingly, these fake degrees can be used to get student visas similar to those purchased by the 9/11 terrorists.

Further, teachers, particularly in higher education, are concerned about the quality of distance education. Several regulatory and advisory bodies, including the congressional Web-based Education Commission, are presently establishing quality standards and guidelines (Carnevale 2001). And although a committee of the American Association of University Professors (AAUP) acknowledged that distance learning may be a "valuable pedagogical tool," it also questioned whether "academic quality, academic freedom, intellectual property rights and instructor's workloads and compensation" will be compromised (Arenson 1998, A14; AAUP 2003).

Nonetheless, distance education continues to grow, in part, because it is a moneymaker—a multibillion-dollar industry. Even one-time critic William Bennett, former U.S. Secretary of Education, has recently thrown his hat into the corporate ring opening K12, a company that markets elementary and secondary school courses to home schoolers and to parents who want their children to have academic help outside of the classroom (Wildavsky 2001). Additionally, many commercial sites now offer "educational" courses and, alternatively, educational sites increasingly carry advertising banners, consumer discounts, product photographs, and the like (Guernsey 2000).

Will distance learning solve all the problems facing education today? The answer is clearly no. Although not the technological fix some are looking for, distance education, from digital libraries to "virtual" charter schools, does provide a provocative and financially lucrative alternative to traditional education providers.

Sources: AAUP (Association of University Professors). 2003. "Distance Education." http://aaup.org/issues/DistanceEd; Laren Arenson. 1998. "More Colleges Plunging into the Uncharted Waters of On-Line Courses." *New York Times,* November 2, A14; William M. Bulkeley. 1998. "Education: Kaplan Plans a Law School via the Web." *Wall Street Journal,* September 16, B1; Dan Carnevale. 2001. "Commission Says Federal Rules on Distance Education Must Be Updated." *The Chronicle of Higher Education,* January 5, A46; CBS News. 2003. "Cracking Down on Diploma Mills." http://www.cbsnews. com/stories/2003; Andy Carvin. 1997. EdWeb: Exploring Technology and School Reform. http://edweb.gsn.org; Lisa Guernsey. 2000. "Education: Web's New Come-On." *New York Times,* March 16, C1, C7; James V. Koch. 1998. "How Women Actually Perform in Distance Education." *The Chronicle of Higher Education* 45:A60; Joe Rudich. 1998. "Internet Learning." *Link-Up* 15:23–25; NCES (National Center for Education Statistics). 2003. "Distance Learning." http://nces.ed.gov/fastfacts; NCES (National Center for Education Statistics). 2002. "Student Participation in Distance Education." http://nces.ed.gov/ programs.coe

School Choice

Traditionally, children have gone to school in the district where they live. School vouchers, charter schools, home schooling, and private schools provide parents with alternative school choices for their children. **School vouchers** are tax credits that are transferred to the public or private school parents select for their child. President Bush's plan calls for federally funded vouchers for students attending failing schools that don't improve test scores for two consecutive years. In 2002, 8,600 schools were identified as failing state standards (Ed.gov 2003). The vouchers can be used for tuition at a private school, out-of-class tutoring, or other supplemental services.

Proponents of the voucher system argue that it reduces segregation and increases the quality of schools because they must compete for students to survive. Opponents argue that vouchers increase segregation because white parents use the vouchers to send their children to private schools with few minorities. Research by Saporito (2003) supports this contention:

> Findings show that white families avoid schools with higher percentages of non-white students. The tendency of white families to avoid schools with higher percentages of non-whites can not be accounted for by other school characteristics such as test scores, safety, or poverty rates. I also find that wealthier families avoid schools with higher poverty rates. The choices of whites and wealthier students lead to increased racial and economic segregation in the neighborhood schools that these children leave. (p. 181)

Vouchers, opponents argue, are also unfair to economically disadvantaged students who are not able to attend private schools because of the high tuition. Further, they argue, the use of vouchers for religious schools violates the constitutional guarantee of separation of church and state. However, the U.S. Supreme Court, in reviewing the voucher program in Cleveland, Ohio, held that the use of tax dollars for enrollment in religious schools is not unconstitutional (Morse 2002d).

Vouchers can also be used for charter schools. **Charter schools** originate in contracts, or charters, which articulate a plan of instruction that must be approved by local or state authorities. Nearly 40 states now have charter school laws (NCLB 2003). Although charter schools can be funded by foundations, universities, private benefactors, and entrepreneurs, many are supported by tax dollars (Mollison 2001). The Department of Education estimates that more than 2,000 charter schools are operating in the United States today, up from just two a decade ago. Charter schools, like school vouchers, were designed to expand schooling options and increase the quality of education through competition. Recent evidence suggests that "charter schools produce student learning gains comparable to those of conventional schools, despite resource limitations" (Rand 2003, 1).

Some parents are choosing not to send their children to school at all but to teach them at home. More than 1 million students in the United States are home-schooled, their numbers increasing by 15 percent annually (Winters 2001). For some parents, **home schooling** is part of a fundamentalist movement to protect children from perceived non-Christian values in the public schools. Other parents are concerned about the quality of their children's education and their safety. How does being schooled at home instead of attending public school affect children? Some evidence suggests that home-schooled children perform as well as or better than their institutionally schooled counterparts (Webb 1989; Winters 2001).

Another choice parents may make is to send their children to a private school. About 10 percent of all students are enrolled in private elementary or private high schools (U.S. Census 2003). The primary reason parents send their children to private schools is for religious instruction. The second most common reason is the belief that private schools are superior to public schools in terms of academic achievement. Evidence may support this belief. Students in private schools generally perform higher on achievement tests than their public school counterparts (NCES 2003g). Parents also choose private schools for their children in order to have greater control over school policy, to avoid busing, or to obtain a specific course of instruction such as dance or music.

Understanding Problems in Education

"In order for all children to enter elementary school prepared to learn, we will have to eliminate child poverty in this country and provide every child who needs them not only with adequate health and social services but also with an early childhood education program well beyond anything envisioned by Head Start."

Evans Clinchy
Educational reformer

What can we conclude about the educational crisis in the United States? Any criticism of education must take into account that just over a century ago the United States had no systematic public education system at all. Many American children did not receive even a primary school education. Instead, they worked in factories and on farms to help support their families. Whatever education they received came from the family or the religious institution. In the mid-1800s, educational reformer Horace Mann advocated at least five years of mandatory education for all U.S. children. Mann believed that mass education would function as the "balanced wheel of social machinery" to equalize social differences among members of an immigrant nation. His efforts resulted in the first compulsory education law in 1852, which required 12 weeks of attendance by school-age children each year. By World War I, every state mandated primary school education, and by World War II, secondary education was compulsory as well.

Public schools are supposed to provide all U.S. children with the academic and social foundations necessary to participate in society in a productive and meaningful way. But as conflict theorists note, for many children the educational institution perpetuates an endless downward cycle of failure, alienation, and hopelessness. As yet unsettled, *Williams v. California* is a class action suit against the state of California on behalf of over 1 million low-income students. Lawyers for the plaintiffs argue that the students are permanently disadvantaged as a result of the schools they attend—outdated books, unqualified teachers, and a general lack of resources. The students and their families want minimum standards to be set and upheld just as academic standards are now in place (Asimov 2003).

Breaking the downward cycle requires providing adequate funding for teachers, school buildings, equipment, and educational materials, a task that has become increasingly difficult as student enrollments swell and diversify, a consequence of the "baby-boom echo" and changing immigration patterns. Between 1989 and 2009, elementary school enrollment will have increased 12 percent, high school enrollment 19 percent, and full-time college student enrollment 11 percent (U.S. Department of Education 2000).

The public is, however, supportive of government spending on education. In a national survey, 77 percent favored paying teachers more; 81 percent favored placing more computers in classrooms; and 92 percent favored fixing run-down schools (NPR 2003). Legislation such as the 1998 amendments to the Higher Education Act, which allocated $300 billion to teacher preparation and recruitment, reduced interest on college student loans, and increased the maximum in Pell grants, is essential. But even with financial support, as functionalists argue, education alone cannot bear the burden of improving our schools.

Jobs must be provided for those who successfully complete their education. If not, students will have few incentives to exert any effort. Rosenbaum (2002, 488) explains:

> What is missing from current practices is a mechanism for creating and conveying signals that tell students the value of their present actions in achieving desirable career goals. Other countries produce such signals with linkage mechanisms. . . . The German system provides a clear mechanism that makes the relationship between school performance and career option totally obvious [through an apprenticeship program]. . . . students know that apprenticeships lead to respected occupations, and that school grades affect selection into apprenticeships.

Finally, "if we are to improve the skills and attitudes of future generations of workers, we must also focus attention and resources on the quality of the lives children lead outside the school" (Murnane 1994, 290). We must provide support to families so children grow up in healthy, safe, and nurturing environments. Children are the future of our nation and of the world. Whatever resources we provide to improve the lives and education of children are sure to be wise investments in our collective future.

Chapter Review

- **Do all countries educate their citizens?**
  No. Many societies have no formal mechanism for educating the masses. As a result, worldwide, over 30 percent of females and 17 percent of males are illiterate. The problem of illiteracy is greater in developing rather than developed nations; in some less-developed countries the adult illiteracy rate is as high as 80 percent.
- **According to structural-functionalist, what are the functions of education?**
  There are four major functions. The first is instruction—that is, teaching students knowledge and skills. The second is socialization that, for example, teaches students to respect authority. The third is sorting individuals into statuses by providing them with credentials. The fourth function is custodial care—a babysitting agency of sorts.
- **What is a self-fulfilling prophecy?**
  A self-fulfilling prophecy occurs when people act in a manner consistent with the expectations of others.
- **What variables predict school success?**
  Three variables tend to predict school success. Socioeconomic status predicts school success: the higher the socioeconomic status, the higher the likelihood of school success. Race predicts school success, with nonwhites and Hispanics having more academic difficulty than whites and non-Hispanics. Gender also predicts success although it varies by grade level.
- **What were the conclusions of the study summarized in *Social Problems Research Up Close*?**
  The authors concluded that (1) carrying a weapon to school is a rare event; (2) age, race, and sex were unrelated to carrying a weapon; (3) prior victimization and fear of crime were unrelated to gun carrying; (4) drug dealing or using were related to carrying a weapon to school; and (5) having a best friend that carries a gun to school was a strong predictor of weapon carrying.
- **What are some of the problems associated with the American school system?**
  One of the main problems is the lack of student achievement in our schools—particularly when you compare U.S. data with data from other industrialized countries. Minority dropout rates are high and school violence continues to be a threat. School facilities are in need of repair and renovations, and personnel, including teachers, have been found to be deficient.
- **What are the arguments for and against school choice?**
  Proponents of school choice programs argue that they reduce segregation and that schools, now having to compete with one another, will be of a higher quality. Opponents argue that school choice programs increase segregation and treat disadvantaged students unfairly. Low-income students can't afford, even with vouchers, to go to private schools. Further, those opposed to school choice are quick to note that using government vouchers to help pay for religious schools is unconstitutional.

Critical Thinking

1. Clearly, home schooling has both advantages and disadvantages. After making a list of each, would you want your child to be home-schooled? Why or why not?
2. As discussed in Chapter 9, the proportion of elderly in the United States is increasing dramatically as we move into the twenty-first century. Because the elderly are unlikely to have children in public schools, how will the allocation of necessary school funds be affected by this demographic trend?
3. Student violence is one of the most pressing problems in U.S. public schools. One response is defensive, that is, having police patrol halls, using metal detectors, and so on. Other than such defensive tactics, what violence prevention techniques should be instituted?

Key Terms

- alternative certification programs
- bilingual education
- character education
- charter schools
- cultural imperialism
- distance learning
- functionally illiterate
- Head Start
- home schooling
- integration hypothesis
- multicultural education
- privatization
- school vouchers
- self-fulfilling prophecy
- social promotion

Taking a Stand

Should schools be privatized?

Given the less than stellar performance of many U.S. schools, some have argued that what is needed is privatization. **Privatization** entails states hiring corporations to operate local institutions. For example, 45 of Philadelphia's worst-performing schools are now being run by seven independent contractors including Edison Schools—the largest of the for-profit corporations (Winters 2002). Edison is hoping to help failing schools and failing students while making a return for investors. While proponents argue that market-driven education will result in a better "product," critics are ideologically opposed to companies making a profit from public school children.

Use Wadsworth's exclusive online resources—InfoTrac College Edition, MicroCase Online, and OVRC—to formulate a position on this topic.

The Wadsworth's Sociology Online Resources and Writing Companion will help you get started. This valuable guide will show you how to use Wadsworth's exclusive online resources when studying social problems. It will also help you to build essential research and writing skills. InfoTrac College Edition, MicroCase Online, OVRC, and an electronic copy of portions of this companion are available at http://sociology.wadsworth.com/mooney_knox_schacht/problems4e, the companion Web site for *Understanding Social Problems,* Fourth Edition.

Media Resources

The Companion Web Site for *Understanding Social Problems,* Fourth Edition

http://sociology.wadsworth.com/mooney_knox_schacht/problems4e

Supplement your review of this chapter by going to the companion Web site to take one of the Tutorial Quizzes, use the flash cards to master key terms, and check out the many other study aids you'll find there. You'll also find special features such as *Wadsworth's Sociology Online Resources and Writing Companion,* GSS Data, and Census 2000 information, data, and resources at your fingertips to help you complete that special project or do some research on your own.

Interactions CD-ROM

Go to the Interactions CD-ROM for *Understanding Social Problems,* Fourth Edition, to access additional interactive learning tools, such as in-depth review materials, corresponding practice quizzes, and other engaging resources and activities to help you study the concepts in this chapter.

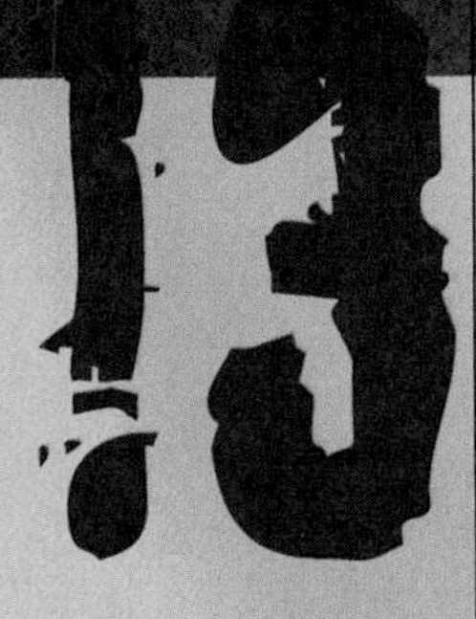

"Population may be the key to all the issues that will shape the future: economic growth; environmental security; and the health and well-being of countries, communities, and families." *Nafis Sakik, Executive Director, UN Population Fund*

Population and Urbanization

Since 1999, world population has exceeded 6 billion. If you have difficulty conceiving a number so large, you are not alone. Many people cannot imagine the meaning of 1 billion, let alone 6 billion. To help you appreciate how much one billion is, consider the following (U.S. Geological Survey 2003; Urban Legends Reference Page 2003):

- *If you were to count to 1 billion, saying one number per second without stopping, it would take you 31.7 years.*
- *1 billion minutes equals 1,901 years, so in a billion minutes from the year 2005, it will be the year 3906.*
- *A billion hours ago, our ancestors were living in the Paleolithic age.*

Although thousands of years passed before the world's population reached 1 billion, in less than three hundred years the population exploded from 1 billion to 6 billion. Affecting virtually every aspect of social life, "population growth is the single most important set of events ever to occur in human history" (Weeks 2002, 1). In this chapter, we provide an overview of the world's population situation, noting trends in population size and growth. As nearly half of the world's population live in urban areas, this chapter ties together the overlapping concerns of population growth and urbanization, focusing on problems of population growth and urbanization and strategies to alleviate these problems.

The Global Context: A World View of Population Growth and Urbanization

We begin this chapter with a brief overview of the history of world population growth, current population trends, and future population projections. We also review the development of urbanization and note the current state of urbanization throughout the world.

World Population: History, Current Trends, and Future Projections

Humans have existed on this planet for at least 200,000 years. During 99 percent of human history, population growth was restricted by disease and limited food supplies. Around 8000 B.C., the development of agriculture and the domestication of animals led to increased food supplies and population growth, but even then harsh living conditions and disease still put limits on the rate of growth. This pattern continued until the mid-eighteenth century when the Industrial Revolution improved the standard of living for much of the world's population. The improvements included better food, cleaner drinking water, and improved housing and sanitation as well as advances in medical technology such as antibiotics and vaccinations against infectious diseases; all contributed to rapid increases in population.

Population Doubling Time. Population **doubling time** is the time required for a population to double from a given base year if the current rate of growth continues. It took several thousand years for the world's population to double from 4 million to 8 million, a few thousand years to double from 8 million to 16 million, about a thousand to double from 16 to 32, and less than a thousand to double to 64. The next doubling of the world's population occurred between the European Renaissance and the Industrial Revolution in a time span of about 400 years. But from 1750, the doubling of world population took little more than 100 years, and the next doubling took less. The most recent doubling (from 3 billion in 1960 to 6 billion in 1999) took only about 40 years (Weeks 2002).

A look at current population trends and future projections suggests that although world population will continue to grow in the coming decades, it may never double in size again. In fact, because fertility rates have dropped around the world, "we're finally at a point where it's possible that a child born today will live to see the stabilization of world population" (Nelson, 2004, 17).

"World population grew from 2.5 billion in 1950 to 6.1 billion in 2000. The growth during those 50 years exceeded that during the 4 million years since we emerged as a distinct species."

Lester Brown
Worldwatch Institute

Current Population Trends and Future Projections. In 2003, the world's population was growing at an annual rate of 1.2 percent, resulting in the addition of 77 million people per year. Ninety-seven percent of world population growth is in developing countries; the only developed country that is growing is the United States, which has steady immigration as well as a relatively high birth rate for a developed country (Population Reference Bureau 2003).

Higher population growth in developing countries is largely due to higher **fertility rates**—the average number of births per woman in a population. In 2003, fertility rates ranged from an average of 1.1 children per woman in the former Soviet republics of Georgia and Ukraine to 8.0 children per woman in the West African country of Niger. The average for the United States was 2.0 and the world average was 2.8—a significant improvement over 1990's world fertility rate of 3.4. If this trend continues, the average world fertility rate could possibly drop to 2.1, the first step toward population stabilization (McFalls 2003; Nelson 2004). However, even if every country in the world achieved replacement-level fertility rates (an average of 2.1 births per woman), populations would continue to grow for several decades because of **population momentum**—continued population growth as a result of past high fertility rates that have resulted in large numbers of young women who are currently entering their childbearing years.

Projections of future population growth suggest that world population will grow from 6.3 billion in 2003 to 8.9 billion in 2050 (United Nations 2003). Most of the growth in world population is expected to occur in Africa and Asia (see Figure 13.1). However, in the four African countries most affected by HIV/AIDS—

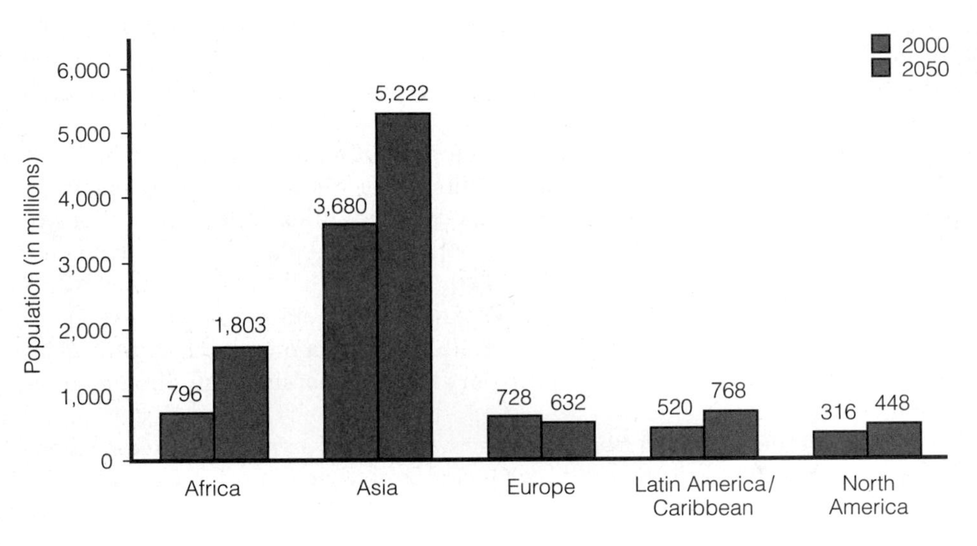

Figure 13.1
Population in Major World Regions, 2000 and Projections for 2050

Source: United Nations. 2003. *World Population Prospects: The 2002 Revision* (medium projection series).

Botswana, Lesotho, South Africa, and Swaziland—population is expected to decline in the coming decades (United Nations 2003). In these countries the fertility rate—the average number of children born to each woman—has fallen below 2.1, the replacement level required to maintain the population.

In 2003, 67 countries had fertility rates below the replacement level, including every country in Europe (Population Reference Bureau 2003). The declining birthrate in many countries, in addition to increasing life span, is contributing to the changing age structure of the world. In more developed regions of the world, the population aged 60 and over is expected to increase from 19 percent in 2000 to 32 percent in 2050. In less-developed regions, the proportion of population aged 60 and over will increase from 8 percent in 2000 to about 20 percent in 2050 (United Nations 2003). Aging societies expect labor shortages and possibly cultural clashes as they resort to imported labor. Furthermore, as the proportions of elderly people grow throughout the world, societies must find ways to meet the increasing health care and pension needs of this population (*Popline* 2003a).

United States Population Growth. During the colonial era (c. 1650), the U.S. population included about 50,000 colonists and 750,000 Native Americans. By 1859, disease and warfare had reduced the American Indian population to 250,000, but the European population had increased to 23 million. The U.S. population continued to increase into the 1900s. From the mid-1940s to the late 1960s the U.S. birthrate increased significantly. This period of high birthrates, commonly referred to as the baby boom, peaked in 1957 when the crude birthrate (number of live births per 1,000 people) reached 25.3. Although the birthrate dropped to 15.3 in 1986, the U.S. population continues to increase. U.S. total population was 292 million in 2003 and is expected to reach more than 422 million by 2050 (Population Reference Bureau 2003). The population growth of 32.7 million people between 1990 and 2000 represents the largest census-to-census increase in U.S. history (Perry & Mackun 2001). The United States is the third largest national population in the world, behind China and India (see Table 13.1).

An Overview of Urbanization Worldwide and in the United States

As early as 5000 B.C., cities of 7,000 to 20,000 people existed along the Nile, Tigris-Euphrates, and Indus River valleys. But not until the Industrial Revolution in the nineteenth century did **urbanization,** the transformation of a society from a rural to an urban one, spread rapidly.

As population has increased, so has the proportion of people living in urban areas. An **urban area** is a spatial concentration of people whose lives are centered around nonagricultural activities. Although countries differ in their definitions of "urban," most countries designate places with 2,000 people or more as being urbanized (Population Reference Bureau 2003). According to the U.S. census definition, an "urban population" consists of persons living in cities or towns of 2,500 or more inhabitants. An "urbanized area" refers to one or more places and the adjacent densely populated surrounding territory that together have a minimum population of 50,000.

The share of the global population living in urban areas has increased from 30 percent in 1950, to 47 percent in 2000, and is expected to reach 60 percent by

Table 13.1
World's Most Populated Countries: 2003 and 2050

World's Most Populated Countries in 2003			World's Most Populated Countries in 2050		
Rank	Country	Population (millions)	Rank	Country	Population (millions)
1	China	1,289	1	India	1,628
2	India	1,069	2	China	1,394
3	United States	292	3	United States	422
4	Indonesia	220	4	Pakistan	349
5	Brazil	176	5	Indonesia	316
6	Pakistan	149	6	Nigeria	307
7	Bangladesh	147	7	Bangladesh	255
8	Russia	146	8	Brazil	221
9	Nigeria	134	9	Congo, Dem. Rep. of	181
10	Japan	128	10	Ethiopia	173

Source: Population Reference Bureau. 2003. *2003 World Population Data Sheet.* Washington DC: Population Reference Bureau.

Table 13.2
Percentage of Population in Urban Areas, by Year

	1950	1975	2000	2030 (projected)
World	29.7	37.9	47.0	60.3
More-developed regions	54.9	70.0	76.0	83.5
Less-developed regions	17.8	26.8	39.9	56.2

Source: United Nations Population Division. 2001. *World Urbanization Prospects: The 1999 Revision.* New York: United Nations.

2030. In more developed regions, 76 percent of the population already live in urban areas (see Table 13.2). Virtually all the population growth expected between 2000 and 2030 will be concentrated in the urban areas of the world. Most of this urban population growth will be in less-developed regions whose urban population will likely increase from 1.9 billion in 2000 to 3.9 billion in 2030. In contrast, the urban population of the more-developed regions is expected to increase only slightly, from 0.9 billion in 2000 to 1 billion in 2030 (United Nations Population Division 2001).

The number of **megacities**—urban areas with 10 million residents or more—is also increasing. In 1950 there was one megacity: New York. In 1975, there were five megacities, increasing to 19 in 2000 and a projected 23 in 2015 (United Nations Population Division 2001). Table 13.3 presents the 10 most populated cities in the world in 2000 and projected in 2015.

Table 13.3
Ten Most Populated Cities: 2000 and 2015

	City	2000		City	2015
1	Tokyo	26.4	1	Tokyo	26.4
2	Mexico City	18.1	2	Bombay	26.1
3	Bombay	18.1	3	Lagos	23.2
4	Sao Paulo	17.8	4	Dhaka	21.1
5	New York	16.6	5	Sao Paulo	20.4
6	Lagos	13.4	6	Karachi	19.2
7	Los Angeles	13.1	7	Mexico City	19.2
8	Calcutta	12.9	8	New York	17.4
9	Shanghai	12.9	9	Jakarta	17.3
10	Buenos Aires	12.6	10	Calcutta	17.3

Source: United Nations Population Division, 2001. *World Urbanization Prospects: The 1999 Revision.* New York: United Nations.

In 1950, New York was the only megacity in the world. In 2000, there were 19 megacities worldwide, and this number is expected to reach 23 by 2015.

Increasing urbanization results about equally from births in urban areas and from migration of people from rural areas to the cities (United Nations Population Fund 1999). Rural dwellers migrate to urban areas to flee from war or natural disasters or to find employment. As foreign corporate-controlled commercial agriculture displaces traditional subsistence farming in poor rural areas, peasant farmers flock to the city looking for employment. Some rural dwellers migrate to urban areas in search of a better job—one that has higher wages and better working conditions. Governments have also stimulated urban growth by spending more to improve urban infrastructures and services while neglecting the needs of rural areas (Clark 1998).

History of Urbanization in the United States. Urbanization of the United States began as early as the 1700s, when most major industries were located in the most populated areas, including New York City, Philadelphia, and Boston. Unskilled laborers, seeking manufacturing jobs, moved into urban areas as industrialization accelerated in the nineteenth century. The "pull" of the city was not the only reason for urbanization, however. Technological advances were making it possible for fewer farmers to work the same amount of land. Thus "push" factors were also involved—making a living as a farmer became more and more difficult as technology, even then, replaced workers.

Urban populations continued to multiply as a large influx of European immigrants in the late 1800s and early 1900s settled in U.S. cities. This influx was followed by a major migration of southern rural blacks to northern urban areas. People were lured to the cities by the promise of employment and better wages and such urban amenities as museums, libraries, and entertainment. Immigrants are also attracted to cities with large ethnic communities that provide a familiar cultural environment in which to live and work.

The growth of the U.S. urban population over the last 200 years has been dramatic. Between 1800 and 2000, the percentage of the U.S. population that is urban grew from 6.1 percent to 79 percent. Urban populations include residents of inner cities and their surrounding suburbs.

Suburbanization. In the late nineteenth century, railroad and trolley lines enabled people to live outside the city and still commute into the city to work. As more and more people moved to the **suburbs**—urban areas surrounding central cities—America underwent **suburbanization.** As city residents left the city to live in the suburbs, cities lost population and experienced **deconcentration,** or the redistribution of the population from cities to suburbs and surrounding areas.

Many factors have contributed to suburbanization and deconcentration. Following World War II, many U.S. city dwellers moved to the suburbs out of concern for the declining quality of city life and the desire to own a home on a spacious lot. Suburbanization was also spurred by racial and ethnic prejudice, as the white majority moved away from cities that, because of immigration, were becoming increasingly diverse. Mass movement into suburbia was encouraged by the federal interstate highway system—financed by the government under the guise of ensuring national defense—the affordability of the automobile, and the dismantling of metropolitan mass transit systems (Lindstrom & Bartling 2003). In the 1950s, Veterans Administration (VA) and Federal Housing Administration (FHA) loans made housing more affordable, enabling many city dwellers to move to the suburbs. Suburb dwellers who worked in the central city could commute to work or work in a satellite branch in suburbia that was connected to the main downtown office. As increasing numbers of people moved to the suburbs, so did businesses and jobs. Without a strong economic base, city services and the quality of city public schools declined, which furthered the exodus from the city. Social problems associated with suburbanization are discussed later in this chapter.

"Cities have been ignored because they are blacker, browner, poorer, and more female than the rest of the nation."

Julianne Malveaux
Writer/Scholar

U.S. Metropolitan Growth and Urban Sprawl. Simply defined, a **metropolitan area** is a densely populated core area, together with adjacent communities. The largest city in each metropolitan area is designated a **central city.** Another term for metropolitan area is **metropolis,** from the Greek meaning "mother city."

Metropolises have grown rapidly in the United States. As new areas reach the minimum required city or urbanized area population, and as adjacent towns, cities, and counties satisfy the requirements for inclusion in metropolitan areas, both the number and size of metropolitan areas have grown. In 2003, 80 percent of Americans lived in one of the 362 metropolitan areas in the nation (Scommega 2003). One U.S. state—New Jersey—is entirely occupied by metropolitan areas as designated by the U.S. census.

The growth of metropolitan areas is often referred to as **urban sprawl**—the ever-increasing outward growth of urban areas. Urban sprawl results in the loss of green, open spaces; the displacement and endangerment of wildlife; traffic congestion and noise; and pollution liabilities—problems that we discuss later in this chapter. In a Gallup poll of U.S. adults, 73 percent said they worried "a great deal" or "a fair amount" about urban sprawl and the loss of open spaces (Dunlap & Saad 2001).

Those who enjoy the conveniences and amenities of urban life, yet find large metropolitan areas undesirable, may choose to live in a **micropolitan area**—a small city (between 10,000 and 50,000 people) located beyond congested metropolitan areas. These areas are large enough to attract jobs, restaurants, community organizations, and other benefits, yet small enough to elude traffic jams, high crime rates, and high costs of housing and other living expenses associated with larger cities. In 2003, 10 percent of Americans lived in micropolitan areas.

Sociological Theories of Population and Urbanization

The three main sociological perspectives—structural-functionalism, conflict theory, and symbolic interactionism—may be applied to the study of population and urbanization.

Structural-Functionalist Perspective

Structural-functionalism focuses on how changes in one aspect of the social system affect other aspects of society. For example, the **demographic transition theory** of population describes how industrialization has affected population growth. According to this theory, in traditional agricultural societies, high fertility rates are necessary to offset high mortality and to ensure continued survival of the population. As a society becomes industrialized and urbanized, improved sanitation, health, and education lead to a decline in mortality. The increased survival rate of infants and children, along with the declining economic value of children, leads to a decline in fertility rates. About one-third of the world's countries have completed the demographic transition—the progression from a population with short lives and large families to one in which people live longer and have smaller families (Cincotta, Engelman, & Anastasion 2003).

Urbanization plays a significant role in the demographic transition. Because health care delivery is more cost-effective in cities than in rural areas, governments prioritize urban health clinics. With greater access to health care, urban dwellers are the first to experience declines in infant mortality and fertility. A study of contraceptive use in Kenya found that women who lived in urban areas were more likely to have used contraception (Kimuna & Adamchak 2001). Recent declines in fertility throughout Africa are mostly an urban phenomenon (Cincotta et al., 2003).

Structural-functionalists view the development of urban areas as functional for societal development. Although cities initially served as centers of production and distribution, today they are centers of finance, administration, education, health care, and information.

Urbanization is also dysfunctional, as it leads to increased rates of anomie, or normlessness, as the bonds between individuals and social groups become weak (see also Chapters 1 and 4). Whereas in rural areas social cohesion is based on shared values and beliefs, in urban areas social cohesion is based on interdependence created by the specialization and social diversity of the urban population. Anomie is linked to higher rates of deviant behavior, including crime, drug addiction, and alcoholism. Overcrowding, poverty, rapid spread of infectious disease, and environmental destruction are also considered dysfunctions associated with urbanization.

Conflict Perspective

The conflict perspective focuses on how wealth, power, or the lack thereof affect population problems. In 1798, Thomas Malthus predicted that population would grow faster than the food supply and that masses of people were destined to be poor and hungry. According to Malthusian theory, food shortages would lead to war, disease, and starvation that would eventually slow population growth. However, conflict theorists argue that food shortages result primarily from inequitable distribution of power and resources (Livernash & Rodenburg 1998).

Conflict theorists also note that population growth results from pervasive poverty and the subordinate position of women in many less-developed countries. Poor countries have high infant and child mortality rates. Hence, women in many poor countries feel compelled to have many children to increase the chances that some will survive into adulthood. Their subordinate position prevents many women from limiting their fertility. For example, in 14 countries around the world, a woman must get her husband's consent before she can receive any contraceptive services (United Nations Population Fund 1997). Thus, according to conflict theorists, population problems result from continued economic and gender inequality.

Power, wealth, and the profit motive also affect the development and operations of urban areas. According to the conflict perspective, capitalism requires that the production and distribution of goods and services be centrally located, thus, at least initially, leading to urbanization. Today, global capitalism and corporate multinationalism, in search of new markets, cheap labor, and raw materials, have largely spurred urbanization of the developing world. Capitalism also contributes to migration from rural areas into cities as peasant farmers who have traditionally produced goods for local consumption are being displaced by commercial agriculture that is geared to producing fruits, flowers, and vegetables for export to the developed world. Displaced from their traditional occupations, peasant farmers are flocking to cities to find employment (Clark 1998).

The conflict perspective also focuses on how individuals and groups with wealth and power influence decisions that affect urban populations. For example, according to citizens' groups working to stop urban sprawl in Central and Eastern Europe, city officials may be bribed to approve a new shopping mall or other development project (Sheehan 2001). And in the United States, "bribery" has taken the form of financial contributions to political parties and candidates. In the 1998 congressional election, industries with a stake in transportation and land use decisions contributed $128 million to political parties and candidates. During the 2000 presidential election campaign, a construction industry lobbyist in favor of increased highway building told the *Wall Street Journal* that he looked forward to meeting with the president if the candidate he supported with a large contribution won: "I'm assuming, since we've supported him throughout his gubernatorial career, that we'd have an opportunity to go visit if we wanted to" (quoted in Sheehan 2001, 120). (Increased highway building encourages urban sprawl and diverts funds away from improvements in public transportation that are needed in urban areas.)

Symbolic Interactionist Perspective

The symbolic interactionist perspective focuses on how meanings, labels, and definitions learned through interaction affect population problems. For example, many societies are characterized by **pronatalism**—a cultural value that promotes having children. Throughout history, many religions have worshiped fertility and recognized it as being necessary for the continuation of the human race. In many countries, religions prohibit or discourage birth control, contraceptives, and abortion. Women in pronatalistic societies learn through interaction with others that deliberate control of fertility is socially unacceptable. Women who use contraception in communities where family planning is not socially accepted face ostracism by their community, disdain from relatives and friends, and even divorce and abandonment by their husbands (Women's Studies Project 2003). However, once some women learn new definitions of fertility control, they become role models and influence the attitudes and behaviors of others in their personal networks (Bongaarts & Watkins 1996). This chapter's *Human Side* feature presents quotes from women and men around the world, speaking of family planning and what it means to them.

The symbolic interaction perspective can also be applied to understanding how urban life affects interaction patterns and social relationships. The classical and modern views represent different observations of how urban living affects social relationships.

Classical Theoretical View. Cities have the reputation of being cold and impersonal. George Simmel observed that urban living involved an overemphasis on punctuality, individuality, and a detached attitude toward interpersonal relationships (Wolff 1978). It is not difficult to find evidence to support Simmel's observations: witness New Yorkers pushing each other to get onto the subway during rush hour, hear motorists curse each other in Los Angeles traffic jams, watch Chicago residents ignore the homeless man asleep on the sidewalk.

Louis Wirth (1938), a second-generation student of Simmel, argued that urban life is disruptive for both families and friendships. He believed that because of the heterogeneity, density, and size of urban populations, interactions become segmented and transitory, resulting in weakened social bonds. Wirth held that as social solidarity weakens, people exhibit loneliness, depression, stress, and antisocial behavior.

The Human Side

Voices from Around the World: Women and Men Talk About Family Planning

The Women's Studies Project, which is described in this chapter's *Social Problems Research Up-Close* feature, involves numerous studies in 14 countries that assess the impact of family planning on women's lives. The following excerpts from focus group discussions and in-depth interviews provide glimpses into the family planning experiences and attitudes of women and men around the world.

If family planning had been available earlier, my future would have been different. That is my life-long regret. Because I had too many children, I had to quit [teaching].

Woman in South Jiangsu, China

Yes, people are happy with family planning. They see that their family is in harmony, their children are big enough to take care of themselves, while the mother can take care of herself.

Woman in rural North Sumatra, Indonesia

If I had [had] access to the method of preventing pregnancy, I wouldn't have been pregnant . . . and I would be working somewhere in town, and maybe I would be having a better life than this one.

Zimbabwean woman

Without family planning and the consequent child spacing and limitation, there is no quality of life. As a woman, you cannot get enough time to give love to your children and your husband if you have many children.

Zimbabwean woman

My parents had eight children. My father died when I was 20—there was no money for the doctor. Some siblings were given to other families. Having too many children—not only do the parents suffer, but also the children, with bad nutrition and bad housing conditions.

Woman in China

Having four children nearly made me crazy. I couldn't give them food and clothes. They wandered from door to door and were driven away like dogs. One day my son asked, "Why did you give me birth if you can't feed me?"

Woman in Bangladesh

Because you have free time to take care of your husband, you can see the affection is reborn.

Contraceptive user in Mali

The man of the house never likes it if the woman can't work. He says, "Did I marry you to keep you as a pet? I married you to work in my house! If you sit around, who will look after the children, and who will do all of the chores?" This is why I stopped taking the [birth control] pills.

Woman in Bangladesh who experienced negative side effects from using the pill

A woman who lived with us, she used family planning. She fell ill and even had two operations. She has not had any more children. . . . When I saw her experience, I was afraid.

Mali woman, explaining why she does not use contraception

People said many things about my having the operation [sterilization]. . . . "Don't you stand next to us. Stay away! Even to look at you is a sin!" I would just weep when people said those things to me.

Woman from rural Bangladesh

I talked secretly with eight or ten women about this. Some of the women said, "If the elders find out about anyone having this operation, they will not let her live in the village anymore. No one will eat food cooked by a woman who has been operated on."

Woman from rural Bangladesh who sought sterilization

The husband always has the final say. What happens is that women are limited in their thinking, and if you do not show your dominance, you will have problems.

Zimbabwean man who believes that men should control family planning decisions

If my wife makes the decision to use family planning without my consent, I would divorce her.

Malian man

If I want four children and my wife wants six; she has to listen to me because I am the one who supports the family financially. If I decide to have five children, this is because I know I can look after them. The husband is the head of the family, and the wife can never tell me the number of children she wants to have.

A husband in Zimbabwe

The phrase "many children, more economic future" is out-of-date. Today, many children means lots of problems, lots of responsibility.

Man from Ujung Pandang, Indonesia

Source: Women's Studies Project. 2003. *Women's Voices, Women's Lives: The Impact of Family Planning.* Family Health International. http://www.fhi.org/

Modern Theoretical View. In contrast to Wirth's pessimistic view of urban areas, Herbert Gans ([1962] 1984) argued that cities do not interfere with the development and maintenance of functional and positive interpersonal relationships. Among other communities, he studied an Italian urban neighborhood in Boston called the West End and found such neighborhoods to be community oriented and marked by close interpersonal ties. Rather than finding the social disorganization described by Wirth, Gans observed that kinship and ethnicity helped bind people together. Intimate small groups with strong social bonds characterized these enclaves. Thus Gans saw the city as a patchwork quilt of different neighborhoods or urban villages, each of which helped individuals deal with the pressures of urban living.

A Theoretical Synthesis. Fisher (1982) interviewed more than 1,000 respondents in various urban areas and found evidence for both the classical and the modern theoretical views of urbanism. From the classical perspective, he found that heterogeneity (the diversity among urban residents) does make community integration and consensus difficult—community cohesion is less tight. Ties that do exist are less often kin-related than in nonurban areas and are more often based on work relationships, memberships in voluntary and professional organizations, and proximity to neighbors.

Fisher also found, however, that the diversity of urban populations facilitates the development of subcultures that have a sense of community ties. For example, large urban areas include such diverse groups as gays, ethnic and racial minorities, and artists. These individuals find each other and develop their own unique subcultures.

Other research has also found support for a synthesis of the classical and modern views of urban life. Tittle (1989) found that the larger the size of the community, the weaker will be a respondent's social bonds and the higher his or her anonymity, tolerance, alienation, and reported incidence of deviant behavior. Tittle also found that racial and ethnic groups created their own sense of community in urban neighborhoods.

Social Problems Related to Population Growth and Urbanization

Some of the most urgent social problems today are related to population growth. They include poor maternal and infant health, increased global food and water requirements, and urban crowding. Other social problems related to urbanization include poverty and unemployment, dilapidated and unaffordable housing, inadequate schools, transportation and traffic problems, and sprawl. Environmental problems, also associated with population growth and urbanization, are discussed in Chapter 14.

Poor Maternal and Infant Health

As noted in Chapter 2, maternal deaths (deaths related to pregnancy and childbirth) are the leading cause of mortality for reproductive-age women in the developing world. Having several children at short intervals increases the chances of premature birth, infectious disease, and death for the mother or the baby. Childbearing at young ages (teens) has been associated with anemia and hemorrhage, obstructed and prolonged labor, infection, and higher rates of infant mortality (Zabin & Kiragu 1998). In developing countries, one in four children are born unwanted, increasing the risk of neglect and abuse. In addition, the more children a woman has, the fewer parental resources (parental income and time, and maternal nutrition) and social resources (health care and education) are available to each child. The adverse health effects of high fertility on women and children are, in themselves, compelling reasons for providing women with family planning services. "Reproductive health and choice are often the key to a woman's ability to stay alive, to protect the health of her children and to provide for herself and her family" (Catley-Carlson & Outlaw 1998, 241).

Worldwide, pregnancy is the leading cause of death for young women aged 15 to 19. Most (95%) maternal deaths occur in Africa and Asia. This woman in sub-Saharan Africa has a 1 in 16 risk of dying in pregnancy or childbirth, compared with a 1 in 2,800 chance for a woman in a developed country.

Increased Global Food and Water Requirements

World food supplies must double in the next 50 years to adequately meet the food needs of the growing population (*Popline* 2003b). This presents a challenge, especially for countries that have not been able to meet the food needs of their current populations.

As global food requirements increase with population growth, so do demands on the environment. Pesticides and fertilizers used in agriculture contaminate soil and water. Agricultural activities contribute to the destruction of forests and the species that inhabit them and deplete water supplies, as agriculture consumes 70 percent of the fresh water used by humans (primarily for irrigation). The world's water usage has tripled since 1950 (Brown 2002). Meeting the water needs of a growing population is an urgent priority around the world; 25 countries currently experience water stress or scarcity, and by 2025, nearly two-thirds of the world's population is expected to be living in countries with significant water stress (*Popline* 2003c; Population Action International 2003a).

Urban Crowding and the Spread of Disease

Population growth contributes to urban crowding and high population density. India has one-third the land area of the United States but more than three times the population. Imagine tripling the U.S. population. Then imagine that this tripled population all lived in the eastern third of the United States. That will give you an idea of how crowded countries like India are.

Without economic and material resources to provide for basic living needs, urban populations in developing countries often live in severe poverty. Urban poverty in turn produces environmental problems such as unsanitary disposal of waste. In

Nigerian urban ghettos, for example, the mounds of garbage and human waste that litter gutters, schools, roads, market places, and town squares have been accepted as the way of life (Agbese 1995). The World Health Organization estimates that half the people in the world do not have access to a decent toilet. Unsanitary disposal of human waste contaminates water supplies. Half the people in the developing world suffer from diseases caused by poor sanitation. Diarrhea caused by many of these diseases is the leading killer of children today (Gardner 1998).

Densely populated urban areas facilitate the spread of disease among people. Infectious diseases cause more than one-third of all deaths worldwide. Crowded conditions in urban areas provide the ideal environment for the culture and spread of diseases such as cholera and tuberculosis (Pimentel et al. 1998).

Urban Poverty and Unemployment

In many countries in developing regions, one in four urban residents is living in absolute poverty (National Research Council 2003). In some of the world's poorest developing countries, half the urban population lives in conditions of extreme deprivation. In the United States, the highest rates of poverty are in the central cities (see Table 13.4).

Urban poverty is reflected in the high numbers of homeless and hungry people in urban areas. In a survey of 25 U.S. cities, the average demand for emergency shelter increased by 13 percent from 2002 to 2003, and requests for emergency food assistance increased by an average of 17 percent (U.S. Conference of Mayors 2003). High poverty rates in cities are related to unemployment, the lack of jobs, and the decrease in high-paying jobs. Many central cities face unemployment rates that are much higher than the national average.

In the United States and other industrialized countries, urban unemployment and poverty are partly the results of deindustrialization, or the loss and/or relocation of manufacturing industries. Since the 1970s, many urban factories have closed or relocated, forcing blue-collar workers into unemployment. When prospects of finding decent employment are low, resulting feelings of frustration and worthlessness can lead to drug use, crime, and violence (Rodriguez 2000).

Table 13.4
U.S. Poverty Rates by Residence: 2002

Residence	Poverty Rate
Inside Central Cities	16.7%
Suburbs	8.9%
Rural Areas	14.2%

Source: Proctor, Bernadette, and Joseph Dalaker. 2003. *Poverty in the United States: 2002.* U.S. Census Bureau, Current Population Reports P60–222.

The United Nations estimates that 56% of urban dwellers in Africa live in slums.

© David Turnley/CORBIS

Urban Housing Problems

Urban housing problems include lack of affordable housing, substandard housing, and housing segregation. Many cities are experiencing a housing crisis, as the number of low-income renters has increased while the number of low-cost rental units has dropped. Housing that is available and affordable is often substandard, characterized by outdated plumbing and wiring, overcrowding, rat infestations, toxic lead paint, and fire hazards (see also Chapter 10).

Low-income housing tends to be concentrated in inner-city areas of extreme poverty; but jobs are increasingly moving to the suburbs, and central-city residents often lack transportation to get to these jobs. The more affluent suburbs restrict development of affordable housing to keep out "undesirables" and maintain their high property values. Suburban zoning regulations that require large lot sizes, minimum room sizes, and single-family dwellings serve as barriers to low-income development in suburban areas (Orfield 1997).

Concentrated areas of poverty and poor housing in urban neighborhoods are called **slums.** In the United States, slums that are occupied primarily by African-Americans are known as **ghettos,** and those occupied primarily by Latinos are called **barrios.** The United Nations estimates that 56 percent of the urban population in Africa, 37 percent in Asia and Oceania, and 26 percent in Latin America and the Caribbean live in slums (Sheehan 2003).

Inadequate Schools

When jobs and middle-class residents left the city for the suburbs in mass exodus, local revenues decreased, and the quality of city schools, particularly in inner-city neighborhoods, declined. With some exceptions, "schools in poor inner-city neighborhoods . . . are dangerous, overcrowded, and ineffective. Moreover, children in those schools, seeing the unemployment and poor earnings of the adults in their neighborhood, may conclude that education has little value" (Jargowsky 1997, 110). Inner-city schools face a chronic shortage of qualified teachers and often hire underprepared and inexperienced teachers, many of whom are hired to teach subjects outside their areas of preparation (Wilson 1996). The poor quality of inner-city schools contributes further to the exodus of middle-class residents from the city and deters businesses and potential residents from moving to those neighborhoods. As a result, property values continue to decline, the tax base decreases, schools further erode, and the cycle repeats itself.

Transportation and Traffic Problems

Urban areas are often plagued with transportation and traffic problems. In a survey of 2,000 U.S. households, the majority (79%) cited heavier traffic as the most negative aspect of urban growth (National Association of Home Builders 1999).

In a telephone survey of randomly selected registered voters, the majority (89%) of respondents agreed that traffic congestion has worsened nationwide (U.S. Conference of Mayors 2001). When asked if traffic had gotten better, worse, or stayed the same in their areas over the past five years, 79 percent said traffic had gotten worse, and one out of two respondents said that traffic was currently "much worse" than it was five years ago.

Many public roads in urban areas are afflicted with what some call *autosclerosis*—defined as "clogged vehicular arteries that slow rush hour traffic to a crawl or a stop, even when there are no accidents or construction crews ahead" ("The Bridge to the 21st Century" 1997, 1). According to Lundberg (in Jensen 2001), "the average vehicle speed for crosstown traffic in New York City is less than six miles per hour—slower than it was in the days of horse-drawn buggies" (p. 6). And "traffic jams in Atlanta have been so entangled that babies have been born in traffic standstills, and some desperate drivers have had to leave their cars to relieve themselves behind roadside bushes" (Shevis 1999, 2). In some foreign cities such as Sao Paulo, Brazil; Bangkok, Thailand; and Cairo, Egypt; traffic jams are even worse than those in the United States. According to one report, "it sometimes takes so long to reach the Bangkok airport from downtown—from 3 to 6 traffic-paralyzed hours—that roadside entrepreneurs sell minitoilet kits for use by desperate riders in traffic-jammed cars" ("The Bridge to the 21st Century" 1997, 1). Traffic congestion creates stress on drivers, which sometimes leads to aggressive driving and violent reactions to other drivers—a phenomenon known as **road rage** (see this chapter's *Self and Society* feature).

In the United States, private automobiles, often carrying only one person, are the dominant mode of travel. Cars and light trucks are the largest single source of air pollution (Union of Concerned Scientists 1999). Air pollution and traffic congestion have been major forces driving residents and businesses away from densely populated urban areas (Warren 1998). Health problems associated with congested traffic include stress, respiratory problems, and death. More than 20 million people are severely injured or killed on the world's roads each year (World Health Organization 2003). Indeed, far more people are killed and injured in automobile accidents than by violent crime. Therefore, it has been argued that despite higher crime rates in the inner city, the suburbs are the more dangerous place to live, because suburbanites "drive three times as much, and twice as fast, as urban dwellers" (Durning 1996, 24).

Dependence on cars has been encouraged by a number of factors: unwillingness to tax gasoline commensurate with its cost to society, free and tax-free parking typically provided by corporations to their employees outside of the city, the glamorization of automobiles—largely perpetuated by the automobile industry—and federal subsidies that favor building highways over investing in public transit. The conversion of the United States to an automobile-based system of transportation was heavily influenced by industries that profit from automobile use. In the 1930s, National City Lines, a company backed by the three major automakers, major oil companies, tire manufacturers, and the trucking and construction industries, succeeded in systematically buying and closing down more than 100 electric trolley lines in 45 U.S. cities. Although National City Lines was convicted of this conspiracy in 1949, the dismantling of the rail transit system had already been accomplished (Warren 1998).

"Everybody says that living in the inner city is dangerous, but the truth is that, if you take car crashes into account, the suburbs are statistically far more dangerous places to live."

Jan Lundberg
Director of the Alliance for a Paving Moratorium

Sprawl and the Displacement and Endangerment of Wildlife

The spread of urban and suburban areas increasingly is replacing natural habitats with pavement, buildings, and human communities. The loss of open green space, trees, and plant life not only affects the quality of life for humans, but also the animals whose homes are turned into parking lots, shopping centers, office buildings, and housing developments. The majority of U.S. adults (81%) in a Gallup poll reported that they worried "a great deal" or "a fair amount" about the loss of natural habitat for wildlife (Dunlap & Saad 2001).

Evidence of wildlife displacement resulting from sprawl is found across the nation. Coyotes, normally found only in the West and in Appalachia, are now being sighted in every state (Shevis 1999). Other displaced species include bear, Canada geese, and deer. "With no place to go, they bound into suburban backyards in search of food and water and across highways, frequently injuring themselves and causing harm to drivers. . . . Deer cause an estimated half-million vehicle accidents a year, killing 100 people and injuring thousands more" (Shevis 1999, 2–3). About 1 million animals are killed on U.S. roads every day (Lundberg, in Jensen 2001).

In Chapter 14, we look at the problem of species extinction and the loss of biodiversity. According to the U.S. Fish and Wildlife Service (USFWS), habitat loss resulting from urban and suburban sprawl is the number one reason that wildlife species are becoming increasingly endangered (Shevis 1999).

> In various parts of the Western United States, the survival of bears, mountain lions, coyotes and black-tailed deer is the most pressing problem; in the Eastern and Midwestern parts of the country the most visible problems are with white-tailed deer and Canada geese. In every case, their enemy is man, relentlessly taking over their territory with backhoes and earthmovers. (Shevis 1999, 3)

Strategies for Action: Responding to Problems of Population and Urbanization

Most strategies related to population problems involve attempts to slow population growth by providing access to family planning services and contraceptive methods, improving the status of women, increasing economic development and improving health status, and imposing governmental regulations and policies. As populations increasingly live in urban areas, strategies that address population problems also

Road Rage Survey

Increased congestion of urban highways has led to a phenomenon called road rage. Please answer the questions below to assess your level of road rage and then compare your answers to those of a national sample of respondents.

1. Which of the following do you ever do?

Category I (check all that apply)

Going 5 to 10 mph over the speed limit ☐
Making rolling stops ☐
Making illegal turns ☐
Lane hopping without signaling ☐
Following very close as a habit ☐
Going through red lights ☐
Denying right of way (failure to yield) ☐
Swearing, cursing, name calling ☐
Combined, how regularly do you do the things in Category I (circle one)?
Never 1 2 3 4 5 6 7 8 9 10 Quite regularly

Category II (check all that apply)

Going 15 to 25 mph over the speed limit ☐
Yelling at another driver ☐
Honking in protest ☐
Revving the engine ☐
Making an insulting gesture ☐
Tailgating dangerously ☐
Shining the headlights to retaliate ☐
Cruising in the passing lane (forcing others to pass on the right) ☐
Combined, how regularly do you do the things in Category II (circle one)?
Never 1 2 3 4 5 6 7 8 9 10 Quite regularly

Category III (check all that apply)

Braking suddenly to punish a tailgater ☐
Deliberately cutting someone off to retaliate ☐
Using the car to block the way ☐
Using the car as a weapon to threaten someone ☐
Chasing another car in pursuit ☐
Getting into a physical fight with another driver ☐
Other things ☐
Combined, how regularly do you do the things in Category III (circle one)?
Never 1 2 3 4 5 6 7 8 9 10 Quite regularly

2. Which of the following emotions do you experience when driving?

Anger or rage behind the wheel
Never 1 2 3 4 5 6 7 8 9 10 Regularly
Enjoying fantasies of violence
Never 1 2 3 4 5 6 7 8 9 10 Regularly
Experiencing fear for self or family in the car
Never 1 2 3 4 5 6 7 8 9 10 Regularly
Feeling compassion for another driver
Never 1 2 3 4 5 6 7 8 9 10 Regularly
Feeling competitive with other drivers
Never 1 2 3 4 5 6 7 8 9 10 Regularly

Feeling impatient and the constant urge to rush

Never 1 2 3 4 5 6 7 8 9 10 Regularly

Wanting to drive dangerously

Never 1 2 3 4 5 6 7 8 9 10 Regularly

Feeling level-headed and calm

Never 1 2 3 4 5 6 7 8 9 10 Regularly

The results of a national survey of 761 U.S. drivers, below, indicate the percentage of people who have engaged in the behavior, ±3 percent.

Category I	**Percentage**
Going 5 to 10 mph over the speed limit	92
Making rolling stops	50
Making illegal turns	15
Lane hopping without signaling	25
Following very close as a habit	14
Going through red lights	10
Denying right of way (failure to yield)	8
Swearing, cursing, name calling	58
Mean regularity rating	6.0
Category II	**Percentage**
Going 15 to 25 mph over the speed limit	39
Yelling at another driver	32
Honking in protest	35
Revving the engine	11
Making an insulting gesture	24
Tailgating dangerously	12
Shining the headlights to retaliate	18
Cruising in the passing lane (forcing others to pass on the right)	11
Mean regularity rating	3.9
Category III	**Percentage**
Braking suddenly to punish a tailgater	32
Deliberately cutting someone off to retaliate	15
Using the car to block the way	18
Using the car as a weapon to threaten someone	2
Chasing another car in pursuit	8
Getting into a physical fight with another driver	2
Other things	8
Mean regularity rating	2.7
Emotions	**Mean Regularity Rating**
Anger or rage behind the wheel	4.5
Enjoying fantasies of violence	2.4
Experiencing fear for self or family in the car	4.0
Feeling compassion for another driver	4.9
Feeling competitive with other drivers	4.9
Feeling impatient and the constant urge to rush	5.7
Wanting to drive dangerously	2.7
Feeling level-headed and calm	6.7

Source: James, Leon. 1998. "World Road Rage Survey" as it appeared at http://www.drdriving.org. Reprinted by permission.

© Dave G. Houser/CORBIS

Recognizing that men play a crucial role in family planning decisions, family planning programs are making efforts to include men in family planning education and services.

address, indirectly, problems of urbanization. Additional strategies that seek to alleviate urban problems include those designed to alleviate poverty and stimulate economic development in inner cities, implement "smart growth" and "new urbanism" into development, improve transportation and alleviate traffic congestion, and curb urban growth in developing countries.

Provide Access to Family Planning Services

Since the 1950s, governments and nongovernmental organizations such as the International Planned Parenthood Federation have sought to lower fertility through family planning programs that provide reproductive health services and access to contraceptive information and methods. Such programs, along with developments in contraceptive technology, have achieved the desired result: Globally, the average number of children born to each woman has fallen from five in 1960 to less than 2.8 in 2003, as more women today want to limit family size and are using modern methods of birth control to control the number and spacing of births (Engelman et al. 2000; Population Reference Bureau 2003). Today, more than half of married women in less-developed countries use some form of modern contraception, compared with 10 percent in 1960 (Population Reference Bureau 2000a; 2003).

Yet, there is still an unmet need for contraception; more than 100 million women in less-developed countries—about 17 percent of married women—would prefer to avoid pregnancy, but are not using any form of family planning (Ashford 2003). While 72 percent of married women in the United States use modern contraception, in 28 countries, the figure is 10 percent or less. The highest rate is in China, where 83 percent of married women use modern contraception (Population Reference Bureau 2003). As discussed in this chapter's *Social Problems Research Up Close* feature, the benefits of family planning go beyond slowing population.

In some countries, family planning personnel often refuse, or are forbidden by law or policy, to make referrals for contraceptive and abortion services for unmarried women. Furthermore, many women throughout the world do not have access to legal, safe abortion. Without access to contraceptives, many women who experience unwanted pregnancy resort to abortion—even under illegal and unsafe conditions. More than half of the nearly 80 million unintended pregnancies that occur worldwide every year end in abortion (Population Action International 2003b). Research in 12 countries of Central Asia and Eastern Europe has found that increased contraception use has resulted in significant declines in the rate of abortion (Leahy 2003).

In 1994, delegations from 179 UN member countries met in Cairo, Egypt, to develop a plan for the future of population and development. This meeting resulted in a declaration called the Programme of Action that stressed four goals to achieve by 2015: (1) universal education, (2) reduction in infant and child mortality, (3) reduction in maternal mortality, and (4) access to reproductive and sexual health services, including family planning. Two-thirds of the funding for the Programme of Action was to come from developing countries, and developed donor countries were to supply one-third. Only three donor countries, Denmark, Norway, and the Netherlands, have met their promised contribution levels (Goodrich 2004).

Cuts in U.S. assistance to international family planning programs are largely the result of opposition to abortion practices in some countries. On his first day of office in 2001, George W. Bush reinstated the "global gag rule," which denies U.S. international family planning assistance to organizations that use their own privately raised funds to counsel women on the availability of abortion, advocate change in abortion laws, or provide abortion services. The gag rule has cost women's health programs millions of dollars, resulting in the closure of thousands of family planning clinics in developing countries (*Popline* 2003d). However, Congress has since voted to repeal the global gag rule.

Improve the Status of Women

Throughout the developing world, the primary status of women is that of wife and mother. Women in developing countries traditionally have not been encouraged to seek education or employment, but rather to marry early and have children.

Improving the status of women is vital to curbing population growth. Education plays a key role in improving the status of women and reducing fertility rates. Educated women are more likely to delay their first pregnancies, to use safe and effective contraception, and to limit and space their children. In Africa, Western Asia, and Latin America and the Caribbean, women with secondary or higher education have three fewer children on average than those with no education (*Popline* 2003e). Providing employment opportunities for women is also important to slow population growth; high levels of female labor force participation and higher wages for women are associated with smaller family size (Population Reference Bureau 2000b).

Increase Economic Development and Improve Health

Although fertility reduction may be achieved without industrialization, economic development may play an important role in slowing population growth. Families in poor countries often rely on having many children to provide enough labor and income to support the family. Economic development decreases the economic value of children and is also associated with more education for women and greater gender equality. As previously noted, women's education and status are related to fertility levels.

Economic development tends to result in improved health status of populations. Reductions in infant and child mortality are important for fertility decline, as couples no longer need to have many pregnancies to ensure that some children survive into adulthood. Finally, the more developed a country is, the more likely are women to be exposed to meanings and values that promote fertility control through their interaction in educational settings and through media and information technologies (Bongaarts & Watkins 1996).

Social Problems Research Up Close

Women's Voices, Women's Lives: The Impact of Family Planning on Women's Lives

The Women's Studies Project (WSP) is a five-year research project to study the impact of family planning on women's lives. This research examines how women's family planning experiences—their contraceptive use and nonuse, their pregnancies and childbearing, and their experiences with family planning and reproductive health programs—affect other aspects of their lives, including their roles as individuals, mothers, and wives; their participation in the workforce; their self-esteem; and their marital and sexual satisfaction. Some studies in the Women's Studies Project interviewed women's husbands or partners, parents, and in-laws to determine how family interactions and power dynamics influence contraceptive experience and use.

Sample and Methods

The Women's Studies Project involved 26 field studies in 14 countries, including Bolivia, Brazil, Egypt, Indonesia, the Philippines, Zimbabwe, China, South Korea, Jamaica, and Mali. Most of the field studies used both qualitative and quantitative methods. Quantitative methods (e.g., longitudinal and cross-sectional surveys, inventories, and secondary analysis of existing data) measure occurrences, trends, and relationships; generalize findings to larger populations; and essentially ask, "How many? How often? How is one thing related to another?" Qualitative approaches (e.g., focus group discussions, in-depth interviews, and case studies) seek depth, insight, and understanding; explore the perceptions of individuals and groups; and ask, "Why? How? Under what circumstances?"

In many cases, quantitative findings validated qualitative findings, and vice versa. For example, in Zimbabwe, both quantitative and qualitative research findings showed that women typically use contraceptives only after they have proven their fertility. In other studies, qualitative and quantitative data were sometimes contradictory. For example, in China, the majority of women responding to a survey said they were satisfied with their current contraceptive method. Yet in focus group discussions, many of these women said they worried about contraceptive failure.

This study exemplifies the benefits of combining qualitative and quantitative approaches. "Although the use of multiple methods can be costly in terms of efficiency and time, the advantages far outweigh the disadvantages for capturing the complexity of women's lives while simultaneously describing more general patterns of behavior and experience."

Findings and Conclusions

Some of the general findings of the Women's Studies Project include the following:

- Most women and men believe that family planning and having smaller families provide economic and health benefits.
- Gender norms strongly influence women's family planning experiences. "Gender shapes family planning experience by determining who has access to reproductive health information, who holds the power to negotiate contraceptive use or to withhold sex, who decides on family size, and who controls the economic resources to obtain health services."
- Some women and men say that family planning offers freedom from fear of unplanned pregnancy and can improve sexual and partner relations.
- Where jobs are available, family planning users are more likely than nonusers to participate in the labor force. While

Restore Urban Prosperity

A number of strategies have been proposed and implemented to restore prosperity to U.S. cities and well-being to their residents, businesses, and workers, including strategies to attract new businesses, create jobs, and repopulate cities. The economic development and revitalization of cities also involves improving affordable housing options (see Chapter 10); alleviating urban problems related to HIV/AIDS, addiction, and crime (discussed in other chapters); and reducing problems of traffic and transportation, which we address later in this chapter.

labor force participation provides more income to the family, and often more self-esteem and status for the woman, it can also mean increased burdens for women who get no relief from domestic responsibilities.

- Contraceptive side effects—real or perceived—are a serious concern for many women and men.
- When husbands, partners, or others in the community are opposed to family planning, using contraceptives can increase a woman's vulnerability to domestic violence, ostracism from the community, and divorce and abandonment by her husband.
- Men often have a dominant role in family decisions but tend to be neglected by family planning programs.

One of the main purposes of the WSP is to use the research findings to improve the quality of women's reproductive health services. Each of the findings in this study generated suggestions for family planning policies and programs. Some of these suggestions include the following:

- Because side effects play a central role in women's decisions to start using family planning, in their choice of methods, and their decisions to stop, providers must address these concerns. Providers need training in how to manage side effects and how to counsel women about side effects.
- Family planning may be promoted as a form of health insurance and as a vehicle to help women obtain an education and earn income for themselves and their families.
- Family planning programs should inform clients that many couples report enhanced marital and sexual relationships once the fear of unplanned pregnancy is reduced. This knowledge may help overcome resistance to using contraception, especially to using male condoms, which are viewed as diminishing sexual pleasure.
- Although men play a central role in family planning decisions, they often do not have access to information and services that would empower them to make informed decisions about contraceptive use. "Men's lack of involvement in family planning programs discourages them from becoming effective contraceptive users or supporting their partners' contraceptive use." Therefore, family planning programs need to direct educational programs and health services to men. For example, men need education about the health risks to women when pregnancies are spaced too closely, or when pregnancies occur before age 20 and after age 40.

The Women's Studies Project makes a significant contribution to the research literature on family planning. This research shows that while women perceive numerous benefits of family planning, they also face significant barriers to using family planning, such as family disapproval and method side effects. "By understanding the intricate realities of women's lives and the factors that affect their reproductive health behaviors, family planning programs can offer services that match women's needs and ultimately . . . improve the quality of women's lives."

Source: Women's Studies Project. 2003. *Women's Voices, Women's Lives: The Impact of Family Planning.* Family Health International. http://www.fhi.org/

Empowerment Zone/Enterprise Community Program. The federal Empowerment Zone/Enterprise Community Initiative, or EZ/EC program, provides tax incentives, grants, and loans to businesses to create jobs for residents living within various designated zones or communities, many of which are in urban areas. Federal money provided to Empowerment Zones and Enterprise Communities is also used to train and educate youth and families and to improve child care, health care, and transportation. The program provides grant funding so communities can design local solutions that empower residents to participate in the revitalization of their neighborhoods.

In Chelsea, Massachusetts, a 178-room hotel has been constructed on a redeveloped brownfield. Across the state of Massachusetts, nearly 175 brownfield redevelopment projects are underway.

AP/Wide World Photos

"Increasing urbanization has the potential for improving human life or increasing human misery. The cities can provide opportunities or frustrate their attainment; promote health or cause disease; empower people to realize their needs and desires or impose on them a simple struggle for basic survival."

United Nations, *1996 State of the World Population Report*

Infrastructure Improvements. Urban revitalization often involves making improvements in the **infrastructure**—the underlying foundation that enables a city to function. Infrastructure includes such things as water and sewer lines, phone lines, electricity cables, sidewalks, streets, curbs, lighting, and storm drainage systems. Improving infrastructure may help attract business to an area. Infrastructure improvements in urban areas also increase property values and renew residents' sense of pride in their neighborhood (Cowherd 2001).

Brownfield Redevelopment. Brownfields are abandoned or undeveloped sites that are located on contaminated land. Cleaning up and redeveloping brownfields is not only an important environmental measure but is also a key component of urban revitalization as it provides jobs, increases tax revenues, and potentially attracts more businesses, residents, and tourists. More than 900 urban brownfield sites across the country have been successfully redeveloped into residential, business, and recreational property. But in 205 U.S. cities, there are nearly 25,000 brownfields awaiting redevelopment. In a survey of U.S. mayors, the most commonly cited obstacle to redeveloping brownfields is lack of funding (U.S. Conference of Mayors 2003).

Gentrification and Incumbent Upgrading. Gentrification is a type of neighborhood revitalization in which middle- and upper-income persons buy and rehabilitate older homes in a depressed neighborhood. They may live there or sell or rent to others. The city provides tax incentives for investing in old housing with the goal of attracting wealthier residents back into these neighborhoods and increasing the tax base. However, low-income residents are often forced into substandard housing as less and less affordable housing is available. In effect, gentrification often displaces the poor and the elderly (Johnson 1997).

An alternative to gentrification is **incumbent upgrading,** in which aid programs help residents of depressed neighborhoods buy or improve their homes and stay in the community. Both gentrification and incumbent upgrading improve decaying neighborhoods, attracting residents as well as businesses.

Community-Based Urban Renewal Efforts. Some residents of deteriorating urban areas have begun grassroots programs to improve living conditions in their neighborhoods. Community-based development programs involve small-scale developers and volunteers working with a small professional staff who are concerned with improving the community. For example, 26,000 volunteers spent a Saturday cleaning Detroit streets, parks, and playgrounds during the fourth annual spring Clean Sweep campaign (Archer 1998). In Detroit's annual Paint the Town event, individuals, community organizations, and corporate volunteers fix and paint the homes of the poor and elderly.

Improve Transportation and Alleviate Traffic Congestion

An important strategy for reducing traffic congestion involves increasing the use of public transit, such as buses, trains, and subways. A national survey found that the majority of Americans (80%) support building more rail systems serving cities, suburbs, and entire regions to give them the option of not driving their cars (U.S. Conference of Mayors 2001).

A resurgence in public transportation may be under way: over a recent six-year period, ridership on the nation's public transportation systems grew by 22 percent (American Public Transportation Association 2003). The 1998 Transportation Equity Act for the 21st Century (TEA21) provided $169.5 billion for highway improvements and $42 billion for mass transit over six years. At the time of this writing, Congress is considering the reauthorization of TEA21.

Transportation planners increasingly recognize that building more roads does not necessarily ease traffic problems. The U.S. public agrees: in a poll of randomly selected registered voters the majority (66%) said they do not think that traffic congestion will be eased if more roads are built (U.S. Conference of Mayors 2001). More important, concern is growing over the social and environmental problems related to road building and motor vehicle use. According to Jan Lundberg, director of a grassroots group called the Alliance for a Paving Moratorium, nearly half of all urban space is paved; more land is devoted to cars than to housing, and every year, nearly 100,000 people are displaced by highway construction (Jensen 2001). The Alliance for a Paving Moratorium advocates a halt to road building. "In the Alliance's view, a paving moratorium would limit the spread of population, redirect investment from suburbs to inner cities, and free up funding for mass transportation and maintenance of existing roads" (Jensen 2001, 6).

"Adding highway capacity to solve traffic congestion is like buying larger pants to deal with your weight problem."

Michael Replogle
Transportation Specialist
Environmental Defense

Another way to ease traffic congestion is to encourage means of transportation other than motor vehicles. This chapter's *Focus on Technology* looks at the use of bicycles as transportation vehicles. Finally, the development of communities that enable residents to walk to schools, shops, and other locations can help relieve

Focus on Technology

A Return to Simpler Technology: Bicycles as a Solution to Urban Problems

The application of technology to solving modern social problems often involves relatively current, state-of-the art technologies. However, sometimes an effective solution to a social problem can be found in an older, simpler technology.

In cities around the world, bicycles are emerging as a solution to some of today's urban problems. "For safer streets, less congestion, and cleaner air, the bicycle is poised to become an integral part of urban transportation systems in the 21st century" (Gardner 1999, 23). The use of bicycles can also result in improved health and lower health care costs—benefits associated with cleaner air, reduced noise, and more exercise. Studies in the United Kingdom have found that the health benefits of cycling, which include decreased risk of heart disease and diabetes, far outweigh the risks of bicycle accidents (Sheehan 2001). Increased bicycle use could also help alleviate the problem of noise pollution and its associated adverse health effects. Noise, which is perceived by many urban residents as one of the greatest problems associated with road traffic, contributes to stress disturbances, cardiovascular disease, and hearing loss (Sheehan 2001). Increased bicycle use would also result in fewer people being injured and killed by cars. Nearly a million people are killed on the world's roads each year, most of whom are pedestrians. And replacing motor vehicle trips with bike trips could improve health by reducing vehicle emissions and improving air quality. Sheehan (2001) cites a report by the World Health Organization stating: "In some parts of the world, vehicular air pollution actually kills more people than traffic accidents do" (110).

Although bicycles could contribute to improving urban life by reducing traffic congestion, reducing pollution-causing vehicle emissions, reducing noise pollution, and improving public health, their use in urban areas is not widespread. A number of factors discourage widespread bicycle use, including unsafe roads and a lack of safe bike parking. Ironically, air pollution, largely caused by vehicle emissions, discourages the use of bicycles, thus hindering a mode of transportation that would help to alleviate the air pollution.

However, bicycles have become a major mode of transportation in some European countries. In the past 20 years, the Netherlands has doubled the length of its bikeways and Germany has tripled its bikeway networks. Cycling accounts for about 12 percent of all trips in Germany, and for 27 percent in the Netherlands, compared with less than 1 percent in the United States (Gardner 2003).

Encouraging the use of bicycles requires changes in urban design and policy to make cycling a safer, more viable, and/or necessary option. Some European cities, such as Munich, Vienna, and Copenhagen, have commercial centers that restrict vehicle traffic to ambulances, delivery trucks, and cars owned by local residents. About 20 car-free communities are in various stages of development in Germany (Sheehan 2001). Programs in Lima, Peru, help low-income residents buy bicycles, and a program in Copenhagen, Denmark, provides bikes for public use (Gardner 1999). The United Kingdom has built an 8,000-kilometer National Cycle Network that will pass within four kilometers of half of the country's population (Brown 2000). It is hoped that the accessibility of safe cycling paths will induce people to shift from cars to bicycles on short trips.

To encourage bike use for longer trips, cities can establish convenient connections between cycling and public transit. "Bicycles and transit can complement each other when people are able to carry their bikes aboard buses or trains, or park them at stations" (Sheehan 2001, 17). Urban planning based on "mixed use" designs can also facilitate bike use by combining housing, public facilities (such as schools, parks, and libraries), and commercial sites (such as grocery stores and banks) in the same neighborhood.

Even if cities are designed to promote safe bicycling as a mode of transportation, people in the United States and other industrialized countries are not likely to trade their car keys for a bicycle helmet. People throughout the developed world have acquired a love for the automobile and its images of freedom, power, adventure, and sexiness. These images are perpetuated by the automobile industry, which spends more money on advertising than any other industry in the United States and worldwide (Sheehan 2001). Until our love for cleaner air and better health outweighs our love of the automobile, most Americans will continue to leave their bicycles (if they own them) in the closet or garage.

Sources: Brown, Lester. 2000. "Overview: The Acceleration of Change." In *Vital Signs: The Environmental Trends that Are Shaping Our Future,* ed. Linda Starke, pp. 17–29. New York: W. W. Norton; Gardner, Gary. 2003. "Bicycle Production Seesaws." In *Vital Signs,* eds. M. Renner and M.O. Sheehan, pp. 58–59. New York: W. W. Norton; Gardner, Gary. 1999. "Cities Turning to Bicycles to Cut Costs, Pollution, and Crime." *Public Management* 81(1):23; Jensen, Derrick. 2001. "Road to Ruin: An Interview with Jan Lundberg." *The Sun* 302(4–13); Sheehan, Molly O'Meara. 2001. "Making Better Transportation Choices." In *State of the World 2001,* ed. Linda Starke, pp. 103–122. New York: W. W. Norton.

traffic congestion. This strategy is part of the "smart growth" and "new urbanism" movement discussed in the next section.

Responding to Urban Sprawl: Growth Boundaries, Smart Growth, and New Urbanism

Some cities have tried to manage urban sprawl by establishing growth boundaries. Since 1973, Oregon has required each city to draw a growth boundary based on its estimate of economic development and community needs in the next 20 years (Geddes 1997). A number of cities in California have also enacted urban growth boundaries (Froehlich 1998). Rather than simply put a limit on urban growth, another approach to managing urban sprawl is to develop land according to principles known as **smart growth.** A smart growth urban development plan entails the following principles (Froehlich 1998; Rees 2003; Smart Growth Network 2004):

- **Mixed-use land,** which allows homes, jobs, schools, shops, workplaces, and parks to be located within close proximity of each other
- Ample sidewalks, encouraging residents to walk to jobs and shops
- Compact building design
- Housing and transportation choices
- Distinctive and attractive community design
- Preservation of open space, farmland, natural beauty, and critical environmental areas
- Redevelopment of existing communities, rather than letting them decay and building new communities around them
- Regional planning and collaboration among businesses, private residents, community groups, and policy makers on development/redevelopment issues

Smart growth is very similar to another movement in urban planning called **New Urbanism.** The goals and methods of New Urbanism are similar to those of smart growth, but the impetus for these movements is slightly different. Smart growth approaches the idea of sustainable urban communities with the primary goal of stopping sprawl. The New Urbanism approach is to raise the quality of life for all those in the community by creating compact communities with a sustainable infrastructure. Smart growth and New Urbanism are both impeded by most local zoning codes that mandate large housing setbacks, wide streets, and separation of residential and commercial areas (Pelley 1999).

Regionalism

Addressing the various social problems facing urban areas may best be achieved through **regionalism**—a form of collaboration among central cities and suburbs that encourages local governments to share common responsibility for common problems. Central cities, declining inner suburbs, and developing suburbs are often in conflict over the distribution of government-funded resources, zoning and land use plans, transportation and transit reform, and development plans. Rather than compete with each other, regional government provides a mechanism for achieving the interests of an entire region. A metropolitan-wide government would handle the inequities and concerns of both suburban and urban areas. As might be expected, suburban officials resist regionalization because they believe it will hurt their neighborhoods economically by draining off money for the cities.

Strategies for Reducing Urban Growth in Developing Countries

In developing countries, limiting population growth is essential for alleviating social problems associated with rapidly growing urban populations. Another strategy for minimizing urban growth in less-developed countries involves redistributing the population from urban to rural areas. Such redistribution strategies include the following: (1) promote agricultural development in rural areas, (2) provide incentives to industries and businesses to relocate from urban to rural areas, (3) provide incentives to encourage new businesses and industries to develop in rural areas, (4) develop the infrastructure of rural areas, including transportation and communication systems, clean water supplies, sanitary waste disposal systems, and social services. Of course, these strategies require economic and material resources, which are in short supply in less-developed countries.

Understanding Problems of Population and Urbanization

What can we conclude from our analysis of population and urbanization? First, although fertility rates have declined significantly in recent years, world population will continue to grow. This growth will largely occur in urban areas in developing regions. Given the problems associated with population growth, such as poor maternal and infant health, depletion of natural resources and pollution of the environment, and urban crowding, most governments recognize the value of controlling population size and support family planning programs. But efforts to control population must go beyond providing safe, effective, and affordable methods of birth control. Slowing population growth necessitates interventions that change the cultural and structural bases for high fertility rates. These interventions include increasing economic development and improving the status of women, which includes raising their levels of education, their economic position, and their (and their children's) health. Addressing problems associated with population growth also requires the willingness of wealthier countries to commit funds to providing reproductive health care to women, improving the health of populations, and providing universal education for people throughout the world (see Table 13.5).

Attention to urban problems and issues is increasingly important as the United States and the rest of the world are becoming increasingly and rapidly urbanized. Aside from population concerns, problems affecting urban residents include poverty and unemployment, inadequate schools, inferior or unaffordable housing, pollution, and traffic and transportation problems.

The social forces affecting urbanization in industrialized countries are different from those in developing countries. Countries such as the United States have experienced urban decline as a result of deindustrialization, deconcentration, and the shift to a service economy in which jobs that pay well and come with full benefits are scarce. At the same time, developing countries have experienced rapid urban growth as a result of industrialization, fueled in part by a global economy in which corporate transnationals locate industry in developing countries to gain access to cheap labor, raw materials, and new markets.

One of the shadows lingering over cities throughout the world is cast by environmental problems, as urban populations consume the largest share of the world's natural resources and contribute most of the pollution and waste that compromise the health of our planet. Global environmental problems, discussed in the next

Table 13.5
Annual Expenditures on Luxury Items Compared with Funding Needed to Meet Selected Basic Needs

Product	Annual Expenditure	Social or Economic Goal	Additional Annual Investment Needed to Achieve Goal
Makeup	$18 billion	Reproductive health care for all women	$12 billion
Pet food in Europe and United States	$17 billion	Elimination of hunger and malnutrition	$19 billion
Perfumes	$15 billion	Universal literacy	$5 billion
Ocean cruises	$14 billion	Clean drinking water for all	$10 billion
Ice cream in Europe	$11 billion	Immunizing every child	$1.3 billion

Source: World Watch Institute Press Release. 2004 (January 7). "State of the World 2004: Consumption by the Numbers." http://www.worldwatch.org

chapter, have many urban roots. "Key global environmental problems have their roots in cities—from the vehicular exhaust that pollutes and warms the atmosphere, to the urban demand for timber that denudes forests and threatens biodiversity, to the municipal thirst that heightens tensions over water" (Sheehan 2003, 131). As we ponder ways to improve cities, and as we strive to meet the needs of growing populations, we must include the larger environment in the equation.

Chapter Review

- **Where is most of the world population growth occurring?**
  Ninety-seven percent of world population growth is in developing countries. Most of this growth is occurring in urban areas.
- **What is "urban sprawl" and why is it a problem?**
  Urban sprawl refers to the ever-increasing outward growth of urban areas. Urban sprawl results in the loss of green, open spaces, the displacement and endangerment of wildlife, traffic congestion and noise, and pollution liabilities.
- **What is the demographic transition?**
  The demographic transition is the progression from a population with short lives and large families to one in which people live longer and have smaller families. About one-third of countries have completed the demographic transition.
- **What are slums, ghettos, and barrios?**
  Slums are concentrated areas of poverty and poor housing in urban neighborhoods. In the United States, slums that are occupied primarily by African-Americans are known as ghettos, and those occupied primarily by Latinos are called barrios.
- **Globally, what was the average number of children born to each woman in 1960? In 2003?**
  Globally, the average number of children born to each woman has fallen from 5 in 1960 to less than 2.8 in 2003.
- **What are brownfields?**
  Brownfields are abandoned or undeveloped sites that are located on contaminated land. Cleaning up and redeveloping brownfields is a key component of urban revitalization as it provides jobs, increases tax revenues, and potentially attracts more businesses, residents, and tourists.
- **What does the term "mixed-use land" refer to? What are the benefits of mixed-use land?**
  The use of mixed-use land is a strategy of the smart growth movement whereby homes, jobs, schools, shops, workplaces, and parks are located within close proximity of each other. Mixed-use land encourages walking as a means of transportation and minimizes the use of cars.

Critical Thinking

1 Today, one of the most significant forces in the United States and throughout the developed world is the aging of the population. As the population ages, what issues must cities confront in order to accommodate the needs of their elderly residents?
2 What could be done in your community to encourage the use of bicycles as an alternative to motor vehicles?

Key Terms

barrio
brownfields
central city
deconcentration
demographic transition theory
doubling time
fertility rate
gentrification
ghetto
incumbent upgrading
infrastructure
megacities
metropolis
metropolitan area
micropolitan area
mixed-use land
New Urbanism
population momentum
pronatalism
regionalism
road rage
slum
smart growth
suburbanization
suburbs
urban area
urbanization
urban sprawl

Taking a Stand

Do you think safe abortion should be part of reproductive health care in developing countries?

The majority of U.S. adults favor U.S. economic aid for family planning programs in developing countries, but only half favor U.S. economic aid to provide voluntary, safe abortion as part of reproductive health care in developing countries that request it (DaVanzo, Adamson, Belden, & Patterson 2000). How would you vote on this issue?

Use Wadsworth's exclusive online resources—InfoTrac College Edition, MicroCase Online, and OVRC—to formulate a position on this topic.

The Wadsworth's Sociology Online Resources and Writing Companion will help you get started. This valuable guide will show you how to use Wadsworth's exclusive online resources when studying social problems. It will also help you to build essential research and writing skills. InfoTrac College Edition, MicroCase Online, OVRC, and an electronic copy of portions of this companion are available at http://sociology.wadsworth.com/mooney_knox_schacht/problems4e, the companion Web site for *Understanding Social Problems,* Fourth Edition.

Media Resources

The Companion Web Site for *Understanding Social Problems,* Fourth Edition

http://sociology.wadsworth.com/mooney_knox_schacht/problems4e

Supplement your review of this chapter by going to the companion Web site to take one of the Tutorial Quizzes, use the flash cards to master key terms, and check out the many other study aids you'll find there. You'll also find special features such as *Wadsworth's Sociology Online Resources and Writing Companion,* GSS Data, and Census 2000 information, data, and resources at your fingertips to help you complete that special project or do some research on your own.

Interactions CD-ROM

Go to the Interactions CD-ROM for *Understanding Social Problems,* Fourth Edition, to access additional interactive learning tools, such as in-depth review materials, corresponding practice quizzes, and other engaging resources and activities to help you study the concepts in this chapter.

"Just as we have the capacity to hasten the degradation and destruction of our planet, so, too, do we have the capacity to preserve, build, and improve the quality of life on our planet. The choice is ours." *Hal Burdett, Population Institute*

Environmental Problems

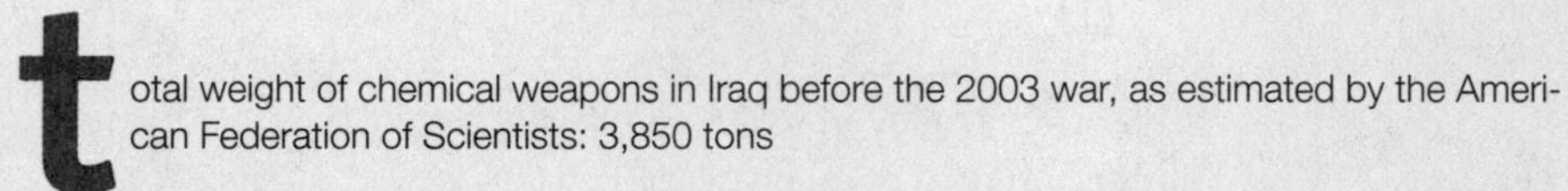

total weight of chemical weapons in Iraq before the 2003 war, as estimated by the American Federation of Scientists: 3,850 tons

Total weight of just six of the most dangerous pesticides at large in the global environment: 7,000,000 tons

People killed by Iraqi chemical weapons in the six-year period preceding the 2003 war: 0

People killed by pesticides, as estimated by the World Health Organization, during the same six-year period: over 1,000,000 ("Matters of Scale: Chemical Warfare" 2003).

After the terrorist attacks of September 11, 2001, the "war on terrorism" and the search for chemical weapons in Iraq commanded much attention and concern, and diverted massive resources into military operations, "homeland security" measures, and the rebuilding of Iraq. Comparisons like those described in the opening of this chapter invite a reconsideration of priorities in the quest to make the world a safer place. In this chapter, we focus on environmental problems that threaten the lives and well-being of people, plants, and animals all over the world—today and in future generations.

After examining how globalization affects environmental problems, we view environmental issues through the lens of structural-functionalism, conflict theory, ecofeminist theory, and symbolic interactionism. Then we overview major environmental problems, examining their social causes and exploring strategies that attempt to reduce or alleviate them.

The Global Context: Globalization and the Environment

In 1992, leaders from across the globe met at the first Earth Summit in Rio de Janeiro to forge agreements to protect the planet's environment and at the same time alleviate world poverty. When world leaders met a decade later in 2002 for the second Earth Summit in Johannesburg, the overall state of the environment had deteriorated and poverty had deepened. How is it that the combined efforts of leaders who met at the first Earth Summit were so ineffectual in achieving their goals? A large part of the answer lies in the increasing globalization of the last two decades. Three aspects of globalization that have affected the environment include (1) the permeability of international borders to pollution and environmental problems, (2) cultural and social integration spurred by communication and information technology, and (3) growth of transnational corporations.

Permeability of International Borders

Environmental problems such as global warming and destruction of the ozone layer (discussed later in this chapter) demonstrate that environmental problems extend far beyond their source to affect the entire planet and its inhabitants. A striking example of the permeability of international borders to pollution is the spread of toxic chemicals (such as PCBs) from the southern hemisphere into the Arctic. In as little as five days, chemicals from the tropics can evaporate from the soil, ride the winds north thousands of miles, condense in the cold air, and fall on the Arctic in the form of toxic snow or rain (French 2000a). This phenomenon was discovered in the mid-

1980s, when scientists found high levels of PCBs (toxic chemicals) in the breast milk of Inuit women in the Canadian Arctic.

Another environmental problem involving permeability of borders is **bioinvasion:** the emergence of organisms into regions where they are not native. Exotic species travel in the ballast water of ships, in packing material, in shipments of crops and other goods, and in many other ways. Invasive species may compete with native species for food, start an epidemic, or prey on natives, threatening not only their immediate victims but the entire ecosystem in which the victims lives. Bioinvasion is largely a product of the growth of global trade.

Cultural and Social Integration

As mass media infiltrate the world, people across the globe aspire to consume the products and mimic the materially saturated lifestyles portrayed in movies, television, and advertising. As patterns of consumption in developing countries increasingly follow those in wealthier Western nations, so do the problems associated with overconsumption: depletion of natural resources, pollution, and global warming.

On the positive side, the Internet and other forms of mass communication have helped to integrate the efforts of diverse environmental groups across the globe. The globalization of the environmental movement has created new opportunities for environmental groups to join forces, share information, and educate the public through mass communication.

The Growth of Transnational Corporations

As discussed in Chapter 11, the world's economy is dominated by transnational corporations. The World Trade Organization (WTO) and free trade agreements such as NAFTA and the FTAA provide transnational corporations with privileges to pursue profits, expand markets, use natural resources, and exploit cheap labor in developing countries while, at the same time, weakening the ability of governments to protect natural resources or to implement environmental legislation. Transnational corporations have influenced the world's most powerful nations to institutionalize an international system of governance that values commercialism, corporate rights, and "free" trade above environment, human rights, worker rights, and human health (Bruno & Karliner 2002).

Many transnational corporations are implicated in environmentally destructive activities—from mining to dumping of toxic waste. One case in point: In the city of Mexicali, near the California border, foreign-owned manufacturing plants known as *maquiladoras* were found to be surrounded by toxic waste. "Chemicals known to cause cancer, birth defects, and brain damage were being emptied into ditches that ran through the shantytowns around the factories" (French 2000a, 85). One reason that U.S.-owned factories set up shop in Mexico and other developing countries is the weak environmental laws that exist in these countries.

Sociological Theories of Environmental Problems

The three main sociological theories—structural functionalism, conflict, and symbolic interactionism each provide insights into social causes of and responses to environmental problems. We also include a brief overview of ecofeminism in the following discussion of sociological theories of environmental problems.

Structural-Functionalist Perspective

Structural-functionalism emphasizes the interdependence between human beings and the natural environment. From this perspective, human actions, social patterns, and cultural values affect the environment, and in turn, the environment affects social life. For example, population growth affects the environment, as more people utilize natural resources and contribute to pollution. However, the environmental impact of population growth varies tremendously according to a society's patterns of economic production and consumption (Hunter 2001).

Structural-functionalism focuses on how changes in one aspect of the social system affect other aspects of society. For example, in the two years after the terrorist attacks of September 11, 2001, public concern about most environmental problems declined sharply, most likely due to increasing concern about the economy and terrorism over the same period (Saad 2002). "Perhaps environmental problems look less serious by contrast with the economy and the terrorist threat, or the environment has been squeezed out of media headlines, or both" (Saad 2002, 2).

The structural-functional perspective is concerned with latent functions—consequences of social actions that are unintended and not widely recognized. For example, the more than 840,000 dams worldwide provide water to irrigate farmlands and supply 17 percent of the world's electricity. Yet dam building has had unintended negative consequences for the environment, including the loss of wetlands and wildlife habitat, the emission of methane (a gas that contributes to global warming) from rotting vegetation trapped in reservoirs, and the altering of river flows downstream killing plant and animal life ("A Prescription for Reducing the Damage Caused by Dams" 2001). Dams have also displaced millions of people from their homes. As philosopher Kathleen Moore points out, "Sometimes in maximizing the benefits in one place, you create a greater harm somewhere else. . . . While it might sometimes seem that small acts of cruelty or destruction are justified because they create a greater good, we need to be aware of the hidden systematic costs" (Jensen 2001, 11). Being mindful of latent functions means paying attention to the unintended and often hidden environmental consequences of human activities.

Conflict Perspective

The conflict perspective focuses on how wealth, power, and the pursuit of profit underlie many environmental problems. Wealth is related to consumption patterns that cause environmental problems. Wealthy nations have higher per capita consumption of petroleum, wood, metals, cement, and other commodities that deplete the earth's resources, emit pollutants, and generate large volumes of waste. The wealthiest 20 percent of the world's population are responsible for 86 percent of total private consumption (Bright 2003). The capitalistic pursuit of profit encourages making money from industry regardless of the damage done to the environment. Further, to maximize sales, manufacturers design products intended to become obsolete. As a result of this **planned obsolescence,** consumers continually throw away used products and purchase replacements. Industry profits at the expense of the environment, which must sustain the constant production and absorb ever-increasing amounts of waste. Industries also use their power and wealth to influence politicians' environmental policies. A report by the Natural Resources Defense Council found that in 2002 the Bush administration made numerous policy decisions that benefited industry at the expense of the environment and public health (Perks & Wetstone 2003). For example, under the Bush administration the Environmental

Protection Agency made changes to the Clean Air Act that exempt the nation's oldest and dirtiest power plants and refineries from installing modern pollution controls when they upgrade or expand their facilities in ways that increase emissions.

Ecofeminist Perspective

Ecological feminism, or **ecofeminism,** began in 1974 when French feminist Francoise d'Eaubonn coined the term "ecological feminisme" to call attention to women's potential to energize an ecological revolution (Warren 2000). Ecofeminists view environmental problems as resulting from human domination of the environment and see connections between the domination of women, people of color, children, and the poor and the domination of nature. Throughout the world, and in developing countries in particular, men are dominant in deciding how natural resources are used. Men are dominant in positions of government and corporate leadership and "own" most of the land. By some estimates, women around the world hold title to less than 2 percent of the land that is owned (MacDonald & Nierenberg 2003). In contrast to a male-oriented view of natural resources as a means to an end—a means to profit and power—ecofeminists often embrace a spiritual approach to addressing environmental problems, drawing on pagan, Native American, New Age, and Eastern religious traditions that emphasize the close connection between women and nature (Mother Earth) (Schaeffer 2003).

"Women must see that there can be no liberation for them and no solution to the ecological crisis within a society whose fundamental model of relationships continues to be one of domination."

Karen J. Warren
Ecofeminist philospher

Symbolic Interactionist Perspective

The symbolic interactionist perspective focuses on how meanings, labels, and definitions learned through interaction and through the media affect environmental problems. Whether an individual recycles, car-pools, or joins an environmental activist group is influenced by the meanings and definitions of these behaviors that the individual learns through interaction with others.

Large corporations and industries commonly utilize marketing and public relations strategies to construct favorable meanings of their corporation or industry. The term **greenwashing** refers to the way environmentally and socially damaging companies portray their corporate image and products as being "environmentally friendly" or socially responsible. Greenwashing is commonly used by public relations firms that specialize in damage control for clients whose reputations and profits have been hurt by poor environmental practices. Philip Morris, the infamous cigarette and food producer, donated $60 million to charity in 1999, but spent another $108 million in advertising to tell the world about their generosity ("Corporate Spotlight" 2001). DuPont, the biggest private generator of toxic waste in the United States, attempted to project a "green" image by producing a TV ad showing seals clapping, whales and dolphins jumping, and flamingos flying. In an Earth Day event on the National Mall in Washington, D.C., the National Association of Manufacturers (NAM) highlighted renewable and energy-efficient technologies. Yet members of NAM have spent millions of dollars lobbying against use of these very same technologies (Karliner 1998). A logging company facing opposition from environmentalists in New Zealand described their activities as "sustainable harvesting of indigenous production forests"—a phrase that sounds more environmentally friendly than "logging of old growth forests" (Hager & Burton 2000).

"Despite their eco-friendly rhetoric, for most corporations in the world green is nothing more than the color of money."

Kenny Bruno and Joshua Karlinger
Authors of *Earthsummit.biz: The Corporate Takeover of Sustainable Development*

Greenwashing can cause controversy and divisions among environmentalists. Consider Earth Day, which has brought environmentalists together to celebrate the earth and to increase public awareness of environmental issues every April 22 since

At an Earth Day 2004 rally in Madison, Wisconsin, "The Raging Grannies" sing original songs in support of environmental and social reform.

it began in 1970. In recent years, Earth Day events have been sponsored by corporations such as Office Depot, Texas Instruments, Raytheon Missile systems, and Waste Management—a Houston-based company responsible for numerous hazardous waste sites. Some environmentalists, including Earth Day's founder—former Senator Gaylord Nelson—consider the participation of corporations in Earth Day evidence of the celebration's success. But other environmentalists oppose corporate sponsorship of Earth Day events, accusing corporations of using their financial support of Earth Day as a public relations "greenwashing" strategy. John Stauber, a critic of corporations that hide their damaging environmental practices behind green advertising and marketing, says that "Waste Management sponsoring Earth Day is similar to Enron sponsoring a seminar on corporate responsibility" (quoted in Cappiello 2003). In recent years, environmentalists have withdrawn from and protested Earth Day events in cities across the country because the events were sponsored by corporations.

Although greenwashing involves manipulation of public perception to maximize profits, many corporations make genuine and legitimate efforts to improve their operations, packaging, or overall sense of corporate responsibility toward the environment. For example, in 1990, McDonald's announced it was phasing out foam packaging and switching to a new, paper-based packaging that is partially degradable. But many environmentalists are not satisfied with what they see as token environmentalism, or as Peter Dykstra of Greenpeace suggests, 5 percent of environmental virtue to mask 95 percent of environmental vice (Hager & Burton 2000).

Environmental Problems: An Overview

Environmental problems include air, land, and water pollution; global warming; depletion of natural resources; threats to biodiversity; environmental illness; environmental injustice; and disappearing livelihoods. Because many of these environmental problems are related to the ways humans produce and consume energy, we begin this section with an overview of global energy use. Before reading further, you may want to assess your knowledge about energy issues and problems by taking the Energy IQ Test in this chapter's *Self and Society* feature.

Energy Use Worldwide: An Overview

Most of the world's energy—77 percent in 2000—comes from fossil fuels, which include oil, coal, and natural gas (Sawin 2003a) (see Figure 14.1). The next most common sources of energy (15%) are hydropower and traditional biomass (e.g., fuel wood, crops, animal wastes), which are primarily used by poor populations in developing countries. Nuclear energy generated by nuclear power plants contributes 6 percent of the world's energy, followed by new renewables—the least common source of energy (2%). The most common form of new renewables are wind energy and photovoltaics (also known as solar power).

Worldwide, the most common source of electricity is fossil fuels (64%), followed by nuclear power (17%) and hydropower (17%), with the least comomon source being other renewables (2%) (see Figure 14.2). As you continue reading this chapter, notice that the major environmental problems facing the world today—air, land, and water pollution; destruction of habitats; biodiversity loss; global warming; and environmental illness are linked to the production and use of fossil fuels.

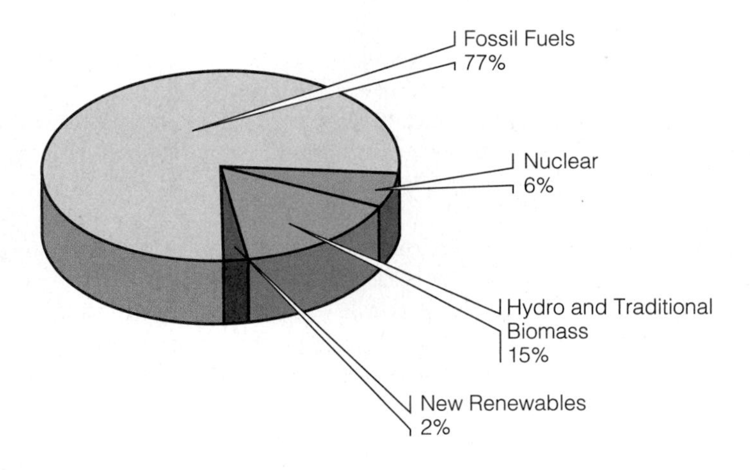

Figure 14.1
World Energy Consumption, by Source: 2000

Source: From *Vital Signs 2003: The Trends That Are Shaping Our Future* by Michael Renner, et al. Copyright © 2003 by WorldWatch Institute. Used by permission of W. W. Norton & Company, Inc.

Fossil Fuels
64%
Nuclear
17%
Hydropower
17%
Other Renewables
2%

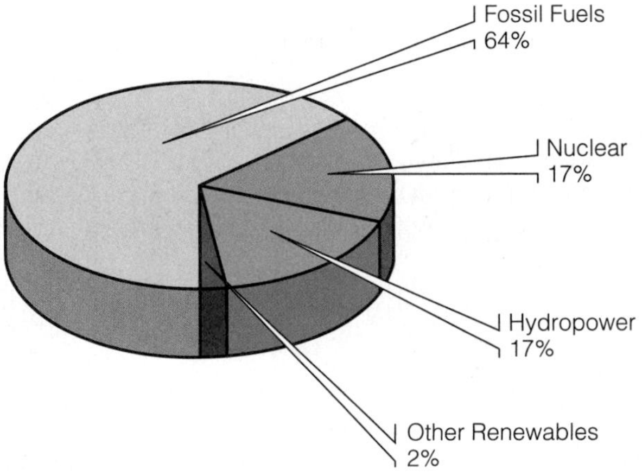

Figure 14.2
World Electricity Generation by Type: 2000

Source: From *Vital Signs 2003: The Trends That Are Shaping Our Future* by Michael Renner, et al. Copyright © 2003 by WorldWatch Institute. Used by permission of W. W. Norton & Company, Inc.

Self and Society

Energy IQ Test

DIRECTIONS: *After answering each of the following items, check your answers for items 2 through 11 using the answer key provided. For items 2 through 11, calculate the total number of items you answered correctly. Compare your score to the results of a national survey.*

1. In general, how much do you feel you know about energy issues and problems—would you say you know a lot, a fair amount, only a little, or practically nothing?
   a. A lot
   b. A fair amount
   c. Only a little
   d. Practically nothing
2. How is most of the electricity in the United States generated?
   a. By burning oil, coal, and wood
   b. With nuclear power
   c. Through solar energy
   d. At hydroelectric power plants
   e. Don't know
3. Which of the following uses the most energy in the average home?
   a. Lighting rooms
   b. Heating water
   c. Heating and cooling rooms
   d. Refrigerating food
   e. Don't know
4. Which of the following sectors of the U.S. economy consumes the greatest percentage of the nation's petroleum?
   a. The residential sector
   b. The commercial sector
   c. The transportation sector
   d. The industrial sector
   e. Don't know
5. Which fuel is used to generate the most energy in the United States?
   a. Petroleum
   b. Coal
   c. Natural gas
   d. Nuclear
6. Though the United States has only 4 percent of the world's population, what percentage of the world's energy does it consume?
   a. 5 percent
   b. 15 percent
   c. 20 percent
   d. 25 percent
   e. Don't know
7. In the last 10 years, which of the following industries in the U.S. economy has increased its energy demands the most?
   a. The food industry
   b. The transportation industry
   c. The computer and technology industry
   d. The health care industry
   e. Don't know
8. In the past 10 years, has the average miles per gallon of gasoline used by vehicles in the United States

Being mindful of environmental problems means seeing the connections between energy use and our daily lives.

> Everything we consume or use—our homes, their contents, our cars and the roads we travel, the clothes we wear, and the food we eat—requires energy to produce and package, to distribute to shops or front doors, to operate, and then to get rid of. We rarely consider where this energy comes from or how much of it we use—or how much we truly need. (Sawin 2004, 25)

Depletion of Natural Resources

Population growth combined with consumption patterns are depleting natural resources such as forests, water, minerals, and fossil fuels. For example, coal and oil—the world's main source of energy—are being depleted, with global oil pro-

a. Increased
b. Remained the same
c. Gone down
d. Not been tracked
e. Don't know

9. Scientists have not determined the best solution for disposing of nuclear waste. In the United States, what do we do with it now?
   a. Use it as nuclear fuel
   b. Sell it to other countries
   c. Dispose of it in landfills
   d. Store and monitor the waste
   e. Don't know
10. The United States currently uses oil from both domestic and foreign sources. What percentage of the oil is imported?
   a. 10 percent
   b. 20 percent
   c. 35 percent
   d. 50 percent
   e. Don't know
11. Scientists say the fastest and most cost-effective way to address our energy needs is to
   a. Develop all possible sources of domestic oil and gas
   b. Build nuclear power plants
   c. Develop more hydroelectric plants
   d. Promote more energy conservation
   e. Don't know

Scoring and Comparison Data from a National Sample

Use the answer key below to score your answers to items 2–11.
ANSWER KEY: 2a, 3c, 4c, 5a, 6d, 7b, 8c, 9d, 10d, 11d

When a national sample of U.S. adults took the *Energy IQ Test,* three in four (75%) rated themselves as having "a lot" (12%) or "a fair amount" (63%) of knowledge about energy issues and problems. Yet most of the participants (76%) failed the test, scoring 5 or fewer items correctly (see table below). This sample scored an average of 4.1 out of 10 items correctly.

Grade	Percent of Total Sample Receiving Grade
A (9 or 10 correct)	1
B (8 correct)	3
C (7 correct)	8
D (6 correct)	13
F (5 or fewer correct)	76

Source: Based on *2001 NEETF/Roper Report Card.* 2001. The National Environmental Education & Training Foundation & Roper Starch Worldwide. Used by permission.

duction expected to peak during the lifetime of the current generation of young adults. Yet, in the next two decades, energy consumption demands are expected to increase nearly 60 percent due to continued population growth and urbanization, and economic and industrial expansion (Sawin 2003a).

Fresh water resources are also being consumed by agriculture, by industry, and for domestic use. More than 1 billion people lack access to clean water, and by 2025, 2.3 billion people—two-thirds of the world's population—will be threatened by water scarcity (Population Institute 2003).

The world's forests are also being depleted. The demand for new land, fuel, and raw materials has resulted in **deforestation**—the conversion of forest land to non-forest land (Intergovernmental Panel on Climate Change 2000b). Global forest cover has been reduced by as much as 50 percent since preagricultural times (Bright 2003). Between the 1960s and the 1990s, one-fifth of the world's tropical forests were cut or burned (Youth 2003). The major causes of deforestation are the expansion of agricultural land, human settlements, commercial logging, and road building.

Deforestation displaces people and wild species from their habitats; soil erosion caused by deforestation can cause severe flooding; and as we explain later in this chapter, deforestation contributes to global warming. Deforestation also contributes to **desertification**—the degradation of semiarid land, which results in the expansion of desert land that is unusable for agriculture. Overgrazing by cattle and other herd animals also contributes to desertification. The problem of desertification is most severe in Africa (Reese 2001). As more land turns into desert, populations can no longer sustain a livelihood on the land, and so they migrate to urban areas or other countries, contributing to social and political instability.

Air Pollution

Transportation vehicles, fuel combustion, industrial processes (such as the burning of coal and the processing of minerals from mining), and solid waste disposal have contributed to the growing levels of air pollutants, including carbon monoxide, sulfur dioxide, nitrogen dioxides, and lead. Air pollution levels are highest in areas with both heavy industry and traffic congestion, such as Los Angeles, New Delhi, Jakarta, Bangkok, Tehran, Beijing, and Mexico City. In the mid-1990s, breathing the air in Mexico City was like smoking two packs of cigarettes a day (Weiner 2001). Although air quality in the United States has improved on average in recent decades, 160 million tons of pollution are emitted into the air each year in the United States and more than half of the U.S. population lives in counties where air pollutants exceed federal levels at times during the year (Environmental Protection Agency 2003).

Indoor air pollution is a serious problem in developing countries. As this woman in Nepal, India, cooks food for her family, she is exposed to harmful air contaminants from the fumes.

© Ted Spiegel/CORBIS

Indoor Air Pollution. When we hear the phrase "air pollution," we typically think of industrial smokestacks and vehicle exhausts pouring gray streams of chemical matter into the air. But indoor air pollution is also a major problem, especially in poor countries. About half the world's population and up to 90 percent of rural households in developing countries rely on wood and unprocessed biomass (dung and crop residues) for cooking and heating fuel (Bruce, Perez-Padilla, & Albalak 2000). These fuels are typically burned indoors in open fires or poorly functioning stoves, producing hazardous emissions such as soot particles, carbon monoxide, nitrous oxides, sulphur oxides, and formaldehyde. Long-term exposure to the smoke contributes to respiratory illness, lung cancer, tuberculosis, and blindness, and an estimated 2 million deaths in developing countries each year result from exposure to indoor air pollution. According to the World Health Organization, indoor air pollution ranks fifth as a risk factor for ill health worldwide—

behind malnutrition, AIDS, tobacco use, and poor water and sanitation (Bruce et al. 2000; Mishra, Retherford, & Smith 2002).

Even in afffluent countries, much air pollution is invisible to the eye and exists where we least expect it—in our homes, schools, workplaces, and public buildings. Common household, personal, and commercial products contribute to indoor pollution. Some of the most common indoor pollutants include carpeting (which emits nearly 100 different chemical gases), mattresses (which may emit formaldehyde and aldehydes), drain cleaners, oven cleaners, spot removers, shoe polish, dry-cleaned clothes, paints, varnishes, furniture polish, potpourri, mothballs, fabric softener, and caulking compounds. Air fresheners, deodorizers, and disinfectants emit the pesticide paradichlorobenzene. Potentially harmful organic solvents are present in numerous office supplies, including glue, correction fluid, printing ink, carbonless paper, and felt-tip markers.

Destruction of the Ozone Layer. The ozone layer of the earth's atmosphere absorbs most of the harmful ultraviolet-B radiation from the sun and completely screens out lethal ultraviolet-C radiation. The ozone layer is thus essential to life on earth. Yet the use of certain chemicals has weakened the ozone layer. The depletion of the ozone layer allows hazardous levels of ultraviolet rays to reach the earth's surface and is linked to increases in skin cancer and cataracts, weakened immune systems, reduced crop yields, damage to ocean ecosystems and reduced fishing yields, and adverse effects on animals.

In 2000, the size of the ozone hole reached a record 29 million square kilometers (more than three times the size of the United States) and in 2003 was measured at 25 million square kilometers (11.1 million square miles) (UNEP 2003). Previously, this ozone hole exposed only ocean and barren land in Antarctica, but now the ozone hole exposes Puntas Arenas, Chile, a populous city near the southern tip of South America.

Ninety-six chemicals have been identified as being harmful to the ozone layer. These include chlorofluorocarbons (CFCs) (used in refrigerators, air conditioners, spray cans, and other applications); hydrochlorofluorocarbons (HCFCs) (developed as a replacement for CFCs), halons (used in fire extinguishers), and methyl bromide (used as a fumigant for crops, pest control, and quarantine treatment of agricultural exports). These and other ozone-damaging chemicals remain in the atmosphere for various lengths of time ranging from about one year to 1,700 years (UNEP 2003).

Acid Rain. Air pollutants, such as sulfur dioxide and nitrogen oxide, mix with precipitation to form **acid rain.** Polluted rain, snow, and fog contaminate crops, forests, lakes, and rivers. Due to the effects of acid rain, 140 Minnesota lakes are totally devoid of fish (Miller 2000). Because pollutants in the air are carried by winds, industrial pollution in the midwest falls back to earth as acid rain on southeast Canada and the northeast New England states. Acid rain is not just a problem in North America; it decimates plant and animal species around the globe.

Global Warming and Climate Change

In 2002, an Antarctic ice shelf the size of Rhode Island melted through the month of February following unusually warm regional temperatures (Saad 2002). This event contributes to growing evidence of **global warming.** Globally, the 1990s was the warmest decade on record, with 1998 being the hottest year since record keeping began in the late 1800s. In 2002, global average temperature reached the second highest level ever recorded (Sheehan 2003). Average global air temperature rose by 0.6 degrees C over the twentieth century and, between 1990 and 2100, is expected to rise another 1.4 to 5.8 degrees C (Intergovernmental Panel on Climate Change 2001a). Over the twentieth century, the average annual U.S. temperature has risen 1 degree F and is expected to rise by 5 to 9 degrees F over the next 100 years (National Assessment Synthesis Team 2000).

Causes of Global Warming. Although the cause of global warming remains a disputed topic, the prevailing view is that **greenhouse gases**—primarily carbon dioxide, methane, and nitrous oxide—accumulate in the atmosphere and act like the glass in a greenhouse, holding heat from the sun close to the earth. Atmospheric concentrations of carbon dioxide, the main greenhouse gas, have increased by 31 percent since 1750; methane by 151 percent, and nitrous oxide by 17 percent (Intergovernmental Panel on Climate Change 2001a). In the year 2002 alone, human activity released more than 6 billion tons of carbon dioxide into the air—a 1 percent increase over the previous year (Sheehan 2003).

The primary source of carbon dioxide emissions is fossil fuel burning. With less than 5 percent of the world's population, the United States produces 24 percent of the world's carbon emissions from fossil fuel burning (Renner & Sheehan 2003). U.S. carbon emissions per person are roughly double that of other major industrialized countries, and 17 times that of India (see Figure 14.3).

Deforestation also contributes to increasing levels of carbon dioxide in the atmosphere. Trees and other plant life utilize carbon dioxide and release oxygen into the air. As forests are cut down, fewer trees are available to absorb the carbon dioxide.

Effects of Global Warming. Numerous effects of global warming have been observed and are anticipated in the future (Intergovernmental Panel on Climate Change 2001b; National Assessment Synthesis Team 2000). As temperature increases, some areas will experience heavier rain, while other regions will get drier. Whereas some regions will experience increased water availability and crop yields, other regions, particularly tropical and subtropical regions, are expected to experience decreased water availability and a reduction of crop yields. Global warming

Figure 14.3
Carbon Emissions Per Person in Selected Countries

Source: From *Vital Signs 2003: The Trends That Are Shaping Our Future* by Michael Renner, et al. Copyright © 2003 by WorldWatch Institute. Used by permission of W. W. Norton & Company, Inc.

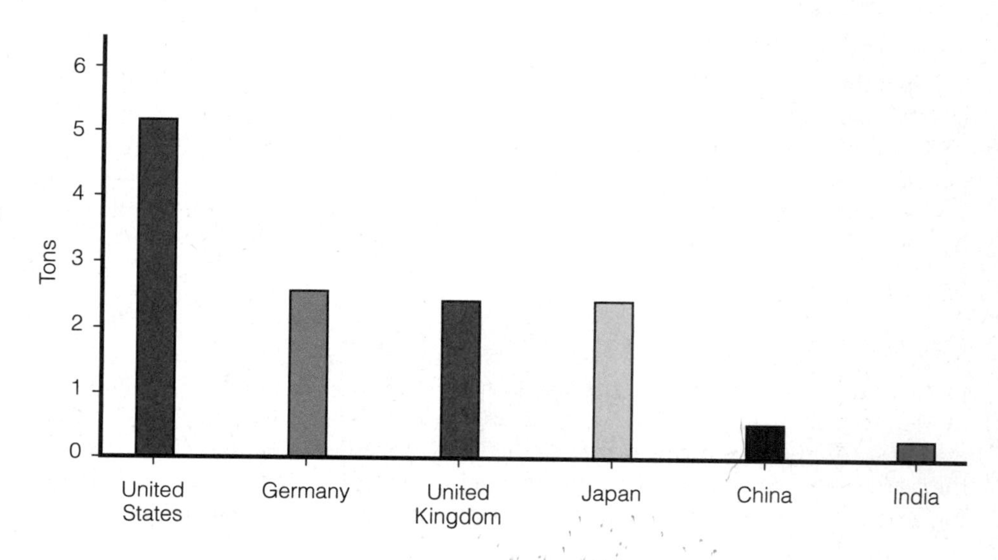

results in shifts of plant and animal habitats and the increased risk of extinction of some species. Regions that experience increased rainfall as a result of increasing temperatures may face increases in waterborne diseases and diseases transmitted by insects. Over time, climate change may produce opposite effects on the same resource. For example, in the short term, forest productivity is likely to increase in response to higher levels of carbon dioxide in the air, but over the long term, forest productivity is likely to decrease because of drought, fire, insects, and disease.

Global warming also threatens to melt glaciers and permafrost, resulting in a rise in sea level. Global average sea levels rose 0.1 to 0.2 meters during the twentieth century and are expected to rise by 0.09 to 0.88 meters from 1990 to 2100. As seas levels rise, some island countries, as well as some barrier islands off the U.S. coast are likely to disappear and low-lying coastal areas will become increasingly vulnerable to storm surges and flooding. Although U.S. coastal counties (excluding Alaska) constitute only 11 percent of U.S. land area, they are home to 53 percent of the population (Hunter 2001).

In urban areas, flooding can be a problem where storm drains and waste management systems are inadequate. Increased flooding associated with global warming is expected to result in increases in drownings, and diarrheal and respiratory diseases. An increase in the number of people exposed to insect and water-related diseases, such as malaria and cholera, are also expected. Even if greenhouse gases are stabilized, global air temperature and sea level are expected to continue to rise for hundreds of years. When asked how much they worry about "the greenhouse effect or global warming," more than half of Americans (58%) worry either a fair amount or a great deal, while 40 percent worry only a little or not at all (Saad 2002).

Land Pollution

About 30 percent of the world's surface is land, which provides soil to grow the food we eat. Increasingly, humans are polluting the land with nuclear waste, solid waste, and pesticides. In 2002, 1,289 hazardous waste sites were on the National Priority List (also called Superfund sites) (*Statistical Abstract* 2003). States with the most Superfund sites include New Jersey (115 sites), California (98 sites), Pennsylvania (96 sites), New York (91 sites), Michigan (69 sites), and Florida (52 sites).

Nuclear Waste. Nuclear waste, resulting from both nuclear weapons production and nuclear reactors or power plants, contains radioactive plutonium—a substance linked to cancer and genetic defects. Radioactive plutonium has a half-life of 24,000 years, meaning that it takes 24,000 years for the radioactivity to be reduced by half (Mead 1998). Thus, nuclear waste in the environment remains potentially harmful to human and other life for thousands of years.

Under the 1982 Nuclear Waste Policy Act, beginning in 1998 the Department of Energy has handled the storage of nuclear waste from commercial nuclear power plants. Scientists have proposed several options for disposing of nuclear waste, including burying it in rock formations deep below the earth's surface or under Antarctic ice, injecting it into the ocean floor, or hurling it into outer space. Each of these options is risky and costly. Accidents at nuclear power plants, such as the one at Chernobyl, and the potential for nuclear reactors to be targeted by terrorists add to the actual and potential dangers of nuclear power plants. Recognizing the hazards of nuclear power plants and their waste, Germany became the first country to order all of its 19 nuclear power plants shut down by 2020 ("Nukes Rebuked" 2000). But other countries are building new reactors. In 2002, seven new nuclear reactors became operational (in China, South Korea, and the Czech Republic), bringing the total number of functioning nuclear reactors worldwide to 437 (Lenssen 2003).

Solid Waste. In 1960, each U.S. citizen generated 2.7 pounds of garbage on average every day. In 2001, this figure had increased to 4.4 pounds (*Statistical Abstract* 2003). This figure does not include mining, agricultural, and industrial waste; demolition and construction wastes; junked autos; or obsolete equipment wastes. Some solid waste is converted into energy by incinerators; more than half is taken to landfills. The availability of landfill space is limited, however. Some states have passed laws that limit the amount of solid waste that can be disposed of; instead, they require that bottles and cans be returned for a deposit or that lawn clippings be used in a community composting program. This chapter's *Focus on Technology* feature examines the hazards of disposing of computers in landfills (and other environmental and health hazards of computers).

Pesticides. Pesticides are used worldwide for crops and gardens; outdoor mosquito control; the care of lawns, parks, and golf courses; and indoor pest control. Pesticides contaminate food, water, and air and can be absorbed through the skin, swallowed, or inhaled. Many common pesticides are considered potential carcinogens and neurotoxins. Recognizing the health risks associated with exposure to the pesticide Dursban, the Environmental Protection Agency issued a ban on the popular pesticide in 2000 (Kaplan & Morris 2000). Even when a pesticide is found to be hazardous and is banned in the United States, other countries may continue to use it. Although DDT is banned in the United States, the pesticide is used in many other countries from which we import food. About one-third of the foods purchased by U.S. consumers have detectable levels of pesticides; from 1 to 3 percent of the foods have pesticide levels above the legal level (Pimentel & Greiner 1997).

Water Pollution

When a Gallup poll asked Americans how much they worry about various environmental issues, pollution of drinking water, rivers, lakes, and reservoirs ranked at the top of the list, with more than half of Americans worried "a great deal" about these issues (Saad 2002). And Americans have good reason to worry. According to the Environmental Protection Agency, 40 percent of all U.S. waters are not safe for fishing or swimming (Smith, Coequyt, & Wiles 2000).

Our water is being polluted by a number of harmful substances, including pesticides, industrial and agricultural waste, acid rain, and oil spills. In 2000, more than 8,000 oil spills occurred in and around U.S. waters alone, totaling nearly one and one-half million gallons of spilled oil (*Statistical Abstract* 2003). Approximately 35,000 river miles in the United States are polluted by runoff from livestock operations ("New Rules for Feedlots" 1998).

Water pollution is most severe in developing countries, where more than 1 billion people lack access to clean water. In developing nations, as much as 95 percent of untreated sewage is dumped directly into rivers, lakes, and seas that are also used for drinking and bathing (Pimentel et al. 1998). Mining operations, located primarily in developing countries are notoriously damaging to the environment. Modern gold-mining techniques use cyanide to extract gold from low-grade ore. Cyanide is extremely toxic: a teaspoon containing a 2 percent cyanide solution can kill an adult (cyanide was used to kill Jews in Hitler's gas chambers). In 2000, a dam

Focus on Technology

Environmental and Health Hazards of Computer Manufacturing, Disposal, and Recycling

More than 1 billion computers worldwide have been sold, and the 2 billion mark may be reached as early as 2008 (Thibodeau 2002). The toxic materials (see Table 1) used in computer manufacturing pose significant hazards for workers in the production of computers, and the toxic **e-waste** (waste from electronic equipment) makes the disposal and recycling of computers a human and environmental hazard.

Table 1
Toxic Materials in Computer Manufacturing

- lead and cadmium in computer circuit boards
- lead oxide and barium in cathode ray tubes of computer monitors
- mercury in switches and flat screens
- brominated flame retardants on printed circuit boards, cables, and plastic casing
- polychlorinated biphenyls (PCBs) present in older capacitors and transformers
- polyvinyl chloride (PVC)-coated copper cables and plastic computer casings

Hazards of Computer Manufacturing

The manufacturing of semiconductors, printed circuit boards, disk drives, and monitors involves hazardous chemicals. Numerous reports have documented higher than normal rates of cancer and birth defects among computer manufacturing workers who are regularly exposed to carcinogenic and other toxic chemicals (Silicon Valley Toxics Coalition 2001b). The health hazards involved in computer manufacturing received heightened media attention after workers at IBM and their families initiated lawsuits against IBM for knowingly exposing workers to hazardous materials. As of this writing, more than 200 such lawsuits against IBM are pending.

Hazards of E-Waste Disposal

When consumers upgrade to faster and sleeker PCs with more memory and new features, what happens to their old machines? Most unwanted, obsolete computers are destined for landfills, incinerators, or hazardous waste exports; less than 10 percent of unwanted computers are recycled (Computer Take Back Campaign 2003).

HAZARDS OF LANDFILL DUMPING AND INCINERATING E-WASTE

The main concern in regard to dumping e-waste in landfills is that hazardous substances in landfills can leach and contaminate the soil and groundwater. In many city landfills, cathode ray tubes from computers and TVs are the largest source of lead (Motavalli 2001). Older computer monitors (14- and 15-inch size) contain an average of 5 to 8 pounds of lead (Fisher 2000). Lead has toxic effects on humans as well as plants, animals, and microorganisms. The leaching of mercury from landfilled e-waste is also a major concern. Just 1/70th of a teaspoon of mercury can contaminate 20 acres of a lake, making the fish unfit for consumption.

Some computer waste is disposed of through incineration (burning). Incineration of scrap computers generates extremely toxic dioxins that are released in air emissions and leaves high concentrations of toxic heavy metals in the flue gas residues, fly ash, and filter cake.

HAZARDS OF RECYCLING COMPUTERS

Computers and printers can be recycled, and the plastic materials and precious metals, such as copper and gold, can be recovered. Humans used to handle the entire recycling process, but new recycling technologies have largely replaced human labor with machines. In modern recycling facilities, workers wearing safety gear first pull out parts such as batteries and toner cartridges. Printers, desktops, and monitors are then processed by machines that smash them to bits and recover materials that can be spun into precious metals and plastic containers.

The presence of toxic polybrominated flame retardants in computers makes recycling difficult and dangerous. High concentrations of these substances have been found in the blood of workers in recycling plants. In addition, a Swedish study found that when computers, fax machines, or other electronic equipment are recycled, dust containing toxic flame retardants is spread in the air (Silicon Valley Toxics Coalition 2001b).

THE EXPORTATION OF HAZARDOUS E-WASTE

An estimated 50 percent to 80 percent of the United States' electronic waste that is collected for recycling is exported to developing countries where labor costs are cheap and worker safety and environmental regulations are lax compared to U.S. law. One pilot program that collected electronic scrap in San Jose, California, estimated that shipping computer monitors to China for recycling was 10 times cheaper than recycling them in the United States (Silicon Valley Toxics Coalition 2001b).

When investigators visited a waste site in Guiyu, China, they saw men, women, and children wearing little or no protective gear smashing, picking apart, and burning computers, exposing themselves and their surroundings to toxic substances (Silicon Valley Toxics Coalition 2002). The groundwater

continued

continued

Much of the electronic waste from the United States is shipped to India. At this scrap shop in New Delhi, mercury and lead from electronic waste pose environmental and health hazards.

EPA Photo/EPA/Harish Tyagi/© AP/Wide World Photos

near the site is so polluted that drinking water is trucked in from 18 miles away. A nearby river water sample contained 190 times the pollution levels allowed by the World Health Organization.

A 1989 treaty known as the Basel Convention restricts the export of hazardous waste from rich countries to poor countries—even when the waste is destined for recycling. But the United States has not ratified this treaty and has lobbied Asian governments to establish bilateral trade agreements to allow continued exporting of hazardous waste.

Solutions to the Problem of Toxic E-Waste

LEGISLATIVE SOLUTIONS

At least 26 states have considered legislation concerning the disposal of e-waste. In 2000, Massachusetts became the first state to ban dumping of video monitors (computer screens and televisions) (Massachusetts Department of Environmental Protection 2001). Maine and Minnesota also prohibit dumping cathode ray tubes (found in monitors). And in 2003, California passed the Electronic Waste Recycling Act, becoming the first state to pass legislation to require recycling. This act requires retailers to add a $6 to $10 fee to certain electonic products to

holding cyanide-laced waste at a Romanian gold mine broke, dumping 22 million gallons of cyanide-laced waste into the Tisza River, which flowed into Hungary and Serbia. Some have called this event the worst environmental disaster since the 1986 Chernobyl nuclear explosion. Today, in West Papua, Indonesia, a U.S.-owned gold mine dumps 120,000 tons of cynanide-laced waste into local rivers every day (Ayres 2004). Most gold is used to make jewelry, hardly a necessity warranting the environmental degradation that results from gold mining. To make matters worse, about 300,000 tons of waste are generated for every ton of marketable gold—which translates roughly into 3 tons of waste per gold wedding ring (Sampat 2003).

Focus on Technology

cover the costs of collection and recycling programs, and requires recycling programs to use environmentally sound methods. Electronics manufacturers favor regulating e-waste recycling through federal legislation so that all manufacturers abide by the same rules.

THE EUROPEAN UNION'S WEEE DIRECTIVE

In 2002, the European Union (25 European nations as of May 2004) passed legislation on waste from electrical and electronic equipment (the WEEE Directive) that makes producers responsible for taking back their old products and for phasing out the use of certain toxic materials in computer manufacturing; the legislation also encourages cleaner product design and less waste generation. This strategy, known as **Extended Producer Responsibility,** is based on the principle that producers bear a degree of responsibility for all the environmental impacts of their products. This includes impacts arising from choice of materials and from the manufacturing process as well as the use and disposal of products. When manufacturers are financially responsible for waste management of their products, they will have a financial incentive to design those products with less hazardous and more recyclable materials.

INDUSTRY INITIATIVES

A number of industries have taken steps to alleviate the environmental and health hazards of computers. Some computer manufacturers such as Dell and Gateway lease out their products, thereby ensuring that they get them back to further upgrade and lease out again. In 1998 IBM introduced the first computer that uses 100 percent recycled resin in all its major plastic parts. Hewlett-Packard (HP) has developed a safe cleaning method for chips, using carbon dioxide for cleaning as a substitute for hazardous solvents. To encourage recycling of e-waste, HP's Planet Partners program allows people to have their old computers and printers picked up at their door for a fee of $17 to $31 to cover transportation and recycling costs. In exchange, they receive a $50 coupon toward the purchase of an HP product. Toshiba is working on a modular upgradable and customizable computer to reduce product obsolescence. Pressures to eliminate halogenated flame retardants and design products for recycling have led to the use of metal shielding in computer housings and a range of lead-free solders is now available.

As governments and industries continue to develop strategies for reducing environmental hazards associated with computer manufacturing and disposal, consumers can also make a difference by choosing manufacturers who practice "Product Stewardship"—making products that are less toxic, conserve natural resources, and reduce waste. *A Guide to Environmentally Preferable Computer Purchasing* is available from the Northwest Product Stewardship Council at http://www.govlink.org/nwpsc/

Sources: Bowman, Lisa M. 2003 (June 5). "Calif. Senate Advances 'E-waste' Bill." CNET News.com. http://news.com.com/; Computer Take Back Campaign. 2003. *Fourth Annual Computer Report Card.* http://www.computertakeback.com; Fisher, Jim. 2000 (September 18). "Poison PCs." http://www.salon.com/tech/feature/2000/09/18/toxic_pc/; Massachusetts Department of Environmental Protection. 2001. "TV and Computer Reuse and Recycling." http://www.state.ma.us/dep/recycle/crt/crthome.htm; Motavalli, Jim. 2001 (March). "Is There an Afterlife for Your Computer? Grappling with America's Techno-Trash Dilemma." *Environmental Defense* 32(2):6; Northwest Product Stewardship Council (Computer Subcommittee). 2000 (October). *A Guide to Environmentally Preferable Computer Purchasing.* http://www.govlink.org/nwpsc/; Silicon Valley Toxics Coalition. 2002. "Exporting Harm: The High-Tech Trashing of Asia." http://www.svtc.org; Silicon Valley Toxics Coalition. 2001a. "Why Focus on Computers?" http://www.svtc.org/cleancc/focus/htm; Silicon Valley Toxics Coalition. 2001b. *Just Say No to E-Waste: Background Document on Hazards and Waste from Computers.* http://www.svtc.org; Thibodeau, Patrick. 2002. "Handling E-Waste." *Computerworld* 36(47):46.

Environment and Health Problems

Human exposure to pollution and toxic substances in the environment contributes to illness, disability, and death. Next, we present just a few of the many health problems associated with the environment.

Health Effects of Air Pollution. Air pollution contributes to respiratory and cardiovascular diseases. In the United States, about 80 million people are exposed to levels of air pollution that can impair health, and more than 2 percent of all deaths annually in the United States are attributed to air pollution (UNEP 2002). As noted earlier, indoor air pollution from burning wood and biomass for heating and cooking contributes to respiratory illness, lung cancer, blindness, and death.

Chemicals and Carcinogens. When scientists tested journalist Bill Moyers' blood as part of a documentary on the chemical industry, they found traces of 84 of the 150 chemicals they had tested (PBS 2001). Sixty years ago, when Moyers was 6 years old, only one of these chemicals—lead—would have been present in his blood. In the United States, more than 62,000 chemical substances are in commercial use, and 1,500 new chemicals are introduced each year. But complete data on the health and environmental effects are known for only 7 percent of chemicals produced in high volume (UNEP 2002).

Although the health effects of most synthetic chemicals are unknown, more than 200 chemical substances are "known to be human carcinogens" or "reasonably anticipated to be human carcinogens," meaning that they are linked to cancer. In the *10th Report on Carcinogens,* steroidal estrogens (used in estrogen replacement therapy and oral contraceptives) and broad spectrum ultraviolet radiation (from the sun or from artificial sources) were added to the official list of "known" human carcinogens (U.S. Department of Health and Human Services 2002). Many of the chemicals we are exposed to in our daily lives can cause not only cancer but other health problems such as infertility, birth defects, and a number of childhood developmental and learning problems (Fisher 1999; Kaplan & Morris 2000; McGinn 2000). Some chemicals, such as persistent organic pollutants (POPs), accumulate in the food chain and persist in the environment, taking centuries to degrade.

Substances found in common household, personal, and commercial products can result in a variety of temporary acute symptoms such as drowsiness, disorientation, headache, dizziness, nausea, fatigue, shortness of breath, cramps, diarrhea, and irritation of the eyes, nose, throat, and lungs. Long-term exposure can affect the nervous system, reproductive system, liver, kidneys, heart, and blood. Fragrances, which are found in many consumer products, may produce sensory irritation, pulmonary irritation, decreases in expiratory airflow velocity, and possible neurotoxic effects (Fisher 1998). Fragrance products can cause skin sensitivity, rashes, headache, sneezing, watery eyes, sinus problems, nausea, wheezing, shortness of breath, inability to concentrate, dizziness, sore throat, cough, hyperactivity, fatigue, and drowsiness (DesJardins 1997). More and more businesses are voluntarily limiting fragrances in the workplace or banning them altogether to accommodate employees who experience ill effects from them (see also this chapter's *Taking a Stand* feature). This chapter's *Social Problems Research Up Close* feature describes the *Second National Report on Human Exposure to Environmental Chemicals.*

Vulnerability of Children. The World Health Organization estimates that 5,500 children die each day from diseases linked to polluted water, air, and food (Mastny 2003). In the United States, an estimated one in 200 children suffers developmental or neurological deficits due to exposure to toxic substances during pregnancy or after birth (UNEP 2002). Asthma, the number one childhood illness in the United States, is linked to air pollution. Children are more vulnerable than adults to the harmful effects of most pollutants for a number of reasons. For instance, children drink more fluids, eat more food, and inhale more air per unit of body weight than do adults; and crawling and a tendency to put their hands and other things in their mouths provide more opportunities to ingest chemical or heavy metal residues.

Multiple Chemical Sensitivity Disorder. **Multiple chemical sensitivity** (MCS), also known as "environmental illness," is a condition whereby individuals experience adverse reactions when exposed to low levels of chemicals found in everyday substances (vehicle exhaust, fresh paint, housecleaning products, perfume and other fragrances, synthetic building materials, and numerous other petrochemical-based products). Symptoms of MCS include headache, burning eyes, difficulty breathing, stomach distress/nausea, loss of mental concentration, and dizziness. The onset of MCS is often linked to acute exposure to a high level of chemicals or to chronic long-term exposure. Sufferers of MCS often avoid public places and/or wear a protective breathing filter to avoid inhaling the many chemical substances in the environment. Some individuals with MCS build houses made from materials that do not contain chemicals that are typically found in building materials. Although estimates of the prevalence of chemical sensitivity vary, one recent study reported that in the Atlanta, Georgia, metropolitan area, 12.6 percent of a population sample reported an unusual sensitivity to common chemicals, and 3.1 percent had been diagnosed as having MCS or environmental illness (Caress, Steinemann, & Waddick 2002).

Environmental Injustice

Although environmental pollution and degradation and depletion of natural resources affect us all, some groups are more affected than others. **Environmental injustice,** also referred to as **environmental racism,** refers to the tendency for socially and politically marginalized groups to bear the brunt of environmental ills.

Environmental Injustice in the United States. In the United States, polluting industries, industrial and waste facilities, and transportation arteries (that generate vehicle emissions pollution) are often located in minority communities (Bullard 2000; Bullard & Johnson 1997). U.S. communities that are predominantly African-American, Hispanic, or Native American are disproportionately affected by industrial toxins, contaminated air and drinking water, and the location of hazardous waste treatment and storage facilities. One study found, for example, that hog industries—and the associated environmental and health risks associated with hog waste—in eastern North Carolina tend to be located in communities with high black populations, low voter registration, and low incomes (Edwards & Ladd 2000). Another study of 370 communities in Massachusetts found that compared with communities with higher median household incomes and lower percentages of people of color, communities with lower household incomes and higher percentages of people of color contained significantly higher levels of chemical emissions from industrial facilities and had more hazardous waste sites (Leutwyler 2001). Another study of air lead concentrations in 3,111 U.S. counties found that counties with the largest proportion of black youth under age 16 have more than 7 percent more lead in the air than counties with no black youth, and counties with the largest proportion of white youth have nearly 10 percent less lead in the air than counties with the smallest proportion of white youth (Stretesky 2003). These findings help explain racial inequality, as the medical literature concludes that lead exposure is associated with deficits in intellectual and academic functioning. Stretesky (2003) reports that nearly 11 percent of black children suffer from lead poisoning, compared with 2 percent of white children.

Social Problems Research Up Close

Second National Report on Human Exposure to Environmental Chemicals

In 2001, the Centers for Disease Control published a landmark study that reported measures of human exposure to 27 environmental chemicals, 24 of which had never been measured before in a nationally representative sample of the U.S. population. The *Second National Report on Human Exposure to Environmental Chemicals* (Centers for Disease Control and Prevention 2003) is a continuation of the ongoing assessment of the U.S. population's exposure to environmental chemicals. This *Second Report* presents exposure data on the 27 chemicals listed in the first report and 89 new chemicals: 116 environmental chemicals total.

Methods and Sample

The *Second National Report on Human Exposure to Environmental Chemicals* is based on 1999–2000 data from the National Center for Health Statistics National Health and Nutrition Examination Survey (NHANES). The NHANES is a series of surveys designed to collect data on the health and nutritional status of the U.S. population.

The 1999–2000 NHANES selected a representative sample of the noninstitutionalized, civilian U.S. population, with an oversampling of African-Americans, Mexican-Americans, adolescents (aged 12–19 years), older Americans (aged 60 years or older), and pregnant women to produce more reliable estimates for these groups. In 2000, targeted sampling of low-income whites was also included. The sample did not specifically target people who were believed to have high or unusual exposures to environmental chemicals.

Assessing research participants' exposure to environmental chemicals involved measuring the chemicals or their metabolites (breakdown products) in the participants' blood or urine—a method known as *biomonitoring.* Blood was obtained by venipuncture for participants aged 1 year and older, and urine specimens were collected for people aged 6 years and older.

The 116 chemicals measured in the research participants' blood and urine are grouped into the following categories:

1. Metals (including barium, cobalt, lead, mercury, uranium, and others)
2. Cotinine (tracks human exposure to tobacco and tobacco smoke)
3. Organophosphate, organochlorine, and carbamate pesticides
4. Phthalate metabolites (tracks chemicals commonly used in such consumer products as soap, shampoo, hair spray, nail polish, and flexible plastics)
5. Polycyclic aromatic hydrocarbons (PAHs)
6. Dioxins, furans, and polychlorinated biphenyls (PCBs)
7. Phytoestrogens
9. Herbicides
10. Insect repellents and disinfectants

The researchers conducting this study had several objectives, including the following:

- To determine which selected environmental chemicals are getting into the bodies of Americans and at what concentrations
- For chemicals that have a known toxicity level, to determine the prevalence of people in the U.S. population with levels above those toxicity levels
- To establish reference ranges that can be used by physicians and scientists to determine whether a person or group has an unusually high level of exposure
- To assess the effectiveness of public health efforts to reduce the exposure of Americans to specific environmental chemicals
- To track, over time, trends in the levels of exposure of the U.S. population to environmental chemicals
- To determine whether exposure levels are higher among minorities, children, women of childbearing age, or other potentially vulnerable goups

Findings and Conclusions

Selected findings of the *Second National Report on Human Exposure to Environmental Chemicals* include the following:

- *Establishment of reference ranges.* The levels of chemicals found in the blood and urine of the research participants provide reference ranges (also known as background exposure levels) for all 116 of the environmental chemicals for which participants were tested. Reference ranges are extremely helpful to physicians and health researchers because levels above the reference range indicate high exposure to a particular chemical. For example, if a physician was concerned about a patient's exposure to mercury and measured the level of mercury in the patient's urine, the results could be compared with the population reference range in the *Second Report.* A mercury level much higher than those found in the *Report* would indicate that the patient may have had an unusual exposure to mercury worthy of further investigation.

- *Reduced exposure of the U.S. population to environmental tobacco smoke.* Cotinine is a metabolite of nicotine that tracks exposure to environmental tobacco smoke (ETS) among nonsmokers; higher continine levels reflect more exposure to ETS. Results from the *Second Report* show that from 1991 to 1994, cotinine levels in nonsmokers decreased by more than 50 percent for children and adolescents, and 75 percent for adults. These findings reflect a dramatic reduction in exposure of the general U.S. population to environmental tobacco smoke. However, in 1999–2000, cotinine levels in children were more than twice those of adults. Although efforts to reduce environmental tobacco smoke exposure in the 1990s were sucessful, ETS exposure remains a major public health concern.
- *Decline in blood lead levels among children since 1991–1994.* Since 1976, CDC has measured blood lead levels as part of the NHANES surveys. Results in the *First Report* showed a decrease in average blood lead levels for children aged 1–5 years compared with levels in 1991–1994, but still found that 4.4 percent of children had elevated blood lead levels. In the *Second Report* the percentage of children with elevated blood lead levels had decreased to 2.2 percent. This significant decrease reflects the success of public health efforts to decrease the exposure of children to lead. Nevertheless, children at high risk for lead exposure (e.g., those living in homes containing lead-based paint or lead-contaminated dust) remain a major public health concern.
- *Evidence that chemicals in consumer products are absorbed by humans.* This study found that seven phthalates—compounds found in such products as soap, shampoo, hair spray, many types of nail polish, and flexible plastics—are absorbed by humans. Surprisingly, human exposure levels to the two phthalates that are produced in greatest quantity were relatively low, while exposure levels to other phthalates produced in much lower quantities were high. Hence, future research should focus on the health effects of phthalates that are least common in the environment because their levels are much higher in the general population.

As the presence of an environmental chemical in a person's blood or urine does not necessarily mean that the chemical will cause disease, more studies are needed to determine which levels of these chemicals in people result in disease. Research has already documented adverse health effects of lead and environmental tobacco smoke. However, for many environmental chemicals, few studies are available, and more research is needed to assess health risks associated with different blood or urine levels of these chemicals. Future *Reports,* expected to be released every two years, will include more than the 116 chemicals in the *Second Report.* Future *Reports* will include data from studies of groups of people in high exposure situations (e.g., pesticide applicators, people living near hazardous waste sites, people working in lead smelters). Data published in future *Reports* will also address the following questions: (1) Are chemical exposure levels increasing or decreasing over time? (2) Are public health efforts to reduce chemical exposure effective? (3) Do certain groups of people have higher levels of chemical exposure than others?

Source: Centers for Disease Control and Prevention, National Center for Environmental Health. 2003. *Second National Report on Human Exposure to Environmental Chemicals.* Atlanta, GA: Centers for Disease Control and Prevention.

Margie Eugene Richard, a resident of the predominantly African-American community of Norco, Louisiana, and president of Concerned Citizens of Norco, describes what it is like living next to Norco's Shell Oil refinery and chemical plant:

> Norco is situated between Shell Oil Refinery on the east and Shell Chemical plant on the west. The entire town of Norco is only half the size of the oil refineries. Nearly everyone in the community suffers from health problems caused by industry pollution. The air is contaminated with bad odors from carcinogens, and benzene, toluene, sulfuric acid, ammonia, xylene and propylene—run-off and dumping of toxic substances also pollute land and water. . . . My youngest daughter and her son suffer from severe asthma; my mother has breathing problems and must use a breathing machine daily. Many of the residents suffer from sore muscles, cardiovascular diseases, liver, blood and kidney toxicants. Many die prematurely from poor health caused by pollution from toxic chemicals. . . . Daily, we smell foul odors, hear loud noises, and see blazing flares and black smoke that emanates from those foul flares. . . . Norco and many other communities of color across our nation suffer the same ills. (quoted in Bullard 2000, 157)

Environmental Injustice Around the World. Environmental injustice affects marginalized populations around the world, including minority groups, indigenous peoples, and other vulnerable and impoverished communities such as peasants and nomadic tribes (Renner 1996). These groups are often powerless to fight against government and corporate powers that sustain environmentally damaging industries. For example, in the early 1970s, a huge copper mine began operations on the South Pacific island of Bougainville. While profits of the copper mine benefited the central government and foreign investors, the lives of the island's inhabitants were being destroyed. Farming and traditional hunting and gathering suffered as mine pollutants covered vast areas of land, destroying local crops of cocoa and bananas, contaminating rivers and their fish. Indigenous groups in Nigeria, such as the Urhobo, Isoko, Kalabare, and Ogoni, are facing environmental threats caused by oil production operations run by transnational corporations. Oil spills, natural gas fires, and leaks from toxic waste pits have polluted the soil, water, and air and compromised the health of various local tribes. "Formerly lush agricultural land is now covered by oil slicks, and much vegetation and wildlife has been destroyed. Many Ogoni suffer from respiratory diseases and cancer, and birth defects are frequent" (Renner 1996, 57). The environmental injustices experienced by Bougainville and the Ogoni reflect only the tip of the iceberg. Renner (1996) warns that "minority populations and indigenous peoples around the globe are facing massive degradation of their environments that threatens to irreversibly alter, indeed destroy, their ways of life and cultures" (p. 59).

> "People everywhere must come to understand that they depend completely on biodiversity— for food, much medicine, and many other products; for global sustainability and stability; and for spiritual renewal. Acting on this realization will put human beings in their proper place, as only one among millions of kinds of organisms, with responsibility for all our fellow humans and for the future of the planet as a whole."
>
> **Peter Raven**
> **Missouri Botanical Garden**

Threats to Biodiversity

About 1.7 million species of life have been identified on earth, but there may be 10 times that number and perhaps many times more if we include microbial diversity (Chivian & Bernstein 2004). This enormous diversity of life, known as **biodiversity,** provides food, fibers, and fuel; purifies air and fresh water; pollinates crops and vegetation; and makes soils fertile. Many species of plants and animals have been used in biomedical research; about 25 percent of drugs prescribed in the United States include chemical compounds derived from wild species (Tuxill 1998).

In recent decades, we have witnessed mass extinction rates of diverse life forms. On average, one species of plant or animal life becomes extinct every 20

Table 14.1

Threatened and Endangered Species* United States and Worldwide: 2002

Item	Mammals	Birds	Reptiles	Amphibians	Fishes	Snails	Clams	Crustaceans	Insects	Arachnids	Plants
Total Listings	342	273	115	30	126	33	72	21	48	12	749
Endangered species,											
total	316	253	78	20	82	22	64	18	39	12	600
United States	65	78	14	12	71	21	62	18	35	12	99
Threatened species,											
total	26	20	37	10	44	11	8	3	9	—	49
United States	9	14	22	9	44	11	8	3	9	—	47

* An endangered species is one in danger of becoming extinct throughout all or a significant part of its natural range. A threatened species is one likely to become endangered in the foreseeable future.

Source: *Statistical Abstract of the United States 2003,* 123rd ed. 2003. Washington DC: U.S. Bureau of the Census.

minutes (Levin & Levin 2002). Unlike the extinction of the dinosaurs millions of years ago, humans are the primary cause of disappearing species today. Air, water, and land pollution; deforestation; disruption of native habitats; and overexploitation of species for their meat, hides, horns, or medicinal or entertainment value threaten biodiversity and the delicate balance of nature. A poacher, for example, can earn several hundred dollars for a single rhinoceros horn—equivalent to a year's salary in many African countries. That same horn, ground up and used as a medicinal product, can sell for half a million dollars in Asia (Schmidt 2004).

The primary cause of species decline is human-induced habitat destruction (Hunter 2001). Biologists expect that at least half of the roughly 10 million species alive today will become extinct during the next few centuries from habitat loss alone; more could disappear as a result of other causes, including pollution, overharvesting, and global warming (Cincotta & Engelman 2000). Table 14.1 lists the numbers of species worldwide and in the United States that are endangered or threatened.

Environmental Problems and Disappearing Livelihoods

Environmental problems threaten the ability of the earth to sustain its growing population. Agriculture, forestry, and fishing provide 50 percent of all jobs worldwide and 70 percent of jobs in sub-Saharan Africa, East Asia, and the Pacific (World Resources Institute 2000). As croplands become scarce or degraded, as forests shrink, and as marine life dwindles, millions of people who make their living from these natural resources must find alternative livelihoods. In Canada's maritime provinces, collapse of the cod fishing industry in the 1990s from overfishing left 30,000 fishers dependent on government welfare payments and decimated the economies of hundreds of communities (World Resources Institute 2000).

Global estimates suggest that as many as 25 million people may be **environmental refugees**—individuals who have migrated because they can no longer secure a livelihood because of deforestation, desertification, soil erosion, and other environmental problems. As individuals lose their source of income, so do nations. In one-quarter of the world's nations, crops, timber, and fish contribute more to the nation's economy than industrial goods (World Resources Institute 2000).

Social Causes of Environmental Problems

Various structural and cultural factors have contributed to environmental problems. These include population growth, industrialization and economic development, and cultural values and attitudes such as individualism, materialism, and militarism.

Population Growth

Global population exceeds 6.2 billion—more than twice what it was in 1950—and is expected to rise to between 7.9 billion and 10.9 billion by 2050 (Bright 2003). Population growth places increased demands on natural resources and results in increased waste. As Hunter (2001) explains:

> Global population size is inherently connected to land, air, and water environments because each and every individual uses environmental resources and contributes to environmental pollution. While the scale of resource use and the level of wastes produced vary across individuals and across cultural contexts, the fact remains that land, water, and air are necessary for human survival. (p. 12)

However, population growth itself is not as critical as the ways in which populations produce, distribute, and consume goods and services. Recall from Chapter 13 that wealthy, industrialized countries have the slowest rates of population growth, yet it is these countries that consume the most natural resources and contribute most to pollution, global warming, and other environmental problems.

Industrialization and Economic Development

Many of the environmental problems confronting the world, including global warming and the depletion of the ozone layer, are associated with industrialization and economic development. Industrialized countries, for example, consume more energy and natural resources and contribute more pollution to the environment than poor countries. But the relationship between level of economic development and environmental pollution is curvilinear rather than linear. For example, industrial emissions are minimal in regions with low levels of economic development, and are high in the middle-development range as developing countries move through the early stages of industrialization. However, at more advanced industrial stages, industrial emissions ease because heavy polluting manufacturing industries decline and "cleaner" service industries increase, and because rising incomes are associated with a greater demand for environmental quality and cleaner technologies. However, a positive linear correlation has been demonstrated between per capita income and national carbon dioxide emissions (Hunter 2001).

In less-developed countries, environmental problems are largely the result of poverty and the priority of economic survival over environmental concerns. Vajpeyi (1995) explains:

> Policymakers in the Third World are often in conflict with the ever-increasing demands to satisfy basic human needs—clean air, water, adequate food, shelter, education—and

to safeguard the environmental quality. Given the scarce economic and technical resources at their disposal, most of these policymakers have ignored long-range environmental concerns and opted for short-range economic and political gains. (p. 24)

The ways we measure economic development and the "health" of economies also influence environmental outcomes. Two primary measures of economic development and the health of economies are gross domestic product (GDP) and consumer spending. But these measures overlook the social and environmental costs of the production and consumption of goods and services. Until definitions and measurements of "economic development" and "economic health" reflect these costs, the pursuit of economic development will continue to contribute to environmental problems.

Cultural Values and Attitudes

Cultural values and attitudes that contribute to environmental problems include individualism and materialism. Individualism, which is a characteristic of U.S. culture, puts individual interests over collective welfare. Even though recycling is good for our collective environment, many individuals do not recycle because of the personal inconvenience involved in washing and sorting recyclable items. Similarly, individuals often indulge in countless behaviors that provide enjoyment and convenience at the expense of the environment: long showers, use of dishwashing machines, recreational boating, meat eating, and use of air conditioning, to name just a few.

"In America we have two dominant religions: Christianity and accumulation."

Brian Swimme
California Institute of Integral Studies

Materialism, or the emphasis on worldly possessions, also encourages individuals to continually purchase new items and throw away old ones. The media bombard us daily with advertisements that tell us life will be better if we purchase a particular product. After the terrorist attacks of September 11, 2001, President Bush advised Americans that it was their patriotic duty to go to the malls and spend money. Materialism contributes to pollution and environmental degradation by supporting polluting and resource-depleting industries and contributing to waste.

The cultural value of militarism also contributes to environmental degradation. This issue is discussed in detail in Chapter 16.

Strategies for Action: Responding to Environmental Problems

One strategy for alleviating environmental problems, discussed in detail in Chapter 13, is to lower fertility rates and slow population growth. Responses to environmental problems also include energy conservation and innovation, environmental activism, environmental education, and government regulations and legislation. Sustainable economic development and international cooperation and assistance also play important roles in alleviating environmental problems.

"Man is here for only a limited time, and he borrows the natural resources of water, land and air from his children who carry on his cultural heritage to the end of time. . . . One must hand over the stewardship of his natural resources to the future generations in the same condition, if not as close to the one that existed when his generation was entrusted to be the caretaker."

Delano Saluskin
Yakima Indian Nation

Environmental Activism

Individuals who are concerned about environmental problems join (or form) environmental activist groups, which exert pressure on government and private industry to initiate or intensify actions related to environmental protection. Environmentalist groups also design and implement their own projects, and disseminate

The Human Side

Excerpts from an Interview with West Virginia Activist Julia Bonds

Julia Bonds, a grandmother and native of West Virginia, was awarded one of six Goldman Environmental Prizes for her activism in fighting mountaintop-removal mining—a practice in which mountain peaks are literally sliced off so that coal can be extracted from within the mountain. In the following excerpts from an interview published in *Grist Magazine,* Julia Bonds describes the devastating effects of mountaintop-removal mining and her activism experiences.

Grist: Can you describe the effects of mountaintop removal mining on communities and the environment?

Julia Bonds: What I've seen happening in the coalfields is the . . . complete annihilation of the communities and culture of Appalachia. . . . The wonderful and valuable hardwood forests are being destoyed and they will not return for over 600 years, if ever. Our beautiful mountain streams have been devastated. . . . The blasts from the mine damage homes . . . and the air quality where they're blasting and mining is the worst anyone can imagine. . . . The worst devastation comes from the flooding. . . . Once you remove soil and vegetation from the top of a very steep mountain valley, you're going to increase runoff from rain. . . .

Grist: . . . You and your family were the last residents to evacuate from your hometown of Marfork Hollow. What happened to the town?

AP/Wide World Photos

This mountain-top removal is just one of the mining sites on Black Mountain in Kentucky.

information to the public about environmental issues. This chapter's *Human Side* feature contains excerpts from an interview with a grandmother who has become nationally known for her activism against the environmentally devastating practice of mountaintop-removal mining.

In the United States, environmentalist groups date back to 1892 with the establishment of the Sierra Club, followed by the Audubon Society in 1905. Other environmental groups include the National Wildlife Federation, World Wildlife Fund, Environmental Defense Fund, Friends of the Earth, Union of Concerned Scientists, Greenpeace, Environmental Action, Natural Resources Defense Council, World

The Human Side

Julia Bonds: I'm the seventh generation to live in that hollow, and my grandson is the ninth. Massey Coal [Company] moved in there around 1994. Now, I'm used to coal mining—I'm from a coal-mining family—but I was not prepared for what Massey brought down on our heads in Marfork. The reserves they're mining now are not the clean reserves they were mining in the '40s, '50s, and '60s. These reserves create more waste than coal. The air pollution, the coal dust, is unbearable in that little community. My grandson now has asthma, and my home and my neighbors' homes were damaged by coal dust.

. . . The thing that really sticks in my mind is a 6-year-old child, my grandson, standing in a stream full of dead fish and asking, "What's wrong with these fish?" I looked down at the water and screamed. My family, for generations, has enjoyed that stream, but we never went back in the river again. We also witnessed several blackwater spills [of coal waste]. Those are so thick they're like pea soup, with big black chunks in it. I knew people were going to have to drink that crap.

In Marfork, there's a huge earthen dam for coal waste—it's eventually going to be 924 feet tall and will hold 7 billion gallons of waste—that sat three miles above my home. I was sitting out on the front porch with my grandson, and he told me he had picked out an escape route in case the dam failed. I knew in my heart there was really no escape. How do you tell a child that his life is a sacrifice for corporate greed? . . .

Grist: What was the first step you took as an activist?

Julia Bonds: When I saw the fish kill, I called the neighbor that lived above me. . . . The neighbor said, "Here's the Department of Environmental Protection's number, call them." Two weeks later I noticed a flyer on a window that said there was a rally against irresponsible mining. I went to that rally, I went to one meeting, and I never looked back.

Grist: Can you tell me about some of the threats you've experienced because of your work?

Julia Bonds: You really haven't been intimidated until you see a 60-ton coal truck swerve at you on a narrow road, when there's a rock cliff on one side and a 100-foot drop-off on the other. . . . The coal companies pack permit hearings with their men, and they brainwash the men, telling them, "These are the people who are going to take your jobs away." So there's foul language, threats, phone calls. . . . Once I heard someone say at a permit hearing, "If I were these ladies, I'd be afraid to go home tonight."

Grist: What's been your greatest victory so far?

Julia Bonds: I think it's watching or reading when an oppressed Appalachian person stands up with a protest sign or writes a lettter to the editor. When people empower themselves, that's the greatest victory.

Source: Michelle Niihuis. 2003 (April 14). "Coal-Miner's Slaughter." *Grist Magazine.* http://www.gristmagazine.com. Used by permission.

Watch Institute, National Recycling Coalition, World Resources Institute, and Rainforest Alliance, to name a few.

Online Activism. The Internet and e-mail provide important tools for environmental activism. For example, the organization Environmental Defense sends Member Action Alerts by e-mail to more than 650,000 activists, informing them when Congress and other decision makers threaten the health of the environment ("Online Activism Lives Up to Its Promise" 2001). These members can then send e-mails and faxes to Congress, the president, and business leaders. "Our online activists pressed EPA for strict controls on diesel trucks and buses and helped persuade former President Clinton to place one-third of our national forests off limits to road building, saving them from oil exploration and logging" (p. 8).

The Role of Industry in the Environmental Movement. Industries are major contributors to environmental problems and often fight against environmental efforts that threaten their profits. However, some industries are joining the environmental movement for a variety of reasons, including pressure from consumers and environmental groups, the desire to improve their public image, and genuine concern for the environment. For example, in 2003, FedEx, in collaboration with the environmental organization Environmental Defense, introduced a new hybrid electric delivery truck that is 50 percent more fuel-efficient and produces lower emissions (Environmental Defense 2003).

Ecoterrorism. An extreme form of environmental activism is **ecoterrorism,** defined as any crime intended to protect wildlife or the environment that is violent, puts human life at risk, or results in damages of $10,000 or more (Denson 2000). The best-known ecoterrorist groups are the Earth Liberation Front (ELF) and the Animal Liberation Front (ALF). The ALF and the ELF are international underground movements consisting of autonomous individuals and small groups who engage in "direct action" to (1) inflict economic damage on those profiting from the destruction and exploitation of the natural environment, (2) save animals from places of abuse (e.g., laboratories, factory farms, fur farms), and (3) reveal information and educate the public on atrocities committed against the earth and all the species that populate it. In recent years, direct actions of the ALF and ELF have included setting fire to a ski resort at Vail, Colorado, to express opposition to the resort's proposed expansion into the forest habitat of the Canada lynx, setting fire to gas-guzzling Hummers and other sport-utility vehicles (SUVs) at car dealerships, vandalizing SUVs parked on streets and in driveways, setting fire to construction sites (in opposition to the development), and smashing windows at the homes of fur retailers. The FBI estimates that between 1996 and 2002, the ALF and ELF have committed more than 600 criminal acts in the United States, resulting in damages in excess of $43 million (Jarboe 2002). Because the direct actions of the ELF and ALF are illegal, activists work anonymously in small groups, or individually, and do not have any centralized organization or coordination. Although critics of ecoterrorism cite the damage done by such tactics, members and supporters of ecoterrorist groups claim that the real terrorists are corporations that plunder the earth.

Environmental Education

One goal of environmental organizations and activists is to educate the public about environmental issues and the seriousness of environmental problems. Some Americans may underestimate the seriousness of environmental concerns because they lack accurate information about environmental issues. In a national survey of U.S. adults, most (69%) rated themselves as having either "a lot" (10%) or "a fair amount" (59%) of knowledge about environmental issues and problems (National Environmental Education & Training Foundation & Roper Starch Worldwide 1999). Yet, on seven of ten questions asked about emerging environmental issues, more Americans gave incorrect answers than correct ones. The average was 3.2 correct answers out of 10 questions. Even those who rated themselves as having "a lot" of environmental knowledge, as well as those with a college degree, scored only four out of ten items correctly. The authors of the study concluded:

> Increased knowledge is the key to changing attitudes and behaviors on issues critical to our environmental future. If Americans can answer an average of only 3 of 10 simple knowledge questions, there is a clear need to provide environmental information in a form that the American public can easily digest and act upon. (National Environmental Education & Training Foundation & Roper Starch Worldwide 1999, 41)

A main source of information about environmental issues for most Americans is the media. However, because media are owned by corporations and wealthy individuals with corporate ties, unbiased information about environmental impacts of corporate activities may not readily be found in mainstream media channels. Indeed, the public must consider the source in interpreting information about environmental issues. Propaganda by corporations sometimes comes packaged as "environmental education." Hager and Burton (2000) explain: "Production of materials for schools is a growth area for public relations companies around the world. Corporate interests realise the value of getting their spin into the classrooms of future consumers and voters" (p. 107).

Energy Conservation and Innovation

As noted earlier, many of the environmental problems facing the world today are linked to energy use, particularly the use of fossil fuels. Clean, renewable sources of energy include wind power, solar power, geothermal power, ocean thermal power, tidal power, and energy from fuel cells. Renewable resources pose significantly lower social, environmental, and health costs than do conventional energy fuels and technologies.

The fastest developing clean energy is wind power, which tripled between 1998 and 2003 and could supply 10 percent of the world's electricity by 2020

Wind energy is harnessed by turbines such as those pictured in this photo of a wind farm in Alamont Pass, California.

(Flavin 2000; Sawin 2003a). Due to strong commitment to pursuing renewable energy sources, Europe has more than 70 percent of global wind capacity. In terms of megawatts of wind energy installations, Germany ranks first in the world, followed by Spain, the United States, and Denmark (Sawin 2003b). One disadvantage of wind power is that wind turbines have been known to result in bird mortality. However, this problem has been mitigated in recent years through the use of painted blades, slower rotational speeds, and careful placement of wind turbines.

In the United States, leadership to promote clean energy has come not from the federal level, but from states and local governments. More than a dozen states have mandated that a portion of the state's overall electricity consumption come from clean, renewable sources such as solar, wind, and geothermal power. For example, California set a target of 20 percent clean energy use by 2017. Several nations have similar policies, including Japan and the entire European Union (Makower, Pernick, & Wilder 2003).

In addition to clean forms of energy, efforts are being made to develop products that use energy more efficiently. In 1992 the U.S. Environmental Protection Agency introduced the Energy Star label for products that meet EPA energy efficiency standards. Originally applied to personal computers and monitors, the Energy Star program has expanded to include new homes and a variety of electronics, home appliances, and office equipment.

Unfortunately, efforts to raise national mandated fuel-efficiency standards in transportation vehicles failed in the mid-1990s and again in 2002. Nevertheless, several car manufacturers have developed eco-friendly cars powered by nonpolluting electric motors or "hybrid" cars that have gasoline-electric motors. For example, hybrid cars made by Toyota and Honda have low emissions and high fuel economy, getting from 48 to nearly 70 miles per gallon of gas ("Hybrid Cars" 2002–2003).

Modifications in Consumer Behavior

Increasingly, consumers are making choices in their behavior and purchases that reflect concern for the environment. For example, consumers are increasingly buying energy-efficient compact fluorescent lamps (CFLs) (Scholand 2000). CFLs last about 10 times longer than the more commonly used incandescent light bulb and they use 75 percent less electricity. Recycling has also increased. Between 1975 and 1997, the share of paper used that is recycled globally increased from 38 percent to more than 43 percent and is expected to reach 45 percent by 2010 (Abramovitz 2000). A national survey found that 60 percent of Americans say they frequently recycle newspapers, cans, and glass; 89 percent report that they frequently turn off lights and electrical appliances when not in use, 65 percent lower the thermostat in the winter to conserve energy (National Environmental Education & Training Foundation & Roper Starch Worldwide 2001).

From credit card and long-distance telephone companies that donate a percentage of their profits to environmental causes, to socially responsible investment services, consumers are increasingly voting with their dollars to support environmentally responsible practices. However, many Americans continue to disregard the environmental impact of their consumer behavior. One of the more environmentally damaging U.S. consumer trends in recent years is the increased demand for sports utility vehicles (SUVs), minivans, and light trucks which have lower fuel economy and higher emissions of carbon and smog-forming pollutants than passenger cars.

Government Policies, Regulations, and Funding

Worldwide, governments spend about $1 trillion (U.S.) per year in subsidies that allow the prices of fuel, timber, metals, and minerals (and products using these materials) to be much lower than they otherwise would be, encouraging greater consumption (Renner 2004). Thus, governmental policies can contribute to environmental problems. But government policies, regulations, and funding can also play a role in protecting and restoring the environment. At the local, state, and federal levels, governments have implemented policies and regulations affecting the production and use of pollutants as well as the preservation of deserts, forests, reefs, wetlands, and endangered plant and animal species.

Some environmentalists propose that governments use taxes to discourage environmentally damaging practices and products (Brown & Mitchell 1998). In the 1990s, a number of European governments implemented a strategy to reduce consumption of natural resources known as "ecological tax shifting," whereby taxes on environmentally harmful activities and products are increased while taxes on income and labor are decreased. In the European Union, environmental tax revenues—mostly from taxes on gasoline, diesel, fuel, and motor vehicles—more than quadrupled between 1980 and 2001 (Renner 2004).

International Cooperation and Assistance

Global environmental concerns such as global warming, destruction of the ozone layer, and loss of biodiversity call for global solutions forged through international cooperation and assistance. For example, the 1987 Montreal Protocol on Substances that Deplete the Ozone Layer forged an agreement made by 70 nations to curb the production of CFCs (which contribute to ozone depletion and global warming). Under the Montreal Protocol, CFC emissions have dropped nearly 90 percent (French 2000b). The 1992 Earth Summit in Rio de Janeiro brought together heads of state, delegates from more than 170 nations and nongovernmental organizations, and participants to discuss an international agenda for both economic development and the environment. The 1992 Earth Summit resulted in the Rio Declaration—"a nonbinding statement of broad principles to guide environmental policy, vaguely committing its signatories not to damage the environment of other nations by activities within their borders and to acknowledge environmental protection as an integral part of development" (Koenig 1995, 15).

In 1997, delegates from 160 nations met in Kyoto, Japan, and forged the **Kyoto Protocol**—the first international agreement to place legally binding limits on greenhouse gas emissions from developed countries. As of February 2003, 188 nations had ratified the Kyoto Protocol, including the European Union and Japan. The United States, which produces one-quarter of the world's greenhouse gas emissions, withdrew from the Kyoto negotiations in 2001—a major blow to international efforts to thwart global warming.

In another international environmental action, representatives from 122 countries drafted a legally binding agreement to phase out a group of dangerous chemicals known as persistent organic pollutants (POPs) (Cray 2001). However, "even as the number of treaties climbs, the condition of the biosphere continues to deteriorate. . . . The main reason that many environmental treaties have not yet turned around the environmental trends they were designed to address is because the governments that created them permitted only vague commitments and lax enforcement" (French 2000b, 135).

In addition, some countries do not have the technical or economic resources necessary for implementing the requirements of environmental treaties. Wealthy, industrialized countries can help less-developed countries address environmental concerns through economic aid. Because industrialized countries have more economic and technological resources, they bear primary responsibility for leading the nations of the world toward environmental cooperation. Jan (1995) emphasizes the importance of international environmental cooperation and the role of developed countries in this endeavor:

> Advanced countries must be willing to sacrifice their own economic well-being to help improve the environment of the poor, developing countries. Failing to do this will lead to irreparable damage to our global environment. Environmental protection is no longer the affair of any one country. It has become an urgent global issue. Environmental pollution recognizes no national boundaries. No country, especially a poor country, can solve this problem alone. (pp. 82–83)

Sustainable Economic Development

Achieving global cooperation on environmental issues is difficult, in part, because developed countries (primarily in the Northern Hemisphere) have different economic agendas from those of developing countries (primarily in the Southern Hemisphere). The northern agenda emphasizes preserving wealth and affluent lifestyles whereas the southern agenda focuses on overcoming mass poverty and achieving a higher quality of life (Koenig 1995). Southern countries are concerned that northern industrialized countries—having already achieved economic wealth—will impose international environmental policies that restrict the economic growth of developing countries just as they are beginning to industrialize. Global strategies to preserve the environment must address both wasteful lifestyles in some nations and the need to overcome overpopulation and widespread poverty in others.

Development involves more than economic growth; it involves sustainability—the long-term environmental, social, and economic health of societies. **Sustainable development** involves meeting the needs of the present world without endangering the ability of future generations to meet their own needs. "The aim here is for those alive today to meet their own needs without making it impossible for future generations to meet theirs. . . . This in turn calls for an economic structure within which we consume only as much as the natural environment can produce, and make only as much waste as it can absorb" (McMichael, Smith, & Corvalan 2000, 1067).

Understanding Environmental Problems

Environmental problems are linked to a number of interrelated features of social life, including corporate globalization, rapid and dramatic population growth, expanding world industrialization, patterns of excessive consumption, and reliance on fossil fuels for energy. Growing evidence of the irreversible effects of global warming and loss of biodiversity, and increased concerns about the health effects of toxic waste and other forms of pollution suggest that we cannot afford to ignore environmental problems. Many Americans believe in a "technological fix" for the environment—that science and technology will solve environmental problems. Paradoxically, the same environmental problems that have been caused by technological progress may be solved by technological innovations designed to clean up pollution, preserve natural resources and habitats, and provide clean forms of energy. But the direction of technical innovation is largely in the hands of big corporations that place profits over environmental protection. Unless the global community challenges the power of transnational corporations to pursue profits at the expense of environmental and human health, corporate behavior will continue to take a heavy toll on the health of the planet and its inhabitants. As oil has been implicated in political and military conflicts involving the Middle East (see Chapter 16), such conflicts are likely to continue as long as oil plays the lead role in providing the world's energy.

Global cooperation is also vital to resolving environmental concerns but is difficult to achieve because rich and poor countries have different economic development agendas: developing poor countries struggle to survive and provide for the basic needs of their citizens; developed wealthy countries struggle to maintain their wealth and relatively high standard of living. Can both agendas be achieved without further pollution and destruction of the environment? Is sustainable economic development an attainable goal? The answer must be yes. But sustainable development will not occur on its own or as the inevitable outcome of current trends. If it is to happen, we must, collectively, make it happen. The motivation for finding and implementing sustainable economies may come from understanding the consequences if we don't.

Since September 11, 2001, world leaders, the media, and citizens in countries across the globe have been preoccupied with terrorism. Lester Brown, of the Earth Policy Institute, warns against the environmental dangers of such preoccupation:

> Terrorism is certainly a matter of concern, but if it diverts us from the environmental trends that are undermining our future until it is too late to reverse them, Osama Bin Laden and his followers will have achieved their goal of bringing down Western civilization in a way they could not have imagined. (Brown, 2003, 5)

Chapter Review

- **In what ways does corporate globalization pose a threat to environmental protection?**
  The rules of corporate globalization, set out by the World Trade Organization (WTO) and free trade agreements such as NAFTA and the FTAA, provide transnational corporations with privileges to pursue profits, expand markets, use natural resources, and exploit cheap labor in developing countries while, at the same time, weakening the ability of governments to protect natural resources or to implement environmental legislation. Corporate globalization is based on values of commercialism, corporate rights, and "free" trade rather than on values of environmental protection, health and safety, and social justice.

- **How did the terrorist attacks of September 11, 2001, affect public concern for environmental issues in the two years following the attacks?**
  Gallup poll surveys found that in the two years after the terrorist attacks of September 11, 2001, public concern about most environmental problems declined sharply, most likely due to increasing concern about the economy and terrorism over the same period. Perhaps environmental problems look less serious by contrast with the economy and the terrorist threat, or the environment did not receive as much media attention, as media stories focused on terrorism and war and economic issues.

- **Where does most of the world's energy and electricity come from?**
  Most of the world's energy and electricity comes from fossil fuels, which include oil, coal, and natural gas. This is significant because many of the serious environmental problems in the world today including pollution, biodiversity loss, and global warming stem from the use of fossil fuels.

- **What are some examples of common household, personal, and commercial products that contribute to indoor pollution?**
  Some common indoor air pollutants include carpeting, mattresses, drain cleaners, oven cleaners, spot removers, shoe polish, dry-cleaned clothes, paints, varnishes, furniture polish, potpourri, mothballs, fabric softener, caulking compounds, air fresheners, deodorizers, disinfectants, glue, correction fluid, printing ink, carbonless paper, and felt-tip markers.

- **When a 2002 Gallup poll asked Americans how much they worry about various environmental issues, what environmental problem ranked at the top of the list?**
  According to a 2002 Gallup poll, the environmental problem that Americans are most concerned about is pollution of drinking water, rivers, lakes, and resevoirs, as more than half of Americans worried "a great deal" about these issues. This concern is not unwarranted, as the Environmental Protection Agency reports that 40 percent of all U.S. waters are polluted to the extent that they are not safe for fishing or swimming.
- **What is the primary cause of global warming?**
  The prevailing view on what causes global warming is that greenhouse gases—primarily carbon dioxide, methane, and nitrous oxide—accumulate in the atmosphere and act like the glass in a greenhouse, holding heat from the sun close to the earth. The primary greenhouse gas is carbon dioxide, which is released into the atmosphere in the emissions of burning fossil fuels.
- **What is the relationship between level of economic development and environmental pollution?**
  There is a curvilinear relationship between level of economic development and environmental pollution. In regions with low levels of economic development, industrial emissions are minimal, but emissions rise in countries that are in the middle economic development range as they move through the early stages of industrialization. However, at more advanced industrial stages, industrial emissions ease because heavy polluting manufacturing industries decline and "cleaner" service industries increase, and because rising incomes are associated with a greater demand for environmental quality and cleaner technologies.
- **What is the first environmental group established in the United States?**
  The first environmental group in the United States is the Sierra Club, established in 1892.
- **What is the fastest growing source of "clean," renewable energy?**
  Wind power, which tripled between 1998 and 2003, is the fastest developing alternative source of energy. In terms of megawatts of wind energy installations, Germany ranks first in the world, followed by Spain, the United States, and Denmark.

Critical Thinking

1. In Chapters 6 and 8, we discussed hate crimes, noting that hate crime laws impose harsher penalties on the perpetrator of a crime if the motive for that crime was hate or bias. Should motives be considered in imposing penalties on persons who are convicted of acts of ecoterrorism? For example, should a person who sets fire to a business to protest that business's environmentally destructive activities receive a lighter penalty than a person who sets fire to a business for some other reason?
2. Consumers who want to make environmentally friendly purchasing decisions are sometimes faced with difficult choices. For example, suppose one lives in an area where locally grown organic produce is not available. In this case, is it better to purchase (1) organic produce that has been trucked from a distant state (no pesticides, but the transportation contributes to fossil-fuel emissions), or (2) locally grown produce from a farm that uses pesticides?

Key Terms

acid rain
biodiversity
bioinvasion
deforestation
desertification
ecofeminism
ecoterrorism
environmental injustice
environmental racism
environmental refugees
e-waste
Extended Producer Responsibility
global warming
greenhouse gases
greenwashing
Kyoto Protocol
multiple chemical sensitivity
planned obsolescence
sustainable development

Taking a Stand

Should the needs of individuals who react adversely to perfumes and other fragrances be accommodated by banning such products from public schools and workplaces?

Consider the case of 16-year-old Kristian Childers who was sprayed in the face with perfume by a classmate, causing a severe asthma attack that required hospitalization. Kristian's family petitioned the school board to ban students from bringing perfume, cologne, and other spray fragrances to school, but the school board denied the proposal (CNN.com 2002). A school board member explained, "You can't make everyone else's lives miserable just to accommodate one child. . . . If you have a . . . young lady and you tell her after gym class that she can't use deodorant to freshen herself up, you're going to have some problems with her and her parents." Kristian's mother, Doris Childers disagrees: "My child has had to fight all her life just to breathe. . . . She should not have to fear for her life on a daily basis just to go to public school." What position would you take if you were a school board member deciding this issue?

Use Wadsworth's exclusive online resources—InfoTrac College Edition, MicroCase Online, and OVRC—to formulate a position on this topic.

The Wadsworth's Sociology Online Resources and Writing Companion will help you get started. This valuable guide will show you how to use Wadsworth's exclusive online resources when studying social problems. It will also help you to build essential research and writing skills. InfoTrac College Edition, MicroCase Online, OVRC, and an electronic copy of portions of this companion are available at http://sociology.wadsworth.com/mooney_knox_schacht/problems4e, the companion Web site for *Understanding Social Problems,* Fourth Edition.

Media Resources

The Companion Web Site for *Understanding Social Problems,* Fourth Edition

http://sociology.wadsworth.com/mooney_knox_schacht/problems4e

Supplement your review of this chapter by going to the companion Web site to take one of the Tutorial Quizzes, use the flash cards to master key terms, and check out the many other study aids you'll find there. You'll also find special features such as *Wadsworth's Sociology Online Resources and Writing Companion,* GSS Data, and Census 2000 information, data, and resources at your fingertips to help you complete that special project or do some research on your own.

Interactions CD-ROM

Go to the Interactions CD-ROM for *Understanding Social Problems,* Fourth Edition, to access additional interactive learning tools, such as in-depth review materials, corresponding practice quizzes, and other engaging resources and activities to help you study the concepts in this chapter.

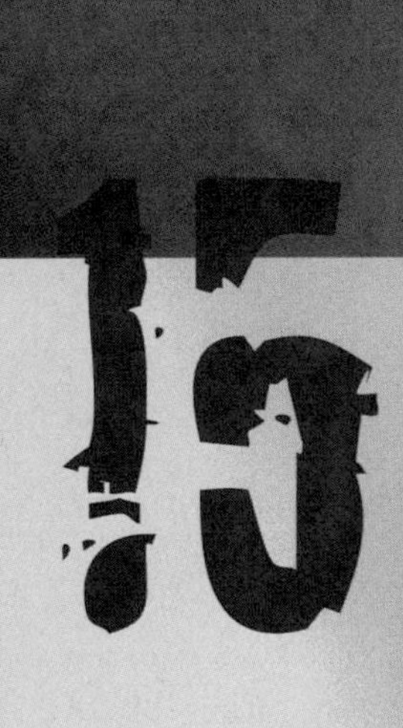

"Most of the consequences of technology that are causing concern at the present time— pollution of the environment, potential damage to the ecology of the planet, occupational and social dislocations, threats to the privacy and political significance of the individual, social and psychological malaise . . . are with us in large measure because it has not been anybody's explicit business to foresee and anticipate them." *Emmanuel Mesthene, Former Director of the Harvard Program on Technology and Society*

Science and Technology

At 19, Erica Robinson lived in a household of "fighters." Her brothers were both athletes—one a star basketball player, the other a wrestler. Her father's strength was summed up in a plaque hanging from the wall of their home: "Don't quit." When Erica found out she was pregnant she was determined to do whatever was best for her baby (Kilen 2003).

But something was terribly wrong. At Erica's six-month checkup the doctor said that the baby wasn't getting enough blood to survive. An emergency Caesarean section was performed and with it E'Maria Robinson-Butler was introduced to the world—10 inches, 11 ounces—one of the tiniest babies ever born. In fact, on any given day in the United States, 1,300 premature babies are born. "Preemies" are more likely than full-term infants to have lasting disabilities such as chronic lung disease, cerebral palsy, mental retardation, and vision and hearing problems. The rate of premature babies has increased 27 percent since 1981 (March of Dimes 2003).

As soon as E'Maria was born, a tube was placed in her windpipe; she was hooked to a respirator, fed intravenously, and given medicine to help her tiny lungs develop. During her five-month stay in the neonatal unit, E'Maria had three surgeries to correct her eyesight, heart, and a hernia. Erica was finally able to bring her little girl home but not without a heart monitor, multiple medications, and an oxygen tank. There were still problems, but the prognosis is good.

Twenty years ago, E'Maria, born at 25 weeks and weighing little more than two sticks of butter, would not have survived. Ten years ago, over half of babies like E'Maria would not have survived. Today, due to medical technology, 85 percent of babies born at 25 weeks' gestation survive. Moreover, the limit of viability is being pushed back one week every five years (Kilen 2003).

"She always keeps her fist balled up," said Grandpa Robinson. "She's a little fighter."

Many of the medical technologies available today seem futuristic. But such technologies, like virtual reality, cloning, and teleportation, are no longer just the stuff of popular sci-fi movies. Virtual reality is now used to train workers in occupations as diverse as medicine, engineering, and professional football. The ability to genetically replicate embryos has sparked worldwide debate over the ethics of reproduction, and California Institute of Technology scientists have transported a ray of light from one location to another. Just as the telephone, the automobile, television, and countless other technological innovations have forever altered social life, so will more recent technologies (see this chapter's *Focus on Technology* feature).

Science and technology go hand in hand. **Science** is the process of discovering, explaining, and predicting natural or social phenomena. A scientific approach to understanding AIDS, for example, might include investigating the molecular structure of the virus, the means by which it is transmitted, and public attitudes about AIDS. **Technology,** as "a form of human cultural activity that applies the principles of science and mechanics to the solution of problems," is intended to accomplish a specific task—in this case, the development of an AIDS vaccine.

Societies differ in their level of technological sophistication and development. In agricultural societies, which emphasize the production of raw materials, the use of tools to accomplish tasks previously done by hand, or **mechanization,** dominates. As societies move toward industrialization and become more concerned with the mass production of goods, automation prevails. **Automation** involves the use of self-operating machines, as in an automated factory where autonomous robots assemble automobiles. Finally, as a society moves toward post-industrialization, it emphasizes service and information professions (Bell 1973). At this stage, technology

Focus on Technology

New Technological Inventions

Here are some of the most innovative inventions of recent years:

- **The Earth Simulator.** A group of Japanese engineers have created a supercomputer capable of 35 trillion calculations a second—five times faster than the next fastest computer. At a cost of $350 million, the computer is capable of forecasting the weather for the entire planet far into the future. It has already calculated global ocean temperatures for the next 50 years.
- **Tendon-activated Hand.** Prosthetics are generally clumsy and not very responsive to fine-tuned actions. However, a new "tendon-activated hand" connects sensors in the artificial hand to remnant tendons providing flexible and natural movement in the prosthetic. One recent accident victim has even returned to playing the piano!
- **Lego Mindstorms.** Researchers at the Massachusetts Institute of Technology have developed computer-controlled building blocks. Described as half robot and half Lego, the custom-designed software helps children write computer programs by piecing together instructions on a computer screen that are transmitted to a robot. The robot then carries out the instructions—whether it be dealing a hand of gin rummy or playing a game of hide-and-seek. Besides being fun, children learn the elementary principles of programming.
- **Talking Lights.** Imagine you are visually impaired and in search of the "Main Street" exit of a building. As you walk through a door the badge you are wearing says, "This is the Main Street exit." Designed by electrical engineers at MIT, talking lights transform ordinary fluorescent light bulbs into "global positioning satellites" that can be used to guide people around malls, senior centers, hospitals, and the like. The lights simply emit information to a decoder that then translates it to verbal messages. The unit is wireless, inexpensive, and consumes no more energy than a regular light bulb.
- **Recodable Lock.** One of the biggest concerns of the twenty-first century is computer security—Internet privacy, virus-free computer environments, secure e-mail, and so on. Although software security continues to be a multibillion dollar industry, two researchers may have discovered a way to secure a computer from the inside out. This new microscopic lock is almost impossible to crack and may do to electronic snoopers what software programs have failed to do—put a lock on cyberspace.
- **Microsystems *Jini.*** All of us have spent hours poring over the manuals of our latest high-tech purchases trying to figure out the operating instructions of each new gadget. Jini, however, is a new software package that once installed allows your computer to talk to other "intelligent" appliances installed into the Jini system. From the Jini web page simply click on an icon to warm your coffee, retrieve pictures from a digital camera, or toast a Pop Tart.
- **Teeth Telephone.** Two British developers have created a phone tooth. The device is a miniature telephone that is embedded in a tooth, usually a molar, where phone calls are received. The signals are then transmitted, through vibrations, from the tooth, to the skull, and then to the inner ear. Unfortunately, at this time, the talking tooth is really a listening tooth—there is no way to talk to the caller.
- **Virtual Touch.** Using a computer is a multisensory experience—bright lights, moving images, and real life audio, but until recently, no tactile stimuli. Researchers at MIT have now created what they call a haptic interface, that is, a joystick, which gives the computer user a sense of touch. The advantages of such an innovation are unlimited, allowing doctors to actually feel tissue during cybersurgery or engineers to experience the goodness of fit of parts just designed.
- **Breaking Through.** That all too familiar sound of a jackhammer may be a thing of the past. In trying to design a quiet way to break up concrete, researchers have developed RAPTOR, a lightweight "gun" that fires penny nails at speeds of 5,000 feet per second. When the nails are fired into the concrete in consecutive lines, stress fractures occur and even the thickest of concrete begins to crumble. Because RAPTOR can be fitted with a silencer, the firing of nails not only breaks up the concrete, it does so quietly and with much less effort on the part of the operator.

Sources: Joseph D'Agnese. 2000. "The 11th Annual Discover Awards." *Discover* 21(7). http://www.discover.com/jul_00featawards.html; Maryanne Buechner, Lev Grossman, and Anita Hamilton. 2002. "The Coolest Invention of 2002." *Time,* November 18, 73–81; "1999 Emerging Technology Finalist." http://www.discover.com/awards/ awards_emerging.html

shifts toward **cybernation,** whereby machines control machines—making production decisions, programming robots, and monitoring assembly performance.

What are the effects of science and technology on humans and their social world? How do science and technology help to remedy social problems and how do they contribute to social problems? Is technology, as Postman (1992) suggests, both a friend and a foe to humankind? This chapter addresses each of these questions.

The Global Context: The Technological Revolution

Less than 50 years ago, traveling across state lines was an arduous task, a long distance phone call was a memorable event, and mail carriers brought belated news of friends and relatives from far away. Today, travelers journey between continents in a matter of hours, and for many, e-mail, faxes, video-conferencing, and electronic fund transfers have replaced conventional means of communication.

The world is a much smaller place than it used to be and will become even smaller as the technological revolution continues. The Internet is projected to have over 945 million users in over 100 countries by the year 2004 (CyberAtlas 2003). Americans constitute the largest share of Internet users (166 million), followed by the Japanese (56 million), Chinese (46 million), Germans (32 million), Italians (19 million), Russians (18 million), and Canadians (17 million). English speakers comprise the largest language group online (37%), but non-English speakers constitute the fastest growing group on the Internet. Although 87 percent of all Internet users live in industrialized countries, there is some movement toward the Internet's becoming a truly global medium as Africans and Latin Americans increasingly "get online" (Sampat 2000; World Employment Report 2001). Table 15.1 displays the Internet activities of the average global user.

The movement toward globalization of technology is, of course, not limited to the use and expansion of the Internet. The world robot market, and America's share of it, continues to expand; Latin America's computer industry is dominated by

Table 15.1
Global Internet Use: August, 2003

Average number of sessions per month	22
Average number of unique domains visited	54
Average pages viewed per month	899
Average pages viewed per surfing session	41
Average time online per month	11 hours, 50 minutes
Average time spent during surfing session	32 minutes
Average duration of a page viewed	47 seconds
Average online population	253,054,814
Current Internet universe	416,339,888

Source: Nielsen/NetRatings. http://cyberatlas.internet.com/big—picture/traffic-patterns

© Bettmann/CORBIS

© Sonda Dawes/The Image Works

The world was made a smaller place in the mid- to late-1800s by the Pony Express. Today, the Internet provides a forum for millions of users worldwide who shop, surf, e-mail, and bank "the net." Globally, the number of Internet users is predicted to grow dramatically over the next several decades.

Texas-based manufacturer Compaq; Microsoft's Internet platform and support products are sold overseas; scientists collect skin and blood samples from remote islanders for genetic research; and a global treaty regulating trade of genetically altered products has been signed by more than a hundred nations.

To achieve such scientific and technological innovations, sometimes called research and development (R&D), countries need material and economic resources. Research entails the pursuit of knowledge; development refers to the production of materials, systems, processes, or devices directed toward the solution of practical problems. In 2002, the United States spent over $292 billion dollars on R&D. As in most other countries, U.S. funding sources were primarily from four sectors: private industry (73%), the federal government (9.9%), colleges and universities (12.9%), and other nonprofit organizations such as research institutes (4%) (NSF 2003).

Scientific discoveries and technological developments also require the support of a country's citizens and political leaders. For example, although abortion has been technically possible for years, millions of the world's citizen's live in countries where abortion is either prohibited or permitted only when the life of the mother is in danger. Thus the degree to which science and technology are considered good or bad, desirable or undesirable, is, as social constructionists argue, a function of time and place.

Postmodernism and the Technological Fix

Many Americans believe that social problems can be resolved through a **technological fix** (Weinberg 1966) rather than through social engineering. For example, a social engineer might approach the problem of water shortages by persuading people to change their lifestyle: use less water, take shorter showers, and wear clothes more than once before washing. A technologist would avoid the challenge of changing people's habits and motivations and instead concentrate on the development of new technologies that would increase the water supply. Social problems may be tackled through both social engineering and a technological fix. In recent years, for example, social engineering efforts to reduce drunk driving have included imposing stiffer penalties for drunk driving and disseminating public service announcements such as "Friends Don't Let Friends Drive Drunk." An example of a technological fix for the same problem is the development of car air bags, which reduce injuries and deaths resulting from car accidents.

Not all individuals, however, agree that science and technology are good for society. **Postmodernism,** an emerging worldview, holds that rational thinking and the scientific perspective have fallen short in providing the "truths" they were once presumed to hold. During the industrial era, science, rationality, and technological innovations were thought to hold the promises of a better, safer, and more humane world. Today, postmodernists question the validity of the scientific enterprise, often pointing to the unforeseen and unwanted consequences of resulting technologies. Automobiles, for example, began to be mass produced in the 1930s in response to consumer demands. But the proliferation of automobiles also led to increased air pollution and the deterioration of cities as suburbs developed, and today, traffic fatalities are the number one cause of accident-related deaths.

"I've got gigabytes. I've got megabytes. I'm voicemailed. I'm e-mailed. I surf the Net. I'm on the Web. I am a Cyber-Man. So how come I feel so out of touch?"

Volkswagen television commercial

Sociological Theories of Science and Technology

Each of the three major sociological frameworks helps us to better understand the nature of science and technology in society.

Structural-Functionalist Perspective

Functionalists view science and technology as emerging in response to societal needs—that "[science] was born indicates that society needed it" (Durkheim [1925] 1973). As societies become more complex and heterogeneous, finding a common and agreed-upon knowledge base becomes more difficult. Science fulfills the need for an assumed objective measure of "truth" and provides a basis for making intelligent and rational decisions. In this regard, science and the resulting technologies are functional for society.

If society changes too rapidly as a result of science and technology, however, problems may emerge. When the material part of culture (i.e., its physical elements) changes at a faster rate than the nonmaterial (i.e., its beliefs and values) a **cultural lag** may develop (Ogburn 1957). For example, the typewriter, the conveyor belt, and the computer expanded opportunities for women to work outside the home. With the potential for economic independence, women were able to remain single or to leave unsatisfactory relationships and/or establish careers. But although new technologies have created new opportunities for women, beliefs about women's roles, expectations of female behavior, and values concerning equality, marriage, and divorce have lagged behind.

Robert Merton (1973), a functionalist and founder of the subdiscipline sociology of science, also argued that scientific discoveries or technological innovations may be dysfunctional for society and create instability in the social system. For example, the development of time-saving machines increases production, but it also displaces workers and contributes to higher rates of employee alienation. Defective technology can have disastrous effects on society. In 1994, a defective Pentium chip was discovered to exist in over 2 million computers in aerospace, medical, scientific, and financial institutions, as well as schools and government agencies. Replacing the defective chip was a massive undertaking but was necessary to avoid thousands of inaccurate computations and organizational catastrophe.

© CORBIS

Motivated by profit, private industry spends more money on research and development than the federal government does.

Conflict Perspective

Conflict theorists, in general, argue that science and technology benefit a select few. For some conflict theorists, technological advances occur primarily as a response to capitalist needs for increased efficiency and productivity and thus are motivated by profit. As McDermott (1993) notes, most decisions to increase technology are made by "the immediate practitioners of technology, their managerial cronies, and for the profits accruing to their corporations" (p. 93). In the United States, private industry spends more money on research and development than the federal government does. The Dalkon Shield and silicone breast implants are examples of technological advances that promised millions of dollars in profits for their developers. However, the rush to market took precedence over thorough testing of the products' safety. Subsequent lawsuits filed by consumers who argued that both products had compromised the physical well-being of women resulted in large damage awards for the plaintiffs.

Science and technology also further the interests of dominant groups to the detriment of others. The need for scientific research on AIDS was evident in the early 1980s, but the required large-scale funding was not made available as long as the virus was thought to be specific to homosexuals and intravenous drug users. Only when the virus became a threat to mainstream Americans were millions of dollars allocated to AIDS research. Hence, conflict theorists argue that granting agencies act as gatekeepers to scientific discoveries and technological innovations. These agencies are influenced by powerful interest groups and the marketability of the product rather than by the needs of society.

Finally, conflict theorists as well as feminists argue that technology is an extension of the patriarchal nature of society that promotes the interests of men and ignores the needs and interests of women. As in other aspects of life, women play a subordinate role in reference to technology in terms of both its creation and its use. For example, washing machines, although time-saving devices, disrupted the communal telling of stories and the resulting friendships among women who gathered together to do their chores. Bush (1993) observes that in a "society characterized by a sex-role division of labor, any tool or technique . . . will have dramatically different effects on men than on women" (p. 204).

Symbolic Interactionist Perspective

Knowledge is relative. It changes over time, over circumstances, and between societies. We no longer believe that the world is flat or that the earth is the center of the universe, but such beliefs once determined behavior as individuals responded to what they thought to be true. The scientific process is a social process in that "truths"—socially constructed truths—result from the interactions between scientists, researchers, and the lay public.

Kuhn (1973) argues that the process of scientific discovery begins with assumptions about a particular phenomenon (e.g., the world is flat). Because unanswered questions always remain about a topic (e.g., why don't the oceans drain?),

Social Problems Research Up Close

The Social Construction of the Hacking Community

Cyberstalking, pornography on the Internet, identity theft are crimes almost unheard of before the computer revolution and the enormous growth of the Internet. One such "high-tech" crime, computer hacking, ranges from childish pranks to deadly viruses that shut down corporations. Below Jordan and Taylor (1998) enter the world of hackers, analyzing the nature of this illegal activity, hackers' motivations, and the social construction of the "hacking community."

Sample and Methods

Jordan and Taylor (1998) researched computer hackers and the hacking community through 80 semistructured interviews, 200 questionnaires, and an examination of existing data on the topic. As is often the case in crime, illicit drug use, and other similarly difficult research areas, a random sample of hackers was not possible. **Snowball sampling** is often the preferred method in these cases; that is, one respondent refers the researcher to another respondent, who then refers the researcher to another respondent, and so forth. Through their analysis, the authors lend insight into this increasingly costly social problem and the symbolic interactionist notion of "social construction"—in this case, of an online community.

Findings and Conclusions

Computer hacking, or "unauthorized computer intrusion," is an increasingly serious problem, particularly in a society dominated by information technologies. Unlawful entry into computer networks or databases can be achieved by several means including (1) guessing someone's password, (2) tricking a computer about the identity of another computer (called "IP spoofing"), or (3) "social engineering," a slang term referring to getting important access information by stealing documents, looking over someone's shoulder, going through their garbage, and so on.

Hacking carries with it certain norms and values for, according to Jordan and Taylor, the hacking community can be thought of as a culture within a culture. The two researchers identified six elements of this socially constructed community:

- *Technology.* The core of the hacking community is the technology that allows it to occur. As one professor interviewed stated, the young today have "lived with computers virtually from the cradle, and therefore have no trace of fear, not even a trace of reverence."
- *Secrecy.* The hacking community must, on the one hand, commit secret acts because their "hacks" are illegal. On the other hand, much of the motivation for hacking requires publicity to achieve the notoriety often sought. Further, hacking is often a group activity that bonds members together. As one hacker stated, hacking "can give you a real kick some time. But it can give you a lot more satisfaction and recognition if you share your experiences with others."
- *Anonymity.* Whereas secrecy refers to the hacking act, anonymity refers to the importance of the hacker's identity remaining unknown. Thus, for example, hackers and hacking groups take on names such as Legion of Doom, the Inner Circle I, Mercury, and Kaos, Inc.
- *Membership Fluidity.* Membership is fluid rather than static, often characterized by high turnover rates, in part, as a response to law enforcement pressures. Unlike more structured organizations, there are no formal rules or regulations.
- *Male Dominance.* Hacking is defined as a male activity; consequently, there are few female hackers. Jordan and Taylor also note, after recounting an incident of sexual harassment, that "the collective identity hackers share and construct . . . is in part misogynist" (p. 768).
- *Motivation.* Contributing to the articulation of the hacking communities' boundaries are the agreed-upon definitions of acceptable hacking motivations, including (1) addiction to computers, (2) curiosity, (3) excitement, (4) power, (5) acceptance and recognition, and (6) community service through the identification of security risks.

Finally, Jordan and Taylor (1998, 770) note that hackers also maintain group boundaries by distinguishing between their community and other social groups, including "an antagonistic bond to the computer security industry (CSI)." Ironically, hackers admit a desire to be hired by the CSI, which would not only legitimize their activities but give them a steady income as well.

The authors conclude that the general fear of computers and of those who understand them underlies the common although inaccurate portrayal of hackers as pathological, obsessed computer "geeks." When journalist Jon Littman asked hacker Kevin Mitnick if he was being demonized because of increased dependence on and fear of information technologies, Mitnick replied, "Yeah. . . . That's why they're instilling fear of the unknown. That's why they're scared of me. Not because of what I've done, but because I have the capability to wreak havoc" (Jordan & Taylor 1998, 776).

Source: Jordan, Tim, and Paul Taylor. 1998. "A Sociology of Hackers." *The Sociological Review,* November, 757–778.

science works to fill these gaps. When new information suggests that the initial assumptions were incorrect (e.g., the world is not flat), a new set of assumptions or framework emerges to replace the old one (e.g., the world is round). It then becomes the dominant belief or paradigm.

Symbolic interactionists emphasize the importance of this process and the impact social forces have on it. Conrad (1997), for example, describes the media's contribution in framing societal beliefs that alcoholism, homosexuality, and racial inequality are genetically determined. Technological innovations are also affected by social forces, and their success is, in part, dependent upon the social meaning assigned to any particular product. As social constructionists argue, reality is socially constructed by individuals as they interpret the social world around them, including the meaning assigned to various technologies. If claims makers can successfully define a product as impractical, cumbersome, inefficient, or immoral, it is unlikely to gain public acceptance. Such is the case with RU486, an oral contraceptive that is widely used in France, Great Britain, and China, but whose availability, although legal in the United States, is opposed by a majority of Americans (Gallup 2000; Gottlieb 2000).

Not only are technological innovations subject to social meaning, but who becomes involved in what aspects of science and technology is also socially defined. Men, for example, outnumber women three to one in earning computer science degrees, and although women make up 46 percent of the general workforce, they make up just 29 percent of the information technology workforce (*Statistical Abstract* 2002; CAW 2003). Societal definitions of men as being rational, mathematical, and scientifically minded and having greater mechanical aptitude than women are, in part, responsible for these differences. This chapter's *Social Problems Research Up Close* feature highlights one of the consequences of the masculinization of technology, as well as the ways in which computer hacker identities and communities are socially constructed.

Technology and the Transformation of Society

A number of modern technologies are considerably more sophisticated than technological innovations of the past. Nevertheless, older technologies have influenced the social world as profoundly as the most mind-boggling modern inventions. Postman (1992) describes how the clock—a relatively simple innovation that is taken for granted in today's world—profoundly influenced not only the workplace but the larger economic institution:

> The clock had its origin in the Benedictine monasteries of the twelfth and thirteenth centuries. The impetus behind the invention was to provide a more or less precise regularity to the routines of the monasteries, which required, among other things, seven periods of devotion during the course of the day. The bells of the monastery were to be rung to signal the canonical hours; the mechanical clock was the technology that could provide precision to these rituals of devotion. . . . What the monks did not foresee was that the clock is a means not merely of keeping track of the hours but also of synchronizing and controlling the actions of men. And thus, by the middle of the fourteenth century, the clock had moved outside the walls of the monastery, and brought a new and precise regularity to the life of the workman and the merchant. . . . In short, without the clock, capitalism would have been quite impossible. The paradox . . . [is] that the clock was invented by men who wanted to devote themselves more rigorously to God; it ended as the technology of greatest use to men who wished to devote themselves to the accumulation of money. (pp. 14–15)

Technology has far-reaching effects not only on the economy but on every aspect of social life. The following sections discuss societal transformations resulting from various modern technologies, including workplace technology, computers, the information highway, and science and biotechnology.

Technology in the Workplace

All workplaces, from doctors' offices to factories and from supermarkets to real estate agencies, have felt the impact of technology. The Office of Technology Assessment of the U.S. Congress estimates that over 7 million U.S. workers are under some type of computer surveillance, which lessens the need for supervisors and makes control by employers easier. Further, technology can make workers more accountable by gathering information about their performance. Through such time-saving devices as personal digital assistants and battery powered store-shelf labels, technology can enhance workers' efficiency. Technology is also changing the location of work. Over 28 million employees **telework**—that is, they complete all or part of their work away from the traditional workplace. By 2003, "22 percent of the labor force will opt for 'flextime, flexplace' arrangements, which allow them to work at home, communicating with the office via computer networks" (Cetron & Davies 2003, 5).

Information technologies are also changing the nature of work. Lilly Pharmaceutical employees communicate via their own "intranet" on which all work-related notices are posted. Federal Express not only created a FedEx network for their 30,000 employees but also allowed customers to enter their package-tracking database, saving the Memphis-based company $2 million a year. The importance of such technologies is empirically documented. In a survey of British office workers, 25 percent reported that information technologies were more important than office managers (Hayday 2003).

Robotic technology, sometimes called computer-aided manufacturing (CAM), has also revolutionized work. Ninety percent of robots work in factories and over

© Adam Lubroth/STONE

Automation means that machines can now perform the labor originally provided by human workers, such as the robots that perform tasks on automobile assembly lines.

half are used in heavy industry such as automobile manufacturing (Robotics 2003). An employer's decision to use robotics depends on direct (e.g., initial investment) and indirect (e.g., unemployment compensation) costs, the feasibility and availability of robots to perform the desired tasks, and the resulting increased rate of productivity. Use of robotics may also depend on whether labor unions resist the replacement of workers by machines.

The Computer Revolution

Early computers were much larger than the small machines we have today and were thought to have only esoteric uses among members of the scientific and military communities. In 1951, only about half a dozen computers existed (Ceruzzi 1993). The development of the silicon chip and sophisticated microelectronic technology allowed tens of thousands of components to be imprinted on a single chip smaller than a dime. The silicon chip led to the development of laptop computers, mini-television sets, cellular phones, electronic keyboards, and singing birthday cards. The silicon chip also made computers affordable. Although the first personal computer (PC) was developed only 20 years ago, today over 56 percent of American homes have a computer (*Statistical Abstract* 2002). Ownership varies by a number of factors (see Table 15.2) but is greatest among those with higher incomes (*Statistical Abstracts* 2002).

Seventy-two million employees use a computer at work—over half the labor force (BLS 2002). The percentage of workers who use computers varies by occupation, with 80 percent of managers and professionals using a computer but only 29 percent of, for example, laborers using a computer. Females, whites, and the more educated have higher rates of computer use. The most common computer activity at work is accessing the Internet or e-mail, followed by word processing, working with spreadsheets or databases, and accessing/updating calendars or schedules.

Not surprisingly, computer education has also mushroomed in the last two decades (see Chapter 12). In 1971, 2,388 U.S. college students earned a bachelor's degree in computer and information sciences; by 2000 that number had increased to 36,195 (*Statistical Abstract 2002*). Universities are moving toward requiring their students to have laptop computers, and college and university spending on hardware and software is at an all-time high.

> "I think there is a world market for maybe five computers."
>
> **Thomas Watson**
> **Chairman of IBM, speaking in 1943**

Computers are also big business, and the United States is one of the most successful producers of computer technology in the world, boasting several of the top companies—IBM, Hewlett-Packard, Dell, Sun Microsystems, and Apple (Plunkett Research 2003). Retail sales of computers exceed $13 billion annually (*Statistical Abstract 2002*). Americans spend over $400 million on educational software alone, and by the year 2006 spending on home computers is predicted to grow tenfold as consumers increasingly define PCs as a necessity rather than a luxury (Tanaka 2000; Klein 1998).

Computer software is also big business and in some states too big. In 2000, a federal judge found that Microsoft Corporation was in violation of antitrust laws, which prohibit unreasonable restraint of trade. At issue were Microsoft's Windows operating system and the vast array of Windows-based applications (e.g., spreadsheets, word processors, tax software)—applications that work *only* with Windows. The court held that the 70,000 programs written exclusively for Windows made "competing against Microsoft impractical" (Markoff 2000). In an agreement reached between the U.S. Department of Justice and Microsoft Corporation, Microsoft is to be divided into two companies—an operating systems company and

Table 15.2
Computer Use from Any Location: 1997 and 2001

	Percentage of Population	
	1997	**2001**
Sex:		
Male	53.8	65.5
Female	53.3	65.8
Age:		
3 to 8 years old	59.0	71.0
9 to 17 years old	85.1	92.6
18 to 24 years old	58.2	71.3
25 to 49 years old	57.7	70.2
50 years and over	27.6	42.5
Race/Ethnicity:		
White, non-Hispanic	57.5	70.0
Black, non-Hispanic	43.6	55.7
Asian-American and Pacific Islander	57.5	71.2
Hispanic	38.0	48.8
Family Income:		
Less than $15,000	29.8	37.3
$15,000 to $24,999	37.4	46.8
$25,000 to $34,999	49.3	57.7
$35,000 to $49,999	60.4	70.0
$50,000 to $74,999	71.7	79.4
$75,000 and over	80.8	88.0
Educational Attainment:		
Less than high school	7.9	17.0
High school diploma/GED	33.5	47.3
Some college	57.8	69.5
Bachelor's degree	74.3	84.9
More than a BA degree	79.1	86.9

Source: *Statistical Abstract of the United States*. 2002. 122nd ed., Table 1134. U.S. Bureau of the Census, Washington, DC: U.S. Government Printing Office.

an applications company. However, nine states have rejected the government's settlement and are continuing their anti-trust suits against Microsoft (AP 2003a).

The Internet

Information technology, or **IT** for short, refers to any technology that carries information. Most information technologies were developed within a 100-year span: photography and telegraphy (1830s), rotary power printing (1840s), the typewriter (1860s), transatlantic cable (1866), telephone (1876), motion pictures (1894), wireless telegraphy (1895), magnetic tape recording (1899), radio (1906), and television

Figure 15.1
U.S. Households with Internet Access, by Race/Ethnicity: 1998 and 2000

Source: National Science Foundation, Science and Engineering Indicators, 2002.

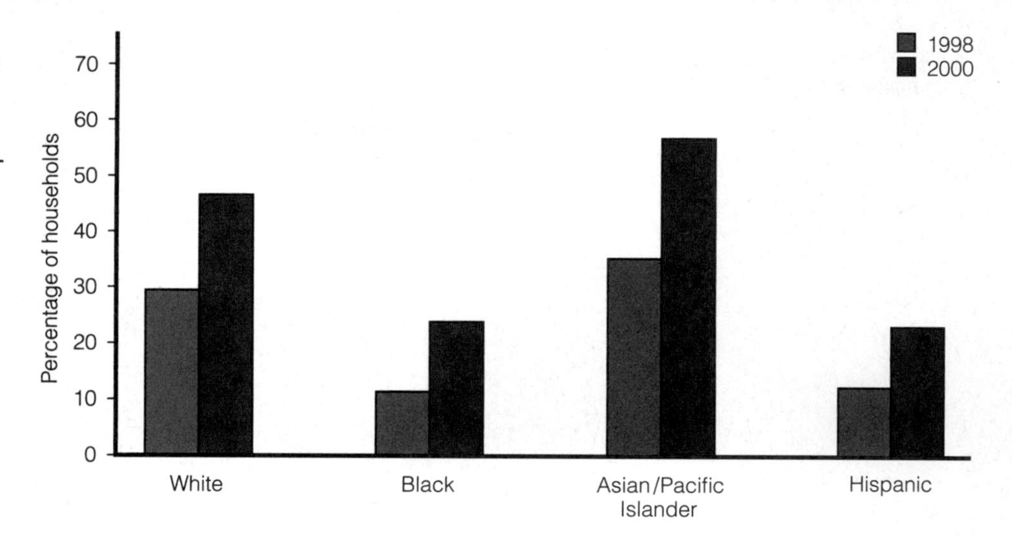

(1923) (Beniger 1993). The concept of an "information society" dates back to the 1950s when an economist identified a work sector he called "the production and distribution of knowledge." In 1958, 31 percent of the labor force was employed in this sector—today more than 50 percent is. When this figure is combined with those in service occupations, more than 78 percent of the labor force is involved in the information society.

The development of a national information infrastructure was outlined in the Communications Act of 1994. An information infrastructure performs three functions (Kahin 1993). First, it carries information, just as a transportation system carries passengers. Second, it collects data in digital form that can be understood and used by people. Finally, it permits people to communicate with one another by sharing, monitoring, and exchanging information based on common standards and networks. In short, an information infrastructure facilitates telecommunications, knowledge, and community integration.

The **Internet** is an international information infrastructure—a network of networks—available through universities, research institutes, government agencies, and businesses. Today over half of all Americans use the Internet. U.S. users are likely to be male, white, and non-Hispanic (see Figure 15.1), and between the ages of 18 and 29. Internet use is higher in the suburbs than in rural or urban areas (Internet Statistics 2003).

Despite dramatic growth, there is some evidence that Internet use in the United States is slowing down. Researchers from the Pew Internet and American Life Project found that 42 percent of those surveyed were nonusers (Pew 2003). The majority of nonusers (56%) said they did not intend to go online, citing cost, difficulty of use, and fear (e.g., computer fraud) as reasons for remaining off line. Nonusers were also less likely to use other technologies (e.g., cell phones) and less likely to be socially content. "Those who are socially content—who trust others, have lots of people to draw on for support, and believe that generally others are fair—are more likely to be wired than those who are less content" (Pew 2003, 1).

E-commerce, or the buying and selling of goods and services over the Internet, continues to grow. Online spending for 8- to 21-year-olds alone is projected to exceed $13 billion by 2006 (Greenspan 2003a). The largest online retailer in the United States is the federal government with annual sales in excess of $3.6 billion

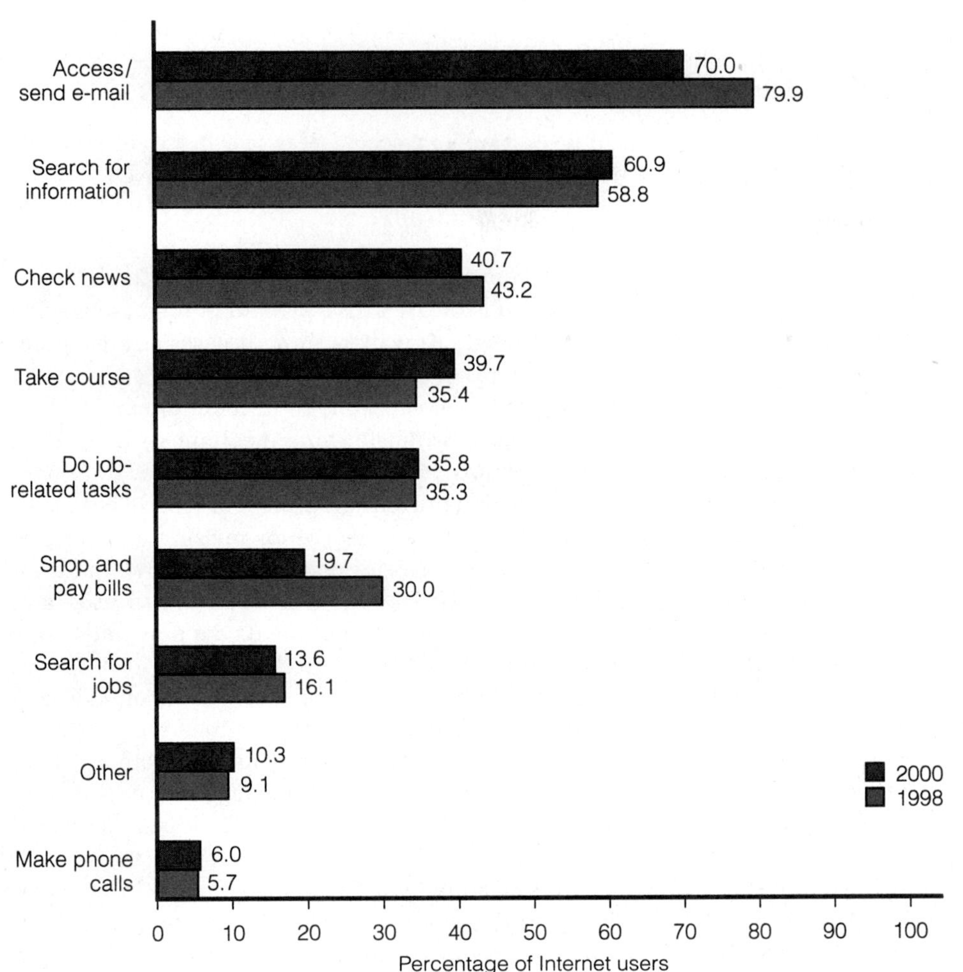

Figure 15.2
Online Activities: 1998 and 2000

Source: National Science Foundation, Science and Engineering Indicators, 2002.

in 2000 (Hasson & Browning 2002). Amazon.com's reported net sales for the same year were $2.8 billion. The government has over 164 sites that sell property or products to the public. The most profitable site is the Treasury Department's "Treasury Direct" page, which sells savings bonds, T-bills, and the like. Figure 15.2 displays the most common online activities. Interestingly, online consumerism may actually help the environment. A team of energy experts reports that given present trends, by 2007 "e-commerce could prevent the annual release of 35 million tons of greenhouse gases by reducing the need for up to 3 billion square feet of energy-consuming office buildings and malls in the United States" (Sampat 2000, 94).

Science and Biotechnology

While recent computer innovations and the establishment of an information highway have led to significant cultural and structural changes, science and its resulting biotechnologies have produced not only dramatic changes but also hotly contested issues. Here we will look at some of the issues raised by developments in genetics and reproductive technology.

Genetics. Molecular biology has led to a greater understanding of the genetic material found in all cells—DNA (deoxyribonucleic acid)—and with it the ability for **genetic screening.**

> If you could uncoil a strip of DNA, it would reach 6 feet in length, a code written in words of four chemical letters: A, T, G and C. Fold it back up, and it shrinks to trillionths of an inch, small enough to fit in any one of our 100 trillion cells, carrying the recipe for how to create human beings from scratch. (Gibbs 2003, 42)

Currently, researchers are trying to complete genetic maps that will link DNA to particular traits. Already, specific strands of DNA have been identified as carrying such physical traits as eye color and height as well as such diseases as breast cancer, cystic fibrosis, prostate cancer, depression, and Alzheimer's.

The U.S. Human Genome Project, a 13-year effort to decode human DNA, is now complete. Conclusion of the project is transforming medicine—"knowledge about the effects of DNA variations among individuals can lead to revolutionary new ways to diagnosis, treat, and someday prevent thousands of disorders that affect us" (HGP 2003a, 2). The hope is that if a defective or missing gene can be identified, possibly a healthy duplicate can be acquired and transplanted in the affected cell. This is known as **gene therapy** (HGP 2003b). Alternatively, viruses have their own genes that can be targeted for removal. Experiments are now under way to accomplish these biotechnological feats.

Genetic engineering is the ability to manipulate the genes of an organism in such a way that the natural outcome is altered. Genetic engineering is accomplished by splicing the DNA from one organism into the genes of another. Often, however, unwanted consequences ensue. For example, through genetic engineering some plants are now self-insecticiding—that is, the plant itself produces an insect-repelling substance. Ironically, the plant's continual production of the insecticide, in contrast to only sporadic application by farmers, is leading to insecticide-resistant pests (Ehrenfeld 1998). The debate over genetically engineered crops is ongoing as advocates note the prospects for expanded food production and proponents question health and environmental consequences (see Chapter 10).

Reproductive Technologies. The evolution of "reproductive science" has been furthered by scientific developments in biology, medicine, and agriculture. At the same time, however, its development has been hindered by the stigma associated with sexuality and reproduction, its link with unpopular social movements (e.g., contraception), and the feeling that such innovations challenge the natural order (Clarke 1990). Nevertheless, new reproductive technologies have been and continue to be developed.

In **in-vitro fertilization** (IVF), an egg and a sperm are united in an artificial setting such as a laboratory dish or test tube. Although the first successful attempt at IVF occurred in 1944, the first test-tube baby, Louise Brown, was not born until 1978. Today, there are over 400,000 frozen embryos in U.S. clinics—"88 percent set aside for future family building by patients. Only about 3 percent have been earmarked for medical research and just 2 percent for donation to other couples" (Rosenberg 2003, 41). Criticisms of IVF are often based on traditional definitions of the family and the legal complications created when a child can have as many as five potential parental ties—egg donor, sperm donor, surrogate mother, and the two people who raise the child (depending on the situation, IVF may not involve donors and/or a surrogate). Litigation over who are the "real" parents has already occurred.

Perhaps more than any other biotechnology, abortion epitomizes the potentially explosive consequences of new technologies. **Abortion** is the removal of an embryo or fetus from a woman's uterus before it can survive on its own. Since the U.S. Supreme Court's ruling in *Roe v. Wade* in 1973, abortion has been legal in the United States. However, recent Supreme Court decisions have limited the *Roe v. Wade* decision. In *Planned Parenthood of Southeastern Pennsylvania v. Casey,* the Court ruled that a state may restrict the conditions under which an abortion is granted, such as requiring a 24-hour waiting period or parental consent for minors. In recent years, the number of abortions has decreased in the United States even though a majority of respondents in a national survey, 55 percent, support "a women's right to have an abortion in the first three months of pregnancy" (Tumulty & Novak 2003, 39).

Several laws that advocate and protect "fetal rights" are presently being considered. The *Unborn Victims of Violence Act* was passed by the U.S. House of Representatives in 2001. It is expected to pass the Senate and, if signed into law, would protect all children in utero, regardless of stage of development. To date, 28 states have criminalized harm to a fetus. For example, because California defines an 8-week-old fetus as a person, Scott Peterson has been charged with the deaths of his wife Laci *and* his unborn son Conner. Observers note that if in law a fetus is defined as a human being with the same rights as women and men, it will in effect overturn *Roe v. Wade* (Rosenberg 2003).

Most recent debates concern intact dilation and extraction (D&X) abortions. Opponents refer to such abortions as **partial birth abortions** because the limbs and the torso are typically delivered before the fetus has expired. D&X abortions were performed because the fetus has a serious defect, the woman's health is jeopardized by the pregnancy, or both. However, in 2003 a federal ban on partial birth abortions was signed into law (White House 2003). Given that the U.S. Supreme court held that a Nebraska law very similar to the federal law was unconstitutional, opponents of the ban hope that it will also be questioned on constitutional grounds (Fagan 2003).

Feminists strongly oppose the ban, arguing that it is just one step closer to making abortions illegal. They are also quick to note that the bill was not supported by "the American Medical Association, the American College of Obstetricians and Gynecologists, the American Medical Women's Association, the American Nurses Association or the American Public Health Association" (U.S. Newswire 2003, 1). Says National Organization for Women (NOW) president Kim Gandy, "Try as you might, you won't find the term 'partial birth abortion' in any medical dictionary. That's because it doesn't exist in the medical world—it's a fabrication of the anti-choice machine" (U.S. Newswire 2003, 1).

Abortion is a complex issue for everyone, but especially for women, whose lives are most affected by pregnancy and childbearing. Women who have abortions are disproportionately poor, unmarried minorities who say they intend to have children in the future. Abortion is also a complex issue for societies, which must respond to the pressures of conflicting attitudes toward abortion and the reality of high rates of unintended and unwanted pregnancy. Figure 15.3 illustrates the percentage of Americans who support legal abortions under various circumstances.

Attitudes toward abortion tend to be polarized between two opposing groups of abortion activists—pro-choice and pro-life. Advocates of the pro-choice movement hold that freedom of choice is a central human value, that procreation choices must be free of government interference, and that because the woman must bear the burden of moral choices, she should have the right to make such decisions. Alternatively, pro-lifers hold that the unborn fetus has a right to live and be protected, that abortion is immoral, and that alternative means of resolving an unwanted pregnancy should be found. Assess your attitudes toward abortion in this chapter's *Self and Society* feature.

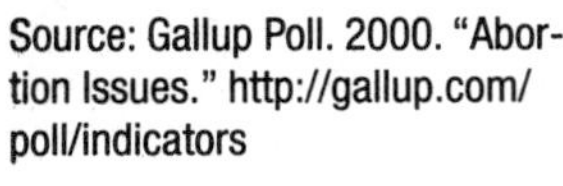

Figure 15.3
Support for Legal Abortions under Specific Circumstances: 2000

Source: Gallup Poll. 2000. "Abortion Issues." http://gallup.com/poll/indicators

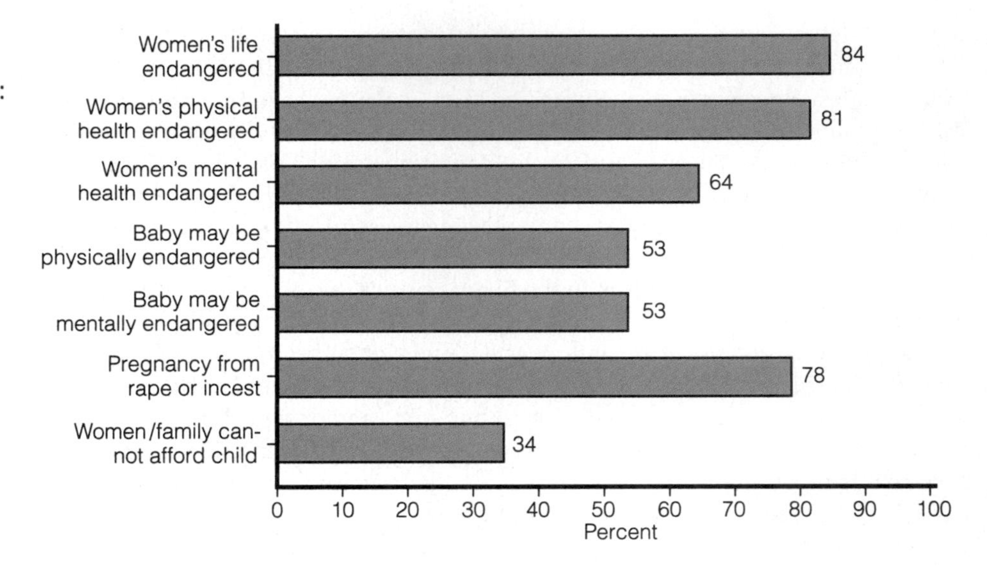

In July of 1996, scientist Ian Wilmut of Scotland successfully cloned an adult sheep named Dolly. To date, cattle, goats, mice, pigs, cats, rabbits, and horses have also been cloned (Weiss 2003). This technological breakthrough has caused worldwide concern about the possibility of human cloning. One argument in favor of developing human cloning technology is its medical value; it may potentially allow everyone to have "their own reserve of therapeutic cells that would increase their chance of being cured of various diseases, such as cancer, degenerative disorders and viral or inflammatory diseases" (Kahn 1997, 54). Human cloning could also provide an alternative reproductive route for couples who are infertile and for those in which one partner is at risk for transmitting a genetic disease.

Arguments against cloning are largely based on moral and ethical considerations. Critics of human cloning suggest that whether used for medical therapeutic purposes or as a means of reproduction, human cloning is a threat to human dignity. For example, cloned humans would be deprived of their individuality, and as Kahn (1997, 119) points out, "creating human life for the sole purpose of preparing therapeutic material would clearly not be for the dignity of the life created." **Therapeutic cloning** uses stem cells from human embryos. Stem cells can produce any type of cell in the human body and thus can be "modeled into replacement parts for people suffering from spinal cord injuries or degenerative diseases, including Parkinson's and diabetes" (Eilperin & Weiss 2003, A06). Because the use of stem cells entails the destruction of human embryos, many conservatives are opposed to the practice. In 2003, the U.S. House of Representatives voted to prohibit the cloning of embryos for reproductive purposes or for medical research.

Despite what appears to be a universal race to the future and the indisputable benefits of such scientific discoveries as the workings of DNA and the technology of IVF, some people are concerned about the duality of science and technology. Science and the resulting technological innovations are often life assisting and life giving (see *The Human Side* feature in this chapter); they are also potentially destructive and life threatening. The same scientific knowledge that led to the discovery of nuclear fission, for example, led to the development of both nuclear power plants and the potential for nuclear destruction. Thus, we now turn our attention to the problems associated with science and technology.

Self and Society

Abortion Attitude Scale

This is not a test. There are no wrong or right answers to any of the statements, so just answer as honestly as you can. The statements ask you to tell how you feel about legal abortion (the voluntary removal of a human fetus from the mother during the first three months of pregnancy by a qualified medical person). Tell how you feel about each statement by circling one of the choices beside each sentence. Respond to each statement and circle only one response.

Strongly Agree	Agree	Slightly Agree	Slightly Disagree	Disagree	Strongly Disagree
5	4	3	2	1	0

1. The Supreme Court should strike down legal abortions in the United States.	5	4	3	2	1	0
2. Abortion is a good way of solving an unwanted pregnancy.	5	4	3	2	1	0
3. A mother should feel obligated to bear a child she has conceived.	5	4	3	2	1	0
4. Abortion is wrong no matter what the circumstances are.	5	4	3	2	1	0
5. A fetus is not a person until it can live outside its mother's body.	5	4	3	2	1	0
6. The decision to have an abortion should be the pregnant mother's.	5	4	3	2	1	0
7. Every conceived child has the right to be born.	5	4	3	2	1	0
8. A pregnant female not wanting to have a child should be encouraged to have an abortion.	5	4	3	2	1	0
9. Abortion should be considered killing a person.	5	4	3	2	1	0
10. People should not look down on those who choose to have abortions.	5	4	3	2	1	0
11. Abortion should be an available alternative for unmarried, pregnant teenagers.	5	4	3	2	1	0
12. Persons should not have the power over the life or death of a fetus.	5	4	3	2	1	0
13. Unwanted children should not be brought into the world.	5	4	3	2	1	0
14. A fetus should be considered a person at the moment of conception.	5	4	3	2	1	0

Scoring and Interpretation

As its name indicates, this scale was developed to measure attitudes toward abortion. It was developed by Sloan (1983) for use with high school and college students. To compute your score, first reverse the point scale for Items 1, 3, 4, 7, 9, 12, and 14. Total the point responses for all items. Sloan provided the following categories for interpreting the results:

70–56 Strong pro-abortion

55–44 Moderate pro-abortion

43–27 Unsure

26–16 Moderate pro-life

15–0 Strong pro-life

Reliability and Validity

The Abortion Attitude Scale was administered to high school and college students, Right to Life group members, and abortion service personnel. Sloan (1983) reported a high total test estimate of reliability (0.92). Construct validity was supported in that Right to Life members' mean scores were 16.2; abortion service personnel mean scores were 55.6; and other groups' scores fell between these values.

Source: "Abortion Attitude Scale" by L. A. Sloan. Reprinted with permission from the *Journal of Health Education* 14(3), May/June, 1983. The *Journal of Health Education* is a publication of the American Allegiance for Health, Physical Education, Recreation and Dance. 1900 Association Drive, Reston, VA 20191.

The Human Side

GROW, CELLS, GROW: One Child's Fight for Survival

© Becky Levine

Tommy Bennett's big brown eyes and sweet demeanor make it that much harder to accept his plight. Just three years old, he blithely endures the constant barrage of drugs, needles, and tests as though he instinctively knows that they are destined to cure him.

Born with a rare, degenerative disease called Sanfilippo syndrome, Tommy lacks a critical enzyme needed for proper organ and brain development. Without enzymes, Tommy will die by adolescence. With the enzymes, Tommy's brain may unlock the potential to allow him to talk, dress, and care for himself.

Such skills have eluded his two affected siblings, four-year-old Hunter and six-year-old Ciara. Ciara had just been diagnosed with Sanfilippo syndrome when their mom became pregnant with Tommy.

Since that time, they have searched desperately for someone willing to take a chance on helping Ciara, Hunter, and Tommy. The Bennetts found hope at Duke University Medical Center, the only program in the country willing to apply the benefits of stem cells—derived from newborn babies' umbilical cords—to treat this disease.

Proof of a Sanfilippo cure remains elusive, and Tommy is only the sixth Sanfilippo patient ever to have received a stem cell transplant. Yet if the transplant is to help, Tommy is a good candidate. He is young enough that the disease has only just begun to wreak havoc on his brain and organs. His siblings have progressed too far to be helped. Still, the sting of disappointment was palpable when doctors deemed Tommy the only viable candidate.

Thankful as they are for the opportunity, the Bennetts have embarked on a costly gamble—financially, emotionally, and physically. The Bennetts uprooted their kids and moved 900 miles away from family and friends to undergo a series of grueling tests before Tommy's transplant could begin.

Then came the real test of endurance. Confined to the hospital unit for four straight weeks, Tommy's small body was ravaged by toxic doses of chemotherapy designed to wipe out his immune system and make way for a new one that might provide the crucial enzyme.

Societal Consequences of Science and Technology

Scientific discoveries and technological innovations have implications for all social actors and social groups. As such, they also have consequences for society as a whole.

Alienation, Deskilling, and Upskilling

As technology continues to play an important role in the workplace, workers may feel there is no creativity in what they do—they feel alienated (see Chapter 11). For example, a study of California's high-tech "white collar factories" found that employees were suffering from isolation, job insecurity, and pressure to update skills (*Technology* 2003). The movement from mechanization, to automation, to

The Human Side

Alicia took on hospital duty, caring for Tommy night and day, and catching a few winks of sleep as time permitted on a pull-down cot. John assumed full-time care for Ciara and Hunter at a rented apartment nearby, no easy task given Ciara's penchant for 3 a.m. awakenings.

The process is clearly daunting, yet the transplant itself is deceptively simple. It takes just fifteen minutes for a bag of red liquid to drip intravenously into a child's bloodstream. Nurses literally squeeze every last drop from the bag, lest they lose a single stem cell that floats amidst a billion blood and supporting cells.

Every parent knows that stem cells hold the key to their child's survival. If they grow, the child has a fighting chance to live. If they do not, the child has probably exhausted his or her last resort at a cure.

Then comes the waiting, and the familiar refrain: "Grow, Cells, Grow." The words resonate within the halls, grace the walls of every room, and are sprinkled throughout cards of love and hope. Parents recite them like a battle cry designed to incite soldiers in action.

Indeed, stem cells are like tiny soldiers who descend upon bone marrow and rescue it from near-certain demise. So powerful are stem cells that it takes only ten to a hundred of them to restore a child's entire blood-forming and immune system—in Tommy's case providing the missing enzymes. Moreover, they know exactly where to go and what function to perform.

Yet such remarkable power is not without its drawbacks. Stem cells can attack the last remnants of the child's immune system, a complication called graft-versus-host disease. Stem cells take time to grow and mature, leaving the child's developing immune system vulnerable to minor infections that could prove deadly.

Children also suffer mightily from the dangerously high levels of chemotherapy needed to wipe out their immune system. Often, their mucous linings literally slough off from within causing severe diarrhea and vomiting. Nausea, painful sores, fatigue, and stomach pains also plague the children as the chemo exerts its effects.

Luckily, Tommy endured far less of the usual symptoms of his transplant, but only time will tell if the new cells have become his own. A year must pass before his new immune system will be running at full force. A lot can happen in that time, but hope, prayer, and a will to overcome will be on their side.

Sadly, Tommy Bennett died just prior to the publication of this book.

cybernation increasingly removes individuals from the production process, often relegating them to flipping a switch, staring at a computer monitor, or entering data at a keyboard. Many low-paid employees, often women, sit at computer terminals for hours entering data and keeping records for thousands of businesses, corporations, and agencies. The work that takes place in these "electronic sweatshops" is monotonous, solitary, and almost completely devoid of autonomy for the worker.

Not only are these activities routine, boring, and meaningless, they promote **deskilling**—that is, "labor requires less thought than before and gives them [workers] fewer decisions to make" (Perrolle 1990, 338). Deskilling stifles development of alternative skills and limits opportunities for advancement and creativity as old skill sets become obsolete. To conflict theorists, deskilling also provides the basis for increased inequality for "throughout the world, those who control the means of producing information products are also able to determine the social organization of the 'mental labor' which produces them" (Perrolle 1990, 337).

Technology in some work environments, however, may lead to **upskilling.** Unlike deskilling, upskilling reduces alienation as employees find their work more rather than less meaningful and have greater decision-making powers as information becomes decentralized. Futurists argue that upskilling in the workplace could lead to a "horizontal" work environment in which "employees do not so much what they are told to do, but what their expansive knowledge of the entire enterprise suggests to them needs doing" (GIP 1998).

Social Relationships and Social Interaction

Technology affects social relationships and the nature of social interaction. The development of telephones has led to fewer visits with friends and relatives; with the coming of VCRs and cable television, the number of places where social life occurs (e.g., movie theaters) has declined. Even the nature of dating has changed as computer networks facilitate cyberdates and "private" chat rooms. As technology increases, social relationships and human interaction are transformed.

Technology also makes it easier for individuals to live in a cocoon—to be self-sufficient in terms of finances (e.g., Quicken), entertainment (e.g., pay-per-view movies), work (e.g., telework), recreation (e.g., virtual reality), shopping (e.g., e-Bay), communication (e.g., Internet, e-cards), and many other aspects of social life. Ironically, although technology can bring people together, it can also isolate them from each other. Children who use a home computer "spend much less time on sports and outdoor activities than non-computer users" (Attewell, Suazo-Garcia, & Battle 2003, 277). Some technological innovations replace social roles—an answering machine may replace a secretary, a computer-operated vending machine may replace a waitperson, an automatic teller machine may replace a banker, and closed circuit television, a teacher. These technologies may improve efficiency, but they also reduce necessary human contact.

Loss of Privacy and Security

Schools, employers, and the government are increasingly using technology to monitor individuals' performance and behavior. A 2003 survey found that 52 percent of U.S. companies monitor employees' e-mails and 22 percent have terminated an employee for an e-mail violation (AMA 2003). Today, the legality of monitoring e-mails is under scrutiny—1 in 20 companies has been sued for e-mail-related surveillance. In addition to e-mail monitoring, high-tech machines monitor countless other behaviors including counting a telephone operator's minutes online, videotaping a citizen walking down a city street, or tracking the whereabouts of a student or faculty member on campus. One group of privacy advocates, under the Freedom of Information Act, has filed for "technical information about a network of video cameras that has been established in their city" (Markoff 2002).

Employers and schools may subject individuals to drug testing technology (see Chapter 3) and in 2002, the number of identity thefts doubled from the previous year, with this crime representing the most frequent complaint to the Federal Trade Commission (Lee 2003). Through computers, individuals can obtain access to someone's phone bills, tax returns, medical reports, credit histories, bank account balances, and driving records. Unauthorized disclosure is potentially devastating. If a person's medical records indicate that he or she is HIV-positive, for example,

"People aren't aware that mouse clicks can be traced, packaged, and sold."

Larry Irving
U.S. Commerce Department

that person could be in danger of losing his or her job or health benefits. If DNA testing of hair, blood, or skin samples reveals a condition that could make the person a liability in the insurer's or employer's opinion, the individual could be denied insurance benefits, medical care, or even employment. In response to the possibility of such consequences, Brin (1998), author of *The Transparent Society,* argues that because it is impossible to prevent such intrusions, "reciprocal transparency," or complete openness, should prevail. If organizations can collect the information, then citizens should have access to it and to its uses.

The development of new technologies has produced new forms of work and new demands for highly skilled workers in certain segments of the labor market.

Our privacy is also disturbed by the intrusion in our e-mail inboxes of unwanted mail called spam. Between 2002 and 2003 the amount of junk mail sent on the Internet increased by 85 percent to a total of 4.9 trillion spam messages. Motivated in part by the "promise of an easy profit, spammers have gone from pests to an invasive species of parasite that threatens to clog the inner workings of the Internet" (Taylor 2003). Recently, spammers have targeted cell phones, sending thousands of unwanted text messages to unsuspecting subscribers (CNN 2004).

Technology has created threats not only to the privacy of individuals but also to the security of entire nations. Computers and modems can be used (or misused) in terrorism and warfare to cripple the infrastructure of a society and tamper with military information and communication operations (see Chapter 16).

Unemployment

Some technologies replace human workers—robots replace factory workers, word processors displace secretaries and typists, and computer-assisted diagnostics reduce the need for automobile mechanics. Unemployment rates can also increase when companies "outsource" their information technology operations to lower-wage countries. For example, U.S. accounting firm Ernst and Young has offices in India that prepare 2 percent of the firm's total tax returns processed. According to one expert, "more than 300 of the Fortune 500 firms do business with Indian information-technology-services companies" (O'Meara 2003, 32).

Not surprisingly, unemployment rates for information technology (IT) workers is at an all time high—5.23 percent in 2002 (Chabrow 2003). When layoffs occur, older IT workers are more likely to remain unemployed, being less likely than younger workers to take a job outside of the field. Downsizing of corporations, business failures, and an economic recession in addition to "outsourcing" have contributed to disappearing high-tech jobs (McNair 2003).

Technology also changes the nature of work and the types of jobs available. For example, fewer semiskilled workers are needed because many of their jobs are now being done by machines. The jobs that remain, often white-collar jobs, require more education and technological skills. Technology thereby contributes to the split labor market as the pay gulf between skilled and unskilled workers continues to grow.

For example, Addison, Fox, and Ruhm (2000) report that employees who use computers at work are at a lower risk of losing their jobs than are non-computer users.

The Digital Divide

One of the most significant social problems associated with science and technology is the increased division between the classes. As Welter (1997, 2) notes,

> it is a fundamental truth that people who ultimately gain access to, and who can manipulate, the prevalent technology are enfranchised and flourish. Those individuals (or cultures) that are denied access to the new technologies, or can not master and pass them on to the largest number of their offspring, suffer and perish.

The fear that technology will produce a "virtual elite" is not uncommon. Several theorists hypothesize that as technology displaces workers, most notably the unskilled and uneducated, certain classes of people will be irreparably disadvantaged—the poor, minorities, and women. There is even concern that biotechnologies will lead to a "genetic stratification," whereby genetic screening, gene therapy, and other types of genetic enhancements are available only to the rich.

The wealthier the family, for example, the more likely the family is to have a computer. Of American families with an income of $75,000 a year or more, 89 percent have at least one computer in the household. However, only 19 percent of households with incomes between $5,000 and $9,999 own a computer (*Statistical Abstract* 2002). Further, 61.1 percent of white Americans own a computer compared to 37.1 percent of African-Americans.

Racial disparities also exist in Internet access—33.6 percent of white children have access to the Internet at home compared with 12.8 percent of Hispanic children and 14.7 percent of black children (*Statistical Abstract* 2002). Inner-city neighborhoods, disproportionately populated by racial and ethnic minorities, are simply less likely to be "wired," that is, to have the telecommunications hardware necessary for schools to access online services. In fact, cable and telephone companies are less likely to lay fiber optics in these areas—a practice called "information apartheid" or "electronic redlining." Students who live in such neighborhoods are technologically disadvantaged and may never catch up to their middle-class counterparts (Welter 1997). A bill is presently being considered in Congress that, if it becomes law, "could provide $1.25 billion over five years to help minority-serving institutions upgrade their computers and communication systems" (Dervarics 2003, 6).

The cost of equalizing such differences is enormous, but the cost of not equalizing them may be even greater. Employees who are technologically skilled have higher incomes than those who are not—up to 15 percent higher (Hancock 1995; World Employment Report 2001). Further, technological disparities exacerbate the structural inequities perpetuated by the split labor force and the existence of primary and secondary labor markets.

Mental and Physical Health

Some new technologies have unknown risks. Biotechnology, for example, has promised and, to some extent, has delivered everything from life-saving drugs to hardier pest-free tomatoes. Biotechnologies have also, however, created **technology-induced diseases** such as those experienced by Chellis Glendinning (1990, 15).

Glendinning, after using the "Pill" and, later, the Dalkon Shield IUD, became seriously ill.

> Despite my efforts to get help, medical professionals did not seem to know the root of my condition lay in immune dysfunction caused by ingesting artificial hormones and worsened by chronic inflammation. In all, my life was disrupted by illness for twenty years, including six years spent in bed. . . . For most of the years of illness, I lived in isolation with my problem. Doctors and manufacturers of birth control technologies never acknowledged it or its sources.

Other technologies that pose a clear risk to a large number of people include nuclear power plants, DDT, automobiles, X rays, food coloring, and breast implants.

The production of new technologies may also place manufacturing employees in jeopardy. For example, the electronics industry uses thousands of hazardous chemicals:

> The semiconductor industry uses large amounts of toxic chemicals to manufacture the components that make up a computer, including disk drives, circuit boards, video display equipment, and silicon chips themselves, the basic building blocks of computer devices. The toxic materials needed to make the 220 billion silicon chips manufactured annually are staggering in amount and include highly corrosive hydrochloric acid; metals such as arsenic, cadmium, and lead; volatile solvents such as methyl chloroform, toluene, benzene, acetone, and trichloroethylene; and toxic gases such as arsine. Many of these chemicals are known or probable human carcinogens. (Chepesiuk 1999, 1)

Finally, technological innovations are, for many, a cause of anguish and stress, particularly when the technological changes are far-reaching (Hormats 2001). Nearly 60 percent of workers report being "technophobes," that is, fearful of technology, and as many as 10 percent of Internet users are "addicted" to being online (Papadakis 2000; Boles & Sunoo 1998). Says Hilarie Cash, a psychologist for Internet/Computer Addiction Services, those who are addicted "lose sleep, spend more and more time online, [and] neglect all areas of their lives" (Markovich & Brahm 2003, 1).

The Challenge to Traditional Values and Beliefs

Technological innovations and scientific discoveries often challenge traditionally held values and beliefs, in part because they enable people to achieve goals that were previously unobtainable. Before recent advances in reproductive technology, for example, women could not conceive and give birth after menopause. Technology that allows postmenopausal women to give birth challenges societal beliefs about childbearing and the role of older women. Macklin (1991) notes that the techniques of egg retrieval, in vitro fertilization, and gamete intrafallopian transfer (GIFT) make it possible for two different women to each make a biological contribution to the creation of a new life. Such technology requires society to reexamine its beliefs about what a family is and what a mother is. Should family be defined by custom, law, or the intentions of the parties involved?

Medical technologies that sustain life lead us to rethink the issue of when life should end. The increasing use of computers throughout society challenges the traditional value of privacy. New weapons systems make questionable the traditional idea of war as something that can be survived and even won. And cloning causes us to wonder about our traditional notions of family, parenthood, and individuality. Toffler (1970) coined the term **future shock** to describe the confusion resulting from rapid scientific and technological changes that unravel our traditional values and beliefs.

Strategies for Action: Controlling Science and Technology

As technology increases, so does the need for social responsibility. Nuclear power, genetic engineering, cloning, and computer surveillance all increase the need for social responsibility: "technological change has the effect of enhancing the importance of public decision making in society, because technology is continually creating new possibilities for social action as well as new problems that have to be dealt with" (Mesthene 1993, 85). In the following section, various aspects of the public debate are addressed, including science, ethics and the law, the role of corporate America, and government policy.

Science, Ethics, and the Law

Science and its resulting technologies alter the culture of society through the challenging of traditional values. Public debate and ethical controversies, however, have led to structural alterations in society as the legal system responds to calls for action. For example, Rhode Island has recently passed several technology relevant laws. These laws do the following (Rhode Island Department of Health 2003):

- Establish DNA testing for convicted offenders
- Require physician screening of newborn babies for certain diseases
- Prohibit hospitals from releasing genetic information to insurance companies without the patient's permission
- Forbid employers to force employees to reveal whether they've had a genetic test and, if so, the results of the test
- Ban human cloning

"Prohibiting scientific and medical activities would also raise troubling enforcement issues. . . . Would they [FBI] raid research laboratories and universities? Seize and read the private medical records of infertility patients? Burst into operating rooms with their guns drawn? Grill new mothers about how their babies were conceived?"

Mark Eibert
Attorney

Are such regulations necessary? In a society characterized by rapid technological and thus social change—a society where custody of frozen embryos is part of the divorce agreement—many would say yes. Cloning, for example, is one of the most hotly debated technologies in recent years. Bioethicists and the public vehemently debate the various costs and benefits of this scientific technique. Despite such controversy, however, the chairman of the National Bioethics Advisory Commission has said that human cloning will be "very difficult to stop" (McFarling 1998). At present, nine states have laws pertaining to human cloning, some prohibiting cloning for reproductive purposes, some prohibiting therapeutic cloning, and still others, both (NCSL 2003).

Should the choices that we make, as a society, be dependent upon what we can do, or what we should do? Whereas scientists and the agencies and corporations who fund them often determine the former, who should determine the latter? Although such decisions are likely to have a strong legal component, that is, they must be consistent with the rule of law and the constitutional right of scientific inquiry, legality or the lack thereof often fails to answer the question, what should be done? *Roe v. Wade* (1973) did little to squash the public debate over abortion and, more specifically, the question of when life begins. Thus it is likely that the issues surrounding the most controversial of technologies will continue into the twenty-first century and with no easy answers (see Figure 15.4).

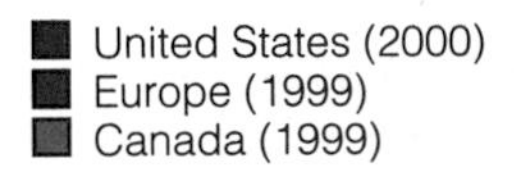

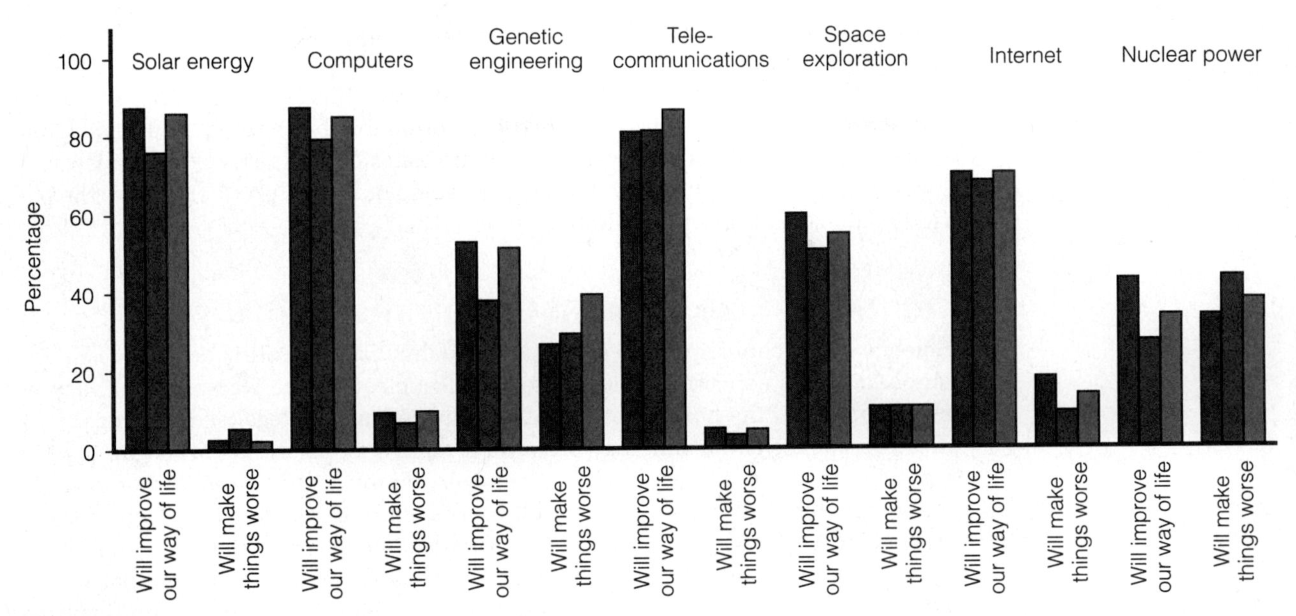

Figure 15.4
Public Attitudes Toward Selected Technologies in the United States, Europe, and Canada
Source: National Science Foundation, Science and Engineering Indicators, 2002.

Technology and Corporate America

As philosopher Jean-Francois Lyotard notes, knowledge is increasingly produced to be sold (Powers 1998). The development of genetically altered crops, the commodification of women as egg donors, and the harvesting of regenerated organ tissues are all examples of potentially market-driven technologies. Like the corporate pursuit of computer technology, profit-motivated biotechnology creates several concerns.

First is the concern that only the rich will have access to such life-saving technologies as genetic screening and cloned organs. Such fears are justified. Companies with obscure names such as Progenitor, Millennium Pharmaceuticals, Darwin Molecular, and Myriad Genetics have been patenting human life. Millennium Pharmaceutical holds the patent on the melanoma gene and the obesity gene, Darwin Molecular controls the premature aging gene, Progenitor controls the gene for schizophrenia, and Myriad Genetics has nine patents on the breast/ovarian cancer gene (Shand 1998; Mayer 2002).

These patents result in **gene monopolies,** which could lead to astronomical patient costs for genetic screening and treatment. One company's corporate literature candidly states that its patent of the breast cancer gene will limit competition and lead to huge profits (Shand 1998, 47). The biotechnology industry argues that such patents are the only way to recoup research costs that, in turn, lead to further innovations. Since 1980, 20,000 gene-related patents have been issued in the United States alone (Doll 2002).

The commercialization of technology causes several other concerns, including issues of quality control and the tendency for discoveries to remain closely guarded secrets rather than collaborative efforts (Rabino 1998; Lemonick & Thompson 1999; Mayer 2002). Further, industry involvement has made government control more difficult as researchers depend less and less on federal funding. Over 75 percent of research and development in the United States is supported by private industry using their own company funds (NSF 2003).

Finally, although there is little doubt that profit acts as a catalyst for some scientific discoveries, other less commercially profitable but equally important projects may be ignored. As biologist Isaac Rabino states, "Imagine if early chemists had thrown their energies into developing profitable household products before the periodic table was discovered" (Rabino 1998, 112).

Runaway Science and Government Policy

Science and technology raise many public policy issues. Policy decisions, for example, address concerns about the safety of nuclear power plants, the privacy of electronic mail, the hazards of chemical warfare, and the legality of surrogacy. In creating science and technology, have we created a monster that has begun to control us rather than the reverse? What controls, if any, should be placed on science and technology? And are such controls consistent with existing law? Consider the use of Napster software to download music files (the question of intellectual property rights and copyright infringement); a proposed federal regulation that would require schools to use blocking software on computers or lose federal funding (free speech issues); and a court decision ordering investigation of *Carnivore*, an FBI surveillance program that can search every message that passes through an Internet service provider (Fourth Amendment privacy issues) (White 2000; Kaplan 2000; Schaefer 2001).

"As president, I will prohibit genetic discrimination, criminalize identity theft, and guarantee the privacy of medical and sensitive financial records."

George W. Bush
U.S. President

The government, often through regulatory agencies and departments, prohibits the use of some technologies (e.g., assisted-suicide devices) and requires others (e.g., seat belts). The Food and Drug Administration, the Environmental Protection Agency, and the Agriculture Department recently investigated the use of genetically altered corn—corn that had been approved only for animal feed—in the making of Taco Bell taco shells (Brasher 2000). A treaty has been adopted by 130 nations that permits countries to "bar imports of genetically altered seeds, microbes, animals and crops that they deem a threat to their environment" (Pollack 2000, 1). Despite the agreement, developing countries are under enormous pressure from corporations to permit such imports (Mayer 2002).

The federal government has instituted several initiatives dealing with technology-related crime. For example, the U.S. Senate Judiciary Committee has approved the *Internet Integrity and Critical Infrastructure Protection Act,* which clarifies the federal role in prosecuting computer hackers and establishes a National Cyber Crime Technical Support Center (Johnson 2000). A government report notes that "many of the attributes of this [Internet] technology—low cost, ease of use, and anonymous nature, among others—make it an attractive medium for fraudulent scams, child sexual exploitation, and . . . cyberstalking" (U.S. Department of Justice 1999, 1). Of late, the issue of online pornography has come to the forefront.

A Federal Bureau of Investigation (FBI) report states that "computer telecommunications have become one of the most prevalent techniques used by pedophiles to share illegal photographic images of minors and to lure children into illicit sexual relationships" (FBI 2002, 1). In fact, there has been a 1,800 percent rise in pornography Web pages in the last five years, increasing from 14 million to 260 million (Greenspan 2003b). In response to such concerns, the *Innocent Images National Initiative* was created. The charge of this initiative is to "identify, investigate and prosecute" sexual predators, "establish a law enforcement presence" on the Internet, and "identify and rescue" child victims (FBI 2002). In 2003, the U.S. Supreme Court agreed to hear arguments on the constitutionality of the *Child Online Protect Act* which, if held constitutional, would limit adult material on the net and require an adults-only screening system (AP 2003b).

Finally, the government has several science and technology boards including the National Science and Technology Council, the Office of Science and Technology Policy, and the President's Council of Advisors on Science and Technology. These agencies advise the President on matters of science and technology including research and development, implementation, national policy, and coordination of different initiatives.

Understanding Science and Technology

What are we to understand about science and technology from this chapter? As functionalists argue, science and technology evolve as a social process and are a natural part of the evolution of society. As society's needs change, scientific discoveries and technological innovations emerge to meet these needs, thereby serving the functions of the whole. Consistent with conflict theory, however, science and technology also meet the needs of select groups and are characterized by political components. As Winner (1993) notes, the structure of science and technology conveys political messages including "power is centralized," "there are barriers between social classes," "the world is hierarchically structured," and "the good things are distributed unequally" (Winner 1993, 288).

The scientific discoveries and technological innovations that are embraced by society as truth itself are socially determined. Research indicates that science and the resulting technologies have both negative and positive consequences—a **technological dualism.** Technology saves lives and time and money; it also leads to death, unemployment, alienation, and estrangement. Weighing the costs and benefits of technology poses ethical dilemmas as does science itself. Ethics, however, "is not only concerned with individual choices and acts. It is also and, perhaps, above all concerned with the cultural shifts and trends of which acts are but the symptoms" (McCormick & Richard 1994, 16).

Thus, society makes a choice by the very direction it follows. These choices should be made on the basis of guiding principles that are both fair and just, such as those listed here (Winner 1993; Goodman 1993; Eibert 1998; Buchanan, Brock, Daniels, & Wikler 2000; Murphie & Potts 2003):

1. Science and technology should be prudent. Adequate testing, safeguards, and impact studies are essential. Impact assessment should include an evaluation of the social, political, environmental, and economic factors.
2. No technology should be developed unless all groups, and particularly those who will be most affected by the technology, have at least some representation "at a very early stage in defining what that technology will be" (Winner 1993, 291). Traditionally, the structure of the scientific process and the development of technologies have been centralized (that is, decisions have been in the hands of a few scientists and engineers); decentralization of the process would increase representation.
3. Means should not exist without ends. Each new innovation should be directed toward fulfilling a societal need rather than the more typical pattern in which a technology is developed first (e.g., high-definition television) and then a market is created (e.g., "you'll never watch a regular TV again!"). Indeed, from the space program to research on artificial intelligence, the vested interests of scientists and

engineers, whose discoveries and innovations build careers, should be tempered by the demands of society.

What the twenty-first century will hold, as the technological transformation continues, may be beyond the imagination of most of society's members. Technology empowers; it increases efficiency and productivity, extends life, controls the environment, and expands individual capabilities. According to a National Intelligence Council report, "Life in 2015 will be revolutionized by the growing effort of multi-disciplinary technology across all dimensions of life: social, economic, political, and personal" (NIC 2003, 1).

As we proceed into the first computational millennium, one of the great concerns of civilization will be the attempt to reorder society, culture, and government in a manner that exploits the digital bonanza, yet prevents it from running roughshod over the checks and balances so delicately constructed in those simpler pre-computer years.

Chapter Review

- **What are the three types of technology?**
The three types of technology, escalating in sophistication, are mechanization, automation, and cybernation. Mechanization is the use of tools to accomplish tasks previously done by hand. Automation involves the use of self-operating machines, and cybernation is the use of machines to control machines.

- **What are some Internet global trends?**
Globally, English speakers are the largest language group online, but non-English speakers constitute the fastest growing group on the Internet. The clear majority of Internet users live in industrialized countries, although there is some movement toward the Internet's becoming truly global as those in developing countries "get online."

- **According to Kuhn, what is the scientific process?**
Kuhn describes the process of scientific discovery as occurring in three steps. First are assumptions about a particular phenomenon. Next, because unanswered questions always remain about a topic, science works to start filling in the gaps. Then, when new information suggests that the initial assumptions were incorrect, a new set of assumptions or framework emerges to replace the old one. It then becomes the dominant belief or paradigm, until it is questioned and the process repeats.

- **What is meant by the computer revolution?**
The silicon chip made computers affordable. Today, over 56 percent of American homes have a computer. Further, over half the labor force uses a computer at work. The most common computer activity at work is accessing the Internet or e-mail, followed by word processing, working with spreadsheets or databases, and accessing/updating calenders or schedules.

- **What is the Human Genome Project?**
The U.S. Human Genome Project is an effort to decode human DNA. The 13-year-old project is now complete, allowing scientists to "transform medicine" through early diagnosis and treatment, as well as possibly preventing disease through gene therapy. Gene therapy entails identifying a defective or missing gene and then replacing it with a healthy duplicate that is transplanted to the affected area.

- **How are some of the problems of the Industrial Revolution similar to the problems of the technological revolution?**
The most obvious example is in unemployment. Just as the Industrial Revolution replaced many jobs with technological innovations, so too has the technological revolution. Further, research indicates that many of the jobs created by the Industrial Revolution, such as working on a factory assembly line, were characterized by high rates of alienation. Rising rates of alienation are also a consequence of increased estrangement as high-tech employees work in "white-collar factories."

- **What is the digital divide?**
The digital divide is the tendency for technology to be most accessible to the wealthiest and most educated. For example, some fear that there will be "genetic stratification" whereby the benefits of genetic screening, gene therapy, and other genetic enhancements are available to only the very richest segments of society.

- **What is meant by the commercialization of technology?**
The commercialization of technology refers to profit-motivated technological innovations. Whether it be the isolation of a particular gene, genetically modified organisms, or the regeneration of organ tissues, where there's a possibility of profit, private enterprise will be there.

Critical Thinking

1. Use of the Internet by neo-Nazi and white supremacist groups has recently increased. Despite such increases, the U.S. Supreme Court has strengthened First Amendment protections of Internet material (Whine 1997). Should such groups have the right to disseminate information about their organizations and recruit members through the Internet?
2. What currently existing technologies have had more negative than positive consequences for individuals and for society?
3. Some research suggests that productivity actually declines with the use of computers (Rosenberg 1998). Assuming this "paradox of productivity" is accurate, what do you think causes the reduction in efficiency?

Key Terms

abortion
automation
cultural lag
cybernation
deskilling
e-commerce
future shock
gene monopolies
gene therapy
genetic engineering
genetic screening
Internet
in-vitro fertilization
IT
mechanization
partial birth abortion
postmodernism
science
snowball sampling
technological dualism
technological fix
technology
technology-induced diseases
telework
therapeutic cloning
upskilling

Taking a Stand

Has the Internet improved the average citizen's life?

Ease of communication, the development of online communities, individual empowerment, and cyberactivism are just a few of the benefits of the Internet. On the other hand, much of the information on the Internet is either wrong, misleading, or outdated, and far too many people are dependent on the Internet as their sole source of information. There are also concerns over Internet privacy, security, and abuses including online pornography, identity theft, and e-commerce fraud.

Use Wadsworth's exclusive online resources—InfoTrac College Edition, MicroCase Online, and OVRC—to formulate a position on this topic.

The Wadsworth's Sociology Online Resources and Writing Companion will help you get started. This valuable guide will show you how to use Wadsworth's exclusive online resources when studying social problems. It will also help you to build essential research and writing skills. InfoTrac College Edition, MicroCase Online, OVRC, and an electronic copy of portions of this companion are available at http://sociology.wadsworth.com/mooney_knox_schacht/problems4e, the companion Web site for *Understanding Social Problems,* Fourth Edition.

Media Resources

The Companion Web Site for *Understanding Social Problems,* Fourth Edition

http://sociology.wadsworth.com/mooney_knox_schacht/problems4e

Supplement your review of this chapter by going to the companion Web site to take one of the Tutorial Quizzes, use the flash cards to master key terms, and check out the many other study aids you'll find there. You'll also find special features such as *Wadsworth's Sociology Online Resources and Writing Companion,* GSS Data, and Census 2000 information, data, and resources at your fingertips to help you complete that special project or do some research on your own.

Interactions CD-ROM

Go to the Interactions CD-ROM for *Understanding Social Problems,* Fourth Edition, to access additional interactive learning tools, such as in-depth review materials, corresponding practice quizzes, and other engaging resources and activities to help you study the concepts in this chapter.

"Every gun that is made, every warship launched, every rocket fired, signifies in the final sense a theft from those who hunger and are not fed, those who are cold and not clothed. The world in arms is not spending money alone. It is spending the sweat of its laborers, the genius of its scientists, and the hopes of its children." *Dwight D. Eisenhower, Former U.S. President/Military Leader*

Conflict, War, and Terrorism

Todd was just one of those great guys. As a teenager he loved sports, playing basketball, football, and soccer at Christian High School in Wheaton, Illinois. He and his family moved to California where Todd finished his senior year of high school and began attending Fresno State, but soon after, he returned to Illinois finishing his bachelor's degree at Wheaton College, a coed Christian school. He worked on his MBA at DePaul University in Chicago and in 1994 married his college sweetheart. They moved to New Jersey, where Todd had taken a job with Oracle Corporation, and began a family (Beamer 2001).

On September 11, 2001, Todd Beamer, 32, left his home at 6:15 A.M. and boarded Flight 93 to San Francisco. At 9:45 A.M. he contacted GTE operator Lisa Jefferson and told her there were three highjackers aboard, each with knives, and one with a bomb tied to his mid-section. The highjackers had separated the passengers, 27 into first class and 10 plus five flight attendants in the back of the plane. Todd didn't know the location or condition of the pilot or co-pilot. The operator told him of the World Trade Center attacks (Beamer 2001).

The two talked for about 15 minutes. He told Jefferson of plans to foil the highjackers' attack and that he knew he wouldn't survive. He asked her to call his wife and kids, David, 4, and Drew, 2, and tell them how much he loved them. He would never see his yet unborn third child (McKinnin 2001). He prayed with her—"Though I walk through the valley of the shadow of death I shall fear no evil." He dropped the phone leaving the line open. It was then Jefferson heard him say, "Are you guys ready? Let's roll."

The events of September 11, 2001, led to the war on terrorism and the war with Iraq. **War,** the most violent form of conflict, refers to organized armed violence aimed at a social group in pursuit of an objective. Wars have existed throughout human history and continue in the contemporary world.

War is one of the great paradoxes of human history. It both protects and annihilates. It creates and defends nations, but also destroys them. Whether war is just or unjust, defensive or offensive, it involves the most horrendous atrocities known to humankind. This chapter focuses on the causes and consequences of conflict, war, and terrorism. Along with population and environmental problems, conflict, war, and terrorism are among the most serious of all social problems in their threat to the human race and life on earth.

The Global Context: Conflict in a Changing World

As societies have evolved and changed throughout history, the nature of war has also changed. Before industrialization and the sophisticated technology that resulted, war occurred primarily between neighboring groups on a relatively small scale. In the modern world, war can be waged between nations that are separated by thousands of miles, as well as between neighboring nations. In the following sections, we examine how war has changed our social world and how our changing social world has affected the nature of war in the industrial and post-industrial information age.

"The significance of wars is not just that they led to major changes during the period of hostilities and immediately after. They produced transformations which have turned out to be of enduring significance."

Anthony Giddens
Sociologist

War and Social Change

The very act that now threatens modern civilization—war—is largely responsible for creating the advanced civilization in which we live. Before large political states existed, people lived in small groups and villages. War broke the barriers of autonomy between local groups and permitted small villages to be incorporated into

© Bob Mahoney/The Image Works

Industrialization may decrease a society's propensity to war, but it also increases the potential destructiveness of war because, with industrialization, warfare technology becomes more sophisticated and lethal.

larger political units known as "chiefdoms." Centuries of warfare between chiefdoms culminated in the development of the state. The **state** is "an apparatus of power, a set of institutions—the central government, the armed forces, the regulatory and police agencies—whose most important functions involve the use of force, the control of territory and the maintenance of internal order" (Porter 1994, 5–6). The creation of the state in turn led to other profound social and cultural changes:

> And once the state emerged, the gates were flung open to enormous cultural advances, advances undreamed of during—and impossible under—a regimen of small autonomous villages. . . . Only in large political units, far removed in structure from the small autonomous communities from which they sprang, was it possible for great advances to be made in the arts and sciences, in economics and technology, and indeed in every field of culture central to the great industrial civilizations of the world. (Carneiro 1994, 14–15)

Industrialization and technology could not have developed in the small social groups that existed before military action consolidated them into larger states. Thus war contributed indirectly to the industrialization and technological sophistication that characterize the modern world. Industrialization, in turn, has had two major influences on war. Cohen (1986) calculated the number of wars fought per decade in industrial and pre-industrial nations and concluded that "as societies become more industrialized, their proneness to warfare decreases" (p. 265). For example, in 2002, there were 21 major armed conflicts the majority of which were in less-developed countries in Africa and Asia (SIPRI 2003).

Although industrialization may decrease a society's propensity to war, it also increases the potential destruction of war. With industrialization, military technology became more sophisticated and more lethal. Rifles and cannons replaced the clubs, arrows, and swords used in more primitive warfare and in turn were replaced by tanks, bombers, and nuclear warheads. This chapter's *Focus on Technology* feature looks at how information technology is transforming cyberwar capabilities.

The Economics of Military Spending

The increasing sophistication of military technology has commanded a large share of resources totaling, worldwide, $794 billion in 2002. Military spending worldwide has been increasing since 1998 with a dramatic increase in 2002 as a consequence of U.S. post-September 11 expenditures. For example, prior to September 11, 2001, only 20 percent of active duty personnel were abroad; by 2003, that number was 50 percent. The majority of these troops are in Iraq at a cost of $1 billion a week (Thompson & Duffy 2003).

Focus on Technology

Cyberwarfare

In the post-industrial information age, computer technology has revolutionized the nature of warfare and future warfare capabilities. Today, a "whole range of new technologies are offered for the next generations of weapons and military operations" (BICC 1998, 3) including the use of high-performance sensors, information processors, directed-energy technologies, precision-guided munitions, and worms and viruses (O'Prey 1995; BICC 1998; PBS 2003). With the increasing proliferation and power of computer technology, military strategists and political leaders are exploring the horizons of information warfare often called **cyberwarfare.** Essentially, cyberwarfare utilizes information technology to attack or manipulate the military and civilian infrastructure and information systems of an enemy. For example, cyberwarfare capabilities include the following, from least to most serious (National Security 2003):

- *Web vandalism*—"deactivate or deface" a government Web page
- *Disinformation campaigns*—use of misinformation to confuse enemy
- *Gathering secret data*—intercepting of and tampering with classified information
- *Disruption in the field*—interfering with military activities including blocking vital communication and intercepting commands
- *Attacking critical infrastructure*—electronic attacks on America's infrastructures including transportation, electricity, water, fuel, and finances.

Concerns over cyberwarfare are warranted. U.S. officials stumbled across a "two-year pattern of probing of computer systems in the Pentagon, NASA, the Energy Department, and university and research labs" (PBS 2003, 2). The attacks were traced to a mainframe in Russia.

At the Naval Postgraduate School students from diverse backgrounds, some military, some civilian, come together for cyberspace maneuvers—mock Internet warfare with a computer, modem, software, and keyboard (Howe 2003). Further, Air Force "battlelabs" have been established across the United States (Sietzen 2000). Because "the ability of American forces to deny access to space by any enemy of the United States or its allies" is paramount, battlelab personnel are conducting research on space control technologies. Today, the concept of space control includes controlling "everyday communications moving through space: voice, e-mail, paging signals, computer data and weather projections, . . . reconnaissance images of enemy forces and basic military communications between forces, fleets and communication centers" (Sietzen 2000, 2).

There are two concerns, however, about the use of cyberwarfare technologies. First, although the United States is developing a cyberwarfare strategy, other countries, including Russia, China, and Israel, are further along in their information-warfare capabilities (Messmer 2000). In fact, a House committee, reviewing cyber-security of federal agencies, gave an overall "F" to the 24 agencies reviewed (McDonald 2003).

Another problem is that cyberwarfare is relatively inexpensive and readily available (McDonald 2003). With a computer, a modem, and some rudimentary knowledge of computers, anyone could initiate an information attack. For example, after the fall of the Taliban and Al-Qaeda, computers were seized and prisoners interrogated. One computer contained:

> software and connections to a programming site where the users had been pulling specific information about digital switches on power and water company system infrastructures. It showed how Al Qaeda was doing research through open, available resources to learn about U.S. critical infrastructure and how to exploit it.

In 2003, the government spent $4.2 billion on cyber-security and President Bush released his National Strategy to Secure Cyberspace (McDonald 2003; PBS 2003). Given that "85 to 95 percent of cyberspace is owned and managed by the private sector," at the heart of the national strategy lies a partnership between the government and private industry (PBS 2003). It is estimated that securing cyberspace will cost tens of billions of dollars over many years.

Sources: BIC (Bonn International Center for Conversion). 1998. "Chapter Six." *Conversion Survey, 1998.* Bonn, Germany: BIC; Kevin Howe. 2003. "War Games at Navy School." April 18. http://www.montereyherald.com; Tim McDonald. 2002. "U.S.: Cyber Strike Could Earn Military Response." *NewsFactor Network,* February 14. http://www.newsfactor.com; Ellen Messier. 2000. "U.S. Army Kick-Starts Cyberwar Machine." November 20. http://www.nwfusion.com; National Security. 2003. "Special Focus: Cyberwarfare." http://www.tecsoc.org/natsec; Kevin P. O'Prey. 1995. *The Arms Export Challenge: Cooperative Approaches to Export Management and Defense Conversion.* Washington, DC: The Brookings Institution; PBS (Public Broadcastiing System). 2003. "FAQ: Cyber-security." http://www.pbs.org/wgbh; Frank Sietzen Jr. 2000. "'Battlelabs' Beef Up Space Defense." http://www.msnbc.com/ews/4/2426

Figure 16.1
U.S. Military Spending vs. the World ($ Billions), 2003

Source: World Military Spending, Center for Defense Information, March 2003.

Note: "Allies" refers to the NATO countries, Australia, Japan, and South Korea. "Rogues" refers to Cuba, Iran, Iraq, Libya, North Korea, Sudan, and Syria.

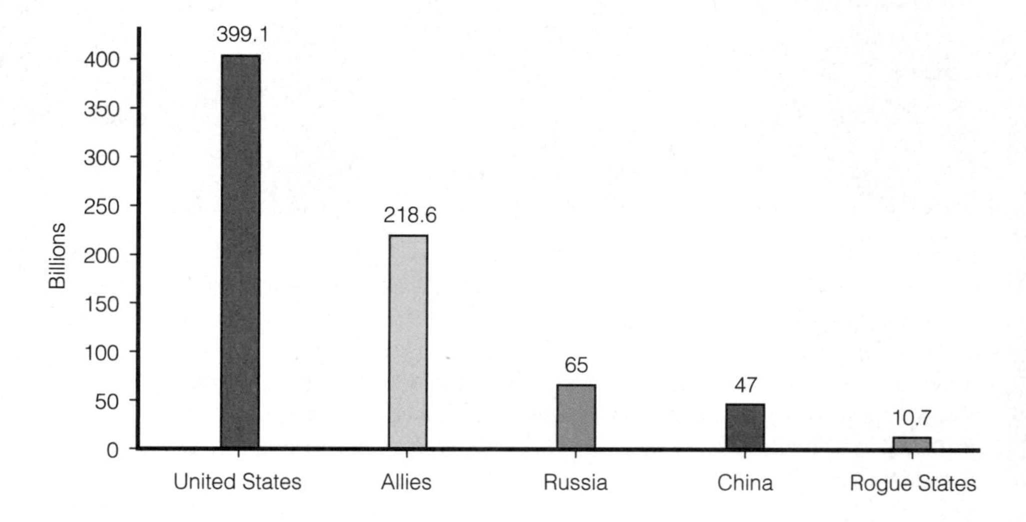

The **Cold War**, the state of political tension and military rivalry that existed between the United States and the former Soviet Union, provided justification for large expenditures on military preparedness. However, the end of the Cold War, along with the rising national debt, resulted in cutbacks in the U.S. military budget in the 1990s. Today, military spending has once again reached Cold War levels. The United States accounts for 46 percent of the world's military spending—the largest single percentage of any nation (see Figure 16.1). The next highest military spenders are Japan, England, France, and China, which, with the United States, account for 62 percent of the world's military budget (SIPRI 2003). U.S. military spending, funded at $400 billion for fiscal year 2004, includes expenditures for salaries of military personnel, research and development, weapons, veterans' benefits, and other defense-related expenses. In addition to these expenditures, the newly created Department of Homeland Security was funded at a total of $37.6 billion for fiscal year 2004 (White House 2003).

The U.S. government not only spends money on its own military and defense but also sells military equipment to other countries, either directly or by helping U.S. companies sell weapons abroad. Although the purchasing countries may use these weapons to defend themselves from hostile attack, foreign military sales may pose a threat to the United States by arming potential antagonists. For example, the United States, which provides almost half of the world's arms exports, supplied weapons to Iraq to use against Iran. These same weapons were then used against Americans in the Gulf and Iraq Wars. Between 1999 and 2003 there were no weapon deliveries to the Middle East by the United States, Russia, China, or any European nation. However, North Korea, increasingly a global threat, is thought to have provided 60 surface-to-surface missiles and 10 anti-ship missiles to Middle Eastern countries (Shanker 2003).

Sociological Theories of War

Sociological perspectives can help us understand various aspects of war. In this section, we describe how structural-functionalism, conflict theory, and symbolic interactionism may be applied to the study of war.

Structural-Functionalist Perspective

Structural-functionalism focuses on the functions war serves and suggests that war would not exist unless it had positive outcomes for society. We have already noted that war has served to consolidate small autonomous social groups into larger political states. An estimated 600,000 autonomous political units existed in the world around the year 1000 B.C. Today, that number has dwindled to fewer than 200.

Another major function of war is that it produces social cohesion and unity among societal members by giving them a "common cause" and a common enemy. Unless a war is extremely unpopular, military conflict promotes economic and political cooperation. Internal domestic conflicts between political parties, minority groups, and special interest groups dissolve as they unite to fight the common enemy. During World War II, U.S. citizens worked together as a nation to defeat Germany and Japan.

In the short term, war also increases employment and stimulates the economy. The increased production needed to fight World War II helped pull the United States out of the Great Depression. The investments in the manufacturing sector during World War II also had a long-term impact on the U.S. economy. Hooks and Bloomquist (1992) studied the effect of the war on the U.S. economy between 1947 and 1972 and concluded that the U.S. government "directed, and in large measure, paid for a 65 percent expansion of the total investment in plant and equipment" (p. 304).

As structural-functionalists argue, a major function of war is that it produces unity among societal members. War provides a common cause and a common identity. Societal members feel a sense of cohesion, and they work together to defeat the enemy.

Another function of war is the inspiration of scientific and technological developments that are useful to civilians. Research on laser-based defense systems led to laser surgery, for example, and research in nuclear fission and fusion facilitated the development of nuclear power. The airline industry owes much of its technology to the development of air power by the Department of Defense, and the Internet was created by the Pentagon for military purposes. Other **dual-use technologies,** a term referring to defense-funded innovations with commercial and civilian applications, include SLICE, a high speed, twin hull water vessel originally made for the Office of Naval Research. SLICE has a variety of commercial applications "including its use as a tour or sport fishing boat, oceanographic research vessel, oil spill response ship, and high-speed ferry" (Hawaii 2000).

Finally, war serves to encourage social reform. After a major war, members of society have a sense of shared sacrifice and a desire to heal wounds and rebuild normal patterns of life. They put political pressure on the state to care for war victims, improve social and political conditions, and reward those who have sacrificed lives, family members, and property in battle. As Porter (1994) explains, "Since . . . the lower economic strata usually contribute more of their blood in battle than the wealthier classes, war often gives impetus to social welfare reforms" (p. 19).

Conflict Perspective

Conflict theorists emphasize that the roots of war are often antagonisms that emerge whenever two or more ethnic groups (e.g., Bosnians and Serbs), countries (United States and Vietnam), or regions within countries (the U.S. North and South) struggle for control of resources or have different political, economic, or religious ideologies. Further, conflict theory suggests that war benefits the corporate, military, and political elites. Corporate elites benefit because war often results in the victor taking control of the raw materials of the losing nations, thereby creating a bigger supply of raw materials for its own industries. Indeed, many corporations profit from defense spending. Under the Pentagon's bid and proposal program, for example, corporations can charge the cost of preparing proposals for new weapons as overhead on their Defense Department contracts. Also, Pentagon contracts often guarantee a profit to the developing corporations. Even if the project's cost exceeds initial estimates, called a cost overrun, the corporation still receives the agreed-upon profit. In the late 1950s, President Dwight D. Eisenhower referred to this close association between the military and the defense industry as the **military-industrial complex.** Conflict theorists would be quick to note that "many former Republican officials and political associates of the Bush administration are associated with the Carlyle Group, an equity investment firm with billions of dollars in military and aerospace assets" (Knickerbocker 2002, 2).

The military elite benefit because war and the preparations for it provide prestige and employment for military officials. For example, Military Professional Resources, Inc. (MPRI), in Virginia, boasts that it can "perform any task or accomplish any mission requiring defense related expertise, military skills short of combat operations" and lists such capabilities as war gaining, anti-terrorism/force protection, and peacekeeping (MPRI 2003).

War also benefits the political elite by giving government officials more power. Porter (1994) observed that "throughout modern history, war has been the lever by which . . . governments have imposed increasingly larger tax burdens on increasingly broader segments of society, thus enabling ever-higher levels of spending to be sustained, even in peacetime" (p. 14). Political leaders who lead their country to a military victory also benefit from the prestige and hero status conferred on them.

Finally, feminists argue that war and other conflicts are often justified using the "language of feminism" (Viner 2002). For example, the attack on Afghanistan in 2001 was, in part, to liberate women who had been subjugated by the Taliban regime. Ironically, the position of women in Afghanistan has improved little in the last several years, leading many Muslim women to reject "Western-style feminism" and embrace Muslim feminism.

> Muslim women deplore misogyny just as western women do, and they know that Islamic societies also oppress them; why wouldn't they? But liberation for them does not encompass destroying their identity, religion, or culture, and many of them want to retain the veil. (p. 2)

Other differences also exist. Muslim feminism is based in the teachings of Islam, is pro-family, and rejects the concept of patriarchy (McElory 2003).

Symbolic Interactionist Perspective

The symbolic interactionist perspective focuses on how meanings and definitions influence attitudes and behaviors regarding conflict and war. The development of attitudes and behaviors that support war begins in childhood. American children learn to glorify and celebrate the Revolutionary War, which created our nation. Movies romanticize war, children play war games with toy weapons, and various video and computer games glorify heroes conquering villains. Indeed, from 1938 to 1942 a series of "Horrors of Wars" cards were manufactured and distributed in the United States and collected by millions of American youth, much like baseball cards (Nelson 1999).

Symbolic interactionism helps to explain how military recruits and civilians develop a mind-set for war by defining war and its consequences as acceptable and necessary. The word "war" has achieved a positive connotation through its use in various phrases—the "war on drugs," the "war on poverty," and the "war on crime." Positive labels and favorable definitions of military personnel facilitate military recruitment and public support of armed forces. Military personnel wear uniforms that command public respect, and they earn badges and medals that convey their status as "heroes."

Many government and military officials convince the masses that the way to ensure world peace is to be prepared for war. Most world governments preach peace through strength rather than strength through peace. Governments may use propaganda and appeals to patriotism to generate support for war efforts and motivate individuals to join armed forces. Salladay (2003), for example, notes that those in favor of the war on Iraq have commandeered the language of patriotism, making it difficult but necessary for peace activists to use the same symbols or phrases.

"Why do we kill people who are killing people to show that killing people is wrong?"

Holly Near
Singer-songwriter

To legitimize war, the act of killing in war is not regarded as "murder." Deaths that result from war are referred to as "casualties." Bombing military and civilian targets appears more acceptable when nuclear missiles are "peacekeepers" that are equipped with multiple "peace heads." Killing the enemy is more acceptable when derogatory and dehumanizing labels such as Gook, Jap, Chink, and Kraut convey the attitude that the enemy is less than human.

Such labels are often socially constructed as images, often through the media, and are presented to the public. Social constructionists, like symbolic interactionists in general, emphasize the social aspects of "knowing." Thus, Li and Izard (2003) used content analysis to analyze newspaper and television coverage of the World Trade Center and Pentagon bombings on September 11. The researchers examined the first eight hours of coverage of the attacks presented on CNN, ABC, CBS, NBC, and FOX as well as in eight major U.S. newspapers (e.g., *Los Angeles Times, New York Times, Washington Post*). Results of the analysis indicate that newspaper articles tended to have a "human interest" emphasis whereas television coverage was more often "guiding and consoling." Other results suggest that both media relied most heavily on government sources, newspaper and network were equally factual, and networks were more homogeneous in their presentation than newspapers. One indication of the importance of the media lies in President Bush's creation of the Office of Global Communications—"a huge production company, issuing daily scripts on the Iraq war to U.S. spokesmen around the world, auditioning generals to give media briefings, and booking administration stars on foreign news shows" (Kemper 2003, 1).

Causes of War

The causes of war are numerous and complex. Most wars involve more than one cause. The immediate cause of a war may be a border dispute, for example, but religious tensions that have existed between the two combatant countries for decades may also contribute to the war. The following section reviews various causes of war.

Conflict over Land and Other Natural Resources

Nations often go to war in an attempt to acquire or maintain control over natural resources, such as land, water, and oil. Michael Klare, author of *Resource Wars: The New Landscape of Global Conflict* (2001), predicts that wars will increasingly be over resources as supplies of the most needed diminish. Disputed borders are one of the most common motives for war. Conflicts are most likely to arise when borders are physically easy to cross and are not clearly delineated by natural boundaries such as major rivers, oceans, or mountain ranges.

Water is another valuable resource that has led to wars. At various times the empires of Egypt, Mesopotamia, India, and China all went to war over irrigation rights. In 1998, five years after Eritrea gained independence from Ethiopia, forces clashed over control of the port city Assab and with it, access to the Red Sea.

Not only do the oil-rich countries in the Middle East present a tempting target in themselves, but war in the region can threaten other nations that are dependent on Middle Eastern oil. Thus, when Iraq seized Kuwait and threatened the supply of oil from the Persian Gulf, the United States and many other nations reacted militarily in the Gulf War. In a document prepared for the Center for Strategic and International Studies, Starr and Stoll (1989) warn that soon:

> water, not oil, will be the dominant resource issue of the Middle East. According to World Watch Institute, "despite modern technology and feats of engineering, a secure water future for much of the world remains elusive." The prognosis for Egypt, Jordan, Israel, the West Bank, the Gaza Strip, Syria, and Iraq is especially alarming. If present consumption patterns continue, emerging water shortages, combined with a deterioration in water quality, will lead to more competition and conflict. (p. 1)

As predicted, Israelis and Palestinians continue to fight over control of the Jordan river basin.

Conflict over Values and Ideologies

Many countries initiate war not over resources, but over beliefs. World War II was largely a war over differing political ideologies: democracy versus fascism. The Cold War involved the clash of opposing economic ideologies: capitalism versus communism. Wars over differing religious beliefs have led to some of the worst episodes of bloodshed in history, in part, because some religions are partial to martyrdom—the idea that dying for one's beliefs leads to eternal salvation. The Shiites (one of the two main sects within Islam) in the Middle East represent a classic example of holy warriors who feel divine inspiration to kill the enemy.

Conflicts over values or ideologies are not easily resolved. The conflict between secularism and Islam has lasted for 14 centuries. Conflict over values and ideologies are less likely to end in compromise or negotiation because they are fueled by people's convictions. For example, when a representative sample of American Jews was asked, "Do you agree or disagree with the following statement? 'The goal of Arabs is not the return of occupied territories but rather the destruction of Israel,'" 82 percent agreed, 15 percent disagreed, and 4 percent were unsure (AJC 2003).

If ideological differences can contribute to war, do ideological similarities discourage war? The answer seems to be yes; in general, countries with similar ideologies are less likely to engage in war with each other than countries with differing ideological values (Dixon 1994). Democratic nations are particularly disinclined to wage war against one another (Doyle 1986).

Racial, Ethnic, and Religious Hostilities

Racial, ethnic, and religious groups vary in their cultural beliefs, values, and traditions. Thus, conflicts between racial, ethnic, and religious groups often stem from conflicting values and ideologies. Such hostilities are also fueled by competition over land and other scarce natural and economic resources. Gioseffi (1993) notes that "experts agree that the depleted world economy, wasted on war efforts, is in great measure the reason for renewed ethnic and religious strife. 'Haves' fight with 'have-nots' for the smaller piece of the pie that must go around" (p. xviii). Racial, ethnic, and religious hostilities are also perpetuated by the wealthy majority to divert attention away from their exploitations and to maintain their own position of power (see Chapter 6).

As described by Paul (1998), sociologist Daniel Chirot argues that the recent worldwide increase in ethnic hostilities is a consequence of "retribalization," that is, the tendency for groups, lost in a globalized culture, to seek solace in the "extended family of an ethnic group" (p. 56). Chirot identifies five levels of ethnic conflict: (1) multiethnic societies without serious conflict (e.g., Switzerland), (2) multiethnic societies with controlled conflict (e.g., United States, Canada), (3) societies with ethnic conflict that has been resolved (e.g., South Africa), (4) societies with serious ethnic conflict leading to warfare (e.g., Sri Lanka), and (5) societies with genocidal ethnic conflict including "ethnic cleansing" (e.g., Kosovo).

Religious differences as a source of conflict have recently come to the forefront. An Islamic *Jihad,* or holy war, has been blamed for the September 11 attacks on the World Trade Center and Pentagon as well as bombings in Kashmir, Sudan, the Philippines, Kenya, Tanzania, Saudi Arabia, and Chechnya. Some claim that Islamic beliefs in and of themselves have led to recent conflicts (Feder 2003, 21).

> If Islam is so mellow, why are the most contemptible crimes regularly committed in its name? There is no United Methodists Jihad. Suicide bombers don't quote the book of Mormon. Individuals aren't given the choice of conversion to Judaism or death.

Others contend that religious fanatics, not the religion itself, are responsible for violent confrontations. For example, Islamic leader Osama bin Laden claims that unjust U.S. Middle East policies are responsible for "dividing the whole world into two sides–the side of believers and the side of infidels" (Williams 2003, 18).

Defense Against Hostile Attacks

The threat or fear of being attacked may cause the leaders of a country to declare war on the nation that poses the threat. The threat may come from a foreign country or from a group within the country. After Germany invaded Poland in 1939, Britain and France declared war on Germany out of fear that they would be Germany's next victims. Germany attacked Russia in World War I, in part out of fear that Russia had entered the arms race and would use its weapons against Germany. Japan bombed Pearl Harbor hoping to avoid a later confrontation with the U.S. Pacific fleet, which posed a threat to the Japanese military. In 2001, a U.S.-led coalition bombed Afghanistan in response to the September 11 terrorist attacks.

Revolution

Revolutions involve citizens warring against their own government and often result in significant political, economic, and social change. A revolution may occur when a government is not responsive to the concerns and demands of its citizens and when strong leaders are willing to mount opposition to the government (Renner 2000; Barkan & Snowden 2001).

The birth of the United States resulted from colonists revolting against British control. Contemporary examples of civil war include Sri Lanka, where the Tamils, a separatist group living in the northern region of the country, have been at war with the Sri Lankan government for over 18 years. The war has resulted in more than 64,000 deaths and, recently, "has been confined to jungle skirmishes, government air strikes and ambushes by the Liberation of Tamil Tigers (LTTE) which wants to carve a separate Tamil state out of Sri Lanka's north and east" (Reuters 2001).

Similarly, Liberia's 14-year on-again off-again civil war pits Liberians United for Reconciliation and Democracy rebels against government officials. In 2003, under heavy pressure from the United States, the president of Liberia resigned. His exile has led to talks between warring factions, the arrival of much needed U.S. supplies, and the beginnings of an interim government (Sengupta 2003). Civil wars have also erupted in newly independent republics created by the collapse of communism in Eastern Europe, as well as in Rwanda, Sierra Leone, Chile, Uganda, and Nepal.

"Those who make peaceful evolution impossible, make violent revolution inevitable."

John F. Kennedy
Former U.S. President

Nationalism

Some countries engage in war in an effort to maintain or restore their national pride. For example, Scheff (1994) argues that "Hitler's rise to power was laid by the treatment Germany received at the end of World War I at the hands of the victors" (p. 121). Excluded from the League of Nations, punished by the Treaty of Versailles, and ostracized by the world community, Germany turned to nationalism as a reaction to material and symbolic exclusion. Further, some observers note that despite the official end of the war in Iraq and, with it, Saddam Hussein's rule, the dictator has "become a symbol of nationalist resistance to forces opposing the U.S.-led occupation of Iraq" (Morahan 2003, 1).

In the late 1970s, Iranian fundamentalist groups took hostages from the American Embassy in Iran. President Carter's attempt to use military forces to free the hostages was not successful. That failure intensified doubts about America's ability to use military power effectively to achieve its goals. The hostages in Iran were eventually released after President Reagan took office, but doubts about the strength and effectiveness of America's military still called into question America's status as a world power. Subsequently, U.S. military forces invaded the small island of Grenada because the government of Grenada was building an airfield large enough to accommodate major military armaments. U.S. officials feared that this airfield would be used by countries in hostile attacks on the United States. From one point of view, the large scale and "successful" attack on Grenada functioned to restore faith in the power and effectiveness of the American military.

Terrorism

Terrorism is the premeditated use, or threatened use, of violence by an individual or group to gain a political or social objective (INTERPOL 1998; Barkan & Snowden 2001; Brauer 2003). Terrorism may be used to publicize a cause, promote an ideology, achieve religious freedom, attain the release of a political prisoner, or rebel against a government. Terrorists use a variety of tactics, including assassinations, skyjackings, suicide bombings, armed attacks, kidnapping and hostage taking,

Table 16.1
U.S. Policy Responses to Terrorism

Diplomacy and Constructive Engagement—Using diplomacy, including verbal contact and direct negotiations, to create a "global anti-terror coalition."

Economic Sanctions—Banning or threatening to ban, for example, trade and investment relations with a nation as a means of control. Other economic sanctions include freezing bank accounts, imposing trade embargos, and suspending foreign aid.

Economic Inducements—Developing assistance programs to reduce poverty and illiteracy in countries that are breeding grounds for terrorist activity.

Covert Action—Infiltrating terrorist groups, military operations, and intelligence gathering. Also included in this category are sabotaging weapons facilities and capturing wanted terrorists.

Rewards for Information Programs—Exchanging money for information on or the capture of a wanted terrorist. For example, the reward for capturing Osama bin Laden is $25 million.

Extradition/Law Enforcement Cooperation—Enlisting the worldwide cooperation of law enforcement agencies including international extradition of terrorists.

Military Force—Using military force in fighting terrorism. For example, military force was used to successfully overthrow the Taliban in Afghanistan.

International Conventions—Entering into agreements with other countries that obligate the signatories to conform to the articles of the convention including, for example, prosecution and extradition of terrorists.

Source: Perl, Raphael. 2003. "Terrorism, the Future, and U.S. Policy." Congressional Research Service. Washington, DC: Library of Congress.

threats, and various forms of bombings. Despite President Bush's declaration of a "war on terrorism," unlike war where there is a winner and a loser, terrorism is unlikely to be completely defeated (Simon 2002). Table 16.1 highlights recent U.S. responses to terrorist activity.

Types of Terrorism

Terrorism can be either transnational or domestic. **Transnational terrorism** occurs when a terrorist act in one country involves victims, targets, institutions, governments, or citizens of another country. The 1988 bombing of Pan-Am Flight 103 in Lockerbie, Scotland, exemplifies transnational terrorism. The incident took the lives of 270 people and, after a 10-year investigation, resulted in the life sentence of a Libyan intelligence agent (CNN 2001a). In 2003, the Libyan government agreed to pay $2.7 billion in compensation to victims' families (Smith 2004). Other examples of transnational terrorism include attacks on American embassies in Kenya and Tanzania and the bombing of the naval ship, the USS *Cole,* moored in Aden Harbor, Yemen. The 2001 bombings of the World Trade Center, the Pentagon, and Flight 93 are also examples of transnational terrorism. The September 11 bombings, linked to Al-Qaeda and its leader Osama Bin Laden, are the most devastating acts of terrorism in U.S. history.

Domestic terrorism, sometimes called insurgent terrorism (Barkan & Snowden 2001), is exemplified by the 1995 truck bombing of a nine-story federal office building in Oklahoma City, resulting in 168 deaths and the injury of more than 200. Gulf War veteran Timothy McVeigh, who, along with Terry Nichols, was convicted of

© CORBIS

Saddam Hussein, former President of Iraq, was captured by U.S. forces on December 14, 2003.

the crime, is reported to have been a member of a paramilitary group that opposes the U.S. government. In 1997, McVeigh was sentenced to death for his actions and was executed in 2001 (Barnes 2004).

Patterns of Global Terrorism

A 2003 report by the U.S. State Department describes patterns of terrorism around the world (U.S. State Department 2003). In 2002 (see Figure 16.2),

- There were 199 transnational acts of terrorism—44 percent fewer than in 2001—resulting in the deaths of 725 people.
- The number of terrorist attacks dropped to a 30-year all-time low.
- The number of anti-U.S. attacks was 77, 65 percent less than in the previous year.
- Thirty U.S. citizens were killed in terrorist attacks.
- Asia had the highest rate of terrorism followed by Latin America and the Middle East.
- Asia had the highest number of casualties, 1,281, followed by the Middle East East (772) and Eurasia (615).

Perhaps reflecting the overall decrease in terrorist acts, a national survey of U.S. adults found that more Americans were worried about the economy than terrorism (Gibbs 2003).

The Roots of Terrorism

Walter Laqueur's book, *No End to War: Terrorism in the 21st Century* (2003), dispels the myths that poverty and political oppression give birth to terrorist activity. As Laqueur notes, almost no terrorism has occurred in the world's poorest countries. Further, the most repressive regimes of the twenty-first century, Hitler's Germany, Russia under Stalin, and Mussolini's fascist Italy, were relatively free of terrorism.

Laqueur's research also suggests that terrorism flourishes in democracies. Similarly, a study of suicide bombers found that most suicide bombings take place in

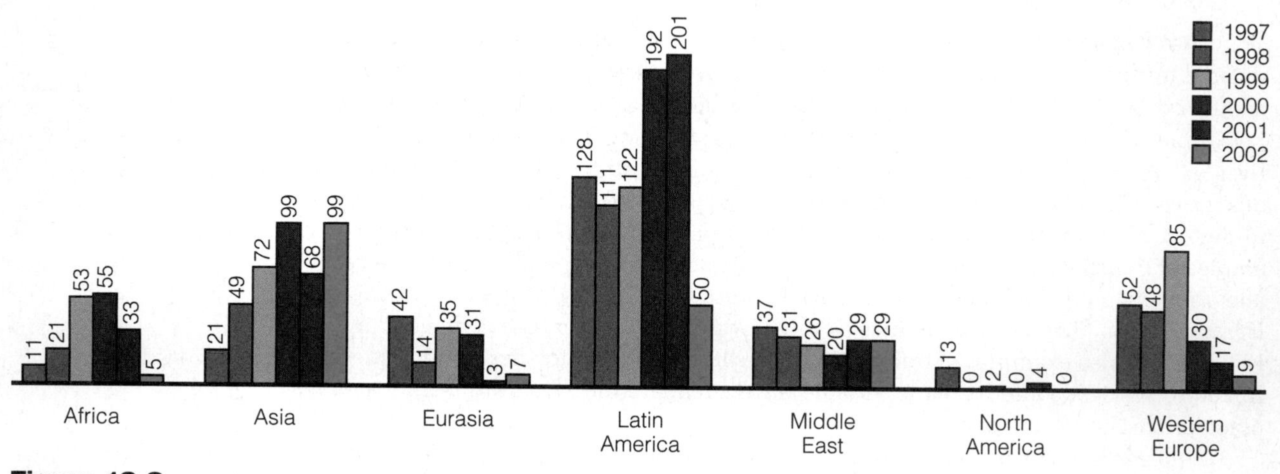

Figure 16.2
Total International Terrorist Attacks by Region: 1997–2002
Source: U.S. Department of State. "2002 Patterns of Global Terrorism Report." http://www.state.gov

democratic countries such as the United States, France, Israel, and Turkey (Pape 2003). Suicide bombings are also directed toward a specific objective and are rarely the acts of fanatics:

> From Lebanon to Israel to Sri Lanka to Kashmir to Chechnya, the sponsors of every campaign have been terrorist groups trying to establish or maintain political self-determination by compelling a democratic power to withdraw from the territories they claim. (p. 2)

But what are the causes of suicide bombings or, for that matter, terrorist attacks in general? In 2003, a panel of terrorist experts came together in Oslo, Norway, to address the causes of terrorism (Bjorgo 2003). Although not an exhaustive list, several causes emerged from the conference:

- A failed or weak state, which is unable to control terrorist operations
- Rapid modernization when, for example, a country's sudden wealth leads to rapid social change
- Extreme ideologies—religious or secular
- A history of political violence, civil wars, and revolutions
- Repression by a foreign occupation (i.e., invaders to the inhabitants)
- Large-scale racial or ethnic discrimination
- The presence of a charismatic leader

Note that Iraq, part of what President Bush calls the "axis of evil," has several of the characteristics listed above, including rapid modernization (e.g., oil reserves), extreme ideologies (e.g., Islamic fundamentalism), a history of violence (e.g., invasion of Kuwait), and a charismatic leader (e.g., Saddam Hussein).

America's Response to Terrorism

A government can use both defensive and offensive strategies to fight terrorism. Defensive strategies include using metal detectors and X-ray machines at airports and strengthening security at potential targets, such as embassies and military

AP/Wide World Photos

On March 11, 2004, ten bombs exploded on four commuter trains in Madrid killing 191 people and injuring at least 1,800. A video, which claimed that Al-Qaeda was responsible for the bombings, explained that the attacks were revenge for Spain's collaboration with Bush and his allies in the war in Iraq and Afghanistan.

command posts. The newly created Department of Homeland Security, made up of 22 domestic agencies (e.g., U.S. Coast Guard, Immigration and Naturalization, the Secret Service) and over 170,000 employees, coordinates such defensive tactics. The new department's first priority is to "protect the nation against further terrorist attacks" as "component agencies analyze threats and intelligence, guard our borders and airports, protect our critical infrastructure, and coordinate the response of our nation for future emergencies" (DHS 2003, 1)

Offensive strategies include retaliatory raids such as the U.S. bombing of terrorist facilities in Afghanistan, group infiltration, and preemptive strikes. New legislation facilitates such offensive tactics. In October 2001, the USA Patriot Act (Uniting and Strengthening America by Providing Appropriate Tools Required to Intercept and Obstruct Terrorism) was signed into law. The act increases police powers both domestically and abroad. Critics hold that the act poses a danger to civil liberties. For example, the act provides for the *indefinite* detention of immigrants if the immigrant group is defined as a "danger to national security" (Romero 2003). Because of the potential harm to civil liberties, several cities across the nation have enacted resolutions condemning the anti-terrorist legislation, and a federal lawsuit has been filed challenging the act on constitutional grounds (Schabner 2002; Lichtblau 2003). Despite such condemnations, less than 39 percent of Americans agree with the statement that "the federal government has become so large and powerful that it poses an immediate threat to the rights and freedoms of ordinary citizens" (Saad 2003, 1).

"The collars used on prisoners, the dogs, and the cameras did not suddenly appear out of thin air. . . . These acts of abuse were not the spontaneous actions of lower-ranking enlisted personnel who lacked the proper supervision."

Carl Levin
Senator from Michigan, on abuse of Iraqi prisoners

Advocates of the legislation note that during war, some restrictions of civil liberties are necessary. Moreover, the legislation is not "a substantive shift in policy but a mere revitalization of already established precedents" (Smith 2003, 25). Finally, many of the provisions of the act are set to expire within three years, making the critics' predictions of a 1984 Orwellian doom and gloom scenario absurd.

Combating terrorism is difficult, and recent trends will make it increasingly problematic (Zakaria 2000; Strobel, Kaplan, Newman, Whitelaw, & Grose, 2001). First, data stored on computers can be easily acquired by hackers who illegally gain access to classified information. Interlopers obtained the fueling and docking schedules of the USS *Cole*. Second, the Internet permits groups with similar interests, once separated by geography, to share plans, fund-raising efforts, recruitment strategies, and other coordinated efforts. Worldwide, thousands of terrorists keep in touch through Hotmail.com Internet accounts. Third, globalization contributes to terrorism by providing international markets where the tools of terrorism—explosives, guns, electronic equipment, and the like—can be purchased.

Finally, fighting terrorism under guerrilla warfare-like conditions is increasingly a concern. Unlike terrorist activity, which targets civilians and may be committed by lone individuals, **guerrilla warfare** is committed by organized groups opposing a domestic or foreign government and its military forces. Guerrilla warfare often involves small groups who use elaborate camouflage and underground tunnels to hide until they are ready to execute a surprise attack. In the fall of 2003, there were an estimated 5,000 Iraqi guerrillas fighting American forces in postwar Baghdad (Wilkinson 2003).

The possibility of terrorists using weapons of mass destruction is the most frightening scenario of all and the motivation for the 2003 war with Iraq. Weapons of mass destruction **(WMD)** include chemical, biological, and nuclear weapons. Anthrax, for example, although usually associated with diseases in animals, is a highly deadly disease in humans and although preventable by vaccine, has a "lethal lag time." In a hypothetical city of 100,000 people, delaying a vaccination program one day would result in 5,000 deaths; 6 days, 35,000 deaths. In 2001, trace amounts of anthrax were found in several letters sent to media and political figures resulting in five deaths and the inspection and closure of several postal facilities (Baliunas 2004).

Other examples of the use of WMD exist. On at least eight occasions, Japanese terrorists dispersed aerosols of anthrax and botulism in Tokyo (Inglesby et al. 1999) and in 2000, a religious cult, hoping to disrupt elections in an Oregon county, "contaminated local salad bars with salmonella, infecting hundreds" (Garrett 2001, 76). Government officials have initiated a variety of laws, policies, and technological innovations designed to combat WMD. For example, federal funds have been used to develop an experimental medicine tentatively called BCTP. The drug has successfully protected mice from injections of anthrax-like bacteria (Stipp 2004).

Social Problems Associated with War and Terrorism

Social problems associated with war and terrorism include death and disability; rape, forced prostitution, and displacement of women and children; social-psychological costs; diversion of economic resources; and destruction of the environment.

Death and Disability

More than 7 million people, most of them civilians, have lost their lives in the 45 wars and conflicts that are currently active (Worldwatch Institute 2003). Many American lives have been lost in wars, including over 53,000 in World War I, 292,000 in World War II, 34,000 in Korea, and 47,000 in Vietnam (*Statistical Abstract* 2002). Over 4,000 Americans lost their lives to terrorism in the last three years (U.S. State Department 2003). Globalization and sophisticated weapons technology combined with increased population density has made it easier to kill large numbers of people in a short amount of time. When the atomic bomb was dropped on the Japanese cities of Hiroshima and Nagasaki during World War II, 250,000 civilians were killed.

The impact of war and terrorism extends far beyond those who are killed. Many of those who survive war incur disabling injuries as well as diseases. For example, more than a quarter of a million people worldwide have been disabled by landmines—a continuing problem in the aftermath of the war with Iraq. In 1997, the Landmine Ban Treaty, requiring that governments destroy stockpiles within four years and clear landmine fields within 10 years, became international law. To date, 141 countries have signed the agreement; 53 countries remain (Landmines 2003a). This chapter's *Self and Society* feature tests your knowledge of the subject.

War-related deaths and disabilities also deplete the labor force, create orphans and single-parent families, and burden taxpayers who must pay for the care of orphans and disabled war veterans.

Individuals who participate in experiments for military research may also suffer physical harm. U.S. Representative Edward Markey of Massachusetts identified 31 experiments dating back to 1945 in which U.S. citizens were subjected to harm from participation in military experiments. Markey charged that many of the experiments used human subjects who were captive audiences or populations considered "expendable," such as the elderly, prisoners, and hospital patients. Eda Charlton of New York was injected with plutonium in 1945. She and 17 other

Self and Society

The Landmine Knowledge Quiz

Carefully read each of the following statements and select the correct answer. When finished, compare your answers to those provided and rank your performance using the scale below.

1. The most heavily mined regions of the world are in Southeast Asia and Central America. True or False?
2. Worldwide, more than 300 different types of landmines exist. True or False?
3. Each landmine costs between
   (a) $1000–2000
   (b) $500–1000
   (c) $3–30
   (d) $100–500
4. What percentage of landmine victims are civilians?
   (a) 10 percent
   (b) 60 percent
   (c) less than 1 percent
   (d) 75 percent
5. Landmines remain active long after their intended use, killing or injuring innocent people up to 10 years after they have been deployed. True or False?
6. It takes about a year and a half to train a mine detection dog. True or False?
7. Metal detectors are effective in identifying buried mines. True or False?
8. Landmines kill or maim 10,000 civilians every year. True or False?
9. In Cambodia, one out of every 1,236 people is an amputee from a landmine. True or False?
10. Today, approximately 10 million landmines lie in thousands of minefields around the world. True or False?

ANSWERS: 1. False (Africa and the Middle East are the most heavily mined); 2. True; 3. c; 4. d; 5. False (up to 75 years); 6. True; 7.False (mines are increasingly plastic); 8. True; 9. False (1 out of every 236); 10. False (there are over 60–70 million active landmines).

Rank your Performance—Number Correct

9 or 10	Excellent
7 or 8	Good
5 or 6	Average
3 or 4	Fair
0, 1 or 2	Poor

Source: Adapted from information at (2003) at http://www.landmines.org. Used by permission.

patients did not learn of their poisoning until 30 years later. Her son, Fred Shultz, said of his deceased mother:

> I was over there fighting the Germans who were conducting these horrific medical experiments . . . at the same time my own country was conducting them on my own mother. (Miller 1993, 17)

Rape, Forced Prostitution, and Displacement of Women and Children

"The use of rape in conflict reflects the inequalities women face in their everyday lives in peacetime. . . . Women are raped because their bodies are seen as the legitimate spoils of war."

Amnesty International

Half a century ago, the Geneva Convention prohibited rape and forced prostitution in war. Nevertheless, both continue to occur in modern conflicts.

Before and during World War II, the Japanese military forced an estimated 100,000 to 200,000 women and teenage girls into prostitution as military "comfort women." These women were forced to have sex with dozens of soldiers every day in "comfort stations." Many of the women died as a result of untreated sexually transmitted disease, harsh punishment, or indiscriminate acts of torture.

Since 1998, Congolese government forces have fought Uganda and Rwanda rebels. Women have paid a high price for this civil war where gang rape is "so violent, so systematic, so common . . . that thousands of women are suffering from vaginal fistula, leaving them unable to control bodily functions and enduring ostracism and the threat of debilitating health problems" (Wax 2003, 1). Rapes of Albanian women by Serbian and Yugoslavian paramilitary soldiers were, according to a Human Rights Watch report, "used deliberately as an instrument to terrorize the civilian population, extort money from families, and push people to flee their homes" (quoted in Lorch & Mendenhall 2000, 2). Feminist analysis of wartime rape emphasizes the practice as reflecting not only a military strategy but ethnic and gender dominance as well. In 2001, three Serbian soldiers accused of mass rape and forced prostitution were convicted by a United Nations tribunal and sentenced to a combined total of 60 years in prison (CNN 2001b).

War and terrorism also force women and children to flee to other countries or other regions of their homeland. For example, an estimated 20 million children have been forced to leave their homeland because of conflict and human rights violations (UNICEF 2003). Refugee women and female children are particularly vulnerable to sexual abuse and exploitation by locals, members of security forces, border guards, or other refugees. In refugee camps, women and children may also be subjected to sexual violation. A 2003 report by *Save the Children* examined the treatment of women and children in 40 conflict zones. The use of child soldiers was reported in 70 percent of the zones, and trafficking of women and girls in 85 percent of the zones. The most dangerous zones for women and children were Angola, Burundi, Sierra Leone, the Democratic Republic of the Congo, and Afghanistan, where the lives of 4 million women and 6 million children are endangered (Save the Children 2003).

Social-Psychological Costs

Terrorism, war, and living under the threat of war interfere with social-psychological well-being as well as family functioning (see this chapter's *Social Problems Research Up Close* feature). In a study of 269 Israeli adolescents, Klingman and Goldstein (1994) found a significant level of anxiety and fear, particularly among younger females, in regard to the possibility of nuclear and chemical warfare. Similarly, Myers-Brown, Walker, and Myers-Walls (2000) report that Yugoslavian children suffer from depression, anxiety, and fear as a response to recent conflicts in that region. Children, as well as adults, had similar emotional responses to the September 11 bombing of the World Trade Center and Pentagon (NASP 2003). Whether it be war or terrorism, "virtually every aspect of a child's development is damaged in such circumstances" (Bellamy 2003).

Civilians who are victimized by war and military personnel who engage in combat may experience a form of psychological distress known as **post-traumatic stress disorder** (PTSD), a clinical term referring to a set of symptoms that may result from any traumatic experience, including crime victimization, rape, or war. Symptoms of PTSD include sleep disturbances, recurring nightmares, flashbacks, and poor concentration (PTSD 2003). For example, Canadian Lt. General Romeo Dallaire, head of the UN peacekeeping mission in Rwanda, witnessed horrific acts of genocide. Four years after his return he continues to have images of "being in a valley at sunset, waist deep in bodies, covered in blood" (quoted in Rosenberg 2000, 14). PTSD is also associated with other personal problems, such as alcoholism, family violence, divorce, and suicide.

One study estimates that about 30 percent of male veterans of the Vietnam War have experienced PTSD, and about 15 percent continue to experience it (Hayman

Social Problems Research Up Close

Family Adjustment to Military Deployment

Research indicates that long and/or frequent absences from home by a spouse may create a variety of family problems including feelings of isolation, depression, and marital conflict. Despite recent trends toward downsizing the military, deployment of military personnel has increased, and as a result, separations of military families have increased. The present investigation by Rohall, Segal, and Segal (1999) examines the impact of military deployment on family functioning.

Sample and Methods

A survey was administered to 518 enlisted personnel stationed at air bases in Dsuwon, Kunson, and Osan, South Korea. All respondents were from one of two battalions. Battalion A personnel were separated from their families for a longer period of time (19 months) than those from Battalion B (7 months), and had been deployed more frequently. Thus, the first independent variable is *length and frequency of separation* from family. Respondents were also questioned about their *morale* (e.g., "How satisfied are you with your current job?"), perceived *organizational support* for family and self (e.g., support from commanding officer, chaplain, etc.), and satisfaction with *resources to communicate* with family members. The dependent variable is the *soldiers' assessment of family adjustment* to their absence. The authors' (1999) research question is thus, "Are soldiers' perceptions of family adjustment affected by (1) time and frequency of separation from family, (2) morale, (3) perceptions of organizational support, and/or (4) satisfaction with resources to communicate with family members?

Findings and Conclusions

As hypothesized, soldiers in Battalion B were more likely to report lower levels of family functioning than soldiers in Battalion A. Thus, longer and more frequent separations are associated with poorer family adjustment or, at least, the perception of poorer family adjustment. Morale was also positively correlated with family adjustment—the higher a soldier's reported morale, the greater his or her perception of positive family functioning. Similarly, respondents' ratings of organizational support and satisfaction with resources to communicate with family members were directly associated with higher family functioning ratings.

Although the results of the study indicate that each of the four independent variables is related to family adjustment, the relative importance of each is unclear. Further analysis by Rohall, Segal, and Segal, (1999) indicates that morale is the best single predictor of perceived family adjustment, followed by organizational support. Surprisingly, length and frequency of separation as measured by unit, that is, Battalion A versus Battalion B, contributed only minimally to variation in family adjustment.

As the authors note, the results of the present study may, in part, reflect what is called selection bias. For example, soldiers who believe that their families are not adjusting to military life have low morale, feel little organizational support, and/or are dissatisfied with communication resources may be more likely to drop out of the military. If that is the case, the present research, by only measuring *active* duty personnel, sampled those *most* satisfied with military life and thus used a biased sample. In this case, however, the possibility of selection bias does not negate the significance of the research results. If concern over family adjustment leads to soldiers leaving the military, keeping the length and frequency of separation low, morale high, organizational support strong, and communication resources satisfactory should reduce dropout rates. In a voluntary army when concerns over the quality of military personnel and recruitment shortfalls are high, insight into retention is invaluable.

Source: David E. Rohall, Mady Wechsler Segal, and David R. Segal. 1999. "Examining the Importance of Organizational Supports on Family Adjustment to Army Life in a Period of Increasing Separation." *Journal of Political and Military Sociology* 27:49–65.

& Scaturo 1993). Another study of 215 Army National Guard and Army Reserve troops who served in the Gulf War and who did not seek mental health services upon return to the states reports that 16 to 24 percent exhibited symptoms of PTSD (Sutker, Uddo, Brailey, & Allain 1993). Compared to other civilians, refugees have higher rates of PTSD and depression owing to stressors common to refugee camps (De Jong, Scholte, Koeter, and Hart 2000). Finally, research on PTSD in children reveals that females are generally more symptomatic than males and that PTSD may disrupt the normal functioning and psychological development of children (Pfefferbaum 1997; Cauffman, Feldman, Waterman, & Steiner 1998; PTSD 2003).

Diversion of Economic Resources

As discussed earlier, maintaining the military and engaging in warfare require enormous financial capital and human support. In 2002, worldwide military expenditures approached $800 billion (SIPRI 2003). This amount exceeds the combined government research expenditures on developing new energy technologies, improving human health, raising agricultural productivity, and controlling pollution. Although military spending by rogue states (Cuba, Iraq, Iran, Libya, North Korea, Sudan, and Syria) was just $14.4 billion in 2002, U.S. allies weighed in at $212 billion (CDI 2003).

Money that is spent for military purposes could be allocated for social programs. The decision to spend $567 million for one Trident II D-5 missile, equal to the operating cost of the Smithsonian Institution (CDI 2003), is a political choice. Similarly, allocating $2.3 billion for a "Virginia" Attack Submarine while our schools continue to deteriorate is also a political choice. Nonetheless, when a national survey of adults were asked about defense spending, 70 percent responded that the government was spending "too little" or "about the right amount" (Roper 2003). As Figure 16.3 indicates, the projected 2006 U.S. budget allocates more

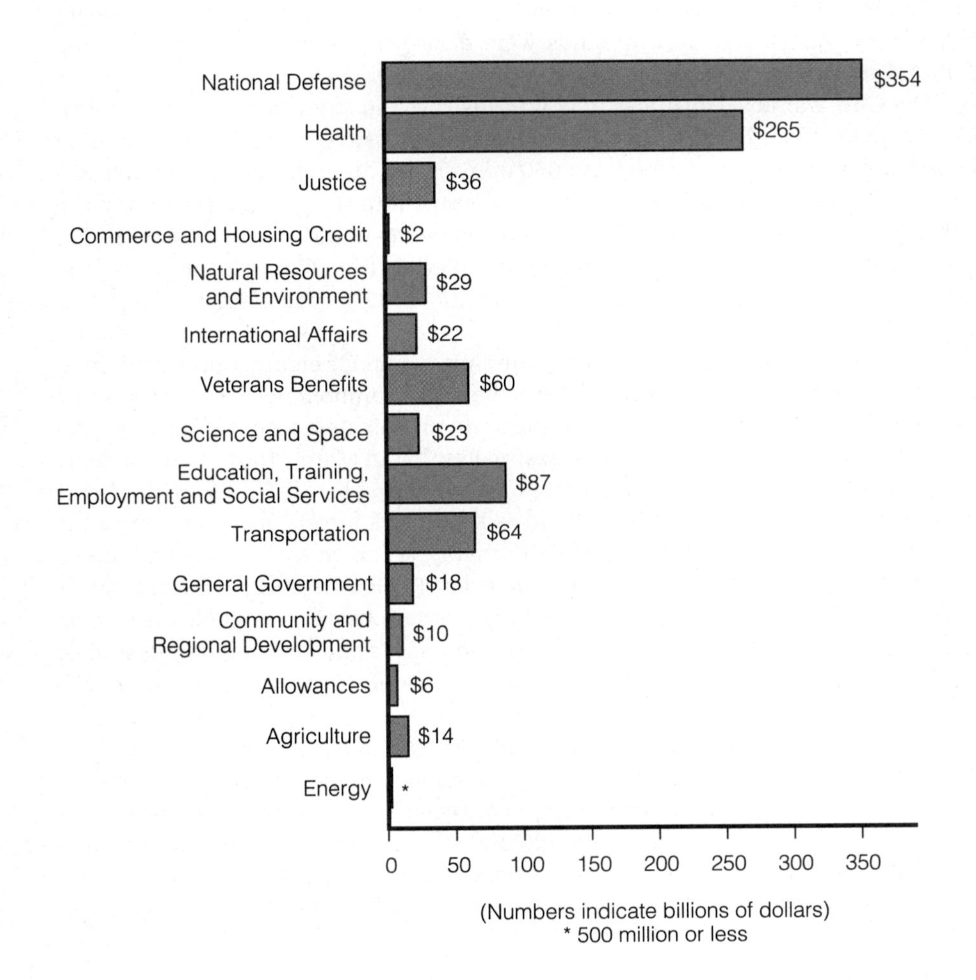

Figure 16.3 Selected Federal U.S. Outlays: 2006 (projected)

Source: Office of Management and Budget. 2003.

money for national defense than for justice, transportation, veterans' benefits, and natural resources and the environment combined (Office of Management and Budget 2003).

Destruction of the Environment

Traditional definitions of and approaches to national security have assumed that political states or groups constitute the principal threat to national security and welfare. This assumption implies that national defense and security are best served by being prepared for war against other states. The environmental costs of military preparedness are often overlooked or minimized in such discussions. Annually, for example, the U.S. Navy Public Works Center in San Diego produces more than 12,000 tons of paint cans. The depleted cans, enough to fill 12 semi-trucks, contain volatile chemicals making their disposal or any proposed recycling methods problematic (Lifsher 1999). A seemingly trivial matter—paint cans, in one year, at one facility. Imagine the environmental problems generated by the military, worldwide.

"My God . . . what have we done?"

Robert Lewis
co-pilot of the Enola Gay, after dropping the atomic bomb on Hiroshima

Destruction of the Environment During War. The environmental damage that occurs during war continues to devastate human populations long after war ceases. As the casings for landmines erode, for example, poison substances–often carcinogenic—leak into the ground (Landmines 2003b). In Vietnam, 13 million tons of bombs left 25 million bomb craters. Nineteen million gallons of herbicides, including Agent Orange, were spread over the countryside. An estimated 80 percent of Vietnam's forests and swamp lands were destroyed by bulldozing or bombing (Funke 1994).

The Gulf War also illustrates how war destroys the environment (Funke 1994; Renner 1993; EMS 2002). In 6 weeks, 1,000 air missions were flown, dropping 6,000 bombs. In 1991, Iraqi troops set 650 oil wells on fire, releasing oil, which still covers the surface of the Kuwaiti desert and seeps into the ground, threatening to poison underground water supplies. The estimated 6–8 million barrels of oil that spilled into the Gulf waters are threatening marine life. This spill is by far the largest in history—25–30 times the size of the 1989 Exxon Valdez oil spill in Alaska.

The clouds of smoke that hung over the Gulf region for eight months contained soot, sulfur dioxide, and nitrogen oxides—the major components of acid rain—and a variety of toxic and potentially carcinogenic chemicals and heavy metals. The U.S. Environmental Protection Agency estimates that in March 1991 about 10 times as much air pollution was being emitted in Kuwait as by all U.S. industrial and power-generating plants combined. Acid rain destroys forests and harms crops. It also activates several dangerous metals normally found in soil, including aluminum, cadmium, and mercury. Further, the over 900,000 depleted uranium munitions fired at Kuwait and Iraq are thought to have serious environmental as well as health consequences. This chapter's *The Human Side* feature poignantly describes the destructive course of Gulf War Syndrome—one of the many diseases thought to be associated with this radioactive material.

The ultimate environmental catastrophe facing the planet is a massive exchange thermonuclear war. Aside from the immediate human casualties, poisoned air, poisoned crops, and radioactive rain, many scientists agree that the dust storms and concentrations of particles would block vital sunlight and lower temperatures in the Northern Hemisphere, creating a **nuclear winter.** In the event of large-scale nuclear war, most living things on earth would die. The fear of nuclear war has

The Human Side

The Death of a Gulf War Veteran

The following letter to American Legion Magazine *is from the parents of a Gulf War veteran.*

Our son may be unique. We really don't know. We're desperate for information that is extremely hard to come by. We are very frustrated that what happened to us may have already happened to other families—or will—before something is done.

This is for Scott, our son, who said, "Go for it, Mom," when I told him I'd never quit trying to find out what made him so ill.

He was sent to the Persian Gulf with the 1133rd Transportation Co., National Guard, of Mason City, Iowa. He left for Saudi as a very healthy young man and he returned in much the same physical condition. Just tired, but not unusual considering where he'd been and what he'd seen.

. . . Two years after he returned home in the summer of 1993, he developed a rash on his torso. It would erupt, disappear and then come back again. It didn't always look the same. We thought it was a heat rash, something he ate, new laundry soap, etc.

In the fall of 1993 he developed sores in his mouth. He could eat very little and quickly lost about 40 pounds. He went from specialist to specialist, who sent tests many places to try and find a cause. He was put on steroids and depending on the dosage, it would get a little better and then flare up again. By winter he could only eat pureed foods and liquids. He could not use a straw as it hurt too much. Many doctors, many tests, many different medications. Nothing helped very much. And no concrete diagnosis.

In May 1994, he was examined by the VA hospital in Des Moines. They were not able to find the cause for the terrible sores in his mouth and the rash that now also affected his feet, hands and arms. He made many trips to Des Moines—a three-hour drive.

In early August 1994, the VA Hospital in Des Moines came up with a diagnosis of lupus. My heart just broke when I heard those words, but he was thrilled to know there was finally a name and treatment for his symptoms.

By mid-August he was hardly able to walk, his feet swollen and so extremely sore from the rash that seemed to get worse by the day. He went to the VA Hospital on Friday, August 19 . . . and was admitted. The rash had become blisters about the size of a 50-cent piece and were breaking and bleeding. He was running a fever . . .

He was transferred to University [hospital] in Iowa City and taken to surgery . . . [where] they removed all his skin and replaced it with what is called pig skin. Out of surgery, bandaged from head to toe, he was given a five percent chance of survival.

Our family gathered together to give him all the love and support we could. Ten days later the pig skin was removed. Infection had been found. It was too risky to take him back to surgery, so it was removed in a sterile room close to his room.

The next seven weeks are a blur. We had many ups—a good day for Scott—and many downs—a bad day for him. He had to endure burn baths every day. Water jets removed sloughed skin. He'd grow a tiny patch of skin only to lose it in a day or so later. He was fed through a tube. Many antibiotics were given in the hopes of warding off infection; morphine for the tremendous pain. We read to him—he'd correctly pronounce words we missed. His sense of humor never left. He worried about all of us.

Scott's life ended at 3 A.M. on October 15, 1994 . . .

We are convinced beyond anything we've ever felt as parents that Persian Gulf Syndrome killed our only son. We want someone to tell us the truth. We won't quit until we have believable answers to our question.

What do we tell his oldest son, now 11, who has nothing but pictures of his daddy . . . ?

What do we say to a three-year-old whose wish is to build a rocket so he can go get daddy and make everyone happy?

What do you say to a son who was born two weeks after his daddy died . . . ?

What do you say to a wife who longs only for her husband's love, strength and support?

If chemicals and gases were used over there, tell us the truth. It kills innocent people and destroys innocent families. We are real people with real feelings and we deserve the truth. This was not easy to write. We did it in hopes it may help someone. Only then will Scott's death make sense to us. He would have wanted it that way. He was that special.

Ardie and Rollie Siefken, Plainfield, Iowa

Source: "The Sad Death of a Gulf War Veteran." *American Legion Magazine,* August 6, 1996. Copyright lifted for distribution. http://www.ascension-research.org/part-5.html

Oil smoke from the 650 burning oil wells left in the wake of the Gulf War contains soot, sulfur dioxide, and nitrogen oxides, the major components of acid rain, along with a variety of toxic and potentially carcinogenic chemicals and heavy metals

greatly contributed to the military and arms buildup, which, ironically, also causes environmental destruction even in times of peace.

Destruction of the Environment in Peacetime. Even when military forces are not engaged in active warfare, military activities assault the environment. For example, modern military maneuvers require large amounts of land. The use of land for military purposes prevents other uses, such as agriculture, habitat protection, recreation, and housing. More important, military use of land often harms the land. In practicing military maneuvers, the armed forces demolish natural vegetation, disturb wildlife habitats, erode soil, silt up streams, and cause flooding. Military bombing and shooting ranges leave the land pockmarked with craters and contaminate the soil and groundwater with lead and other toxic residues. Further,

> mined areas can restrict access to large areas of agricultural land, forcing populations to use small tracts of land to earn their livelihoods. The limited productive land that is available is over-cultivated. which contributes to long term underproduction, as minerals are depleted from the soil, and the loss of valuable vegetation. (Landmines 2003b, 1)

Bombs exploded during peacetime leak radiation into the atmosphere and groundwater. From 1945 to 1990, 1,908 bombs were tested—that is, exploded—at more than 35 sites around the world. Although underground testing has reduced radiation, some still escapes into the atmosphere and is suspected of seeping into groundwater. Similarly, in Russia decommissioned submarines have been found to emit radioactive pollution (Editorial 2000).

Finally, although arms-control and disarmament treaties of the last decade have called for the disposal of huge stockpiles of weapons, no completely safe means of disposing of weapons and ammunition exist. Many activist groups have called for placing weapons in storage until safe disposal methods are found. Unfortunately, the longer weapons are stored, the more they deteriorate, increasing the likelihood of dangerous leakage. In 2003, a federal court judge gave permission, despite objections by environmentalists, to incinerate over 2,000 tons of nerve agents and mustard gas left from the Cold War era. Although the Army says that it is safe to dispose of the weapons, they have issued protective gear in case of an "accident" to the nearly 20,000 residents who live nearby (CNN 2003).

Strategies for Action: In Search of Global Peace

Various strategies and policies are aimed at creating and maintaining global peace. These include the redistribution of economic resources, the creation of a world government, peacekeeping activities of the United Nations, mediation and arbitration, and arms control.

Redistribution of Economic Resources

Inequality in economic resources contributes to conflict and war as the increasing disparity in wealth and resources between rich and poor nations fuels hostilities and resentment. Therefore, any measures that result in a more equal distribution of economic resources are likely to prevent conflict. John J. Shanahan (1995), retired U.S. Navy vice admiral and director of the Center for Defense Information, suggests that wealthy nations can help reduce social and economic roots of conflict by providing economic assistance to poorer countries. Nevertheless, U.S. military expenditures for national defense far outweigh U.S. economic assistance to foreign countries.

As we discussed in Chapter 14, strategies that reduce population growth are likely to result in higher levels of economic well-being. Funke (1994) explains that "rapidly increasing populations in poorer countries will lead to environmental overload and resource depletion in the next century, which will most likely result in political upheaval and violence as well as mass starvation" (p. 326). Although achieving worldwide economic well-being is important for minimizing global conflict, it is important that economic development does not occur at the expense of the environment.

Finally, UN Secretary General Kofi Annan in an address to the United Nations concludes that it is not poverty per se that leads to conflict but rather the "inequality among domestic social groups" (Deen 2000). Referencing a research report completed by the Tokyo-based UN University, Annan argues that "inequality . . . based on ethnicity, religion, national identity or economic class . . . tends to be reflected in unequal access to political power that too often forecloses paths to peaceful change" (Deen 2000).

World Government

Some analysts have suggested that world peace might be attained through the establishment of a world government. The idea of a single world government is not new. In 1693, William Penn advocated a political union of all European monarchs, and in 1712, Jacques-Henri Bernardin de Saint-Pierre of France suggested an all-European Union with a "Senate of Peace." Proposals such as these have been made throughout history.

From the 1950s until the early 1990s, world power was primarily divided between the United States and the Soviet Union. With the fall of communism in the early 1990s, the United States became the world's leader, causing observers to debate the proper role of the United States in global politics. Many argued that with the demise of the Soviet Union the world no longer needed a watchdog—a role the United States had historically played. Still others argued that "it is both the moral duty and in the strategic interest of the United States to become involved in regions where there is unrest, and stand as a leading force in international organizations, especially the United Nations" (SSRC 2003,1).

Although some commentators are pessimistic about the likelihood of a new world order, Lloyd (1998) identifies three global trends that, he contends, signify the "ghost of a world government yet to come" (p. 28). First is the increasing tendency for countries to engage in "ecological good behavior," indicating a concern for a global rather than national well-being. Second, despite several economic crises worldwide, major world players such as the United States and Great Britain have supported rather than abandoned nations in need. And, finally, an International Criminal Court has been created with "powers to pursue, arraign, and condemn

On September 11, 2001, the most devastating acts of terror ever killed thousands. On September 20, President Bush announced that all the evidence we have gathered points to a collection of loosely affiliated terrorist organizations known as Al-Qaeda.

© Reuters/CORBIS

those found guilty of war and other crimes against humanity" (p. 28). On the basis of these three trends, Lloyd concludes, "Global justice, a wraith pursued by peace campaigners for over a century, suddenly seems achievable" (p. 28).

The United Nations

The United Nations (UN), whose charter begins, "We the people of the United Nations—Determined to save succeeding generations from the scourge of war . . ." has engaged in over 55 peacekeeping operations since 1948 (UN 2003). The UN Security Council can use force, when necessary, to restore international peace and security.

> United Nations peacekeepers—military personnel in their distinctive blue helmets or blue berets, civilian police and a range of other civilians—help implement peace agreements, monitor cease fires, create buffer zones, or support complex military and civilian functions essential to maintain peace and begin reconstruction and institution-building in societies devastated by war. (UN 2003, 1).

Recently, the United Nations has been involved in overseeing multinational peacekeeping forces in Afghanistan, East Timor, Sierra Leone, Kosovo, Congo, and Iraq. However, the 2003 bombing of UN headquarters in Iraq led to the immediate although temporary withdrawal of UN officials from that country (UNDP 2003).

In the last few years, the United Nations has come under heavy criticism. First, in recent missions, developing nations have supplied more than 75 percent of the troops while developed countries—United States, Japan, and Europe—have contributed 85 percent of the finances. As one UN official commented, "you can't have a situation where some nations contribute blood and others only money" (quoted in Vesely 2001, 8). Second, a recent review of UN peacekeeping operations noted several failed missions including an intervention in Somalia in which 44 American marines were killed (Lamont 2001). Third, as typified by the debate over the disarming of Iraq, the UN cannot take sides but must wait for a consensus of its members which, if not forthcoming, undermines the strength of the organization (Goure 2003).

Finally, the concept of the United Nations is that its members represent individual nations, not a region or the world. And because nations tend to act in their own best economic and security interests, UN actions performed in the name of world peace may be motivated by nations acting in their own interests. Despite such criticisms, the proposed 2004–2005 budget is an increase over the previous year totaling over $3 billion (UN News Service 2003).

Mediation and Arbitration

Most conflicts are resolved through nonviolent means (Worldwatch Institute 2003). Mediation and arbitration are just two of the nonviolent strategies used to resolve conflicts and stop or prevent war. In mediation, a neutral third party intervenes and facilitates negotiation between representatives or leaders of conflicting groups. Mediators do not impose solutions but rather help disputing parties generate options for resolving the conflict (Conflict Research Consortium 2003). Ideally, a mediated resolution to a conflict meets at least some of the concerns and interests of each party to the conflict. In other words, mediation attempts to find "win-win" solutions in which each side is satisfied with the solution. Although mediation is used to resolve conflict between individuals, it is also a valuable tool for resolving international conflicts. For example, in 2000, mediators were used to resolve the escalating violence between Israeli and Palestinian troops. Using mediation as a means of resolving international conflict is often difficult given the complexity of the issues.

Arbitration also involves a neutral third party who listens to evidence and arguments presented by conflicting groups. Unlike mediation, however, in arbitration the neutral third party arrives at a decision or outcome that the two conflicting parties agree to accept. As Sweet and Brunell (1998) note, **triad dispute resolution,** that is, resolution that involves two disputants and a negotiator, performs "profoundly political functions including the construction, consolidation, and maintenance of political regimes" (p. 64).

"War doesn't determine who's right, just who's left."

George Carlin
Comedian

Arms Control and Disarmament

In the 1960s, the United States and the Soviet Union led the world in an arms race, each competing to build a more powerful military arsenal than its adversary. If either superpower were to initiate a full-scale war, the retaliatory powers of the other nation would result in the destruction of both nations. Thus, the principle of **mutually assured destruction (MAD)** that developed from nuclear weapons capabilities transformed war from a win-lose proposition to a lose-lose scenario. If both sides would lose in a war, the theory goes, neither side would initiate war.

Because of the end of the Cold War and the growing realization that current levels of weapons literally represented "overkill," governments have moved in the direction of arms control, which involves reducing or limiting defense spending, weapons production, and armed forces. Recent arms-control initiatives include SALT (Strategic Arms Limitation Treaty), START (Strategic Arms Reduction Treaty), SORT (Strategic Offensive Reduction Treaty), and NPT (Nuclear Nonproliferation Treaty).

Strategic Arms Limitation Treaty. Under the 1972 SALT agreement (SALT I), the United States and the Soviet Union agreed to limit both their defensive weapons and their land-based and submarine-based offensive weapons. Also in 1972, Henry Kissinger drafted the Declaration of Principles, known as **detente,** which means "negotiation rather than confrontation." A further arms limitation agreement (SALT II) was reached in 1979, but was never ratified by Congress because of the Soviet invasion of Afghanistan. Subsequently, the arms race continued with the development of new technologies and an increase in the number of nuclear warheads.

Strategic Arms Reduction and Strategic Offensive Reduction Treaties. Strategic arms talks resumed in 1982 but made relatively little progress for several years. During this period, President Reagan proposed the Strategic Defense Initiative, more commonly known as "Star Wars," which purportedly would be able to block missiles launched by another country against the United States. Although some research was conducted on the system, Star Wars was never actually built. Today, research on the development of a national missile defense system continues with proposed funding for 2001–2005 reaching $10.4 billion.

In 1991, the international situation changed dramatically. The communist regime in the Soviet Union had fallen, the Berlin Wall had been dismantled, and many Eastern European and Baltic countries were under self-rule. SALT was renamed START (Strategic Arms Reduction Treaty) and was signed in 1991. A second START agreement, signed in 1993 and ratified by the U.S. Senate, signaled the end of the Cold War. START II called for the reduction of nuclear warheads to 3,500 by the year 2003, but was superseded by SORT (Strategic Offensive Reduction Treaty), which requires both the United States and Russia to reduce their nuclear arsenals to 1,700–2,200 warheads by December 31, 2012 (Arms Control Association 2003a).

Nuclear Nonproliferation Treaty. The 1970 Nuclear Nonproliferation Treaty (NPT) was renewed in 2000, adopted by 187 countries. Only India, Israel, and Pakistan have not signed the agreement (Arms Control Association 2003b). The treaty holds that countries without nuclear weapons will not to try to get them; in exchange, the nuclear-weapon countries (the United States, United Kingdom, France, China, and Russia) agree they will not provide nuclear weapons to countries that did not have them. In January of 2003, North Korea, under suspicion of secretly producing nuclear weapons, announced that it was withdrawing from the treaty effective immediately (Arms Control Association 2003b).

> "I hate war as only a soldier who has lived it can, only as one who has seen its brutality, its futility, its stupidity."
>
> **Dwight D. Eisenhower**
> **Former U.S. President, military leader, hero**

However, even if military superpowers honor agreements to limit arms, the availability of black market nuclear weapons and materials presents a threat to global security. For example, Kyl and Halperin (1997) note that U.S. security is threatened more by nuclear weapons falling into the hands of a terrorist group than by a nuclear attack from an established government such as Russia. The authors further conclude that if Russia were to launch missiles directed at the United States "the odds are overwhelming that it [would] be a Russian missile fired by accident or without authority" (p. 28).

In 2003, the United States accused Iran of operating a covert nuclear weapons program. When Iranian officials responded that their nuclear program was solely for the purpose of generating electricity, the U.S. ambassador to the investigating agency "accused Iran of 'stalling and stonewalling' the true goal of its nuclear activities" (AP 2003, 1).

Understanding Conflict, War, and Terrorism

As we come to the close of this chapter, how might we have an informed understanding of war and terrorism? Each of the three theoretical positions discussed in this chapter reflects the realities of global conflict. As functionalists argue, war offers societal benefits—social cohesion, economic prosperity, scientific and technological developments, and social change. Further, as conflict theorists contend, wars often occur for economic reasons as corporate elites and political leaders benefit from the spoils of war—land and water resources and raw materials. The symbolic interactionist perspective emphasizes the role that meanings, labels, and definitions play in creating conflict and contributing to acts of war.

The September 11 attacks on the World Trade Center and the Pentagon and the aftermath—the battle against terrorism, the bombing of Afghanistan, and the war with Iraq—forever changed the world we live in. For some theorists, these events were inevitable. Political scientist Samuel P. Huntington argues that such conflict represents a **clash of civilizations.** In *The Clash of Civilizations and the Remaking of World Order* (1996, 28), Huntington argues that in the new world order

> the most pervasive, important and dangerous conflicts will not be between social classes, rich and poor, or economically defined groups, but between people belonging to different cultural entities . . . the most dangerous cultural conflicts are those along the fault lines between civilizations . . . the line separating peoples of Western Christianity, on the one hand, from Muslim and Orthodox peoples on the other.

Although not without critics, the clash of civilizations hypothesis has some support. In an interview of almost 10,000 people from nine Muslim states representing half of all Muslims worldwide, only 22 percent had favorable opinions toward the United States (CNN 2003). Even more significantly, 67 percent saw the September 11 attacks as "morally justified" and the majority of respondents found the United States to be overly materialistic, secular, and having a corrupting influence on other nations.

Ultimately, we are all members of one community—Earth—and have a vested interest in staying alive and protecting the resources of our environment for our own and future generations. But, as we have seen, conflict between groups is a feature of social life and human existence that is not likely to disappear. What is at stake—human lives and the ability of our planet to sustain life—merits serious attention. World leaders have traditionally followed the advice of philosopher Karl von Clausewitz: "If you want peace, prepare for war." Thus, nations have sought to protect themselves by maintaining large military forces and massive weapons systems. These strategies are associated with serious costs, particularly in hard economic times. In diverting resources away from other social concerns, defense spending undermines a society's ability to improve the overall security and well-being of its citizens. Conversely, defense-spending cutbacks, although unlikely in the present climate, could potentially free up resources for other social agendas, including lowering taxes, reducing the national debt, addressing environmental concerns, eradicating hunger and poverty, improving health care, upgrading educational services, and improving housing and transportation. Therein lies the promise of a "peace dividend." The hope is that future dialogue on the problems of war and terrorism will redefine national and international security to encompass social, economic, and environmental well-being.

Chapter Review

- **What is the relationship between war and industrialization?**
  War indirectly impacts industrialization and technological sophistication as military research and development advances civilian-used technologies. Industrialization, in turn, has had two major influences on war: the more industrialized a country, the lower the rate of conflict, and if conflict occurs, the higher the rate of destruction.

- **In general, how do feminists view war?**
  Feminists are quick to note that wars are part of the patriarchy of society. Although women and children may be used to justify a conflict (e.g., improving women's lives by removing the repressive Taliban in Afghanistan), the basic principles of male dominance and control are realized through war.

- **What are some of the causes of war?**
  The causes of war are numerous and complex. Most wars involve more than one cause. Some of the causes of war are conflict over values or ideologies; racial, ethnic, and religious hostilities; defense against hostile attacks; revolution; and nationalism.

- **What is terrorism and what are the different types of terrorism?**
  Terrorism is the premeditated use, or threatened use, of violence by an individual or group to gain a political or social objective. Terrorism can be either transnational or domestic. Transnational terrorism occurs when a terrorist act in one country involves victims, targets, institutions, governments, or citizens of another country. Domestic terrorism involves only one nation such as the 1995 truck bombing of a nine-story federal office building in Oklahoma City.

- **How has the United States responded to the threat of terrorism?**
  The United States has used both defensive and offensive strategies to fight terrorism. Defensive strategies include using metal detectors and X-ray machines at airports and strengthening security at potential targets, such as embassies and military command posts. The newly created Department of Homeland Security coordinates such defensive tactics. Offensive strategies include retaliatory raids such as the U.S. bombing of terrorist facilities in Afghanistan, group infiltration, and preemptive strikes.

- **What is meant by "diversion of economic resources"?**
  Worldwide, the billions of dollars used on defense could be channeled into social programs dealing with, for example, education, health, and poverty. Thus, defense monies are economic resources diverted from other needy projects.

- **What are some of the criticisms of the United Nations?**
  First, in recent missions, developing nations have supplied more than 75 percent of the troops. Second, several recent UN peacekeeping operations have failed. Third, the UN cannot take sides but must wait for a consensus of its members which, if not forthcoming, undermines the strength of the organization. Finally, the concept of the United Nations is that its members represent individual nations, not a region or the world. Because nations tend to act in their own best economic and security interests, UN actions performed in the name of world peace may be motivated by nations acting in their own interests.

Critical Thinking

1. Certain actions constitute "war crimes." Such actions include the use of forbidden munitions such as biological weapons, purposeless destruction, killing civilians, poisoning waterways, and violation of surrender terms. In addition to those listed above, what other actions should constitute "war crimes"?
2. Selecting each of the five major institutions in society, what part could each play in attaining global peace?
3. Make a list of famous war movies (e.g., *Schindler's List*). With specific movies in mind, list media sounds and images of war. Has the portrayal of war in movies changed over time? If so, how and why?

Key Terms

arbitration
clash of civilizations
Cold War
cyberwarfare
detente
domestic terrorism
dual-use technologies
guerrilla warfare
military-industrial complex
mutually assured destruction (MAD)
nuclear winter
post-traumatic stress disorder
state
terrorism
transnational terrorism
triad dispute resolution
war
WMD

Taking a Stand

Does anti-terrorism legislation threaten civil liberties?

After September 11, 2001 the Bush administration passed several laws in the fight against terrorism. Some of these measures allow for covert government surveillance, detention of immigrant groups, tracking of e-mail, and roving wiretaps. Although many civil libertarians believe that the new laws violated citizens' rights, advocates argue that during war, measures must be taken to protect national security.

Use Wadsworth's exclusive online resources—InfoTrac College Edition, MicroCase Online, and OVRC—to formulate a position on this topic.

The Wadsworth's Sociology Online Resources and Writing Companion will help you get started. This valuable guide will show you how to use Wadsworth's exclusive online resources when studying social problems. It will also help you to build essential research and writing skills. InfoTrac College Edition, MicroCase Online, OVRC, and an electronic copy of portions of this companion are available at http://sociology.wadsworth.com/mooney_knox_schacht/problems4e, the companion Web site for *Understanding Social Problems,* Fourth Edition.

Media Resources

The Companion Web Site for *Understanding Social Problems,* Fourth Edition

http://sociology.wadsworth.com/mooney_knox_schacht/problems4e

Supplement your review of this chapter by going to the companion Web site to take one of the Tutorial Quizzes, use the flash cards to master key terms, and check out the many other study aids you'll find there. You'll also find special features such as *Wadsworth's Sociology Online Resources and Writing Companion,* GSS Data, and Census 2000 information, data, and resources at your fingertips to help you complete that special project or do some research on your own.

Interactions CD-ROM

Go to the Interactions CD-ROM for *Understanding Social Problems,* Fourth Edition, to access additional interactive learning tools, such as in-depth review materials, corresponding practice quizzes, and other engaging resources and activities to help you study the concepts in this chapter.

> "Never doubt that a small group of thoughtful, committed citizens can change the world. Indeed, it's the only thing that ever has."
> *Margaret Mead, Anthropologist*

Epilogue

Today, there is a crisis—a crisis of faith: faith in the ideals of equality and freedom, faith in political leadership, faith in the American dream, and, ultimately, faith in the inherent goodness of humankind and the power of one individual to make a difference. To some extent, faith is shaken by texts such as this one. Drug use is up; marriages down; political corruption is everywhere; bigotry's on the rise; the environment is killing us—if we don't kill it first. Social problems are everywhere, and what's worse, many solutions only seem to create more problems.

The transformation of American society in recent years has been dramatic. With the exception of the Industrial Revolution, no other period in human history has seen such rapid social change. The structure of society, forever altered by such macro-sociological processes as multi-nationalization, de-industrialization, and globalization, continues to be characterized by social inequities—in our schools, in our homes, in our cities, and in our salaries.

> "A pessimist sees the difficulty in every opportunity; an optimist sees the opportunity in every difficulty."
>
> **Sir Winston Churchill**
> **British statesman**

The culture of society has also undergone rapid change, leading many politicians and lay persons alike to call for a return to traditional values and beliefs and to emphasize the need for moral education. The implication is that somehow things were better in the "good old days," and if we could somehow return to those times, things would be better again. Some things were better—for some people.

Fifty years ago, there were fewer divorces and less crime. AIDS and crack cocaine were unheard of, and violence in schools was almost nonexistent. At the same time, however, in 1950, the infant mortality rate was more than three times what it is today; racial and ethnic discrimination flourished in an atmosphere of bigotry and hate, and millions of Americans were routinely denied the right to vote because of the color of their skin; more than half of all Americans smoked cigarettes; and persons over the age of 25 had completed a median of 6.8 years of school.

The social problems of today are the cumulative result of structural and cultural alterations over time. Today's problems are not necessarily better or worse than those of generations ago—they are different and, perhaps, more diverse as a result of the increased complexity of social life. But, as surely as we brought the infant mortality rate down, prohibited racial discrimination in education, housing, and employment, increased educational levels, and reduced the number of smokers, we can continue to meet the challenges of today's social problems. But how does positive social change occur? How does one alter something as amorphous as society? The answer is really quite simple. All social change takes place because of the acts of individuals. Every law, every regulation and policy, every social movement and media exposé, and every court decision began with one person.

> "The day will come when nations will be judged not by their military or economic strength, nor by the splendor of their capital cities and public buildings, but by the well-being of their peoples; by their levels of health, nutrition and education; by their opportunities to earn a fair reward for their labours; by their ability to participate in the decisions that affect their lives; by the respect that is shown for their civil and political liberties; by the provision that is made for those who are vulnerable and disadvantaged; and by the protection that is afforded to the growing minds and bodies of their children."
>
> **UNICEF The Progress of Nations, 2000**

Sociologist Earl Babbie (1994) recounts how the behavior of one person—Rosa Parks—made a difference. Rosa Parks was a seamstress in Montgomery, Alabama, in the 1950s. Like almost everything else in the South in the 1950s, public transportation was racially segregated. On December 1, 1955, Rosa Parks was on her way home from work when the "white section" of the bus she was riding became full. The bus driver told black passengers in the first row of the black section to relinquish their seats to the standing white passengers. Rosa Parks refused.

She was arrested and put in jail, but her treatment so outraged the black community that a boycott of the bus system was organized by a new minister in town—Martin Luther King, Jr. The Montgomery bus boycott was a success. Just 11 months later, in November of 1956, the U.S. Supreme Court ruled that racial segregation of public facilities was unconstitutional. Rosa Parks had begun a process that in time would echo her actions—the civil rights movement, the March on Washington, the 1963 Equal Pay Act, the 1964 Civil Rights Act, the 1965 Voting Rights Act, regulations against discrimination in housing, and affirmative action.

Was social change accomplished? In 1960, just 20 percent of blacks 25 years old and older had completed high school compared to 40 percent of whites. By 2000, 79 percent of blacks 25 years old and older had completed high school compared to 85 percent of whites (*Statistical Abstract* 2002). While many would point out that such changes and thousands like them have created other problems that need to be addressed, who among us would want to return to the "good old days" of the 1950s in Montgomery, Alabama?

> "The only thing necessary for the triumph of evil is for good men [and women] to do nothing."
>
> **Edmund Burke**
> **English political writer/orator**

Millions of individuals make a difference daily. Chuck Beattie and Bret Byfield, two social workers in Minneapolis, began Phoenix Group in 1991. Purchasing houses with a few grants and some private donations, they hired "street people" to renovate the houses and then let them move in. Today, Phoenix Group has more than 39 properties, 300 residents, and 11 businesses. Susan Brotchie, deserted by her husband and unable to get child support, began Advocates for Better Child Support, which has helped more than 9,000 people establish and collect child-support payments. Pedro José Greer was an intern in Miami when he treated his first homeless patient. Appalled by the fatal incidence of tuberculosis, a curable disease, Dr. Greer opened a medical clinic in a homeless shelter. Today, he heads the largest provider of medical care for the poor in Florida—Camillus Health Concern—annually serving over 4,500 patients (Chinni et al. 1995, 34). After suffering chronic illness from exposure to air pollutants from a sewage plant and home renovation materials, Mary Lamielle founded the National Center for Environmental Health Strategies (NCEHS), a national nonprofit organization dedicated to the development of creative solutions to environmental health problems (Lamielle 1995). Through her work as director of the NCEHS, Lamielle has influenced policy development and research and has provided support and advocacy to sufferers of environmental pollution. Julie Posey, a victim of childhood sexual abuse, helps thousands of people annually. She is the director of Pedowatch.com, a Web site that provides information on child sexual abuse.

> "Some people see things that are and say, 'Why?' He saw things that never were and said, 'Why not?'"
>
> **Ted Kennedy**
> **Of his brother, Bobby Kennedy**

College students have also worked to bring about social change. College students have prompted the adoption of multi-cultural curriculums, helped the homeless, made their schools more environmentally accountable, and organized against sweatshop labor (Loeb 1995; Alvarado 2000). College students also influenced universities to rid themselves of South African investments and played an important role in building the international movement that helped end apartheid.

While only a fraction of the readers of this text will occupy social roles that directly influence social policy, one need not be a politician or member of a social reform group to make a difference. We, the authors of this text, challenge you, the reader, to make individual decisions and take individual actions to make the world a more humane, just, and peaceful place for all. Where should we begin? Where Rosa Parks and others like her began—with a simple individual act of courage, commitment, and faith.

Appendix A
Methods of Data Analysis

There are three levels of data analysis: description, correlation, and causation. Data analysis also involves assessing reliability and validity.

Description

Qualitative research involves verbal descriptions of social phenomena. Having a homeless and single pregnant teenager describe her situation is an example of qualitative research.

Quantitative research often involves numerical descriptions of social phenomena. Quantitative descriptive analysis may involve computing the following: (1) means (averages), (2) frequencies, (3) mode (the most frequently occurring observation in the data), (4) median (the middle point in the data; half of the data points are above, and half are below the median), and (5) range (the highest and lowest values in a set of data).

Correlation

Researchers are often interested in the relationship between variables. *Correlation* refers to a relationship among two or more variables. The following are examples of correlational research questions: What is the relationship between poverty and educational achievement? What is the relationship between race and crime victimization? What is the relationship between religious affiliation and divorce?

If there is a correlation or relationship between two variables, then a change in one variable is associated with a change in the other variable. When both variables change in the same direction, the correlation is positive. For example, in general, the more sexual partners a person has, the greater the risk of contracting a sexually transmissible disease. As variable A (number of sexual partners) increases, variable B (chance of contracting an STD) also increases. Similarly, as the number of sexual partners decreases, the chance of contracting an STD decreases. Notice that in both cases, the variables change in the same direction, suggesting a positive correlation (see Figure A.1).

When two variables change in opposite directions, the correlation is negative. For example, there is a negative correlation between condom use and contracting STDs. In other words, as condom use increases, the chance of contracting an STD decreases (see Figure A.2).

The relationship between two variables may also be curvilinear, which means that they vary in both the same and opposite directions. For example, suppose a re-

Figure A-1
Positive Correlation

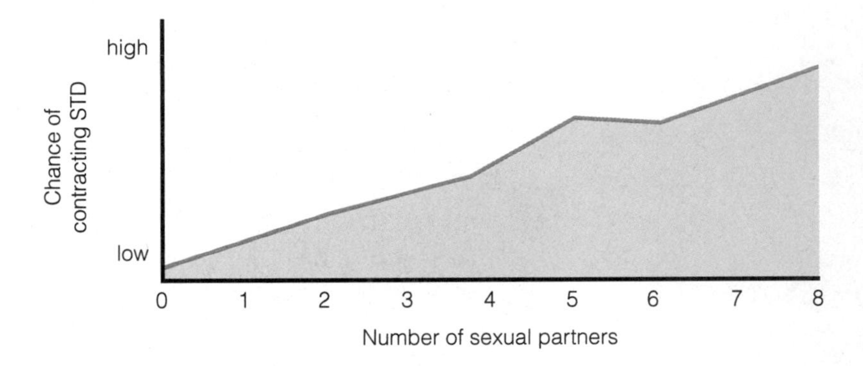

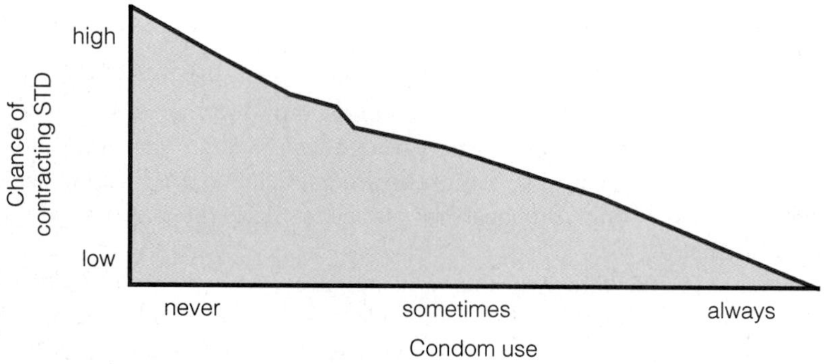

high

Chance of contracting STD

low

never sometimes always

Condom use

Figure A-2
Negative Correlation

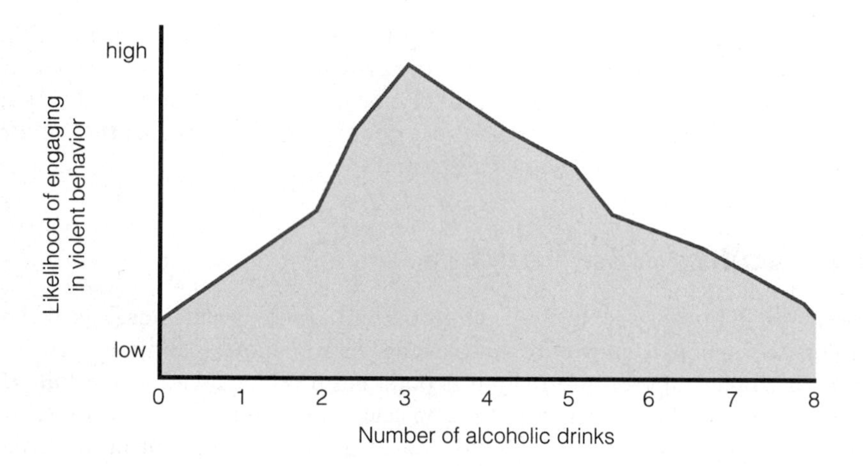

Figure A-3
Curvilinear Correlation

searcher finds that after drinking one alcoholic beverage, research participants are more prone to violent behavior. After two drinks, violent behavior is even more likely, and this trend continues for three and four drinks. So far, the correlation between alcohol consumption and violent behavior is positive. After the research participants have five alcoholic drinks, however, they become less prone to violent behavior. After six and seven drinks, the likelihood of engaging in violent behavior decreases further. Now the correlation betweeen alcohol consumption and violent behavior is negative. Because the correlation changed from positive to negative, we say that the correlation is curvilinear (the correlation may also change from negative to positive) (see Figure A.3).

A fourth type of correlation is called a spurious correlation. Such a correlation exists when two variables appear to be related, but the apparent relationship occurs only because they are both related to a third variable. When the third variable is controlled through a statistical method in which the variable is held constant, the apparent relationship between the variables disappears. For example, blacks have a lower average life expectancy than whites do. Thus, race and life expectancy appear to be related. However, this apparent correlation exists because both race and life expectancy are related to socioeconomic status. Because blacks are more likely than whites to be impoverished, they are less likely to have adequate nutrition and medical care.

Causation

If the data analysis reveals that two variables are correlated, we know only that a change in one variable is associated with a change in another variable. We cannot assume, however, that a change in one variable causes a change in the other variable unless our data collection and analysis are specifically designed to assess causation. The research method that best assesses causality is the experimental method (discussed in Chapter 1).

To demonstrate causality, three conditions must be met. First, the data analysis must demonstrate that variable A is correlated with variable B. Second, the data analysis must demonstrate that the observed correlation is not spurious. Third, the analysis must demonstrate that the presumed cause (variable A) occurs or changes prior to the presumed effect (variable B). In other words, the cause must precede the effect.

It is extremely difficult to establish causality in social science research. Therefore, much social research is descriptive or correlative, rather than causative. Nevertheless, many people make the mistake of interpreting a correlation as a statement of causation. As you read correlative research findings, remember the following adage: "Correlation *does not* equal causation."

Reliability and Validity

Assessing reliability and validity is an important aspect of data analysis. *Reliability* refers to the consistency of the measuring instrument or technique; that is, the degree to which the way information is obtained produces the same results if repeated. Measures of reliability are made on scales and indexes (such as those in the *Self and Society* features in this text) and on information-gathering techniques, such as the survey methods described in Chapter 1.

Various statistical methods are used to determine reliability. A frequently used method is called the "test-retest method." The researcher gathers data on the same sample of people twice (usually one or two weeks apart) using a particular instrument or method and then correlates the results. To the degree that the results of the two tests are the same (or highly correlated), the instrument or method is considered reliable.

Measures that are perfectly reliable may be absolutely useless unless they also have a high validity. *Validity* refers to the extent to which an instrument or device measures what it intends to measure. For example, police officers administer "Breathalyzer" tests to determine the level of alcohol in a person's system. The Breathalyzer is a valid test for alcohol consumption.

Validity measures are important in research that uses scales or indexes as measuring instruments. Validity measures are also important in assessing the accuracy of self-report data that are obtained in survey research. For example, survey research on high-risk sexual behaviors associated with the spread of HIV relies heavily on self-report data on such topics as number of sexual partners, types of sexual activities, and condom use. Yet how valid are these data? Do survey respondents underreport the number of their sexual partners? Do people who say they use a condom every time they engage in intercourse really use a condom every time? Because of the difficulties in validating self-reports of number of sexual partners and condom use, we may not be able to answer these questions.

Ethical Guidelines in Social Problems Research

Social scientists are responsible for following ethical standards designed to protect the dignity and welfare of people who participate in research. These ethical guidelines include the following (Schutt 1999; Neuman 2000; American Sociological Association 2001):

1. *Freedom from coercion to participate.* Research participants have the right to decline to participate in a research study or to discontinue participation at any time during the study. For example, professors who are conducting research using college students should not require their students to participate in their research.
2. *Informed consent.* Researchers are required to inform potential participants of any aspect of the research that might influence a subject's willingness to participate. After informing potential participants about the nature of the research, researchers typically ask participants to sign a consent form indicating that the participants are informed about the research and agree to participate in it.
3. *Deception and debriefing.* Sometimes the researcher must disguise the purpose of the research in order to obtain valid data. Researchers may deceive participants as to the purpose or nature of a study only if there is no other way to study the problem. When deceit is used, participants should be informed of this deception (debriefed) as soon as possible. Participants should be given a complete and honest description of the study and why deception was necessary.
4. *Protection from harm.* Researchers must protect participants from any physical and psychological harm that might result from participating in a research study. This is both a moral and a legal obligation. It would not be ethical, for example, for a researcher studying drinking and driving behavior to observe an intoxicated individual leaving a bar, getting into the driver's seat of a car, and driving away.

   Researchers are also obligated to respect the privacy rights of research participants. If anonymity is promised, it should be kept. Anonymity is maintained in mail surveys by identifying questionnaires with a number coding system rather than with the participants' names. When such anonymity is not possible, as is the case with face-to-face interviews, researchers should tell participants that the information they provide will be treated as confidential. Although interviews may be summarized and excerpts quoted in published material, the identity of the individual participants is not revealed. If a research participant experiences either physical or psychological harm as a result of participation in a research study, the researcher is ethically obligated to provide remediation for the harm.
5. *Reporting of research.* Ethical guidelines also govern the reporting of research results. Researchers must make research reports freely available to the public. In these reports, a researcher should fully describe all evidence obtained in the study, regardless of whether the evidence supports the researcher's hypothesis. The raw data collected by the researcher should be made available to other researchers who might request it for purposes of analysis. Finally, published research reports should include a description of the sponsorship of the research study, its purpose, and all sources of financial support.

Glossary

abortion The intentional termination of a pregnancy.

absolute poverty The lack of resources that leads to hunger and physical deprivation.

acculturation Learning the culture of a group different from the one in which the learner was originally raised.

achieved status A status assigned on the basis of some characteristic or behavior over which the individual has some control.

acid rain The mixture of precipitation with air pollutants, such as sulfur dioxide and nitrogen oxide.

acquaintance rape Rape that is committed by someone known by the victim.

activity theory A theory claiming that the elderly disengage, in part, because they are structurally segregated and isolated with few opportunities to engage in active roles.

adaptive discrimination Discrimination that is based on the prejudice of others.

affirmative action A broad range of policies and practices to promote equal opportunity and diversity in the workplace and on campuses.

age grading The assignment of social roles to given chronological ages.

ageism The belief that age is associated with certain psychological, behavioral, and/or intellectual traits.

agricultural biotechnology The application of biotechnology to agricultural crops and livestock.

Aid to Families with Dependent Children (AFDC) Before 1996, a cash assistance program that provided single parents (primarily women) and their children with a minimum monthly income.

alienation The concept used by Karl Marx to describe the condition when workers feel powerlessness and meaninglessness as a result of performing repetitive, isolated work tasks. Alienation involves becoming estranged from one's work, the products one creates, other human beings, and/or one's self; it also refers to powerlessness and meaninglessness experienced by students in traditional, restrictive educational institutions.

alternative certification programs Programs that permit college graduates, without education degrees, to be certified to teach school based on job and/or life experiences.

amalgamation The physical blending of different racial and/or ethnic groups resulting in a new and distinct genetic and cultural population; results from the intermarriage of racial and ethnic groups over generations.

androgynous Having both feminine and masculine characteristics.

anomie A state of normlessness in which norms and values are weak or unclear; results from rapid social change and is linked to many social problems, including crime, drug addiction, and violence.

antimiscegenation laws Laws that prohibited interracial marriages.

arbitration Dispute settlement in which a neutral third party listens to evidence and arguments presented by conflicting groups and arrives at a decision or outcome that the two conflicting parties agree to accept.

ascribed status A status that society assigns to an individual on the basis of factors over which the individual has no control.

assimilation The process by which minority groups gradually adopt the cultural patterns of the dominant majority group.

automation The replacement of human labor with machinery and equipment.

aversive racism A subtle, often unintentional form of prejudice that involves (1) feelings of discomfort, uneasiness, disgust, and sometimes fear (as opposed to feelings of hate and hostility); and (2) the presence of prowhite attitudes as opposed to antiblack attitudes.

barrios In the United States, slums that are occupied primarily by Latinos.

behavior-based safety programs A controversial health and safety strategy used by business management that attributes health and safety problems in the workplace to workers' behavior, rather than to work processes and conditions.

beliefs Definitions and explanations about what is assumed to be true.

bias-motivated crime See *hate crime.*

bigamy In the United States, the criminal offense of marrying one person while still legally married to another.

bilingual education Educational instruction provided in two languages—the student's native language and another language. In the United States, bilingual education involves teaching individuals in both English and their non-English native language.

biodiversity The diversity of living organisms on earth.

biofeedback A process in which information that is fed back to the brain enables a person to change his or her biological functioning. Biofeedback treatment can teach a person to influence biological responses such as heart rate, nervous system arousal, muscle contractions, and even brain wave functioning.

bioinvasion The emergence of organisms into regions where they are not native, usually as a result of being carried in the ballast water of ships, in packing material, and in shipments of crops and other goods that are traded around the world. Invasive species may compete with native species for food, start an epidemic, or prey on natives.

biomedicalization The view that medicine can not only control particular conditions, but can also transform bodies and lives.

biphobia Negative attitudes toward bisexuality and people who identify as bisexual.

biomonitoring A method of assessing a person's exposure to environmental chemicals or their metabolites (breakdown products) by testing for the presence of these in blood or urine.

bisexuality A sexual orientation that involves emotional and sexual attraction to members of both sexes.

bonded labor The repayment of a debt through labor.

bourgeoisie The owners of the means of production.

Brady Bill A law that requires a five-day waiting period for handgun purchases so that sellers can screen buyers for criminal records or mental instability.

brain drain The phenomenon in developing countries of many individuals with the highest level of skill and education leaving the country in search of work abroad.

brownfields Abandoned or undeveloped sites that are located on contaminated land.

burden of disease The number of deaths in a population combined with the impact of premature death and disability on that population.

capitalism An economic system in which private individuals or groups invest capital to produce goods and services, for a profit, in a competitive market.

central city The largest city in a metropolitan area.

character education Education that emphasizes the moral and interpersonal aspects of an individual.

charter schools Public schools founded by parents, teachers, and communities, and maintained by school tax dollars.

chattel slavery An old form of slavery whereby slaves are considered property that can be bought and sold.

chemical dependency A condition in which drug use is compulsive, and users are unable to stop because of physical and/or psychological dependency.

child abuse The physical or mental injury, sexual abuse, negligent treatment, or maltreatment of a child under the age of 18 by a person who is responsible for the child's welfare.

child labor Children performing work that is hazardous, that interferes with a child's education, or that harms a child's health or physical, mental, spiritual, or moral development.

civil union A legal status that entitles same-sex couples who apply for and receive a civil union certificate to nearly all the benefits available to married couples.

classic rape A rape committed by a stranger, with the use of a weapon, resulting in serious bodily injury to the victim.

clearance rate The percentage of crimes in a jurisdiction for which the police report and arrest.

club drugs A general term for illicit, often synthetic drugs commonly used at nightclubs or all-night dances called "raves."

Cold War The state of political tension and military rivalry that existed between the United States and the former Soviet Union from the 1950s through the late 1980s.

colonialism When a racial and/or ethnic group from one society takes over and dominates the racial and/or ethnic group(s) of another society.

common couple violence Occasional acts of violence that result from conflict that gets out of hand between persons in a relationship.

community policing A type of policing in which uniformed police officers patrol and are responsible for certain areas of the city as opposed to simply responding to crimes as they occur.

comprehensive primary health care An approach to health care that focuses on the broader social determinants of health, such as poverty and economic inequality, gender inequality, environment, and community development.

compressed workweek Workplace option in which employees work full-time, but in four 10-hour rather than five 8-hour days.

computer crime Any violation of the law in which a computer is the target or means of criminal activity.

conflict perspective A sociological perspective that views society as comprising different groups and interests competing for power and resources.

control theory A theory that a strong social bond between a person and society constrains some individuals from violating norms.

convergence hypothesis The argument that capitalist countries will adopt elements of socialism and socialist countries will adopt elements of capitalism—that is, their systems will converge.

conversion therapy See *reparative therapy.*

cooperative learning Learning in which a heterogeneous group of students, of varying abilities, help one another with either individual or group assignments.

corporal punishment The intentional infliction of pain for a perceived misbehavior.

corporate downsizing The corporate practice of discharging large numbers of employees. Simply put, the term "downsizing" is a euphemism for mass firing of employees.

corporate violence The production of unsafe products and the failure of corporations to provide a safe working environment for their employees.

corporate welfare Laws and policies that favor corporations, such as low-interest government loans to failing businesses and special subsidies and tax breaks to corporations.

covenant marriage A type of marriage offered in Louisiana that permits divorce only under condition of fault or after a marital separation of more than two years.

crack A crystallized illegal drug product made by boiling a mixture of baking soda, water, and cocaine.

crime An act or the omission of an act that is a violation of a federal, state, or local law and for which the state can apply sanctions.

cultural imperialism The indoctrination into the dominant culture of a society; when cultural imperialism exists, the norms, values, traditions, and languages of minorities are systematically ignored.

cultural lag A condition in which the material part of the culture changes at a faster rate than the nonmaterial part.

cultural sexism The ways in which the culture of society perpetuates the subordination of individuals based on their sex classification.

culture of poverty The set of norms, values, and beliefs and self-concepts that contribute to the persistence of poverty among the underclass.

cumulative trauma disorders Also known as *repetitive motion disorders,* the most common type of workplace illness in the United States; includes muscle, tendon, vascular, and nerve injuries that result from repeated or sustained actions or exertions of different body parts.

cybernation The use of machines that control other machines in the production process; characteristic of post-industrial societies that emphasize service and information professions.

cyberwarfare Utilization of information technology to attack or manipulate the military and civilian infrastructure and information systems of an enemy.

cycle of abuse A pattern in abusive relationships whereby a violent or abusive episode is followed by a makeup period when the abuser expresses sorrow and asks for forgiveness and "one more chance." The "honeymoon period" may last for days, weeks, or even months before the next outburst of violence occurs.

date-rape drugs Drugs that are used to render victims incapable of resisting sexual assaults.

debt bondage See *bonded labor.*

de facto segregation Segregation that is not required by law but exists "in fact," often as a result of housing and socioeconomic patterns.

de jure segregation Segregation that is required by law.

deconcentration The redistribution of the population from cities to suburbs and surrounding areas.

decriminalization The removal of criminal penalties for a behavior, as in the decriminalization of drug use.

Defense of Marriage Act Federal legislation that states that marriage is a legal union between one man and one woman and denies federal recognition of same-sex marriage.

deforestation The conversion of forest land to non-forest land.

deindustrialization The loss and/or relocation of manufacturing industries.

deinstitutionalization The removal of individuals with psychiatric disorders from mental hospitals and large residential institutions to outpatient community mental health centers.

demographic transition theory A theory that attributes population growth patterns to changes in birthrates and death rates associated with the process of industrialization. In preindustrial societies, the population remains stable because, although the birthrate is high, the death rate is also high. As a society becomes industrialized, the birthrate remains high, but the death rate declines, causing rapid population growth. In societies with advanced industrialization, the birthrate declines, and this decline, in conjunction with the low death rate, slows population growth.

dependency ratio The number of societal members who are under 18 or are 65 and over compared to the number of people who are between 18 and 64.

dependent variable The variable that the researcher wants to explain. See also *independent variable.*

deregulation The reduction of government control of, for example, certain drugs.

desertification The degradation of semi-arid land, which results in the expansion of desert land that is unusable for agriculture. Desertification is caused by deforestation and overgrazing by cattle and other herd animals.

deskilling The tendency for workers in a post-industrial society to make fewer decisions and for labor to require less thought.

detente The philosophy of "negotiation rather than confrontation" in reference to relations between the United States and the former Soviet Union; put forth by Secretary of State Henry Kissinger's Declaration of Principles in 1972.

deterrence The use of harm or the threat of harm to prevent unwanted behaviors.

devaluation hypothesis The hypothesis that women are paid less because the work they perform is socially defined as less valuable than the work performed by men.

differential association A theory developed by Edwin Sutherland that holds that through interaction with others, individuals learn the values, attitudes, techniques, and motives for criminal behavior.

disability-adjusted life year (DALY) Years lost to premature death and years lived with illness or disability. More simply, 1 DALY equals 1 lost year of healthy life.

discrimination Actions or practices that result in differential treatment of categories of individuals.

disengagement theory A theory that the elderly disengage from productive social roles in order to relinquish these roles to younger members of society. As this process continues, each new group moves up and replaces another, which, according to disengagement theory, benefits society and all of its members.

distance learning Learning in which, by time or place, the student is separated from the teacher.

diversity training Workplace training programs designed to reduce prejudice and discrimination in the workplace, increase employees' awareness of cultural differences in the workplace and how these differences may affect job performance.

divorce law reform Policies and proposals designed to change divorce law. Usually, divorce law reform measures attempt to make divorce more difficult to obtain.

divorce mediation A process in which divorcing couples meet with a neutral third party (mediator) who assists the individuals in resolving such issues as property division, child custody, child support, and spousal support in a way that minimizes conflict and encourages cooperation.

divorce rate The number of divorces per 1,000 population.

domestic partners A status granted to unmarried couples, including gay and lesbian couples, by some states, counties, cities, and workplaces that conveys various rights and responsibilities. These may include coverage under a partner's health and pension plan, rights of inheritance and community property, tax benefits, access to married student housing, child custody and child and spousal support obligations, and mutual responsibility for debts.

domestic terrorism Domestic terrorism, sometimes called insurgent terrorism, occurs when the terrorist act involves victims, targets, institutions, governments, or citizens from one country.

double effect In reference to physician-assisted suicide (PAS), the use of medical interventions to relieve pain and suffering but that also may hasten death.

double jeopardy See *multiple jeopardy.*

doubling time The time required for a population to double in size from a given base year if the current rate of growth continues.

drug Any substance other than food that alters the structure and functioning of a living organism when it enters the bloodstream.

drug abuse The violation of social standards of acceptable drug use, resulting in adverse physiological, psychological, and/or social consequences.

drug addiction See *chemical dependency.*

dual-use technologies Defense-funded technological innovations with commercial and civilian use.

dumbing down The lowering of educational standards or expectations by students and/or teachers.

Earned Income Tax Credit (EITC) A refundable tax credit based on a working family's income and number of children. In addition to the federal EITC, many states also have a state EITC.

ecofeminism A perspective that views environmental problems as resulting from human domination of the environment and sees connections between the domination of women, people of color, children, and the poor and the domination of nature. In contrast to a male-oriented view of natural resources as a means to an end—a means to profit and power—ecofeminists often embrace a spiritual approach to addressing environmental problems and emphasize the close connection between women and nature.

e-commerce The buying and selling of goods and services over the Internet.

economic institution The structure and means by which a society produces, distributes, and consumes goods and services.

ecoterrorism Any crime intended to protect wildlife or the environment that is violent, puts human life at risk, or results in damages of $10,000 or more.

elder abuse The physical or psychological abuse, financial exploitation, or medical abuse or neglect of the elderly.

Employment Nondiscrimination Act (ENDA) A bill that would make it illegal to discriminate based on sexual orientation. ENDA was introduced in Congress in 1994, and, as of 2003, had failed to pass the Senate.

environmental injustice (See also *environmental racism.*) The tendency for socially and politically marginalized groups to bear the brunt of environmental ills.

environmental racism See *environmental injustice.*

environmental refugees Individuals who have migrated because they can no longer secure a livelihood because of deforestation, desertification, soil erosion, and other environmental problems.

epidemiological transition The shift from a society characterized by low life expectancy and parasitic and infectious diseases to one characterized by high life expectancy and chronic and degenerative diseases.

ergonomics The designing or redesigning of the workplace to prevent and reduce cumulative trauma disorders.

ethnicity A shared cultural heritage and/or national origin.

e-waste Waste from electronic equipment.

extended producer responsibility An environmental strategy whereby producers are responsible for taking back their old products, phasing out the toxic materials in manufacturing, and designing cleaner products with less waste generation.

experiment A research method that involves manipulating the independent variable to determine how it affects the dependent variable.

expulsion When a dominant group forces a subordinate group to leave the country or to live only in designated areas of the country.

familism A value system that encourages family members to put their family's well-being above their individual and personal needs.

Family and Medical Leave Act (FMLA) A federal law passed in 1993 that requires all companies with 50 or more employees to provide eligible workers (who work at least 25 hours a week and have been working for at least a year) with up to 12 weeks of job-protected, unpaid leave so they can care for a seriously ill child, spouse, or parent; stay home to care for their newborn, newly adopted, or newly placed child; or take time off when they are seriously ill.

family household As defined by the U.S. Census Bureau, a family household consists of two or more persons related by birth, marriage, or adoption who reside together.

family preservation programs In-home interventions for families who are at risk of having a child removed from the home because of abuse or neglect.

feminism The belief that women and men should have equal rights and responsibilities.

feminization of poverty The disproportionate distribution of poverty among women.

fertility rate The average number of births per woman.

field research A method of research that involves observing and studying social behavior in settings in which it naturally occurs; includes participant observation and nonparticipant observation.

flextime An option in work scheduling that allows employees to begin and end the workday at different times as long as they perform 40 hours of work per week.

folkways The customs and manners of society.

forced labor A form of slavery whereby individuals are lured by the promise of a good job and instead find themselves enslaved.

free trade agreements Pacts between two countries, or a group of countries, that make it easier to trade goods across national boundaries. Free trade agreements reduce or eliminate foreign restrictions on exports, reduce or eliminate tariffs (or taxes) on imported goods, and prevent U.S. technology from being copied and used by competitors through protection of "intellectual property rights."

functionally illiterate Being unable to carry out many of the tasks required of an adult in today's society because of reading, writing, and/or quantitative deficiencies.

future shock The state of confusion resulting from rapid scientific and technological changes that challenge traditional values and beliefs.

gateway drug A drug (e.g., marijuana) that is believed to lead to the use of other drugs (such as cocaine and heroin).

gender The social definitions and expectations associated with being male or female.

gender tourism The recent tendency for definitions of masculinity and femininity to become less clear, resulting in individual exploration of the gender continuum.

gene monopolies Exclusive control over a particular gene as a result of government patents.

gene therapy The transplantation of a healthy gene to replace a defective or missing gene.

genetic engineering The manipulation of an organism's genes in such a way that the natural outcome is altered.

genetic screening The use of genetic maps to detect predispositions to human traits or disease(s).

genetically modified organisms (GMOs) Also known as "genetically engineered foods" (GE foods) and transgenic crops, products that have been created or modified through agricultural biotechnology.

genocide The deliberate, systematic annihilation of an entire nation or people.

gentrification A type of neighborhood revitalization in which middle- and upper-income persons receive tax incentives to buy and rehabilitate older homes in a depressed neighborhood.

geographic steering The practice whereby realtors discourage minorities from moving into certain areas by showing them homes only in minority neighborhoods.

gerontophobia Fear or dread of the elderly.

ghettos In the United States, slums that are occupied primarily by African Americans.

glass ceiling An invisible, socially created barrier that prevents some women and other minorities from being promoted into top corporate positions.

GLBT A term used to refer collectively to gays, lesbians, bisexuals, and transgendered individuals; see also "LGBT."

global economy An interconnected network of economic activity that transcends national borders.

global warming The increasing average global air temperature, caused mainly by the accumulation of various gases (see *greenhouse gases*) that collect in the atmosphere.

globalization The economic, political, and social interconnectedness among societies throughout the world.

greenhouse gases Gases (primarily carbon dioxide, methane, and nitrous oxide) that accumulate in the atmosphere and act like the glass in a greenhouse, holding heat from the sun close to the earth.

greenwashing The way in which environmentally and socially damaging companies portray their corporate image and products as being "environmentally friendly" or socially responsible.

guerrilla warfare Warfare in which organized groups oppose domestic or foreign governments and their military forces; often involves small groups of individuals who use camouflage and underground tunnels to hide until they are ready to execute a surprise attack.

harm reduction A recent public health position that advocates reducing the harmful consequences of drug use for the user as well as society as a whole.

hate crime Also known as a "bias-motivated crime"; and "ethnoviolence," an unlawful violent act motivated by prejudice or bias. Hate crimes include intimidation (e.g., threats), destruction/damage of property, physical assault, and murder.

Head Start A project begun in 1965 by the federal government to help preschool children from disadvantaged families.

health A state of complete physical, mental, and social well-being.

health maintenance organizations (HMOs) Prepaid group plans in which a person pays a monthly premium for comprehensive health care services.

heterosexism The institutional and societal reinforcement of heterosexuality as the privileged and powerful norm. Heterosexism is based on the belief that heterosexuality is superior to homosexuality and results in prejudice and discrimination against homosexuals and bisexuals.

heterosexuality A sexual orientation that involves the predominance of emotional and sexual attraction to persons of the other sex.

home schooling The education of children at home instead of in a public or private school; often part of a fundamentalist movement to protect children from perceived non-Christian values in the public schools.

homophobia Negative attitudes and emotions toward homosexuality and those who engage in same-sex sexual behavior.

homosexuality The predominance of cognitive, emotional, and sexual attraction to persons of the same sex.

HPI-1 The Human Poverty Index for developing countries.

HPI-2 The Human Poverty Index for industrialized countries.

human capital The skills, knowledge, and capabilities of the individual.

human capital hypothesis The hypothesis that female-male pay differences are a function of differences in women's and men's levels of education, skills, training, and work experience.

Human Poverty Index (HPI) A composite measure of poverty based on three measures of deprivation: (1) deprivation of a long, healthy life; (2) deprivation of knowledge; and (3) deprivation in decent living standards.

hypothesis A prediction or educated guess about how one variable is related to another variable.

in-vitro fertilization (IVF) The union of an egg and a sperm in an artificial setting such as a laboratory dish.

incapacitation A criminal justice philosophy that views the primary purpose of the criminal justice system as preventing criminal offenders from committing further crimes against the public by putting them in prison.

incumbent upgrading Aid programs that help residents of depressed neighborhoods buy or improve their homes and stay in the community.

independent variable The variable that is expected to explain change in the dependent variable.

index offenses Crimes identified by the FBI as the most serious, including personal crimes (homicide, rape, robbery, assault) and property crimes (burglary, larceny, car theft, arson).

individual discrimination The unfair or unequal treatment of individuals because of their group membership.

individualism The tendency to focus on one's individual self-interests rather than on the interests of one's family and community.

Industrial Revolution The period between the mid-eighteenth and the early nineteenth century when machines and factories became the primary means for producing goods. The Industrial Revolution led to profound social and economic changes.

infant mortality rate The number of deaths of infants under 1 year of age per 1,000 live births in a year.

infantilizing elders The portrayal of the elderly in the media as childlike in terms of clothes, facial expression, temperament, and activities.

infotech An abbreviation for "information technology"; any technology that carries information.

infowar The utilization of technology to manipulate or attack an enemy's military and civilian infrastructure and information systems.

infrastructure The underlying foundation that enables a city to function. Infrastructure includes water and sewer lines, phone lines, electricity cables, sidewalks, streets, curbs, lighting, and storm drainage systems.

institution An established and enduring pattern of social relationships. The five traditional social institutions are family, religion, politics, economics, and education. Institutions are the largest elements of social structure.

institutional discrimination Discrimination in which the normal operations and procedures of social institutions result in unequal treatment of minorities.

integration hypothesis A theory that the only way to achieve quality education for all racial and ethnic groups is to desegregate the schools.

intergenerational poverty Poverty that is transmitted from one generation to the next.

internalized homophobia A sense of personal failure and self-hatred among some lesbians and gay men due to social rejection and stigmatization.

Internet An international information infrastructure available through many universities, research institutes, government agencies, and businesses; developed in the 1970s as a Defense Department experiment.

intimate partner violence Actual or threatened violent crimes committed against persons by their current or former spouses, boyfriends, or girlfriends.

intimate terrorism Almost entirely perpetrated by men, a form of violence that is motivated by a wish to control one's partner and involves the systematic use of not only violence, but economic subordination, threats, isolation, verbal and emotional abuse, and other control tactics. This form of violence is more likely to escalate over time and to involve serious injury.

Jim Crow laws Laws that separated blacks from whites by prohibiting blacks from using "white" buses, hotels, restaurants, drinking fountains, and restrooms.

job exportation The relocation of U.S. jobs to other countries where products can be produced more cheaply.

job sharing A work option in which two people, often husband and wife, share and are paid for one job.

labeling theory A symbolic interactionist theory that is concerned with the effects of labeling on the definition of a social problem (e.g., a social condition or group is viewed as problematic if it is labeled as such) and with the effects of labeling on the self-concept and behavior of individuals (e.g., the label "juvenile delinquent" may contribute to the development of a self-concept and behavior consistent with the label).

labor unions Worker organizations that originally developed to protect workers and to represent them at negotiations between management and labor. Labor unions have played an important role in fighting for fair wages and benefits, healthy and safe work environments, and other forms of worker advocacy.

latent functions Consequences that are unintended and often hidden, or unrecognized; for example, a latent function of education is to provide schools that function as baby-sitters for employed parents.

law Norms that are formalized and backed by political authority.

legalization Making prohibited behavior legal; for example, legalizing marijuana or prostitution.

lesbigays A collective term sometimes used to refer to lesbians, gays, and bisexuals.

LGBT A term used to refer collectively to lesbians, gays, bisexuals, and transgendered individuals; see also "GLBT."

life chances A term used by Max Weber to describe the opportunity to obtain all that is valued in society, including happiness, health, income, and education.

life expectancy The average number of years that a person born in a given year can expect to live.

living wage laws Laws that require state or municipal contractors, recipients of public subsidies or tax breaks, or, in some cases, all businesses to pay employees wages significantly above the federal minimum, enabling families to live above the poverty line.

looking-glass self The idea that individuals develop their self-concept through social interaction.

macro sociology The study of large aspects of society, such as institutions and large social groups.

MADD Mothers Against Drunk Driving. A social action group committed to reducing drunk driving.

mailbox economy The tendency for a substantial portion of local economies to be dependent on pension and social security checks received in the mail by older residents.

Malthusian theory The theory proposed by Thomas Malthus in which he predicted that population would grow faster than the food supply and that masses of people were destined to be poor and hungry. According to Malthus, food shortages would lead to war, disease, and starvation that would eventually slow population growth.

managed care A type of medical insurance plan that controls costs through monitoring and controlling the decisions of health care providers.

manifest functions Consequences that are intended and commonly recognized; for example, a manifest function of education is to transmit knowledge and skills to youth.

master status The status that is considered the most significant in a person's social identity.

maternal mortality rates A measure of deaths that result from complications associated with pregnancy, childbirth, and unsafe abortion.

means-tested programs Public assistance programs that have eligibility requirements based on income and assets.

mechanization The use of tools to accomplish tasks previously done by workers; characteristic of agricultural societies that emphasize the production of raw materials.

Medicaid A public assistance program designed to provide health care for the poor.

medicalization The tendency to label behaviors and conditions as medical problems in need of medical intervention.

Medicare A national public insurance program created by Title XVIII of the Social Security Act of 1965; originally designed to protect people 65 years of age and older from the rising costs of health care. In 1972, Medicare was extended to permanently disabled workers and their dependents and persons with end-stage renal disease.

Medigap The difference between Medicare benefits and the actual cost of medical care.

megacities Cities with 10 million residents or more.

melting pot The product of different groups coming together and contributing equally to a new, common culture.

mental health The successful performance of mental function, resulting in productive activities, fulfilling relationships with other people, and the ability to adapt to change and to cope with adversity.

mental illness A term that refers collectively to all mental disorders, which are health conditions that are characterized by alterations in thinking, mood, and/or behavior associated with distress and/or impaired functioning, and that meet specific criteria (such as level of intensity and duration) specified in the American Psychiatric Association's classification manual used to diagnose mental disorders: *Diagnostic and Statistical Manual of Mental Disorders.*

metropolis From the Greek meaning "mother city." See also *metropolitan area.*

metropolitan area A densely populated core area and any adjacent communities that have a high degree of social and economic integration with the core; a large city and its surrounding suburbs; also called a "metropolis."

micropolitan area A small city located beyond congested metropolitan areas.

micro-society school A simulation of the "real" or nonschool world where students design and run their own democratic, free-market society within the school.

micro sociology The study of the social psychological dynamics of individuals interacting in small groups.

military-industrial complex A term used by Dwight D. Eisenhower to connote the close association between the military and defense industries.

minority A category of people who are denied equal access to positions of power, prestige, and wealth because of their group membership.

mixed-use land use The location of homes, schools, shops, workplaces, and parks in close proximity to each other. A main purpose of mixed-use land use is to reduce driving distances.

modern racism A subtle form of racism that involves the belief that serious discrimination in America no longer exists, that any continuing racial inequality is the fault of minority group members, and the demands for affirmative action for minorities are unfair and unjustified.

modernization theory A theory claiming that as society becomes more technologically advanced, the position of the elderly declines.

monogamy Marriage between two partners; the only legal form of marriage in the United States.

morbidity Illnesses, symptoms, and impairments they produce.

mores Norms that have moral basis.

mortality Death.

multicultural education A broad range of programs and strategies designed to dispel myths, stereotypes, and ignorance about minorities; promote tolerance and appreciation of diversity; and include minority groups in the school curriculum.

multiple chemical sensitivity (MCS) Also known as "environmental illness," a condition whereby individuals experience adverse reactions when exposed to low levels of chemicals found in everyday substances (vehicle exhaust, fresh paint, house cleaning products, perfume and other fragrances, synthetic building materials, and numerous other petrochemical-based products). Symptoms of MCS include headache, burning eyes, difficulty breathing, stomach distress/nausea, loss of mental concentration, and dizziness. The onset of MCS is often linked to acute exposure to a high level of chemicals or to chronic long-term exposure.

multiple jeopardy The disadvantages associated with being a member of two or more minority groups.

mutual violent control A rare pattern of abuse when two intimate terrorists battle for control. See also *intimate terrorism*.

mutually assured destruction (MAD) A perspective that argues that if both sides in a conflict were to lose in a war, neither would initiate war.

naturalized citizen An immigrant who applied and met the requirements for U.S. citizenship.

needle exchange programs Programs designed to reduce transmission of HIV among injection drug users, their sex partners, and their children, by providing new, sterile syringes in exchange for used, contaminated syringes.

neglect A form of abuse involving the failure to provide adequate attention, supervision, nutrition, hygiene, health care, and a safe and clean living environment for a minor child or a dependent elderly individual.

New Urbanism A movement in urban planning, similar to *smart growth,* that approaches the idea of sustainable urban communities with the goal of raising the quality of life for all those in the community by creating compact communities with a sustainable infrastructure.

no-fault divorce A divorce that is granted based on the claim that there are irreconcilable differences within a marriage (as opposed to one spouse being legally at fault for the marital breakup).

non-family household A household that consists of one person who lives alone, two or more people as roommates, and cohabiting heterosexual or homosexual couples involved in a committed relationship.

norms Socially defined rules of behavior, including folkways, mores, and laws.

nuclear winter The predicted result of a thermonuclear war whereby dust storms and concentrations of particles would block out vital sunlight, lower temperatures in the Northern Hemisphere, and lead to the death of most living things on earth.

objective element of social problems Awareness of social conditions through one's own life experience and through reports in the media.

occupational sex segregation The concentration of women in certain occupations and of men in other occupations.

one drop of blood rule A rule that specified that even one drop of Negroid blood defined a person as black and, therefore, eligible for slavery.

operational definition In research, a definition of a variable that specifies how that variable is to be measured (or was measured) in the research.

organized crime Criminal activity conducted by members of a hierarchically arranged structure devoted primarily to making money through illegal means.

overt discrimination Discrimination that occurs because of an individual's own prejudicial attitudes.

Parental Alienation Syndrome An emotional and psychological disturbance in which children engage in exaggerated and unjustified denigration and criticism of a parent.

parity The concept of equality between mental health care insurance coverage and other health care insurance.

partial birth abortion Also called an intact dilation and extraction (D&X) abortion, the procedure may entail delivering the limbs and the torso of the fetus before it has expired.

patriarchal Literally, rule by father; today, connotes rule by males.

patriarchal terrorism A form of abuse in which husbands control their wives by the systematic use of not only violence, but also economic subordination, threats, isolation, and other control tactics.

patriarchy A tradition in which families are male dominated.

perinatal transmission The transmission of a virus (such as HIV) from an infected mother to a fetus or newborn.

Personal Responsibility and Work Opportunity Reconciliation Act (PRWORA) The 1996 legislation that affected numerous public assistance programs, primarily in the form of cutbacks and eligibility restrictions. This law ended Aid to Families with Dependent Children (AFDC) and replaced it with Temporary Assistance to Needy Families (TANF).

pink-collar jobs Jobs that offer few benefits, often have low prestige, and are disproportionately held by women.

planned obsolescence The manufacturing of products that are intended to become inoperative or outdated in a fairly short period of time.

pluralism A state in which racial and/or ethnic groups maintain their distinctness but respect each other and have equal access to social resources.

polyandry The concurrent marriage of one woman to two or more men.

polygamy A form of marriage in which one person may have two or more spouses.

polygyny The concurrent marriage of one man to two or more women.

population momentum Continued population growth that occurs even if a population achieves replacement-level fertility (2.1 births per woman) due to past high fertility rates which have resulted in large numbers of young women who are currently entering their childbearing years.

population transfer See *expulsion.*

post-industrialization The shift from an industrial economy dominated by manufacturing jobs to an economy dominated by service-oriented, information-intensive occupations.

postmodernism A worldview that questions the validity of rational thinking and the scientific enterprise.

post-traumatic stress disorder A set of symptoms that may result from any traumatic experience, including crime victimization, war, natural disasters, or abuse.

poverty The lack of resources necessary for material well-being—most importantly food and water, but also housing, land, and health care.

poverty line An annual level of personal or household income below which individuals or families are considered officially poor by the government.

preferred provider organizations (PPOs) Health care organizations in which employers who purchase group health insurance agree to send their employees to certain health care providers or hospitals in return for cost discounts.

prejudice Negative attitudes and feelings toward or about an entire category of people.

primary aging Biological changes associated with aging that are due to physiological variables such as cellular and molecular variation (e.g., gray hair).

primary assimilation The integration of different groups in personal, intimate associations such as friends, family, and spouses.

primary deviance Deviance committed before the person is caught and labeled as an offender; i.e., the act is defined as deviant.

primary group Small groups characterized by intimate and informal interaction.

primary prevention strategies Family violence prevention strategies that target the general population.

privatization A practice in which states hire businesses to provide services or operate local institutions, replacing government employees who formerly carried out these functions.

proletariat Workers, often exploited by the bourgeoisie.

pronatalism A cultural value that promotes having children.

public assistance A general term referring to some form of support by the government to citizens who meet certain established criteria.

public housing An assistance program that provides federal subsidies for low-income housing units built, owned, and operated by local public housing authorities; also known as *subsidized housing.*

race A category of people who share distinct physical characteristics that are deemed socially significant.

racial profiling The law enforcement practice of targeting suspects based upon race.

racism The belief that certain groups of people are innately inferior to other groups of people based on their racial classification. Racism serves to justify discrimination against groups that are perceived as inferior.

redlining The practice of mortgage companies to deny loans for the purchase of houses in minority neighborhoods, arguing that the financial risk is too great.

refugees Immigrants who apply from abroad for admission on the basis of persecution or fear of persecution for their political or religious beliefs.

regionalism A form of collaboration among central cities and suburbs that encourages local governments to share responsibility for common problems.

registered partnerships Federally recognized same-sex relationships that convey most but not all of the rights of marriage (some countries also offer registered partnerships to opposite-sex couples).

regressive taxes Taxes that absorb a much higher proportion of the incomes of lower-income households than of higher-income households.

rehabilitation A criminal justice philosophy that views the primary purpose of the criminal justice system as changing the criminal offender through such programs as education and job training, individual and group therapy, substance abuse counseling, and behavior modification.

relative poverty A deficiency in material and economic resources compared with some other population.

reparative therapy Various therapies that are aimed at changing homosexuals' sexual orientation.

restorative justice A philosophy primarily concerned with reconciling conflict between the offender, the community, and the victim.

road rage Aggressive and violent driving behavior.

role A set of rights, obligations, and expectations associated with a status.

sample In survey research, the portion of the population selected to be questioned.

sanctions Social consequences for conforming to or violating norms. Types of sanctions include positive, negative, formal, and informal.

sandwich generation The generation that has the responsibility of simultaneously caring for their children and their aging parents.

school vouchers Tax credits that are transferred to the public or private school of a parent's choice.

science The process of discovering, explaining, and predicting natural or social phenomena.

scientific apartheid The growing gap between the industrial and developing countries in the rapidly evolving knowledge frontier.

second shift The household work and child care that employed parents (usually women) do when they return home from their jobs.

secondary aging Biological changes associated with aging that can be attributed to poor diet, lack of exercise, and increased stress.

secondary assimilation The integration of different groups in public areas and in social institutions, such as neighborhoods, schools, the workplace, and in government.

secondary deviance Deviance that results from being caught and labeled as an offender; i.e., the actor is defined as deviant.

secondary group A group characterized by impersonal and formal interaction.

secondary prevention strategies Family violence prevention strategies that target groups thought to be at high risk for family violence.

Section 8 housing A federal low-income housing program in which federal rent subsidies are provided either to tenants (in the form of certificates and vouchers) or to private landlords.

segregation The physical and social separation of categories of individuals, such as racial or ethnic groups.

selective primary health care An approach to health care that uses technocratic solutions to target a specific health problem, such as using immunization and oral rehydration therapy to promote child survival.

self-fulfilling prophecy A concept referring to the tendency for people to act in a manner consistent with the expectations of others.

senescence The biology of aging.

serial monogamy A succession of marriages in which a person has more than one spouse over a lifetime but is legally married to only one person at a time.

sex A person's biological classification as male or female.

sexism The belief that there are innate psychological, behavioral, and/or intellectual differences between females and males and that these differences connote the superiority of one group and the inferiority of another.

sex slavery A form of forced labor in which girls are forced into prostitution by their own husbands, fathers, and brothers to earn money to pay family debts, or they are lured by offers of good jobs and then forced to work in brothels under the threat of violence.

sexual aggression Sexual interaction that occurs against one's will through the use of physical force, threat of force, pressure, use of alcohol/drugs, or use of position of authority.

sexual harassment An employer's requiring sexual favors in exchange for a promotion, salary increase, or any other employee benefit and/or the existence of a hostile environment that unreasonably interferes with job performance, as in the case of sexually explicit remarks or insults being made to an employee.

sexual orientation The identification of individuals as heterosexual, bisexual, or homosexual, based on their emotional and sexual attractions, relationships, self-identity, and lifestyle.

shaken baby syndrome A form of child abuse whereby the caretaker shakes a baby to the point of causing the child to experience brain or retinal hemorrhage. Shaken baby syndrome most often occurs in response to a baby, who typically is younger than 6 months, who won't stop crying.

single-payer system A single tax-financed public insurance program that replaces private insurance companies.

slavery The loss of free will, where a person is forced through violence or the threat of violence to give up the ability to sell freely his/her labor power.

slums Concentrated areas of poverty and poor housing in urban areas.

smart growth A strategy for managing urban sprawl that serves the economic, environmental, and social needs of communities.

social class Group of people who share a similar position or social status within the stratification system.

social group Two or more people who have a common identity, interact, and form a social relationship. Institutions are made up of social groups.

social problem A social condition that a segment of society views as harmful to members of society and in need of remedy.

social promotion The passing of students from grade to grade even if they are failing.

socialism An economic ideology that emphasizes public rather than private ownership. Theoretically, goods and services are equitably distributed according to the needs of the citizens.

socialized medicine National health insurance systems in countries such as Canada, Great Britain, Sweden, Germany, and Italy.

sociological imagination A term coined by C. Wright Mills to refer to the ability to see the connections between our personal lives and the social world in which we live.

sodomy Oral and anal sexual acts.

soft money Money that flows through a loophole in the law to provide political parties, candidates, and contributors a means to evade federal limits on political contributions.

state The organization of the central government and government agencies such as the armed forces, police force, and regulatory agencies.

status A position a person occupies within a social group.

stereotypes Exaggerations or generalizations about the characteristics and behavior of a particular group.

stigma Any personal characteristic associated with social disgrace, rejection, or discrediting.

strain theory A theory that when legitimate means of acquiring culturally defined goals are limited by the structure of society, the resulting strain may lead to crime or other deviance.

structural-functionalism A sociological perspective that views society as a system of interconnected parts that work together in harmony to maintain a state of balance and social equilibrium for the whole; focuses on how each part of society influences and is influenced by other parts.

structural sexism The ways in which the organization of society, and specifically its institutions, subordinate individuals and groups based on their sex classification.

subcultural theories A set of theories arguing that certain groups or subcultures in society have values and attitudes that are conducive to crime and violence.

subculture The distinctive lifestyles, values, and norms of discrete population segments within a society.

subjective element of social problems The belief that a particular social condition is harmful to society, or to a segment of society, and that it should and can be changed.

subsidized housing See *public housing.*

suburbanization The development of urban areas outside central cities, and the movement of populations into these areas.

suburbs The urbanlike areas surrounding central cities.

survey research A method of research that involves eliciting information from respondents through questions; includes interviews (telephone or face-to-face) and written questionnaires.

sustainable development Societal development that meets the needs of current generations without threatening the future of subsequent generations.

sweatshops Work environments that are characterized by less than minimum wage pay, excessively long hours of work often without overtime pay, unsafe or inhumane working conditions, abusive treatment of workers by employers, and/or the lack of worker organizations aimed at negotiating better work conditions.

symbol Something that represents something else.

symbolic interactionism A sociological perspective emphasizing that human behavior is influenced by definitions and meanings that are created and maintained through symbolic interaction with others.

technological dualism A term referring to the tendency for technology to have both positive (e.g., time saving) and negative (e.g., unemployment) consequences.

technological fix The use of scientific principles and technology to solve social problems.

technology Activities that apply the principles of science and mechanics to the solution of specific problems.

technology-induced diseases Diseases that result from the use of technological devices, products, and/or chemicals.

telemedicine Using information and communication technologies to deliver a wide range of health care services, including diagnosis, treatment, prevention, health support and information, and education of health care workers.

telework A form of work that allows employees to work part- or full-time at home or at a satellite office.

Temporary Assistance to Needy Families (TANF) The welfare program that resulted from 1996 welfare reform legislation. Under the TANF program, which replaced Aid to Families with Dependent Children (AFDC), after two consecutive years of receiving aid, welfare recipients are required to work at least 20 hours per week or to participate in a state-approved work program (few exceptions are made). A lifetime limit of five years is set for families to receive benefits.

terrorism The premeditated use or threatened use of violence by an individual or group to gain a political objective.

tertiary prevention strategies Family violence prevention strategies that target families who have experienced family violence.

therapeutic cloning Use of stem cells from human embryos to produce body cells that can be used to grow needed organs or tissues.

therapeutic communities Organizations in which approximately 35 to 100 individuals reside for up to 15 months to abstain from drugs, develop marketable skills, and receive counseling.

tracking An educational practice in which students are grouped together on the basis of similar levels of academic achievement and abilities.

traditional family Families in which the husband is the breadwinner and the wife is a homemaker.

transgendered individuals Includes a range of people whose gender identities do not conform to traditional notions of masculinity and femininity. Transgendered individuals include homosexuals, bisexuals, cross-dressers, transvestites, and transsexuals.

transnational corporations Also known as *multinational corporations,* corporations that have their home base in one country and branches, or affiliates, in other countries.

transnational crime Crime that, directly or indirectly, involves more than one country.

transnational terrorism Terrorism that occurs when a terrorist act in one country involves victims, targets, institutions, governments, or citizens of another country.

triad dispute resolution Dispute resolution that involves two disputants and a negotiator.

triangulation The use of multiple methods and approaches to study a social phenomenon.

triple jeopardy See *multiple jeopardy.*

underclass A persistently poor and socially disadvantaged group that disproportionately experiences early sexual activity, unmarried parenthood, joblessness, reliance on public assistance, illegitimate income-producing activities (e.g., selling drugs), and substance use.

underemployment Unemployed workers as well as (1) those working part-time but who wish to work full-time ("involuntary" part-timers); (2) those who want to work but have been discouraged from searching by their lack of success ("discouraged" workers), and (3) others who are neither working nor seeking work but who indicate that they want and are available to work and have looked for employment in the last 12 months.

unemployment To be currently without employment, actively seeking employment, and available for employment, according to measures of U.S. unemployment. Unemployment figures do not include discouraged workers, who have given up on finding a job and are no longer looking for employment.

union density The percentage of workers who belong to unions.

universal health care See *socialized medicine.*

upskilling The opposite of de-skilling; upskilling reduces employee alienation and increases decision-making powers.

urban area A spatial concentration of people whose lives are centered around nonagricultural activities.

urban sprawl The ever-increasing outward growth of urban areas.

urbanization The transformation of a society from a rural to an urban one.

values Social agreements about what is considered good and bad, right and wrong, desirable and undesirable.

variable Any measurable event, characteristic, or property that varies or is subject to change.

victimless crimes Illegal activities, such as prostitution or drug use, that have no complaining party; also called "vice crimes."

violent resistance Acts of violence by a partner that are committed in self-defense. Violent resistance is almost exclusively perpetrated by women against a male partner.

virtual reality Computer-generated three-dimensional worlds that change in response to the movements of the head or hand of the individual; a simulated experience of people, places, sounds, and sights.

war Organized armed violence aimed at a social group in pursuit of an objective.

wealth The total assets of an individual or household, minus liabilities.

wealthfare Governmental policies and regulations that economically favor the wealthy.

white-collar crime Includes both occupational crime, in which individuals commit crimes in the course of their employment, and corporate crime, in which corporations violate the law in the interest of maximizing profit.

WMD Weapons of mass destruction including chemical, biological, and nuclear weapons.

working poor Individuals who spend at least 27 weeks a year in the labor force (working or looking for work), but whose income falls below the official poverty line.

References

Chapter 1

Blumer, Herbert. 1971. "Social Problems as Collective Behavior." *Social Problems* 8(3):298–306.

Caldas, Stephen, and Carl L. Bankston III. 1999. "Black and White TV: Race, Television Viewing, and Academic Achievement." *Sociological Spectrum* 19:39–61.

Canedy, Dana. 2003. "Lifting Veil for Photo ID Goes Too Far, Driver Says." *New York Times,* June 27. http://www.nytimes.com/2003/06/27/national

Dordick, Gwendolyn A. 1997. *Something Left to Lose: Personal Relations and Survival Among New York's Homeless.* Philadelphia: Temple University Press.

Eitzen, Stanley, and Maxine Baca Zinn. 2000. *Social Problems.* Boston: Allyn and Bacon.

Engel, Robin Shepard, and Robert E. Worden. 2003. "Police Officers' Attitudes, Behavior, and Supervisory Influences: An Analysis of Problem Solving. *Criminology* 41:131–166.

Fleming, Zachary. 2003. "The Thrill of It All." In *In Their Own Words,* ed. Paul Cromwell, pp. 99–107. Los Angeles, CA: Roxbury.

Fordham University. 2003. "The Social Health of the States." Fordham Institute for Innovation in Social Policy. http://www.fordham.edu/images/graduate_school

Hewlett, Sylvia Ann. 1992. *When the Bough Breaks: The Cost of Neglecting Our Children.* New York: Harper Perennial.

Human Development Report. 2003. "2003 Human Development Index Reveals Development Crisis." http://www.undp.org/hdr2003

Jekielek, Susan M. 1998. "Parental Conflict, Marital Disruption and Children's Emotional Well-Being." *Social Forces* 76:905–935.

Jacobs, Bruce A. 2003. "Researching Crack Dealers." In *In Their Own Words,* ed. Paul Cromwell, pp. 1–11. Los Angeles, CA: Roxbury.

Kmec, Julie A. 2003. "Minority Job Concentration and Wages." *Social Problems* 50:38–59.

Merton, Robert K. 1968. *Social Theory and Social Structure.* New York: Free Press.

Mills, C. Wright. 1959. *The Sociological Imagination.* London: Oxford University Press.

Newport, Frank. 2003. "Public's Satisfaction at Lowest Point in Six Years." Gallup Poll. http://www.gallup.com

Palacios, Wilson R., and Melissa E. Fenwick. 2003. "'E' is for Ecstacy." In *In Their Own Words,* ed. Paul Cromwell, pp. 277–283. Los Angeles, CA: Roxbury.

Reiman, Jeffrey. 2003. *The Rich Get Richer and the Poor Get Prison.* Boston, MA: Addison-Wesley.

———. 2001. *The Rich Get Richer and the Poor Get Prison.* Needham Heights: MA: Allyn and Bacon.

Romer, D., R. Hornik, B. Stanton, M. Black, X. Li, I. Ricardo, and S. Feigelman. 1997. "Talking Computers: A Reliable and Private Method to Conduct Interviews on Sensitive Topics with Children." *The Journal of Sex Research* 34:3–9.

Sax, Linda, Jennifer Lindholm, Alexander Astin, William Korn, and Kathryn Mahoney. 2002. *The American Freshman: National Norms for Fall 2002.* Los Angeles, CA: Higher Education Research Institute.

Schwalbe, Michael. 1998. *The Sociologically Examined Life: Pieces of the Conversation.* Mountain View, CA: Mayfield.

Thomas, W. I. [1931]1966. "The Relation of Research to the Social Process." In *W. I. Thomas on Social Organization and Social Personality,* ed. Morris Janowitz, pp. 289–305. Chicago: University of Chicago Press.

Troyer, Ronald J., and Gerald E. Markle. 1984. "Coffee Drinking: An Emerging Social Problem." *Social Problems* 31:403–416.

Ukers, William H. 1935. *All About Tea,* Vol. 1. *The Tea and Coffee Trade Journal Co.*

Wilson, John. 1983. *Social Theory.* Englewood Cliffs, NJ: Prentice Hall.

Chapter 2

"Access to Free Health Care Means Crime Does Pay for Some Sick Inmates." 2001 (April 20). News, theIndependent.com. http://www.theindependent.com/stories/042001/new-inmates20.html (accessed July 2003).

Allen, P. L. 2000. *The Wages of Sin: Sex and Disease, Past and Present.* Chicago: University of Chicago Press.

Altman, Dennis. 2003. "Understanding HIV/AIDS as a Global Security Issue." In *Health Impacts of Globalization,* ed. Kelley Lee, pp. 33–46. New York: Palgrave MacMillan.

American Psychiatric Association. 2000. *Diagnostic and Statistical Manual of Mental Disorders,* 4th ed. Text Revision DSM–TR. Washington, DC.

Anderson, Robert N. 2002. "Deaths: Leading Causes for 2000." *National Vital Statistics Reports* 50(16): all.

*The authors and Wadsworth acknowledge that some of the Internet sources may have become unstable, that is, they are no longer hot links to the intended reference. In that case, the reader may want to access the article through the search engine or archives of the homepage cited (e.g., fbi.gov, cbsnews.com).

Associated Press. 2000. "Millions of Kids Could Have Insurance." MSNBC, August 9. http://www.msnbc.com/news

Barker, Kristin. 2002. "Self-Help Literature and the Making of an Illness Identity: The Case of Fibromyalgia Syndrome (FMS)." *Social Problems* 49(3):279–300.

Bartley, Mel, Jane Ferrie, and Scott M. Montgomery. 2001. "Living in a High-Unemployment Economy: Understanding the Health Consequences. In *Social Determinants of Health,* eds. M. Marmot and R. G. Wilkinson, pp. 81–104. New York: Oxford University Press.

Bekelman, J. E., Y. Lit, and C. P. Gross. 2003. "Scope and Impact of Financial Conflicts of Interest in Biomedical Research: A Systematic Review." *Journal of the American Medical Association* 289:454–465.

Centers for Disease Control and Prevention. 1997. "Youth Risk Behavior Surveillance: National College Health Risk Behavior Survey—United States, 1995." *Surveillance Summaries* 46(SS-6):1–54.

Cherner, Linda L. 1995. *The Universal Healthcare Almanac.* Phoenix: Silver & Cherner.

Children's Defense Fund. 2000. "The State of America's Children Yearbook 2000." http://www.childrensdefense.org/keyfacts.htm

Clarke, Adele E., Laura Mamo, Jennifer R. Fishman, Janet K. Shim, and Jennifer Ruth Fosket. 2003. "Biomedicalization: Technoscientific Transformations of Health, Illness, and U.S. Biomedicine." *American Sociological Review* 68:161–194.

Cockerham, William C. 1998. *Medical Sociology,* 7th ed. Upper Saddle River, NJ: Prentice Hall.

Collin, Jeff. 2003. "Think Global, Smoke Local: Transnational Tobacco Companies and Cognitive Globalization." In *Health Impacts of Globalization: Towards Global Governance,* ed. K. Lee, pp. 61–85. New York: Palgrave MacMillan.

Conrad, Peter, and Phil Brown. 1999. "Rationing Medical Care: A Sociological Reflection." In *Health, Illness, and Healing: Society, Social Context, and Self,* eds. Kathy Charmaz and Debora A. Paterniti, pp. 582–590. Los Angeles: Roxbury.

Conyers, John. 2003. "A Fresh Approach to Health Care in the United States: Improved and Expanded Medicare for All." *American Journal of Public Health* 93(2):193.

Diamond, Catherine, and Susan Buskin. 2000. "Continued Risky Behavior in HIV-Infected Youth." *American Journal of Public Health* 90(1):115–118.

Durkheim, Emile. [1897] 1951. *Suicide: A Study in Sociology.* Translated by J. A. Spaulding and G. Simpson. Edited by G. Simpson. New York: Free Press.

Egwu, Igbo N. 2002. "Health Promotion Challenges in Nigeria: Globalization, Tobacco Marketing and Policies." In *Health Communication in Africa: Contexts, Constraints and Lessons,* eds. A. O. Alali and B. A. Jinadu, pp. 40–55. Lanham, MD: University Press of America.

Ezzell, Carol. 2003. "Why? The Neuroscience of Suicide." *Scientific American* 288(2):45–51.

Family Care International. 1999. "Safe Motherhood." http://www.safemotherhood.org/init_facts.htm

Feachum, Richard G. A. 2000. "Poverty and Inequality: A Proper Focus for the New Century." *The International Journal of Public Health* (Bulletin of the World Health Organization) 78:1–2.

Feldman, Debra S., Dennis H. Novack, and Edward Gracely. 1998. "Effects of Managed Care on Physician Patient Relationships, Quality of Care, and the Ethical Practice of Medicine: A Physician Survey." *Archives of Internal Medicine* 158:1626–1632.

Gallup Poll Social Series. 2002 (December). *Health & Healthcare.* Princeton, NJ: Gallup Organization.

Garfinkel, P. E., and D. S. Goldbloom. 2000. "Mental Health—Getting Beyond Stigma and Categories." *Bulletin of the World Health Organization* 78(4):503–505.

Goldstein, Michael S. 1999. "The Origins of the Health Movement." In *Health, Illness, and Healing: Society, Social Context, and Self,* eds. Kathy Charmaz and Debora A. Paterniti, pp. 31–41. Los Angeles: Roxbury.

Hadley, Jack, and John Holahan. 2003. "How Much Medical Care Do the Uninsured Use." *Health Affairs* (February 12):all. http://www.healthaffairs.org

Harrell, Joseph, and Olveen Carrasquillo. 2003. "The Latino Disparity in Health Coverage." *The Journal of the American Medical Association* 289(9):1167.

Harrison, Mary M. 2002. "Silence that Promotes Stigma." *Teaching Tolerance* 21:52–53.

Health Care Financing Administration. 2000. "The State Children's Health Insurance Program." http://www.hcfa.gov/facts/fs000224.htm

Hippert, Christine. 2002. "Multinational Corporations, the Politics of the World Economy, and Their Effects on Women's Health in the Developing World: A Review." *Health Care for Women International,* 23:861–869.

Hollander, Dore. 2003. "Having Two Parents Helps." *Perspectives on Sexual and Reproductive Health* 35(2):60.

Hughes, Mary Elizabeth, and Linda J. Waite. 2002. "Health in Household Context: Living Arrangements and Health in Late Middle Age." *Journal of Health and Social Behavior* 43:1–21.

Iglehart, John K. 1999. "The American Health Care System: Expenditures." *New England Journal of Medicine* 340:70–76.

Institute of Medicine. 2004. *Insuring America's Health: Principles and Recommendations.* Washington, DC: National Academies Press.

International Women's Health Coalition. 2003. *Bush's Other War: The Assault on Women's Sexual and Reproductive Health and Rights.* http://www.iwhc.org

Johnson, Tracy L., and Elizabeth Fee. 1997. "Women's Health Research: An Introduction." In *Women's Health Research: A Medical and Policy Primer,* eds. Florence P. Haseltine and Beverly Greenberg Jacobson, pp. 3–26. Washington, DC: Health Press International.

Kaiser Commission on Medicaid and the Uninsured. 2000 (May). "The Uninsured and Their Access to Health Care." Washington, DC: The Kaiser Family Foundation.

Kaiser Family Foundation. 2000. "Preliminary Findings from a New National Survey of Teens on HIV/AIDS, 2000." http://www.kff/org

Kann, Laura, Steven A. Kinchen, Barbara I. Williams, James G. Ross, Richard Lowry, Jo Anne Grunbaum, Lloyd J. Kolbe, and State and Local YRBSS Coordinators. 2000. "Youth Risk Behavior Surveillance—United States, 1999." *Journal of School Health* 70(7):271–285.

Kessler, Ronald C., Katherine A. McGonagle, Shanyang Zhao, Christopher B. Nelson, Michael Hughes, Suzann Eshleman, Hans-Ulrich Wittchen, and Kenneth S. Kendler. 1994. "Life-time and 12-Month Prevalence of DSM-III-R Psychiatric Disorders in the United States." *Archives of General Psychiatry* 51:8–19.

Komiya, Noboru, Glenn E. Good, and Nancy B. Sherrod. 2000. "Emotional Openness as a Predictor of College Students' Attitudes Toward Seeking Psychological Help." *Journal of Counseling Psychology* 47(1)138–143.

Lantz, Paula M., James S. House, James M. Lepkowski, David R. Williams, Richard P. Mero, and Jieming Chen. 1998. "Socioeconomic Factors, Health Behaviors, and Mortality: Results from a Nationally Representative Prospective

Study of U.S. Adults." *Journal of the American Medical Association* 279: 1703–1708.
LaPorte, Ronald E. 1997. "Improving Public Health via the Information Superhighway." http://www.the-scientist.library.upenn.edu/yr1997/August/opin_97018.html
Lee, Kelley. 2003. "Introduction." In *Health Impacts of Globalization,* ed. Kelley Lee, pp. 1–10. New York: Palgrave MacMillan.
Lehmann, Christine. 2003. "Parents Giving Up Custody of Mentally Ill Children." *Psychiatric News* 38(11):13–15.
Lerer, Leonard B., Alan D. Lopez, Tord Kjellstrom, and Derek Yach. 1998. "Health for All: Analyzing Health Status and Determinants." *World Health Statistics Quarterly* 51:7–20.
"Libya Leader: Only Gays Get AIDS." 2003 (July 14). *Planet Out.* http://www.planetout.com
Link, Bruce G., and Jo Phelan. 2001. "Social Conditions as Fundamental Causes of Disease." In *Readings in Medical Sociology,* 2nd ed., eds. William C. Cockerham, Michael Glasser, and Linda S. Heuser, pp. 3–17. Upper Saddle River, NJ: Prentice Hall.
Malatu, Mesfin Samuel, and Carmi Schooler. 2002. "Causal Connections between Socio-economic Status and Health: Reciprocal Effects and Mediating Mechanisms." *Journal of Health and Social Behavior* 43:22–41.
Mathews, T. J., Sally C. Curtin, and Marian F. MacDorman. 2000 (July 20). "Infant Mortality Statistics from the 1998 Period Linked Birth/Infant Death Data Set." *National Vital Statistics Reports* 48(12):1–28.
McGinn, Anne P. 2003. "Combating Malaria." In *State of the World 2003,* ed. L. Starke, pp. 62–83. New York: W. W. Norton.
Mercy, James A., Etienne G. Krug, Linda L. Dahlberg, and Anthony B. Zwi. 2003. "Violence and Health: The United States in a Global Perspective. *American Journal of Public Health* 93(2): 256–261.
Miller, K., and A. Rosenfield. 1996. "Population and Women's Reproductive Health: An International Perspective." *Annual Review of Public Health* 17:359–382.
Mirowsky, John and Catherine E. Ross. 2003. *Social Causes of Psychological Distress,* 2nd ed. New York: Walter de Gruyter.
Monardi, Fred, and Stanton A. Glantz. 1998. "Are Tobacco Industry Campaign Contributions Influencing State Legislative Behavior?" *American Journal of Public Health* 88:918–923.
Mulligan, Kate. 2003. "APA, Advocacy Groups Decry MH Budget Cuts." *Psychiatric News* 38(12):12.
Murphy, Elaine M. 2003. "Being Born Female Is Dangerous for Your Health." *American Psychologist* 58(3):205–210.
Murray, C., and A. Lopez, eds. 1996. *The Global Burden of Disease.* Boston: Harvard University Press.
National Center for Health Statistics. 2000. *Health, United States, 2000 with Adolescent Health Chartbook.* Hyattsville, MD: U.S. Government Printing Office.
———. 2004. *Health, United States, 2003.* Hyattsville, MD: U.S. Government Printing Office.
National Council on Disability. 2002. *The Well-Being of Our Nation: An Inter-Generational Vision of Effective Mental Health Services and Supports.* Washington DC.
National Institutes of Health. 2004 (January). "HIV/AIDS Statistics." www.niaid.nih.gov/factsheets/aidsstat.htm
Nelson, Alan R., Brian D. Smedley, and Adrienne Y. Stith. 2002. *Unequal Treatment: Confronting Racial and Ethnic Disparities in Health Care.* Washington, DC: Institute of Medicine, National Academy Press.
Nelson, Lyle. 2003 (May). *How Many People Lack Health Insurance for How Long?* Congressional Budget Office. http://www.cbo.gov
Ninan, Ann. 2003 (March). "'Without My Consent'—Women and HIV-Related Stigma in India." Population Reference Bureau. http://www.prb.org
Park, Edwin, Melanie Nathanson, Robert Greenstein, and John Springer. 2003 (December 8). "The Troubling Medicare Legislation." Washington, DC: Center on Budget and Policy Priorities. http://www.cbpp.org
Parsons, Talcott. 1951. *The Social System.* New York: Free Press.
Pear, Robert. 2004 (January 9). "Health Spending at Record Rate." New York Times Online. http://www.nytimes.com
Peeno, Linda. 2000 (Spring). "Taking On the System." *Hope* (22):18–21.
PNHP Data Update. 2003 (Spring). Physicians for a National Health Program http://www.pnhp.org.
Public Citizen. 2002 (April 18). "Pharmaceutical Industry Ranks as Most Profitable Industry—Again." http://www.publiccitizen.org
Ransom, Elizabeth I., and Nancy V. Yinger. 2002. *Making Motherhood Safer: Overcoming Obstacles on the Pathway to Care.* Population Reference Bureau. http://www.prb.org
Rustein, Shea O. 2000. "Factors Associated with Trends in Infant and Child Mortality in Developing Countries During the 1990s." *Bulletin of the World Health Organization* 78(10):1256–1270.
Sanders, David, and Mickey Chopra. 2003. "Globalization and the Challenge of Health for All: A View from sub-Saharan Africa." In *Health Impacts of Globalization,* ed. Kelley Lee, pp. 105–119. New York: Palgrave Macmillan.
Save the Children. 2002. *State of the World's Mothers, 2002.* http://www.savethechildren.org.
Schaeffer, Robert K. 2003. *Understanding Globalization: The Social Consequences of Political, Economic, and Environmental Change,* 2nd ed. Lanham, MD: Rowman & Littlefield.
Sidel, Victor W., and Barry S. Levy. 2002. "The Health and Social Consequences of Diversion of Economic Resources to War and Preparation for War." In *War or Health: A Reader,* eds. Ilkka Taipale, P. Helena Makela, Kati Juva, and Vappu Taipale, pp. 208–221. New York: Palgrave.
SmokeFree Educational Services. 2003 (June 5). "5 Million Deaths a Year Worldwide from Smoking." CorpWatch. http://www.corpwatch.org
Stephenson, Joan. 2003. "Growing, Evolving HIV/AIDS Pandemic Is Producing Social and Economic Fallout." *Journal of the American Medical Association* 289(1):31–33.
Stockard, Jean, and Robert M. O'Brien. 2002. Cohort Effects on Suicide Rates: International Variations. *American Sociological Review 67*(6):854–872.
Substance Abuse and Mental Health Services Administration. 2003 (July 21). "Treatment of Adults with Serious Mental Illness." *The NHSDA (National Household Survey on Drug Abuse) Report.* http://www.samhsa.gov
Szasz, Thomas. 1970 (orig. 1961). *The Myth of Mental Illness: Foundations of a Theory of Personal Conduct.* New York: Harper & Row.
Thomas, Caroline. 2003. "Trade Policy, the Politics of Access to Drugs and Global Governance for Health." In *Health Impacts of Globalization,* ed. Kelley Lee, pp. 177–191. New York: Palgrave MacMillan.
Toner, Robin, and Sheryl Gay Stolberg. 2002 (August 11). "Decade After Health Care Crisis, Soaring Costs Bring New Strains." New York Times Online. http://www.nytimes.com
UNICEF. 2003. *The State of the World's Children, 2003.* New York: UNICEF. http://www.unicef.org
United Nations Population Fund. 2000. *The State of World Population Report 2000.* http://www.unfpa.org
———. 2002a. *The State of World Population 2002: People, Poverty, and Possibilities.* New York: United Nations Population Fund.
———. 2002b. "Addressing Obstetric Fistulas." New York: United Nations Population Fund.
U.S. Department of Health and Human Services. 1998. "Needle Exchange Pro-

grams: Part of a Comprehensive HIV Prevention Strategy." http://www.hhs.gov/news/press/1998pres/980420.html

———. 1999. *Mental Health: A Report of the Surgeon General: Executive Summary.* Rockville, MD: U.S. Government Printing Office.

———. 2001. *Mental Health: Culture, Race, and Ethnicity—A Supplement to Mental Health: A Report of the Surgeon General.* Rockville, MD: U.S. Government Printing Office.

Vastag, Brian. 2003. "Mental Health Parity." *Journal of the American Medical Association (JAMA)* 289(1):35.

Ward, Darrell E. 1999. *The AmFAR AIDS Handbook.* New York: W. W. Norton.

Weitz, Rose. 2004. *The Sociology of Health, Illness, and Health Care: A Critical Approach,* 3rd ed. Belmont, CA: Wadsworth.

White, Frank. 2003. "Can International Public Health Law Help to Prevent War?" *Bulletin of the World Health Organization* 81(3):228.

WHO International Consortium in Psychiatric Epidemiology. 2000. "Cross-National Comparisons of the Prevalences and Correlates of Mental Disorders." *Bulletin of the World Health Organization* 78(4):413–426.

Wiegand, Steve. 2002 (January 25). "State Inmate Gets New Heart." United for No Injustice, Oppression, or Neglect (U.N.I.O.N). http://www.geocities.com

Williams, David R. 2003. "The Health of Men: Structured Inequalities and Opportunities." *American Journal of Public Health* 93(5):724–731.

Williams, David R., and Chiquita Collins. 1999. "U.S. Socioeconomic and Racial Differences in Health: Patterns and Explanations." In *Health, Illness, and Healing: Society, Social Context, and Self,* eds. Kathy Charmaz and Debora A. Paterniti, pp. 349–376. Los Angeles: Roxbury.

World Health Organization. 1946. "Constitution of the World Health Organization." New York: World Health Organization Interim Commission.

———. 1999. *The World Health Report 1999.* http://www.who.int

———. 2000. *The World Health Report 2000.* http://www.who.int

———. 2002. *The World Health Report 2002: Reducing Risks, Promoting Healthy Life.* http://www.who.int

———. 2003. *The World Health Report 2003: Shaping the Future.* http://www.who.int

Chapter 3

Affects Child. 2003. "How Does Alcohol Affect the World of a Child." http://www.alcoholfreechildren.org

AHA (American Heart Association). 2003a. "How Does the Tobacco Industry Target Youth?" July 30. http://www.americanheart.org

———. 2003b. "Cigarette Smoking and Cardiovascular Diseases." August 7. http://www.americanheart.org

Alcohol Alert. 2000. "Mechanisms of Addiction." National Institute on Alcohol Abuse and Alcoholism 46 (April), 2.

Alcoholism and Drug Abuse Weekly. 2000. "Hawaii Is Seventh State to Permit Medical Marijuana." June 26, 8.

AP (Associated Press). 1999. "Alcoholism Touches Millions." http://www.abcnews.go.com. December 30.

———. 2000. "MIT pays $4.75M in Drinking Death." *Philadelphia Daily News,* September 14, 11.

BBC (British Broadcasting Company). 2003. "Know About Drugs so You Won't Make a Mistake About Drugs." http://www.bbc.uk/stoke/features. July 29.

Becker, H. S. 1966. *Outsiders: Studies in the Sociology of Deviance.* New York: Free Press.

Belluck, Pam. 2003. "Methadone Grows as Killer Drug." *New York Times,* February 9. NYTimes.com.

Charon, Joel. 2002. *Social Problems: Readings with Four Questions.* Belmont, CA:Wadsworth.

Cloud, John. 2000. "The Lure of Ecstasy." *Time,* June 5, 63–72.

Crittenden, Jules. 2002. "MIT Fraternity Settles Lawsuit in Freshman's Drinking Death." Boston Herald. http://www.groups.yahoo.com/group/fraternalnews

DEA (Drug Enforcement Administration). 2000. "An Overview of Club Drugs." Drug Intelligence Brief, February, 1–10. Washington, DC: U.S. Department of Justice.

Drug Policy Alliance. 2003a. "Drug Policy Around the World: The Netherlands." http://www.drugpolicy.org/global

———. 2003b. "Drug Policy Around the World: England." http://www.drugpolicy.org/global

———. 2003c. "Fuzzy Math in New ONDCP Report." http://www.drugpolicy.org/global

———. 2003d. "Election Results 2002." http://www.drugpolicy.org/statebystate

Duke, Steven, and Albert C. Gross. 1994. *America's Longest War: Rethinking Our Tragic Crusade Against Drugs.* New York: G. P. Putnam.

Easley, Margaret, and Norman Epstein. 1991. "Coping with Stress in a Family with an Alcoholic Parent." *Family Relations* 40:218–224.

Feagin, Joe R., and C. B. Feagin. 1994. *Social Problems.* Englewood Cliffs, NJ: Prentice Hall.

Fields, Richard. 2001. *Drugs in Perspective.* Boston, MA: McGraw-Hill.

Final Report. 2000. "Methamphetamine Interagency Task Force Final Report." Federal Advisory Committee. Washington DC: Department of Justice.

Firshein, Janet. 2003. "The Role of Biology and Genetics." *Moyers on Addiction.* Public Broadcasting System (PBS).

Fletcher, Michael A. 2000. "War on Drugs Sends More Blacks to Prison than Whites." *Washington Post,* June 8, A10.

Gentry, Cynthia. 1995. "Crime Control through Drug Control." In *Criminology,* 2d ed., ed. Joseph F. Sheley, pp. 477–493. Belmont, CA: Wadsworth.

Gusfield, Joseph. 1963. *Symbolic Crusade: Status Politics and the American Temperance Movement.* Urbana: University of Illinois Press.

Hanson, Glen R. 2002. "New Vistas in Drug Abuse Prevention." *NIDA Notes* 16:3–4.

Harris Poll. 2003. IRSS Study Number S12851, Question Number Q605. The Odum Institute, Chapel Hill, NC.

HHS (U.S. Department of Health and Human Services). 2002. "2001 National Household Survey on Drug Abuse." Substance Abuse and Mental Health Service Administration. Washington, DC: U.S. Government Printing Office.

Heroin Drug Conference. 1997. "Administrator's Message." U.S. Department of Justice: Drug Enforcement Administration. http://www.udsdoj.gov/dea/pubs/special/heroin.html

International Narcotics Control Strategy Report. 2000. "Policy and Program Development." Bureau for International Narcotics and Law Enforcement Affairs. Washington, DC: U.S. Department of State.

James, Joni. 2003. "State's Tobacco Money in Danger." April 11. Bradenton.com

Jarvik, M. 1990. "The Drug Dilemma: Manipulating the Demand." *Science* 250:387–392.

Johnson, Lynn M. 2003. "Alcohol, Drugs and Violence: Detrimental Effects on Children." http://www.babyparenting.about.com

Klutt, Edward C. 2000. "Pathology of Drug Abuse." http://www. medlib-utah.edu/webpath

Kornblum, William, and Joseph Julian. 2004. *Social Problems.* Upper Saddle River, NJ: Prentice Hall.

Leonard, K. E., and H. T. Blane. 1992. "Alcohol and Marital Aggression in a National Sample of Young Men." *Journal of Interpersonal Violence* 7:19–30.

Leshner, Alan. 2003. "Using Science to Counter the Spread of Ecstasy Abuse." *NIDA Notes* 16:3–4.

MacCoun, Robert J., and Peter Reuter. 2001. "Does Europe Do It Better? Lessons from Holland, Britain and Switzerland." In *Solutions to Social Problems,* eds. D. Stanley Eitzen and

Craig S. Leedham, pp. 260–264. Boston: Allyn and Bacon.
MADD (Mothers Against Drunk Driving). 2003a. "Alcohol Advertising." http:www.madd.org/stats
———. 2003b. "Did you Know . . . ?" http://www.madd.org/stats
———. 2003c. "The Limiting Factor: Economic Cost of Underage Drinking." http:www. madd.org/activism
———. 2003d. "New 8-Point Plan to Jumpstart Stalled War on Drunk Driving." http://www.madd.org
Mann, Judy. 2000. "Make War on the War on Drugs." *Washington Post,* July 26, C13.
Mayell, Hillary. 1999. "Tobacco on Course to Become World's Leading Cause of Death." *National Geographic News.* http://ngnews/news/1999/12149
McCaffrey, Barry. 1998. "Remarks by Barry McCaffrey, Director, Office of National Drug Control Policy, to the United Nations General Assembly: Special Session on Drugs." Office of National Drug Control Policy. http://www.whitehousedrugpolicy.gov/news/speeches
Moore, Martha T. 1997. "Binge Drinking Stalks Campuses." *USA Today,* October 1, A3.
Morgan, Patricia A. 1978. "The Legislation of Drug Law: Economic Crisis and Social Control." *Journal of Drug Issues* 8:53–62.
Monk, Richard C. 2001. Taking Sides: *Clashing Views on Controversial Issues, Crime and Criminology,* 6th ed. Guilford, CN: Dushkin.
MTF (Monitoring the Future). 2002. "Data from In School Surveys of 8th, 10th, and 12th Grade Students." Ann Arbor: National Institute on Drug Abuse and University of Michigan.
NHSDA (National Household Survey on Drug Abuse) Report. 2003. "Cigarette Brand Preference." http://www.samhsa.gov/oas/2k3 /cigBrands
NHTSA (National Highway Traffic Safety Administration). 2003. "National Survey of Drinking & Driving Attitudes & Behaviors." Washington, DC: U.S. Department of Transportation.
NIDA (National Institute on Drug Abuse). 2000. "Heroin Abuse and Addiction," Research Report Series. NIH Publication No. 00-4165. http://www.nida.nih/gov/ResearchReports/Heroin/Heroin.html
———. 2003. "InfoFacts: Drug Addiction Treatment Methods." http://www.nida.org.
ODCCP (Office of Drug Control and Crime Prevention). 2003. "Global Illicit Drug Trends 2002." New York: United Nations.
ONDCP (Office of National Drug Control Policy). 2000. "The Link Between Drugs and Crime." Chapter II. The National Drug Control Strategy 2000 Annual Report. http://www.whitehousedrugpolicy.gov
———. 2002. "Illegal Drugs Drain $160 Billion a Year from American Economy." http://www.whitehousedrugpolicy.gov
———. 2003a. "Drug Use Trends." http://www.whitehousedrugpolicy.gov/publications/factsht/druguse
———. 2003b. "Drug Facts: Marijuana." http://www.whitehousedrugpolicy.gov/marijuana
———. 2003c. "Drug Facts: Cocaine." http://www.whitehousedrugpolicy.gov/cocaine
———. 2003d. "Drug Facts: Crack." http://www.whitehousedrugpolicy.gov/crack
———. 2003e. "Drug Facts: Club Drugs." http://www.whitehousedrugpolicy.gov/club
———. 2003f. "Drug Facts: LSD." http://www.whitehousedrugpolicy.gov/LSD
———. 2003g. "Fact Sheet: Rohypnol." http://www.whitehousedrugpolicy.gov/publications/factsht/rohypnol
———. 2003h. "Fact Sheet: Heroin." http://www.whitehousedrugpolicy.gov/publications/factsht/heroin
———. 2003i. "Drug Facts: Methamphetamine." http:// www.whitehousedrugpolicy.gov/methamphetamine
———. 2003j. "Juveniles and Drugs." http://www.whitehousedrugpolicy.gov/publications/factsht/juveniles
———. 2003k. "Fact Sheet: Inhalants." http://www.whitehousedrugpolicy.gov/publications/factsht/inhalants
———. 2003l. "Drug Data Summary." http://www.whitehousedrugpolicy.gov/publications/factsht/summary
———. 2003m. "National Drug Control Strategy: Drug Use Consequences." http://www.whitehousedrugpolicy.gov/publications/policy
———. 2003n. "Budget Highlights." http://www.whitehousedrugpolicy.gov/publications/policy/ndsc03
———. 2003o. "National Drug Control Strategy." http://www.whitehousedrugpolicy.gov/publications/policy
———. 2003p. "Multimedia Leader YO-GI-OH and White House Drug Policy Office Announce Partnership to Prevent Drug Use." Media Campaign Press Release. June 10.
Pew. 2002. Illegal Drugs." Pew Research Center for the People and Press Survey. http://www.pollingreport.com/drugs
Rorabaugh, W. J. 1979. *The Alcoholic Republic: An American Tradition.* New York: Oxford University Press.
Rychtarik, Robert G., Gerald J. Connors, Kurt H. Dermen, and Paul Stasiewicz. 2000. "Alcoholics Anonymous and the Use of Medications to Prevent Relapse." *Journal of Studies on Alcohol* 61:134–141.
SAMHSA (Substance Abuse and Mental Health Administration). 2003. "New Report Estimates that 6 Million Children Lived with Addicted Parents in 2001." http://www.samhsa.gov/news/newsreleases
Sax, Linda, Jennifer Lindholm, Alexander Astin, William Korn, and Kathryn Mahoney. 2002. *The American Freshman: National Norms for Fall 2002.* Los Angeles: University of California, Higher Education Research Institute.
Schemo, Diana Jean. 2002. "Study Calculates the Effects of College Drinking in the U.S." *New York Times,* April 10. http://www.nytimes.com
Sheldon, Tony. 2000. "Cannabis Use Among Dutch Youth." *British Medical Journal* 321:655.
Siegel, Larry J. 2002. *Juvenile Delinquency.* Belmont, CA: Wadsworth.
Snider, Valerie. 2003. "Monitoring Prescription Drug Use Among Older Adults Can Prevent Misuse." http://www.ncadi.samhsa.gov/newsroom
State Legislatures. 2000. "All You Ever Wanted to Know About Drunk Drivers." *State Legislatures* 26:7.
Stein, Joel. 2002. "The New Politics of Pot." *Time Magazine,* November 4.
Straus, Murry, and S. Sweet. 1992. "Verbal/Symbolic Aggression in Couples: Incidence Rates and Relationships to Personal Characteristics." *Journal of Marriage and the Family* 54:346–57.
Sullivan, Thomas J. 2003. *Social Problems.* New York: Allyn and Bacon.
Thio, Alex. 2004. *Deviant Behavior.* Boston, MA: Allyn and Bacon.
Time/CNN. 2002. "The New Politics of Pot." *Time,* November 4, 56–66.
Thompson, Don. 2000. "States Ballot Questions Focus on Drug Rehab Instead of Prison." Excite News. http://news.excite.com/news
Update. 2003. "Update on Nicotine Addiction and Tobacco Research." *NIDA Notes* 15:15.
Van Dyck, C., and R. Byck. 1982. "Cocaine." *Scientific American* 246:128–41.
Van Kammen, Welmoet B., and Rolf Loeber. 1994. "Are Fluctuations in Delinquent Activities Related to the Onset and Offset in Juvenile Illegal Drug Use and Drug Dealing?" *Journal of Drug Issues* 24:9–24.
Weitzman, Elissa, Henry Wechsler, and Toben F. Nelson. 2003. "Environment, not Education, a Stronger Predictor of Binge Drinking Behavior Among College Freshman." Press Release: Harvard School of Public Health, January 21.
Willing, Richard. 2002. "Study Shows Alcohol Is Main Problem for Addicts." *USA Today,* October 3, B4.
Witters, Weldon, Peter Venturelli, and Glen Hanson. 1992. *Drugs and Society,* 3d ed. Boston: Jones & Bartlett.
Wysong, Earl, Richard Aniskiewicz, and David Wright. 1994. "Truth and Dare:

Tracking Drug Education to Graduation and as Symbolic Politics." *Social Problems* 41:448–68.
YRBSS (Youth Risk Behavior Surveillance System). 2002. Center for Disease Control. http://www.appa.nccd.cdc.gov/yrbbs
Zickler, Patrick. 2003a. "Study Demonstrates that Marijuana Smokers Experience Significant Withdrawal." *NIDA Notes* 17: 7, 10.

Chapter 4

ABCNews. 2001. "U.S.–Russia Child Porn Bust." http://www.abcnews.go.com/sections/world/Daily/News/childpornbust_010326.htm
Albanese, Jay. 2000. *Criminal Justice.* Boston: Allyn and Bacon.
Amnesty International. 2002. "Abolitionist and Retentionist Countries." http://www.amnestyusa.org
Anderson, David. 1999. "The Aggregate Burden of Crime." *Journal of Law and Economics* 42:611–642.
AP (Associated Press). 2003, "Video Games Get Updated Rating System." *New York Times,* June 20. http://www.nytimes.com
Bartollas, Clemens. 2003. *Juvenile Delinquency,* 6th ed. Boston: Allyn and Bacon.
Becker, Howard S. 1963. *Outsiders: Studies in the Sociology of Deviance.* New York: Free Press.
Bipartisan Bill. 2003. "Bipartisan Bill to Legalize Marijuana Introduced to Congress." http://www.talkleft.com/archives/May 22.
BJS (Bureau of Justice Statistics). 2003a. "Nation's Prison and Jail Population Exceeds 2 Million Inmates for the First Time." Bureau of Justice Statistics. http://www. ojp.usdoj.gov/bjs/pub
———. 2003b. "Bureau of Justice Statistics Releases Annual Report on Capital Punishment." http://www.ojp.usdoj.gov/bjs/pub/press
Butterfield, Fox. 2002a. "Tight Budgets Force States to Reconsider Crime and Penalties." *New York Times,* January 21. http://www.nytimes.com
———. 2002b. "Study Shows Building Prisons Did Not Prevent Repeat Crimes." June 3. http://www.ojp.usdoj.gov/bjs/pub
Chesney-Lind, Meda, and Randall G. Shelden. 2004. *Girls, Delinquency and Juvenile Justice.* Belmont, CA: Wadsworth.
Chicago Tribune. 2003. "Ford Fined, Ordered to Hand over Van Data." Tribune News Services. http://www.lieffcabraser.com
Conklin, John E. 1998. *Criminology,* 6th ed. Boston: Allyn and Bacon.
COPS. 2003. "What Is Community Policing?" http://www.cops.usdoj.gov
D'Alessio, David, and Lisa Stolzenberg. 2002. "A Multilevel Analysis of the Relationship Between Labor Surplus and Pretrial Incarceration." *Social Problems* 49:178–193.
———. 2003, "Race and the Probability of Arrest." *Social Forces* 81:1381–1397.
Dickerson, Debra. 2000. "Racial Profiling: Are We All Really Equal in the Eyes of the Law?" *Los Angeles Times,* July 16. http://www.latimes.com
Dixon, Travis L., and Daniel Linz. 2000. "Race and the Misrepresentation of Victimization on Local Television News." *Communication Research* 27:547–74.
Dobriansky, Paula. 2001. "The Explosive Growth of Globalized Crime." *Global Issues: Arresting Transnational Crime* Vol. 6 (August):1–3.
Eisenberg, Daniel. 2002. "Jail to the Chiefs?" *Time,* August 12, 23–26.
Eitle, David, Stewart D'Alessio, and Lisa Stolzenberg. 2002. "Racial Threat and Social Control: A Test of the Political, Economic, and Threat of Black Crime Hypothesis." *Social Forces* 81:557–576.
Erikson, Kai T. 1966. *Wayward Puritans.* New York: John Wiley.
Exonerations. 2003. "Death Row Exonerations, 1973–2002." http://www. infor please.com/ipa
(FBI) Federal Bureau of Investigation. 2000. *Crime in the United States, 1999.* Uniform Crime Reports. Washington, DC: U.S. Government Printing Office.
———. 2002. *Crime in the United States, 2001.* Uniform Crime Reports. Washington, DC: U.S. Government Printing Office.
Felson, Marcus. 2002. *Crime and Everyday Life,* 3rd ed. Thousand Oaks, CA: Sage.
Finckenauer, James O. 2000. "Meeting the Challenge of Transnational Crime." *National Institute of Justice Journal,* July, 2–7.
Foundation (Amadou Diallo Foundation, Inc). 2003a. "Amadou's Profile." http://www.amadoudiallofoundationinc.com
———. 2003b. "The Amadou Diallo Foundation, Inc." http://www.amadoudiallo.foundationinc.com
Fletcher, Michael A. 2000. "War on Drugs Sends More Blacks to Prison than Whites." *The Washington Post,* June 8, A10.
Gallup Poll. 2000a. "Most Important Problem." June 22–25. http://www.gallup.com/poll/indicators
———. 2000b. "Crime Tops List of Americans' Local Concerns." June 21. http://www.gallup.com/poll/releases
Garey, M. 1985. "The Cost of Taking a Life: Dollars and Sense of the Death Penalty." *U.C. Davis Law Review* 18: 1221–1273.
Gest, Ted, and Dorian Friedman. 1994. "The New Crime Wave." *U.S. News and World Report,* August 29, 26–28.
Harris Poll. 2001. IRSS Study Number S13955, Question Number Q520. The Odum Institute, Chapel Hill, NC.
———. 2002. IRSS Study Number S15938, Question Numbers Q405, Q415 . The Odum Institute, Chapel Hill, NC.
Haughney, Christine. 2001. "U.S. Refuses to Charge Police in Diallo Slaying." *Washington Post,* February 1, A9.
Heimer, Karen, and Stacy DeCoster. 1999. "The Gendering of Violent Delinquency." *Criminology* (May):377–389.
Herbert, Bob. 2002. "The Fatal Flaws." *New York Times,* February 11. http://www.nytimes.com
Hirschi, Travis. 1969. *Causes of Delinquency.* Berkeley: University of California Press.
HSPH (Harvard School of Public Health). 2002. "American Females at Highest Risk of Murder." Press Release, April 17.
Human Rights Watch. 2000. *Human Rights Watch World Report 2000.* United States. http://www.hrw.org/wr2k/us.html
INTERPOL. 1998. "INTERPOL Warning: Nigerian Crime Syndicate's Letter Scheme Fraud Takes on New Dimension." *Press Releases.* http://www.kenpubs.co.uk/INTERPOL.COM/English/pres/nig.html
Kong, Deborah, and Jon Swartz. 2000 "Experts See Rash of Hack Attacks Coming." *USA Today,* September 27, 1B.
Kubrin, Charis, and Ronald Weitzer. 2003. "Retaliatory Homicide: Concentrated Disadvantage and Neighborhood Culture." *Social Problems* 50:157–180.
Laub, John, Daniel S. Nagan, and Robert Sampson. 1998. "Trajectories of Change in Criminal Offending: Good Marriages and the Desistance Process." *American Sociological Review* 63 (April):225–238.
Lee, Jennifer. 2002. "Some States Track Parolees by Satellite." *New York Times,* January 31. http://www.nytimes.com
———. 2003. "Identity Theft Complaints Double in 2002." *New York Times,* January 22. http://www.nytimes.com
Lichtblau, Eric. 2003. "Panel Clears Harsher Terms in Corporate Crime Cases." *New York Times,* January 9. http://www.nytimes.com
Liptak, Adam. 2003. "Death Penalty Found More Likely When Victim Is White." *New York Times,* January 8. http://www.nytimes.com
Lott, John R. Jr. 2003. "Guns Are an Effective Means of Self-Defense." In *Gun Control,* ed. Helen Cothran, pp. 86–93. Farmington Hills, MI: Greenhaven Press.
MAD DADS. 2003. "MAD DADs National: Who are MAD DADS?" http://www.maddadsnational.com/whoweare
Madriz, Esther. 2000. "Nothing Bad Happens to Good Girls." In *Social Prob-*

lems of the Modern World, ed. Frances Moulder, pp. 293–297. Belmont, CA: Wadsworth.

Major Rulings. 2003. "Major Rulings of the 2002–2003 Term." *USA Today,* June 27, 4A.

Merton, Robert. 1957. *Social Theory and Social Structure.* Glencoe, IL: Free Press.

Mosk, Matthew. 2003. "Sniper Families Fight Bill Aiding Gunmakers." *Washington Post,* July 17, B01.

Murray, Mary E., Nancy Guerra, and Kirk Williams. 1997. "Violence Prevention for the Twenty-First Century." In *Enhancing Children's Awareness,* ed. Roger P. Weissberg, Thomas Gullota, Robert L. Hampton, Bruce Ryan, and Gerald Adams, pp. 105–128. Thousand Oaks, CA: Sage.

Myths and Facts About the Death Penalty. 1998. "Death Penalty: Focus on California." http://www.members.aol.com/Dpfocus/facts.htm

Napolitano, Jo. 2004. "Top Illinois Court Upholds Total Amnesty of Death Row." http://www.nytimes.com

National Research Council. 1994. *Violence in Urban America: Mobilizing a Response.* Washington, DC: National Academy Press.

NNO (National Night Out). 2003. "National Night Out 2002 Is Largest Ever." http://www.nationaltownwatch.org/nno/topstory

PBS (Public Broadcasting System). 2000. "Police Divide." *Online News-Hour,* February 28. http://www.pbs.org/newshour

Pertossi, Mayra. 2000. "ANALYSIS—Argentine Crime Rate Soars." September 27. http://www.news.excite.com

Pew Research Center. 2000. "Respondents' Percepticn of Safety." The Pew Research Center for the People and the Press. May 12. http://www.peoplepress.org/april00rpt.htm

Philips, Julie, 2002. "White, Black and Latino Homicide Rates: Why the Difference?" *Social Problems* 49:349–374.

Pickler, Nedra. 2000. "Documents Point to Tire Problem." September 6. http://www.news.excite.com

Reid, Sue Titus. 2003. *Crime and Criminology,* 10th ed. Boston: McGraw-Hill.

Reiman, Jeffery. 2001. *The Rich Get Richer and the Poor Get Prison.* Boston, MA:Allyn and Bacon.

Richtel, Matt. 2002. "Credit Card Theft Thrives Online as Global Market." *New York Times.* http://www.nytimes.com

———. 2003, "Mayhem, and Far from the Nicest Kind." *New York Times,* February 10. http://www.nytimes.com

Ripley, Amanda. 2003. "The Night Detective." *Time,* January 6, 45–50.

Rosoff, Stephen, Henry Pontell, and Robert Tillman. 2002. "White Collar Crime." In *Social Problems: Readings with Four Questions,* ed. Joel M. Charon, pp. 339–350. Belmont, CA: Wadsworth.

Rubin, Paul H. 2002. "The Death Penalty and Deterrence." *Forum* (Winter): 10–12.

Sealey, Geraldine. 2002. "No Second Chances?" December 10. http://www.abcnews.com

Sherman, Lawrence. 2003. "Reasons for Emotion." *Criminology* 42:1–37.

Sieber, Samantha. 2003. "Senators Push Hate Crime Legislation." Knight Rider/Tribune News Service. http://www.bayarea.com/mid/mercurynews/politics

Siegel, Larry. 2000. *Criminology.* Belmont, CA: Wadsworth.

Sileo, Chi Chi. 2000. "Crime Fighters Get Streetwise." In *Social Problems 00/01,* ed. Kurt Finsterbusch, pp. 189–191. Guilford, CN: Dushkin/McGraw-Hill.

State Department. 2003. "Victims of Trafficking and Violence Protection Act 2000." *Global Issues* 6 (August). http://wwwusinfo.state.gov/journals

Sullivan, Thomas. 2003. *Social Problems.* Boston, MA: Allyn and Bacon.

Surgeon General. 2002. "Cost-Effectiveness." *Youth Violence: A Report of the Surgeon General.* http://www.mentalhealth.org/youthviolence/surgeongeneral

Sutherland, Edwin H. 1939. *Criminology.* Philadelphia: Lippincott.

The Situation in the Netherlands. 2003. "Prostitution in Holland." http://www.ex.ac.uk/politics/por_data

Thio, Alex. 2004. *Deviant Behavior.* Boston, MA: Allyn and Bacon

Three Strikes. 2003. "The California Corporate 3 Strikes Campaign." http://www. corporate.3strikes.org

Travis, Jeremy, and Michelle Waul. 2002. *Reflections on the Crime Decline: Lessons for the Future.* Washington, DC: The Urban Institute, Justice Policy Center.

United Nations (UN). 1997. "Crime Goes Global." Document No. DPI/1518/SOC/CON/30M. New York: United Nations.

———. 2003. "Creating Guidelines for Restorative Justice Progrmmes." http://www.restorativejustice.org/rjs/feature/2003

U.S. Department of Justice (USDOJ). 2003a. "Global Crime Issues." *National Institute of Justice: International Center Global Crimes Issues.* Washington, DC: U.S. Government Printing Office.

———. 2003b. "Half of all Violent Crimes . . ." Press Release. March 9. Bureau of Justice Statistics, Office of Justice Programs. http://www.ojp.usdoj.gov/bjs/pub/press

———. 2003c. "Preliminary Uniform Crime Report." Press Release. June 16. Federal Bureau of Investigation. http://www.fbi.gov/pressrel03

Vander Ven, Thomas, Francis Cullen, Mark Carrozza, and John Wright. 2001. "Home Alone: The Impact of Maternal Employment on Delinquency." *Social Problems* 48:236–257.

(VORP) Victim-Offender Reconciliation Program. 2003. "About Victim-Offender Mediation and Reconciliation." http://www.vorp.com

Walker, Samuel, Cassia Spohn, and Miriam Delone. 1996. *The Color of Justice: Race, Ethnicity, and Crime in America.* Belmont, CA: Wadsworth.

Warner, Barbara, and Pamela Wilcox Rountree. 1997. "Local Social Ties in a Community and Crime Model." *Social Problems* 4(4):520–536.

Weed and Seed. 2003. "Operation Weed and Seed." Executive Office. http://www.ojp.usdoj.gov/eows/nutshell.htm

Williams, Linda. 1984. "The Classic Rape: When Do Victims Report?" *Social Problems* 31:459–467.

Wilgoren, Jodi. 2003. "Governor Empties Illinois Death Row." *New York Times,* January 12. http://www.nytimes.com

Chapter 5

Aassve, A. 2003. "The Impact of Economic Resources on Premarital Childbearing and Subsequent Marriage Among Young American Women." *Demography* 40:105–126.

Administration for Children and Families. 2003. *Prevention Pays: The Costs of Not Preventing Child Abuse and Neglect.* U.S. Dept. of Health and Human Services. http://www.acf.hhs.gov

Amato, Paul. 1999. "The Postdivorce Society: How Divorce Is Shaping the Family and Other Forms of Social Organization." In *The Postdivorce Family: Children, Parenting, and Society,* eds. R. A. Thompson and P. R. Amato, pp. 161–190. Thousand Oaks, CA: Sage.

——— 2003. "The Consequences of Divorce for Adults and Children." In *Family in Transition,* 12th ed., eds. Arlene S. Skolnick and Jerome H. Skolnick, pp. 190–213. Boston: Allyn and Bacon.

"American Bar Association Supports Equal Protections for Children of Same-Sex Parents." 2003 (Aug. 13). Human Rights Campaign. http://www.hrc.org

American Council on Education and University of California. 2002. *The American Freshman: National Norms for Fall, 2002.* Los Angeles: Los Angeles Higher Education Research Institute.

Anderson, Kristin L. 1997. "Gender, Status, and Domestic Violence: An Integration of Feminist and Family Violence Approaches." *Journal of Marriage and the Family* 59:655–669.

Applewhite, Ashton. 2003. "Covenant Marriage Would Not Benefit the Family." In *The Family: Opposing Viewpoints,* ed. Auriana Ojeda, pp. 189–195. Farmington Hill, MI: Greenhaven Press.

Babcock, J. C., S. A. Miller, and C. Siard. 2003. "Toward a Typology of Abusive Women: Differences Between Partner-Only and Generally Violent Women in the Use of Violence." *Psychology of Women Quarterly* 27:153–161.

Bachrach, C., M. J. Hindin, and E. Thomson. 2000. "The Changing Shape of Ties That Bind: An Overview and Synthesis." In *The Ties That Bind,* ed. Linda J. Waite, pp. 3–16. New York: Aldine de Gruyter.

Bachu, Amara. 1999. "Trends in Premarital Childbearing." *Current Population Reports,* pp. 23–197. Washington, DC: U.S. Bureau of the Census.

Barth, Richard P. 2003. "Abusive and Neglecting Parents and the Care of Their Children." In *All Our Families,* 2nd ed., eds. Mary Ann Mason, Arlene Skolnick, and Stephen D. Sugarman, pp. 265–284. New York: Oxford University Press.

Beitchman, J. H., K. J. Zuker, J. E. Hood, G. A. daCosta, D. Akman, and E. Cassavia. 1992. "A Review of the Long-Term Effects of Child Sexual Abuse." *Child Abuse and Neglect* 16:101–119.

Block, Nadine. 2003. "Disciplinary Spanking Should Be Banned." In *Child Abuse: Opposing Viewpoints,* ed. L. I. Gerdes, pp. 182–190. Farmington Hills, MI: Greenhaven Press.

Browning, Christopher R., and Edward O. Laumann. 1997. "Sexual Contact Between Children and Adults: A Life Course Perspective." *American Sociological Review* 62:540–60.

Browning, Don S. 2003. *Marriage and Modernization: How Globalization Threatens Marriage and What to Do About It.* Grand Rapids, MI: William B. Eerdmans.

Carrington, Victoria. 2002. *New Times: New Families.* Dorderecht, the Netherlands: Kluwer Academic.

"Child Abuse and Neglect National Statistics." 2000 (April). National Clearinghouse on *Child Abuse and Neglect* Information. 330 C St., SW, Washington, DC 20447.

Cole, Charles L., Anna L. Cole, and Jessica G. Gandolfo. 2000. "Marriage Enrichment for Newlyweds: Models for Strengthening Marriages in the New Millennium." Poster Presentation at the 62nd Annual Conference of the National Council on Family Relations, Minneapolis, MN, November 10–13.

Coltrane, Scott, and Randall Collins. 2001. *Sociology of Marriage and the Family: Gender, Love, and Property,* 5th ed. Belmont CA: Wadsworth.

Coontz, Stephanie. 2000. "Marriage: Then and Now." *Phi Kappa Phi Journal* 80:10–15.

Daley, S. 2000. "French Couples Take Plunge That Falls Short of Marriage." *New York Times,* April 18, A1, A4.

Demian. 2003. "Legal Marriage Report: Global Status of Legal Marriage." Partners Task Force for Gay & Lesbian Couples. http://www.buddybuddy.com

Demo, David H. 1992. "Parent-Child Relations: Assessing Recent Changes." *Journal of Marriage and the Family* 54: 104–117.

———. 1993. "The Relentless Search for Effects of Divorce: Forging New Trails or Tumbling Down the Beaten Path?" *Journal of Marriage and the Family* 55:42–45.

Demo, David H., Mark A. Fine, and Lawrence H. Ganong. 2000. "Divorce as a Family Stressor." In *Families & Change: Coping with Stressful Events and Transitions,* 2nd ed., eds. P. C. McKenry and S. J. Price, pp. 279–302. Thousand Oaks, CA: Sage.

DiLillo, D., G. C. Tremblay, and L. Peterson. 2000. "Linking Childhood Sexual Abuse and Abusive Parenting: The Mediating Role of Maternal Anger." *Child Abuse and Neglect* 24:767–769.

"Domestic Violence Fact Sheet." 1999 (January). Department of Health and Human Services, Administration for Children and Families. http://www.acf.dhhs.gov/p. . .pa/facts/domsvio.htm

Drummond, Tammerlin. 2000. "Mom on Her Own." *Time* (August 28), pp. 54–55.

Edin, Kathryn. 2000. "What Do Low-Income Single Mothers Say About Marriage?" *Social Problems* 47(1): 112–133.

Edin, Kathryn, and Laura Lein. 1997. *Making Ends Meet: How Single Mothers Survive Welfare and Low-Wage Work.* New York: Russell Sage Foundation.

Edleson, J. L., L. F. Mbilinyi, S. K. Beeman, and A. K. Hagemeister. 2003. "How Children Are Involved in Adult Domestic Violence." *Journal of Interpersonal Violence* 18:18–32.

Edwards, Tamala M. 2000. "Flying Solo." *Time,* August 28, pp.49–53.

Elliott, D. M., and J. Briere. 1992. "The Sexually Abused Boy: Problems in Manhood." *Medical Aspects of Human Sexuality* 26:68–71.

Eltahawy, Mona. 2000. "Giving Wives a Way Out." *U.S. News & World Report,* March 6, 35.

Emery, Robert E. 1999. "Postdivorce Family Life for Children: An Overview of Research and Some Implications for Policy." In *The Postdivorce Family: Children, Parenting, and Society,* eds. R. A. Thompson and P. R. Amato, pp. 3–27. Thousand Oaks, CA: Sage.

Family Court Reform Council of America. 2000. "Parental Alienation Syndrome." 31441 Santa Margarita Parkway, Suite A184. Rancho Santa Margarita, CA 92688.

Fass, Paula S. 2003. "A Sign of Family Disorder? Changing Representations of Parental Kidnapping." In *All Our Families,* 2nd ed. eds. Mary Ann Mason, Arlene Skolnick, and Stephen D. Sugarman, pp. 170–195. New York: Oxford University Press.

Federal Interagency Forum on Child and Family Statistics. 2003. *America's Children: Key National Indicators.* Washington, DC: U.S. Government Printing Office.

Fields, Jason. 2003. "Children's Living Arrangements and Characteristics: March 2002" *Current Population Reports,* P20–547. Washington, DC: U.S. Census Bureau.

Fisher, Bonnie S., Francis T. Cullen, and Michael G. Turner. 2000. *The Sexual Victimization of College Women.* National Institute of Justice and Bureau of Justice Statistics. Washington, DC: U.S. Department of Justice.

Fogle, Jean M. 2003. "Domestic Violence Hurts Dogs, Too." *DogFancy,* April, 12.

Forum on Child and Family Statistics. 2000. *America's Children: Key National Indicators of Well-Being, 2000.* http://www.childstats.gov

Gardner, Richard A. 1998. *The Parental Alienation Syndrome,* 2nd ed. Cresskill, NJ: Creative Therapeutics.

Gelles, Richard J. 2000. "Violence, Abuse, and Neglect in Families." In *Families & Change: Coping with Stressful Events and Transitions,* 2nd ed., eds. P. C. McKenry and S. J. Price, pp. 183–207. Thousand Oaks, CA: Sage.

———. 1993. "Family Violence." In *Family Violence: Prevention and Treatment,* eds. Robert L. Hampton, Thomas P. Gullotta, Gerald R. Adams, Earl H. Potter III, and Roger P. Weissberg, pp. 1–24. Newbury Park, CA: Sage.

Gilbert, Neil. 2003. "Working Families: Hearth to Market." In *All Our Families,* 2nd ed., eds. M. A. Mason, A. Skolnick, and S. D. Sugarman, pp. 220–243. New York: Oxford University Press.

Global Study of Family Values. 1998. The Gallup Organization. http://198.175.140.8/Special_Reports/family.htm

Goldstein, Joshua R., and Catherine T. Kenney. 2001. "Marriage Delayed or Marriage Forgone? New Cohort Forecasts of First Marriage for U.S. Women." *American Sociological Review,* 66:506–519.

Gore, Al, and Tipper Gore. 2002. *Joined at the Heart.* New York: Henry Holt.

Hacker, Andrew. 2003. *Mismatch: The Growing Gulf Between Women and Men.* New York: Scribner.

Hackstaff, Karla B. 2003. "Divorce Culture: A Quest for Relational Equality in Marriage." In *Family in Transition,* 12th ed., eds. Arlene S. Skolnick and Jerome H. Skolnick, pp. 178–190. Boston: Allyn and Bacon.

Harrington, Donna, and Howard Dubowitz. 1993. "What Can Be Done to Prevent Child Maltreatment?" In *Family Violence: Prevention and Treatment,* eds. Robert L. Hampton, Thomas P. Gullotta, Gerald R. Adams, Earl H. Potter III, and Roger P. Weissberg, pp. 258–280. Newbury Park, CA: Sage.

Hendy, H. M., D. Eggen, C. Gustitus, K. C. McLeod, and P. Ng. 2003. "Decision to Leave Scale: Perceived Reasons to Stay in or Leave Violent Relationships." *Psychology of Women Quarterly* 27:162–173.

Henry, Ronald K. 1999. "Child Support at a Crossroads: When the Real World Intrudes upon Academics and Advocates." *Family Law Quarterly* 33(1): 235–264.

Hewlett, Sylvia Ann, and Cornel West. 1998. *The War Against Parents: What We Can Do for Beleaguered Moms and Dads.* Boston: Houghton Mifflin.

Heyman, R. E., and A. M. S. Slep. 2002. "Do Child Abuse and Interpersonal Violence Lead to Adult Family Violence?" *Journal of Marriage and Family* 64:864–870.

Hochschild, Arlie Russell. 1997. *The Time Bind: When Work Becomes Home and Home Becomes Work.* New York: Henry Holt.

———. 1989. *The Second Shift: Working Parents and the Revolution at Home.* New York: Viking.

Hogan, D. P., R. Sun, and G. T. Cornwell. 2000. "Sexual and Fertility Behaviors of American Females Aged 15–19 Years: 1985, 1990, and 1995. *American Journal of Public Health* 90:1421–1425.

Jackson, Shelly, Lynette Feder, David R. Forde, Robert C. Davis, Christopher D. Maxwell, and Bruce G. Taylor. 2003. *Batterer Intervention Programs: Where Do We Go From Here?"* U.S. Department of Justice, June. http://www.usdoj.gov

Janofsky, Michael. 2001. "Utah Man Is Sentenced to 5 Years in Polygamy Case." *New York Times,* August 25. http://www.nytimes.com

Jasinski, J. L., L. M. Williams, and J. Siegel. 2000. "Childhood Physical and Sexual Abuse as Risk Factors for Heavy Drinking Among African-American Women: A Prospective Study. *Child Abuse and Neglect* 24:1061–1071.

Jekielek, Susan M. 1998. "Parental Conflict, Marital Disruption and Children's Emotional Well-Being." *Social Forces* 76:905–935.

Johnson, Michael P. 2001. "Patriarchal Terrorism and Common Couple Violence: Two Forms of Violence Against Women." In *Men and Masculinity: A Text Reader,* ed. T. F. Cohen, pp. 248–260. Belmont, CA: Wadsworth.

Johnson, Michael P., and Kathleen Ferraro. 2003. "Research on Domestic Violence in the 1990s: Making Distinctions." In *Family in Transition,* 12th ed., eds. A. S. Skolnick and J. H. Skolnick, pp. 493–514. Boston: Allyn and Bacon.

Jorgensen, Stephen R. 2000. "Adolescent Pregnancy Prevention: Prospects for 2000 and Beyond." Presidential Address at the National Council on Family Relations, 62nd Annual Conference, Minneapolis, MN, November 11.

Kaufman, Joan, and Edward Zigler. 1992. "The Prevention of Child Maltreatment: Programming, Research, and Policy." In *Prevention of Child Maltreatment: Developmental and Ecological Perspectives,* ed. Diane J. Willis, E. Wayne Holden, and Mindy Rosenberg, pp. 269–295. New York: John Wiley.

Kitzmann, K. M., N. K. Gaylord, A. R. Holt, and E. D. Kenny. 2003. "Child Witnesses to Domestic Violence: A Meta-analytic Review." *Journal of Clinical and Consulting Psychology* 71:339–352.

Knox, David (with Kermit Leggett). 1998. *The Divorced Dad's Survival Book: How to Stay Connected with Your Kids.* New York: Insight Books.

Knutson, John F., and Mary Beth Selner. 1994. "Punitive Childhood Experiences Reported by Young Adults over a 10-Year Period." *Child Abuse and Neglect* 18:155–166.

Krug, Ronald S. 1989. "Adult Male Report of Childhood Sexual Abuse by Mothers: Case Description, Motivations, and Long-Term Consequences." *Child Abuse and Neglect* 13:111–119.

Lachs, Mark S., Christianna Williams, Shelley O'Brien, Leslie Hurst, and Ralph Horwitz. 1997. "Risk Factors for Reported Elder Abuse and Neglect: A Nine-Year Observational Cohort Study." *Gerontologist* 37:469–474.

Lanz, Jean B. 1995. "Psychological, Behavioral, and Social Characteristics Associated with Early Forced Sexual Intercourse Among Pregnant Adolescents." *Journal of Interpersonal Violence* 10:188–200.

Leite, Randy W., and Patrick C. McKenry. 2000. "Aspects of Father Status and Post-Divorce Father Involvement with Children." Poster session at the National Council on Family Relations 62nd Annual Conference, Minneapolis, MN, November 10–13.

Lewin, Tamar. 2000. "Fears for Children's Well-Being Complicates a Debate over Marriage." *New York Times on the Web,* November 4. http://www.nytimes.com/2000/11/04/arts/04MARR.html

Lloyd, Sally A. 2000. "Intimate Violence: Paradoxes of Romance, Conflict, and Control." *National Forum* 80(4):19–22.

Lloyd, Sally A., and Beth C. Emery. 2000. *The Dark Side of Courtship: Physical and Sexual Aggression.* Thousand Oaks, CA: Sage.

Luker, Kristin. 1996. *Dubious Conceptions: The Politics of Teenage Pregnancy.* Cambridge, MA: Harvard University Press.

Magdol, L., T. E. Moffitt, A. Caspi, and P. A. Silva. 1998. "Hitting Without a License: Testing Explanations for Differences in Partner Abuse Between Young Adult Daters and Cohabitors." *Journal of Marriage and the Family* 60:41–55.

Marlow, L., and S. R. Sauber. 1990. *The Handbook of Divorce Mediation.* New York: Plenum.

Mason, Mary Ann. 2003. "The Modern American Step-Family: Problems and Possibilities." In *All Our Families,* 2nd ed., eds. Mary Ann Mason, Arlene Skolnick, and Stephen D. Sugarman, pp. 96–116. New York: Oxford University Press.

Mason, Mary Ann, Arlene Skolnich, and Stephen D. Sugarman. 2003. "Introduction." In *All Our Families,* 2nd ed., eds. Mary Ann Mason, Arlene Skolnick, and Stephen D. Sugarman, pp. 1–13. New York: Oxford University Press.

Mauldon, Jane. 2003. "Families Started by Teenagers." In *All Our Families,* 2nd ed., eds. Mary Ann Mason, Arlene Skolnick, and Stephen D. Sugarman, pp. 40–65. New York: Oxford University Press.

Mercy, James A., Etienne G. Krug, Linda L. Dahlberg, and Anthony B. Zwi. 2003. "Violence and Health: The United States in a Global Perspective." *American Journal of Public Health,* 93(2): 256–261.

Mindel, Charles H., Robert W. Habenstein, and Roosevelt Wright, Jr. 1998. *Ethnic Families in America: Patterns and Variations.* Upper Saddle River, NJ: Prentice Hall.

Monson, C. M., G. R. Byrd, and J. Langhinrichsen-Rohling. 1996. "To Have and to Hold: Perceptions of Marital Rape." *Journal of Interpersonal Violence* 11:410–424.

Moore, David W. 2003. "Family, Health Most Important Aspects of Life." *Gallup News Service,* January 3. http://www.gallup.org

National Center for Injury Prevention and Control. 2000. "Intimate Partner Violence Fact Sheet." National Center for Injury Prevention and Control. Mailstop K60, 4770 Buford Highway NE, Atlanta, GA 30341-3724.

National Mental Health Association. 2003. "Effective Discipline Techniques for Parents: Alternatives to Spanking." Strengthening Families Fact Sheet. http://www.nmha.org

National Parenting Association. 1996. *What Will Parents Vote For? Findings of the First National Survey of Parent Priorities.* New York: Author.

Nelson, B. S., and K. S. Wampler. 2000. "Systemic Effects of Trauma in Clinic Couples: An Exploratory Study of Secondary Trauma Resulting from Childhood Abuse." *Journal of Marriage and Family Counseling* 26:171–184.

Nielsen, L. 1999. "College Aged Students with Divorced Parents: Facts and Fiction." *College Student Journal* 33:543–572.

Nock, Steven L. 1995. "Commitment and Dependency in Marriage." *Journal of Marriage and the Family* 57:503–514.

Ozer, Elizabeth, M. Jane Park, Tina Paul, Claire D. Brindis, and Charles E. Irwin, Jr. 2003. *America's Adolescents: Are They Healthy?* San Francisco: University of California; San Francisco, National Adolescent Health Information Center.

Parker, Marcie R., Edward Bergmark, Mark Attridge, and Jude Miller-Burke. 2000. "Domestic Violence and Its Effect on Children." *National Council on Family Relations Report* 45(4):F6–F7.

Pasley, Kay, and Carmelle Minton. 2001. "Generative Fathering After Divorce and Remarriage: Beyond the 'Disappearing Dad.'" In *Men and Masculinity: A Text Reader,* ed. T. F. Cohen, pp. 239–248. Belmont CA: Wadsworth.

Peterson, Richard R. 1996. "A Reevaluation of the Economic Consequences of Divorce." *American Sociological Review* 61:528–536.

Popenoe, David. 1993. "Point of View: Scholars Should Worry About the Disintegration of the American Family." *Chronicle of Higher Education,* April 14, A48.

———. 1996. *Life Without Father.* New York: Free Press.

Rand, M. R. 2003. "The Nature and Extent of Recurring Intimate Partner Violence Against Women in the United States." *Journal of Comparative Family Studies* 34:137–146.

Rennison, Callie M. 2003. "Intimate Partner Violence, 1993–2001." Bureau of Justice Statistics. Crime Data Brief. Washington, DC: U.S. Department of Justice.

Rennison, Callie M., and Sarah Welchans. 2000. "Intimate Partner Violence." U.S. Department of Justice. Office of Justice Programs. Washington, DC: Bureau of Justice Statistics.

Resnick, Michael, Peter S. Bearman, Robert W. Blum, Karl E. Bauman, Kathleen M. Harris, Jo Jones, Joyce Tabor, Trish Beubring, Renee E. Sieving, Marcia Shew, Marjore Ireland, Linda H. Berringer, and J. Richard Udry. 1997. "Protecting Adolescents from Harm." *Journal of the American Medical Association* 278(10):823–832.

Ricci, L., A. Giantris, P. Merriam, S. Hodge, and T. Doyle. 2003. "Abusive Head Trauma in Maine Infants: Medical, Child Protective, and Law Enforcement Analysis." *Child Abuse and Neglect* 27:271–283.

Rubin, D. M., C. W. Christian, L. T. Bilaniuk, K. A. Zaxyczny, and D. R. Durbin. 2003. "Occult Head Injury in High-risk Abused Children." *Pediatrics* 111:1382–1386.

Russell, D. E. 1990. *Rape in Marriage.* Bloomington: Indiana University Press.

Schacht, Thomas E. 2000. "Protection Strategies to Protect Professionals and Families Involved in High-Conflict Divorce." *UALR Law Review* 22(3): 565–592.

Scott, K. L., and D. A. Wolfe. 2000. "Change Among Batterers: Examining Men's Success Stories." *Journal of Interpersonal Violence* 15:827–842.

Shapiro, Joseph P., and Joannie M. Schrof. 1995. "Honor Thy Children." *U.S. News and World Report,* February 27, 39–49.

Sigle-Rushton, W., and S. McLanahan. 2002. "The Living Arrangements of New Unmarried Mothers." *Demography* 39:415–433.

Simmons, T., and M. O'Connell. 2003. "Married-Couple and Unmarried Partner Households: 2000." Census 2000 Special Reports. http://www.census.gov

Simonelli, C. J., T. Mullis, A. N. Elliott, and T. W. Pierce. 2002. "Abuse by Siblings and Subsequent Experiences of Violence Within the Dating Relationship." *Journal of Interpersonal Violence* 17:103–121.

Singh, Susheela, and Jacqueline E. Darroch. 2000. "Adolescent Pregnancy and Childbearing: Levels and Trends in Developed Countries." *Family Planning Perspectives* 32(1):14–23.

Smith, J. 2003. "Shaken Baby Syndrome." *Orthopaedic Nursing.* 22:196–205.

Spiegel, D. 2000. "Suffer the Children: Long-Term Effects of Sexual Abuse." *Society* 37:18–20.

Stanley, Scott M., Howard J. Markman, Michelle St. Peters, and B. Douglas Leber. 1995. "Strengthening Marriage and Preventing Divorce: New Directions in Prevention Research." *Family Relations* 44:392–401.

Statistical Abstract of the United States: 2003. 2003. 123rd ed. U.S. Bureau of the Census. Washington, DC: U.S. Government Printing Office.

Stets, J. E., and M. A. Straus. 1989. "The Marriage as a Hitting License: A Comparison of Assaults in Dating, Cohabiting, and Married Couples." In *Violence in Dating Relationships,* eds. M. A. Pirog-Good and J. E. Stets, pp. 33–52. New York: Greenwood Press.

Stock, J. L., M. A. Bell, D. K. Boyer, and F. A. Connell. 1997. "Adolescent Pregnancy and Sexual Risk-Taking Among Sexually Abused Girls." *Family Planning Perspectives* 29:200–203.

Straus, Murray. 2000. "Corporal Punishment and Primary Prevention of Physical Abuse." *Child Abuse and Neglect* 24:1109–1114.

Sugarman, Stephen D. 2003. "Single-Parent Families." In *All Our Families,* 2nd ed., eds. Mary Ann Mason, Arlene Skolnick, and Stephen D. Sugarman, pp. 14–39. New York: Oxford University Press.

Thakkar, R. R., P. M. Gutierrez, C. L. Kuczen, and T. R. McCanne. 2000. "History of Physical and/or Sexual Abuse, and Current Suicidality in College Women." *Child Abuse and Neglect* 24:1345–1354.

Thompson, Ross A., and Paul R. Amato. 1999. "The Postdivorce Family: An Introduction to the Issues." In *The Postdivorce Family: Children, Parenting, and Society,* eds. R. A. Thompson and P. R. Amato, pp. xi–xxiii. Thousand Oaks, CA: Sage.

Thompson, Ross A., and Jennifer M. Wyatt. 1999. "Values, Policy, and Research on Divorce." In *The Postdivorce Family: Children, Parenting, and Society,* eds. R. A. Thompson and P. R. Amato, pp. 191–232. Thousand Oaks, CA: Sage.

Ulman, A. 2003. "Violence by Children Against Mothers in Relation to Violence Between Parents and Corporal Punishment by Parents." *Journal of Comparative Family Studies* 34:41–56.

Umberson, D., K. L. Anderson, K. Williams, and M. D. Chen. 2003. "Relationship Dynamics, Emotion State, and Domestic Violence: A Stress and Masculine Perspective." *Journal of Marriage and the Family* 65:233–247.

United Nations Development Programme. 2000. *Human Development Report 2000.* Cary, NC: Oxford University Press.

United States Conference of Mayors—Sodexho. 2003. *Hunger and Homelessness Survey.* http://www.usmayors.org

U.S. Bureau of the Census. 1998. "Poverty in the United States." *Current Population Reports* P60–201. Washington, DC: U.S. Government Printing Office.

———. 2000 (September). "Money Income in the U.S." *Current Population Reports.* Washington, DC: U.S. Government Printing Office.

U.S. Department of Health and Human Services. 2000 (June 17). "HHS Fatherhood Initiative." http://www.hhs.gov/news/press/2000pres/20000617.html

———. Administration on Children, Youth and Families. 2003. *Child Maltreatment 2001.* Washington, DC: U.S. Government Printing Office.

U.S. Department of Justice. 1998 (March 16). "Murder by Intimates Declined 36 Percent Since 1976, Decrease Greater for Male than for Female Victims." Washington, DC. http://www.ojp.usdoj.gov/bjs/pub/press/vi.pr

Ventura, Stephanie J., and Christine A. Bachrach. 2000 (October 18). "Nonmarital Childbearing in the United States, 1940–99." *National Vital Statistics Report* 48(16), entire issue.

Ventura, Stephanie J., Sally C. Curtin, and T. J. Mathews. 2000 (April 24). "Variations in Teenage Birth Rates, 1991–1998: National and State Trends." *National Vital Statistics Report* 48(6), entire issue.

Viano, C. Emilio. 1992. "Violence Among Intimates: Major Issues and Approaches." In *Intimate Violence: Interdisciplinary Perspectives,* ed. C. E. Viano, pp. 3–12. Washington, DC: Hemisphere.

Walker, Alexis J. 2001. "Refracted Knowledge: Viewing Families Through the Prism of Social Science." In *Understanding Families into the New Millennium: A Decade in Review,* ed. Robert M. Milardo, pp. 52–65. Minneapolis: National Council on Family Relations.

Wallerstein, Judith S. 2003. "Children of Divorce: A Society in Search of Policy." In *All Our Families,* 2nd ed., eds. Mary Ann Mason, Arlene Skolnick, and Stephen D. Sugarman. pp. 66–95. New York: Oxford University Press.

Whiffen, V. E., J. M. Thompson, and J. A. Aube. 2000. "Mediators of the Link Between Childhood Sexual Abuse and Adult Depressive Symptoms." *Journal of Interpersonal Violence* 15: 1100–1120.

Whitehead, Barbara Dafoe, and David Popenoe. 2003. *The State of Our Unions: The Social Health of Marriage in America.* The National Marriage Project. Rutgers, State University of New Jersey. http://marriage.rutgers.edu

Chapter 6

American Council on Education and American Association of University Professors. 2000. *Does Diversity Make a Difference? Three Research Studies on Diversity in College Classrooms.* Washington, DC: American Council on Education and American Association of University Professors.

American Council on Education and University of California. 2003. *The American Freshman: National Norms for Fall 2003.* Los Angeles Higher Education Research Institute.

Arner, Mildred L. 2003. "Membership of the 108th Congress: A Profile." Congressional Research Service. http://www.senate.gov

Beeman, Mark, Geeta Chowdhry, and Karmen Todd. 2000. "Educating Students About Affirmative Action: An Analysis of University Sociology Texts." *Teaching Sociology* 28(2):98–115.

Cohen, Mark Nathan. 1998. "Culture, Not Race, Explains Human Diversity." *Chronicle of Higher Education* 44(32): B4–B5.

Conley, Dalton. 1999. *Being Black, Living in the Red: Race, Wealth, and Social Policy in America.* Berkeley: University of California Press.

———. 2002 (October 13). "The Importance of Being White." *Newsday, Inc.* http://www.newsday.com

Day, Jennifer Cheeseman, and Eric Newburger, 2002. "The Big Payoff: Educational Attainment and Synthetic Estimates of Work-Life Earnings." Washington, DC: U.S. Census Bureau.

Dees, Morris. 2000 (December 28). Personal correspondence. Morris Dees, cofounder of the Southern Poverty Law Center. 400 Washington Avenue, Montgomery, AL 36104.

"Details on the Huge Advantage for Legacy Applicants." 2003. *The Journal of Blacks in Higher Education.* http://www.jbhe.com

EEOC Press Release. 2003 (July 17). "Muslim Pilot Fired Due to Religion and Appearance, EEOC Says in Post-9/11 Backlash Discrimination Suit." U.S. Equal Employment Opportunity Commission. http://www.EEOC.gov

Etzioni, Amitai. 1997. "New Issues: Rethinking Race." *The Public Perspective* (June–July):39–40. http://www.ropercenter.unconn.edu/pubper/pdf/!84b.htm

Federal Bureau of Investigation. 2003. *Hate Crime Statistics 2001.* http://www.fbi.gov

Fix, Michael E., and Randolph Capps. 2002. "The Dispersal of Immigrants in the 1990s." Washington DC: Urban Institute.

"The Forgotten." 2002 (Spring). *Intelligence Report* 105:n.p. http://www.splcenter.org/intelligenceproject

Foust, Dean, Brian Grow, and Aixa M. Pascual. 2002. "The Changing Heartland." *BusinessWeek,* September 9, 80–84.

Frey, William H. 2003. "Charticle." *The Milken Institute Review,* 3rd Quarter:7–10.

Gaertner, Samuel L., and John F. Dovidio. 2000. *Reducing Intergroup Bias: The Common Ingroup Identity Model.* Philadelphia: Taylor & Francis Group.

Gardyn, Rebecca, and John Fetto. 2000 (June). "Demographics . . . It's All the Rage!" *American Demographics.* http://www.demographics.com/publications/htm

Glaser, Jack, Jay Dixit, and Donald P. Green. 2002. "Studying Hate Crime with the Internet: What Makes Racists Advocate Racial Violence?" *Journal of Social Issues* 58(1):177–193.

Goldstein, Joseph. 1999 (January). "Sunbeams." *The Sun* 277:48.

Goodnough, Abby. 2001 (January 11). "New York City Is Short-Changed in School Aid, State Judge Rules." *New York Times on the Web.* http://www.nytimes.com/2001/01/11/nyregion/11SCHO.html

Greenberg, Daniel S. 2003 (March 1). "Supreme Court Sets Showdown on Affirmative Action." *The Lancet* 361:762. http://www.thelancet.com

Greenhouse, Steven. 2003 (February 9). "Suit Claims Discrimination Against Hispanics on Job." *New York Times.* http://www.nytimes.com

Grieco, Elizabeth M., and Rachel C. Cassidy. 2001 (March). "Overview of Race and Hispanic Origin: Census 2000 Brief." U.S. Census Bureau. http://www.census.gov/prod/2001pubs/cenbr01-1.pdf

Guinier, Lani. 1998. Interview with Paula Zahn. CBS Evening News, July 18.

Gurin, Patricia. 1999 (Spring). "New Research on the Benefits of Diversity in College and Beyond: An Empirical Analysis." *Diversity Digest,* 5–15. Washington, DC: Association of American Colleges and Universities.

Halton, Beau. 1998 (March 26). "City's Housing Bias Called 'Abysmal.'" http://www.jacksonville.c...98/met_2blhousi.html

"Hate on Campus." 2000 (Spring). *Intelligence Report* 98:6–15.

Healey, Joseph F. 1997. *Race, Ethnicity, and Gender in the United States: Inequality, Group Conflict, and Power.* Thousand Oaks, CA: Pine Forge Press.

Hodgkinson, Harold L. 1995. "What Should We Call People? Race, Class, and the Census for 2000." *Phi Delta Kappa,* October, 173–179.

Holzer, Harry, and David Neumark. 2000. "Assessing Affirmative Action." *Journal of Economic Literature* 38(3): 483–568.

Hooks, Bell. 2000. *Where We Stand: Class Matters.* New York: Routledge.

Humphreys, Debra. 1999. "Diversity and the College Curriculum: How Colleges and Universities Are Preparing Students for a Changing World." *DiversityWeb.* http://www.inform.umd.edu/EdRes/Topic/Di...Leadersguide/CT/curriculum_briefing.html

———. 2000 (Fall). "National Survey Finds Diversity Requirements Common

Around the Country." *Diversity Digest.* http://www.diversityweb.org/Digest/F00/survey.html
Immigration and Naturalization Service. 2001. "General Naturalization Requirements." http://www.ins.usdoj.gov/natz/general.html
"Intelligence Briefs." 2000 (September). *SPLC Report* 30(3):3.
Jensen, Derrick. 2001 (April). "Saving the Indigenous Soul: An Interview with Martin Prechtel." *The Sun,* 304:4–15.
Keita, S. O. Y., and Rick A. Kittles. 1997. "The Persistence of Racial Thinking and the Myth of Racial Divergence." *American Anthropologist* 99(3): 534–544.
King, Joyce E. 2000 (Fall). "A Moral Choice." *Teaching Tolerance,* 18: 14–15.
Kozol, Jonathan. 1991. *Savage Inequalities: Children in America's Schools.* New York: Crown.
Landau, Elaine. 1993. *The White Power Movement: America's Racist Hate Groups.* Brookfield, CT: Millbrook Press.
Lawrence, Sandra M. 1997. "Beyond Race Awareness: White Racial Identity and Multicultural Teaching." *Journal of Teacher Education* 48(2):108–117.
Levin, Jack, and Jack McDevitt. 1995. "Landmark Study Reveals Hate Crimes Vary Significantly by Offender Motivation." *Klanwatch Intelligence Report,* August, 7–9.
Lollock, Lisa. 2001 (March). "The Foreign-Born Population in the United States: March 2000." *Current Population Reports* P20-534. Washington DC: U.S. Bureau of the Census.
Ludwig, Jack. 2000 (February 28). "Perceptions of Black and White Americans Continue to Diverge Widely on Issues of Race Relations in the U.S." Gallup Organization, Poll Releases. http://www.gallup.com/poll/releases/pr000228.asp
Massey, Douglas, and Nancy Denton. 1993. *American Apartheid: Segregation and the Making of an American Underclass.* Cambridge, MA: Harvard University Press.
McLemore, S. Dale, Harriet D. Romo, and Susan Gonzalez Baker. 2001. *Racial and Ethnic Relations in America,* 6th ed. Needham Heights, MA: Allyn and Bacon.
Miller, Patti, McCrae A. Parker, Eileen Espejo, and Sarah Grossman-Swenson. 2002 (May). *Fall Colors: Prime Time Diversity Report 2001–02.* Oakland, CA: Children Now & the Media Program.
Molnar, Stephen. 1983. *Human Variation: Races, Types, and Ethnic Groups,* 2nd ed. Englewood Cliffs, NJ: Prentice-Hall.
Morrison, Pat. 2002 (September 6). "September 11: A Year Later—American Muslims Are Determined Not to Let Hostility Win." *National Catholic Reporter,* 38(38):9–10.
Motavalli, Jim, and Christina Zarrella. 2004. "The Numbers Game." *E Magazine: The Environmental Magazine,* 15(1):26–34.
NAACP Press Release. 2000 (November 11). "NAACP Voting Irregularities Public Hearing." National Association for the Advancement of Colored People. http://www.NAACP.org
———. 2001 (January 10). "NAACP National Civil Rights Groups File Florida Voting Rights Lawsuit to Eliminate Unfair Voting Practices." National Association for the Advancement of Colored People. http://www.NAACP.org
Nash, Manning. 1962. "Race and the Ideology of Race." *Current Anthropology* 3:258–288.
National Coalition on Black Civic Participation Press Release. 2000 (November 10). "National Coalition's Efforts Lead to Upsurge in Black Voter Turnout." http://www.bigvote.org
Navarro, Mireya. 2003 (April 28). "For New York's Black Latinos, a Growing Racial Awareness." *New York Times.* http://www.nytimes.com
Newburger, Eric C., and Andrea Curry. 2000 (March). "Educational Attainment in the United States." *Current Population Reports,* p. 20–528. Washington, DC: U.S. Census Bureau.
Olson, James S. 2003. *Equality Deferred: Race, Ethnicity, and Immigration in America Since 1945.* Belmont CA: Wadsworth/Thomson.
Orfield, Gary. 2001 (July). *Schools More Separate: Consequences of a Decade of Resegregation.* Cambridge, MA: Harvard University, the Civil Rights Project.
Padgett, Tim, and Frank Sikora. 2003. "Color-Blind Love." *Time,* May 12, n.p.
Pager, Devah. 2003. "The Mark of a Criminal Record." *American Journal of Sociology* 108(5):937–975.
Parsons, Sharon, William Simmons, Frankie Shinhoster, and John Kilburn. 1999. "A Test of the Grapevine: An Empirical Examination of Conspiracy Theories Among African-Americans." *Sociological Spectrum* 19(2):201–222.
Paul, Pamela. 2003 (May). "Attitudes Toward Affirmative Action." *American Demographics* 25(4):18–19.
Pew Research Center. 2003 (May 13). "Conflicted Views of Affirmative Action." http://www.people-press.org
Plous, S. 2003. "Ten Myths About Affirmative Action." In *Understanding Prejudice and Discrimination,* ed. S. Plous, pp. 206–212. New York: McGraw-Hill.
Polling Report. 2003. http://www.pollingreport.com/race/htm
Purdum, Todd S. 2001 (March 29). "California Census Confirms Whites Are in Minority." *New York Times on the Web.* http://www.nytimes.com/2001/03/30/national/30CALI.html
Race Relations Reporter. 1999 (October 15). Vol. 7, no. 8. New York: CH II Publishers, 200 West 57th St., New York, NY 10019.
Ramirez, Roberto R., and G. Patricia de la Cruz. 2003 (June). *The Hispanic Population in the United States: March 2002,* Current Population Reports, P20-545. Washington, DC: U.S. Census Bureau.
Reardon-Anderson, Jane, Randolph Capps, and Michael E. Fix. 2002. "The Health and Well-Being of Children in Immigrant Families." Washington, DC: Urban Institute.
Rothenberg, Paula S. 2002. *White Privilege.* New York: Worth.
Schaefer, Richard T. 1998. *Racial and Ethnic Groups,* 7th ed. New York: HarperCollins.
Schiller, Bradley R. 2004. *The Economics of Poverty and Discrimination,* 9th ed. Upper Saddle River, NJ: Pearson Education.
Schmidley, Dianne. 2003. *The Foreign-Born Population in the United States: March 2002.* Current Population Reports, P20-539. Washington DC: U.S. Bureau of the Census.
Schmitt, Eric. 2001 (March 31). "Blacks Split on Disclosing Multiracial Roots." *New York Times on the Web.* http://www.nytimes.com/2001/03/31/national/31RACE.html
———. 2001 (April 4). "Analysis of Census Finds Segregation Along with Diversity." *New York Times on the Web.* http://www.nytimes.com/2001/04/04/national/04CENS.html
Schuman, Howard, and Maria Krysan. 1999. "A Historical Note on Whites' Beliefs About Racial Inequality." *American Sociological Review* 64:847–855.
Schuman, Howard, Charlotte Steeh, Lawrence Bobo, and Maria Krysan. 1997. *Racial Attitudes in America: Trends and Interpretations.* Cambridge, MA: Harvard University Press.
Shipler, David K. 1998 (March 15). "Subtle vs. Overt Racism." *Washington Spectator* 24(6):1–3.
SPLC Report 2003 (March). "Hate Group Numbers Rise, but New-Nazis in Disarray."33(1):3. Southern Poverty Law Center. 400 Washington Ave. Montgomery, AL 36104.
"Study Finds Benefits from Immigration." 1997. Minneapolis *Star Tribune,* May 18, 4A.
Teaching Tolerance. 2000 (Fall). "Hear & Now," p. 5.
Tolbert, Caroline J., and John A. Grummel. 2003. "Revisiting the Racial Threat Hypothesis: White Voter Support for California's Proposition 209. *State Politics & Policy Quarterly* 3(2):183–202; 215–216.

Turner, Margery Austin, and Felicity Skidmore. 1999. *Mortgage Lending Discrimination: A Review of Existing Evidence.* Washington, DC: The Urban Institute.

Turner, Margery Austin, Stephen L. Ross, George Galster, and John Yinger. 2002. *Discrimination in Metropolitan Housing Markets.* Washington, DC: The Urban Institute.

U.S. Department of Labor. 2002. "Facts on Executive Order 11246 Affirmative Action." http://www.dol.gov

Van Ausdale, Debra, and Jor R. Feagin. 2001. *The First R: How Children Learn Race* and *Racism.* Lanham; MD: Rowman & Littlefield.

Wheeler, Michael L. 1994. *Diversity Training: A Research Report.* New York: The Conference Board.

"White Power Bands." 2002 (January). *Hate in the News.* Southern Poverty Law Center. http://www.tolerance.org

Williams, Eddie N., and Milton D. Morris. 1993. "Racism and Our Future." In *Race in America: The Struggle for Equality,* eds. Herbert Hill and James E. Jones Jr., pp. 417–424. Madison: University of Wisconsin Press.

Willoughby, Brian. 2003 (June 13). "Hate on Campus." *Tolerance in the News,* http://www.tolerance.org

Wilson, William J. 1987. *The Truly Disadvantaged: The Inner City, the Underclass and Public Policy.* Chicago: University of Chicago Press.

Winter, Greg. 2003a (January 21). "Schools Resegregate, Study Finds." *New York Times.* http://www.nytimes.com.

———. 2003b (August 29). "U. of Michigan Alters Admissions Use of Race." *New York Times.* http://www.nytimes.com

"The Year in Hate." 2001 (Spring). *Intelligence Report,* Issue 101. http://www.splcenter.org/intelligenceproject/ip-index.html

Zack, Naomi. 1998. *Thinking About Race.* Belmont, CA: Wadsworth.

Zinn, Howard. 1993. "Columbus and the Doctrine of Discovery." In *Systemic Crisis: Problems in Society, Politics, and World Order,* ed. William D. Perdue, pp. 351–357. Fort Worth: Harcourt Brace Jovanovich.

Chapter 7

Abernathy, Michael. 2003. "Male Bashing on TV." *Tolerance in the News.* http://www.tolerance.org/news

Anderson, Margaret L. 1997. *Thinking About Women.* 4th ed. New York: Macmillan.

Alvarez, Lizette. 2003. "Norway v. Glass Ceiling." *New York Times.* June 14. http://www.nytimes.com

Athreya, Bama. 2003. "Trade Is a Women's Issue." *ATTAC* February 20. http://www.globalpolicy.org/socecon

Austin, Jonathan D. 2000. "U.N. Report: Women's Unequal Treatment Hurts Economies." CNN.com. September 20. http://www.cnn.com/2000/world/europe/09/20.un. population.report

Baker, Robin, Gary Kriger, and Pamela Riley. 1996. "Time, Dirt and Money: The Effects of Gender, Gender Ideology, and Type of Earner Marriage on Time, Household Task, and Economic Satisfaction Among Couples with Children." *Journal of Social Behavior and Personality* 11:161–77.

Banister, Judith. 2003. "Shortage of Girls in China: Causes, Consequences, International Comparisons and Solutions." Population Reference Bureau online. http://www.prb.org

Bannon, Lisa. 2000. "Why Girls and Boys Get Different Toys." *Wall Street Journal,* February 14, B1.

Barko, Naomi. 2003. "Equal Pay for Equal Work." In *Women's Rights,* ed. Shasta Gaughen, pp. 43–48. Farmington, MA: Greenhaven Press.

Basow, Susan A. 1992. *Gender: Stereotypes and Roles,* 3rd ed. Pacific Grove, CA: Brooks/Cole.

Begley, Sharon. 2000. "The Stereotype Trap." *Newsweek,* November 6, 66–68.

Belkachla, Said. 2003. "Monitoring Gender Equality in Framework of Education for All." UNESCO. ECE Work Session on Gender Statistics, Geneva, Switzerland, September 23–25.

Beutel, Ann M., and Margaret Mooney Marini. 1995. "Gender and Values." *American Sociological Review* 60:436–48.

Bianchi, Susanne M., Melissa A. Milkie, Liana C. Sayer, and John Robinson. 2000. "Is Anyone Doing the Housework? Trends in the Gender Division of Household Labor." *Social Forces* 79:191–228.

Bittman, Michael, and Judy Wajcman. 2000. "The Rush Hour: The Character of Leisure Time and Gender Equity." *Social Forces* 79:165–189.

BLS (Bureau of Labor Statistics). 2000. *Report on the Youth Labor Force.* U.S. Department of Labor. Washington, DC.

———. 2002. "Highlights of Women's Earnings in 2001." U.S. Department of Labor. http://www.bls.gov

———. 2003. "Characteristics of Minimum Wage Workers." U.S. Department of Labor. http://www.bls.gov/cps/minwage

Budig, Michelle. 2003. "Male Advantage and the Gender Composition of Jobs: Who Rides the Glass Escalator?" *Social Problems* 49:258–277.

Burger, Jerry M., and Cecilia H. Solano. 1994. "Changes in Desire for Control over Time: Gender Differences in a Ten-Year Longitudinal Study." *Sex Roles* 31:465–72.

CBSNews. 2003a. "Air Force Cadets Claim Rape Cover Up." February 17. http://www.cbsnews.com/stories/2003

———. 2003b "Changes Planned at A. F. Academy." March 11. http://www.cbsnews.com/stories/2003

CEDAW (Committee on the Elimination of Discrimination Against Women). 2003. "Welcome New Legislation to Foster Gender Equality." Press Release. August 7.

Cejka, Mary Ann, and Alice Eagly. 1999. "Gender Stereotypic Images of Occupations Correspond to the Sex Segregation of Employment." *Personality and Social Psychology Bulletin* 25: 413–423.

Chavez, Linda. 2000. *The Color Bind.* Berkeley: University of California Press.

Civil Rights Monitor. 2000. "Sexual Harassment Decisions, Supreme Court 1997–1998 Term." http://www.civilrights.org/crlibrary/monitor/winter_spring1999

CNN.com. 2003. "More Air Force Cadets Speak Out." March 7. http://www.cnn.com

Cohen, Philip, and Matt Huffman. 2003. "Individuals, Jobs, and Labor Markets: The Devaluation of Women's Work: *American Sociological Review* 68:443–463.

Cohen, Theodore. 2001. *Men and Masculinity.* Belmont, CA: Wadsworth.

Cullen, Lisa Takeuchi. 2003. "I Want Your Job, Lady!" *Time,* May 12, 52–56.

Dittrich, Liz. 2002a. "About-face Facts on the Children and the Media." http://www.about-face.org/r/facts/children media

———. 2002b. "About-face Facts on the Media." http://www.about-face.org/r/facts/childrenmedia

EEOC (Equal Employment Opportunity Commission). 2003. "Women of Color Make Gains in Employment and Job Status." July 31. http://www.eeoc.org/pres

Evans, Lorraine, and Kimberly Davies. 2000. "No Sissy Boys Here." *Sex Roles* (February):255–271.

Faludi, Susan. 1991. *Backlash: The Undeclared War Against American Women.* New York: Crown.

FGC (Female Genital Cutting). 2003. "Female Genital Cutting (FGC): An Introduction." http://www.fgcnetwork.org/intro

Fitzgerald, Louise F., and Sandra L. Shullman. 1993. "Sexual Harassment: A Research Analysis and Agenda for the '90s." *Journal of Vocational Behavior* 40:5–27.

Fitzpatrick, Catherine. 2000. "Modern Image of Masculinity Changes with Rise of New Celebrities." *Detroit News,* June 24. http://www.detnews.com/2000/religion/0006/24

Gandy, Kim. 2003. "It's Not About Golf: Feminists Blast Discrimination at Home of Masters Tournament." http://www.now/org/press

GEM (Gender Empowerment Measure). 2003. Human Development Index. http://www.undp.org

Gupta, Sanjay. 2003. "Why Men Die Young." *Time,* May 12, 84.

Heyzer, Noeleen. 2003. "Enlisting African Women to Fight AIDS." *Washington Post,* July 8. http://www.globalpolicy.org/socecon/inequal

IWRP (International Women's Right's Project). 2000. "The First CEDAW Impact Study." http://www.yorku.ca/iwrp/cedawReport

Jones, Del. 2003. "Few Women Hold Top Executive Jobs, Even when CEOs Are Female." *USA Today,* May 22. http://www.usatoday.com

Kilbourne, Barbara S., Georg Farkas, Kurt Beron, Dorothea Weir, and Paula England. 1994. "Returns to Skill, Compensating Differentials, and Gender Bias: Effects of Occupational Characteristics on the Wages of White Women and Men." *American Journal of Sociology* 100:689–719.

Kopelman, Loretta M. 1994. "Female Circumcision/Genital Mutilation and Ethical Relativism." *Second Opinion* 20:55–71.

Leeman, Sue. 2000. "The More Things Change . . ." September 20. http://www.abcnews.go.com/sections/living/Daily/News/women_unreport00920.html

Long, J. Scott, Paul D. Allison, and Robert McGinnis. 1993. "Rank Advancement in Academic Careers: Sex Differences and the Effects of Productivity." *American Sociological Review* 58:703–22.

Lorber, Judith. 1998. "Night to His Day." In *Reading Between the Lines,* eds. Amanda Konradi and Martha Schmidt, pp. 213–220. Mountain View, CA: Mayfield.

Major Rulings. 2003. "Major Rulings of the 2002-2003 Term." *Daily Reflector,* June 27, 4A

Marini, Margaret Mooney, and Pi-Ling Fan. 1997. "The Gender Gap in Earnings at Career Entry." *American Sociological Review* 62:588–604.

Martin, Patricia Yancey. 1992. "Gender, Interaction, and Inequality in Organizations." In *Gender, Interaction, and Inequality,* ed. Cecilia Ridgeway, pp. 208–231. New York: Springer-Verlag.

Mattingly, Marybeth, and Suzanne Bianchi. 2003. "Gender Differences in the Quantity and Quality of Free Time: The U.S. Experience." *Social Forces* 81:999–1030.

Mazure, C. M., G. P. Keita, and M. Blehar. 2002. *Summit on Women and Depression: Proceedings and Recommendations.* Washington, DC: American Psychological Association.

McBrier, Debra Branch. 2003, "Gender and Career Dynamics Within a Segmented Professional Labor Market: The Case of Law Academic." *Social Forces* 81:1201–1266.

McCammon, Susan, David Knox, and Caroline Schacht. 1998. *Making Choices in Sexuality.* Pacific Grove, CA: Brooks/Cole.

McGregor, Liz. 2003. "Women Bear Brunt of AIDS Toll." http://www.globalpolicy.org/socecon/develop

Mensch, Barbara, and Cynthia Lloyd. 1997. "Gender Differences in the Schooling Experiences of Adolescents in Low-Income Countries: The Case of Kenya." Policy Research Working Paper no. 95. New York: Population Council.

Mitchell, Melissa. 2002. "U.S. Moving Closer to Approving International Women's Rights Bill." News Bureau, October 1, University of Illinois at Urbana.

Moen, Phyllis, and Yan Yu. 2000. "Effective Work/Life Strategies: Working Couples, Working Conditions, Gender and Life Quality." *Social Problems* 47:291–326.

Morin, Richard, and Megan Rosenfeld. 2000. "The Politics of Fatigue." In *Annual Editions: Social Problems,* ed. Kurt Finsterbusch, pp. 152–154. Guilford, CT: Dushkin/McGraw-Hill.

NCFM (National Coalition of Free Men). 1998. "Historical." Manhasset, NY: http://www.ncfm.org

Nichols-Casebolt, Ann, and Judy Krysik. 1997. "The Economic Well-Being of Never and Ever-Married Mother Families." *Journal of Social Service Research* 23(1):19–40.

Orecklin, Michelle. 2003. "Now She's Got Game." *Time,* March 3, 57–59.

Parker, Kathleen. 2000. "It's Time for Women to Get Angry but Not at Men." *Greensboro News Record,* March 14.

Poe, Marshall. 2004. "The Other Gender GAP." *The Atlantic Online.* January/February.

Pollack, William. 2000a. *Real Boys' Voices.* New York: Random House.

———. 2000b. "The Columbine Syndrome." *National Forum* 80:39–42.

Popline. 2003. "Gender Disparities in Transition Nations." September–October. World Population News Service. Volume 24.

———. 1999. "World Population: More than Just Numbers." Washington, DC. http://www.prb.org

Purcell, Piper, and Lara Stewart. 1990. "Dick and Jane in 1989." *Sex Roles* 22:177–185.

Quist-Areton, Ofeibea. 2003. "Fighting Prejudice and Sexual Harassment of Girls in Schools." *All Africa,* June 12. http://www.globalpolicy.org/socecon

Rabin, Sarah. 2000. "Feminists Take CEDAW into Our Own Hands." National Organization for Women. http://www.63111.42.146/cgs/gs_article.asp?ArticleD-1863

Reid, Pamela T., and Lillian Comas-Diaz. 1990. "Gender and Ethnicity: Perspectives on Dual Status." *Sex Roles* 22:397–408.

Renzetti, Claire, and Daniel Curran. 2003. *Women, Men and Society.* Boston: Allyn and Bacon.

Reskin, Barbara, and Debra McBrier. 2000. "Why Not Ascription? Organizations' Employment of Male and Female Managers." *American Sociological Review* 65:210–233.

Robinson, John P., and Suzanne Bianchi. 1997. "The Children's Hours." *American Demographics,* December, 1–6.

Rosenberg, Janet, Harry Perlstadt, and William Phillips. 1997. "Now That We Are Here: Discrimination, Disparagement, and Harassment at Work and the Experience of Women Lawyers." In *Workplace/Women's Place,* ed. Dana Dunn, pp. 247–259. Los Angeles: Roxbury.

Rubenstein, Carin. 1990. "A Brave New World." *New Woman* 20(10):158–64.

Saad, Lydia. 2000. "Most Working Women Deny Gender Discrimination in Their Pay." Gallup Poll. http://www.gallup.poll/releases/pr000207.asp

Sadker, Myra, and David Sadker. 1990. "Confronting Sexism in the College Classroom." In *Gender in the Classroom: Power and Pedagogy,* eds. S. L. Gabriel and I. Smithson, pp. 176–187. Chicago: University of Illinois Press.

Sapiro, Virginia. 1994. *Women in American Society.* Mountain View, CA: Mayfield.

Sax, Linda, Jennifer Lindholm, Alexander Astin, William Korn, and Kathryn Mahoney. 2002. *The American Freshman: National Norms for Fall 2002.* Los Angeles, CA: Higher Education Research Institute.

Schneider, Margaret, and Susan Phillips. 1997. "A Qualitative Study of Sexual Harassment of Female Doctors by Patients." *Social Science and Medicine* 45:669–676.

Schroeder, K. A., L. L. Blood, and D. Maluso. 1993. "Gender Differences and Similarities Between Male and Female Undergraduate Students Regarding Expectations for Career and Family Roles." *College Student Journal* 27:237–249.

Schwalbe, Michael. 1996. *Unlocking the Iron Cage: The Men's Movement, Gender Politics, and American Culture.* New York: Oxford University Press.

Sheehan, Molly. 2000. "Women Slowly Gain Ground in Politics." In *Vital Signs: The Environmental Trends That Are Shaping Our Future,* ed. Linda Starke., pp. 152–153. New York: W.W. Norton.

Smith, Shellee. 2003. "New Report on Academy Sex Assault." NSNBC.com. August 29. http://www.msnbc.com/m/pt/printms

Smolken, Rachael. 2000. "Girls' SAT Scores Still Lag Boys'." *Post Gazette.* http://www.post-gazette.com/headlines/20000830sat2.asp

Sommers, Christina. 2000. *The War Against Boys.* New York: Simon and Schuster.

Statistical Abstract of the United States: 2002. 122nd ed. U.S. Bureau of the Census. Washington, DC: U.S. Government Printing Office.

Tam, Tony. 1997. "Sex Segregation and Occupational Gender Inequality in the United States: Devaluation or Specialized Training?" *American Journal of Sociology* 102(6):1652–1692.

Thomas, Cathy Booth. 2003. "Conduct Unbecoming." *Time,* March 10, 46–47.

Thomas, Karen. 1999. "Equality Is for the Daughters." *USA Today,* February 17, 7A.

United Nations. 2000a. *The World's Women 2000: Trends and Statistics.* New York: United Nations Statistics Division.

———. 2000b. "Convention on the Elimination of All Forms of Discrimination Against Women." http://www.un.org/womenwatch/daw/cedaw

UNFPA (United Nations Population Fund). 2003. *State of the World Population, 2002.* "Women and Gender Inequality." http://www.unfpa.org/swp

UNICEF. 2003. "Gender Equality: The Big Picture." http://www.unicef.org/gender

U.S. Census. 2003. "Women Closing the Gap with Men in Some Measures, According to Census Bureau." Press Release, March 24. http://www.census.gov/Press-Release/www/2003

Urban Institute. 2003. "Gender Gap in Higher Education Focus of New Urban Institute Research." September 2. http://www.urban.org

Van Willigen, Marieke, and Patricia Drentea. 1997. "Benefits of Equitable Relationships: The Impact of Sense of Failure, Household Division of Labor, and Decision-Making Power on Social Support." Presented at the American Sociological Association, Toronto, Canada, August.

WGI (World Global Issues). 2003. "Women." http://www.osearth.cpm/reources

WHO (World Health Organization). 2001. "Prevalence Rates for FGM." http://www.who.int/frh-whd/FGM

Williams, Christine L. 1995. *Still a Man's World: Men Who Do Women's Work.* Berkeley: University of California Press.

Williams, John E., and Deborah L. Best. 1990. *Sex and Psyche: Gender and Self Viewed Cross-Culturally.* London: Sage Publications.

WIN *(Women's International Network)* News. 2000. "Reports from Around the World." Autumn, 50–58.

Winfield, Nicole. 2000. "Activists Give U.S. Mixed Rating for Efforts on Gender." *Boston Globe,* June 8, A21.

Witt, S. D. 1996. "Traditional or Androgynous: An Analysis to Determine Gender Role Orientation of Basal Readers." *Child Study Journal* 26:303–318.

World Bank. 2003 (April). "Gender Equality and the Millennium Development Goals." *Gender and Development Group.* World Bank.

Yamaguchi, Mari. 2000. "Female Government Workers Face Harassment." http://www.news.excite.com/news/ap/001227/05/int.Japan

Yoder, Janice D., and Patricia Aniakudo. 1997 "Outsiders Within the Firehouse: Subordination and Difference in the Social Interactions of African American Women Firefighters." *Gender and Society* 11(3):324–341.

Yumiko, Ehara (2000). "Feminism's Growing Pains." *Japan Quarterly* 47:41–48.

Zimmerman, Marc A., Laurel Copeland, Jean Shope, and T. E. Dielman. 1997. "A Longitudinal Study of Self-Esteem: Implications for Adolescent Development." *Journal of Youth and Adolescence* 26(2):117–141.

Chapter 8

"Adoption/Foster Care Laws in the U.S." (2002). National Gay and Lesbian Task Force. http://www.ngltf.org

Bayer, Ronald. 1987. *Homosexuality and American Psychiatry: The Politics of Diagnosis,* 2nd ed. Princeton, NJ: Princeton University Press.

Besen, Wayne. 2000. "Introduction." In *Feeling Free: Personal Stories: How Love and Self-Acceptance Saved Us from "Ex-Gay" Ministries.* Human Rights Campaign, p. 7. Washington, DC: Human Rights Campaign Foundation.

Black, Dan, Gary Gates, Seth Sanders, and Lowell Taylor. 2000 (May). "Demographics of the Gay and Lesbian Population in the United States: Evidence from Available Systematic Data Sources." *Demography* 37(2):139–154.

Bobbe, Judith. 2002. "Treatment with Lesbian Alcoholics: Healing Shame and Internalized Homophobia for Ongoing Sobriety." *Health & Social Work* 27(3): 218–223.

Bontempo, Daniel E., and Anthony R. D'Augelli. 2002. "Effects of At-School Victimization and Sexual Orientation on Lesbian, Gay, or Bisexual Youths' Health Risk Behavior." *Journal of Adolescent Health,* 30:364–374.

Bradford, Judith, Kirsten Barrett, and Julie A. Honnold. 2002. *The 2000 Census and Same-Sex Households: A User's Guide.* New York: The National Gay and Lesbian Task Force Policy Institute, the Survey and Evaluation Research Laboratory, and the Fenway Institute. http://www.ngltf.org

Button, James W., Barbara A. Rienzo, and Kenneth D. Wald. 1997. *Private Lives, Public Conflicts: Battles over Gay Rights in American Communities.* Washington, DC: CQ Press.

Cahill, Sean, and Kenneth T. Jones. 2003. "Child Sexual Abuse and Homosexuality: The Long History of the 'Gays as Pedophiles' Fallacy." National Gay and Lesbian Task Force. http://www.ngltf.org

Cahill, Sean, Mitra Ellen, and Sarah Tobias. 2002. *Family Policy: Issues Affecting Gay, Lesbian, Bisexual, and Transgender Families.* National Gay and Lesbian Task Force Policy Institute. http://www.ngltf.org

Cahill, Sean, and Samuel Slater. 2004. *Marriage: Legal Protections for Families and Children.* Policy Brief. Washington, DC: National Gay and Lesbian Task Force Policy Institute.

Chase, Bob. 2000. "NEA President Bob Chase's Historic Speech from 2000 GLSEN Conference." http://www.glsen.org

Cianciotto, Jason, and Sean Cahill. 2003. *Educational Policy: Issues Affecting Lesbian, Gay, Bisexual, and Transgender Youth.* Washington, DC: National Gay & Lesbian Task Force Policy Institute.

"Constitutional Protection." 1999. GayLawNet. http://www.nexus.net.au/~dba/news.html#top

Curtis, Christopher. 2003 (October 7). "Poll: U.S. Public is 50-50 on Gay Marriage." PlanetOut.com. http://www.planetout.com

"Custody and Visitation." 2000. Human Rights Campaign FamilyNet. http://familynet.hrc.org

D'Emilio, John. 1990. "The Campus Environment for Gay and Lesbian Life." *Academe* 76(1):16–19.

Diamond, Lisa M. 2003. "What Does Sexual Orientation Orient? A Biobehavioral Model Distinguishing Romantic Love and Sexual Desire." *Psychological Review* 110 (1):173–192.

Dozetos, Barbara. 2001 (March 7). "School Shooter Taunted as 'Gay.'" PlanetOut.com. http://www.planetout.com

Durkheim, Emile. 1993. "The Normal and the Pathological." Originally published in *The Rules of Sociological Method,* 1938. In *Social Deviance,* ed. Henry N. Pontell, pp. 33–63. Englewood Cliffs, NJ: Prentice-Hall.

Esterberg, K. 1997. *Lesbian and Bisexual Identities: Constructing Communities, Constructing Selves.* Philadelphia: Temple University Press.

Federal Bureau of Investigation. 2003. *Hate Crime Statistics 2002.* http://www.fbi.gov

Firestein, B. A. 1996. "Bisexuality as Paradigm Shift: Transforming Our Disciplines." In *Bisexuality: The Psychology and Politics of an Invisible Minority,* ed. B. A. Firestein, pp. 263–291. Thousand Oaks, CA: Sage.

Fone, Byrne. 2000. *Homophobia: A History.* New York: Henry Holt.

Frank, Barney. 1997. Foreword to *Private Lives, Public Conflicts: Battles over Gay Rights in American Communities,* by J. W. Button, B. A. Rienzo, and K. D. Wald. Washington DC: CQ Press.

Frank, David John, and Elizabeth H. McEneaney. 1999. "The Individualization of Society and the Liberalization of State Policies on Same-Sex Relations, 1984–1995." *Social Forces* 77(3): 911–944.

Franklin, Karen. 2000. "Antigay Behaviors Among Young Adults." *Journal of Interpersonal Violence* 15(4):339–362.

Garnets, L., G. M. Herek, and B. Levy. 1990. "Violence and Victimization of Lesbians and Gay Men: Mental Health Consequences." *Journal of Interpersonal Violence* 5:366–383.

Gilman, Stephen E., Susan D. Cochran, Vickie M. Mays, Michael Hughes, David Ostrow, and Ronald C. Kessler. 2001. "Risk of Psychiatric Disorders Among Individuals Reporting Same-Sex Sexual Partners in the National Comorbidity Survey." *American Journal of Public Health,* 91(6):933–939.

Goode, Erica E., and Betsy Wagner. 1993. "Intimate Friendships." *U.S. News and World Report,* July 5, 49–52.

Hallett, Vicky. 2003. "Who Do You Love?" *U.S. News & World Report,* July 14, 38.

"Hate Crime Laws in the U.S." 2004 (February). National Gay and Lesbian Task Force. http://www.ngltg.org

Homophobia 101: Teaching Respect for All. 2000. The Gay, Lesbian, and Straight Education Network. http://www.glsen.org

Human Rights Campaign. 2000. *Feeling Free: Personal Stories: How Love and Self-Acceptance Saved Us from "Ex-Gay" Ministries.* Washington, DC: Human Rights Campaign Foundation.

——— 2003. *The State of the Workplace for Lesbian, Gay, Bisexual and Transgendered Americans, 2002.* Washington DC: Human Rights Campaign. http://www.hrc.org

Human Rights Watch. 2001. *World Report 2001.* http://www.hrw.org

International Gay and Lesbian Human Rights Commission. 1999. "Antidiscrimination Legislation." http://www.iglhrc.org/news/factsheets/990604-antidis.html

———. 2003a. "Where Having Sex Is a Crime: Criminalization and Decriminalization of Homosexual Acts." http://www.iglhrc.org

———. 2003b. "Where You Can Marry: Global Summary of Registered Partnership, Domestic Partnership, and Marriage Laws." http://www.iglhrc.org

———. 2004 (January 30). "IGLHRC Calls for Global Mobilization to Help Pass the United Nations Resolution on Sexual Orientation and Human Rights."

Johnston, Eric. 2003a (October 1). "Kentucky State Senator Reveals He Is Gay." PlanetOut.com. http://www.planetout.com

———. 2003b (October 14). "Clinton Blasts Bush's Gay Marriage Attack." Planetout.com.

Kaiser Family Foundation. 2000. *Sex Education in America: A View from Inside the Nation's Classrooms.* Menlo Park, CA: Henry J. Kaiser Family Foundation.

Kinsey, A. C., Pomeroy, W. B., and Martin, C. E. 1948. *Sexual Behavior in the Human Male.* Philadelphia: W. B. Saunders.

Kinsey, A. C., Pomeroy, W. B., Martin, C. E., and Gebhard, P. H. 1953. *Sexual Behavior in the Human Female.* Philadelphia: W. B. Saunders.

Kirkpatrick, R. C. 2000. "The Evolution of Human Sexual Behavior." *Current Anthropology* 41(3):385.

Kite, M. E., and B. E. Whitley, Jr. 1996. "Sex Differences in Attitudes Toward Homosexual Persons, Behavior and Civil Rights: A Meta-analysis." *Personality and Social Psychology Bulletin* 22:336–352.

Kosciw, Joseph G., and M. K. Cullen. 2002. *The 2001 National School Climate Survey.* Gay, Lesbian, and Straight Education Network. http://www.glsen.org

Landis, Dan. 1999 (February 17). "Mississippi Supreme Court Made a Tragic Mistake in Denying Custody to Gay Father, Experts Say." American Civil Liberties Union, News. http://www.aclu.org

LAWBriefs. 2003 (Spring). "Recent Developments in Sexual Orientation and Gender Identity Law." Vol. 6, No. 1.

———. 2000 (Fall). "Recent Developments in Sexual Orientation and Gender Identity Law." Vol. 3, No. 3.

Lever, Janet. 1994. "The 1994 Advocate Survey of Sexuality and Relationships: The Men." *The Advocate,* August 23, 16–24.

Loftus, Jeni. 2001. "America's Liberalization in Attitudes Toward Homosexuality, 1973 to 1998." *American Sociological Review* 66:762–782.

Louderback, L. A., and B. E. Whitley. 1997. "Perceived Erotic Value of Homosexuality and Sex-Role Attitudes as Mediators of Sex Differences in Heterosexual College Students' Attitudes Toward Lesbians and Gay Men." *Journal of Sex Research* 34:175–182.

Mathison, Carla. 1998. "The Invisible Minority: Preparing Teachers to Meet the Needs of Gay and Lesbian Youth." *Journal of Teacher Education* 49:151–155.

McCammon, Susan, David Knox, and Caroline Schacht. 2004. *Choices in Sexuality,* 2nd ed. Cincinnati: Atomic Dog Publishing.

Michael, Robert T., John H. Gagnon, Edward O. Laumann, and Gina Kolata. 1994. *Sex in America: A Definitive Survey.* Boston: Little, Brown.

Miller, Patti, McCrae A. Parker, Eileen Espejo, & Sarah Grossman-Swenson. 2002. *Fall Colors: Prime Time Diversity Report 2001–02.* Oakland CA: Children Now & The Media Program.

Moore, David W. 1993. "Public Polarized on Gay Issue." *Gallup Poll Monthly* 331 (April):30–34.

National Gay & Lesbian Task Force. 2003. "GLBT Civil Rights Laws in the U.S.–August 2003." http://www.ngltf.org

"NCLR Wins Equal Tax Benefits for Nonbiological Lesbian Mother." 2000 (Fall). *NCLR Newsletter* (National Center for Lesbian Rights), pp. 1, 10. http://www.nclrights.org/index.html

Newport, Frank. 2002 (September). "In-Depth Analysis: Homosexuality." The Gallup Organization. http://www.gallup.com

Page, Susan. 2003. "Gay Rights Tough to Sharpen into Political 'Wedge Issue.'" *USA Today,* July 28, 10A.

Patterson, Charlotte J. 2001. "Family Relationships of Lesbians and Gay Men." In *Understanding Families into the New Millennium: A Decade in Review,* ed. Robert M. Milardo, pp. 271–288. Minneapolis, MN: National Council on Family Relations.

Patton, Clarence. 2003. *Anti-LGBT Violence in 2002.* New York: The National Coalition of Anti-Violence Programs. 240 West 35th St., Suite 200. New York, NY 10001.

Paul, J. P. 1996. "Bisexuality: Exploring/Exploding the Boundaries." In *The Lives of Lesbians, Gays, and Bisexuals: Children to Adults,* eds. R. Savin-Williams and K. M. Cohen, pp. 436–461. Fort Worth: Harcourt Brace.

Platt, Leah. 2001. "Not Your Father's High School Club." *The American Prospect* 12(1):A37–A39.

Plugge-Foust, C. 2001. "Homophobia, Irrationality, and Christian Ideology: Does a Relationship Exist? *Journal of Sex Education and Therapy,* 25: 240–244.

"Post-Election Analysis." 2000. Human Rights Campaign. http://www.hrc.org

Price, Jammie, and Michael G. Dalecki. 1998. "The Social Basis of Homophobia: An Empirical Illustration." *Sociological Spectrum* 18:143–159.

Ricks, Thomas E. 2000. "Pentagon Vows to Enforce 'Don't Ask.'" *Washington Post,* July 22, A01.

Rosin, Hanna, and Richard Morin. 1999 (January 11). "In One Area, Americans Still Draw a Line on Acceptability." *Washington Post National Weekly Edition* 16(11):8.

Rostow, Ann. 2004 (January 22). "More States Join Marriage-Amendment Frenzy." *PlanetOut.* http://www.planetout.com
Safe Zone Resources. 2003. National Consortium of Directors of LGBT Resources in Higher Education. http://www.lgbtcampus.org
Sanday, Peggy. R. 1995. "Pulling Train." *In Race, Class, and Gender in the United States,* 3d ed., ed. P. S. Rothenberg, pp. 396–402. New York: St. Martin's Press.
Schellenberg, E. Glenn, Jessie Hirt, and Alan Sears. 1999. "Attitudes Toward Homosexuals Among Students at a Canadian University." *Sex Roles* 40(1/2):139–152.
"Second-Parent/Step-parent Adoption in the U.S." 2002. National Gay and Lesbian Task Force. http://www.ngltf.org
Sexuality Information and Education Council of the United States (SIECUS). 2000. "Fact Sheets: Sexual Orientation and Identity." 130 West 42nd St., Suite 350. NY, NY 10036-7802.
Simmons, Tavia, and Martin O'Connell. 2003 (February). *Married-Couple and Unmarried-Partner Households: 2000.* Washington, DC: U.S. Census Bureau.
Simon, A. 1995. "Some Correlates of Individuals' Attitudes Toward Lesbians." *Journal of Homosexuality* 29:89–103.
Singh, Daniel P., and Heather D. Wathington. 2003. "Valuing Equity: Recognizing the Rights of the LGBT Community." *Diversity Digest,* 7 (1, 2):8–9.
Smith, D. M, and G. J. Gates. 2001. *Gay and Lesbian Families in the United States: Same-Sex Unmarried Partner Households: Preliminary Analysis of 2000 U.S. Census Data.* Human Rights Campaign. http://www.hrc.org
Sullivan, A. 1997. "The Conservative Case." In *Same-Sex Marriage: Pro and Con,* ed. A. Sullivan, pp. 146–154. New York: Vintage Books.
———. 2003. "Legalizing Same-Sex Marriage Would Strengthen Marriage." In *The Family: Opposing Viewpoints,* ed. A. Ojeda, pp. 196–200. Farmington Hill, MI: Greenhaven Press.
Thompson, Cooper. 1995. "A New Vision of Masculinity." In *Race, Class, and Gender in the United States,* 3d ed., ed. P. S. Rothenberg, pp. 475–481. New York: St. Martin's Press.
Tobias, Sarah, and Sean Cahill. 2003. "School Lunches, the Wright Brothers, and Gay Families." National Gay and Lesbian Task Force. http://www.ngltf.org
The United Methodist Church and Homosexuality. 1999. http://www.religioustolerance.org/hom_umc.htm
Vogel, Steve. 2003 (January 23). "U.S. Awards 9/11 Benefits for Loss of a Gay Partner." *Washington Post,* p. B1.
Wilcox, Clyde, and Robin Wolpert. 2000. "Gay Rights in the Public Sphere: Public Opinion on Gay and Lesbian Equality." In *The Politics of Gay Rights,* eds. Craig A. Rimmerman, Kenneth D. Wald, and Clyde Wilcox, pp. 409–432. Chicago: University of Chicago Press.
Yang, Alan. 1999. *From Wrongs to Rights 1973 to 1999: Public Opinion on Gay and Lesbian Americans Moves Toward Equality.* New York: The Policy Institute of the National Gay and Lesbian Task Force.

Chapter 9

AARP 2000. "Baby Boomers Envision Their Retirement: An AARP Segmentation Analysis." Executive Summary Part I. http://www.research.aarp.org/econ/boomer_seg_1.html
———2003. "Fighting Ageism." *AARP Magazine,* July and August. http://www.aarpmagazine.org
African Eye News. 2003. "Elderly Become 'Muti' Targets." April 15.
AOA (Administration on Aging). 2000a. "A Profile of Older Americans" http://www.aoa.gov/STATS/profile/default.html
———. 2000b. "Older Women" http://www.aoa.dhhs.gov/may200/factsheets/olderwomen.html
———. 2001. "The Older Americans Act." http://www.aoa.gov/may2001/factsheets/OAA.htm
———. 2003a. "Statistics." http://www.aoa.gov
———. 2003b. "A Profile of Older Americans: 2002." http://www.aoa.gov/prof
———. 2003c. "Aging into the 21st Century." http://www.aoa.gov/prof/statistics/future_growth/aging21
———. 2003d. "Did you Know? Income and Poverty Among the Elderly." http://www.ada.gov/press/did_you_know
———. 2003e. "HHS Secretary Tommy G. Thompson, Americans' Doctors Team Up for Better Benefits, More Choices in Medicare." Press Release. http://www.aoa.gov/press
———. 2003f. "Fact Sheets Older Americans Act." http://www.aoa.gov/press/fact
Atchley, Robert C. 2000. *Social Forces and Aging.* Belmont, CA: Wadsworth.
BBC (British Broadcasting Company). 2003. "Children of Conflict." BBC World Service. http://www.bbc.co.uk/worldservice/people/features
Becker, J., 2004 "Children as Weapons of War." Human Rights watch. http://www.hrw.org/wrzk4
Brazzini, D. G., W. D. McIntosh, S. M. Smith, S. Cook, and C. Harris. 1997. "The Aging Woman in Popular Film: Underrepresented, Unattractive, Unfriendly, and Unintelligent." *Sex Roles* 36:531–543.
Brooks-Gunn, Jeanne, and Greg Duncan. 1997. "The Effects of Poverty on Children." *Future of Children* 7(2):55–70.
Burtner, Kent. 2003. "AARP, National Council on the Aging: Americans Working Longer." OregonLive News, April 1. http://www.globalaging.org/elderrights
Carpenter, MacKenzie, and Ginny Kopas. 1999. "Casualties of Custody Wars: Special Report." *Pittsburgh Post-Gazette.* http://www.post-gazette.com/custody
CASA Alianza. 2003. "Living in the Streets." http://www.casa-alianza.org
CASA (Center on Addiction and Substance Abuse). 2000. "Back to School, 2000." The CASA National Survey of American Attitudes on Substance Abuse: Teens and Their Parents." http://www.casacolumbia. org/usr_doc17645.pdf
CBS News. 2003. "AIDS Creating Global 'Orphans Crisis.'" July 10. http://www.cbsnews.com/stories/2002
CBS/New York Times Poll. 2000. "Priorities." September 9–11. http://www.pollingreport.com/prioritl/htm
CDF (Children's Defense Fund). 2000. "Child Care Now!" http://www.childrensdefense.org/childcare/cc_polls.html
———. 2003a. "Basic Facts on Poverty." http://www. childrensdefense.org
———. 2003b. "June Jobless Rate Among America's Teens Highest in 55 Years." http://www.childrensdefense.org/release
———. 2003c. "Among Industrialized Countries, the United States Ranks . . ." http://www.childrensdefense.org/factsfigures
———. 2003d. "Fact Sheet: Media Violence." http://www.childrensdefense.org
———. 2003e. "Senate Tax Bill Costly Enough to Pay for Essential Protections for Children." May 14. http://www.childrensdefense.org/releases
CNN. 2004. "Lionel Tate Released." http://www.cnn.com. January 27
Costa, Dora L. 2000. "A Century of Retirement." *TIAA-CREF Participant.* November, pp.12–13.
DeAngelis, Tori. 1997. "Elderly May Be Less Depressed Than the Young." *APA Monitor,* October. http://www.apa.org/monitor/oct97/elderly.html
Dorman, Peter. 2001. "Child Labour in the Developed Economies." Geneva: International Labour Office.
Duncan, Greg, W. Jean Yeung, Jeanne Brooks-Gunn, and Judith Smith. 1998. "How Much Does Childhood Poverty Affect the Life Chance of Children?" *American Sociological Review* 63:402–423.
Ennis, Dave. 2000. "Survey Finds Retirement Plans Often Include a Job." *Morning Star,* November 27, 1D, 3D.
EurekAlert. 2003. "Children's Sterotypes of Aging Start Early." April 22. http://www.globalaging.org

FBI (Federal Bureau of Investigation). 2002. *Crime in the United States, 2001.* Uniform Crime Reports. Washington, DC: U.S. Government Printing Office.

Gallup Poll. 2000. "Children and Violence." http://www.gallup.com/poll/indicators/indchild_violence.asp

Generations United. 2003. "About GU." http://www.gu.org

Goldberg, Beverly. 2000. *Age Works.* New York: Free Press.

Green, Shane. 2003. "Hidden Abuse of Elderly Emerging Problem for Japan." *Sydney Morning Herald,* June 21. http://www.globalaging.org/elderrights

Harris, Diana K. 1990. *Sociology of Aging.* New York: Harper & Row.

Harris, Kathleen, and Jeremy Marmer. 1996. "Poverty, Paternal Involvement and Adolescent Well-Being." *Journal of Family Issues* 17(5):614–640.

HHS (Health and Human Services). 1999. "Children's Health Insurance Program National Back-to-School Kick Off." HHS News, September 22. http://www.hcfa.gov/init/9909922wh.htm

Hooyman, Nancy R., and H. Asuman Kiyak. 1999. *Social Gerontology: A Multidisciplinary Perspective,* 2nd ed. Boston: Allyn and Bacon.

Hounsell, Cindy, and Pat Humphlett. 2003. "Older Minority Women Need Retirement Help Now." *WomenENews,* May 29. http://www. globalaging.org

HRW (Human Rights Watch). 2000. *Fingers to the Bone: United States Failure to Protect Child Farmworkers.* Human Rights Watch. 350 Fifth Ave. 34th Floor. NY, NY 10118-3299.

———. 2003a. "Child Farmworkers." http://www.hrw.org/campaigns

———. 2003b. "Orphans and Abandoned Children." http://www.hrw.org/children

Jackson, Maggie. 2003. "More Sons Are Juggling Jobs and Care for Parents." *New York Times,* June 15. http://www.globalaging.org/elderrights

Jacobson, Linda. 2000. "Children's Early Needs Seen as Going Unmet." *Washington Post,* October 3. http://washingtonpost.com/wp-dyn/articles/A4014-2000Oct5.html

Livni, Ephrat. 2000. "Exercise, the Anti-Drug." September 21. http://abcnews.go.com/sections/living/Daily/News/Depression_elderly00921.html

Mack, Brandy, and Kathi Jones. 2003 (August 5). "Elder Abuse: Identification and Prevention." *ASHA Leader 8:10–12A.*

Matras, Judah. 1990. *Dependency, Obligations, and Entitlements: A New Sociology of Aging, the Life Course, and the Elderly.* Englewood Cliffs, NJ: Prentice-Hall.

McCoy, Kevin. 2003. "Study Offers First Assisted-Living Guidelines." *USA Today,* April 29. http://www.usatoday.com

Miner, Sonia, John Logan, and Glenna Spitze. 1993. "Predicting Frequency of Senior Center Attendance." *The Gerontologist* 33:650–657.

National Center for Victims of Crime. 2003. "Federal Legislation: Children and Teens." http://www.ncvc.org/print

NCCP (National Center for Children in Poverty). 2003. "Low Income Children in the United States: Overview." http://www.nccp.org

NCHS (National Center for Health Statistics). 2003. "Fast Stats: Health of the Elderly." http://www.cdc.gov/nchn

NCSC (National Council of Senior Citizens). 2000a. "Our Issues: Poverty." http://www.ncscinc.org/issues/poverty.htm

———. 2000b. "Affordable Housing." http://www.ncscinc.org /issues/affordhouse.htm

Newman, Cathy. 2000. "Older, Healthier and Wealthier." *Washington Post,* August 10, A03.

Nicholson, Trish. 2003. "Age Complaints Surge as Midlife Workers Find the Going Harder. *AARP Bulletin.* March. http://www.globalaging.org/elderrights/us

NIH (National Institutes of Health). 2003. "New Prevalence Study Suggests Dramatically Rising Numbers of People with Alzheimer's Disease." http://www.nih.gov/news

NIMH (National Institute of Mental Health). 2003. "Older Adults: Depression and Suicide Facts." http://www.mental-health-matters.com/articles

NMHA (National Mental Health Association). 2003. "Children's Mental Health Statistics." http://www.nimh.org/children/prevent

Pear, Robert. 2002. "9 out of 10 Nursing Homes Lack Adequate Staff, Study Finds." *New York Times,* February 18. http://www.globalaging.org/elderrights

Peterson, Karen. 2003. "Kids Are Better Off than Adults Believe." *USA Today,* July 2. http://www.usatoday.com

Ramirez, Eddy. 2002. "Ageism in the Media Is Seen as Harmful to Health of Elderly." Los Angeles Times, September 5. http://www.globalaging.org

Reio, Thomas G., and Joanne Sanders-Reto. 1999. "Combating Workplace Ageism." *Adult Learning* 11:10–13.

Riley, Matilda White. 1987. "On the Significance of Age in Sociology." *American Sociological Review* 52 (February):1–14.

Riley, Matilda W., and John W. Riley. 1992. "The Lives of Older People and Changing Social Roles." In *Issues in Society,* eds. Hugh Lena, William Helmreich, and William McCord, pp. 220–231. New York: McGraw-Hill.

Seeman, Teresa E., and Nancy Adler. 1998. "Older Americans: Who Will They Be?" *National Forum,* Spring, 22–25.

"Snapshots of the Elderly." 2000. *Washington Post.* http://www.washingtonpost.com/wp-srv/health/images/elderly.html

Statistical Abstract of the United States 2002, 122nd ed. 2002. U.S. Bureau of the Census. Washington, DC: U.S. Government Printing Office.

Thompson, Tommy. 2003. "On the Trafficking Victims Protection Reauthorization Act of 2003." *HHS News,* December 22. http://www.acf.dhhs.gov/news

Thurow, Lester C. 1996. "The Birth of a Revolutionary Class." *The New York Times Magazine,* May 19, 46–47.

Tirrito, Terry. 2003. *Aging in the New Millennium.* Columbia: University of South Carolina Press.

Torres-Gil, Fernando. 1990. "Seniors React to Medicare Catastrophic Bill: Equity or Selfishness?" *Journal of Aging and Social Policy* 2(1):1–8.

"Underage and Unprotected: Child Labor in Egypt's Cotton Fields." 2001 (January). *Human Rights Watch* 13(1).

UN (United Nations). 2003. "Indicators of Youth and Elderly Populations." United Nations: Statistics Division. http://www.unstat.un.org/unsd/demographics/social

———. 1998. "The First Nearly Universally Ratified Human Rights Treaty in History." *Status.* Washington, DC: UNICEF. http://www.unicef.org/crc/status.html

———. 2003a. "Introduction to Convention on the Rights of the Child." http://www.unicef.org/crc/introduction

———. 2003b. "Facts on Children: Child Protection." http://www.unicef.org/media

———. 2003c. "The State of the World's Children." New York: United Nations Children's Fund.

U.S. Department of Labor. 1995. *By the Sweat and Toil of Children. Vol. 2, The Use of Child Labor in U.S. Agricultural Imports and Forced and Bonded Child Labor.* Washington, DC: U.S. Department of Labor, Bureau of International Labor Affairs.

Welch, William. 2003. "AARP Balks at Drug Plan." *USA Today,* July 16. http://www.usatoday.com

Willson, Andrea, and Melissa Hardy. 2002. "Racial Disparities in Income Security for a Cohort of Aging American Women." *Social Forces* 80:1283–1306.

Zhong Xin Net. 2003. "China Becomes an Aging Society." http://www.globalaging.org/elderrights/world

Chapter 10

Administration for Children and Families. 2002. "Early Head Start Benefits Children and Families." U.S. Department of Health and Human Services. http://www.acf.hhs.gov

Albelda, Randy, and Chris Tilly. 1997. *Glass Ceilings and Bottomless Pits: Women's Work, Women's Poverty.* Boston, MA: South End Press.

Alex-Assensoh, Yvette 1995. "Myths about Race and the Underclass." *Urban Affairs Review* 31:3–19.

Anderson, Sarah, John Cavanagh, Chuck Collins, Chris Hartman, and Scott Klinger. 2003. *Executive Excess: Tenth Annual CEO Compensation Survey.* Boston: Institute for Policy Studies and United for a Fair Economy.

Anderson, Sarah, John Cavanagh, Chuck Collins, Chris Hartman, and Felice Yeskel. 2000. *Executive Excess: Seventh Annual CEO Compensation Survey.* Boston: Institute for Policy Studies and United for a Fair Economy.

Andrews, Edmund. 2003 (January 23). "Economic Inequality Grew in 90's Boom, Fed Reports." *New York Times Online.* http://www.nytimes.com

Associated Press. 2003. "More U.S. Families Hungry or Too Poor to Eat, Study Says." *NY Times.com.* http://www.nytimes.com

Barlett, Donald L., and James B. Steele. 1998. "The Empire of the Pigs." *Time,* November 30, 52–64.

Briggs, Vernon M. Jr. 1998. "American-Style Capitalism and Income Disparity: The Challenge of Social Anarchy." *Journal of Economic Issues* 32(2): 473–481.

Bureau of Labor Statistics. 2002. "2001 National Occupational Employment and Wage Estimates. Personal Care and Service Occupations." http://www.bls.gov

———. 2003 (August). "Characteristics of Minimum Wage Workers: 2002." http://stats.bls.gov

Children's Defense Fund. 2002. *Low-Income Families Bear the Burden of State Child Care Cuts.*" http://www.childrensdefense.org

———. 2003. "Children in the United States." http://www.childrensdefense.org

Chossudovsky, Michel. 1998. "Global Poverty in the Late 20th Century." *Journal of International Affairs* 52(1):293–303.

Citizens for Tax Justice. 2002 (April 17). "Surge in Corporate Tax Welfare Drives Corporate Tax Payments Down to Near Record Low." http://www.ctj.org

Corcoran, Mary, and Terry Adams. 1997. "Race, Sex, and the Intergenerational Transmission of Poverty." In *Consequences of Growing Up Poor,* eds. Greg J. Duncan and Jeanne Brooks-Gunn, pp. 461–517. New York: Russell Sage Foundation.

Davis, Kingsley, and Wilbert Moore. 1945. "Some Principles of Stratification." *American Sociological Review* 10:242–249.

Deen, Thalif. 2000. "NGOs Call for UN Poverty Eradication Fund." Global Policy Forum. http://www.globalpolicy.org

Deng, Francis M. 1998. "The Cow and the Thing Called 'What': Dinka Cultural Perspectives on Wealth and Poverty." *Journal of International Affairs* 52(1): 101–115.

Duncan, Greg J., and Jeanne Brooks-Gunn. 1997. "Income Effects Across the Life Span: Integration and Interpretation." In *Consequences of Growing Up Poor,* eds. Greg J. Duncan and Jeanne Brooks-Gunn, pp. 596–610. New York: Russell Sage Foundation.

Duncan, Greg J., and P. Lindsay Chase-Lansdale. 2001. "Welfare Reform and Child Well-Being." Paper presented at the Blank/Haskins conference, "The New World of Welfare Reform," Washington, DC, February 1–2.

Economic Policy Institute. 2000. "Issue Guide to the Minimum Wage." http://www.epinet.org

Food Research and Action Center. 2004 (January 6). "Food Stamp Participation Increases in October 2003 to More than 23.3 Million Persons." *Current News & Analysis.* http://www.frac.org

Fremstad, Sean. 2004 (January 30). "Recent Welfare Reform Research Findings: Implications for TANF Reauthorization and State TANF Policies." Center on Budget and Policy Priorities. http://www.cbpp.org

Gans, Herbert J. 1972. "The Positive Functions of Poverty." *American Journal of Sociology* 78 (September):275–388.

Goesling, Brian. 2001. "Changing Income Inequalities Within and Between Nations: New Evidence." *American Sociological Review* 66:745–761.

Hill, Lewis E. 1998. "The Institutional Economics of Poverty: An Inquiry into the Causes and Effects of Poverty." *Journal of Economic Issues* 32(2): 279–286.

Hooks, Bell. 2000. *Where We Stand: Class Matters.* New York: Routledge.

HUD (U.S. Department of Housing and Urban Development). 2000 (April). "Gun-Related Violence: The Costs to Public Housing Communities." *Recent Research Results,* pp. 1–2. Rockville, MD.

Human Development Report 1997. 1997. United Nations Development Programme. New York: Oxford University Press.

Human Development Report 2003. 2003. United Nations Development Programme. New York: Oxford University Press.

Independent Sector. 2001. *Giving and Volunteering in the United States, 2001.* http://www.independentsector.org

Jargowsky, Paul A. 1997. *Poverty and Place: Ghettos, Barrios, and the American City.* New York: Russell Sage Foundation.

Johnson, Nicholas, Joseph Llobrera, and Bob Zahradnik. 2003. "A HAND UP: How State Earned Income Tax Credits Helped Working Families Escape Poverty in 2003." Center on Budget and Policy Priorities. http://www.cbpp.org

Kennedy, Bruce P., Ichiro Kawachi, Roberta Glass, and Deborah Prothrow-Stith. 1998. "Income Distribution, Socioeconomic Status, and Self-Rated Health in the U.S.: Multilevel Analysis." *British Medical Journal* 317 (7163):917–921.

Kingsley, G. Thomas, and Kathryn L. S. Pettit. 2003 (May). "Concentrated Poverty: A Change in Course." Urban Institute. http://www.urban.org/nnip

Knickerbocker, Brad. 2000. "Nongovernmental Organizations Are Fighting and Winning Social, Political Battles." Global Policy Forum. http://www.globalpolicy.org/ngos

Kraut, Karen, Scott Klinger, and Chuck Collins. 2000. *Choosing the High Road: Businesses that Pay a Living Wage and Prosper.* Boston: United for a Fair Economy.

Leventhal, Tama, and Jeanne Brooks-Gunn. 2003. "Moving to Opportunity: An Experimental Study of Neighborhood Effects on Mental Health." *American Journal of Public Health 93*(9): 1576–1585.

Levitan, Sar A., Garth L. Mangum, and Stephen L. Mangum. 1998. *Programs in Aid of the Poor,* 7th ed. Baltimore: Johns Hopkins University Press.

Lewis, Oscar. 1966. "The Culture of Poverty." *Scientific American* 2(5):19–25.

———. 1998. "The Culture of Poverty: Resolving Common Social Problems." *Society* 35(2):7–10.

Living Wage Resource Center. 2004. "Living Wage Successes." Living Wage Resource Center, Boston, MA.

Luker, Kristin. 1996. *Dubious Conceptions: The Politics of Teenage Pregnancy.* Cambridge, MA: Harvard University Press.

Malatu, Mesfin Samuel, and Carmi Schooler. 2002. "Causal Connections Between Socio-economic Status and Health: Reciprocal Effects and Mediating Mechanisms." *Journal of Health and Social Behavior* 43:22–41.

Mann, Judy. 2000 (May 15). "Demonstrators at the Barricades Aren't Very Subtle, but They Sometimes Win." *The Washington Spectator* 26(10):1–3.

Massey, D. S. 1991. "American Apartheid: Segregation and the Making of the American Underclass." *American Journal of Sociology* 96:329–357.

Mayer, Susan E. 1997. *What Money Can't Buy: Family Income and Children's Life Chances.* Cambridge, MA: Harvard University Press.

McIntyre, Robert. 2003 (June 18). "Testimony of Robert S. McIntyre, Director,

Citizens for Tax Justice Before the Committee on the Budget, United States House of Representatives, Concerning 'Waste, Fraud, [and] Abuse in Federal Mandatory Programs.'" Citizens for Tax Justice. http://www.ctj.org
McIntyre, Robert, and T. D. Coo Nguyen. 2000 (October). "Corporate Income Taxes in the 1990s." Institute on Taxation and Economic Policy. 1311 L Street, NW. Washington, DC 20005.
Michel, Sonya. 1998. "Childcare and Welfare (In)justice." *Feminist Studies* 24: 44–54.
Mishel, Lawrence, Jared Bernstein, and Heather Boushey. 2003. *The State of Working America 2002–2003.* Ithaca, NY: Cornell University Press.
Narayan, Deepa. 2000. *Voices of the Poor: Can Anyone Hear Us?* New York: Oxford University Press.
National Center for Children in Poverty. 2003. "The Effects of Parental Education on Income." http://www.nccp.org
National Law Center on Homelessness and Poverty in America. 2002. "Homelessness and Poverty in America." http://www.nlchp.org
Newman, Katherine S. 1999. *No Shame in My Game: The Working Poor in the Inner City.* New York: Alfred A. Knopf and The Russell Sage Foundation.
Parenti, Michael. 1998. "The Super Rich Are Out of Sight." *Dollars and Sense* 217(May–June):36–37.
Parisi, Domenico, Steven Michael Grice, and Michael Taquino. 2003. "Poverty and Inequality in the Context of Welfare Reform." *Social Problems Forum: The SSSP Newsletter,* 34(2):19–21.
Paul, James A. 2000. "NGOs and Global Policy-Making." Global Policy Forum. http://www.globalpolicy.org/ngos
Pressman, Steven. 1998. "The Gender Poverty Gap in Developed Countries: Causes and Cures." *The Social Science Journal* 35(2):275–287.
Proctor, Bernadette D., and Joseph Dalaker. 2003. *Poverty in the United States: 2002.* Current Population Reports P60–222. U.S. Census Bureau. Washington, DC: U.S. Government Printing Office.
"Reducing Poverty Is Key to Global Stability." 2003 (May–June). *Popline* (World Population News Service): p. 4.
Schiller, Bradley R. 2004. *The Economics of Poverty and Discrimination,* 9th ed. Upper Saddle River, NJ: Pearson Prentice Hall.
Seccombe, Karen. 2001. "Families in Poverty in the 1990s: Trends, Causes, Consequences, and Lessons Learned." In *Understanding Families into the New Millennium: A Decade in Review,* ed. Robert M. Milardo, pp. 313–332. Minneapolis, MN: National Council on Family Relations.
Sorensen, Elaine. 2003. "Child Support Gains Some Ground." Urban Institute. http://www.urbaninstitute.org
Speth, James Gustave. 1998. "Poverty: A Denial of Human Rights." *Journal of International Affairs* 52(1):277–286.
Streeten, Paul. 1998. "Beyond the Six Veils: Conceptualizing and Measuring Poverty." *Journal of International Affairs* 52(1):1–8.
United Nations. 1997. *Report on the World Social Situation, 1997.* New York: United Nations.
UNDP (United Nations Development Programme). 1997. *Human Development Report 1997.* New York: Oxford University Press.
———. 2000. *Human Development Report 2000.* New York: Oxford University Press.
U.S. Census Bureau. 2003. "Historical Poverty Tables–People." http://www.census.gov/hhes/poverty/histpov
United States Conference of Mayors. 2003 (December). *Hunger and Homelessness Survey: A Status Report on Hunger and Homelessness in America's Cities.* Washington, DC: United States Conference of Mayors.
USDA (U.S. Department of Agriculture). 2003. "Characteristics of Food Stamp Households: Fiscal Year 2002." http://www.fns.usda.gov
U.S. Department of Health and Human Services. 2003a. *Fiscal Year 2001: Characteristics and Financial Circumstances of TANF Recipients.* http://www.acf.dhhs.gov
———. 2003b. (September 3). "HHS Releases Data Showing Continuing Decline in Number of People Receiving Temporary Assistance." New Release. http://www.hhs.gov/news
U.S. Department of Labor. 2004 (January). "Minimum Wage Laws in the United States." http://www.dol.gov/esa
Van Kempen, Eva T. 1997. "Poverty Pockets and Life Chances: On the Role of Place in Shaping Social Inequality." *American Behavioural Scientist* 41(3):430–450.
Vleminckx, Koen, and Timothy M. Smeeding (Eds.). 2001. *Child Well-Being, Child Poverty and Child Policy in Modern Nations.* University of Bristol: The Policy Press.
Wagstaff, Adam. 2003. "Child Health on a Dollar a Day: Some Tentative Cross-Country Comparisons." *Social Science & Medicine* 57:1529–1538.
Washburn, Jennifer. 2004. "The Tuition Crunch." *The Atlantic Monthly* 293(1):140.
Wilson, William J. 1987. *The Truly Disadvantaged: The Inner City, the Underclass, and Public Policy.* Chicago: University of Chicago Press.
———. 1996. *When Work Disappears: The World of the New Urban Poor.* New York: Alfred A Knopf.
World Bank. 2001. *World Development Report: Attacking Poverty, 2000/2001.* Herndon, VA: World Bank and Oxford University Press.
World Health Organization. 2002. *The World Health Report 2002.* http://www.who.int/pub/en
Zedlewski, Sheila R. 2003. "Work and Barriers to Work Among Welfare Recipients in 2002." Urban Institute. http://www.urban.org
Zedlewski, Sheila R., Sandi Nelson, Kathryn Edin, Heather L. Koball, and Kate Roberts. 2003. "Families Coping Without Earnings or Government Cash Assistance." Urban Institute: http://www.urbaninstitute.org

Chapter 11

AFL-CIO. 2002. *Death on the Job: The Toll of Neglect,* 11th ed. http://www.aflcio.org
Ambrose, Soren. 1998. "The Case Against the IMF." *Campaign for Human Rights Newsletter,* no. 12. http://www.summersault.co...wsletter/news12.html
"Attitudes in the American Workplace IX." 2003. The Marlin Company, 100 Kenna Drive, North Haven, CT 06473.
Austin, Colin. 2002. "The Struggle for Health in Times of Plenty." In *The Human Cost of Food: Farmworkers' Lives, Labor, and Advocacy,* eds. C. D. Thompson, Jr., and M. F. Wiggins, pp. 198–217. Austin: University of Texas Press.
Bales, Kevin. 1999. *Disposable People: New Slavery in the Global Economy.* Berkeley: University of California Press.
Bales, Kevin, and Peter T. Robbins. 2001. "No One Shall Be Held in Slavery or Servitude: A Critical Analysis of International Slavery Agreements and Concepts of Slavery." *Human Rights Review* 2(2):18–45.
Barlett, Donald L., and James B. Steele. 1998. "Corporate Welfare: First in a Series." *Time* 152(19):36–39.
Barstow, David, and Lowell Bergman. 2003 (January 10). "Deaths on the Job, Slaps on the Wrist." *New York Times OnLine.* http://www.nytimes.com
Bassi, Laurie J., and Jens Ludwig. 2000 (January). "School-to-Work Programs in the United States: A Multi-Firm Case Study of Training, Benefits, and Costs." *Industrial and Labor Relations Review* 53(2):219.
Bavendam, James. 2000. "Managing Job Satisfaction." Special Reports, Vol. 6:1–2. Bavendam Research Incorporated. http://www.bavendam.com

Bello, Walden. 2001 (January/Febuary). "Lilliputians Rising: 2000: The Year of Global Protest Against Corporate Globalization." *Multinational Monitor* 22(Nos. 1 & 2). http://www.essential.org/monitor

Benjamin, Medea. 1998. "What's Fair About Fair Labor Association (FLA)?" *Sweatshop Watch.* http://www.sweatshopwatch.org

"Big Business for Reform." 2000 (November). *Multinational Monitor* 21(11). http://www.essential.org/monitor

Bivens, Josh, Robert Scott, and Christian Weller. 2003. "Mending Manufacturing: Reversing Poor Policy Decisions Is the Only Way to End Current Crisis." Economic Policy Institute. http://www.epinet.org

Bond, James T., Cindy Thompson, Ellen Galinsky, and David Prottas. 2002. *Highlights of the National Study of the Changing Workforce.* Families and Work Institute. http://www.familiesandwork.org

Bond, James T., Ellen Galinsky, and Jennifer E. Swanberg. 1997. *The 1997 National Study of the Changing Workforce.* New York: Families and Work Institute.

Bronfenbrenner, Kate. 2000 (December). "Raw Power: Plant-Closing Threats and the Threat to Union by Organizing." *Multinational Monitor* 21(12). http://www.essential.org/monitor

Bureau of Labor Statistics. 2002. "Workplace Injuries and Illnesses in 2001." Washington, DC: U.S. Department of Labor.

———. 2003a. "Workplace Injuries and Illnesses in 1999." Washington DC: U.S. Department of Labor. http://stats.bls.gov/oshhome.htm

———. 2003b. "National Census of Fatal Occupational Injuries, 2002." Washington DC: U.S. Department of Labor

———. 2003c. "Employment Characteristics of Families in 2002." http://www.bls.gov

Cantor, David, Jane Waldfogel, Jeffrey Kerwin, Mareena McKinley Wright, Kerry Levin, John Rauch, Tracey Hagerty, and Martha Stapelton Kudela. 2001. "Balancing the Needs of Families and Employers: The Family and Medical Leave Surveys, 2000 Update." U.S. Department of Labor. http://www.dol.gov/dol

Caston, Richard J. 1998. *Life in a Business-Oriented Society: A Sociological Perspective.* Boston: Allyn and Bacon.

Cernasky, Rachel. 2002 (December). "Slavery: Alive and Thriving in the World Today—The Satya Interview with Kevin Bales." http://www.satyamag.com

Clarke, Tony. 2002. "Twilight of the Corporation." In *Social Problems, Annual Editions 02/03,* 30th ed., Kurt Finsterbusch, ed., pp. 41–45. Guilford, CT: McGraw-Hill/Dushkin.

Cockburn, Andrew. 2003. "21st Century Slaves." *National Geographic,* September, pp. 2–11, 18–24.

Common Cause. 2000. "Top Soft-Money Donors: January 1, 1999 through December 31, 1999." http://www.commoncause.org

———. 2001 (February 7). "National Parties Raise Record $463 Million in Soft Money During 1999–2000 Election Cycle." *Common Cause News.* http://commoncause.org

Durkheim, Emile. [1893] 1966. *On the Division of Labor in Society,* trans. G. Simpson. New York: Free Press.

Franco, Lynn. 2003 (September). "Job Satisfaction Continues to Wither." *Executive Action* 69:1–5. The Conference Board. http://www.conference-board.org

Frederick, James, and Nancy Lessin. 2000 (November). "Blame the Worker: The Rise of Behavior-Based Safety Programs." *Multinational Monitor* 21(11). http://www.essential.org/monitor

Galinsky, Ellen, and James T. Bond. 1998. *The 1998 Business Work-Life Study.* New York: Families and Work Institute.

Galinsky, Ellen, Stacy S. Kim, and James T. Bond. 2001. *Feeling Overworked: When Work Becomes Too Much.* New York: Families and Work Institute.

Gallup Poll. 2002. "The Racial Divide: Job Satisfaction." http://www.gallup.com/poll

"The Garment Industry." 2001. Sweatshop Watch. http://www.sweatshopwatch.org/swatch/industry

George, Kathy. 2003 (December 1). "Myanmar: Unocal Faces Landmark Trial over Slavery." *CorpWatch.* http://www.corpwatch.org/news

Gordon, David M. 1996. *Fat and Mean: The Corporate Squeeze of Working Americans and the Myth of Managerial "Downsizing."* New York: Free Press.

Greenhouse, Steven. 2000 (January 26). "Anti-Sweatshop Movement Is Achieving Gains Overseas." *New York Times on the Web.* http://www.nytimes.com

———. 2001 (January 21). "Unions Hit Lowest Point in 6 Decades." *New York Times on the Web.* http://www.nytimes.com

Hargis, Michael J. 2001 (January/February). "Bangladesh: Garment Workers Burned to Death." *Industrial Worker* 1630, 98(1). http://www.parsons.www.org/~iw/jan2001/stories/intl.html

Hochschild, Arlie Russell. 1997. *The Time Bind: When Work Becomes Home and Home Becomes Work.* New York: Henry Holt and Company.

Human Rights Watch. 2000. *Unfair Advantage: Workers' Freedom of Association in the United States Under International Human Rights Standards.* http://www.hrw.org/reports/2000/uslabor

———. 2001. *World Report 2001.* http://www.hrw.org

Inglehart, Ronald. 2000. "Globalization and Postmodern Values." *Washington Quarterly* (Winter):215–228.

International Confederation of Free Trade Unions. 2003. *Annual Survey of Violations of Trade Union Rights: 2003.* http://www.icflu.org

International Labor Organization. *2003. Global Employment Trends.* Geneva: International Labour Office.

———. 2004. *Global Employment Trends: 2004.* Geneva: International Labour Office.

Kenworthy, Lane. 1995. *In Search of National Economic Success.* Thousand Oaks, CA: Sage.

Koch, Kathy. 1998. "High-Tech Labor Shortage." *CQ Researcher* 8(16): 361–384.

"Labor's 'Female Friendly' Agenda." 1998. *Labor Relations Bulletin* no. 690, 2.

Lazarus, David. 2003 (January 3). "USA: Shooting the Messenger—Report on Layoffs Killed." *Corpwatch.* http://www.corpwatch.org

Lenski, Gerard, and J. Lenski. 1987. *Human Societies: An Introduction to Macrosociology,* 5th ed. New York: McGraw-Hill.

Leonard, Bill. 1996 (July). "From School to Work: Partnerships Smooth the Transition." *HR Magazine* (Society for Human Resource Management). http://www.shrm.org/hrmag...articles/0796cov.htm

Levitan, Sar A., Garth L. Mangum, and Stephen L. Mangum. 1998. *Programs in Aid of the Poor,* 7th ed. Baltimore: Johns Hopkins University Press.

Lovell, Vicky, and Hedieh Rahmanou. 2000 (November). "Paid Family and Medical Leave: Essential Support for Working Women and Men." Institute for Women's Policy Research Publication #A124. http://www.iwpr.org

Miers, Suzanne. 2003. *Slavery in the Twentieth Century: The Evolution of a Global Problem.* Walnut Creek, CA: AltaMira Press.

Mishel, Lawrence, Jared Bernstein, and Heather Boushey. 2003. *The State of Working America, 2002/2003.* An Economic Policy Institute Book. Ithaca, NY: ILR Press, an imprint of Cornell University Press.

Moen, Phyllis. 2003. "Epilogue: Toward a Policy Agenda." In *It's About Time: Couples and Careers,* ed. P. Moon, pp. 333–337. Ithaca, NY: Cornell University Press.

Mokhiber, Russell, and Robert Weissman. 1998. "Focus on the Corporation." *Multinational Monitor.* http://www.essential.org

———. 2001 (January 4). "The Corporate Conservative Administration." *Focus on the Corporation.* http://www.essential.org

Moore, David W. 2002. "Public Support for Unions Remains Strong." *Gallup News Service.* http://www.gallup.com

Multinational Monitor. 2000 (November). "Editorial: What Is Society Willing to Spend on Human Beings?" *Multinational Monitor* 21(11). http://www.essential.org/monitor

National Labor Committee. 2001 (January 16). "Nightmare at J.C. Penney Contractor." http://www.nlcnet.org

"New OSHA Policy Relieves Employees." 1998. *Labor Relations Bulletin* no. 687, 8.

Newport, Frank. 2002. "Most Workers Satisfied with Their Jobs." *Gallup News Service.* http://www.gallup.com

Report on the World Social Situation. 1997. New York: United Nations.

Schaeffer, Robert K. 2003. *Understanding Globalization: The Social Consequences of Political, Economic, and Environmental Change,* 2nd ed. Lanham, MD: Rowman & Littlefield.

Scott, Robert E. 2003 (November). "The High Price of Free Trade." Briefing Paper #147. Economic Policy Institute. http://www.epinet.org

"Sex Trade Enslaves Millions of Women, Youth." 2003. *Popline* 25:6.

Statistical Abstract of the United States: 2002. 2002. 122nd ed. U.S. Bureau of the Census. Washington, DC: U.S. Government Printing Office.

Still, Mary C., and David Strang. 2003. "Institutionalizing Family-Friendly Policies." In *It's About Time: Couples and Careers,* ed. Phyllis Moen, pp. 288–309. Ithaca, NY: Cornell University Press.

Thompson, Charles D., Jr. 2002. "Introduction." In *The Human Cost of Food: Farmworkers' Lives, Labor, and Advocacy,* eds. C. D. Thompson, Jr., and M. F. Wiggins, pp. 2–19. Austin: University of Texas Press.

Uchitelle, Louis. 2003 (October 5). "A Missing Statistic: U.S. Jobs that Went Overseas." *New York Times Online.* http://www.nytimes.com

"Unions Forge Global Network." 2002 (April 22). *CorpWatch.* http://www.corpwtch.org

U.S. Department of Labor. 2001 (January 16). "OSHA Statement: Statement of Assistant Secretary Charles N. Jeffress on Effective Date of OSHA Ergonomic Standard." http://www.osha.gov

Vandivere, Sharon, Kathryn Tout, Martha Zaslow, Julia Calkins, and Jeffrey Cappizzano. 2003. "Unsupervised Time: Family and Child Factors Associated with Self-Care." Assessing the New Federalism Occasional Paper No. 71. Urban Institute. http://www.urban-institute.org

Watkins, Marilyn. 2002. "Building Winnable Strategies for Paid Family Leave in the United States." Economic Opportunity Institute. http://www.econop.org

Went, Robert. 2000. *Globalization: Neoliberal Challenge, Radical Responses.* Sterling, VA: Pluto Press.

Werhane, Patricia H., Tara J. Radin, & Norman E. Bowie. 2004. *Employment and Employee Rights.* Malden, MA: Blackwell.

"Workers at Risk." 2003. *Multinational Monitor* 24(6). http://www.essential.org/monitor

Wright, Carter. 2001 (January/February). "A Clean Sweep: Justice for Janitors." *Multinational Monitor* 22(1 & 2). http://www.essential.org/monitor/mm2001/01jan-feb/corp3.html

Chapter 12

Addington, Lynne, Sally Ruddy, Amanda Miller, Jill Devoe, and Kathryn Chandler. 2004. "Are America's Schools Safe? Students Speak Out." In *Schools and Society,* eds. Jeanne Ballantine and Joan Spade, pp. 161–164. Belmont, CA: Thomson Wadsworth.

Asimov, Nanette. 2003. "Bitter Battle over Class Standards." *San Francisco Chronicle.* http://www.sfgate.com

Associated Press, 2004. "Group Pushes Fying Teacher Pay to Progress of Students." *USA Today,* January 14.

Baker, David P., and Deborah P. Jones. 1993. "Creating Gender Equality: Cross-National Gender Stratification and Mathematical Performance." *Sociology of Education* 66:91–103.

Bankston, Carl, and Stephen Caldas. 1997. "The American School Dilemma: Race and Scholastic Performance." *Sociological Quarterly* 8(3):423–429.

BBC (British Broadcasting Company). 2003. "School Drop Outs 'Global Problem.'" BBC News, November 21. http://www.news.bbs.co.uk

Bowman, Darcia. 2002. "Lethal School Shootings Resemble Workplace Rampages, Report Says." *Education Week,* May 29. http://www.edweek.com

Briefing Paper. 2003. "Gender and Education." http://www.eurostep.org/pubs

Bushweller, Kevin. 1995. "Turning Our Backs on Boys." *Education Digest,* January, 9–12.

Call, Kathleen, Lorie Grabowski, Jeylan Mortimer, Katherine Nash, and Chaimun Lee. 1997. "Impoverished Youth and the Attainment Process." Presented at the annual meeting of the American Sociological Association, Toronto, Canada, August.

CDF (Children's Defense Fund). 2003a. "Key Facts About Education." http://www.childrensdefense.org/keyfacts_education

———. 2003b. "Administration Proposal to Test Head Start Children." http://www.childrensdefense.org/hs_test_proposal

———. 2003c. "New TV Ad Tells Congress: 'Head Start's Not Broken, Don't Break it.'" http://www.childrensdefense.org/release

Cochran-Smith, Marilyn. 2000. "Teacher Education at the Turn of the Century." *Journal of Teacher Education* 51:163–165.

Coleman, James S., J. E. Campbell, L. Hobson, J. McPartland, A. Mood, F. Weinfield, and R. York. 1966. *Equality of Educational Opportunity.* Washington, DC: U.S. Government Printing Office.

Crampton, Faith, and David Thompson. 2002. "The Condition of American Schools: A National Disgrace." *School Operations and Facilities.* http://www.asbointl.org

Day, Sherri. 2003. "Sizing Up Snapple's Drink Deal with New York City." *New York Times,* September 12, C2.

Digest of Education Statistics. 2001. Chapters 1–6. National Center for Educational Statistics. http://www.nces.ed.gov/pubs2001/digest

Dillon, Sam. 2003. "Cameras Watching Students, Especially in Biloxi." *New York Times,* September 24. http://www.nytimes.com

ED.gov. 2003. "Paige Releases Number of Schools in School Improvement in Each State." http://www.ed.gov/news/pressreleases

Elam, Stanley M., Lowell C. Rose, and Alec M. Gallup. 1994. "The 26th Annual Phi Delta Kappa/Gallup Poll of the Public's Attitudes Toward the Public Schools." *Phi Delta Kappan,* September, 41–56.

Evans, Lorraine, and Kimberly Davies. 2000. "No Sissy Boys Here." *Sex Roles* (February):255–271.

Fact Sheet. 2003. "Fact Sheet on the Major Provisions of the Conference Report to H.R. 1, the No Child Left Behind Act." http://www.ed/gov/print/nclb

FAQ (Frequently Asked Questions). 2003. "What Is Character Education?" Boston College: Center for the Advancement of Ethics and Character.

Fletcher, Robert S. 1943. *History of Oberlin College to the Civil War.* Oberlin, OH: Oberlin College Press.

Flexner, Eleanor. 1972. *Century of Struggle: The Women's Rights Movement in the United States.* New York: Atheneum.

Frankenberg, Erika, and Chungmei Lee. 2002. *Race in American Public Schools: Rapidly Resegregating School Districts.* Harvard University: The Civil Rights Project.

Goldberg, Carey. 1999. "After Girls Get Attention, Focus Is on Boys' Woes." In *Themes of the Times: New York Times,* p. 6. Upper Saddle River, NJ: Prentice-Hall.

Grant, Eerald, and Cristine Murray. 2004. "Teaching in America: The Slow Revolution." In *Schools and Society,* eds.

Jeanne Ballantine and Joan Spade, pp. 93–101. Belmont, CA: Thomson Wadsworth.

Harris Poll. 2003. "As Economy Grows the Public's Priorities for Growth Are Health, Education and Defense." http://www.harrisinteractive.com

HHS (Department of Heath and Human Services). 2003. "Head Start Program Fact Sheet." Administration for Children and Families. http://www.acf.hhs.gov

Independent Sector. 2002. "Engaging Youth in Lifelong Service." Newsroom. http://www.independentsector.org

International Education Report. 2003a. "International Education Report: U.S. Students Are Average." http://www.ed.gov/print/news

———. 2003b. "New International Study Compares Fourth Grade Reading Literacy in U.S. and Thirty Other Countries." http://www.ed.gov/print/news

Indicators. 2003. "Elementary and Secondary Education: IT in Schools." http://www.nsf.gov

Jacobs, Joanne. 1999. "Gov. Davis to Make Volunteering Mandatory for Students." *Jose Mercury News,* April 23.

Jencks, Christopher, and Meredith Phillips. 1998. "America's Next Achievement Test: Closing the Black-White Test Score Gap." *The American Prospect,* September/October, 44–53.

Kanter, Rosabeth Moss. 1972. "The Organization Child: Experience Management in a Nursery School." *Sociology of Education* 45:186–211.

Kozol, Jonathan. 1991. *Savage Inequalities: Children in America's Schools.* New York: Crown.

Kronholz, June. 2003. "Education Plan Falling Short: States Struggle to Meet Standards of No Child Left Behind." *The Wall Street Journal,* September 17, A4.

Lareau, Annette. 1989. *Home Advantage: Social Class and Parental Intervention in Elementary Education.* Philadelphia: Falmer Press.

Lareau, Annette, and Erin Horvat. 2004. "Moments of Social Inclusion and Exclusion." In *Schools and Society,* eds. Jeanne Ballantine and Joan Spade, pp. 276–286. Belmont, CA: Thomson Wadsworth.

Leo, John. 1998. "Dumbing Down Teachers." *U.S. News and World Report,* August 3, 15.

Lewin, Tamar. 2003. "Writing in Schools Is Found Both Dismal and Neglected." *New York Times,* April 26. http://www.nytimes.com

Levinson, Arlene. 2000. "Study Evaluates Higher Education." Excite.News. http://www.excite.com/news/ap/00130/08/grading

Literacy. 2003. "Facts on Literacy in America." Literacy Volunteers of America. http://www.literacyvolunteers.org

Merton, Robert K. 1968. *Social Theory and Social Structure.* New York: Free Press.

Mollison, Andrew. 2001. "Boom in Charter Schools Continues Despite Mixed Results." *The Atlanta Journal Constitution,* September 24. http://www.accessatlanta.com/partners/ajc/epaper

Morse, Jodie. 2002a. "Learning While Black." *Time,* May 27, 50–52.

———. 2002b. "Flunking Lunch." *Time,* December 2, 74–75.

———. 2002c. "Anything to Avoid an F." *Time,* September 23, 22.

———. 2002d. "A Victory for Vouchers." Time, July 8, 52–53.

Muller, Chandra, and Katherine Schiller. 2000. "Leveling the Playing Field?" *Sociology of Education* 73:196–218.

Murnane, Richard J. 1994. "Education and the Well-Being of the Next Generation." In *Confronting Poverty: Prescriptions for Change,* eds. Sheldon H. Danziger, Gary D. Sandefur, and Daniel H. Weinberg, pp. 289–307. New York: Russell Sage Foundation.

NAAL (National Assessment of Adult Literacy). 2000. "FAQ." http://nces.ed.gov/naal/faq

Nash, Madeleine. 2003. "Obesity Goes Global." *Time,* August 25, p. 53.

Natriello, Gary. 1995. "Dropouts: Definitions, Causes, Consequences, and Remedies." In *Transforming Schools,* eds. Peter W. Cookson, Jr. and Barbara Schneider, pp. 107–128. New York: Garland.

NCES (National Center for Education Statistics). 2000. "Teacher Trends." http://www.nces.ed.gov/fastfacts

———. 2002. "Drop Out Rates in the United States." http://www.nces.ed.gov/pubs2002

———. 2003a. "Comparative Indicators of Education in the United States and Other G8 Countries." National Center for Educational Statistics. http://nces.ed.gov

———. 2003b. "Fast Fact: Family Reading." http://www.nces.ed.gov/fastfacts

———. 2003c. "Average Grade." http://www.nces.ed.gov/pubs2003

———. 2003d. "Mathematics 2000 Major Results." http://www.nces.ed.gov/nations

———. 2003e. "Writing 2002 Major Results" http://www.nces.ed.gov/nations report card

———. 2003e. "Academic Background of College Graduates Who Enter and Leave Teaching." *Contexts of Elementary and Secondary Education.* http://nces.ed.gov/programs

———. 2003f. "Young Children's Access to Computers in the Home and at School in 1999 and 2000: Executive Summary." http://www.nces.ed.gov

———. 2003g. "Special Analysis 2002: Private Schools: A Brief Portrait." http://www.nces.ed.gov/programs

NCJRS (National Criminal Justice Reference Service). 2003. "School Safety Resources—Facts and Figures." http://www.ncjrs.org/school_safety

NCLB (No Child Left Behind). 2003. "The Facts About Supporting Charter Schools." http://www.nclb.gov/start/facts

NPR (National Public Radio). 2003. "NPR/Kaiser/Kennedy School Education Survey." http://www.npr.org/programs/specials

Orfield, Gary. 2002. *Schools More Separate: Consequences of a Decade of Resegregation.* Cambridge, MA: Harvard University, the Civil Rights Project.

Pollock, William. 2000. *Real Boys' Voices.* New York: Random House.

PRB (Population References Bureau). 2002. "2002 Women of Our World." http://www.prb.org

———. 2003. "The Changing Age Structure of U.S. Teachers." http://www.prb.org

Rand, 2003. "Rand Study Finds California Charter Schools Produce Achievement Gains Similar to Conventional Public Schools." http://www.rand.org

Ramierz-Valles, and Amanda Brown. 2003. "Latinos' Community Involvement in HIV/AIDS: Organizational and Individual Perspectives on Volunteering." *AIDS Education and Prevention* 15: 90–104.

Riehl, Carolyn. 2004. "Bridges to the Future: Contributions of Qualitative Research to the Sociology of Education." In *Schools and Society,* eds. Jeanne Ballantine and Joan Spade, pp. 56–72. Belmont, CA: Thomson Wadsworth.

Rodriguez, Richard. 1990. "Searching for Roots in a Changing World." In *Social Problems Today,* ed. James M. Henslin, pp. 202–213. Englewood Cliffs, NJ: Prentice-Hall.

Roscigno, Vincent. 1998. "Race and the Reproduction of Educational Disadvantage." *Social Forces* 76(3):1033–1060.

Rosenbaum, James. 2002. "Beyond College for All: Career Paths for the Forgotten Half." In *Schools and Society,* eds. Jeanne Ballantine and Joan Spade, pp. 485–490. Belmont, CA: Thomson Wadsworth.

Rosenthal, Robert, and Lenore Jacobson. 1968. *Pygmalion in the Classroom: Teacher Expectations and Pupils' Intellectual Development.* New York: Holt, Rinehart & Winston.

Rumberger, Russell W. 1987. "High School Dropouts: A Review of Issues and Evidence." *Review of Educational Research* 57:101–121.

Saporito, Salvatore. 2003. "Private Choices, Public Consequences: Magnet School Choice and Segregation by Race and Poverty." *Social Problems* 50: 181–203.

Sadovnik, Alan. 2004. "Theories in the Sociology of Education." In *Schools and Society,* eds. Jeanne Ballantine and Joan Spade, pp. 7–26. Belmont, CA: Thomson Wadsworth.
Schemo, Diana. 2002. "Neediest Schools Receive Less Money, Report Finds." *New York Times,* August 9. http://www.nytimes.com
Selingo, Jeffery. 2003. "What Americans Think About Higher Education." *Chronicle of Higher Education* May, A10–A17.
Silber, John. 1998. "The Correct Answer: Too Many Can't Teach." *The News & Observer,* July 8, A2.
Slobogin, Kathy. 2002. "Survey: Many Students Say Cheating's OK." CNN.news. http://www.cnn.com
Sommers, Christina. 2000. *The War Against Boys.* New York: Simon and Schuster.
Spade, Joan. 2004. "Gender and Education in the United States." In *Schools and Society,* eds. Jeanne Ballantine and Joan Spade, pp. 287–295. Belmont, CA: Thomson Wadsworth.
Steptoe, Sonja. 2003. "Taking the Alternative Route." *Time,* January 13, 50–51.
Summary Report. 2001. "Building Their Futures." http://www2.acf.dhhs.gov/programs/hsb/EHS
Teachman, Jay D., Kathleen Paasch, and Karen Carver. 1997. "Social Capital and the Generation of Human Capital." *Social Forces* 75(4):1343–1359.
Toppo, Geg, 2004. "Low-Income College Students Are Increasingly Left Behind." *USA Today,* January 14. http://www.usatoday.com
UNDP (United Nations Development Fund). 2003. "Human Development Indicators." http://www.undp.org
UNESCO. 2003. "Literacy." http://portal.unesco.org
U.S. Census. 2003. "Facts and Figures: Back to School." http://www.census.gov/Press-Release.
U.S. Department of Education. 2000. "The Baby Boom Echo: No End in Sight: Figure 2." http://www.ed.gov/pubs/bbecho00/figure2
Wakefield, Julie, 2002. "Learning the Hand Way." *Environmental Health Perspectives* 110(6):1.
Webb, Julie. 1989. "The Outcomes of Home-Based Education: Employment and Other Issues." *Educational Review* 41:121–133.
Weiner, Rebecca. 2000. "Industry Group's Education Study Draws Conclusions and Critics." *New York Times on the Web.* http://www.nytimes.com
Wenglinsky, Harold. 2004. "How Money Matters: The Effect of School District Spending on Academic Achievement." In *Schools and Society,* eds. Jeanne Ballantine and Joan Spade, pp. 213–219. Belmont, CA: Thomson Wadsworth.
White House Fact Sheet. 2002. "A Quality Teacher in Every Classroom." http://www.ed.gov/print/news/pressrelease
Wilgoren, Jodi. 2001. "Calls for Change in the Scheduling of the School Day." *New York Times on the Web,* January 10. http://www.nytimes.com
Winters, Rebecca. 2001. "From Home to Harvard." *Time,* September 11. http://www.time.com/time/magazine/articles
———. 2003. "The Philadelphia Experiment." *Time,* October 21, 64–69.
Zigler, Edward, Sally Styfco, and Elizabeth Gilman. 2004. "The National Head Start Program for Disadvantaged Preschoolers." In *Schools and Society,* eds. Jeanne Ballantine and Joan Spade, pp. 341–346. Belmont, CA: Thomson Wadsworth.

Chapter 13

Agbese, Pita Ogaba. 1995. "Nigeria's Environment: Crises, Consequences, and Responses." In *Environmental Policies in the Third World: A Comparative Analysis,* eds. O. P. Dwivedi and Dhirendra K. Vajpeyi, pp. 125–144. Westport, CT: Greenwood Press.
American Public Transportation Association. 2003. "Public Transportation Facts." http://www.apta.com
Archer, Dennis W. 1998. "The Lesson of Detroit: Never Underestimate a City." *Vital Speeches* 64(11):340–343.
Ashford, Lori. 2003. "Unmet Need for Family Planning: Recent Trends and Their Implications for Programs." Population Reference Bureau, Policy Brief. http://www.prb.org
Bongaarts, John, and Susan Cotts Watkins. 1996. "Social Interactions and Contemporary Fertility Transitions." *Population and Development Review* 22(4): 639–682.
"The Bridge to the 21st Century Leads to Gridlock in and Around Decaying Cities." 1997. *The Washington Spectator* 23(12). Washington, DC: The Public Concern Foundation, Inc.
Brown, Lester. 2002. "Assessing the Food Prospect." In *The Earth Policy Reader,* eds. L. Brown, J. Larsen, and B. Fischlowitz-Roberts, pp. 29–58. New York: W. W. Norton.
Catley-Carlson, Margaret, and Judith A. M. Outlaw. 1998. "Poverty and Population Issues: Clarifying the Connections." *Journal of International Affairs* 52(1): 233–243.
Cincotta, Richard, Robert Engelman, and Daniele Anastasion. 2003. *The Security Demographic: Population and Civil Conflict After the Cold War.* Washington DC: Population Action International.
Clark, David. 1998. "Interdependent Urbanization in an Urban World: An Historical Overview." *The Geographical Journal* 164(1):85–96.
Cowherd, Phil. 2001. "What Is the Business Case for Investing in Inner-City Neighborhoods?" *Public Management* 83(1):12–14.
DaVanzo, Julie, David M. Adamson, Nancy Belden, and Sally Patterson. 2000. *How Americans View World Population Issues: A Survey of Public Opinion.* Santa Monica, CA: Rand Corporation.
Dunlap, Riley E., and Lydia Saad. 2001. "Only One in Four Americans Are Anxious About the Environment." Gallup News Service. http://www.gallup.com
Durning, Alan. 1996. *The City and the Car.* Northwest Environment Watch. Seattle: Sasquatch Books.
Engleman, Robert, Richard P. Cincotta, Bonnie Dye, Tom Gardner-Outlaw, and Jennifer Wisnewski. 2000. *People in the Balance: Population and Natural Resources at the Turn of the Millennium.* Washington, DC: Population Action International.
Fisher, Claude. 1982. *To Dwell Among Friends: Personal Networks in Town and City.* Chicago: University of Chicago Press.
Froehlich, Maryann. 1998. "Smart Growth: Why Local Governments Are Taking a New Approach to Managing Growth in Their Communities." *Public Management* 80(5):5–9.
Gans, Herbert. [1962] 1984. *The Urban Villagers,* 2nd ed. New York: Free Press. (First published 1962).
Gardner, Gary. 1998. "Sanitation Access Lacking." In *Vital Signs 1998,* eds. Lester R. Brown, Michael Renner, and Christopher Flavin, pp. 70–71. New York: W. W. Norton.
Geddes, Robert. 1997. "Metropolis Unbound: The Sprawling American City and the Search for Alternatives." *The American Prospect* 35 (November–December):40–46.
Goodrich, Catherine. 2004. "Women's Empowerment, Population, and Development: An Update on the Cairo Conference." *The Reporter* Winter:1–15. Population Connection. http://www.popconnect.org
Jargowsky, Paul A. 1997. *Poverty and Place: Ghettos, Barrios, and the American City.* New York: Russell Sage Foundation.
Jensen, Derrick. 2001. "Road to Ruin: An Interview with Jan Lundberg." *The Sun* 302:4–13.
Johnson, William C. 1997. *Urban Planning and Politics.* Chicago: American Planning Association, Planners Press.
Kimuna, Sitawa R., & Adamchak, Donald J. 2001. "Gender Relations: Husband-Wife Fertility and Family Planning Decisions in Kenya." *Journal of Biosocial Science* 33:13–23.
Leahy, Elizabeth. 2003. "As Contraceptive Use Rises, Abortions Decline." *Popline,* November–December, 3, 8.

Lindstrom, Matthew J., and Hugh Bartling. 2003. "Introduction." In *Suburban Sprawl: Culture, Theory, and Politics,* eds. M. J. Lindstrom and H. Bartling, pp. 93–114. Lanham, MD: Rowman & Littlefield.

Livernash, Robert, and Eric Rodenburg. 1998. "Population Change, Resources, and the Environment." *Population Bulletin* 53(1):1–36.

McFalls, Joseph A., Jr. 2003. "Population: A Lively Introduction, 4th ed." *Population Bulletin* 58(4):all.

National Association of Home Builders. 1999. "Smart Growth: Building Better Places to Live, Work and Play." Washington, DC: National Association of Home Builders. http://www.smartgrowth.org/

National Research Council. 2003. *Cities Transformed: Demographic Change and Its Implications in the Developing World. Panel on Urban Population Dynamics.* Mark R. Montgomery, Richard Stren, Barney Cohen, and Holly Reed (eds.), Committee on Population. Washington DC: National Academy Press.

Nelson, Mara. 2004. "Silence & Myths: A Response to the 'Birth Dearth.'" *The Reporter* (Winter):17–21, 24. Population Connection. http://www.popconnect.org

Orfield, Myron. 1997. *Metropolitics: A Regional Agenda for Community and Stability.* Washington, DC: Brookings Institution Press and Cambridge, MA: Lincoln Institute of Land Policy.

Pelley, Janet. 1999. "Building Smart-Growth Communities." *Environmental Science & Technology News* 33(1): 28A–32A.

Perry, Marc J., and Paul J. Mackun. 2001 (April). "Population Change and Distribution." U.S. Census Bureau. Washington, DC: U.S. Government Printing Office.

Pimentel, David, Maria Tort, Linda D'Anna, Anne Krawic, Joshua Berger, Jessica Rossman, Fridah Mugo, Nancy Doon, Michael Shriberg, Erica Howard, Susan Lee, and Jonathan Talbot. 1998. "Ecology of Increasing Disease: Population Growth and Environmental Degradation." *BioScience* 48 (October): 817–827.

Popline. 2003a. "Aging Trend Continuing in Japan." May–June, 4.

———. 2003b. "World Food Output Must Double by 2050." March–April, 1.

———. 2003c. "Up to 2.5 Billion May Face Water Shortages in Developing World." November–December, 8.

———. 2003d. "Senate Upholds Gag Rule Repeal." July–August, 1.

———. 2003e. "Children Worldwide Denied Access to School." November–December, 6.

Population Action International. 2003a. "Fact Sheet: Why Population Matters to Natural Resources." Washington, DC: Population Action International.

———. 2003b. "Fact Sheet: How Family Planing and Reproductive Health Services Affect the Lives of Women, Men, and Children." Washington, DC: Population Action International.

Population Reference Bureau. 2003. *2003 World Population Data Sheet.* Washington DC: Population Reference Bureau.

———. 2000a. "How Does Family Planning Influence Women's Lives?" *MEASURE Communication: Reports: Women 2000 Policy Briefs.* Washington, DC: Population Reference Bureau.

———. 2000b. "Is Education the Best Contraceptive?" *MEASURE Communication: Reports: Women 2000 Policy Briefs.* Washington, DC: Population Reference Bureau.

Rees, Amanda. 2003. "New Urbanism: Visionary Landscapes in the Twenty-First Century." In *Suburban Sprawl: Culture, Theory, and Politics,* eds. M. J. Lindstrom and H. Bartling, pp. 93–114. Lanham, MD: Rowman & Littlefield.

Rodriquez, Luis. 2000. "Urban Renewal: The Resurrection of an Ex-Gang Member" (in an interview with Derrick Jensen). *The Sun,* April, pp 4–13.

Scommega, Paola. 2003 (June). "Census Bureau to Track Both Metropolitan and 'Micropolitan' Areas." Population Reference Bureau. http://www.prb.org

Sheehan, M. O. 2001. "Making Better Transportation Choices." In *State of the World 2001,* ed. L. Starke, pp. 103–122. New York: W. W. Norton.

———. 2003. "Uniting Divided Cities." In *State of the World 2003,* ed. L. Starke, pp. 130–151. New York: W. W. Norton.

Shevis, Jim. 1999. "More Affluent Than Their Inner-City Neighbors, Suburbanites Still Have Growth Problems." *The Washington Spectator* 25(19):1–3).

Smart Growth Network. 2004. "About Smart Growth." http://www.smartgrowth.org

Tittle, Charles. 1989. "Influences on Urbanism: A Test of Predictions from Three Perspectives." *Social Problems* 36(3):270–288.

Union of Concerned Scientists. 1999. "The Hidden Costs of Transportation." http://www.ucsusa.org/transportation/hidden.html

United Nations. 2003. *World Population Prospects: The 2002 Revision.* United Nations Population Division. http://www.un.org

United Nations Population Division. 2001. *World Urbanization Prospects: The 1999 Revision.* New York: United Nations.

United Nations Population Fund. 1997. *1997 State of the World Population.* New York: United Nations.

———. 1999. *The State of World Population 1999.* New York: United Nations.

United States Conference of Mayors. 2001. *Traffic Congestion and Rail Investment.* Washington, DC: Global Strategy Group.

———. 2003. *Recycling America's Land: A National Report on Brownfields Redevelopment,* Vol. IV. Washington DC: U.S. Conference of Mayors.

United States Conference of Mayors. 2003. *Hunger and Homelessness Survey.* http://www.usmayors.org

Urban Legends Reference Page. 2003. "How Much Is a Billion?" http://www.snopes.com

U.S. Geological Survey. 2003. "How Much Is One Billion?" National Park Service. Department of the Interior. http://wrgis.wr.usgs.gov

Warren, Roxanne. 1998. *The Urban Oasis: Guideways and Greenways in the Human Environment.* New York: McGraw-Hill.

Weeks, John R. 2002. *Population: An Introduction to Concepts and Issues,* 8th ed. Belmont, CA: Wadsworth/Thomson.

Wilson, William Julius. 1996. *When Work Disappears: The World of the New Urban Poor.* New York: Alfred A. Knopf.

Wirth, Louis. 1938. "Urbanism as a Way of Life." *American Journal of Sociology* 44:8–20.

Wolff, Kurt H. 1978. *The Sociology of George Simmel.* Toronto: Free Press.

Women's Studies Project. 2003. *Women's Voices, Women's Lives: The Impact of Family Planning.* Family Health International. http://www.fhi.org

World Health Organization. 2003. *World Health Report 2003: Shaping the Future.*

Zabin, L. S., and K. Kiragu. 1998. "The Health Consequences of Adolescent Sexual and Fertility Behavior in Sub-Saharan Africa." *Studies in Family Planning* 2 (June 29):210–232.

Chapter 14

Abramovitz, Janet N. 2000. "Paper Recycling Remains Strong." In *Vital Signs 2000,* eds. Lester R. Brown, Michael Renner, and Brian Halweil, pp. 132–133. New York: W. W. Norton.

Ayres, Ed. 2004. "The Hidden Shame of the Global Industrial Economy." *World Watch Magazine,* January–February, pp. 18–29.

Bright, Chris. 2003. "A History of Our Future." In *State of the World 2003,*ed. Linda Starke, pp. 3–13. Worldwatch Institute. New York: W. W. Norton.

Brown, Lester R. 2003. *Plan B—Rescuing a Planet Under Stress and a Civi-*

lization in Trouble. New York: W. W. Norton.

Brown, Lester R., and Jennifer Mitchell. 1998. "Building a New Economy." In *State of the World 1998,* eds. Lester R. Brown, Christopher Flavin, and Hilary French, pp. 168–187. New York: W. W. Norton.

Bruce, Nigel, Rogelio Perez-Padilla, and Rachel Albalak. 2000. "Indoor Air Pollution in Developing Countries: A Major Environmental and Public Health Challenge." *Bulletin of the World Health Organization,* 78(9):1078–1092.

Bruno, Kenny, and Joshua Karliner. 2002. *Earthsummit.biz: The Corporate Takeover of Sustainable Development.* CorpWatch and Food First Books. http://www.corpwatch.org

Bullard, Robert D. 2000. *Dumping in Dixie: Race, Class, and Environmental Quality,* 3rd ed. Boulder, CO: Westview Press.

Bullard, Robert D., and Glenn S. Johnson. 1997. "Just Transportation." In *Just Transportation: Dismantling Race and Class Barriers to Mobility,* eds. Robert D. Bullard and Glenn S. Johnson, pp. 1–21. Stony Creek, CT: New Society Publishers.

Cappiello, Dina. 2003 (April 22). "Corporations Co-opt Earth Day." http://www.houstonchronicle.com

Caress, Stanley, Anne C. Steinemann, and Caitlin Waddick. 2002. "Symptomatology and Etiology of Multiple Chemical Sensitivities in the Southeastern United States." *Archives of Environmental Health* 57(5):429–436.

Chivian, Eric, and Aaron S. Bernstein. 2004. "Embedded in Nature: Human Health and Biodiversity." *Environmental Health Perspectives,* 112(1). http://ehp.niehs.nih.gov

Cincotta, Richard P., and Robert Engelman. 2000. *Human Population and the Future of Biological Diversity.* Washington, DC: Population Action International.

CNN.com. 2002 (October 3). "Parents Fight to Ban Perfume, Aerosols in Schools." http://www.cnn.com

"Corporate Spotlight." 2001 (March/April). *Adbusters* 34:38.

Cray, Charlie. 2001. "Taking on Toxics I: Stopping POPs." *Multinational Monitor* 22(1 & 2).

Denson, Bryan. 2000. "Shadowy Saboteurs." *The IRE Journal* (Investigative Reporters and Editors, Inc.) 23 (May–June):12–14.

DesJardins, Andrea. 1997. "Sweet Poison: What Your Nose Can't Tell You About the Dangers of Perfume." http://members.aol.com/enviroknow/perfume/sweet–poison.htm

Edwards, Bob, and Anthony Ladd. 2000. "Environmental Justice, Swine Production and Farm Loss in North Carolina." *Sociological Spectrum* 20(3):263–290.

Environmental Defense. 2003. *Environmental Defense Annual Report 2003.* http://www.environmentaldefense.org

Environmental Protection Agency. 2003. "Air Trends: 2002 Highlights." http://www.epa.gov

Fisher, Brandy E. 1998 (December). "Scents and Sensitivity." *Environmental Health Perspectives* 106(12). http://ehpnet1.niehs.nih.gov/docs/1998/106-12/focus-abs.html

———. 1999 (January). "Focus: Most Unwanted." *Environmental Health Perspectives* 107(1). http://ehpnet1.niehs.nih.gov/docs/1999/107-1/focus-abs.html

Flavin, Christopher. 2000. "Wind Power Booms." In *Vital Signs 2000,* eds. Lester R. Brown, Michael Renner, and Brian Halweil, pp. 56–57. New York: W. W. Norton.

French, Hilary. 2000a. *Vanishing Borders: Protecting the Planet in the Age of Globalization.* New York: W. W. Norton.

———. 2000b. "Environmental Treaties Gain Ground." In *Vital Signs 2000,* eds. Lester R. Brown, Michael Renner, and Brian Halweil, pp. 134–135. New York: W. W. Norton.

Hager, Nicky, and Bob Burton. 2000. *Secrets and Lies: The Anatomy of an Anti-Environmental PR Campaign.* Monroe, ME: Common Courage Press.

Hunter, Lori M. 2001. *The Environmental Implications of Population Dynamics.* Santa Monica, CA: Rand Corporation.

"Hybrid Cars." 2002–2003. *Earthwise* (Updates from the Union of Concerned Scientists), 5(1):1, 4.

Intergovernmental Panel on Climate Change. 2000. *Land Use, Land-Use Change, and Forestry.* http://www.ipcc.ch

———. 2001a. *Climate Change 2001: The Scientific Basis.* United Nations Environmental Programme and the World Meteorological Organization. http://www.ipcc.ch

———. 2001b. *Climate Change 2001: Impacts, Adaptation, and Vulnerability.* United Nations Environmental Programme and the World Meteorological Organization. http://www.ipcc.ch

Jan, George P. 1995. "Environmental Protection in China." *Environmental Policies in the Third World: A Comparative Analysis,* ed. O. P. Dwivedi and Dhirendra K. Vajpeyi, pp. 71–84. Westport, CT: Greenwood Press.

Jarboe, James F. 2002 (February 12). "The Threat of Eco-Terrorism." Congressional Statement. Federal Bureau of Investigation. http://www.fbi.gov

Jensen, Derrick. 2001. "A Weakened World Cannot Forgive Us: An Interview with Kathleen Dean Moore." *The Sun* 303(March):13.

Kaplan, Sheila, and Jim Morris. 2000. "Kids at Risk." *U.S. News & World Report,* June 19:47–53.

Karliner, Joshua. 1998. "Corporate Greenwashing." *Green Guide* 58 (August): 1–3.

Koenig, Dieter. 1995. "Sustainable Development: Linking Global Environmental Change to Technology Cooperation." *Environmental Policies in the Third World: A Comparative Analysis,* eds. O. P. Dwivedi, and Dhirendra K. Vajpeyi, pp. 1–21. Westport, CT: Greenwood Press.

Lenssen, Nicholas. 2003. "Nuclear Power Rises." In *Vital Signs 2003,* eds. Michael Renner and Molly O. Sheehan, pp. 36–37. Worldwatch Institute. New York: W. W. Norton.

Leutwyler, Kristin. 2001. "The Poor Face More Environmental Hazards." *Scientific American,* January 8. http://www.sciam.com/news/010801/3.html

Levin, Phillip S., and Donald A. Levin. 2002. "The Real Biodiversity Crisis." *American Scientist* 90(1):6

MacDonald, Mia, with Danielle Nierenberg. 2003. "Linking Population, Women, and Biodiversity." In *State of the World 2003,* ed. Linda Starke, pp. 38–61. Worldwatch Institute. New York: W. W. Norton.

Makower, Joel, Ron Pernick, and Clint Wilder. 2003. *Clean Energy Trends 2003.* Clean Edge. http://www.cleanedge.com

Mastny, Lisa. 2003. "State of the World: A Year in Review." In *State of the World 2003,* ed. Linda Starke, pp. xix–xxiii. Worldwatch Institute. New York: W. W. Norton.

"Matters of Scale: Chemical Warfare." 2003 (July–August). World Watch Institute. http://www.worldwatch.org

McGinn, Anne Platt. 2000. "Endocrine Disrupters Raise Concern." In *Vital Signs 2000,* eds. Lester R. Brown, Michael Renner, and Brian Halweil, pp. 130–131. New York: W. W. Norton.

McMichael, Anthony J., Kirk R. Smith, and Carlos F. Corvalan. 2000. "The Sustainability Transition: A New Challenge." *Bulletin of the World Health Organization* 78(9):1067.

Mead, Leila. 1998. "Radioactive Wastelands." *The Green Guide* 53(April 14):1–3.

Miller, Norman. 2000 (May). "Rains of Terror." *Geographical* 75(5):90.

Mishra, Vinod, Robert D. Retherford, and Kirk R. Smith. 2002. "Indoor Air Pollution: The Quiet Killer." *AsiaPacific Issues* 63:1–8.

National Assessment Synthesis Team. 2000. *Climate Change Impacts on the United States: The Potential Consequences of Climate Variability and Change.* Washington, DC: U.S. Global Change Research Program.

National Environmental Education and Training Foundation and Roper Starch Worldwide. 1999. *1999 NEETF/Roper Report Card.* Washington, DC: National

Environmental Education and Training Foundation.
———. 2001. *2001 NEETF/Roper Report Card.* Washington, DC: National Environmental Education and Training Foundation.
"New Rules for Feedlots." 1998. *Environmental Health Perspectives* 106(12). http://ehpnet1.niehs.nih.gov/docs/1998/106-12/forum.html
"Nukes Rebuked." 2000 (July 1). *The Washington Spectator* 26(13):4.
"Online Activism Lives Up to Its Promise." 2001 (March). *Environmental Defense* 32(2):8.
PBS. 2001. "Trade Secrets: A Moyers Report." http://www.pbs.org/tradesecrets/program/program.html
Perks, Robert, and Gregory Wetstone. 2003. *Rewriting the Rules, Year-End Report 2002: The Bush Administration's Assault on the Environment.* Natural Resources Defense Council. http://www.nrdc.org
Pimentel, David, and Anthony Greiner. 1997. "Environmental and Socio-Economic Costs of Pesticide Use." In *Techniques for Reducing Pesticide Use,* ed. D. Pimentel, pp. 50–78. New York: John Wiley.
Pimentel, David, Maria Tort, Linda D'Anna, Anne Krawic, Joshua Berger, Jessica Rossman, Fridah Mugo, Nancy Don, Michael Shriberg, Erica Howard, Susan Lee, and Jonathan Talbot. 1998. "Ecology of Increasing Disease: Population Growth and Environmental Degradation." *BioScience* 48(October): 817–27.
Population Institute. 2003. *Water: The 21st Century Crisis.* United Nations Population Fund. http://www.populationinstitute.org
"A Prescription for Reducing the Damage Caused by Dams." 2001 (March). *Environmental Defense* 32(2).
Reese, April. 2001 (February). "Africa's Struggle with Desertification." Population Reference Bureau. http://www.prb.org/regions/africa/africadesertification.html
Renner, Michael. 1996. *Fighting for Survival: Environmental Decline, Social Conflict, and the New Age of Insecurity.* New York: W. W. Norton.
———. 2004. "Moving Toward a Less Consumptive Economy." In *State of the World 2004,* ed. Linda Starke, pp. 96–119. Worldwatch Institute. New York: W. W. Norton.
Renner, Michael, and Molly O. Sheehan. 2003. "Overview." In *Vital Signs 2003,* eds. M. Renner and M. O. Sheehan, pp. 17–24. Worldwatch Institute. New York: W. W. Norton.
Saad, Lydia. 2002. "Americans Sharply Divided on Seriousness of Global Warming." *Gallup News Service.* http://www.gallup.org
Sampat, Payal. 2003. "Scrapping Mining Dependence." In *State of the World 2003,* ed. Linda Starke, pp. 110–129. Worldwatch Institute. New York: W. W. Norton.
Sawin, Janet. 2003a. "Charting a New Energy Future." In *State of the World 2003,* ed. Linda Starke, pp. 85–109. Worldwatch Institute. New York: W. W. Norton.
———. 2003b. "Wind Power's Rapid Growth Continues." In *Vital Signs 2003,* eds. M. Renner and M. O. Sheehan, pp. 38–39. Worldwatch Institute. New York: W. W. Norton.
———. 2004. "Making Better Energy Choices." In *State of the World 2004,* ed. Linda Starke, pp. 24–43. Worldwatch Institute. New York: W. W. Norton.
Schaeffer, Robert K. 2003. *Understanding Globalization: The Social Consequences of Political Economic, and Environmental Change,* 2nd ed. Lanham, MD: Rowman & Littlefield.
Schmidt, Charles W. 2004. "Environmental Crimes: Profiting at the Earth's Expense." *Environmental Health Perspectives* 112(2):A96–A103.
Scholand, Michael 2000. "Compact Fluorescents Light Up the Globe." In *Vital Signs 2000,* eds. Lester R. Brown, Michael Renner, and Brian Halweil, pp. 60–61. New York: W. W. Norton.
Sheehan, Molly O. 2003. "Carbon Emissions and Temperatures Climb." In *Vital Signs, 2003,* eds. M. Renner and M. O. Sheehan, pp. 40–41. Worldwatch Institute. New York: W. W. Norton.
Smith, Velma, John Coequyt, and Richard Wiles. 2000. *Clean Water Report Card.* Washington, DC: Environmental Working Group.
Statistical Abstract of the United States: 2003, 123rd ed. 2003. U.S. Bureau of the Census. Washington, DC: U.S. Government Printing Office.
Stretesky, Paul. 2003. "The Distribution of Air Lead Levels Across U.S. Counties: Implications for the Production of Racial Inequality." *Sociological Spectrum* 23:91–118.
Tuxill, John. 1998. *Losing Strands in the Web of Life: Vertebrate Declines and the Conservation of Biological Diversity.* Worldwatch Paper 141. Washington, DC: Worldwatch Institute.
United Nations Environment Programme (UNEP). 2002. *North America's Environment: A Thirty-Year State of the Environment and Policy Retrospectives.* http://www.na.unep.net
———. 2003 (August 3). "Basic Facts and Data on the Science and Politics of Ozone Protection." http://www.unep.org/ozone
U.S. Department of Health and Human Services. 2002. *Tenth Report on Carcinogens.* Washington, DC: Public Health Service.
Vajpeyi, Dhirendra K. 1995. "External Factors Influencing Environmental Policymaking: Role of Multilateral Development Aid Agencies." *Environmental Policies in the Third World: A Comparative Analysis,* eds. O. P. Dwivedi and Dhirendra K. Vajpeyi, pp. 24–45. Westport, CT: Greenwood Press.
Warren, Karen J. 2000. *Ecofeminist Philosophy: A Western Perspective on What It Is and Why It Matters.* Lanham, MD: Rowman & Littlefield.
Weiner, Tim. 2001. "Terrific News in Mexico City: Air Is Sometimes Breathable." *The New York Times on the Web,* January 5. http://www.nytimes.com/2001/01/05/world/05MEXI.html
World Resources Institute. 2000. *World Resources 2000–2001: People and Ecosystems: The Fraying Web of Life.* Washington DC: World Resources Institute.
Youth, Howard. 2003. "Watching Birds Disappear." In Linda Starke (Ed.), *State of the World 2003,* ed. Linda Starke, pp. 85–109. Worldwatch Institute. New York: W. W. Norton.

Chapter 15

Addison, John T., Douglas Fox, and Christopher Ruhm. 2000. "Technology, Trade Sensitivity, and Labor Displacement." *Southern Economic Journal* 66:682–699.
AMA (American Management Association). 2003 (May 14). "2003 E-Mail Rules, Policies, and Practices." American Management Association in conjunction with The ePolicy Institute and Clearswift.
AP (Associated Press). 2003a. "Bill Gates Takes the Stand." *CNNMoney,* April 22. http://www.cnnmoney.com
———. 2003b. "Supreme Court Again to Review Online Pornography Law." *USA Today,* October 14. http://www.usatoday.com
Attewell, Paul, Belkis Suazo-Garcia, and Juan Battle. 2003. "Computers and Young Children: Social Benefit or Social Problem?" *Social Forces* 82: 277–296.
Bell, Daniel. 1973. *The Coming of Post-Industrial Society: A Venture in Social Forecasting.* New York: Basic Books.
Beniger, James R. 1993. "The Control Revolution." In *Technology and the Future,* ed. Albert H. Teich, pp. 40–65. New York: St. Martin's Press.
BLS (Bureau of Labor Statistics). 2002. "Computer and Internet Use at Work in 2001 Summary." http://www.bls.gov
Boles, Margaret, and Brenda Sunoo. 1998. "Do Your Employees Suffer from Techno Phobia?" *Workforce* 77(1):21.

Brasher, Philip. 2000. "Government Probes Biotech Corn Allegations." Excite. News. http://www.news.excite.com/news/ap/000918

Brin, David. 1998. *The Transparent Society: Will Technology Force Us to Choose Between Privacy and Freedom?* Reading, MA: Addison Wesley.

Buchanan, Allen, Dan Brock, Norman Daniels, and Daniel Wikler. 2000. *From Chance to Choice: Genetics and Justice.* New York: Cambridge University Press.

Bush, Corlann G. 1993. "Women and the Assessment of Technology." In *Technology and the Future,* ed. Albert H. Teich, pp. 192–214. New York: St. Martin's Press.

CAW (Center for the Advancement of Women). 2003. "Women in Science, Engineering and Technology." http://www.advancewomen.org

Ceruzzi, Paul. 1993. "An Unforseen Revolution." In *Technology and the Future,* ed. Albert H. Teich, pp. 160–174. New York: St. Martin's Press.

Cetron, Marvin, and Owen Davies. 2003. "Fifty Trends Now Changing the World." *Special Report.* Bethesda, MD: World Future Society.

Chabrow, Eric. 2003. "IT Unemployment Hits Record High." *Information Week,* March 18. http://www.informationweek.com

Chepesiuk, Ron. 1999. "Where the Chips Fall: Environmental Health in the Semiconductor Industry." *Environmental Health Perspectives* 107(9). http://www.junkscience.com/aug99/chips.htm

Clarke, Adele E. 1990. "Controversy and the Development of Reproductive Sciences." *Social Problems* 37(1):18–37.

CNN. 2004. "Spam Invasion Targets Mobile Phones." February 5. http//www.cnn.com/2004.

Conrad, Peter. 1997. "Public Eyes and Private Genes: Historical Frames, New Constructions, and Social Problems." *Social Problems* 44:139–154.

CyberAtlas. 2003. "Population Explosion." http://www.cyberatlas.internet.com/big_picture

Dervarics, Charles. 2003. "House Moves Ahead on Digital Divide Help." *Black Issues in Higher Education* 20:6.

Doll, John. 2002. "Talking Gene Patents." *Scientific American,* August 17. http://www.sciam.com

Durkheim, Emile. [1925] 1973. *Moral Education.* New York: Free Press.

Ehrenfeld, David. 1998. "A Techno-Pox upon the Land." *Harper's,* October, 13–17.

Eibert, Mark D. 1998. "Clone Wars." *Reason* 30(2):52–54.

Eilperin, Juliet, and Rick Weiss. 2003. "House Votes to Prohibit All Human Cloning." *Washington Post,* February 28. http://www.washingtonpost.com

Fagan, Amy. 2003. "Partial Birth Abortion Ban Ready for Vote." *Washington Times,* October 1. http://www.dynamic.washtimes.com

FBI (Federal Bureau of Investigation). 2002. "Online Child Pornography: Innocent Images National Initiative." http://www.fbi.gov

Gallup Poll. 2000. "Abortion Issues." http://www.gallup.com/poll/indicators

Gibbs, Nancy. 2003. "The Secret of Life." *Time,* February 17, 42–45.

GIP (Global Internet Project). 1998. "The Workplace." http://www.gip.org/gip2g.html

Glendinning, Chellis. 1990. *When Technology Wounds: The Human Consequences of Progress.* New York: William Morrow.

Goodman, Paul. 1993. "Can Technology Be Humane?" In *Technology and the Future,* ed. Albert H. Teich, pp. 239–255. New York: St. Martin's Press.

Gottlieb, Scott. 2000. "Abortion Pill Is Approved for Sale in United States." *British Medical Journal* 321:851.

Greenspan, Robyn. 2003a. "The Kids Are Alright with Spending." http://www.cyberatlas.internet.com/big_picture

———. 2003b. "Porn Pages Reach 260 Million." http://www.cyberatlas.internet.com/big_picture

Hancock, LynNell. 1995. "The Haves and the Have-Nots." *Newsweek,* February 27, 50–53.

Hasson, Judi, and Graeme Browning. 2002. "The Federal Government Has Become One of the Biggest Online Retailers." Pew Internet and American Life. http://www.pewinternet/reports

Hayday, Graham. 2003. "Technology in the Workplace: More Important than Managers?" http://news.zdnet.co.uk

HGP (Human Genome Project). 2003a. "About the Human Genome Project." http:www.ornl.gov/TechResources/Human_genome

———. 2003b. "Gene Therapy." http://www.ornl.gov/TechResources/Human_genome

Hormats, Robert D. 2001. "Asian Connection." *Across the Board* 38:47–50.

Internet Statistics. 2003. "How Internet Access Has Changed." http://www.infoplease.com

Johnson, Margaret. 2000. "Committee Approves Watered-Down Anti-Hacking Bill." October 6. http://www2.infoworld.com/articles/hn/xml/00/10/06

Kahin, Brian. 1993. "Information Technology and Information Infrastructure." In *Empowering Technology: Implementing a U.S. Strategy,* ed. Lewis M. Branscomb, pp. 135–166. Cambridge, MA: MIT Press.

Kahn, A. 1997. "Clone Mammals . . . Clone Man." *Nature,* March 13, 119.

Kaplan, Carl S. 2000. "The Year in Technology Law." December 22. http://www.nytimes.com/2000/12/22/technology/22CYBERLAW

Kilen, Mike. 2003. "She's a Little Fighter." People and Places. *The Des Moines Register,* February 23. http://www.desmoinesregister.com/news

Klein, Matthew. 1998. "From Luxury to Necessity." *American Demographics* 20(8):8–12.

Kuhn, Thomas. 1973. *The Structure of Scientific Revolutions.* Chicago: Chicago University Press.

Lee, Jennifer. 2003. "Identity Theft Complaints Double in 2002." *New York Times,* January 22. http://www.nytimes.com

Lemonick, Michael, and Dick Thompson. 1999. "Racing to Map Our DNA." *Time Daily,* 153:1–6. http://www.time.com

Macklin, Ruth. 1991. "Artificial Means of Reproduction and Our Understanding of the Family." *Hastings Center Report,* January/February, 5–11.

March of Dimes. 2003. "What We Know and What We Don't." http://www.marchofdimes.com/prematurity

Markoff, John. 2000. "Report Questions a Number in Microsoft Trial." *New York Times on the Web,* August 28. http://www.nytimes.com/library/tech/00/08

———. 2002. "Protesting the Big Brother Lens, Little Brother Turns an Eye Blind." *New York Times,* October 7. http://www.nytimes.com

Markovich, Matt, and Suzanne Brahm. 2003. "Addiction or Compulsion?" ABC News. http://www.abcnews.com

Mayer, Sue. 2002. "Are Gene Patents in the Public Interest?" *BIO-IT World,* November 12. http://www.bio-itworld.com

McCormick, S. J., and A. Richard. 1994. "Blastomere Separation." *Hastings Center Report,* March/April, 14–16.

McDermott, John. 1993. "Technology: The Opiate of the Intellectuals." In *Technology and the Future,* ed. Albert H. Teich, pp. 89–107. New York: St. Martin's Press.

McFarling, Usha L. 1998. "Bioethicists Warn Human Cloning Will Be Difficult to Stop." *Raleigh News and Observer,* November 18, A5.

McNair, James. 2003. "Rising Unemployment Affects even Tech Sector." *Cincinnati Enquirer,* March 26, 1.

Merton, Robert K. 1973. "The Normative Structure of Science." In *The Sociology of Science,* ed. Robert K. Merton. Chicago: University of Chicago Press.

Mesthene, Emmanuel G. 1993. "The Role of Technology in Society." In *Technology and the Future,* ed. Albert H. Teich, pp. 73–88. New York: St. Martin's Press.

Murphie, Andrew, and John Potts. 2003. *Culture and Technology.* New York: Palgrave Macmillan.

NCSL (National Conference of State Legislatures). 2003. "State Human Cloning Laws." http://www.ncsl.org/programs
NIC (National Intelligence Council). 2003. *The Global Technology Revolution.* Preface and Summary. Rand Corporation. http://www.rand.org
NSF (National Science Foundation). 2003. "Science and Engineering Indicators." http://www.nsf.org
Ogburn, William F. 1957. "Cultural Lag as Theory." *Sociology and Social Research* 41:167–174.
O'Meara, Kelly Patricia. 2003. "Cheap Labor at America's Expense." *Insight on the News,* May 27, 32.
Papadakis, Maria. 2000. "Complex Picture of Computer Use in Home Emerges." National Science Foundation, March 31. NSF000-314.
Perrolle, Judith A. 1990. "Computers and Capitalism." In *Social Problems Today,* ed. James M. Henslin, pp. 336–342. Englewood Cliffs, NJ: Prentice-Hall.
Pew. 2003. "Summary of Findings." Pew Internet and American Life. http://www.pewinternet.org/reports
Plunkett Research Ltd. 2003. "Top 15 Hardware Manufacturing Companies, by Sales, 2001." http://www.plunkettresearch.com
Pollack, Andrew. 2000. "Nations Agree on Safety Rules for Biotech Food." *New York Times,* January 30. http://www.nytimes.com/library/national/science
Postman, Neil. 1992. *Technopoly: The Surrender of Culture to Technology.* New York: Alfred A. Knopf.
Powers, Richard. 1998. "Too Many Breakthroughs." Op-Ed. *New York Times,* November 19, 35.
Rabino, Isaac. 1998. "The Biotech Future." *American Scientist* 86(2):110–112.
Rhode Island Department of Health. 2003. "Rhode Island's Genetic Laws." http://www.health.rh.gov/genetics
Robotics. 2003. "Universal Robots: The History and Workings of Robots." http:www.thetech.org/robotics
Rosenberg, Debra. 2003. "The War over Fetal Rights." *Newsweek,* June 9, 40–44.
Rosenberg, Jim. 1998. "Troubles and Technologies." *Editor and Publisher* 131(6):4.
Sampat, Payal. 2000. "Internet Use Accelerates." In *Vital Signs: The Environmental Trends That Are Shaping Our Future,* ed. Linda Starke, pp. 94–94. New York: W. W. Norton.
Schaefer, Naomi. 2001. "The Coming Internet Privacy Scrum." *The American Enterprise* 12:50–51.
Shand, Hope. 1998. "An Owner's Guide." *Mother Jones,* May/June, 46.
Statistical Abstract of the United States. 2002. 122nd ed. U.S. Bureau of the Census. Washington, DC: U.S. Government Printing Office.
Tanaka, Jennifer. 2000. "An Extreme Reaction." *Newsweek,* September 25, 75.
Taylor, Chris. 2003. "Spam's Big Bang." *Time,* June 16, 50–53.
Technology. 2003. "Hi-tech Workplaces no Better than Factories." 2003. BBC News, November 27. http://www.bbc.co.uk
Toffler, Alvin. 1970. *Future Shock.* New York: Random House.
Tumulty, Karen, and Viveca Novak. 2003. "Under the Radar." *Time,* January 27, 38–41.
U.S. Department of Justice. 1999. "1999 Report in Cyberstalking: A New Challenge for Law Enforcement and Industry." Washington, DC: Department of Justice.
U.S. Newswire. 2003. "Feminists Condemn House Passage of Deceptive Abortion Ban, Urge Activists to March on Washington." October 2. http://www.usnewswire.com
Weinberg, Alvin. 1966. "Can Technology Replace Social Engineering?" *University of Chicago Magazine* 59 (October):6–10.
Weiss, Rick. 2003. "First Cloned Horse Announced." *Washington Post,* August 6. http://www.washingtonpost.com
Welter, Cole H. 1997. "Technological Segregation: A Peek Through the Looking Glass at the Rich and Poor in an Information Age." *Arts Education Policy Review* 99(2):1–6.
Whine, Michael. 1997. "The Far Right on the Internet." In *The Governance of Cyberspace,* ed. Brian D. Loader, pp. 209–227. London: Routledge.
White, Lawrence. 2000. "Colleges Must Protect Privacy in the Digital Age." *Chronicle of Higher Education,* June 30, B5–6.
White House. 2003. "President Bush Signs Partial Birth Abortion Ban Act of 2003." November 5. http://www.whitehouse.gov/news/release
Winner, Langdon. 1993. "Artifact/Ideas as Political Culture." In *Technology and the Future,* ed. Albert H. Teich, pp. 283–294. New York: St. Martin's Press.
World Employment Report. 2001. "Digital Divide Is Wide and Getting Wider." Geneva: International Labor Organization. http://www.ilo.org/public/english/bureau/inf/pkits/wer2001

Chapter 16

AJC (American Jewish Committee). 2003. "2002 Annual Survey Americans Jewish Opinion." http://www.ajc.org
AP (Associated Press). 2003. "U.S.: Iran in Noncompliance of Nuclear Nonproliferation Treaty." *Fox News,* September 9. http://www.foxnews.com
Arms Control Association. 2000. "Start III at a Glance." September. http://www.armscontrol.org/FACTS
———. 2003a. "Start II and Its Extension at a Glance." January. http://www.armscontrol.org/factsheets
———. 2003b. "The Nuclear Proliferation Treaty at a Glance." May. http://www.armscontrol.org/factsheets
Baliunas, Sallie. 2004. "Anthrax Is a Serious Threat." In *Biological Warfare,* ed. William Dudley, pp. 53–58. Farmington Hills, MA: Greenhaven Press.
Barkan, Steven, and Lynne Snowden. 2001. *Collective Violence.* Boston: Allyn and Bacon.
Barnes, Steve. 2004. "No Cameras in Bombing Trial." *New York Times,* January 29, 24.
Beamer. 2001. "Passenger: Todd Beamer." *Post-Gazette,* October 28. http://www.post-gazette.com
Bellamy, Carol. 2003. "Children are War's Greatest Victims." Organization for Economic Cooperation and Development. http://www.oecdobserver.org
Bjorgo, Tore. 2003. "Root Causes of Terrorism." International Expert Meeting, June 9–11. The Norwegian Institute of International Affairs, Oslo, Norway.
Brauer, Jurgen. 2003. "On the Economics of Terrorism." *Phi Kappa Phi Forum* Spring:38–41.
Carneiro, Robert L. 1994. "War and Peace: Alternating Realities in Human History." In *Studying War: Anthropological Perspectives,* eds. S. P. Reyna and R. E. Downs, pp. 3–27. Langhorne, PA: Gordon & Breach Science Publishers.
Cauffman, Elizabeth, Shirley Feldman, Jaime Waterman, and Hans Steiner. 1998. "Post-Traumatic Stress Disorder Among Female Juvenile Offenders." *Journal of the American Academy of Child and Adolescent Psychiatry* 37:1209–1217.
CDI (Center for Defense Information). 2003. *Military Almanac.* http://www.cdi.org
CNN. 2001a. "Libyan Bomber Sentenced to Life." January 31. http://www.europe.cnn.com/2001/LAW
———. 2001b. "Rape War Crime Verdict Welcomed." February 23. http://www.cnn.com/2001/WORLD/europe/
———. 2003. "Poll: Muslims Call U.S. 'Ruthless, Arrogant.'" February 26. http://www.cnn.usnews
Cohen, Ronald. 1986. "War and Peace Proneness in Pre- and Post-industrial States." In *Peace and War: Cross-Cultural Perspectives,* eds. M. L. Foster and R. A. Rubinstein, pp. 253–267. New Brunswick, NJ: Transaction Books.
Conflict Research Consortium. 2003. "Mediation." http://www.colorado.edu/conflict/peace
Deen, Thalif. 2000, September 9. "Inequality Primary Cause of Wars, says Annan." http://www.hartford-hwp.com/archives

Dejong, J., W. Scholte, M. Koeter, and A. Hart. 2000. "The Prevalence of Mental Health Problems in Rwandan and Burundese Refugee Camps." *Acta Psychiatrica Scandinavica* 102:171–177.
DHS (Department of Homeland Security). 2003. "Building a Secure Homeland." http://www.dhs.gov
Dixon, William J. 1994. "Democracy and the Peaceful Settlement of International Conflict." *American Political Science Review* 88(1):14–32.
Doyle, Michael. 1986. "Liberalism and World Politics." *American Political Science Review* 80 (December): 1151–1169.
Editorial. 2000. "Foreign Conservationists Under Siege." *New York Times,* April 1, A14.
EMS (Environmental Media Services). 2002, October 7. "Environmental Impacts of War." http://www.ems.org
Feder, Don. 2003. "Islamic Beliefs led to the Attack on America." In *The Terrorist Attack on America,* ed. Mary E. Williams, pp. 20–23. Farmington Hills, MA: Greenhaven Press.
Funke, Odelia. 1994. "National Security and the Environment." In *Environmental Policy in the 1990s: Toward a New Agenda,* 2nd ed., eds. Norman J. Vig and Michael E. Kraft, pp. 323–345. Washington, DC: Congressional Quarterly, Inc.
Garrett, Laurie. 2001. "The Nightmare of Bioterrorism." *Foreign Affairs* 80:76.
Gibbs, Nancy. 2003. "After 9/11 Where Do We Go from Here?" *Time,* September 15, 36–37.
Gioseffi, Daniela. 1993. Introduction to *On Prejudice: A Global Perspective,* ed. Daniele Gioseffi, pp. xi–1. New York: Anchor Books, Doubleday.
Goure, Don. 2003. "First Casualties? NATO, the U.N." *MSNBC News,* March 20. http://www.msnbc.com/news
Hayman, Peter, and Douglas Scaturo. 1993. "Psychological Debriefing of Returning Military Personnel: A Protocol for Post-Combat Intervention." *Journal of Social Behavior and Personality* 8(5): 117–130.
Hawaii, State of. 2000. "Dual-Use Technologies." http://www.state.hi.us/dbedt/ert/key
Hooks, Gregory, and Leonard E. Bloomquist. 1992. "The Legacy of World War II for Regional Growth and Decline: The Effects of Wartime Investments on U.S. Manufacturing, 1947–72." *Social Forces* 71(2):303–337.
Huntington Samuel. 1996. *The Clash of Civilizations and the Remaking of World Order.* New York: Simon and Schuster.
Inglesby, Thomas, Donald Henderson, John Bartlett, Michael Archer, et al. 1999. "Anthrax as a Biological Weapon: Medical and Public Health Management." *Journal of the American Medical Association* 281:1735–1745.
INTERPOL. 1998. "Frequently Asked Questions About Terrorism." http://www.kenpubs.co.uk/INTERPOL.COM/English/faq
Kemper, Bob. 2003. "Agency Wages Media Battle." *Chicago Tribune,* April 7. http://www.chicagotribune.com
Klare, Michael. 2001. *Resource Wars: The New Landscape of Global Conflict.* New York: Metropolitan Books.
Klingman, Avigdor, and Zehara Goldstein. 1994. "Adolescents' Response to Unconventional War Threat Prior to the Gulf War." *Death Studies* 18:75–82.
Knickerbocker, Brad. 2002. "Return of the Military-Industrial Complex?" *Christian Science Monitor,* February 13. http://www.csmonitor.com/2002/0213
Kyl, Jon, and Morton Halperin. 1997. "Q: Is the White House's Nuclear-Arms Policy on the Wrong Track?" *Insight on the News* 42:24–28.
Lamont, Beth. 2001. "The New Mandate for UN Peackeeping." *The Humanist* 61:39–41.
Landmines. 2003a. "Global Landmine Crisis." http://www.landmines.org
———. 2003b. "The Problem: Impact of Landmines." http://www.landmines.org
Laqueur, Walter. 2003. *No End to War: Terrorism in the 21st Century.* New York: Continuum International Publishing Group.
Lichtblau, Eric. 2003. "Suit Challenges Constitutionality of Powers in Antiterrorism Law." *New York Times,* July 31. http://www.nytimes.com
Li, Xigen and Ralph Izard. 2003. "Media in a Crisis Situation Involving National Interest: A Content Analysis of Major U.S. Newspapers' and TV Networks' Coverage of the 9/11 Tragedy." *Newspaper Research Journal* 24:1–16.
Lloyd, John. 1998. "The Dream of Global Justice." *New Statesman* 127:28–30.
Lorch, Donatella, and Preston Mendenhall. 2000. "A War's Hidden Tragedy." *Newsweek.* http://www.msnbc.com/news
McElroy, Wendy. 2003. "Iraq War may Kill Feminism as We Know It." Fox News Channel, March 18. http://www.foxnews.com
McKinnin, Jim. 2001. "The Phone Line from Flight 93 Was Still Open . . ." *Post-Gazette,* September 16. http://www.post-gazette.com
Miller, Susan. 1993. "A Human Horror Story." *Newsweek,* December 27, 17.
Morahan, Lawrence. 2003. "Hussein Threatens to Become Symbol of Nationalism in Iraq." *Townhall,* July 22. http://www.townhall.com/news
MPRI (Military Professional Resources, Inc.). 2003. "MPRI—Taking Expertise Around the World." http://www.mpri.com/channels/about
Myers-Brown, Karen, Kathleen Walker, and Judith A. Myers-Walls. 2000. "Children's Reactions to International Conflict: A Cross-Cultural Analysis." Presented at the National Council of Family Relations, Minneapolis, MN, November 20.
NASP (National Association of School Psychologists). 2003. "Children and Fear of War and Terrorism." http://www.nasponline.org/NEAT/children
Nelson, Murry R. 1999. "An Alternative Medium of Social Education—the 'Horrors of War' Picture Cards." *The Social Studies* 88:100–108.
Office of Management and Budget. 2003.
Pape, Robert A. 2003. "Dying to Kill Us." http://www.nytimes.com
Paul, Annie Murphy. 1998. "Psychology's Own Peace Corps." *Psychology Today* 31:56–60.
Pfefferbaum. Betty. 1997. "Post-Traumatic Stress Disorder in Children: A Review of the Last Ten Years." *Journal of the American Academy of Child and Adolescent Psychiatry* 36:1503–1512.
Porter, Bruce D. 1994. *War and the Rise of the State: The Military Foundations of Modern Politics.* New York: Free Press.
PTSD (Post Traumatic Stress Disorder). 2003. "What Is Post Traumatic Stress Disorder?" National Center for Post Traumatic Stress Disorder. http://www.ncptsd.org/facts
Renner, Michael. 1993. "Environmental Dimensions of Disarmament and Conversion." In *Real Security: Converting the Defense Economy and Building Peace,* eds. Karl Cassady and Gregory A. Bischak, pp. 88–132. Albany: State University of New York Press.
———. 2000. "Number of Wars on Upswing." In *Vital Signs: The Environmental Trends That Are Shaping Our Future,* ed. Linda Starke, pp. 110–111. New York: W. W. Norton.
Reuters. 2001. "Death Toll Climbs Despite Lull in Sri Lanka Conflict." November 26. http://www.in.news.yahoo.com
Romero, Anthony. 2003. "Civil Liberties Should Not Be Restricted During Wartime." In *The Terrorist Attack on America,* ed. Mary Williams, pp. 27–34. Farmington Hills, MA: Greenhaven Press.
Roper. 2003. "U.S. Public Opinion on Defense Spending." The Roper Center for Public Opinion Research, June. http://www.ropercenter.uconn.edu
Rosenberg, Tina. 2000. "The Unbearable Memories of a U.N. Peacekeeper." *New York Times,* October 8, 4, 14.
Saad, Lydia. 2003. "Most Americans Don't Feel Government Threatens Civil Rights." Gallup Organization, October 1. http://www.gallup.com/poll
Salladay, Robert. 2003, April 7. "Anti-War Patriots Find They Need to Reclaim

Words, Symbols, even U.S. Flag from Conservatives." http://www.commondreams.org

Save the Children. 2003. "Report: Women, Children Bear Brunt of War." *USA Today,* May 6. http://www.usatoday.com

Schabner, Dean. 2002. "Patriot Revolution." ABC News. July 1. http://www.abcnews.com

Scheff, Thomas. 1994. *Bloody Revenge.* Boulder, CO: Westview Press.

Sengupta, Somini. 2003. "Taylor Steps Down as President of Liberia." *New York Times,* August 11. http://www.nytimes.com

Shanahan, John J. 1995. "Director's Letter." *The Defense Monitor* 24(6):8. Washington, DC: Center for Defense Information.

Shanker, Thom. 2003. "U.S. Remains Leader on Global Arms Sales, Report Says." September 25. http://www.nytimes.com

Simon, Jeffery. 2002. "The Global Terrorist Threat." *Phi Kappa Phi Forum* (Spring): 10–12.

SIPRI (Stockholm International Peace Research Institute). 2003. *SIPRI Yearbook 2003: Armaments, Disarmament and International Security.* Oxford: Oxford University Press.

Smith, Craig. 2004. "Libya to Pay More to French in '89 Bombing." *New York Times,* January 9, 6.

Smith, Lamar. 2003. "Restricting Civil Liberties During Wartime Is Justifiable." In *The Terrorist Attack on America,* ed. Mary Williams, pp. 23–26. Farmington Hills, MA: Greenhaven Press.

SSRC (Social Science Research Council). 2003. "Introduction to New World Order." http://www.ssrc.org/sept11/essays

Starr, J. R., and D. C. Stoll. 1989. "U.S. Foreign Policy on Water Resources in the Middle East." Washington, DC: Center for Strategic and International Studies.

Statistical Abstract of the United States: 2002. 2002. 122nd ed. U.S. Bureau of the Census. Washington, DC: U.S. Government Printing Office.

Stipp, David. 2004. "The United States Must Spend More on High-Tech Defenses Against Biological Warfare." In *Biological Warfare,* ed. William Dudley, pp. 109–116. Farmington Hills, MA: Greenhaven Press.

Strobel, Warren, David Kaplan, Richard Newman, Kevin Whitelaw, and Thomas Grose. 2001. "A War in the Shadows." *U.S. News and World Report* 130:22.

Sutker, Patricia B., Madeline Uddo, Karen Brailey, and Albert N. Allain, Jr. 1993. "War Zone Trauma and Stress-Related Symptoms in Operation Desert Storm (ODS) Returnees." *Journal of Social Issues* 40(4):33–50.

Sweet, Alec Stone, and Thomas L. Brunell. 1998. "Constructing a Supranational Constitution: Dispute Resolution and Governance in the European Community." *American Political Science Review* 92:63–82.

Thompson, Mark, and Duffy, Michael. 2003. "Is the Army Stretched Too Thin?" *Time,* September 1, 37–43.

UN (United Nations). 2003. "Some Questions and Answers." http://www.unicef.org

UNDP (United Nations Development Programme). 2003. "UNDP's Work in Iraq," October. http://www.iq.undp.org

UNICEF. 2003. "Child Protection: Armed Conflict." http://www.unicef.org

UN News Service. 2003. "Annan Submits $3.05 Billion Budget for UN, Representing .05 percent Growth." http://www.un.org/apps/news

U.S. State Department. 2003. "2002 Patterns of Global Terrorism Report." http://www.state.gov

Vesely, Milan. 2001. "UN Peacekeepers: Warriors or Victims?" *African Business* 261:8–10.

Viner, Katharine. 2002. "Feminism as Imperialism." *The Guardian,* September 21. http://www.guardian.co.uk

Wax, Emily. 2003. "War Horror: Rape Ruining Women's Health." *Miami Herald,* November 3. http://www.miami.com

White House. 2003. "FY 2004 Budget Fact Sheet." http://www.whitehouse.gov/news

Wilkinson, Marian. 2003. "U.S. General Estimate 5000 Iraqi Guerillas." *The Age,* November 15. http://www.theage.com

Williams, Mary E., ed. *The Terrorist Attack on America.* Farmington Hills, MA: Greenhaven Press.

Worldwatch Institute. 2003. *Vital Signs: The Trends That Are Shaping Our Future.* New York: W. W. Norton.

Zakaria, Fareed. 2000. "The New Twilight Struggle." *Newsweek,* October 12. http://www.msnbc.com/news

Photo Credits

Chapter 1: 2, 3 © AP/Wide World Photos; **23** © Paul Fusco/Magnum Photos.

Chapter 2: 38 © AP/Wide World Photos; **40, 41** Photo courtesy of the National Institute of Mental Health; **50** © AP/Wide World Photos; **53** © Mary Kay Denny/Photo Edit.

Chapter 3: 66 © AP/Wide World Photos; **73** © John Kobal Foundation/Hulton/Archive; **80** © Partnership for a Drug-Free America®; **84** © AP/Wide World Photos.

Chapter 4: 99 © Michael Newman/PhotoEdit; **104** © William Campbell/CORBIS Sygma; **107** © A. Ramey/PhotoEdit; **115** © Billy E. Barnes/PhotoEdit.

Chapter 5: 124 (above) © Caroline Schacht; **124** (below) © AFP/CORBIS; **137** Photo by Carol Geddes of Sonoma Image Studios; **155** © AP/Wide World Photos.

Chapter 6: 166 © AP/Wide World Photos; **168** © Reuters NewMedia Inc./CORBIS; **181** © Gamma Press Images/Liaison Agency; **188** © Mark Richards / PhotoEdit; **190** © Reuters NewMedia Inc./CORBIS; **196** Southern Poverty Law Center.

Chapter 7: 203 © David Turnley/CORBIS; **207** © Marjorie Farrell/The Image Works; **226** (left) © Kim Kulish/CORBIS; **226** (right) © Bob Sacha.

Chapter 8: 230 Photo courtesy Lynn Rosenberg; **232** © AP/ Wide World Photos; **234** Courtesy of Palms of Manasota; **236** Courtesy of Tracy St. Pierre; **244, 245** © AP/ Wide World Photos; **247** © Reuters NewMedia Inc./CORBIS; **251** (left) © AP/Wide World Photos; **251** (right) © Kim Kulish/CORBIS; **261** (left to right) Courtesy of the University of Pennsylvania, Courtesy of University of North Carolina at Chapel Hill, Courtesy of Purdue University, Courtesy of Cornell College.

Chapter 9: 269 © Reuters NewMedia Inc./CORBIS; **279** © Tony Freeman/PhotoEdit; **285** © Tom Miner/The Image Works; **290** © Frank Fournier/Contact Press.

Chapter 10: 311 © Elena Rooraid/PhotoEdit; **315** Photo courtesy of the USDA Food Stamp Program; **321** Photo courtesy of the National Student Campaign Against Hunger and Homelessness.

Chapter 11: 340 © AP/Wide World Photos; **346** © CORBIS; **350** Photo by Jesse Kirschner and courtesy of United Students Against Sweatshops; **351, 362** © AP/Wide World Photos.

Chapter 12: 368 © Karen Kasmauski/Woodfin Camp & Associates; **371** (left) © James Wilson/Woodfin Camp & Associates; **371** (right) © Dan Habib; **377** © Michael Newman/PhotoEdit; **386** © Reuters/Kevin Lamarque/Hulton Archive.

Chapter 13: 398 © CORBIS; **405** © Peter Johnson/CORBIS; **407** © David Turnley/CORBIS; **412** © Dave G. Houser/CORBIS; **416** © AP/Wide World Photos.

Chapter 14: 428 © AP/Wide World Photos; **432** © Ted Spiegel/CORBIS; **438** EPA Photo/EPA/Harish Tyagi/AP/Wide World Photos; **447** © AP/Wide World Photos; **451** © Morton Beebe/CORBIS.

Chapter 15: 462 (left) © Bettmann/CORBIS; **462** (right) © Sonda Dawes/The Image Works; **464** © CORBIS; **467** © Adam Lubroth/Stone; **479** © David Sams/Stock Boston.

Chapter 16: 490 © Bob Mahoney/The Image Works; **493** © David Pollack/CORBIS; **500** © CORBIS; **501** © AP/Wide World Photos; **510, 512** © Reuters/CORBIS.

Name Index

Subject Index